D0403519

**DIGITAL
INTEGRATED
ELECTRONICS**

McGRAW-HILL ELECTRICAL AND ELECTRONIC ENGINEERING SERIES

FREDERICK EMMONS TERMAN: Consulting Editor
W. W. HARMAN, J. G. TRUXAL, AND R. A. ROHRER: Associate Consulting Editors

ANGELAKOS AND EVERHART: Microwave Communications
ANGELO: Electronic Circuits
ANGELO: Electronics: BJTs, FETs, and Microcircuits
ASELTINE: Transform Method in Linear System Analysis
BELOVE, SCHACHTER, AND SCHILLING: Digital and Analog Systems, Circuits, and Devices: An Introduction
BENNETT: Introduction to Signal Transmission
BERANEK: Acoustics
BRACEWELL: The Fourier Transform and Its Application
BRENNER AND JAVID: Analysis of Electric Circuits
CARLSON: Communication Systems: An Introduction to Signals and Noise in Electrical Communication
CHEN: The Analysis of Linear Systems
CHEN: Linear Network Design and Synthesis
CHIRLIAN: Analysis and Design of Electronic Circuits
CHIRLIAN: Basic Network Theory
CHIRLIAN: Electronic Circuits: Physical Principles, Analysis, and Design
CHIRLIAN AND ZEMANIAN: Electronics
CLEMENT AND JOHNSON: Electrical Engineering Science
D'AZZO AND HOUPIS: Feedback Control System Analysis and Synthesis
D'AZZO AND HOUPIS: Linear Control System Analysis and Design: Conventional and Modern
ELGERD: Control Systems Theory
ELGERD: Electric Energy Systems Theory: An Introduction
EVELEIGH: Adaptive Control and Optimization Techniques
EVELEIGH: Introduction to Control Systems Design
FEINSTEIN: Foundations of Information Theory
FITZGERALD, HIGGINBOTHAM, AND GRABEL: Basic Electrical Engineering
FITZGERALD, KINGSLEY, AND KUSKO: Electric Machinery
FRANK: Electrical Measurement Analysis
GEHMLICH AND HAMMOND: Electromechanical Systems
GHAUSI: Principles and Design of Linear Active Circuits
GREINER: Semiconductor Devices and Applications
HAMMOND AND GEHMLICH: Electrical Engineering
HANCOCK: An Introduction to the Principles of Communication Theory
HARMAN: Principles of the Statistical Theory of Communication
HAYT: Engineering Electromagnetics
HAYT AND KEMMERLY: Engineering Circuit Analysis
HILL: Electronics in Engineering
JOHNSON: Transmission Lines and Networks
KRAUS: Antennas
KRAUS AND CARVER: Electromagnetics
KUO: Linear Networks and Systems
LEPAGE: Complex Variables and the Laplace Transform for Engineering
LEVI AND PANZER: Electromechanical Power Conversion
LINVILL AND GIBBONS: Transistors and Active Circuits
LYNCH AND TRUXAL: Introductory System Analysis
LYNCH AND TRUXAL: Principles of Electronic Instrumentation
McCLUSKEY: Introduction to the Theory of Switching Circuits
MEISEL: Principles of Electromechanical-energy Conversion
MILLMAN AND HALKIAS: Electronic Devices and Circuits

**McGRAW-HILL
BOOK COMPANY**
New York
St. Louis
San Francisco
Auckland
Bogotá
Düsseldorf
London
Madrid
Mexico
Montreal
New Delhi
Panama
Paris
São Paulo
Singapore
Sydney
Tokyo
Toronto

HERBERT TAUB
DONALD SCHILLING
Professors of Electrical Engineering
The City College of the
City University of New York

Digital Integrated Electronics

This book was set in Times Roman.
The editors were Peter D. Nalle and Madelaine Eichberg;
the cover was designed by Scott Chelius;
the production supervisor was Charles Hess.
The drawings were done by J & R Services, Inc.

Library of Congress Cataloging in Publication Data

Taub, Herbert, date
    Digital integrated electronics.

    (McGraw-Hill electrical and electronic engineering series)
    Includes index.
    1. Digital electronics.   2.   Integrated circuits.
I.   Schilling, Donald L., joint author.   II.   Title.
TK7868.D5T37   621.381   76-4585
ISBN 0-07-062921-8

**DIGITAL
INTEGRATED
ELECTRONICS**

3 4 5 6 7 8 9 0 DODO 8 9 8

To our wives

ESTHER and ANNETTE

CONTENTS

PREFACE

In 1956 the McGraw-Hill Book company published the text "Pulse and Digital Circuits" coauthored by J. Millman and H. Taub. That book, which undertook to present a rather complete account of the state of the art of digital electronics dealt almost exclusively with vacuum-tube circuits. Semiconductor devices and circuits, which had not long before been introduced, appeared in a single final chapter, added at the last moment, while the book was in production. In the decade that followed semiconductor devices completely supplanted tubes in digital circuitry. In response to this development the same authors prepared a replacement volume "Pulse Digital and Switching Waveforms" which appeared in 1965. In the newer volume the overwhelming importance of the semiconductor was appropriately emphasized and vacuum-tube circuits were presented only incidentally. Now, again after about a decade, the advances in integrated circuitry have prompted this present volume. However, this book is intended as a continuation of the 1965 work rather than as a replacement. Here the present authors have undertaken to describe and analyze all the basic integrated-circuit building blocks from which digital circuits and systems are assembled. As reasonably as is feasible in a textbook, the material presented is up to date. As was the case in the earlier volume, the present authors have taken great pain with the style of pedagogy.   We have striven to make the explanations clear and easily understood without sacrificing depth and completeness of presentation.   For this reason, we hope that this work will find a place not only in the classroom but also in a program of self-study for a reader who may want to keep informed about current developments.

The material in the text has been used at the City College of New York in a two-semester course offered to junior and senior electrical engineering students and has been used as well as the basis of two graduate courses.   This material has also been presented in a two-semester course offered to technical staff members of the Bell Laboratories, to engineering personnel at NASA and at Lockheed, and in short courses offered in the continuing-education program at the George Washington University.

It is assumed that the reader already has a background in semiconductor devices and circuits. Nevertheless we find it useful to provide in Chapter 1 a review of certain special matters pertaining to the operation of semiconductor devices in a *switching* mode.   Semiconductors have rather involved and highly nonlinear volt-ampere characteristics.   An exact analysis of semiconductor circuits results in considerable mathematical complexity.   In Chapter 1 we present some convenient simplifications which lead to quite good and useful approximations.

The first part of Chapter 2 discusses operational amplifiers.   Such amplifiers, intended to be operated linearly rather than in a switching mode, are not our proper concern.   Still, in a number of cases we find that operational amplifiers appear as components in what are otherwise digital circuits.   Furthermore, by a rather natural extension, operational amplifiers lead to the discussion, in the second part of the chapter, of comparators which are indeed important switching devices.

Chapter 3 introduces the concept of logical variables, Boolean algebra, and methods of analyzing circuits composed of logical gates.   Karnaugh maps and their various applications are presented.   This chapter is complete in the sense that it presumes no prior acquaintance with the subject and explains all the principles of design and analysis of logical circuits required for an understanding of the entire text. On the other hand, the content of this chapter is inevitably included in a course in logic design and, hence, may be bypassed by readers who have already been exposed to this material.

The electronics of logical gates is begun in Chapter 4.   The first part of this chapter deals with resistor-transistor logic (RTL) while the second part is concerned with integrated-injection logic (IIL).   RTL is not presently used in new design.   Yet there are a number of reasons on account of which it is valuable to consider this family of logic.   Being the first widely used family of IC logic available, there are in operation many installations in which it is incorporated.   Then, again on account of its elegant simplicity, it is an ideal vehicle through which to present many of the basic concepts and principles universally important in the electronics of logical gates.   Finally, it bears an interesting topological relationship to IIL which is one of the most recently developed families of logic. Chapter 5 considers diode-transistor logic (DTL). In the family of DTL we find high-threshold logic (HTL), which finds extensive application in highly noisy environments.

Chapters 6 and 7 discuss transistor-transistor logic (TTL) and emitter-coupled logic (ECL) respectively. At the present time these are the most widely used saturating and nonsaturating logic families. Hence the analysis of these families is rather extensive. In ECL particularly, it turns out that some appreciation of the nature of signal transmission over transmission lines is required. Readers who are unfamiliar with transmission line propagation will find an adequate introductory presentation in Appendix A. A more complete discussion appears in Chapter 3 of "Pulse Digital and Switching Waveforms" referred to above. Metal-oxide semiconductor (MOS) logic and complementary-symmetry (CMOS) logic is presented in Chapter 8.

The various families of logic having been considered (Chapters 4 through 8), we begin in Chapter 9 to consider the basic digital structures which are assembled from these gates. Chapter 9 explores in considerable detail the principles of operation of various types of flip-flops and, in addition, analyzes the electronics of the circuitry of a number of representative commercial units. We take considerable pains to make clear how flip-flops are adapted to circumvent timing problems that would otherwise develop in synchronous systems. Registers and counters are discussed in Chapter 10. Procedures for the design of both synchronous and ripple counters of arbitrary modulo are explained, and the use of registers to generate pseudorandom and other specified sequences is also presented.

Logic circuits for performing arithmetic operations are considered in Chapter 11. Emphasis is placed on the operations of addition (and subtraction) since generally multiplication and division are performed by algorithms involving the operation of addition (or subtraction). We have taken rather more care than is usual to explain clearly how negative numbers are expressed and how subtraction is effected in one's-complement and two's-complement notation through the use of logic circuitry which actually performs addition. The use of saturation logic for overflow correction in addition is presented as is the operation of the *arithmetic logic unit* which is the heart of every *microprocessor*. Semiconductor memories are examined in Chapter 12. We have omitted core memories since it appears that such core memories are in the process of being supplanted by semiconductor systems. This chapter includes sequential memories, read-only memories and random-access dynamic and static memories. The electronics of memories involving field-effect transistors, the CCD and bipolar junction transistors are also described.

In Chapters 13 and 14 we consider the matter of the interface between digital and analog signals. Chapter 13 presents analog gates, analog multiplexers, sample-and-hold circuits, integrate-and-dump circuits, etc. Chapter 14 examines digital-to-analog and analog-to-digital systems. The various analog-to-digital systems considered are reasonably representative of the systems which are in wide use.   Finally, in Chapter 15, timing circuits—the integrated-circuit equivalents of monostable and astable multivibrators—are discussed.

The circuits presented in this text are typical of those encountered in the field.   More than 400 homework problems are provided, ranging from routine exercises to rather sophisticated design problems.   A solutions manual is available which instructors can obtain from the publisher.   An answer book is also available.   The authors will be happy to furnish a set of laboratory experiments currently used at CCNY in conjunction with this text.

We acknowledge gratefully the encouragement given by our colleagues and students.   In particular we thank Mr. T. Apelewicz who prepared the solutions manual, Dr. J. Garodnick to whom we are indebted for a critical review and criticism of much of the text material and Mr. Edward Tynan and Dr. Ronald Schilling through whose kindness we were able to receive a great deal of the very useful technical literature published by the Motorola company.   We express our particular appreciation to Mrs. Joy Rubin for her skillful service in typing the manuscript.

HERBERT TAUB

DONALD SCHILLING

**DIGITAL
INTEGRATED
ELECTRONICS**

ELECTRONIC DEVICES

As with analog circuits, the electronic devices used in digital processing circuits include the diode, the bipolar transistor, and the field-effect transistor. We assume that the reader is familiar with these devices but principally in applications involving analog circuitry, where they are used as linear elements. In digital circuits, these devices are used principally in a nonlinear manner, i.e., in a switching mode, where they are abruptly driven between the extremes of nonconduction and conduction. In this chapter we shall review some matters of interest in connection with these devices with special emphasis on their behavior when used as switches.

1.1 THE IDEAL SEMICONDUCTOR DIODE

For an ideal pn junction diode the current I is related to the voltage V by the equation

$$I = I_0(\epsilon^{V/V_T} - 1) \tag{1.1-1}$$

As indicated in Fig. 1.1-1a, the current I is positive when the current flows

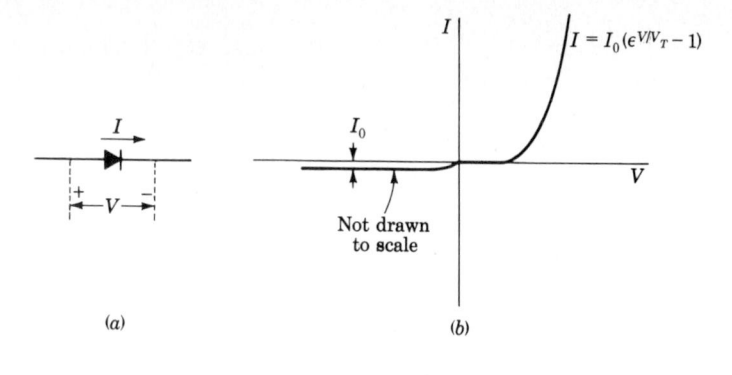

FIGURE 1.1-1
(a) The symbols I and V used in the diode equation (1.1-1), defined. (b) The volt-ampere characteristic of an ideal diode.

from the p side to the n side of the diode. The voltage V is the voltage drop from the p side to the n side. When V is positive, the diode is forward-biased. The symbol V_T stands for the electronvolt equivalent of the temperature and is given by

$$V_T \equiv \frac{kT}{e} \tag{1.1-2}$$

where k = Boltzmann constant = 1.38×10^{-23} J/K
e = electronic charge = 1.602×10^{-19} C
T = absolute temperature, kelvins

Substituting, we find that $V_T = T/11,600$ V and that at room temperature ($T \approx 300$ K) $V_T \approx 25$ mV.

The form, in principle, of the diode volt-ampere characteristic is shown in Fig. 1.1-1b. When the voltage V is positive and several times V_T, the exponential term in Eq. (1.1-1) greatly exceeds unity and the -1 term in the parentheses may be neglected. Consequently, except for a small range in the neighborhood of the origin, the current increases exponentially with voltage. When the diode is reversed-biased and $|V|$ is several times larger than V_T, $|I| \approx I_0$. The reverse current is therefore constant, independently of the applied reverse bias. Accordingly, I_0 is referred to as the *reverse saturation current*. This current is shown in Fig. 1.1-1b using a greatly enlarged scale since the value of I_0 is orders of magnitude less than typical values of I.

As noted, we shall be interested in the operation of diodes (and other elements) as switches. The diode is an *open* switch when back-biased and a *closed* switch when forward-biased. We shall generally find, in circuits of interest to us, that when a diode is called upon to make its presence felt in a circuit as a closed switch, it may typically carry a current of the order of a milliampere, i.e., in the range 0.1 to 10 mA. How large a voltage must be impressed across the diode to produce this nominal forward current depends, of course, on the diode cross section. If a diode yielded a forward current of 1 μA at an applied voltage V, a second diode of cross section 1,000 times larger would yield a current of 1 mA.

When a diode is manufactured, whether as a discrete component or an element in an integrated circuit, it is economical to use a cross section no larger than necessary. Such is particularly the case in integrated circuits (IC). For here, since many circuit elements are included on a single chip, a small increase in the cross section of one element is multiplied many-fold. This may result in an appreciable increase in the size of the silicon chip, or, equivalently, the same size chip will contain fewer diodes. The cross section of a diode will then be selected in part on the basis that with a reasonable margin of safety the diode should be able to dissipate the heat generated within it without an unacceptable increase in temperature. Additionally, the cross-sectional area must be large enough to reduce the ohmic resistance of the diode to an acceptable value.

A diode model When we examine the volt-ampere characteristics of commercial silicon diodes intended for application in low-power electronic circuits, we find that currents of the order of a milliampere correspond to a forward voltage of about 0.75 V. Diodes incorporated into integrated circuits appear to have comparable characteristics, again requiring about 0.75 V for forward currents in the range of a milliampere. Since we shall frequently have occasion to refer to this voltage, we assign to it a symbol $V_\sigma = 0.75$ V. When, then, the forward diode voltage is V_σ, the diode, used as a switch, is in the closed position.

If the diode switch is to be in the open position, it is really not necessary, as a matter of practicality, that the diode be reverse-biased. It is only necessary that the voltage across the diode correspond to a forward current which is negligibly small in comparison with the current corresponding to V_σ. Let us consider that the diode current is negligible when it has been reduced to 1 percent of the current corresponding to V_σ. The diode voltage, corresponding to this reduced current, we call V_γ.

If currents I_σ and I_γ correspond to voltages V_σ and V_γ, then from Eq. (1.1-1), we have

$$I_\sigma = I_0(\epsilon^{V_\sigma/V_T} - 1) \tag{1.1-3a}$$

and
$$I_\gamma = I_0(\epsilon^{V_\gamma/V_T} - 1) \tag{1.1-3b}$$

Since ϵ^{V_σ/V_T} and ϵ^{V_γ/V_T} are each much greater than unity, we have

$$\frac{I_\sigma}{I_\gamma} = 100 = \epsilon^{(V_\sigma - V_\gamma)/V_T} \tag{1.1-4}$$

Hence
$$V_\sigma - V_\gamma = V_T \ln 100 \approx 120 \text{ mV} \tag{1.1-5}$$

Thus, since $V_\sigma \approx 0.75$ V, $V_\gamma \approx 0.63$ V.

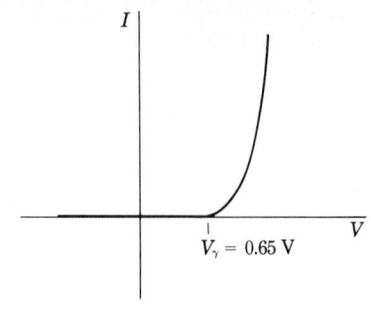

FIGURE 1.1-2
The cut-in at $V_\gamma = 0.65$ V of a physical
silicon diode.

We find it convenient to use the value $V_\gamma = 0.65$ V rather than 0.63 V so that the difference between V_σ and V_γ is a nice round number. With $V_\gamma = 0.65$ V, $V_\sigma - V_\gamma = 0.1$ V, and, as is readily verified, such a 0.1-V voltage change corresponds to a current ratio $I_\sigma/I_\gamma = 55$. However, we shall assume that no current flows for diode voltages $V < V_\gamma$ and therefore the precise value of I_γ is irrelevent. If we draw the diode characteristic using a current scale such that at diode voltage V_γ the current is "zero," the characteristic would appear as in Fig. 1.1-2. V_γ is called the *cut-in* voltage.

It should not be imagined that diodes are carefully tailored in area to make their forward voltage equal to 0.75 V. On the other hand, it is to be noted that because of the exponential nature of the diode characteristic a large relative change in current corresponds to a relativity small change in diode voltage. Suppose, for example, that a particular diode yields a current I_σ at $V_\sigma = 0.75$ V. Suppose now that we require that the same diode, in some application, carry a current $2I_\sigma$. Then, as can be verified (Prob. 1.1-2), V_σ would change only by 17 mV, which is only 2 percent of its value.

We can summarize this entire discussion concerning diode voltages as follows. We shall consider that a silicon diode has a cut-in voltage $V_\gamma = 0.65$ V. Thus, we consider that when the diode voltage V is less than 0.65 V, the diode current is small enough to be neglected and the presence of the diode just begins to make itself felt when V increases beyond 0.65 V. Further, whenever we find a diode in a circuit carrying about as much current as it appears to be intended to carry, we shall consider that the diode voltage is 0.75 V. There may be occasions when the diode appears to be operating in some middle region, and in this case we shall take the diode voltage to be 0.7 V.

The absolute values assigned to these voltages represent a composite experience derived from examining countless circuits, integrated and discrete, and from examining countless pages of manufacturers' specification sheets for moderate-sized diodes. This manner of dealing with diodes gives results with an accuracy which is entirely adequate and avoids the complications that would ensue if in our analysis of electronic circuits we tried to take account of the real volt-ampere characteristic of the diode.

1.2 TEMPERATURE DEPENDENCE OF DIODE CHARACTERISTICS

The characteristics of a diode are temperature-dependent. This dependence appears in Eq. (1.1-1) explicitly in the parameter V_T and implicitly in the fact that the reverse saturation current I_0 is also temperature-dependent. It is found that I_0 is an exponential function of temperature, and, in the neighborhood of room temperature, it doubles, approximately, for each 10°C rise in temperature. Figure 1.2-1 displays the reverse saturation current of a typical discrete silicon diode for several temperatures. The plots also call attention to the fact that in many diodes the reverse current is not constant but increases with increasing reverse voltage. This increase in I_0 results from leakage across the surface of the diode and from the fact that new carriers are generated by collisions in the transition region of the diode junction. It is of interest to note that as a matter of practice, I_0 as measured in the reverse direction may well be several orders of magnitude larger than the value required to give agreement with Eq. (1.1-1) in the forward direction.

In the forward direction, the overall effect of temperature on the diode characteristic is shown in the representative plots of Fig. 1.2-2. Here we note that a temperature change has the principal effect of leaving the general form of the characteristic unchanged but yet translating it along the abscissa. Thus, at a fixed current, an *increase* in temperature ΔT produces a *decrease* in voltage ΔV. It may be estimated theoretically and verified experimentally that

$$\frac{\Delta V}{\Delta T} \approx -2 \text{ mV/°C} \tag{1.2-1}$$

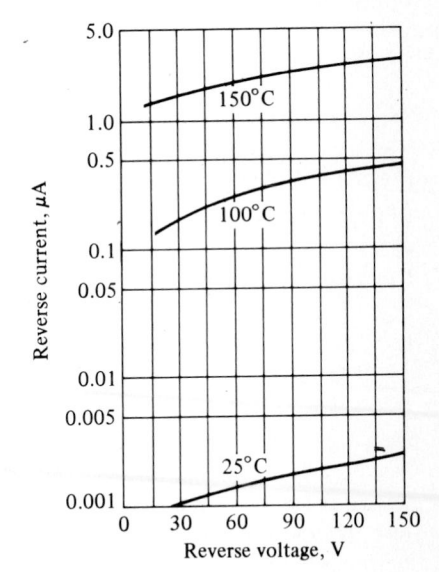

FIGURE 1.2-1
The dependence on temperature and reverse voltage of the reverse saturation current on the type 1N461 silicon diode.

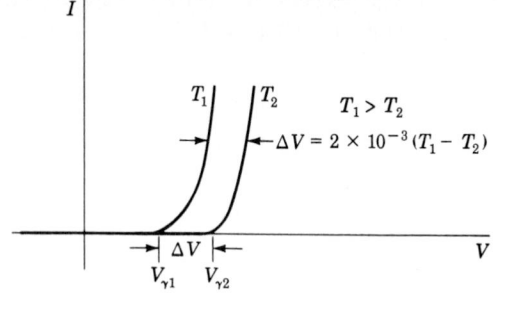

FIGURE 1.2-2
The effect of temperature on the diode volt-ampere characteristic in the forward direction.

Strictly, the temperature sensitivity $\Delta V / \Delta T$ depends on the temperature and also on the operating current. Generally the magnitude of $\Delta V / \Delta T$ decreases with increasing temperature and with increasing current. The value given in Eq. (1.2-1) is appropriate at room temperature (25°C). For simplicity, we shall use the figure of Eq. (1.2-1) quite generally with the comfort of knowing that thereby we shall err on the conservative side.

1.3 DIODE TRANSITION CAPACITANCE

A diode is driven to the reverse-biased condition when it is desired to turn off a current or prevent the transmission of a signal. When diodes are used for such purposes in circuits which handle fast waveforms or high frequencies, we must take account of the capacitance which appears across a reverse-biased junction. This capacitance is called the *barrier* or *transition capacitance* C_T. If this capacitance is large enough, the current which is to be restrained by the low conductance of the reverse-biased diode will flow through the capacitor.

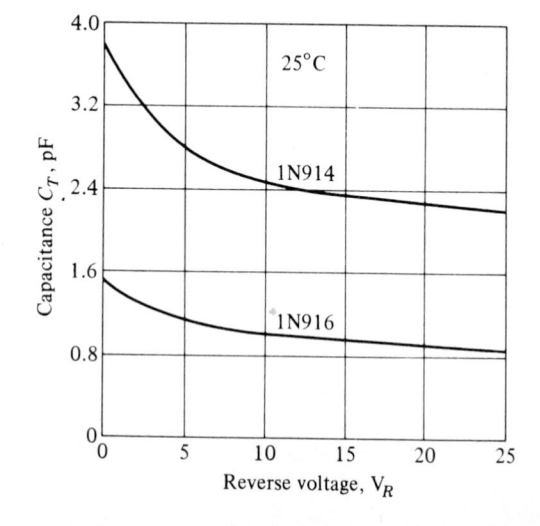

FIGURE 1.3-1
Typical barrier-capacitance variation, with reverse voltage, of silicon diodes 1N914 and 1N916. (*Fairchild Semiconductor Corporation*).

Diodes intended for service with fast waveforms have transition capacitances of the order of 1 to 10 pF. This capacitance decreases with increasing reverse voltage, as illustrated, for two typical diodes, in Fig. 1.3-1. A simple yet good approximation for the transition capacitance is

$$C_T = \frac{C_0}{(1 + V_R)^n} \tag{1.3-1}$$

where V_R is the reverse-biasing voltage and C_0 is the capacitance when $V_R = 0$ and $n = \frac{1}{2}$ or $\frac{1}{3}$ for abrupt or linearily graded junctions, respectively.

1.4 THE ZENER DIODE

As the reverse voltage across a diode is increased, a point is eventually reached where the diode breaks down; i.e., it begins abruptly to allow the flow of large current. The reverse voltage characteristic of the diode, including the breakdown region, is shown in Fig. 1.4-1a. Diodes which are designed with adequate power-dissipation capabilities to operate in the breakdown region may be employed as voltage-reference or constant-voltage sources. Such diodes are known as *avalanche*, *breakdown*, or *zener* diodes. They are used characteristically in the manner indicated in Fig. 1.4-1b. The input source V_i and resistor R are selected so that initially the diode is operating in the breakdown region. Here the diode voltage, which is also the voltage across the load R_L, is V_Z, as in Fig. 1.4-1a, and the diode current is I_Z. The diode will now regulate the load voltage against variations in load current and against variations in input supply voltage V_i because in the breakdown region large changes in diode current produce only small changes in diode voltage. Moreover, as load current or supply voltage changes, the diode current will accommodate itself to these changes to maintain

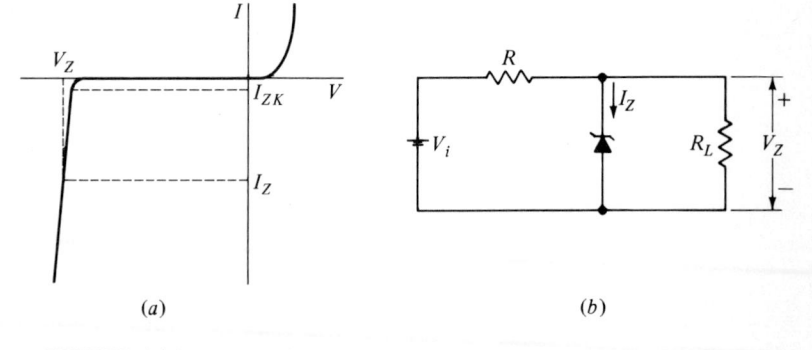

(a) (b)

FIGURE 1.4-1

(a) The volt-ampere characteristic of an avalanche or zener diode; (b) a circuit in which such a diode is used to regulate the voltage across R_L against changes due to variations in load current and supply voltage.

a nearly constant load voltage. The diode will continue to regulate until the circuit operation requires the diode current to fall to I_{ZK}, in the neighborhood of the knee of the diode volt-ampere curve. The upper limit on diode current is determined by the power-dissipation rating of the diode.

Two mechanisms of diode breakdown for increasing reverse voltage are recognized. In one the thermally generated electrons and holes acquire sufficient

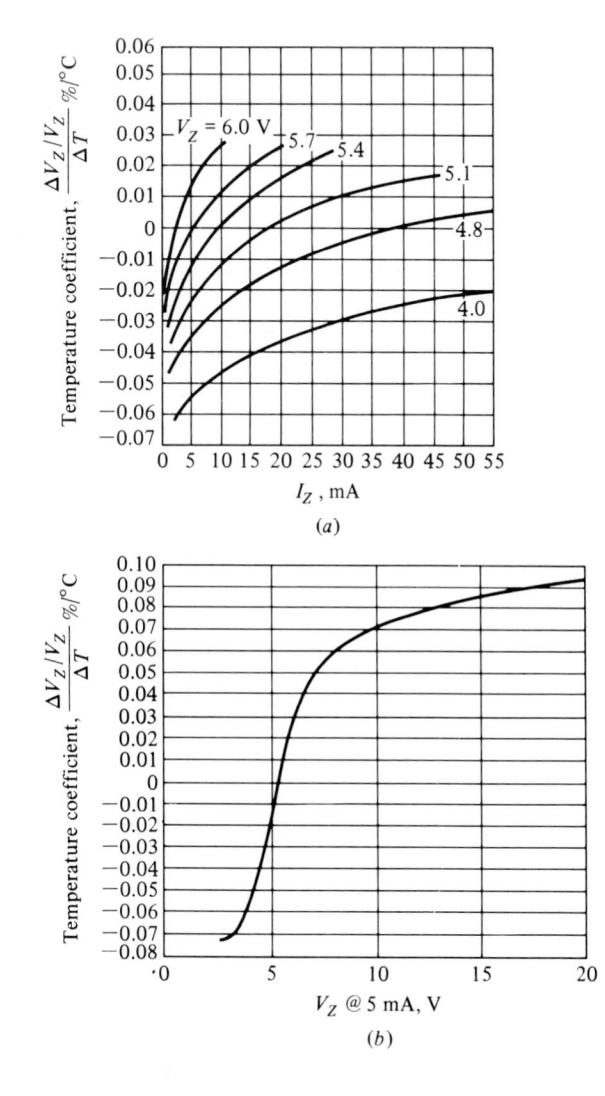

FIGURE 1.4-2
Temperature coefficients for a number of zener diodes having different operating voltages (a) as a function of operating current and (b) as a function of operating voltage. The voltage V_Z is measured at $I_Z = 5$ mA (from 25 to 100°C). (*Pacific Semiconductors, Inc.*)

energy from the applied potential to produce new carriers by removing valence electrons from their bonds. These new carriers, in turn, produce additional carriers, again through the process of disrupting bonds. This cumulative process is referred to as *avalanche multiplication*. It results in the flow of large reverse currents, and the diode finds itself in the region of *avalanche breakdown*. Even if the initially available carriers do not acquire sufficient energy to disrupt bonds, it is possible to initiate breakdown through a direct rupture of the bonds because of the existence of the strong electric field. Under these circumstances the breakdown is referred to as a *zener breakdown*. This zener effect is now known to play an important role only in diodes with breakdown voltages below about 6 V. Nevertheless, the term *zener* is commonly used also for the avalanche breakdown which occurs at higher voltages. Silicon diodes operated in avalanche breakdown are available with maintaining voltages from several volts to several hundred volts and with power ratings up to 50 W.

A matter of interest in connection with zener diodes, as with semiconductor devices generally, is their temperature sensitivity. The temperature dependence of the reference voltage, which is indicated in Fig. 1.4-2, is typical of what may be expected generally. In Fig. 1.4-2a the temperature coefficient of the reference voltage is plotted as a function of the operating current through the diode for various different diodes whose reference voltage at 5 mA is specified. The temperature coefficient is a percentage change in reference voltage per Celsius degree change in diode temperature. Figure 1.4-2b is a plot of the temperature coefficient at a fixed diode current of 5 mA as a function of zener voltage. The data which are used to plot this curve are taken from a series of different diodes of different zener voltages but of fixed dissipation rating. From the curves in Fig. 1.4-2 we note that the temperature coefficients may be positive or negative and normally are in the range ± 0.1 percent/°C. Note that if the reference voltage is above 6 V, where the physical mechanism involved is avalanche multiplication, the temperature coefficient is positive. However, below 6 V, where true zener breakdown is involved, the temperature coefficient is negative.

1.5 DIODES FOR INTEGRATED CIRCUITS

In the fabrication of an integrated circuit, the geometry and the doping of the various layers must be chosen to optimize the characteristic of the transistors which are the most important elements. It is usually not economically feasible to provide additional masking and diffusions to construct diodes. As a result, diodes are generally transistors adapted for diode operation. The five ways in which transistors can be so adapted are shown in Fig. 1.5-1. In Fig. 1.5-1a the diode terminals are the transistor base and emitter, and the collector is left floating. In Fig. 1.5-1b the diode terminals are as in Fig. 1-5-1a except that the collector terminal is connected to the base, etc.

The preferred configuration in integrated circuitry is generally the one shown in Fig. 1.5-1b. It has generally the lowest forward voltage drop for a given

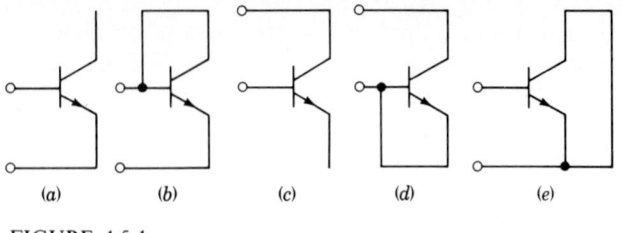

FIGURE 1.5-1
The five ways in which transistors can be adapted for use as diodes.

forward current, the lowest storage time (see Sec. 1.18), and the lowest reverse-bias junction capacitance. Its breakdown voltage, on the other hand, is inclined to be low, of the order of 7 V. However, since supply voltages encountered in digital integrated circuits rarely exceed 5 or 6 V, this limitation is not a serious disadvantage.

1.6 THE TRANSISTOR AS A SWITCH

The transistor in Fig. 1.6-1a is being used as a switch to connect and disconnect the load R_L from the source V_{CC}. Except that the transistor may be operated electrically and may be made to respond more rapidly, it serves the same function as that of the mechanical switch in Fig. 1.6-1b. The mechanical-switch arrangement allows no current to flow when the switch is open, but when the switch is closed, all the voltage V_{CC} appears across the load R_L. Ideally, the transistor switch should have these same properties. In this section we discuss the steady-state characteristics of the circuit of Fig. 1.6-1 corresponding to the cases when the transistor switch is open and when it is closed. Section 1.20 discusses the speed with which a transition between these two states can be made.

When a transistor is used as a switch, it is useful to divide its range of operation into three regions: the *cutoff*, the *active*, and the *saturation regions*. These regions are easily identified on the common-base characteristics of the transistor, as in Fig. 1.6-2. In the cutoff region both the emitter junction and the collector junction are reverse-biased, and only very small reverse saturation currents flow across the junctions. The transistor operates in the region below the characteristic for $I_E = 0$. This characteristic corresponds to a collector current I_{CO}, the reverse collector saturation current. It is almost but not precisely coincident with the axis $I_C = 0$. It is required that the transistor be in the cutoff region at times when it is to act as an open (nonconducting) switch.

When the base-emitter junction is forward-biased and the base-collector junction is reverse-biased, the transistor output current responds almost linearly to an input current. In switching operations, this region is not of great interest because the transistor switches abruptly from the cutoff region to the saturation

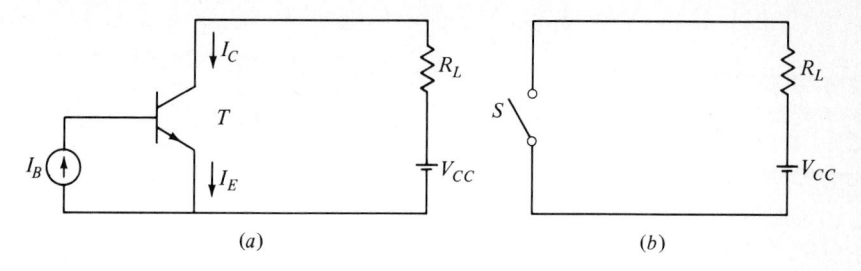

FIGURE 1.6-1

The transistor in (*a*) is being used as a switch. It serves the same function as the switch *S* in (*b*). The positive reference direction for each current is as shown. The symbol V_{CC} is a positive number representing the magnitude of the supply voltage.

region (or vice versa) and spends, ideally, a relatively insignificant time in the active region.

The region to the left of the ordinate $V_{CB} = 0$ and above $I_E = 0$ is the saturation region. Here the emitter junction and the collector junction are both forward-biased. The voltages across the individual junctions or across the combination of junctions are small (less than 1 V). Accordingly, when a transistor switch is required to be in the closed (conducting) condition, it is driven into saturation.

When a transistor is used as a switch in the common-base configuration, the input emitter current required to operate the switch is nominally as large as the collector current being switched. In the common-collector configuration, the

FIGURE 1.6-2

Typical common-base characteristics of an *npn* transistor. The cutoff, active, and saturation regions are indicated. Note the expanded voltage scale in the saturation region.

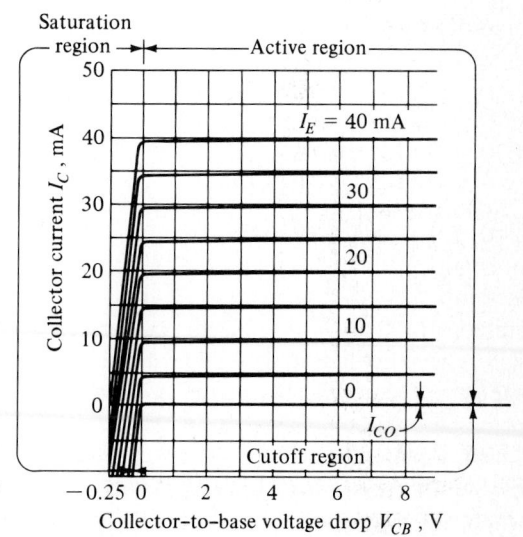

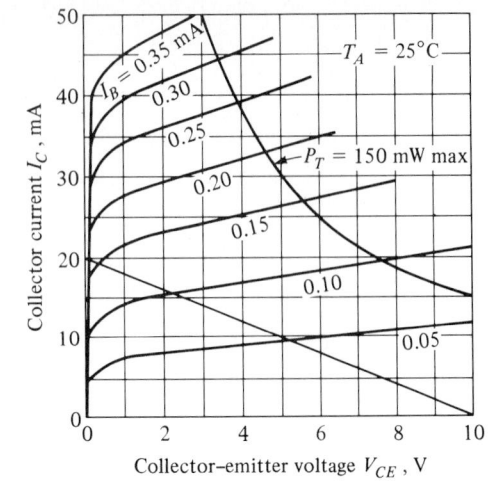

FIGURE 1.6-3
Typical common-emitter characteristics of
a medium-power (150-mW) *npn* transistor.
The load line corresponds to $V_{CC} = 10$ V
and $R_L = 500$ Ω.

input voltage required to operate the switch is nominally as large as the supply voltage. In the common-emitter configuration the input switching signal, current or voltage, is small in comparison with the switched output current or voltage. Hence, the common-emitter configuration is the most generally useful for a transistor switch.

Returning now to the common-emitter switching circuit of Fig. 1.6-1*a*, we display in Fig. 1.6-3 the common-emitter characteristics typical for a medium-power transistor. We have selected a load resistor $R_L = 500$ Ω and a supply voltage $V_{CC} = 10$ V and have superimposed the corresponding load line on the characteristics. The cutoff and saturation regions are not as clearly shown in Fig. 1.6-3 as in Fig. 1.6-2. In the next several sections, we shall discuss details of importance in connection with the transistor operation at cutoff and operation in saturation.

1.7 ANALYTIC EXPRESSIONS FOR TRANSISTOR CHARACTERISTICS[1]

Transistor characteristics which are required for switching-circuit design are normally furnished for each transistor type by the manufacturer. It is nonetheless important that we have some analytic procedure for determining the operating states of a transistor. In the first place, the availability of such an analysis will relieve us of complete dependence on published specifications and plots for each transistor type we may plan to use. Second, it will permit us to arrive at principles which apply to transistors generally rather than to particular types. Third, we shall be able to arrive at estimates of parameters not usually supplied by the manufacturer. Finally, it will help in the analysis of integrated circuits, where the characteristics of the individual transistors are not available.

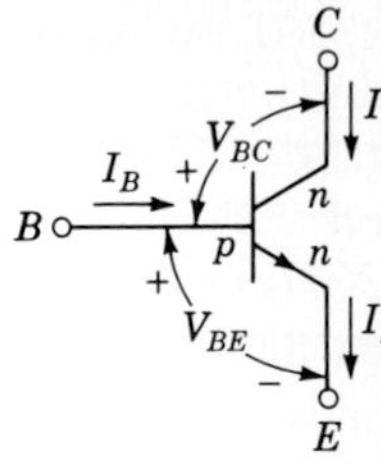

FIGURE 1.7-1
The voltages and currents in the Ebers-
Moll equations defined.

We assume an *npn* transistor, as shown in Fig. 1.7-1, and assume positive directions for the currents as indicated. These are the directions in which the currents normally flow. Collector and base currents I_C and I_B are positive when these currents flow into the transistor, and the emitter current I_E is positive when the emitter current flows out of the transistor. The symbol V_{BC} stands for the voltage drop across the collector junction from the *p*-type base to the *n*-type collector. The collector junction is forward-biased when V_{BC} is positive. Correspondingly V_{BE} is the voltage drop across the emitter junction. The emitter junction is forward-biased when V_{BE} is positive.

The current that crosses a transistor junction has two sources. One such component of the current is the *diode component*. It is due to the voltage V maintained across the junction by external constraints. This component of current is given by the diode equation (1.1-1). The second component is the *transistor component* due to the minority carriers which have crossed the other junction and have diffused across the base. Of the current which crosses the other junction, only a fraction, α, crosses the junction in question. When the transistor junctions are identified as the base-collector and the base-emitter junctions, the corresponding junction currents I_C and I_E are written

$$I_C = \alpha_N I_E - I_{C0}(\epsilon^{V_{BC}/V_T} - 1) \qquad (1.7\text{-}1)$$

$$I_E = \alpha_I I_C + I_{E0}(\epsilon^{V_{BE}/V_T} - 1) \qquad (1.7\text{-}2)$$

Here the first members on the right-hand side of Eqs. (1.7-1) and (1.7-2) represent the transistor component of current and the second terms represent the diode component. A transistor is not normally symmetrical both because of geometry and because of doping. Hence different α's have been used in the two equations. As a result of prejudices that develop naturally when transistors are used as linear elements in the active region, minority carriers which diffuse from emitter to collector are viewed as diffusing in the "normal" direction, while carriers diffusing in the other direction are judged to be diffusing in the inverse or reverse direction. Hence the subscripts N (normal) and I (inverse) on the α's. The currents I_{C0} and I_{E0} are the reverse saturation currents for collector and emitter junction respectively, corresponding to I_0 in Eq. (1.1-1). Equations (1.7-1) and (1.7-2) are known as the *Ebers-Moll equations*. Corresponding equations for a *pnp* transistor are easily written.

If we use Eqs. (1.7-1) and (1.7-2) simultaneously, we shall be assuming tacitly that the response of the transistor to a current injected at the collector junction is independent of any possible current injected at the emitter junction and vice versa. Thus we shall be assuming that superposition applies to the transistor currents, and this assumption implies, in turn, that the transistor is a linear device. This linearity does not apply, of course, to the volt-ampere characteristic of the junction but only to relationships of the junction currents to one another. If the doping of the base region is uniform, such linearity will indeed prevail. If the doping is not uniform, linearity will not apply. However, the range of applicability of these equations is very wide and covers many transistor types.

We can use Eqs. (1.7-1) and (1.7-2) to solve explicitly for the junction currents in terms of the junction voltages as defined in Fig. 1.7-1, with the result that

$$I_E = \frac{I_{E0}}{1 - \alpha_N \alpha_I} (\epsilon^{V_{BE}/V_T} - 1) - \frac{\alpha_I I_{C0}}{1 - \alpha_N \alpha_I} (\epsilon^{V_{BC}/V_T} - 1) \qquad (1.7\text{-}3)$$

$$I_C = \frac{\alpha_N I_{E0}}{1 - \alpha_N \alpha_I} (\epsilon^{V_{BE}/V_T} - 1) - \frac{I_{C0}}{1 - \alpha_N \alpha_I} (\epsilon^{V_{BC}/V_T} - 1) \qquad (1.7\text{-}4)$$

The third current I_B is determined from the condition

$$I_E = I_B + I_C \qquad (1.7\text{-}5)$$

We can solve explicitly for the junction voltages in terms of the currents from Eqs. (1.7-1) and (1.7-2), with the result that

$$V_{BE} = V_T \ln \left(1 + \frac{I_E - \alpha_I I_C}{I_{E0}}\right) \qquad (1.7\text{-}6)$$

$$V_{BC} = V_T \ln \left(1 - \frac{I_C - \alpha_N I_E}{I_{C0}}\right) \qquad (1.7\text{-}7)$$

The parameters α_N, α_I, I_{C0}, and I_{E0} are not independent but are related by the condition

$$\alpha_N I_{E0} = \alpha_I I_{C0} \qquad (1.7\text{-}8)$$

For a transistor operating in its active region, it is useful to have information about the parameter h_{fe}, the incremental current gain; that is, $h_{fe} = \Delta I_C / \Delta I_B$. This parameter is useful in calculations concerning a transistor circuit in which the transistor operates linearly, as in an amplifier. A second related parameter is the dc current gain $h_{FE} \equiv I_C / I_B$. These current-gain parameters can be related to the parameters which appear in the Ebers-Moll equations. In the active region the base-collector junction is reverse-biased, and if we assume that the magnitude of this bias is large in comparison with V_T, Eq. (1.7-1) can be written

$$I_C = \alpha_N I_E + I_{C0} \qquad (1.7\text{-}9)$$

From Eqs. (1.7-9) and (1.7-5) we find

$$I_C = \frac{\alpha_N}{1 - \alpha_N} I_B + \frac{I_{CO}}{1 - \alpha_N} \qquad (1.7\text{-}10)$$

From Eq. (1.7-10) we find

$$h_{fe} = \frac{\Delta I_C}{\Delta I_B} = \frac{\alpha_N}{1 - \alpha_N} \qquad (1.7\text{-}11)$$

Actually since normally I_{CO} is very small in comparison with the transistor currents which flow when a transistor is in the active region, we may ignore the last term in Eq. (1.7-10). In this case we have as well that

$$h_{FE} = \frac{\alpha_N}{1 - \alpha_N} \qquad (1.7\text{-}12)$$

Accordingly, henceforth we shall make no distinction between h_{FE} and h_{fe}. In a typical case we find $\alpha_N = 0.98$, in which case $h_{FE} \approx 50$.

The parameter h_{FE} varies widely over the operating range of a transistor and often varies widely from sample to sample of a given transistor type. It is not unusual to find that h_{FE} varies by as much as 50 percent over a collector-current range of 10. Similarly samples of a particular transistor type may vary in h_{FE} by as much as a factor of 3 or more. The parameter h_{FE} is also temperature-sensitive and may also show a variation by a factor of 3 or more over a temperature range from -50 to $+150°C$. Still it is useful to recognize that it is possible to design transistors, whether integrated or discrete, in such a manner that the average and typical value of h_{FE} becomes a controllable parameter. Thus there are transistors (discrete and integrated) of moderate h_{FE}, by which is meant $h_{FE} \approx 50$, and there are transistors with high h_{FE}, $h_{FE} \approx 200$.

The subscripts on the parameter h_{FE} are derived from the fact that h_{FE} (which we assume equal to h_{fe}) is the *hybrid* parameter which is characterized as the *forward* common-*emitter* current gain. If the transistor is used in the inverse manner, the collector becomes the common terminal, and corresponding to h_{FE} in the normal direction we have, in the inverse direction, the parameter h_{FC}. This parameter, in correspondence with Eq. (1.7-12) is related to α_I by

$$h_{FC} = \frac{\alpha_I}{1 - \alpha_I} \qquad (1.7\text{-}13)$$

The geometry and doping of transistors is generally such that while α_N is normally quite close to unity, α_I is much smaller, being in the range 0.01 to 0.2. Thus h_{FC} will be found in the range 0.01 to 0.25.

In the normal mode of operation and in the active region, $V_{BC} \leq 0$ and $V_{BE} \gg V_T$. When we take account, as well, of the fact that $\alpha_I \ll 1$, Eq. (1.7-3) reduces to the very useful result

$$I_E \approx I_{E0} \epsilon^{V_{BE}/V_T} \qquad (1.7\text{-}14)$$

1.8 THE TRANSISTOR AT CUTOFF

Consider a transistor in a switching circuit as in Fig. 1.6-1a. If the voltage between base and emitter is not sufficiently large to forward-bias the emitter junction adequately, the emitter current (and collector current) will be reduced nominally to zero and the transistor is *cut off*. As in the case of the diode, it is important to know the *cut-in* voltage of the transistor. The Ebers-Moll equations can yield some information of this point. Recognizing that in Fig. 1.6-1, near cut-in, the base-collector junction will be substantially reverse-biased, we have $V_{BC}/V_T \ll 1$. Therefore Eq. (1.7-3) can be written

$$I_E = \frac{I_{E0}}{1 - \alpha_N \alpha_I} (\epsilon^{V_{BE}/V_T} - 1) + \frac{\alpha_I I_{C0}}{1 - \alpha_N \alpha_I} \tag{1.8-1}$$

Using Eq. (1.7-8), we can rewrite Eq. (1.8-1) as

$$I_E = \frac{I_{E0}}{1 - \alpha_N \alpha_I} (\epsilon^{V_{BE}/V_T} - 1 + \alpha_N) \tag{1.8-2}$$

Normally $\alpha_N \approx 1$, so that we have

$$I_E \approx \frac{I_{E0}}{1 - \alpha_I} \epsilon^{V_{BE}/V_T} \tag{1.8-3}$$

As noted, the parameter α_I lies generally in the range 0.01 to 0.2 so that the factor $I_{E0}/(1 - \alpha_I)$ lies in the range $1.25\, I_{E0}$ to about I_{E0}. Assume at first that $I_{E0}/(1 - \alpha_I) \approx I_{E0}$. Then Eq. (1.8-3) becomes identical to the diode equation (1.1-1), I_{E0} being the reverse saturation current of the base-emitter junction. Hence all our previous discussion concerning the cut-in point V_γ of a diode applies in the present case as well, and we have the important result that the transistor has a base-to-emitter cut-in voltage $V_\gamma = 0.65$ V. More generally, Eq. (1.8-3) can be rewritten

$$I_E = I_{E0} \exp \frac{V_{BE} - V_T \ln (1 - \alpha_I)}{V_T} \tag{1.8-4}$$

A plot of Eq. (1.8-4) is identical to a plot of the diode equation except for a displacement in the positive voltage direction by an amount $V_T \ln (1 - \alpha_I)$. Even for α_I as large as 0.2 this shift amounts to only about -0.005, which is negligible in comparison with 0.65 V. Thus, we may assume $V_\gamma \approx 0.65$ V quite independently of α_I.

The reverse collector saturation current I_{CB0} The collector current when the emitter current is zero is designated by I_{CB0}. Two factors cooperate to make $|I_{CB0}|$ larger than $|I_{C0}|$. First, there exists a leakage current which flows not through the junction but around it and across the surfaces. The leakage current is proportional to the voltage across the junction. The second reason why $|I_{CB0}|$ exceeds $|I_{C0}|$ is that new carriers may be generated by collision in the junction transition region, leading to avalanche multiplication of current and eventual breakdown, as discussed in Sec. 1.4. However, even before breakdown is approached, this *multiplication* component of current may attain considerable proportions.

At 25°C, I_{CB0} for a medium-power silicon transistor is in the range of nanoamperes. A rough but useful rule for the temperature sensitivity of I_{CB0} is that it doubles for every 10°C increase in temperature.

In addition to the variability of reverse saturation current with temperature, there is also a wide variability of reverse current with particular samples of a given transistor type. For example, the specification sheet for a Texas Instruments type 2N337 grown diffused-silicon switching transistor indicates that this type number includes units with values of I_{CB0} extending over the extremely large range from 0.2 nA to 0.3 μA. Accordingly, any particular transistor may have an I_{CB0} which differs very considerably from the average characteristic for the type.

1.9   THE TRANSISTOR SWITCH IN SATURATION

We consider again the transistor switch of Fig. 1.6-1. Let the base-emitter junction be forward-biased and the base current I_B be set so that the corresponding collector current I_C causes a drop $R_L I_C$ across R_L which is substantially less than the supply voltage V_{CC}. In this case, while the emitter junction is forward-biased, the collector junction is reverse-biased. The principal mechanism which accounts for the continuity of current between emitter and collector is the following. Minority carriers are injected across the emitter junction. These carriers diffuse across the base and are swept out of the base as they are drawn across the collector junction. The transistor is in its *active* region. An increase in the forward bias of the base-emitter junction increases the injected minority current and hence increases the magnitude of the collector current. It should be pointed out that in this active region there is also a component of current which crosses the base and which is due to minority carriers injected into the base at the collector junction and collected at the emitter junction. Such is the case in spite of the fact that the collector junction is reverse-biased and the emitter junction forward-biased. Of course, in the active region this component of current is relatively so insignificant that it may be neglected.

With continuing increase in base-emitter-junction forward bias and consequent increase in the magnitude of the collector current, a point will eventually be reached where the base-collector-junction voltage becomes zero and finally reverses. The base-collector junction will thereby become forward-biased. Under these circumstances, the injection of minority carriers into the base at the collector junction will become comparable to the injection of such carriers at the emitter. The transistor is now in *saturation*. When the transistor is in saturation, an attempt to increase the magnitude of the collector current by further forward-biasing the base-emitter junction will meet with only limited success, for any increase in the magnitude of collector current will further forward-bias the base-collector junction. The increased injection at the collector junction results in an increased current component in the direction *opposite* to the increase potentially induced by the original increase in forward biasing the base-emitter junction.

There is a point of interest to be noted in connection with the application of the term *saturation* to a transistor. Ordinarily when the term is applied to a device it is intended to convey the idea that the device is furnishing as much current as it can. Note however, that, in the circuit of Fig. 1.6-1, when a base current I_B is supplied, the transistor is able to furnish a current $I_C = h_{FE} I_B$. If the current I_C is actually less than $h_{FE} I_B$, the transistor is said to be in saturation. However, such is the case because of the constraint imposed by the *circuit* and not by the transistor. Hence, strictly, we should speak of a *saturated circuit* and not a *saturated transistor*. Actually the transistor is saturated when it is in the active region since here an increase in collector voltage produces no increase in collector current. Nonetheless, in accord with common practice, we shall use the terminology that a transistor is saturated when both junctions are forward-biased.

We have already noted that for the transistor, as for the diode, cut-in occurs at about 0.65 V. This result is of course most particularly reasonable in integrated circuitry since a diode is generally the base-emitter junction of what would otherwise be a transistor. On the basis of such considerations we may reasonably estimate that when the base-emitter junction is carrying a "substantial" current, as is the case when the transistor is in saturation, the base-emitter voltage will be about 0.75 V, as was the case for the diode. This surmise is well borne out in practice for integrated circuitry as well as for discrete transistors with power-dissipation ratings of the order of 100 mW. In integrated circuitry the substantial current referred to is a current of the order of milliamperes. For, in saturation, when the collector-emitter voltage is very small, the transistor emitter current is approximately V_{CC}/R_L (see Fig. 1.6-1). With V_{CC} typically of the order of 5 V and R_L typically of the order of 5 kΩ, the collector current (which is nearly the same as the emitter current) is 1 mA.

In summary then, we shall consider that cut-in occurs at a base-emitter voltage $V_{BE} = 0.65$ V and saturation at $V_{BE} = 0.75$ V. When we must deal with a transistor which is presumed to be in its active region, we shall assume V_{BE} midway between cut-in and saturation, i.e., at $V_{BE} = 0.70$ V. Again, we must note that these voltage values are typical for transistors operating with currents of about 1 mA. In Chap. 7 in our study of emitter-coupled logic, and for reasons discussed there, we shall use a cut-in voltage of 0.7 V and a saturation voltage of 0.8 V.

In Fig. 1.9-1 we present typical common-emitter characteristics of an *npn* silicon transistor in the range of low collector-emitter voltages (< 1.0 V). The onset of saturation is noted to occur at about 0.35 V. For at this point we note that a further reduction in voltage is accompanied by a reduction in collector current. If we assume that at saturation $V_{BE} = 0.75$, then we calculate that the base-collector junction is forward-biased to the extent $V_{BC} = 0.75 - 0.35 = 0.4$ V.

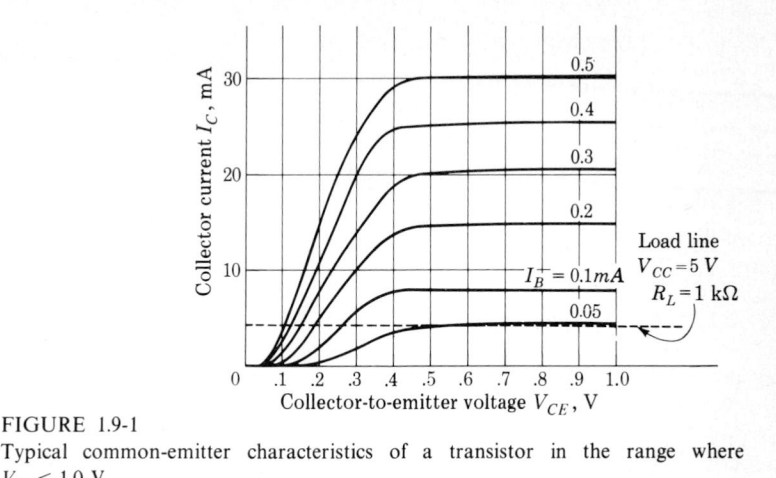

FIGURE 1.9-1

Typical common-emitter characteristics of a transistor in the range where $V_{CE} \leq 1.0$ V.

We had noted that the onset of saturation occurs when the base-collector junction is forward-biased to the extent that appreciable minority current is injected across this junction. On the basis of our earlier discussion we would expect that such a saturation would develop at about $V_{BC} = 0.65$ or higher. Instead we find that $V_{BC} = 0.4$ V. One reason for this discrepancy is the fact that a collector junction has a very much larger cross section than an emitter junction. Hence at the collector junction larger currents are injected at lower voltages than at the emitter junction.

Now, having in mind again the circuit of Fig. 1.6-1, we have drawn on the coordinate axes of Fig. 1.9-1 a load line corresponding to $V_{CC} = 5$ V and $R_L = 1$ kΩ. We drive the transistor in Fig. 1.6-1a into saturation in order to simulate the closing of the switch in Fig. 1.6-1b. We want the voltage across the transistor switch to be as small as feasible, to minimize power dissipation in the switch, and also to maximize the change in voltage across R_L at the closing of the switch. The change in voltage across R_L separates the logic levels, and the wider the separation the better from the point of view of reliability against unpredictable disturbances (noise). When the base current $I_B = 0$, the collector-emitter voltage is $V_{CE} = 5$ V. From Fig. 1.9-1 we note that at $I_B = 0.1$ mA the transistor is in saturation and that V_{CE} has dropped to $V_{CE} = 0.27$ V. A further increase in I_B does not effect an appreciable further reduction in V_{CE}. Still, we would be disinclined to leave the transistor at $I_B = 0.1$ mA because at this point the transistor is too close to the "edge" of saturation. Thus, while a modest increase in I_B would not *reduce* V_{CE} very much, a decrease in I_B might *increase* V_{CE} significantly. Thus, to make allowance for noise and for variations from transistor to transistor we might well drive the transistor more substantially into saturation by increasing I_B by a factor of 2 or even a factor of 3. In this case, as we note from Fig. 1.9-1, we should have $V_{CE}(\text{sat}) \approx 0.2$ V or even 0.1 V. In any event, it is apparent from Fig. 1.9-1 that, depending on the load line and on the enthusiasm with which we drive the transistor into saturation, we shall generally have $V_{CE}(\text{sat})$ in the range 0.1 to 0.3 V. Accordingly, as a useful approximation and generalization we shall often estimate that $V_{CE}(\text{sat}) = 0.2$ V; however, we shall have occasion to keep in mind that a transistor which is brought just to the edge of saturation may have V_{CE} as high as 0.4 V while a transistor driven violently into saturation may have $V_{CE}(\text{sat})$ as low as some tens of millivolts.

1.10   APPLICATION OF EBERS-MOLL EQUATIONS TO SATURATION

When a transistor is in its active region, the collector current and base currents are related by $I_C = h_{FE} I_B$. If a restraint is placed on the maximum possible collector current, then, as the base current is increased, a point will be reached where the transistor saturates. For example, in the transistor switch circuit of Fig. 1.6-1 the collector current at maximum is $I_C \approx V_{CC}/R_L$. The base current required just to bring the collector current to this level is $I_B \approx V_{CC}/h_{FE} R_L$. This base current, then, brings the transistor just to the edge of saturation, and further increases in base current drive the transistor deeper into saturation. Throughout the active region the base current I_B remains equal to I_C/h_{FE}. In saturation I_B is larger than I_C/h_{FE}, and the ratio of I_C/h_{FE} to I_B becomes smaller as the transistor is driven further into saturation. These considerations suggest that the parameter

$$\sigma \equiv \frac{I_C/h_{FE}}{I_B} = \frac{I_C}{h_{FE} I_B} \qquad (1.10\text{-}1)$$

may well serve as a measure of the extent to which the transistor has been driven into saturation. As long as $\sigma = 1$, the transistor is in its active region. The transistor is driven progressively farther into saturation as σ decreases below unity, and we may expect that the saturation collector-emitter voltage $V_{CE}(\text{sat})$ may be closely related to σ.

This expectation is borne out by the Ebers-Moll equations. Assuming that the transistor is in saturation, the junction voltages V_{BC} and V_{BE} in Eqs. (1.7-3) and (1.7-4) are large in comparison with V_T ($=25$ mV). We may then assume that I_E and I_C are exponential in behavior, and we can neglect the -1 terms. The collector-emitter saturation voltage is given by

$$V_{CE}(\text{sat}) = V_{CB} + V_{BE} = V_{BE} - V_{BC} \qquad (1.10\text{-}2)$$

Using Eqs. (1.7-6) and (1.7-7), neglecting the 1 in the argument of the logarithm, and also using Eqs. (1.7-5), (1.7-12), and (1.7-13), we then find that

$$V_{CE}(\text{sat}) = V_T \ln \frac{h_{FE}\, \sigma/h_{FC} + (1 + h_{FC})/h_{FC}}{1 - \sigma} \qquad (1.10\text{-}3)$$

The details of the algebraic manipulation leading to Eq. (1.10-3) are left as a student exercise (see Prob. 1.10-1).

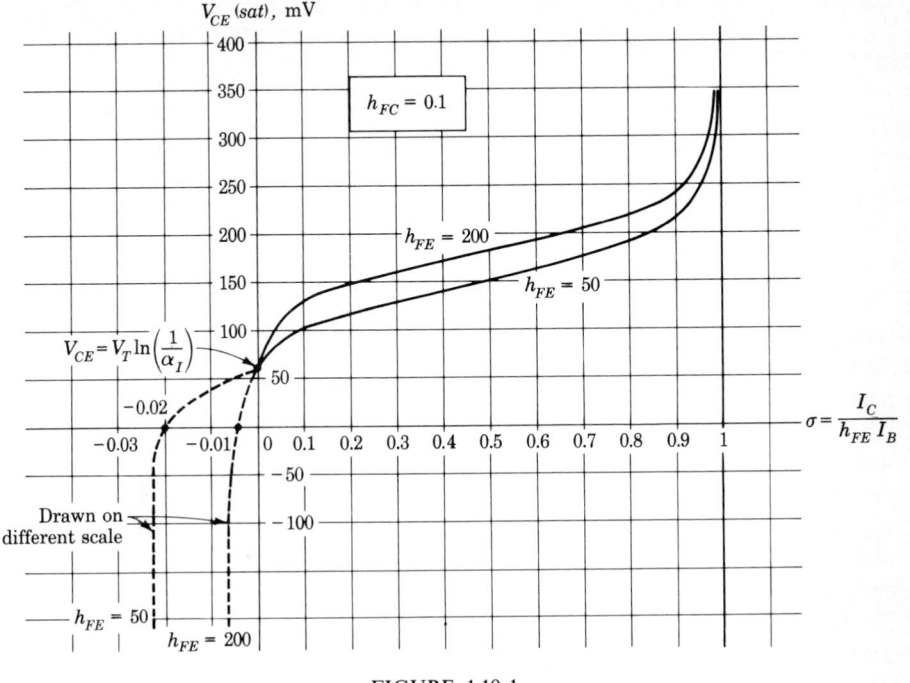

FIGURE 1.10-1
Saturation voltage $V_{CE}(\text{sat})$ as a function of base overdrive.

Equation (1.10-3) is plotted in Fig. 1.10-1 for $h_{FC} = 0.1$ and for a moderate and a high value of h_{FE}. This equation, which was developed from the Ebers-Moll equations, characterizes the transistor alone and is not related to a particular circuit configuration. As a matter of fact, as we shall see, no single simple switch configuration will provide access to the entire range of σ contemplated in Fig. 1.10-1. We have an interest in this matter of accessibility; for we drive the transistor into saturation precisely in order to minimize $V_{CE}(\text{sat})$, and we shall want to know which circuit configuration will give us access to the region of smallest $V_{CE}(\text{sat})$.

In Fig. 1.10-1, if we assume a fixed I_B, the distances marked along the abscissa are proportional to the collector current I_C. Referring to the basic transistor switch circuit of Fig. 1.6-1, we then observe that the plot of Fig. 1.10-1 has the form of the volt-ampere characteristic (Fig. 1.9-1 turned on its side) seen looking into the transistor switch from collector to ground (the emitter). The collector current in the configuration must be positive. Hence the only part of the plot accessible is the part in the first quadrant. Suppose that for fixed I_B ($I_B > 0$ to keep the transistor ON) we increase I_C, say by reducing R_L. Then $V_{CE}(\text{sat})$ will increase. Eventually, as σ approaches unity, the transistor will come out of saturation into the active region. In the active region, to a fair approximation, the collector current is independent of the collector-emitter

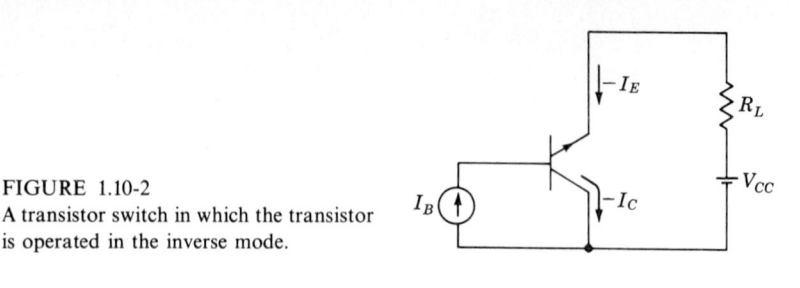

FIGURE 1.10-2

A transistor switch in which the transistor is operated in the inverse mode.

voltage V_{CE}. This independence is evidenced in the plot by the fact that V_{CE} rises vertically at $\sigma = 1$. That is, at $\sigma = 1$ we are free to make V_{CE} anything we please (say by changing V_{CC}) without affecting I_C.

If, on the other hand, we decrease I_C, say by increasing R_L, then V_{CE} will decrease and the minimum accessible V_{CE} occurs at $I_C = 0$. As a matter of practice, I_C will ordinarily be nominally limited to $I_C = V_{CC}/R_L$, and the minimum V_{CE} will be approached by increasing I_B. In such a case we shall never be able to reach $\sigma = 0$ precisely. We shall, however, be able to approach close enough for V_{CE} corresponding to $\sigma = 0$ to be a useful number. We find from Eq. (1.10-3) that with $\sigma = 0$

$$V_{CE}(\text{sat}) = V_T \ln \frac{1 + h_{FC}}{h_{FC}} = V_T \ln \frac{1}{\alpha_I} \qquad (1.10\text{-}4)$$

Consider next the configuration of Fig. 1.10-2. Here the transistor has been turned around so that the junction originally intended as the collector junction is being actually used as the emitter junction and the intended emitter junction is being used as the collector junction. The directions of the actual junction currents are also reversed; that is, I_E and I_C are both negative. The present configuration is described as one in which the transistor is being used in an *inverse mode*. Correspondingly, in the basic configuration of Fig. 1.6-1 the transistor operation is in the *normal mode*.

In the normal mode the asymptote between active and saturation regions occurs when I_C approaches $h_{FE} I_B$, that is, at $\sigma = 1$. In the inverse mode, in comparison, the roles of h_{FE} and h_{FC} are interchanged, and the asymptote occurs at

$$-I_E = h_{FC} I_B \qquad (1.10\text{-}5)$$

We also have, from Fig. 1.10-2, that

$$-I_C = -I_E + I_B \qquad (1.10\text{-}6)$$

Combining Eqs. (1.10-5) and (1.10-6), we find that

$$I_C = -I_B(1 + h_{FC}) \qquad (1.10\text{-}7)$$

Substituting I_C as given in Eq. (1.10-7) into the equation for σ as defined by Eq. (1.10-1), we find that the asymptote between the active and saturation regions occurs at

$$\sigma = -\frac{1 + h_{FC}}{h_{FE}} \tag{1.10-8}$$

For the values of h_{FC} and h_{FE} indicated in Fig. 1.10-1 the values of σ are $\sigma = -0.022\,(h_{FE} = 50)$ and $\sigma = -0.0055\,(h_{FE} = 200)$. At these values, as shown, the plots become vertical.

   In the inverse mode, the voltage across the transistor will be a minimum when $I_E = 0$.   In this case $I_B = -I_C$ and $\sigma = I_C/h_{FE}\,I_B = -1/h_{FE}$. Substituting this value of σ into Eq. (1.10-3), we find that the minimum $V_{CE}(\text{sat})$ is

$$V_{CE}(\text{sat}) = -V_T \ln \frac{1 + h_{FE}}{h_{FE}} = -V_T \ln \frac{1}{\alpha_N} \tag{1.10-9}$$

In this mode, all the accessible range lies in the third quadrant of Fig. 1.10-1. However, not all of the range in the quadrant is accessible. The voltage V_{CE} cannot approach closer to the abcissa than is indicated in Eq. (1.10-9). Note further that the form of the plot in the third quadrant is not the form of the volt-ampere characteristic between emitter and collector (ground). Such is the case because the abcissa is marked off in proportion to the collector and not in proportion to the emitter current.

   Finally, we turn to the configuration of Fig. 1.10-3a. Here we have modified the basic switch circuit of Fig. 1.6-1 by returning the base-drive current source to the collector rather than to the emitter. With increasing I_B the currents I_C and I_E will increase, and the transistor will be driven to saturation. With further increase in I_B a point will be reached where I_C begins, to fall, eventually reaches $I_C = 0$, and thereafter reverses.   As can be verified, I_C will be $I_C = 0$ when

$$I_B = \frac{V_{CC} - V_T \ln\,(1/\alpha_I)}{R_L} \tag{1.10-10}$$

This configuration, then, provides access in Fig. 1.10-1 not only to the range in the first quadrant but also to the range in the second and third quadrants not available to the other two configurations considered.

   In particular, it is possible to arrange that $V_{CE}(\text{sat}) = 0$.   From Eq. (1.10-3), when $V_{CE}(\text{sat}) = 0$,

$$\sigma = \frac{-1}{h_{FE} + h_{FC}} \tag{1.10-11}$$

Unfortunately the base current required to achieve $V_{CE}(\text{sat}) = 0$ turns out to be unreasonably large. As can be verified (Prob. 1.10-4), to make $V_{CE}(\text{sat}) = 0$ requires a base current

$$I_B = \frac{V_{CC}}{R_L}\left(1 + \frac{h_{FE}}{h_{FC}}\right) \tag{1.10-12}$$

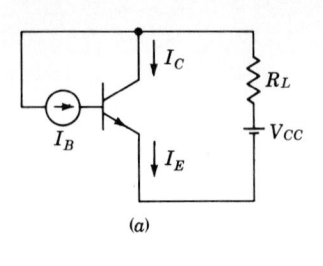

(a)

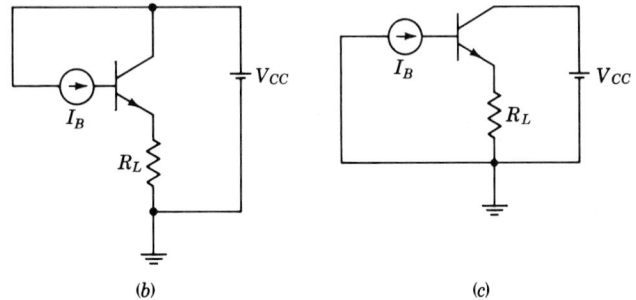

(b)   (c)

FIGURE 1.10-3

(a) Transistor switch in the normal mode with base current source returned to the collector. (b) Circuit in (a) with supply V_{CC} and load R_L interchanged and with a ground connection. (c) Circuit in (b) with the current source returned to the other side of the supply.

With $V_{CC} = 5$ V, $R_L = 1$ kΩ, $h_{FE} = 50$, and $h_{FC} = 0.1$, we find $I_B = 500 V_{CC}/R_L = 2.5$ A. On the other hand, suppose we limit I_B to, say, $I_B = 5 V_{CC}/R_L = 5 I_E$. Then $I_C = I_E - I_B = -4 I_E$, and $\sigma = I_C/h_{FE} I_B = -\frac{4}{5}(1/h_{FE}) = -0.016$ for $h_{FE} = 50$. From Fig. 1.10-1, the corresponding transistor voltage is $V_{CE}(\text{sat}) \approx 25$ mV.

In connection with the last-considered configuration it is to be noted that the circuit is precisely that of an emitter follower. That such is the case is to be seen in Fig. 1.10-3b. Here we have simply interchanged the order of the series combination of V_{CC} and R_L shown in Fig. 1.10-3a and have added a ground connection. In Fig. 1.10-3c we have moved one end of the current source I_B from one side to the other of a fixed voltage source, a change which makes absolutely no difference with respect to the transistor currents and junction voltages.

Comparison of configurations For the common-emitter normal-mode switch we find from Eq. (1.10-4) for $h_{FC} = 0.1$ that $V_{CE}(\text{sat}) = 60$ mV. From Eq. (1.10-9) for the switch with a transistor in the inverse mode we find for $h_{FE} = 50$ that $V_{CE}(\text{sat}) = -0.5$ mV. For the emitter-follower switch we found that, at least in principle, we may have $V_{CE}(\text{sat}) = 0$. If the transistor is operated in the inverse mode as a collector follower, we shall find, interchanging h_{FE} and h_{FC} in Eq. (1.10-12), that $I_B \approx V_{CC}/R_L$ for $V_{CE}(\text{sat}) = 0$ and that hence $V_{CE}(\text{sat}) = 0$ would be more reasonably attainable. Altogether, it would appear that where

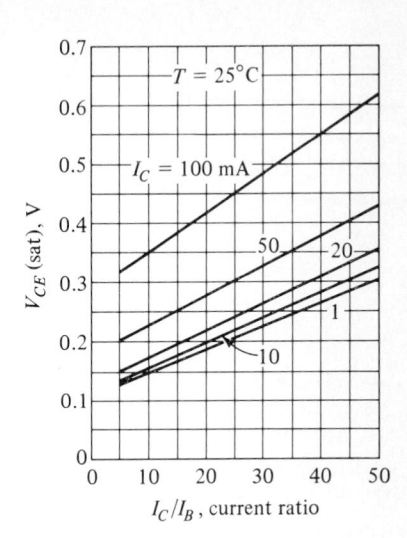

FIGURE 1.10-4
The collector-emitter saturation voltage
V_{CE}(sat) as a function of the ratio I_C/I_B
with I_C as a parameter for a planar
epitaxial passivated silicon transistor.

a premium is to be placed on keeping to a minimum the voltage across a closed transistor switch (as in analog switching circuits, Chap. 13), the advantage lies with using transistors in the inverse mode.

Unfortunately when drive is applied to transistors to turn them ON and OFF, a transistor operating in the inverse mode responds appreciably more slowly than a transistor in the normal mode. Hence the inverse mode finds application only in situations where there is no urgency about speed. In the normal mode the emitter follower has an advantage in the matter of lower voltage. However, the drive current required with an emitter follower may become inconveniently large. Still, as we shall see (Chap. 14), there are occasions when the emitter follower is the configuration of choice.

Manufacturer's specifications at saturation The idea that the saturation collector-to-emitter voltage is largely determined by the extent of the base overdrive is further borne out by examination of plots of V_{CE}(sat) supplied by manufacturers. We generally find that in these plots V_{CE}(sat) is plotted either as a function of $\sigma = I_C/h_{FE} I_B$ or simply as a function of I_C/I_B. A typical plot is shown in Fig. 1.10-4. The transistor shown has $h_{FE} \approx 50$. We may then note that these plots are again consistent with our assumptions that V_{CE}(sat) ≈ 0.2 V. For at $I_C/I_B = 50$ we would just be at the edge of saturation. To be well into saturation we might double I_B, giving $I_C/I_B = 25$. If we assume that I_C is between 1 and 20 mA, we find that $V_{CE} \approx 0.2$ V. We note in Fig. 1.10-4 that V_{CE}(sat) depends also on I_C. This situation results from the ohmic drop across the collector and the emitter ohmic resistance. We have neglected these drops at the small currents (several milliamperes) in which we are interested.

As a further example of the usefulness of the parameter h_{FE}, consider the simple circuit of Fig. 1.10-5. Here we are interested in determining a value of

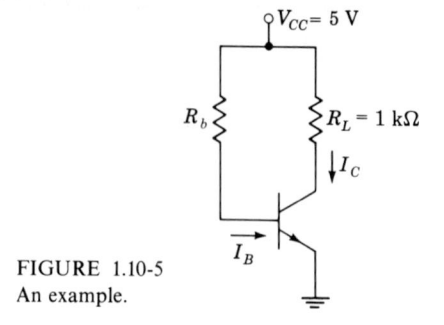

FIGURE 1.10-5
An example.

R_b appropriate to establish the transistor comfortably in saturation. For this purpose we need only to know h_{FE}. For, assuming $V_{CE}(\text{sat}) = 0.2$ V, we have that $I_C = (5 - 0.2)/1$ k$\Omega = 4.8$ mA. Let us take $h_{FE} = 50$. Then to bring the transistor to the edge of saturation we need a base current $I_B = 4.8/50 = 0.096$ mA. Let us double this base current in accordance with the reasons discussed above, so that we take $I_B = 0.192$ mA. Using $V_{BE} = 0.75$, we now find

$$R_b = \frac{5 - 0.75}{0.192 \times 10^{-3}} = 22 \text{ k}\Omega$$

As a final point in connection with the saturation collector-emitter voltage we note that the temperature sensitivity of $V_{CE}(\text{sat})$ is quite small in comparison with the sensitivity of an individual forward-biased junction. This result is, of course, to be expected since $V_{CE}(\text{sat})$ is the *difference* in voltage between two such junctions. In future discussions we shall ignore the temperature sensitivity of $V_{CE}(\text{sat})$.

1.11 THE FIELD-EFFECT TRANSISTOR

Two types of field-effect transistors are available, the *junction field-effect transistor* (JFET) and the *metal-oxide-semiconductor field-effect transistor* (MOSFET). We shall have some interest in JFETs in connection with their use in analog switching circuits (Chap. 13). MOSFETs and, more particularly, *complementary-symmetry* MOSFETs, referred to as *CMOS* devices, find extensive applications in integrated-circuit logic-gate systems.

In principle, a JFET is constructed as in Fig. 1.11-1. Ohmic contacts are affixed at the end of a bar of silicon which has been doped with n-type impurities (p-type doping is also possible). A voltage source V_{DS} connected across the bar causes current to flow. The silicon, being of n type, enables electron carriers to flow across the bar, which has relatively low resistance. The terminal at the end of the bar at which these carriers (electrons) enter the bar is called the *source*. The terminal at which these carriers leave the bar is called the *drain*. Between source and drain, p-type regions, called the *gate*, have been

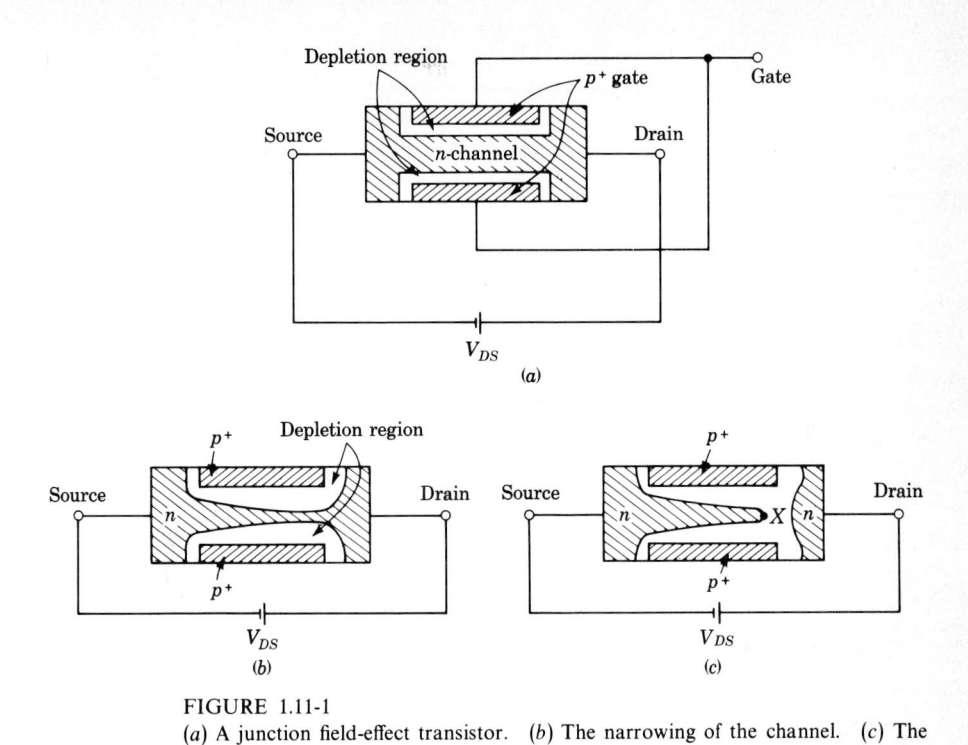

FIGURE 1.11-1
(*a*) A junction field-effect transistor. (*b*) The narrowing of the channel. (*c*) The FET above pinch-off.

diffused as indicated in the figure. At the junction between the *p*- and *n*-type semiconductor there develops a *depletion region*, which is devoid of free current carriers. Nonetheless, as seen in Fig. 1.11-1*a*, between depletion regions there is a *channel* of *n*-type silicon connecting source to drain which serves as a conducting path.

The depletion region can be made to spread further into the channel by further reverse biasing the *pn* junction between gate and channel. With sufficient reverse bias the depletion regions will spread completely across the transistor and the channel will thereby have been *pinched off*.

If the channel is not pinched-off, then in the presence of a voltage between drain and source the depletion regions will have the appearance indicated in Fig. 1.11-1*b*. Near the drain the depletion regions will approach closer to one another than near the source. The reason for the difference is that the drain is at the more positive voltage, and hence, near the drain, the *pn* junction, formed by the gate and the channel, is more back-biased than near the source.

The drain-source voltage V_{DS}, which causes the channel to pinch off, is called the pinch-off voltage V_P. For values of drain-source voltage greater than the pinch-off voltage, the depletion region expands as shown in Fig. 1.11-1*c*. The potential between point X and the source remains at the pinch-off voltage V_P as the drain voltage increases. Hence the voltage across the depletion region

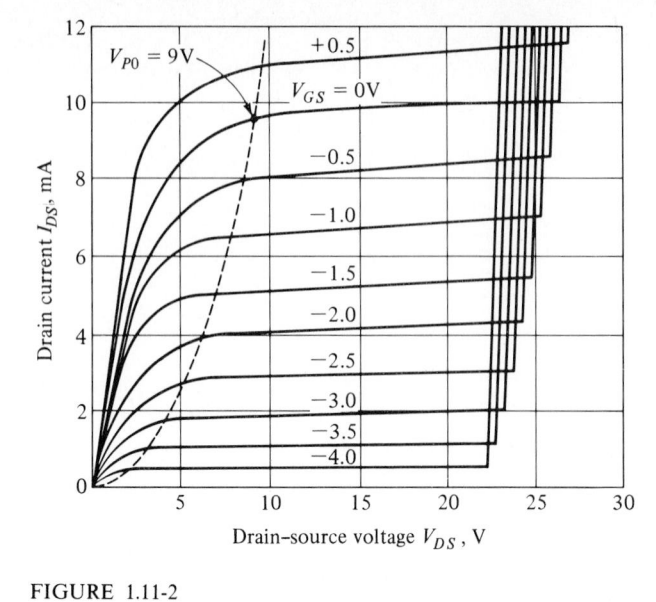

FIGURE 1.11-2
Typical common-source drain characteristics of an n-channel JFET.

increases as V_{DS} increases. The electric field produced by the voltage difference $V_{DS} - V_P$ draws the electrons from the channel through the depletion region to the drain. However, since V_P, the channel voltage, is maintained almost constant, the drain current also remains approximately constant at its pinch-off value.

A typical drain-source volt-ampere characteristic of an n-channel JFET is shown in Fig. 1.11-2. Let us consider an individual plot for some fixed gate-to-source voltage, say $V_{GS} = 0$ V. When the drain-to-source voltage is low, a uniform channel is open between source and drain and the JFET behaves much like a resistor. The volt-ampere characteristic between drain and source is "linear." With increasing drain-to-source voltage the depletion region spreads and constricts the channel. As a result the channel resistance increases, and the rate of increase of drain current with drain-to-source voltage becomes smaller. Finally with further increase in V_{DS} the channel pinches off, and the *drain current saturates;* i.e., the drain current increases hardly at all with further increase in drain-to-source voltage. The drain-to-source voltage (which is not very sharply defined) at which saturation begins when $V_{GS} = 0$ V is called the *pinch-off voltage* V_{PO}. The dashed line in the plot of Fig. 1.11-2 is the locus of V_P for various gate-to-source voltages. Typically, we find that

$$V_P \approx V_{PO} + V_{GS} \qquad (1.11\text{-}1)$$

Note that the drain-source voltage at pinch-off V_P becomes equal to zero when $V_{GS} = -V_{PO}$. At this gate-to-source voltage the FET is cut off. Referring to

Fig. 1.11-2, we see that cutoff occurs when $V_{GS} = -V_{PO} \approx -9$ V. However, at gate-source voltages far less negative than this value the FET can be considered cut off for all intents and purposes.

We observe in Fig. 1.11-2 that with continuing increases in V_{DS} a point is eventually reached where avalanche breakdown occurs between drain and gate and the drain current increases precipitously. Breakdown occurs at a fixed junction voltage, and when the gate supplies a part of this voltage, breakdown occurs at a lower value of V_{DS}. For example, we see in the figure that at breakdown $V_{DS} = 26$ V for $V_{GS} = 0$ whereas $V_{DS} = 22$ V for $V_{GS} = 4$ V. A useful relationship between the breakdown voltage BV_{DS} and the gate-source voltage is then

$$BV_{DS} = BV_{DS0} + V_{GS} \qquad (1.11\text{-}2)$$

where BV_{DS0} is the breakdown voltage when $V_{GS} = 0$ V.

The JFET operates with the gate-channel *pn* junction reverse-biased. As a consequence, the gate current is of the order of nanoamperes, being typically in the range of reverse-bias currents for silicon diodes. A slight gate bias in the forward direction may be allowed since no appreciable gate current will flow until the cut-in voltage (0.65 V) is reached. Thus we note in Fig. 1.11-2 a plot given for $V_{GS} = 0.5$ V.

One of the useful features of a JFET is that, over an appropriately restricted voltage range, it can be used as a linear resistor whose resistance value is controllable by the gate voltage. That such is the case is immediately apparent in examining the plots in Fig. 1.11-2. For we observe that in the neighborhood of small V_{DS}, the $V_{DS} - I_{DS}$ characteristic is a "straight" line of slope determined by the gate voltage V_{GS}. Further, we note that the FET is, in principle, an entirely symmetrical device in which the roles of source and drain may be interchanged. Hence the FET, as a voltage-controllable resistor, allows for drain-source voltages of either polarity. The common-drain characteristic of a typical JFET in the neighborhood of the origin is shown in Fig. 1.11-3. From this figure it appears that if linearity is required, the range of V_{DS} must be restricted to voltages less than ± 0.2 V. Note also how precisely linear are the characteristics over such a small range. Note especially that the characteristics pass exactly through the origin. This feature is of importance when the FET is used in analog gate circuits (see Chap. 13). Observe that in the third quadrant the polarity of the drain voltage is such that with decreasing V_{DS} the gate-channel junction would eventually become forward-biased. We have, however, restricted the range of the plots to avoid that situation.

The field-effect transistor is so called because the mechanism of current control is the effect of the *field* produced by the drain-source and gate-source voltages. In a field-effect transistor current is carried by either electrons (*n* channel) or holes (*p* channel) but not both. Further, these are *majority* carriers, i.e., the carriers in the *n* channel are electrons. Because of this single polarity of carriers, field-effect transistors are called *unipolar* transistors, in distinction to the two-junction transistors. In the two-junction transistors, carriers of both

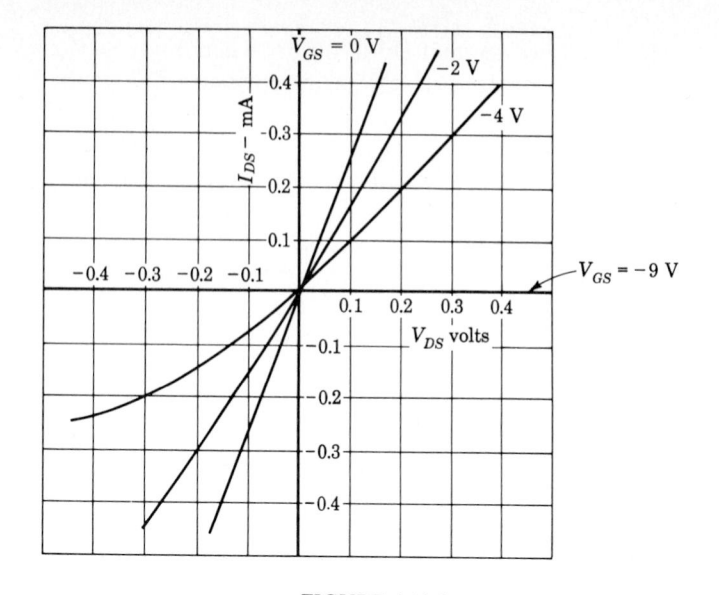

FIGURE 1.11-3
Low-level drain characteristics of a JFET.

polarities are involved, and hence such transistors are characterized as *bipolar*. Further in such bipolar transistors the current crossing the base region (which might be considered to be the "channel") is a current of *minority* carriers. Circuit symbols for *n*-channel and *p*-channel JFET transistors are given in Fig. 1.11-4. Note that in each case the arrow points from the *p* region to the *n* region. It therefore symbolizes the gate-channel "diode" which should be maintained with a reverse bias.

When the reverse bias on the gates of the junction FET is zero, the region between the gates has available throughout, majority carriers (electrons in an *n* channel) which carry current under the influence of an applied field between drain and source. When a reverse bias is applied to the gates, the supply of carriers is *depleted* progressively as the bias is increased. The range of operation of the FET is therefore a range over which the available carriers are depleted

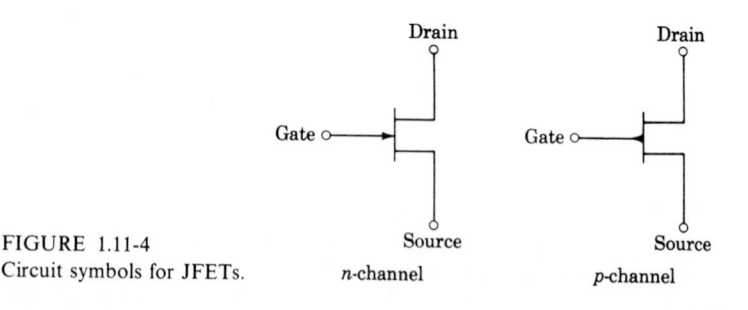

FIGURE 1.11-4
Circuit symbols for JFETs.

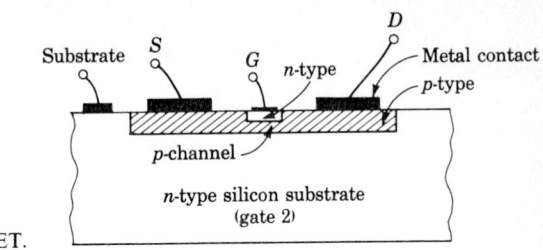

FIGURE 1.11-5
The structure of a planar JFET.

to a greater or lesser extent. For this reason, the JFET is described as a device which operates in the *depletion mode*.

While the geometry suggested in Fig. 1.11-1 is convenient for a discussion of basic operating principles, JFETs are generally manufactured by planar technology in which all diffusions and contacts are made from one side of a silicon slice referred to as the *substrate*. A simplified cross-sectional view of such a planar (*p* channel) FET is shown in Fig. 1.11-5. Construction starts from an *n*-type substrate which is often common to many FETs, resistors, etc., if an integrated circuit is planned. A *p*-type diffusion, as shown, becomes the channel, and finally a heavily doped *n*-type diffusion into the channel becomes the gate. Metal contacts are used to obtain connections for external leads. The substrate is usually connected to the most positive point in the circuit.

1.12 THE METAL-OXIDE-SEMICONDUCTOR FET (MOSFET)

The junction FET of the previous section finds extensive application in linear circuits (amplifiers, etc.) and in a number of nonlinear circuits used for processing analog signals (see Chap. 13). In digital logic circuitry, on the other hand, the MOSFET, now to be described, is the field-effect transistor of choice. In comparison with the bipolar transistor, one immediate advantage of the MOSFET is that it requires a silicon chip area which is only about 15 percent as large as that required for a bipolar transistor. This economy of chip area makes the MOSFET especially attractive in medium- and large-scale integrated circuitry. At the present writing, however, MOSFETs are not yet capable of operating at the speeds of bipolar transistors.

The basic structure of the *n*-channel MOSFET is shown in Fig. 1.12-1*a*. Into a *p*-type substrate are diffused two *n*-type regions. One such region is to become the source, the other the drain. A voltage difference applied between drain and source will not cause a current to flow in the device. For any such current must cross two junctions, one between source and substrate and the other between drain and substrate. The substrate potential is made negative, thereby reverse-biasing the *pn* diode junctions and keeping the "transistor" formed by the source-substrate-drain at cutoff.

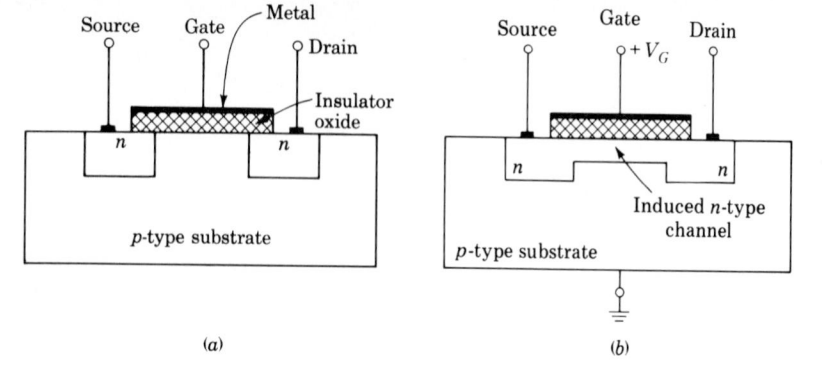

FIGURE 1.12-1

(*a*) The structure of a metal-oxide-semiconductor field effect transistor (MOSFET).
(*b*) The procedure by which a channel is generated.

A metal gate electrode, as appears in the figure, is located above the substrate region which separates source from drain and is insulated from the rest of the MOS structure. The insulator is composed, at least in part, of silicon dioxide. The whole structure is hence appropriately referred to as a *metal-oxide-semiconductor* (MOS) field-effect transistor. A second appropriate name frequently used is *insulated-gate* FET.

If a positive voltage is applied to the gate, the field from the gate will draw electrons into the substrate region surrounding the gate. This is illustrated in Fig. 1.12-1*b*. As a result, a section of the substrate will change from *p*-type to *n*-type silicon, and, as indicated, an *n*-type channel will form. This channel forms a continuity of *n*-type material from drain to source so that current can flow across the transistor. In the MOSFET the application of a gate voltage enhances the availability of electron carriers to carry current from drain to source. Hence, the present type of FET is referred to as operating in an *enhancement* mode, in distinction to the JFET, which operates in a *depletion* mode.

With the gate voltage present, an increase in the drain-source voltage will result in an increase in drain current, producing a resistor-type operation, as before. This is shown in Fig. 1.12-2. As the drain-source voltage continues to increase, the electric field produced under the gate varies, being largest at the source and smallest at the drain since the gate-source voltage is V_{GS} while the gate-drain voltage, $V_{GS} - V_{DS}$, is smaller. Pinch-off occurs when the drain-source voltage is sufficiently large to reduce the field, near the drain, to zero. Above pinch-off the drain current remains relatively constant.

The value of V_{DS} needed for pinch-off is

$$V_{DS} = V_{GS} - V_T \tag{1.12-1}$$

where V_T is a *threshold voltage* having values between 2 and 4 V depending on the manufacturing process. The value of V_T is given by the manufacturer.

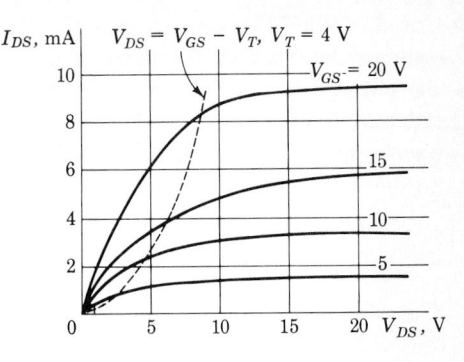

FIGURE 1.12-2
Volt-ampere characteristics of a MOSFET. Drain current as a function of drain-source voltage for various gate-source voltages.

A rather distinctive characteristic of the MOSFET is that the gate current, for either polarity of gate voltage, is extremely small. This feature is, of course, a result of the insulation of the gate. Gate currents are normally of the order of picoamperes.

It is to be noted that it is possible to build a MOSFET which works in the *depletion* mode. Such an FET can be built by arranging an additional diffusion of n-type impurities into the surface of the substrate between the two n-type regions (see Fig. 1.12-1a). With such an n-channel established by diffusion a current would flow between drain and source even in the absence of a positive voltage on the gate. The application of a negative voltage to the gate would control the drain current. The drain current would decrease with increasingly negative gate voltage as the diffused n channel is depleted. On the other hand, the drain current would increase with increasing positive gate voltage as the n channel is enhanced. Thus such a transistor as we here describe would be able to operate in both modes.

The value of V_{DS} needed for pinch-off in a depletion mode FET is given by Eq. (1.12-1), where V_T is negative. A typical value for V_T is $V_T = -6$ V.

A p-channel MOSFET, called a PMOS, can be constructed by reversing the impurity types in the structure of Fig. 1.12-1. At the present writing a PMOS is somewhat easier to fabricate than an NMOS. However the *mobility* of electrons which are the carriers in an NMOS is about 3 times greater than the mobility of hole carriers in a PMOS. Hence an NMOS is faster than a PMOS, and, as a consequence, much effort is being expended to improve NMOS fabrication technology.

For a PMOS we can write an equation corresponding to Eq. (1.12-1), which applies to NMOS. For a PMOS, however, it is more convenient to refer to the source-to-drain and source-to-gate voltages. Hence we write

$$V_{SD} = V_{SG} - V_T \tag{1.12-2}$$

For an enhancement-mode PMOS, V_T is again a positive number; it is a negative number for a depletion-mode PMOS.

In digital circuitry the enhancement-mode transistor is generally preferred since it is a great convenience that the transistor be cut off at zero gate voltage.

Cutoff   When the gate voltage is decreased to a value less than the threshold voltage, the MOSFET is cut off.   In this region the channel is depleted of all the induced free charges, and the device is as shown in Fig. 1.12-1a, where we see that for a positive drain-source voltage an *npn* transistor exists which is operating at cutoff.   Hence, drain current flows which is equivalent to I_{CBO} in an *npn* transistor.   Typical currents are of the order of picoamperes.

A PMOS FET is cut off when the source-to-gate voltage is reduced to a value which is less than the threshold voltage.

The effect of substrate voltage   The conductance of the channel between drain and source will be affected by a bias applied to the substrate.   The substrate thus may serve as a second gate and is occasionally referred to as the *back gate*. Consider that an *n* channel has been formed (Fig. 1.12-1) as a result of the application of a positive voltage to the insulated gate.   Then between this *n* channel and the rest of the *p*-type substrate a *pn* junction is formed.   At this junction there is a depletion region which is devoid of carriers.   And, as in the JFET, the width of the depletion region (and consequently the width of the con-ducting channel) will be influenced by the bias voltage on the back gate. Increasing the reverse bias between the substrate and source decreases the drain current for a fixed voltage on the insulated gate.   As a result pinch-off is affected, and the threshold voltage is therefore made larger by such reverse bias. An approximate rule concerning the threshold voltage is given by the formula

$$\Delta V_T \approx C\sqrt{V_{SB}} \qquad (1.12\text{-}3)$$

where V_{SB} = source-to-substrate voltage

$\quad \Delta V_T$ = corresponding change in threshold voltage

The parameter C, which depends on the doping of the substrate, generally lies in the range 0.5 to 2.0.

Circuit symbols   Circuit symbols for an *n*-channel MOSFET are shown in Fig. 1.12-3.   In Fig. 1.12-3a is shown an *n*-channel enhancement device.   The broken-line connection between source and drain serves to remind us that at zero gate voltage there is no conduction between source and drain.   The source is the terminal located at the right angled bend in the L-shaped gate symbol.   In Fig. 1.12-3b is shown a depletion *n*-channel device.   An alternative symbol for the MOSFET is shown in Fig. 1.12-3c.   Here the substrate has been omitted for simplicity.   This symbol is not intended to give any information concerning enhancement or depletion mode operation.

Observe that the arrow on the substrate lead reminds us of the composition of the substrate.   The diode formed between substrate and source-channel-drain is reflected in the direction of the diode arrow.   PMOS device symbols are identical except that the arrow is reversed.   In integrated-circuit structures all MOS devices are built into a common substrate, which is reverse-biased with respect to the source-channel-drain.   This situation being understood, in circuit diagrams of integrated circuits the substrate connection is often not shown.

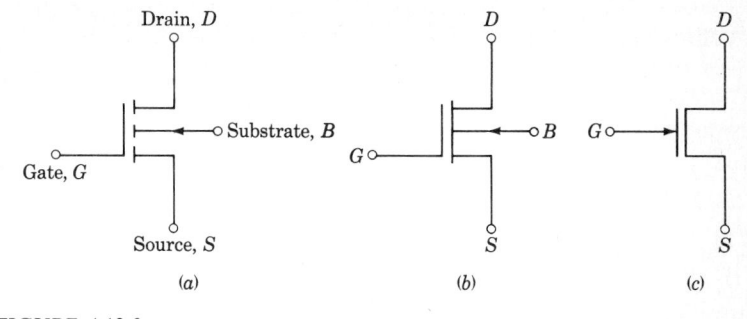

FIGURE 1.12-3
Circuit symbols for the *n*-channel MOSFET. In *p*-channel devices the arrow is reversed. (*a*) An enhancement type; (*b*) a depletion type; (*c*) an alternative symbol which does not identify source and drain and does not show the substrate.

1.13 THE MOS SWITCH

As with the transistor, we are interested in using the MOSFET as a switch. For this purpose we might be inclined, as with the transistor, to use a resistor as the load. In the MOSFET case, however, it turns out to be of great advantage instead to use a second MOSFET as the load. The basic MOSFET switch then appears as in Fig. 1.13-1. The arrangement in Fig. 1.13-1*a* has some slight advantage, inasmuch as the substrate can be connected directly to the source and would normally be the case with discrete devices. Integrated circuits would be restricted to the arrangement in Fig. 1.13-1*b*, since the substrate is common to both $T1$ and $T2$. The lower transistor is the "drive" transistor, and the upper transistor is the "load" transistor. For the sake of generality the drain supply voltage V_{DD} and the load gate bias V_{GG} have been separately indicated. There are occasions when some advantages result from such separation. More generally we find that $V_{GG} = V_{DD}$.

In bipolar-transistor switch circuits we noted the use of load resistors of the order of some thousands of ohms. With such resistors, we found that when a transistor was turned OFF and ON (saturation), the switch output voltage varied between two limits, V_{CC} and $V_{CE}(\text{sat}) \approx 0.2$ V, which were appropriate to turn ON and OFF another transistor switch. We hope to arrange a similar situation with the MOSFET switches. With MOSFET switches, however, it turns out that load resistors are required with resistance of many tens of thousands of ohms, even up to well over 100 kΩ. Conventional diffused resistors as fabricated in integrated circuits occupy areas on the silicon chip which are roughly proportional to the resistance. It is thus not surprising that a MOSFET "load" can be fabricated in an area which is very much smaller than the area required for a diffused resistor. By way of comparison we note that even a 20-kΩ resistor might occupy an area of 20 square mils while a complete MOSFET switch as in Fig. 1.13-1, in which the load MOSFET is roughly equivalent to a 100-kΩ resistor, may occupy an area of only 2 square mils. An incidental advantage

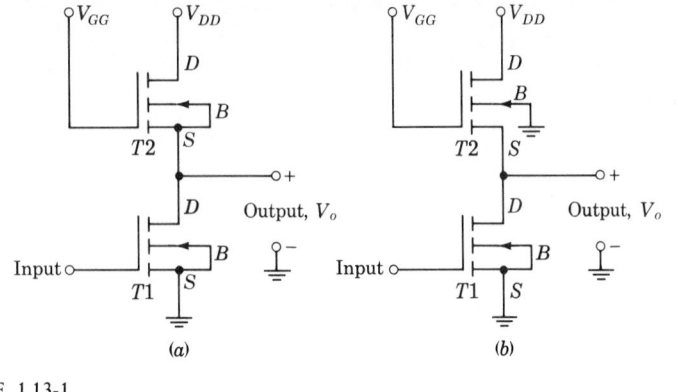

FIGURE 1.13-1
The basic MOSFET switch: (*a*) load substrate returned to load source; (*b*) load substrate returned to ground.

of this reduced area is a corresponding reduction in the distributed stray capacitance from "load" resistor to the substrate.

When the drive transistor in Fig. 1.13-1 is ON, we should like the output voltage to be as close to ground as feasible. For this purpose, the drive transistor should be capable of furnishing a large amount of current while the load transistor should look like a large resistor; i.e., it should be designed to carry relatively small current. For this reason, the drive and load devices are not identical. On the contrary, the drive transistor should have relatively high conductance while the load should have low conductance. Thus we find that in a typical MOSFET switch the driver device can be designed to have a wide and short channel, the width being typically 1 mil while the length is only 0.2 mil, the width-to-length ratio being 5. On the other hand, the channel in the load device may be of just about reversed dimensions, the length being 1 mil and the width 0.2 mil. If, then, the driver device were operating under bias conditions such that the channel resistance from drain to source was 10 kΩ, under the same bias conditions the load-device channel would have a resistance which is $5^2 = 25$ times as large, that is, 250 kΩ. We find 'further, typically, that with a supply voltage V_{DD} (= V_{GG}) of the order of $+15$ V, the voltage at the switch output is about $+0.5$ V when the drive transistor is ON. Assuming a threshold voltage of about 4 V, this 0.5 V output is low enough to assure that when applied to a succeeding switch, that switch will be kept OFF.

With $V_{GG} = V_{DD}$ and if the driver is turned OFF, the output cannot rise to the supply voltage. For when the driver gate voltage is below threshold, there is still some small drain current (the reverse saturation current, similar to I_{CBO}) in the driver. This current must continue to flow through the load transistor. For the load to be able to carry this current, the load transistor itself must be slightly ON. As a consequence, the source of the load transistor (which is the output voltage of the switch) must be negative with respect to the gate by at least the magnitude of the threshold voltage. Thus if in Fig. 1.13-1a $V_{DD} = V_{GG} = +15$ V, and if $V_T = 4$ V, then the output may rise at most to $15 - 4 = 11$ V. Such a voltage is certainly more than adequate when applied to the input of a succeeding switch to turn that switch ON. If it should be necessary for the output to be allowed to rise to V_{DD}, it is only required that V_{GG} be increased in magnitude to $V_{GG} = V_{DD} + V_T = 15 + 4 = +19$ V.

In integrated circuitry the substrate of the load will be at 0 V, as in Fig. 1.13-1b. In this case the substrate is reverse-biased, and the magnitude of the threshold voltage is increased. As a consequence, when the driver transistor is not conducting, the extent to which the output V_o falls short of the gate voltage V_{GG} will be even more pronounced than is the case in the circuit of Fig. 1.13-1a. By way of example, consider again that $V_{GG} = V_{DD} = +15$ V and that the threshold voltage (in the absence of a substrate reverse bias) is $V_T = 4$ V. Assume further that Eq. (1.12-3) applies with $C = 1$. Then it can be verified (Prob. 1.13-1) that with the driver OFF the output voltage is given as the solution of

$$V_o = +11 - \sqrt{V_o} \qquad (1.13\text{-}1)$$

from which we find $V_o \approx +8$ V. Thus we find that with the driver OFF, nearly half of the supply voltage is lost across the load transistor. If the threshold voltage were larger and the parameter C in Eq. (1.12-3) were larger, this voltage loss would be even more pronounced. This situation is in contrast to the situation which prevails in bipolar-transistor circuitry, where the load is a passive resistor. In this case when the transistor is OFF, there is no loss of voltage across the load.

1.14 INPUT-OUTPUT CHARACTERISTICS OF A MOSFET SWITCH

In the bipolar transistor switch of Fig. 1.6-1 the output voltage V_{CE} (the voltage from collector to emitter) goes through its full range from V_{CC} to $V_{CE}(\text{sat}) = 0.2$ V as the input voltage from base to emitter V_{BE} changes from about 0.65 to 0.75 V. Thus a change of only about 0.1 V at the input is adequate to accomplish the switching. The situation is different in the MOSFET switch, and the difference is brought out in Fig. 1.14-1. Here is shown a number of *calculated* plots of output versus input voltage. The vaues of geometrical and doping factors assumed in this calculation are typical of a MOSFET. We do not specify them here since they hold no particular interest for us. The single parameter which is specified is λ, defined as

$$\lambda = \frac{(W/L)_D}{(W/L)_L} \qquad (1.14\text{-}1)$$

where $(W/L)_D$ = ratio of width to length of channel in driver transistor $T1$
$(W/L)_L$ = corresponding ratio for load device $T2$

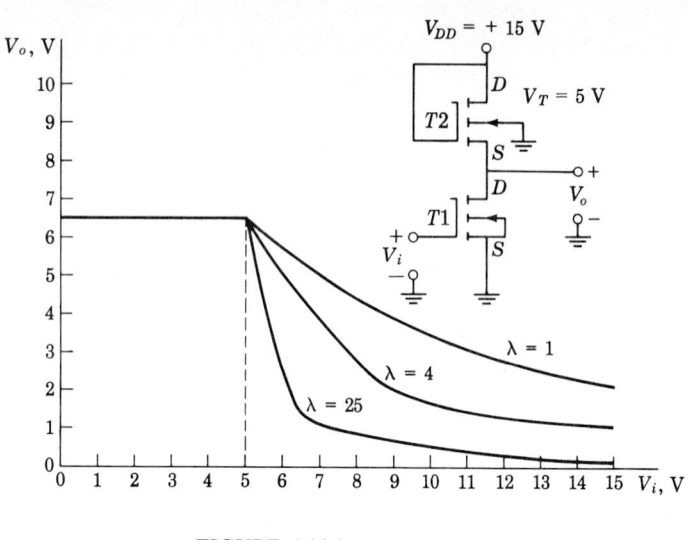

FIGURE 1.14-1
Calculated input-output plots for a MOSFET switch.
(Ref. 3.)

We note that in order to effect a reasonably sharp response of output to input it is necessary that λ be large; i.e., relative to the driver the load channel should be narrow and long. The threshold voltage is $V_T = +5$ V.

We note that, for reasons discussed earlier [see Eqs. (1.12-3) and (1.13-1)] the output voltage goes no higher than about 6.5 V even though a supply voltage of 15 V is employed. Further, we note that even in the case $\lambda = 25$ the input voltage V_i must change by a full 2 V to switch the output from 6.5 to about 1 V. In many a MOSFET even larger input switching voltage changes are required. An input change of 5 V is not uncommon.

1.15 COMPLEMENTARY-SYMMETRY MOSFETs (CMOS)

A complementary-symmetry MOSFET (CMOSFET) is fabricated as indicated in Fig. 1.15-1. The CMOSFET consists of two MOSFETS, one a p-channel type and the second an n-channel type, fabricated side by side on an n-type substrate. To allow the fabrication of the n-channel MOSFET it is necessary at first to diffuse into the n-type substrate a p-type tub or well.

The basic CMOS switch is shown in Fig. 1.15-2. The p-channel and n-channel FETs are joined at their drains, and the series combination is connected across the supply voltage. The positive side of the supply is connected to the source of the p-channel device, and the negative side is connected to the source of the n-channel device and grounded. The output is taken at the common drains. The input is applied in common to both gates. Both transistors operate in the enhancement mode.

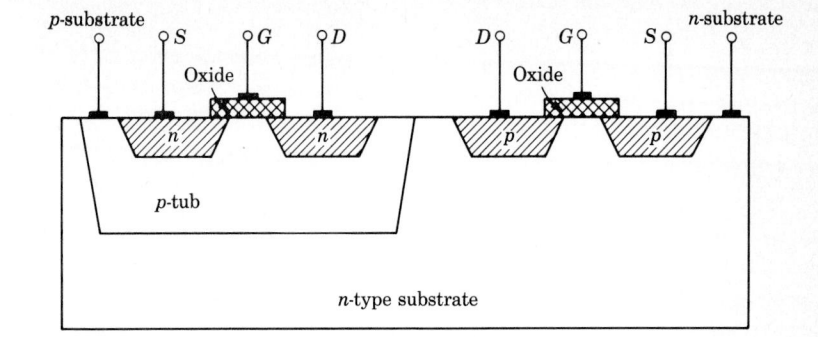

FIGURE 1.15-1
Enhancement-mode CMOSFET pair.

Consider, now, in Fig. 1.15-2 that the input voltage V_i swings over the range of the supply voltage from ground to V_{SS}. When V_i is in the neighborhood of ground, the lower, n-channel device will be OFF since $V_{GS} < V_T$, while the upper device will be ON, since $V_{SG} > V_T$. The output V_o will correspondingly be at V_{SS}. On the other hand, when $V_i = V_{SS}$, $T1$ will be ON and $T2$ will be OFF and V_o will be near ground. Typical input-output voltage plots of a CMOS switch are shown in Fig. 1.15-3 for various supply voltages and for two temperature extremes. In comparing these plots with the corresponding plots for the MOSFET switch (Fig. 1.14-1) we observe a number of relative advantages in the present case. In the CMOS case, V_o equals the full supply voltage when $V_i = 0$, and when V_i approaches V_{SS}, the output voltage drops to a very low value, typically of the order of 10 mV. We observe further that the transition of V_o between its two levels, ground and V_{SS}, is much sharper than in the case of the MOS switch. Note finally the lack of sensitivity of the switch characteristic to temperature.

A great merit of the CMOS switch is that when the switch is at one or the other limit of its range, its power dissipation is nominally zero. For in either

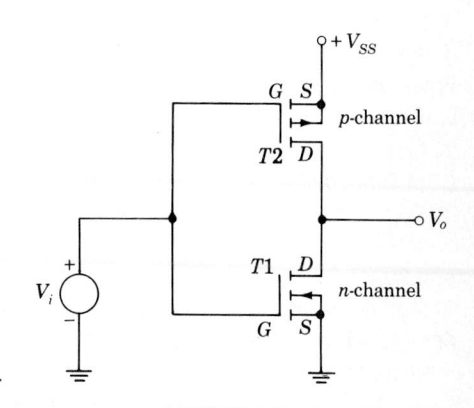

FIGURE 1.15-2
A CMOSFET switch.

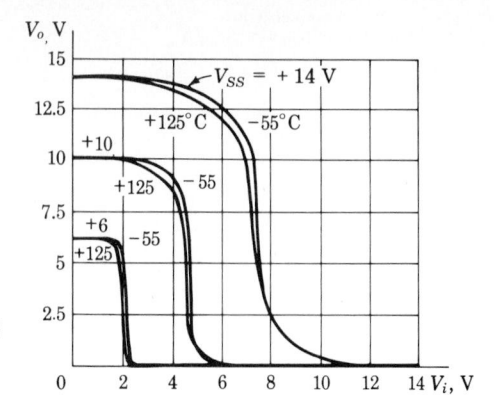

FIGURE 1.15-3
Typical input-output plots for a CMOS switch. The supply voltages are $V_{SS} = +14, +10,$ and $+6$ V.

case one or the other FET is cut off, and the current supplied by the supply voltage is nominally zero. (Actual leakage currents are of the order of 10 nA.) Further, as already noted, the impedance seen looking into the gate of either FET is extremely high (≈ 10 TΩ) so that here again, in the quiescent condition, there is no power dissipation.

On the other hand, when the switch is being driven back and forth between its two states, there is power dissipation. One source of this power dissipation is the inevitable load capacitance C_L (not shown in Fig. 1.15-2), which shunts the output to ground. This capacitance will normally be due principally to the capacitive input impedance of other CMOS switch inputs to which the output will be connected. As the switch swings between its states, this capacitor will charge and discharge and all the energy expended in charging it will be dissipated. This situation is represented in Fig. 1.15-4. Switches $S1$ and $S2$ are closed or open depending on V_i. When $S1$ is open, $S2$ is closed and vice versa. The "resistors" R_1 and R_2 in series with $S1$ and $S2$ are not fixed linear resistors and are intended simply to indicate the availability of a conducting path through the FET when it is ON. With $S1$ open and $S2$ closed the capacitor charges to V_{SS} and has stored in it an energy $\frac{1}{2}C_L V_{SS}^2$. When $S1$ closes and $S2$ opens, all this energy is dissipated in "resistor" R_1. If the frequency of the driving waveform V_i is f, the power dissipated in R_1 is $\frac{1}{2}C_L V_{SS}^2 f$. The dissipation in R_2 while C_L is charging will be same as in R_1 while C_L is discharging. We then have, as an estimate of the total power dissipation,

$$P = C_L V_{SS}^2 f \qquad (1.15\text{-}1)$$

In a typical case, with $C_L = 25$ pF, $V_{SS} = 10$ V, and $f = 10^5$ Hz we find $P = 0.25$ mW.

There is a second mechanism, which does not depend on the presence of a capacitor C_L, by which energy can be dissipated as the switch is carried back and forth between its states. To be specific, assume $V_{SS} = 15$ V and that each of the FETs in Fig. 1.15-2 has a threshold of magnitude $V_T = 3$ V. Then with $V_i = 0$, $T2$ is ON and $T1$ is OFF. When V_i becomes $V_i = 3$ V, $T1$ will turn

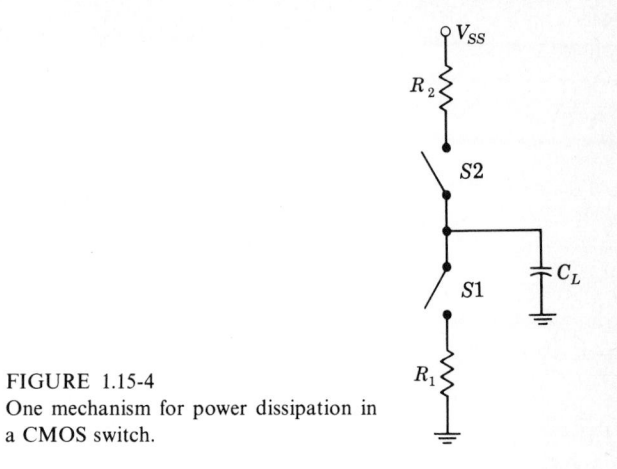

FIGURE 1.15-4
One mechanism for power dissipation in
a CMOS switch.

ON but $T2$ will remain ON. As V_i rises further, both $T1$ and $T2$ will remain ON until V_i rises to 12 V. At that point the threshold of $T2$ will have been reached, and $T2$ will turn OFF. Hence, we see that while V_i is in the range from 3 to 12 V, both FETs are ON and current will be drawn from the supply. The energy drawn from the supply will be dissipated in the two FETs. A typical experimental plot of the supply current drawn as V_i swings through its range is shown in Fig. 1.15-5. In taking data for this plot the input V_i was changed slowly enough to permit the effect of all stray capacitance to be neglected.

The energy dissipated because both FETs are ON will depend on how long they remain ON and hence will depend on how rapidly the switching is accomplished. If V_i swings very abruptly from one level to another, the energy dissipated will be minimal. If V_i swings slowly through the range where both FETs are ON, the energy dissipated will be correspondingly large. In any event, the average power dissipated in a CMOS switch will vary directly with the number of times in a given interval the switching is performed. Thus, like the power dissipation due to the load capacitor C_L, the power dissipation due to the finite speed of switching will increase with frequency.

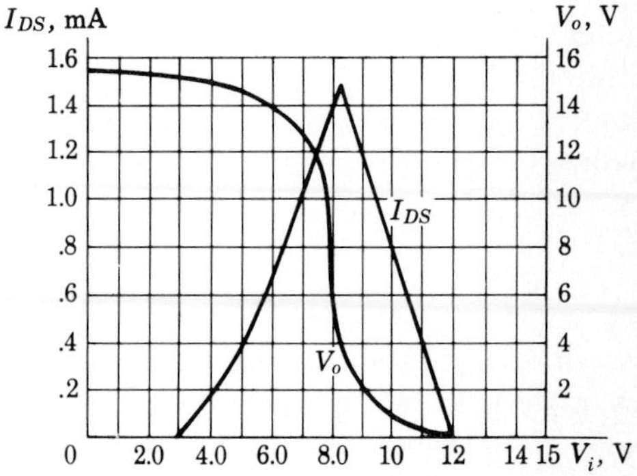

FIGURE 1.15-5
Input-output plot and drain current in a
MOSFET switch during switching.

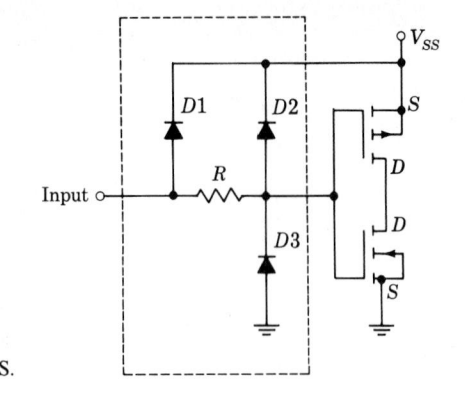

FIGURE 1.16-1
Gate protection in CMOS.

1.16 INPUT PROTECTION

Gate-source breakdown in a MOSFET occurs at about 100 V. However, due to the very high input impedance between gate and source, which is of the order of 1 TΩ, even a low energy source such as "static charge" can cause the breakdown voltage to be exceeded. It is interesting to note in this regard that when a bipolar junction transistor suffers breakdown, it can recover (if not burned out), but when a MOSFET suffers breakdown, the gate oxide is permanently shorted, rendering the device useless.

Gate protection devices are incorporated in almost all MOS integrated circuits to prevent gate-oxide breakdown. A typical protective device employed in conjunction with CMOS gates is shown in Fig. 1.16-1. Diodes $D1$ and $D2$ serve to clamp the input and gate, respectively, to V_{SS}, preventing large positive voltages from reaching the gate. Diode $D3$ clamps the gate to ground, thereby eliminating large negative voltages. Typical resistor values for R vary between 250 and 500 Ω.

The protective circuit shown significantly reduces the input impedance of the gate and results in a typical input current of about 10 pA. This current is negligible for most applications. It is also found that diode capacitors and the resistor R do not slow the speed of the gate down significantly.

Care must be taken when operating an MOS with gate protection in an actual circuit. In normal operation diode $D1$ serves to clip large voltage peaks from the input. However, if the supply voltage is turned down and a low-output-impedance-voltage generator is attached to the input, then a large positive input voltage may burn out D1.

1.17 SWITCHING SPEED OF A DIODE

In our earlier consideration of the bipolar diode we discussed only its volt-ampere characteristics when forward-biased and reversed-biased. Now we must give some thought to the matter of how rapidly the diode can switch from one

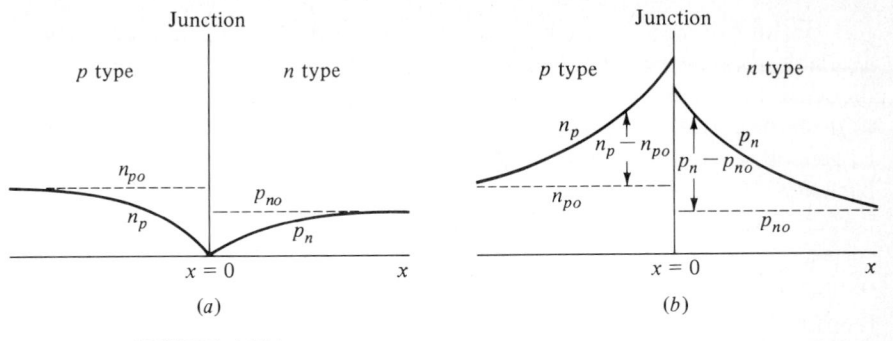

FIGURE 1.17-1
Minority-carrier density distribution as a function of the distance x from a junction. (a) A reverse-biased junction; (b) a forward-biased junction. The injected or excess hole (electron) density is $p_n - p_{no}(n_p - n_{po})$. (The diagram is not drawn to scale since $p_n \gg p_{no}$ and $n_p \gg n_{po}$.)

condition to the other. A principal speed limitation is encountered when we try to switch the diode from ON to OFF. The speed limitation in switching from OFF to ON is generally negligible in comparison, and we shall neglect it.

When an external voltage is impressed across a diode in the direction to reverse-bias it, very little current flows. The reason is that only the minority carriers on each side of the junction (holes on the n side, electrons on the p side) have a charge of the proper sign to carry current across the diode junction. The density of minority carriers in the neighborhood of the junction in the steady state is shown in Fig. 1.17-1a. Here the levels p_{no} and n_{po} are the thermal-equilibrium values of the minority-carrier densities on the two sides of the junction in the absence of an externally applied voltage. When a reverse voltage is applied, the density of minority carriers is as shown by the solid lines marked p_n and n_p. Far from the junctions the minority-carrier density remains unaltered, but as these carriers approach the junction, they are rapidly swept across and the density of minority carriers diminishes to zero at the junction. The current which flows, the reverse saturation current, is small because the density of thermally generated minority carriers is very small.

When the external voltage forward-biases the junction, the steady-state density of minority carriers is as shown in Fig. 1.17-1b. Near the junction the density of minority carriers is very large. These minority carriers have in each case been supplied from the other side of the junction, where, being majority carriers, they are in plentiful supply. With increasing time and as the minority carriers move away from the junction, a progressively larger number of them are lost by recombination with majority carriers. For this reason the density of minority carriers falls off with increasing distance from the junction. The density of majority carriers (not shown) as a function of distance from the junction is the same as the density of minority carriers. Such must be the case because in the body of the semiconductor the net charge must be zero.

In an ordinary ohmic conductor, when current flows, it does so because available charge carriers are pushed by the presence of an electric field. In the diode, in the immediate neighborhood of the junction there is no electric field to push the carriers, and the mechanism that accounts for the current is *diffusion*. In general, a current flows due to diffusion whenever the density of carriers of a particular type is not constant. Thus in Fig. 1.17-1b consider a surface perpendicular to the x axis at, say, some positive value of x. To the left of this surface the density p_n is greater than it is to the right. Hence, simply as a result of the random motion of holes we may well anticipate that purely as a statistical phenomenon more holes will move to the right across the surface than move to the left. Thereby, quite in the absence of an electric field, there will be a net transfer of holes to the right, and such a continuous transfer constitutes a current. As may well be expected, the diffusion current at any point is *proportional to the slope of the plot of density versus distance*. As can be seen in Fig. 1.17-1b, the magnitudes of the slopes of p_n and n_p fall off with increasing distance from the junction. Hence, correspondingly the diffusion current falls off. The total current, however, must remain constant. The difference is made up by the ohmic current, which increases with increasing distance from the junction.

Finally, then, we can now see that if the external voltage is suddenly reversed in a diode circuit which has been carrying current in the forward direction, the current will not immediately fall to its steady-state reverse-voltage value. For the current cannot attain its steady-state value until the minority-carrier distribution, which at the moment of reversal had the form in Fig. 1.17-1b, reduces to the distribution in Fig. 1.17-1a. Until such time as the *injected* or *excess minority-carrier density* $p_n - p_{no}$ (or $n_p - n_{po}$) has dropped nominally to zero, the diode will continue to conduct and the current will be determined by the external resistance in the diode circuit.

1.18 STORAGE AND TRANSITION TIME

The sequence of events which accompanies the reverse biasing of a conducting diode is indicated in Fig. 1.18-1. We consider that the voltage in Fig. 1.18-1b is applied to the diode resistor circuit in Fig. 1.18-1a. For a long time and up to the time t_1, the voltage $V_i = V_F$ has been in the direction to forward-bias the diode. The resistance R_L is assumed large enough for the drop across R_L to be large in comparison with the drop across the diode. Then the current is $I = V_F/R_L = I_F$. At time $t = t_1$ the input voltage reverses abruptly to $-V_R$. For the reasons described above, the current does not drop to zero but instead reverses and remains at the value $I = -V_R/R_L = -I_R$ until the time $t = t_2$. At $t = t_2$ the excess minority-carrier density has fallen to zero. If the diode ohmic resistance is R_d, then at the time t_1 the diode voltage falls slightly [by $(I_F + I_R)R_d$] but does not reverse. At $t = t_2$, when the excess minority carriers in the immediate neighborhood of the junction have been swept back across the junction,

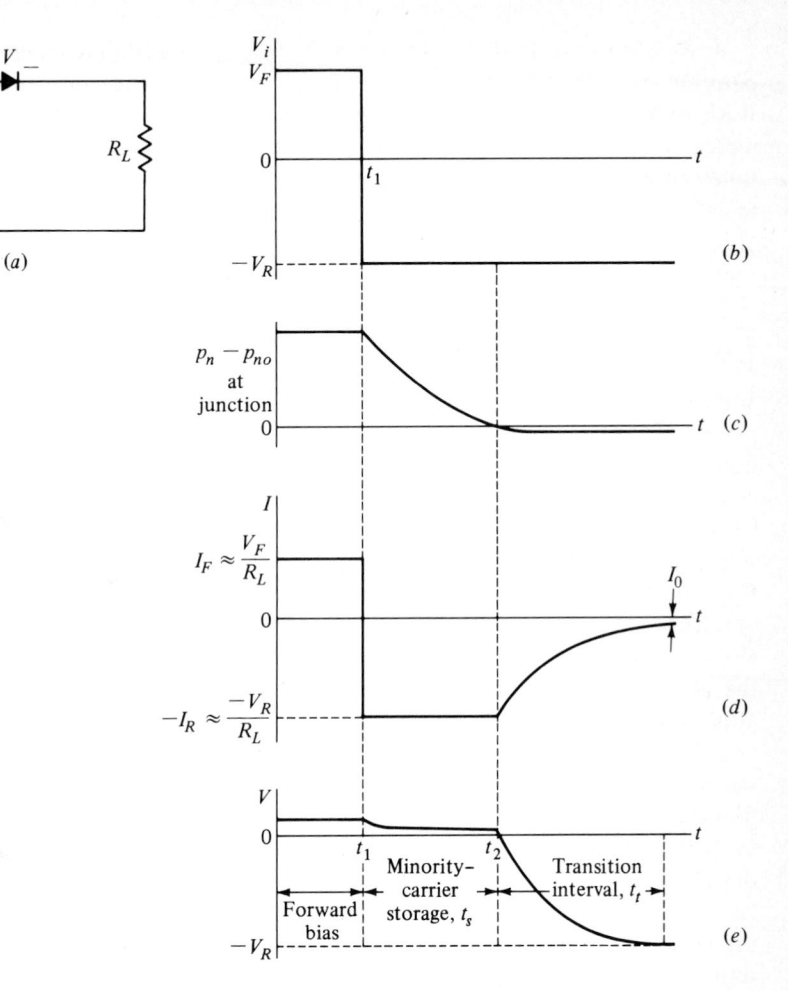

FIGURE 1.18-1
The waveform in (b) is applied to the diode circuit in (a); (c) the excess-carrier density at the junction; (d) the diode current; (e) the diode voltage.

the diode voltage begins to reverse, and the magnitude of the diode current begins to decrease. The interval t_1 to t_2 required for the stored minority charge to become zero is called the *storage time*.

The time which elapses between t_2 and the time when the diode has nominally recovered is called the *transition time* t_t. This recovery interval will be completed when the minority carriers at some distance from the junction have diffused to the junction and crossed it and when, in addition, the junction transition capacitance across the reverse-biased junction has charged through R_L to the voltage $-V_R$.

As is to be anticipated, it is found experimentally that for a fixed reverse current $-I_R$, the storage time increases with increasing forward currents I_F. With fixed forward current the storage time decreases as the reverse current is increased. Commercial switching-type diodes are available with total recovery times (storage plus transition times) in the range from less than a nanosecond to as high as some microseconds in diodes intended for switching large currents.

1.19 THE SCHOTTKY DIODE

We have considered, up to the present, only diodes formed by the junction of p-type and n-type semiconductor materials. Diodes can also be formed by the junction of a semiconductor and a metal. For example, when aluminum makes contact with silicon, the aluminum acts as a p-type impurity. Hence the junction between aluminum and n-type semiconductor constitutes a diode. Such metal-semiconductor diodes, called Schottky diodes, are represented by the symbol of Fig. 1.19-1 and have advantages with respect to speed of operation. The static volt-ampere characteristics of Schottky diodes are entirely similar to the characteristics of semiconductor diodes. However, depending on the metal used, the cut-in voltage of Schottky diodes will be in the range 0.2 to 0.5 V. The aluminum n-type semiconductor diode has a cut-in voltage of about 0.35 V.

In a metal-semiconductor diode, conduction in the forward direction occurs when the metal is biased positively with respect to the semiconductor. Current flows across the junction by virtue of the transport of electrons from semiconductor to metal. Electrons which have entered the metal are not distinguishable from the plentiful electrons already in the metal. Hence, the electrons which have crossed the junction do not constitute a minority charge. Accordingly when the junction voltage is reversed, these transported electrons are no more able to return across the junction than the electrons originally in the metal. Accordingly, Schottky diodes exhibit a negligible storage time. The reverse recovery time of a Schottky diode, typically 50 ps, is at least an order of magnitude smaller than for a pn diode.

We are, of course, aware that every semiconductor device requires connections to external leads, the connections of which must be ohmic and not rectifying. Such junctions are provided by interposing between the metal and the semiconductor an intervening layer in which the doping density is graded. The doping density is very heavy (metal-like) at the metal contact and diminishes toward the semiconductor.

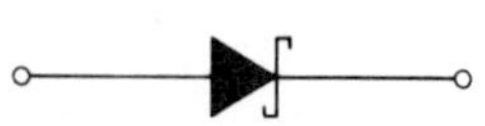

FIGURE 1.19-1
Symbol for a Schottky diode.

1.20 SWITCHING SPEED OF A BIPOLAR TRANSISTOR

We now examine qualitatively the mechanism which accounts for the speed limitations encountered in switching a bipolar transistor. In Fig. 1.20-1*a* is indicated the base region between emitter and collector junctions of an *npn* transistor. When the transistor is turned ON, current flows across the emitter-base junction. This current consists almost entirely of electrons from the emitter which cross the emitter junction into the base. The current across the emitter-base junction due to holes in the base entering the emitter is quite negligible because the doping of the emitter is significantly heavier than the doping of the base. Hence the density of electrons in the base available for current transport across the junction is overwhelmingly larger than density of holes in the base.

The injected electrons are transported across the very narrow base by *diffusion*. As we have noted, a diffusion current is proportional to the slope of the plot of density versus distance. Since the current in the base is constant, the plot of electron density versus distance x across the base is a straight line, as indicated in Fig. 1.20-1*a*. As each electron reaches the collector-base junction, it is immediately swept across the junction because the collector junction is reverse-biased. Hence at the collector the electron density is zero. Actually, at the collector the base current is slightly less than at the emitter because there will be a small but progressive loss of electrons crossing the base due to recombination with holes. Therefore, the current falls off slightly as x increases and the electron-density plot has a slight upward concavity, which, for simplicity, we have not indicated in the figure. (Typically, since $\alpha_N \approx 0.98$, the collector current is only about 2 percent smaller than the emitter current.)

The area under the electron-density plot is proportional to the total excess minority charge in the base Q_B. The total charge in the base is zero; hence

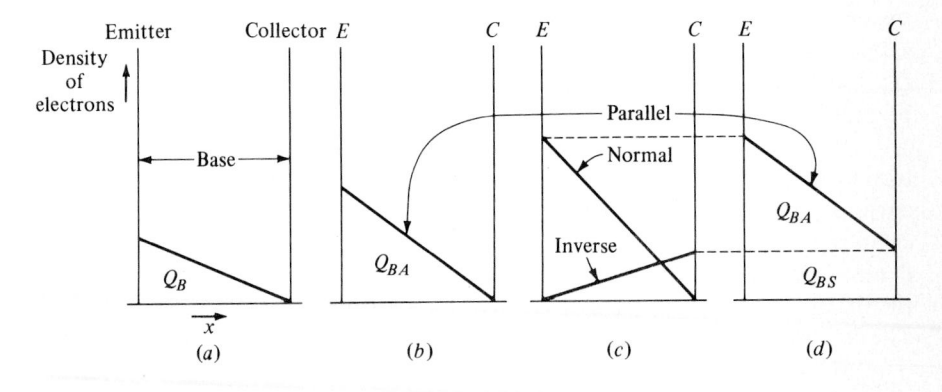

FIGURE 1.20-1

(*a*) Electron-density distribution through the base of a transistor in the active region. (*b*) Here the current through the base is larger than in (*a*); hence $Q_{BA} > Q_B$ and the magnitude of the slope of the density curve is greater. (*c*) In saturation there is a normal and an inverse charge distribution. (*d*) A composite sketch of (*c*).

there must also be an equal positive charge with a distribution throughout the base which matches the electron charge.

When the transistor is OFF, the base charge is nominally zero. When it is turned ON, the equilibrium situation represented in Fig. 1.20-1a cannot establish itself immediately. Instead some finite time must elapse to allow the charge distributions to become established. The negative charge is provided by electrons crossing the emitter junction. The positive charge is furnished by base current. Suppose now that the electrons and holes in the base did not recombine at all. In such a case the positive charge introduced by the base current would not dissipate, and after the base current had served to establish the positive charge, the base current would no longer be required. Actually, of course, because of the recombination of holes and electrons a continuous base current is called for.

In Fig. 1.20-1b the transistor current is larger, and consequently so also is the magnitude of the slope of the electron-density plot. We consider in Fig. 1.20-1b that we have reached the limit of the active range of the transistor circuit (as determined by the supply voltage V_{CC} and collector resistor R_c with which the transistor is operating) and have hence labeled the total base charge Q_{BA}.

Figure 1.20-1c corresponds to the transistor driven into saturation. The externally impressed voltage across the emitter junction has increased, and the number of electrons injected across the emitter junction has correspondingly increased as well. The slope of the electron-density plot (normal) is also higher to provide for more rapid diffusion of the electrons. However, the total current across the base is limited to V_{CC}/R_c. This current limitation is established by virtue of the fact that in saturation the collector junction is forward-biased. As a result there is an inverse injection of electrons from collector into the base, as shown in the figure. The composite of normal and inverse injections gives rise to the situation represented in Fig. 1.20-1d. Since the current in d (saturation) is nominally the same as in b (at the onset of saturation), the slopes of the electron-density plots are the same in the two cases. Observe particularly in d that the accumulated excess minority-carrier charge in d consists of a uniform saturation charge Q_{BS}, which makes no contribution to the current, and a charge Q_{BA}, which establishes the density gradient necessary to sustain the current flow by diffusion.

We now see the mechanism which accounts for the delays involved in turning a transistor ON and OFF. When a transistor is driven ON, its collector current will not attain its final value until the charges Q_B or Q_{BA} have been established. On the other hand, consider that a transistor is ON and in saturation. If we undertake to turn it OFF, we must first remove the charge Q_{BS}. During the interval when Q_{BS} is being removed, the collector current will not diminish. The collector current will begin to fall off only as the charge Q_{BA} begins to dissipate.

We can also see why the inverse-mode transistor switch of Fig. 1.10-2 is slower in operation than the normal-mode switch of Fig. 1.6-1. A transistor has an emitter junction whose cross-sectional area is appreciably smaller than the cross-sectional area of the collector junction. Such geometry is incorporated in a transistor to provide that the angle subtended by the collector junction as viewed from the emitter junction shall be as large as possible. Such a large angle serves to ensure that very nearly every carrier injected into the base at the emitter junction will indeed find its way to the collector. This feature, in turn, yields a common-base current gain α_N very nearly equal to unity, as required in an effective transistor. As can be seen in Fig. 1.20-1*b*, when a transistor is used in the normal mode, the excess minority-carrier charge density is highest where the cross-sectional area is smallest (the emitter) and the density is lowest where the area is largest (the collector). In the inverse mode, the situation is just reversed; the density is largest where the area is largest. Altogether, then, the stored charge in the inverse mode is larger than in the normal mode, and the delays associated with the inverse mode are correspondingly larger.

Figure 1.20-2 shows a transistor switch which is to be driven ON and OFF by an input waveform operating between voltage levels V_1 and V_2 and applied to the base through a resistor R_b. The input waveform V_i is shown in Fig. 1.20-3*a* and carries the transistor from cutoff to saturation and back to cutoff. The collector current I_C is shown in *b*. The time t_d, the *delay time*, is the time required to bring I_C to $0.1I_{CS}$, I_{CS} being the saturation collector current. The time t_r is the *rise time*, defined as the time to carry I_C from $0.1I_{CS}$ to $0.9I_{CS}$. The *storage time* t_s is the time which elapses after the reversal of base current before I_C falls again to $0.9I_{CS}$. Observe particularly, as is to be anticipated in the basis of our previous discussion, that there is an interval during which the collector current does not respond. This is the interval during which the saturation charge Q_{BS} in Fig. 1.20-1*d* is being removed. Finally the *fall time* is t_f. An increase in V_1 will serve to shorten the times t_d and t_r but only at the expense of an increase in the storage time t_s. The waveform of the base current is shown in Fig. 1.20-3*c*. Here for simplicity we have ignored the base-emitter drop of the saturated transistor in calculating I_B. Observe that the base current does not fall to zero until about the time the stored base charge has been dissipated, as evidenced by the fall to zero of the collector current.

The Schottky transistor The speed of a transistor response can be improved by preventing the transistor from going into saturation. One way of doing this is shown in Fig. 1.20-4*a*. A Schottky diode is connected from collector to base. When the transistor is in the active region, diode D does not conduct. The diode will conduct when the base-collector voltage is about 0.4 V and will not allow the collector to fall lower than 0.4 V below the base voltage. We have seen, however, that at saturation with $V_{BE}(\text{sat}) = 0.75$ V and $V_{CE}(\text{sat}) = 0.2$ V

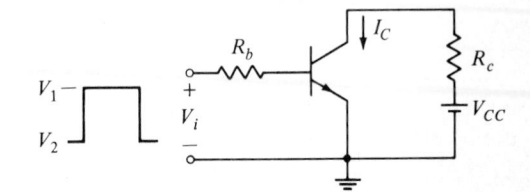

FIGURE 1.20-2
The pulse waveform operating between levels V_2 and V_1 drives the transistor from cutoff to saturation and back again.

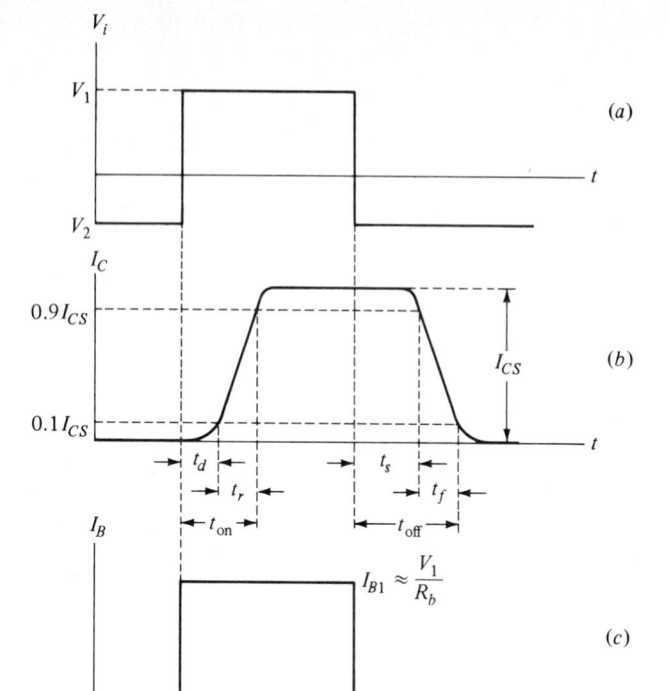

FIGURE 1.20-3
The collector current (b) and the base current (c) in response to the driving pulse in (a).

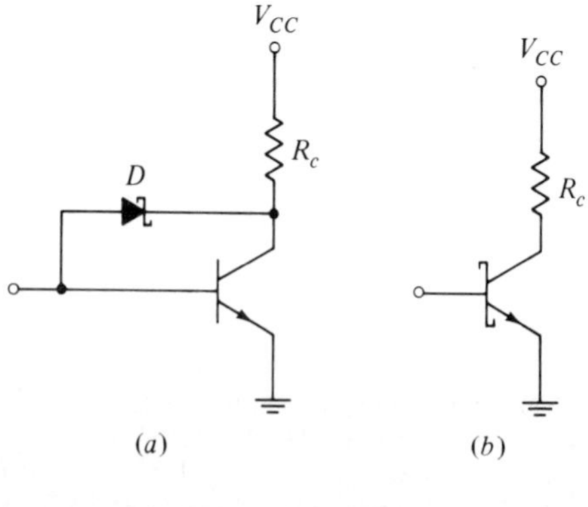

FIGURE 1.20-4
(a) A transistor and a Schottky diode to prevent saturation. (b) Symbol for the Schottky transistor.

that the voltage base-collector voltage must be $0.75 - 0.2 = 0.55$ V. Hence the diode never lets the collector voltage fall low enough to substantially forward-bias the collector junction, and the transistor is thereby restrained from becoming saturated.

It turns out that the Schottky diode can be incorporated with the transistor rather easily during the process of fabrication. Such diode-transistor combinations are called *Schottky transistors* and are represented as in Fig. 1.20-4b.

1.21   SWITCHING SPEEDS IN FET DEVICES

In field-effect devices current is carried by majority rather than minority carriers. It consequently turns out that when such devices are turned ON and OFF, the times required to effect a redistribution of charge is relatively insignificant. Far more significant is the time required to charge and discharge the capacitances associated with the devices and the inevitable stray capacitance associated with the circuitry. The problem of charging these capacitors is exaggerated by virtue of the fact that the resistive impedances encountered in FET circuits, through which the capacitors must charge and discharge, are generally substantially higher than those encountered in bipolar circuitry.

1.22   RISE AND FALL TIMES AND DELAYS

It is not an altogether outrageous exaggeration to say that an array of logical gates used for digital processing consists of a long cascade of switches. Let us therefore give some consideration to the cascade indicated in Fig. 1.22-1.

Consider initially just the first stage in the cascade. Let us also for the moment consider that the first stage is not connected to the input of the second stage, and finally let us assume that the waveform V_i rises and falls with limitless speed. If now V_i turns ON this first stage, the output will fall and there will be associated with this fall a *fall time*. Let us assume that $V_{CC} = 5$ V and that $V_{CE}(\text{sat}) = 0$; let us be somewhat casual about the matter and assume that the output falls linearly with time, making the transition from V_{CC} to zero in a time t_f. This fall time would depend on the characteristics of the transistor through the mechanism described above and would be somewhat slower than anticipated because of the presence of stray capacitances associated with the device and the circuit.

Suppose now that the input V_i itself did not rise absolutely abruptly but had a finite rise time. In general, this finite rise time would have the effect of slowing the fall of the output of the stage. *However, if the input rise time were negligibly small in comparison with the fall time of the stage itself, the effect of the finite rise time would also be negligible.*

Now let us restore the connection of the output of the first stage to the input of the second stage. The input to the second stage does not fall abruptly.

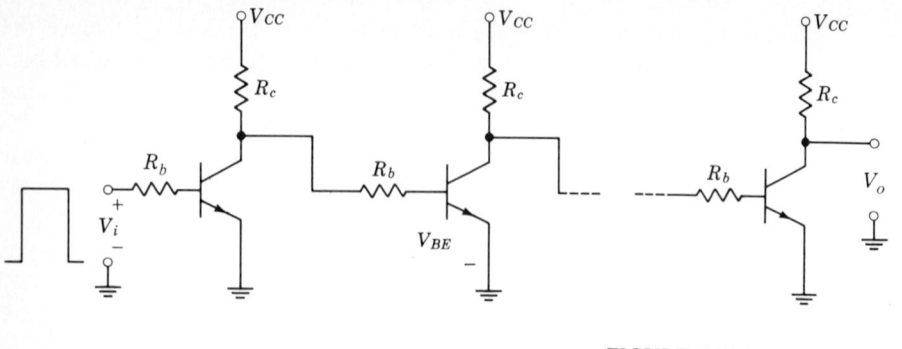

FIGURE 1.22-1
A cascade of transistor switches.

However, the only part of the input waveform of interest is the 0.1 V part of the waveform, which carries the base voltage V_{BE} from 0.75 V (saturation) to 0.65 V (cutoff). Since the fall at the output of the first stage is at a rate 5 V in time t_f, the effective speed of the input to the second stage is 50 times faster. Hence, insofar as the second stage is concerned, its input gating waveform has an effective speed which is so much greater than the speed of the stage itself that the effect of the input speed may be ignored. Hence, altogether, we reach the conclusion that as we proceed down the cascade of Fig. 1.22-1 the rise and fall times will not get progressively longer or will get longer only very slowly with increasing number of stages.

With FET devices the situation is somewhat different because the required input-voltage swing required to turn a device ON and OFF is a larger fraction of the output swing than with bipolar devices. Still, even if we went to the extreme case in which each of the stages in the cascade operated entirely linearly, we would find approximately that the rise and fall times increase only with the square root of the number of stages in the cascade.

Propagation delay time On the other hand, as we proceed down the cascade we would find that there is a progressive delay. Thus, if stage k is carried through its threshold region from cutoff to saturation at time t_k, the next stage will be carried through its threshold region from saturation to cutoff at a later time t_{k+1}. The delay, called the *propagation delay*, is the time $t_{k+1} - t_k$. That there are such delays associated with each edge of the driving waveform is apparent from Fig. 1.20-3. It is also apparent that this delay need not be the same for the rising and falling part of an input driving waveform. Finally, and most important, it is to be noted that the delays are cumulative; i.e., they add up. Because of this cumulative property of propagation delays, these delays are more often of concern than the rise and fall times.

Manufacturers of logical gates and other digital components often specify rise times, fall times, and propagation delays in the manner shown in Fig. 1.22-2. This figure shows the input and output waveforms of a logical gate or other binary

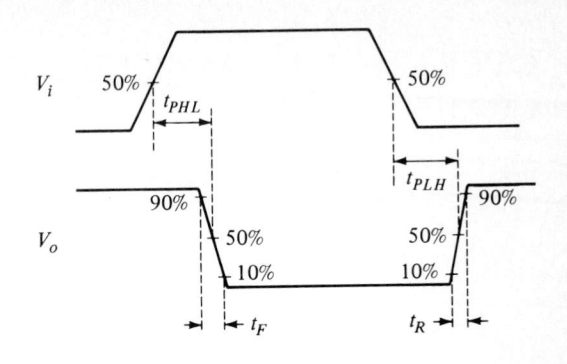

FIGURE 1.22-2
Input and output waveforms to define rise time, fall time, and propagation delay time.

device. The rise (and fall) times are seen to be defined as the time required for V_o to change from the 10 to 90 percent or 90 to 10 percent voltage levels. The propagation delay times t_{PHL} and t_{PLH} are the delay times between input and output when the output goes from the *high* state to the *low* state and vice versa. Note that delay time is measured between the 50 percent voltage levels.

REFERENCES

1 Ebers, J. J., and J. L. Moll, Large Signal Behavior of Junction Transistors, *Proc. IRE*, vol. 42, pp. 1761–1772, December 1954.
2 Millman, J., and H. Taub, "Pulse, Digital, and Switching Waveforms," McGraw-Hill, New York, 1965, chap. 20.
3 Crawford, R. H., "MOSFET in Circuit Design," McGraw-Hill, 1967, chap. 5.

2
OPERATIONAL AMPLIFIERS AND COMPARATORS

There are available, as integrated circuits, high-gain difference-signal amplifiers which are called *operational amplifiers* (*op-amps*). A modification of the op-amp yields a type of amplifier which is called a *comparator*. Comparator amplifiers which have been rendered unstable through the use of positive (regenerative) feedback are called *Schmitt trigger circuits*. Since we shall see that op-amps, comparators, and Schmitt triggers are frequently found as components in digital and switching systems, in this chapter we discuss the features and characteristics of these three circuits which are relevant to their use in digital systems.

2.1 THE OPERATIONAL AMPLIFIER [1,2]

There are commercially available, in integrated-circuit form, amplifiers which are characterized as *operational amplifiers* or more simply *op-amps*. The importance of these devices is borne out by the large number of manufacturers who supply them and by the large number of types available. We shall comment below on the significance of the term "operational" as applied to these amplifiers. These amplifiers are generally intended to operate linearly rather

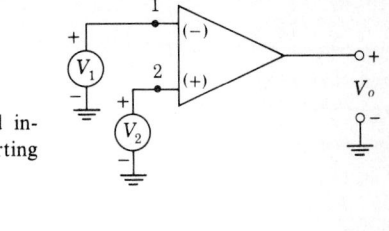

FIGURE 2.1-1

Symbol representing an op-amp and indicating its inverting ($-$) and noninverting ($+$) inputs.

than in a digital switching mode and, in this sense, are not fundamentally of concern to us. However, they do serve as components in sample-and-hold circuits, in analog gate circuits (Chap. 13), and in circuits used to effect conversions between analog and digital systems (Chap. 14). Hence, it is appropriate that we discuss their external characteristics and operating features.

An operational amplifier is represented in Fig. 2.1-1. It has two inputs to accommodate two input signals V_1 and V_2 and a single output, V_o. Ideally, the output voltage is related to the input by the equation

$$V_o = A(V_2 - V_1) \tag{2.1-1}$$

in which A, the voltage gain is (at least at low frequencies) a large real positive number, not uncommonly as large as 5×10^4.

There are three implications of Eq. (2.1-1), which for all their apparent nature, are important enough to merit being spelled out explicitly. We observe first that when an input is applied at input 1, the amplifier provides gain *with inversion*, while when an input is applied at input 2, the amplifier provides gain of equal magnitude *without inversion*. Specifically, with $V_2 = 0$, $V_o = -AV_1$ and with $V_1 = 0$, $V_o = AV_2$. The signs ($-$) and ($+$) at the input terminals in Fig. 2.1-1 are intended to serve as indications of this inversion and noninversion. Next we note from Eq. (2.1-1) that even if V_1 and V_2 make voltage excursions but $V_1 = V_2$, that is, if V_1 and V_2 make *common* voltage excursions, then $V_o = 0$. That is, the amplifier *rejects a signal applied in common* to both inputs. Finally we observe that Eq. (2.1-1) is intended to apply not only on an incremental basis but on an absolute basis as well. Thus the voltages V_o, V_1, and V_2 in Eq. (2.1-1) are not to be interpreted as departures from some initial quiescent operating point but are the actual voltages measured with respect to ground. Specifically, if, say, V_1 and V_2 are dc voltages, each measured to be 2 V, then the output is a dc voltage, $V_o = 0$ V. Needless to say, real physical operational amplifiers fail to reject common-mode signals with absolute precision and similarly fail to maintain $V_o = 0$, when required, with absolute precision and stability.

2.2 THE VIRTUAL GROUND

In Fig. 2.2.-1 we have applied an input V_i to the inverting terminal and have grounded the noninverting terminal. We acknowledge the existence of a finite input impedance R_i seen looking into the amplifier. As viewed from the output,

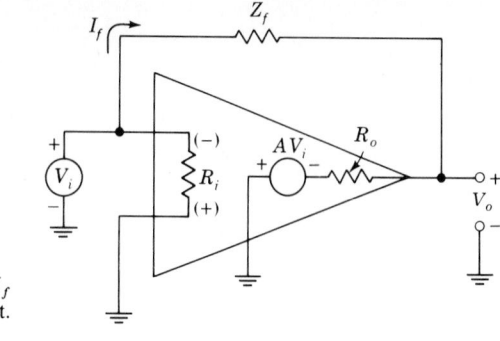

FIGURE 2.2-1
An op-amp with a feedback resistor Z_f
connected from output to inverting input.

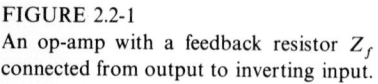

the amplifier is replaced by a voltage generator AV_i in series with a Thevenin equivalent output impedance R_o.   A feedback impedance Z_f has been connected externally from the output back to the input. We are now interested in calculating the impedance which is seen by the source V_i. We find that the current I_f through Z_f is

$$I_f = \frac{V_i + AV_i}{Z_f + R_o} \qquad (2.2\text{-}1)$$

We now define an impedance, $Z_i = V_i/I_f$.   From Eq. (2.2-1) we deduce that

$$Z_i = \frac{V_i}{I_f} = \frac{Z_f + R_o}{1 + A} \qquad (2.2\text{-}2)$$

This quantity Z_i is (neglecting R_i, which is in parallel with Z_i) the impedance seen by the generator V_i. Since A is a very large number, this impedance will be low.   Typically, R_o is of the order of some hundreds of ohms.   Let us take $R_o = 200\ \Omega$.   In many applications Z_f is also resistive and of the order of some thousands of ohms, let us say 5 kΩ.   Then with $A = 5 \times 10^4$ we find $Z_i = 5{,}200/(1 + 5 \times 10^4) \approx 0.1\ \Omega$.   The impedance R_i of the amplifier proper lies in the range of tens of kilohms to tens of megohms.   Hence altogether the input impedance seen by V_i may be taken as Z_i.

It is common to characterize the amplifier configuration of Fig. 2.2-1 as having a *virtual ground* at its input.   The input terminal is grounded in the sense that the impedance between input terminal and ground is very low (in comparison with other impedances in circuits which employ this configuration).   The grounding however is virtual in the sense that the current furnished by V_i flows not to ground but through the impedance Z_f.

2.3   OPERATIONS

In Fig. 2.3-1 an input V_s has been applied to the amplifier configuration of Fig. 2.2-1 through an impedance Z; for simplicity we have not indicated the

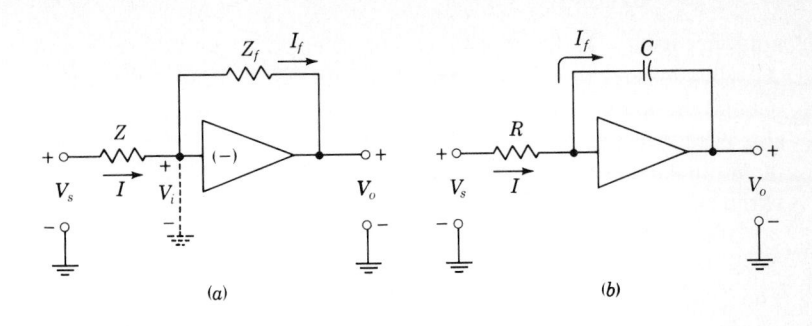

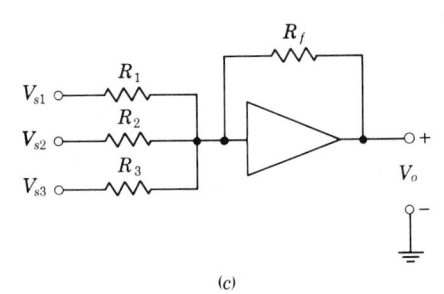

FIGURE 2.3-1
(a) A scale changer when $Z_f = R_f$ and $Z = R$; (b) an integrator; (c) an adder and scale changer.

(grounded) noninverting input terminal. Because of the virtual ground (indicated dashed)

$$I \approx \frac{V_s}{Z} \qquad (2.3\text{-}1)$$

and

$$V_o = -IZ_f \qquad (2.3\text{-}2)$$

Combining Eqs. (2.3-1) and (2.3-2), we find

$$V_o = -\frac{Z_f}{Z} V_s \qquad (2.3\text{-}3)$$

We recognize in Fig. 2.3-1a that the voltage V_i that appears at the amplifier input terminal is a linear combination of the externally impressed voltage and the output voltage. That is, the amplifier now incorporates *feedback*. In the presence of this feedback the overall amplifier gain with feedback A_f is, from Eq. (2.3-3),

$$A_f \equiv \frac{V_o}{V_s} = -\frac{Z_f}{Z} \qquad (2.3\text{-}4)$$

We observe from Eq. (2.3-4) that the gain with feedback depends only on the impedances Z_f and Z and depends in no way on the characteristics of the amplifier itself. If the impedances are selected to be stable components, the gain A_f will similarly be stable. The variability of the gain of the basic amplifier to changes in operating point, temperature, changing of component value with aging, etc., is suppressed by the feedback. Similarly the feedback suppresses the effect of nonlinearities in the basic amplifier, for nonlinearities may be viewed as a variation of device parameters with operating point.

It must be emphasized that the precision with which Eq. (2.3-4) applies depends on how perfect the virtual ground established at the amplifier input is. The quality of this virtual ground depends, in turn, on the gain A of the basic amplifier being very high. Hence high gain (ideally limitless gain) is of merit in an operational amplifier.

We now see that the configuration of Fig. 2.3-1 may be used to perform certain simple linear *mathematical operations* on input waveforms. It is this feature which accounts for the name associated with the amplifier. An early and continuing important use of such mathematical operational amplifier configurations is found in the field of *analog computers*. Strictly the term *operational amplifier* should be applied to the configuration of Fig. 2.3-1 including Z_f and Z. However, it is common practice to apply the term as well to the basic amplifier itself.

If Z_f and Z are resistors R_f and R, then $A_f = -R_f/R$ and the operational amplifier becomes a combined *sign changer* and *scale changer*. If $R_f = R$, the scale of the input is not changed and only sign changing is performed. If $Z = R$, and Z_f becomes capacitive, as in Fig. 2.3-1b, we have $I = V_s/R$ and

$$V_o = -\frac{1}{C}\int I_f \, dt = -\frac{1}{C}\int I \, dt = -\frac{1}{RC}\int V_s \, dt \qquad (2.3\text{-}5)$$

In this case the operational amplifier serves as an *integrator*. As is readily verified, if the resistor and capacitor are interchanged, the operational amplifier becomes a differentiator.

Figure 2.3-1c illustrates the use of an operational amplifier as an *adder* of waveforms. If $R_1 = R_2 = R_3 = R$, we shall have $V_o = -(R_f/R)(V_{s1} + V_{s2} + V_{s3})$. More generally we can combine addition with scale changing. If the input resistors are not equal, we shall have

$$V_o = -\left(\frac{R_f}{R_1} V_{s1} + \frac{R_f}{R_2} V_{s2} + \frac{R_f}{R_3} V_{s3}\right) \qquad (2.3\text{-}6)$$

A most useful feature of the configuration in Fig. 2.3-1c is that on account of the virtual ground there is no interaction between input voltages.

2.4 OUTPUT IMPEDANCE

A frequently useful feature of the operational amplifier is that the feedback incorporated in it serves to lower its output impedance very substantially. This characteristic is useful when low impedance loads and especially capacitive loads must be driven.

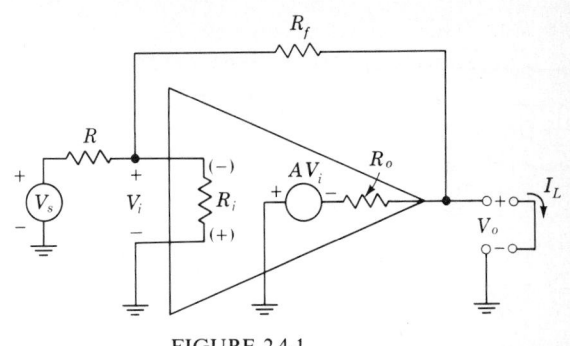

FIGURE 2.4-1
Circuit used to calculate output impedance.

The output impedance can be calculated from Fig. 2.4-1 as the ratio of the open-circuit output voltage V_o to the load current I_L when the load is a short circuit. We have that the open-circuit voltage is

$$V_o = -\frac{R_f}{R} V_s \qquad (2.4\text{-}1)$$

and the short-circuit current is

$$I_L = \frac{V_s}{R + R_f} - \frac{A}{R_o} V_i = \frac{V_s}{R + R_f} - \frac{A}{R_o} \frac{R_f}{R + R_f} V_s \qquad (2.4\text{-}2)$$

In these equations we have ignored the amplifier input impedance R_i. This neglect is certainly justified in connection with Eq. (2.4-1) since when the output is open-circuited, the amplifier provides feedback and there is a virtual ground shunting R_i. In Eq. (2.4-2) we have used the relation $V_i = R_f V_s/(R + R_f)$. For this relationship between V_i and V_s to be valid it is only necessary that R_i be large in comparison with the parallel combination of R and R_f, a requirement which in practice is invariably satisfied.

The second term in Eq. (2.4-2) is overwhelmingly larger than the first term because A is very large. Hence, when the first term is dropped, the output resistance Z_o is

$$Z_o = \frac{V_o}{I_L} = \frac{R_o}{A}\left(1 + \frac{R_f}{R}\right) = \frac{R_o}{A}(1 - A_f) \qquad (2.4\text{-}3)$$

In a typical case, with $A_f = -9$, $R_o = 100\ \Omega$, and $A = 5 \times 10^4$, $Z_o = \frac{1}{50}\ \Omega$.

2.5 ELECTRONICS OF OPERATIONAL AMPLIFIERS

Before proceeding with a discussion of the performance characteristics of operational amplifiers it will serve us to look briefly into the electronics of these devices. The basic circuit used to provide gain in the operational amplifier is

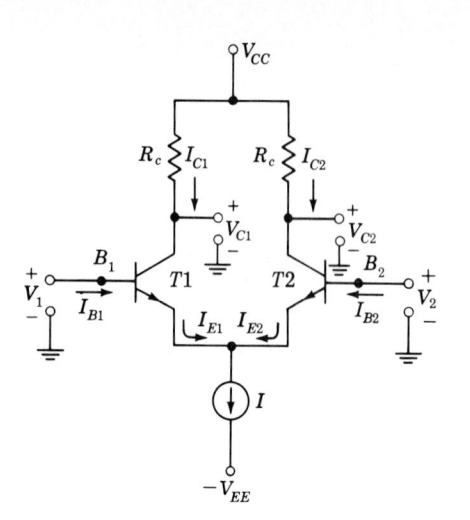

FIGURE 2.5-1
The basic difference amplifier used to
provide gain in the op-amp.

as shown in Fig. 2.5-1. The emitters of the two transistors are joined and connected to a constant current source. (A method of obtaining such a constant current source is given in Prob. 2.5-1. The details of the method are not relevant here.) With $V_1 = V_2$ the collector currents are $I_{C1} = I_{C2} \approx I/2$ (neglecting the base currents). With V_2 fixed, an increase in V_1 will divert a larger fraction of the *fixed* current I into $T1$. Hence V_{C1} will fall, and V_{C2} will rise. A decrease in V_1 will divert a larger part of I into $T2$. Hence, the gain $g \equiv \Delta V_{C1}/\Delta V_1$ from the input of $T1$ to the collector of $T1$ will be negative (inverting) while the gain $\Delta V_{C2}/\Delta V_1$ will be positive (noninverting). Corresponding comments apply to the gain of a signal applied at the base of $T2$.

As long as the current source I is precisely fixed, the change in current in one transistor must be equal and opposite to the change in current in the other transistor. In this case, the various gains must be equal or equal and opposite; i.e.,

$$g \equiv \frac{\Delta V_{C1}}{\Delta V_1} = \frac{\Delta V_{C2}}{\Delta V_2} = -\frac{\Delta V_{C1}}{\Delta V_2} = -\frac{\Delta V_{C2}}{\Delta V_1} \tag{2.5-1}$$

Suppose, then, that starting from an arbitrary initial condition, V_1 and V_2 are changed by arbitrary increments ΔV_1 and ΔV_2. Then the change in the output at the collector of T_1 would be

$$\Delta V_{C1} = \frac{\Delta V_{C1}}{\Delta V_1}\bigg|_{\Delta V_2 = 0} \Delta V_1 + \frac{\Delta V_{C1}}{\Delta V_2}\bigg|_{\Delta V_1 = 0} \Delta V_2 \tag{2.5-2a}$$

$$= g(\Delta V_1 - \Delta V_2) \tag{2.5-2b}$$

from Eq. (2.5-1). Correspondingly $\Delta V_{C2} = g(\Delta V_2 - \Delta V_1)$. It thus appears that if ΔV_1 and ΔV_2 are equal, i.e., a common signal is applied to both inputs, the outputs ΔV_{C1} and ΔV_{C2} will be zero. This feature is described by saying that the amplifier rejects a common-mode signal or by saying that the common-

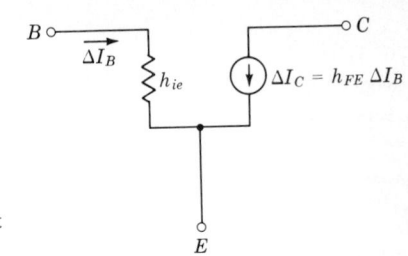

FIGURE 2.5-2
Simplified *h*-parameter equivalent circuit
of a transistor.

mode gain is zero. On the other hand when a *difference* develops between
ΔV_1 and ΔV_2, this difference is amplified. For this reason the circuit is often
referred to as a *difference amplifier*.

To calculate the gain *g* of the difference amplifier we replace the transistors
by the equivalent representation shown in Fig. 2.5-2. This equivalent circuit is a
simplified form of the *h*-parameter circuit. The base-emitter input resistance h_{ie}
can be estimated from the equation [see Eq. (1.7-14)] which relates the emitter
current to the base-emitter voltage, i.e.,

$$I_E = I_{E0} \, \epsilon^{V_{BE}/V_T} \tag{2.5-3}$$

The incremental resistance seen looking between the base and emitter is
dV_{BE}/dI_B. The base current is $I_B = I_E/(h_{FE} + 1)$. Hence the incremental
resistance seen looking into the base is

$$h_{ie} = \frac{dV_{BE}}{dI_B} = (h_{FE} + 1)\frac{dV_{BE}}{dI_E} = \frac{(h_{FE} + 1)V_T}{I_E} \approx \frac{h_{FE} V_T}{I_E} \tag{2.5-4}$$

When the simplified *h*-parameter equivalent circuit of a transistor shown in
Fig. 2.5-2 is used, the incremental equivalent circuit of the difference amplifier
appears as in Fig. 2.5-3. This equivalent circuit takes account only of departures

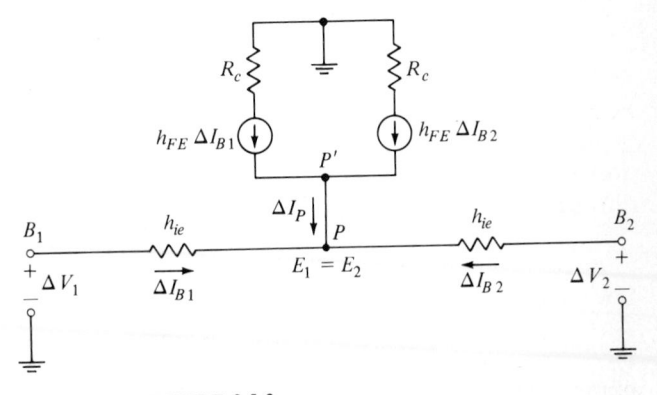

FIGURE 2.5-3
Incremental equivalent circuit of a difference amplifier.

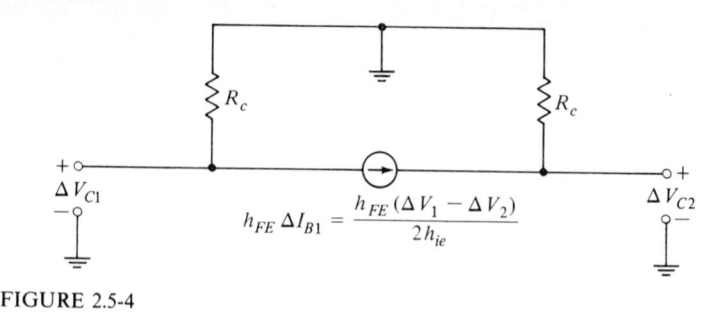

FIGURE 2.5-4
Equivalent circuit of the difference amplifier used to calculate ΔV_{C1} and ΔV_{C2}.

from an initial quiescent operating point; hence the cupply voltage has been replaced by a short circuit to ground and the current source in the emitter has been replaced by an open circuit. The sum of the currents entering node P or P' must equal zero. Hence

$$h_{FE} \, \Delta I_{B1} + \Delta I_{B1} + h_{FE} \, \Delta I_{B2} + I_{B2} = 0 \tag{2.5-5}$$

Therefore

$$\Delta I_{B1} = -\Delta I_{B2} \tag{2.5-6}$$

Consequently the current $\Delta I_P = 0$ and the lead from P' to P may be removed. Finally, the equivalent circuit for the purpose of calculating the currents through the collector resistors R_c is as shown in Fig. 2.5-4. Here the two current sources in series, each carrying the same current, have been replaced by a single current source $h_{FE} \, \Delta I_{B1}$, where

$$\Delta I_{B1} = \frac{\Delta V_1 - \Delta V_2}{2h_{ie}} \tag{2.5-7}$$

The gain of the difference amplifier is, using Eqs. (2.5-2b) and (2.5-7) and Fig. 2.5-4,

$$g = \frac{\Delta V_{C1}}{\Delta V_1 - \Delta V_2} = \frac{-R_c h_{FE}}{2h_{ie}} = -\frac{I_E R_c}{2V_T} \tag{2.5-8}$$

It also follows from Fig. 2.5-3, since $\Delta I_P = 0$, that the resistance from *base to base* is

$$R_i = 2h_{ie} \tag{2.5-9}$$

Typically $h_{ie} \approx 25$ k at $I_E = 0.1$ mA, so that we find the input impedance to be about 50 kΩ. It may be noted that higher input impedances are possible through the use of Darlington input circuits and FET inputs.

2.6 OVERALL AMPLIFIER

Commercial integrated-circuit op-amps generally consists of four stages. The first two stages are cascaded difference amplifiers used to provide the required high gain. A third stage serves to couple the second difference amplifier to a

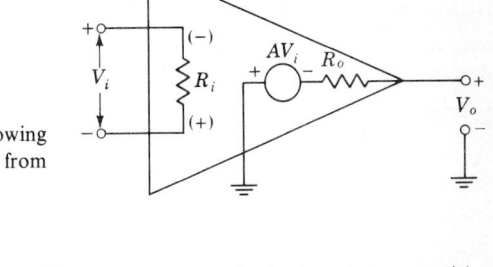

FIGURE 2.6-1

A representation of an op-amp showing the impedance R_i, which is bridged from input to input.

fourth stage, which is the output stage. The output stage is designed to provide a low-impedance output which can swing symmetrically with respect to ground. To allow such symmetrical swing it is necessary to provide for the amplifier both positive and negative supply voltages. We have already noted that the increment in output voltage of the difference amplifier is proportional to the difference in increments of the inputs. However, in an op-amp the dc levels throughout are adjusted so that this proportionality applies not only to increments, as in Eq. (2.5-2), but to the total voltages, as in Eq. (2.1-1). Thus, ideally at least, when $V_1 - V_2 = 0$, so also does $V_o = 0$. One of the functions of the third stage is to allow for shifting of the dc voltage levels to attain this end. Additionally, an op-amp generally incorporates circuitry to provide temperature compensation.

On the basis of our discussions it is now seen to be valid to represent the op-amp as in Fig. 2.6-1. Here we have taken into account that the output V_o depends only on the difference input and not on the inputs individually. We have further included an input resistance R_i which does not appear between individual inputs and ground but is bridged from input to input. Then if one input is indeed grounded, the input resistance seen at the other input is R_i. If, on the other hand, one input is connected to ground through a resistor R_g, the resistance seen looking into the other input is $R_i + R_g$.

2.7 NONINVERTING AMPLIFICATION USING AN OP-AMP

An alternative op-amp configuration is shown in Fig. 2.7-1. Here a signal source is applied to the noninverting input rather than to the inverting input. The voltage fed back to the inverting input is V_1. An exact calculation of the gain V_o/V_s is left as a problem (Prob. 2.7-1). A good approximate result may be obtained from the following simple considerations.

Since R_o is normally small enough to permit us to neglect the drop across it, we have $V_o = -AV_i = -A(V_1 - V_s)$. Since A is very large, the difference $V_1 - V_s$ must be very small in comparison with V_o. We shall make no serious error if we then assume that $V_i = V_1 - V_s \approx 0$, so that $V_1 = V_s$. In this case also, the current I through R_i is zero. We then find that

$$V_1 = \frac{R}{R + R_f} V_o = V_s \qquad (2.7\text{-}1)$$

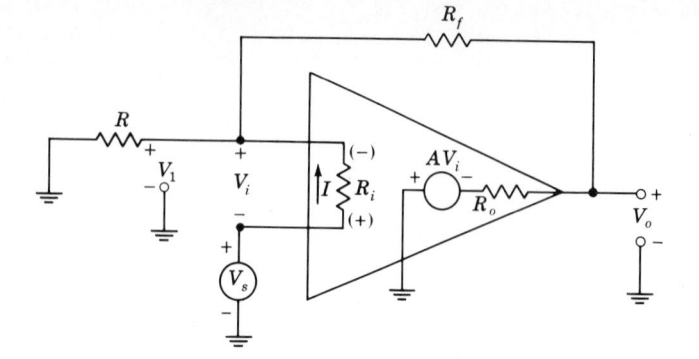

FIGURE 2.7-1
A noninverting op-amp.

from which the gain $A_f = V_o/V_s$ is

$$A_f = 1 + \frac{R_f}{R} \qquad (2.7\text{-}2)$$

Observe, again, that the gain depends only on the external resistors and not on the amplifier.

2.8   IMPEDANCE OF NONINVERTING AMPLIFIER

The input impedance seen by V_s in Fig. 2.7-1 is determined as the ratio of V_s to the current I through R_i. If we persist in our approximation, as above, that $V_1 = V_s$ then $I = 0$ and the input impedance is infinite. We therefore need a somewhat improved calculation for V_1 since the input impedance is clearly not infinite.

We have that

$$V_o = -A(V_1 - V_s) \qquad (2.8\text{-}1)$$

and using the first-order approximation that $I = 0$ (see Prob. 2.8-1) and again neglecting R_o,

$$V_1 = \frac{R}{R + R_f} V_o \qquad (2.8\text{-}2)$$

From Eqs. (2.8-1) and (2.8-2), eliminating V_o, we find that

$$V_1 = \frac{RA}{(1 + A)R + R_f} V_s \qquad (2.8\text{-}3)$$

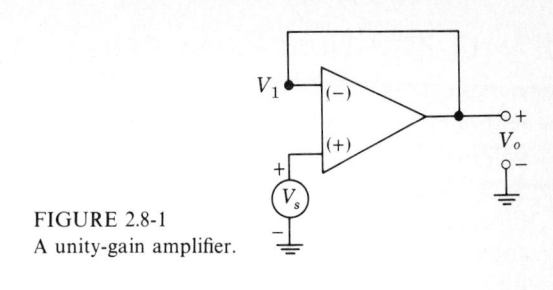

FIGURE 2.8-1
A unity-gain amplifier.

The current through R_i is $I = (V_s - V_1)/R_i$, and the input impedance $Z_i = V_s/I$ is

$$Z_i = R_i \frac{R(1 + A) + R_f}{R + R_f} \tag{2.8-4}$$

Since $R(1 + A) \gg R_f$ and $A \gg 1$, we have finally, using Eq. (2.7-2),

$$Z_i = \frac{R_i A}{1 + R_f/R} = \frac{R_i A}{A_f} \tag{2.8-5}$$

In a typical case with $R_i = 50$ kΩ, $A = 5 \times 10^4$, and $A_f = 10$, $Z_i = 250$ MΩ.

This many-fold multiplication of the input impedance is an example of the effect of *bootstrapping*. We note that when V_s changes the voltage at the $(+)$ side of R_i, the feedback changes the voltage at the $(-)$ side in the *same* direction. As a result, a change in V_s causes a smaller voltage change across the resistor than would be the case if the $(-)$ side were at fixed potential. The current I is thereby smaller and the impedance higher.

The output impedance of the amplifier is the impedance seen looking back into the output under the circumstances that all independent generator voltages are set to zero. Hence the shift of the generator from its location in Fig. 2.4-1 to its location in Fig. 2.7-1 has no effect on the output impedance. Consequently, in the present case, the output impedance, as before, is given by Eq. (2.4-3).

A unity-gain amplifier If, in Fig. 2.7-1 we set $R_f = 0$, R serves no purpose and may be removed. We arrive then at the configuration shown in Fig. 2.8-1. In this case $V_1 = V_o$. Hence the gain of the stage $V_o/V_s = V_1/V_s$ is given by Eq. (2.8-3) as $V_o/V_s = RA/[R(1 + A) + R_f] = A/(A + 1) \approx 1$. This amplifier configuration has many uses as a unity-gain high-input-impedance–low-output-impedance amplifier. Like an emitter follower it provides no gain but serves as an excellent buffer stage.

2.9   A PRACTICAL CONSIDERATION

As noted, the coupling from inputs to output of an op-amp is *direct* and without capacitive intervention. Hence a change in the dc voltage at either input will make itself felt at the output. One source of such dc input voltages is the base currents of the input transistors (see Fig. 2.5-1) which must flow to ground through whatever resistive return connection is available. Consider then the situation as in Fig. 2.4-1. The base current from the $(-)$ input flows to ground through a resistance R_p, which is the parallel combination of R and R_f (neglecting R_o). Current flowing at the $(+)$ input has a zero resistance dc path to ground. Typically the base currents are nominally equal and may be of the order of 1 μA. If R_p has a value of even 1 kΩ, then the $(-)$ input will be at $10^{-6} \times 10^3 = 1$ mV. However the $(+)$ input will remain at zero. Assuming a gain of 10^4, we would find $V_o = 10$ V. This initial offset can be suppressed by introducing an equivalent resistance to ground from the $(+)$ input. In the circuit of Fig. 2.7-1 a resistor R_p should be placed in series with V_s.

2.10   COMPENSATION

The op-amp has frequency limitations due to the stray capacitances which are inevitably present and due to the inherent frequency limitations of the active components. Hence, with increasing frequency the magnitude of the gain falls off, and the output instead of being exactly in phase (or out of phase) with an input suffers a phase shift which is a function of frequency. A feedback amplifier will oscillate if the loop gain is unity, or larger, at some frequency where the loop phase shift is 360°. The feedback op-amp is not immune to such oscillations.

It is therefore necessary in an op-amp to tailor the gain and phase characteristics to one another so that at no frequency is the condition for oscillation satisfied. This tailoring is achieved by introducing additional frequency-sensitive compensating circuits not essential to provide gain. In some op-amps the compensation is built right into the integrated circuit. More generally provision is made to allow some of the elements of the compensation circuitry (generally capacitors or series capacitor-resistor combinations) to be external to the integrated circuit. Such an arrangement is advantageous in two respects. It allows the use of larger capacitors than would be feasible in the integrated-circuit chip itself. Further it allows an adjustment of components to the optimum value, which depends on the gain being used. On a manufacturer's specification sheet recommended values of compensating elements as a function of gain are generally given.

In general, in using op-amps it is well, in order to minimize the risk of oscillations, even with compensation, to make all connections to the op-amp with short leads. This precaution applies not only in connection with the leads to the external compensating elements but also to power-supply and ground leads. Further, all power-supply leads should be well bypassed capacitively.

2.11 COMMON-MODE REJECTION RATIO

Let the input signals to the op-amp be V_1 and V_2, as in Fig. 2.1-1. Instead of specifying the inputs directly by giving V_1 and V_2, we may equivalently specify the inputs in terms of a *difference* signal input V_d and a *common-mode* input V_c, defined in terms of V_1 and V_2 by

$$V_d = V_1 - V_2 \quad \text{and} \quad V_c = \tfrac{1}{2}(V_1 + V_2) \tag{2.11-1}$$

If V_1 and V_2 are equal and opposite, then $V_c = 0$, and if V_1 and V_2 are equal, $V_d = 0$. The signals V_d and V_c measure respectively the difference and average value of the input signal. The signals V_1 and V_2 are uniquely determined in terms of V_d and V_c by the equations

$$V_1 = V_c + \tfrac{1}{2}V_d \quad \text{and} \quad V_2 = V_c - \tfrac{1}{2}V_d \tag{2.11-2}$$

Suppose that, as in an ideal amplifier, the gain A_1 measured with respect to input 1 and the gain A_2 measured with respect to input 2 are equal and opposite. Then if V_1 and V_2 are equal, $V_d = 0$ and the output will be zero even if $V_c \neq 0$. On the other hand, suppose the gains are not equal. Then we would have

$$V_o = A_1 V_1 + A_2 V_2 \tag{2.11-3}$$

When Eq. (2.11-2) is used, Eq. (2.11-3) becomes

$$V_o = \tfrac{1}{2}(A_1 - A_2)V_d + (A_1 + A_2)V_c \tag{2.11-4}$$

In this case we then see that there is not only a gain $A_d = \tfrac{1}{2}(A_1 - A_2)$ for the difference signal V_d but also a gain $A_c = A_1 + A_2$ to the common-mode signal V_c. The relative sensitivity of an op-amp to a difference signal as compared to a common-mode signal is called the *common-mode rejection ratio* (CMRR) and is given by

$$\text{CMRR} = \frac{A_d}{A_c} \tag{2.11-5}$$

For commercial op-amps the CMRR lies in the range 60 to 100 dB. The CMRR is important whenever the op-amp is being used in an application in which the output is required to respond precisely to the *difference* of two independent signals applied at the amplifier inputs.

2.12 CHARACTERISTICS OF OP-AMPS

An ideal op-amp would have infinite gain, infinite input impedance, infinite CMRR, and zero output impedance. As we have noted, physical op-amps fall short of the ideal. Gains are of the order of 5×10^4, input impedance may be in the range 10 kΩ to 100 MΩ, the CMRR lies in the range 60 to 100 dB, and output impedances are in the range of tens to hundreds of ohms. We consider now some additional characteristics of real op-amps. These characteristics are invariably specified on a manufacturer's specification sheet.

Output voltage swing   This is the peak output voltage with respect to zero available at the output without distortion.   It is a function of the supply voltage.   Thus in an amplifier operating between supply voltages of $+6$ and -6 V, the peak-to-peak undistorted output swing might be from -4 to $+4$ V.   (In practice the allowable swing is not always symmetrical.)   The output swing is a function of the supply voltages and may range from about 50 to 80 percent of the supply voltage.

Input common-mode voltage swing   This is the maximum voltage which can be applied to the input without causing abnormal operation or damage to the op-amp.   It may be as high as the supply voltage but often is restricted to one-third or one-half the supply voltage.

Input offset current   If we supplied equal dc currents to the two inputs of an op-amp, then ideally we should have $V_o = 0$.   In a real amplifier, because of lack of perfect symmetry in the input difference amplifier and for other reasons, such is not the case.   We find instead that to establish $V_o = 0$ we need to make the input currents different by an amount I_{io}.   This *input offset current* I_{io} typically lies in the range 20 to 60 nA and is measured under the circumstance that $V_i = 0$.

Input offset voltage   It is found similarly that if equal voltages are applied to the two amplifier inputs, then again, in real amplifier $V_o \neq 0$.   To set $V_o = 0$ requires an input offset voltage V_{io} which typically lies in the range 1 to 4 mV when the input voltages are nominally zero.

Power-supply voltage-rejection ratio   A change in supply voltage ΔV_{CC} will produce a change ΔV_o in the amplifier output.   The *power-supply voltage-rejection ratio* $\Delta V_o/\Delta V_{CC}$ is normally specified for the condition that the difference voltage input $V_i = 0$.   Typically this ratio is in the range 10^{-5} to 7×10^{-5}.

Frequency response   Most op-amps are relatively low-frequency devices.   Rather typically, the open-loop (no feedback) 3-dB bandwidth of general-purpose op-amps will be in the neighborhood of 100 Hz or less.   Bandwidths as low as 10 Hz are not unusual.   On the other hand, the low-frequency (nominally dc) gain of op-amps is so high that the frequency at which the open-loop gain falls to unity may be at 1 MHz or even higher.   The use of negative feedback simultaneously reduces the gain and increases the bandwidth.   Hence it turns out that through the use of feedback it is feasible to provide useful gain (say in the range 1 to 10) over a frequency range extending up to hundreds of kilohertz to some megahertz. There are also available special wide-band op-amps whose open-loop gains extend to several kilohertz and which can be used to provide useful gain up to 10 MHz or even somewhat higher.   Manufacturers generally need to provide fairly extensive information concerning the frequency response because the response depends not only on the nominal gain but also on the capacitive loading of the output and on how the amplifier has been compensated.

Slew rate   The *slew rate* of an op-amp is the maximum possible rate of change of output voltage. It is determined by the capacitances in the amplifier (incidentally present or deliberately introduced) which must be charged and the maximum current available to charge them.   The slew rate is measured as dV_o/dt under the circumstances that the input is a very fast (ideally zero rise time) step of amplitude which is very much larger than the amplifier can handle linearly and which carries the amplifier output from one limiting output to the other.   Since in a slew-rate determination the op-amp operates nonlinearly, the slew rate is not simply related to the frequency response (or equivalently to the rise-time response) of the amplifier restrained to its linear range.   Amplifiers are available with slew rate through the range 50 mV/μs to 50 V/μs and beyond. In measuring slew rate the amplifier is driven abruptly from one limiting output to the other.   Consequently the feedback cannot be effective and the slew rate is rather independent of the amount of feedback used with the op-amp; i.e., the slew rate does not depend much on the gain.

One important use of the slew-rate parameter is that it permits us to estimate the extent to which the linear range of operation of an op-amp is affected by the speed of the input signal (see Prob. 2.12-2).

2.13   THE COMPARATOR

There is frequent need for a device which provides an output V_o which is at one fixed dc voltage level or at another dc level depending on whether the input signal V_s is larger or smaller than the reference voltage V_R.   Such a device is called a *comparator* since it compares two voltages.   Specifically if the outputs are V_{o1} and V_{o2}, then $V_o = V_{o1}$ when $V_s < V_R$ and $V_o = V_{o2}$ when $V_s > V_R$.   When V_s passes through V_R, the output makes an abrupt transition from V_{o1} to V_{o2} or vice versa, depending on the direction of V_s as it passes through V_R.   (We shall comment below on the situation which may prevail in practice if we try to hold V_s at precisely $V_s = V_R$.)

Comparators of the type described here are often required to provide inputs to *logical gates*.   Logical gates are discussed in detail throughout the remainder of the text.   For the present let it suffice to note that in order to serve their function properly logical gates must be supplied at any time with one of two voltages which are separated by some minimum interval.   For example, a TTL gate (Chap. 6) requires an input which is either less than about 0.4 V or is more than about 2.8 V.   Hence a comparator suitable for use with such a gate would provide $V_o < 0.4$ or $V_o > 2.8$ V, depending on whether $V_s < V_R$ or $V_s > V_R$.

Taking into account the fact that any physical amplifier has a limited range over which it can respond linearly to an input signal, a high-gain amplifier may serve as a comparator of the type described above. For suppose we build an amplifier whose output is limited so that at one limit $V_o = 0$ V and at the other limit $V_o = 4$ V. Suppose the gain A is large, say $A = 4{,}000$, and that we have adjusted matters so that $V_o = 0$ V when $V_s = -0.5$ mV. Then we shall have $V_o = 4$ V when V_s *changes* by $4/4{,}000 = 1$ mV to $V_s = +0.5$ mV. Altogether we shall have a comparator suitable for use with a TTL gate. The abrupt transition in a comparator will occur as the input goes from -0.5 to $+0.5$ mV. We would say that our comparator is operating with a reference voltage $V_R = 0$ V and that the comparator range of uncertainty is ± 0.5 mV.

A final feature which we shall have to incorporate into our amplifier used as a comparator is provision for adjusting the reference voltage. This can be achieved by using an amplifier with some of the features of the difference amplifier. The important characteristics of those amplifiers for present interest is that they have two inputs between which, ideally, *the common-mode gain is zero.* Thus we need only apply the input V_s to one amplifier input terminal and the reference voltage V_R to the other input terminal. It is only the *difference* in the two inputs that controls the output. Thus if the reference voltage is changed from V_{R1} to V_{R2}, the input voltage V_s at which the amplifier output swings from one level to another will correspondingly change by amount $V_{R2} - V_{R1}$.

2.14   AN INTEGRATED-CIRCUIT AMPLIFIER COMPARATOR

As an example of the comparator type described in the previous section, we consider the integrated-circuit Fairchild μA 710 comparator whose schematic is shown in Fig. 2.14-1. The input stage is a difference amplifier using transistors $T1$ and $T2$. Two inputs are available, one for the reference voltage and one for the input signal. If the input signal V_s is applied as V_1 (and the reference V_R as V_2) the output V_o will be out of phase with the input signal, while if the signal is applied as V_2 (and the reference as V_1), the output V_o will be in phase with the input signal.

We have noted that to minimize the common-mode gain we require a current source in the common-emitter circuit. One way to effect a current source is to use an extremely large resistor in the common emitter. Such a large resistor would develop across itself a very large voltage due to the quiescent current which must flow through it. The circuit involving $T3$ serves to provide such a large resistor without the correspondingly large voltage. It is left as a student exercise (Prob. 2.14-1) to show that using typical transistor parameters, the incremental resistance seen looking into the collector of $T3$ is of the order of 250 kΩ. We shall see below that the quiescent current through $T3$ is 2 mA. If we really required that 2 mA flow through 250 kΩ, we would develop a voltage $2(250) = 500$ V. In any event, because of this large incremental resistance it can be verified that the common-mode rejection ratio of the input stage is approximately 10^4.

The output signal of the difference amplifier is taken at the collector of $T2$.

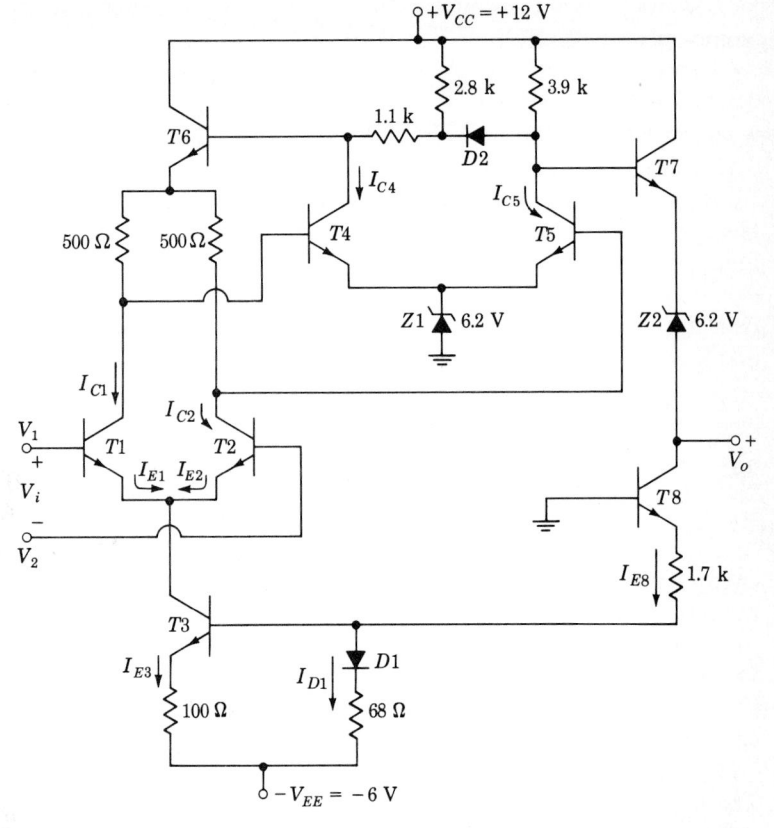

FIGURE 2.14-1
Schematic circuit of the Fairchild type μA 710 amplifier-type voltage comparator.

The stage involving $T5$ provides additional gain, and $T7$ provides an emitter-follower output stage. Zener diodes $Z1$ and $Z2$ provide the necessary dc voltages to allow the dc coupling needed and to arrange proper output voltage levels.

To minimize the range over which the input must swing to carry the output from one limiting voltage to the other, it is advantageous to increase the amplifier gain. One way to increase the gain is to include more stages. A second way to increase the gain is to incorporate *positive feedback*. Positive feedback is not often incorporated in an amplifier which is intended to reproduce an input signal faithfully because positive feedback affects the stability and the linearity of the amplifier adversely and has other disadvantageous characteristics. In the present case of the amplifier used as a comparator we have no interest in preserving an input waveform. Most of the time the comparator will dwell in one limiting position or the other where the gain is zero. And when we do swing the input V_s into the range where the amplifier has gain, it might well be advantageous if the amplifier became somewhat unstable so that the output would swing violently from one limit to the other.

Positive feedback is provided in the comparator by transistors $T4$ and $T6$. To see that such is the case, let V_R be applied as V_2 (to the base of $T2$) and V_s as V_1. If then V_s increases, the collector voltage of $T2$ will also increase. The collector voltage of $T1$ will decrease, as will the base voltage of $T4$. Consequently the collector of $T4$ and the base of $T6$ will rise. Finally since the base of $T6$ rises, the emitter voltage of $T6$ will increase ($T6$ is an emitter follower) and the collector of $T2$ will rise farther. Thus, when an input signal drives the collector of $T2$ in either direction, the feedback serves to push the collector of $T2$ further in the same direction. The feedback is therefore positive, and the gain from input to collector of $T2$ is increased by the feedback.

2.15 CALCULATIONS FOR THE INTEGRATED-CIRCUIT AMPLIFIER COMPARATOR

In this section we shall make some approximate calculations for the integrated-circuit amplifier comparator of Fig. 2.14-1. For simplicity we shall assume that the transistors have current gains which are high enough to permit us to ignore the base currents in comparison with the emitter and collector currents. Correspondingly we assume that collector and emitter currents are equal. As usual, we use $V_{BE} = 0.65$, 0.7 V and 0.75 V for the transitor at cutoff, in the active region, and in saturation, respectively, and $V_{CE} = 0.2$ V at saturation.

The output levels The output voltage V_o limits on one side when $V_1 - V_2$ is sufficiently greater than zero for T_5 to be driven to saturation. Correspondingly V_o is

$$V_o = V_{Z1} + V_{CE5} - V_{BE7} - V_{Z2} = 6.2 + 0.2 - 0.7 - 6.2 = -0.5 \text{ V} \quad (2.15\text{-}1)$$

On the other side, V_o limits if $V_2 - V_1$ is sufficiently positive, so that $T1$ cuts off and the collector of $T2$ decreases to such an extent that $T5$ cuts off. Since $T1$ is OFF, negligible current flows in its 500-Ω collector resistor and the collector voltage of $T4$ is $V_{C4} = V_{BE6} + V_{BE4} + V_{Z1} = 0.7 + 0.7 + 6.2 = 7.6$ V. To find V_o in this case requires calculating V_{C5}. This is easily done using Fig. 2.15-1. A simple ohm's-law analysis of this circuit (see Prob. 2.15-1) yields $V_{C5} = 10$ V. Hence

$$V_o = V_{C5} - V_{BE7} - V_{Z2} = 10 - 0.7 - 6.2 = 3.1 \text{ V} \quad (2.15\text{-}2)$$

The diode $D2$ in Fig. 2.14-1 serves in two ways to establish the level at 3.1 V: (1) by conducting it draws current through the collector resistor of $T5$ even when $T5$ is OFF, thereby limiting the rise in collector voltage; (2) just as the connection of the collector of $T4$ to the base of $T6$ provides positive feedback, so the connection through $D2$ to $T6$ from $T5$ provides negative feedback as soon as $D2$ begins to conduct.

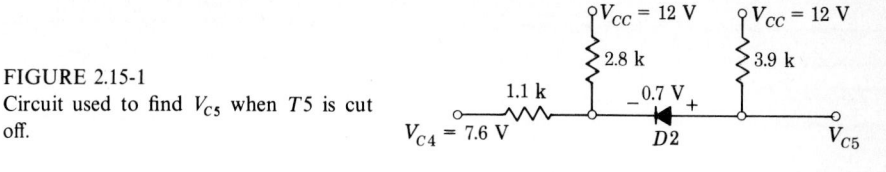

FIGURE 2.15-1
Circuit used to find V_{C5} when $T5$ is cut off.

The two output voltage levels are TTL-compatible (see Chap. 6), and this comparator may be used to generate binary signals for use in TTL logic circuits.

Operating currents and voltages   We shall now estimate some of the currents and voltages when $V_1 = V_2$.   The circuitry associated with $T3$ provides a constant current for the difference amplifier whose transistors are $T1$ and $T2$.   This current I_{E3} can be determined as follows.   (In estimating this current, as in other calculations, we shall neglect base currents for simplicity.)   We add up the voltage drops from the base of $T8$ to the $-V_{EE} = -6$ V supply and find

$$V_{BE8} + 1.7 \times 10^3 I_{E8} + V_{D1} + 68 I_{D1} - 6 = 0 \qquad (2.15\text{-}3)$$

Assuming $V_{BE8} = V_{D1} = 0.7$ V and $I_{E8} = I_{D1}$, we find $I_{E8} = 2.6$ mA and $V_{B3} = -5.1$ V.   The emitter voltage $V_{E3} = V_{B3} - V_{BE3} = -5.1 - 0.7 = -5.8$ V, and $I_{E3} = (0.2 \text{ V})/100 = 2.0$ mA.   Therefore, at least when $V_1 = V_2$ and the current $I_{E1} = I_{E2}$, each of these transistors $T1$ and $T2$ operates at a current of 1 mA.

Next it is of interest to estimate the operating currents of $T4$ and $T5$. For this purpose we add the voltage rises encountered from ground, through $Z1$, through the emitter junction of $T4$, the 500-Ω resistor (carrying 1 mA) in the collector circuit of $T1$, and through the emitter junction of $T6$.   We then find that the collector voltage V_{C4} of $T4$ is

$$V_{C4} = V_{Z1} + V_{BE4} + 500 I_{C1} + V_{BE6} = 6.2 + 0.7 + 500 \times 10^{-3} + 0.7 = 8.1 \text{ V}$$
$$(2.15\text{-}4)$$

The drop across the 3.9-kΩ collector resistor of $T4$ is $12 - 8.1 = 3.9$ V.   Hence the current in $T4$ as in $T5$ is 1 mA and $V_{C5} = V_{C4}$.   (With $V_1 = V_2$, $I_{C1} = I_{C2}$, $I_{C4} = I_{C5}$, and diode $D2$ does not conduct.)

The output voltage when $V_1 = V_2$ is

$$V_o = V_{C5} - V_{BE7} - V_{Z2} = 8.1 - 0.7 - 6.2 = 1.2 \text{ V} \qquad (2.15\text{-}5)$$

which is nearly midway between the limited output voltages, -0.5 and 3.1 V.

The amplifier gain   We have noted that it is advantageous for a comparator to have a high gain.   We shall now estimate the gain of the comparator of Fig. 2.14-1.   With this gain we shall be able to estimate the input swing required to carry the comparator output from one level to the other.   This input swing serves as a measure of the precision with which the comparator accomplishes its purpose.

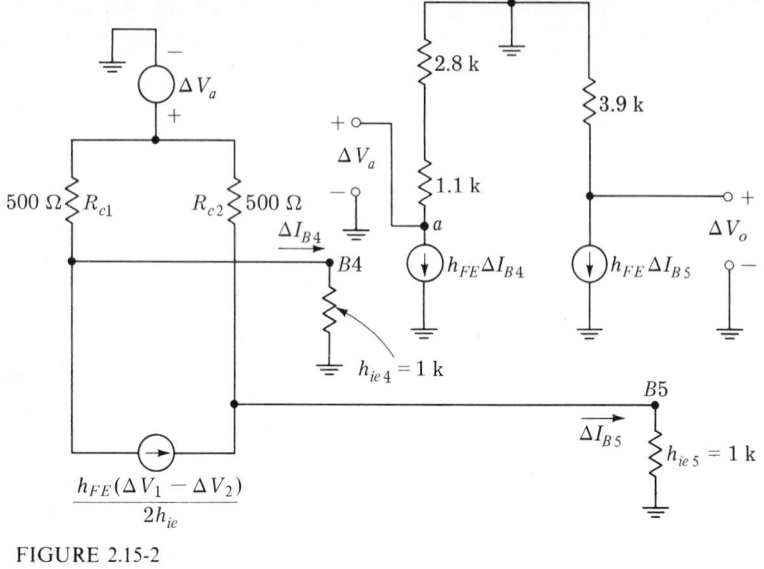

FIGURE 2.15-2

Equivalent circuit for the purpose of calculating the gain of the comparator of Fig. 2.14-1.

An equivalent circuit for calculating the gain is shown in Fig. 2.15-2. Here each transistor has been replaced by its simplified h-parameter equivalent, as in Fig. 2.5-2. Assuming the transistors have $h_{FE} = 40$ and since $I_E = 1$ mA, we find from Eq. (2.5-4) that $h_{ie4} = h_{ie5} = 1$ kΩ. The difference amplifier is replaced by its equivalent circuit as given in Fig. 2.5-4. The transistor $T6$, which provides a feedback path, and $T7$, which is the output transistor, are both emitter followers. We assume each has a gain of unity and each has an input impedance large enough to be ignored. Diode $D2$ has been omitted from the figure since it is reverse-biased when ΔV_1 and ΔV_2 are small.

We therefore have

$$\Delta I_{B4} = \frac{\Delta V_a}{1,500} - \frac{500}{500 + 1,000} \frac{h_{FE}}{2h_{ie}} (\Delta V_1 - \Delta V_2) \qquad (2.15\text{-}6a)$$

$$\Delta I_{B5} = \frac{\Delta V_a}{1,500} + \frac{500}{500 + 1,000} \frac{h_{FE}}{2h_{ie}} (\Delta V_1 - \Delta V_2) \qquad (2.15\text{-}6b)$$

and
$$\Delta V_a = -3,900 h_{FE} \, \Delta I_{B4} \qquad (2.15\text{-}7)$$

When we use Eqs. (2.15-6a) and (2.15-7) and note that

$$\Delta V_o = -3,900 h_{FE} \, \Delta I_{B5} \qquad (2.15\text{-}8)$$

it turns out that the comparator gain $A \equiv \Delta V_o/(\Delta V_1 - \Delta V_2) = -2,030$. Since the total swing at the output is $3.1 - (-0.5) = 3.6$ V, the corresponding input swing is $3.6/2,030 \approx 2$ mV.

A final note is in order to explain the function of transistor $T8$ in Fig. 2.14-1. Suppose we take into account the characteristics of the loads which the comparator will be called upon to drive. It then turns out that when V_o is at its higher level, the *total* current (not just an incremental current) at the output terminal is outward, i.e., *away* from the comparator. When the comparator is thus supplying current, it is described as *sourcing* current. When, on the other hand, the output V_o is at its lower level, the total current at the output is in the direction *into* the comparator. In this case the comparator is *sinking* current. This characteristic of acting as a source of current at one level and as a sink at the other level is a feature not only of comparators but of many of the digital devices we shall consider.

Returning now to Fig. 2.14-1, we note that the presence of $T7$ provides the mechanism through which the comparator can source current. Transistor $T7$ cannot provide for current sinking since its emitter current will not reverse. Transistor $T8$ provides for current sinking. When $V_o = -0.5$ V, the collector junction of $T8$ is forward-biased and the transistor is in saturation. The amount of current which the comparator can sink is then limited principally by the 1.7-kΩ emitter resistor. When $V_o = 3.1$ V, transistor $T8$ is in its active region. On account of the drop across $T8$, the current through the 1.7-kΩ resistor, which now reduces the available source current, is smaller than it would be if $T8$ were not present.

There is no essential reason why the 1.7-kΩ resistor is connected to $D1$. In principle, this resistor could be returned to the -6-V supply. However, it would then be necessary to provide a resistor from $D1$ to V_{CC} to furnish current to keep $D1$ conducting. The present connection eliminates one resistor and somewhat reduces the drain on the power supplies.

Effects of temperature   We now determine the variation of output voltage V_o produced by temperature changes. First consider that $V_o = -0.5$ V as occurs when $T5$ is saturated. From Fig. 2.14-1, we have

$$V_o = V_{Z1} + V_{CE5}(\text{sat}) - V_{BE7} - V_{Z2} \qquad (2.15\text{-}9)$$

Since zener diodes $Z1$ and $Z2$ have the same zener voltage, we shall assume that their voltage variation with temperature is the same. As a matter of fact, as noted in Sec. 1.4, at a zener voltage of 6.2 V, the temperature coefficient of a zener diode is approximately 0. Neglecting the temperature variation of V_{CE5} since $T5$ is saturated, we have

$$\frac{\Delta V_o}{\Delta T} = -\frac{\Delta V_{BE7}}{\Delta T} = +2 \text{ mV/}^\circ\text{C} \qquad (2.15\text{-}10)$$

Thus, over the temperature range from -55 to $+125^\circ$C the output voltage varies from -0.66 to -0.3 V.

Now consider that $V_o = 3.1$ V. In this case $T5$ and $T1$ are cut off. The incremental equivalent circuit used to find ΔV_{C5} due to a temperature change ΔT is shown in Fig. 2.15-3. Note that the 500-Ω collector resistor of $T1$ is

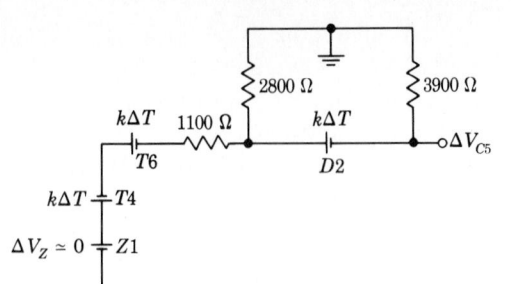

FIGURE 2.15-3
Equivalent circuit for the purpose of calculating the change ΔV_{C5} due to a temperature change ΔT when $V_o = 3.1$ V.

omitted from this circuit since we are assuming that $T1$ is cut off and that
negligible base current flows in $T4$. Also note that in this circuit each base-
emitter voltage drop is represented by its incremental value $-k\,\Delta T$.

Analyzing Fig. 2.15-3 yields (Prob. 2.15-4)

$$\Delta V_{C5} \approx -2k\,\Delta T \qquad (2.15\text{-}11)$$

Thus, from Fig. 2.14-1,

$$\Delta V_o = -\Delta V_{BE7} + \Delta V_{C5} \approx k\,\Delta T - 2k\,\Delta T$$
$$= -k\,\Delta T \qquad (2.15\text{-}12a)$$

and

$$\frac{\Delta V_o}{\Delta T} = -2 \text{ mV/}^\circ\text{C} \qquad (2.15\text{-}12b)$$

Note at this higher output ($V_o = 3.1$ V) the temperature dependence is equal in
magnitude but opposite in direction to the temperature sensitivity at the lower
output level ($V_o = -0.5$ V).

Figure 2.15-4 displays the experimentally determined transfer characteristic,
as supplied by the manufacturer for the comparator under discussion. Observe
that both in temperature sensitivity and range of input and output swings the
plot generally confirms our expectations.

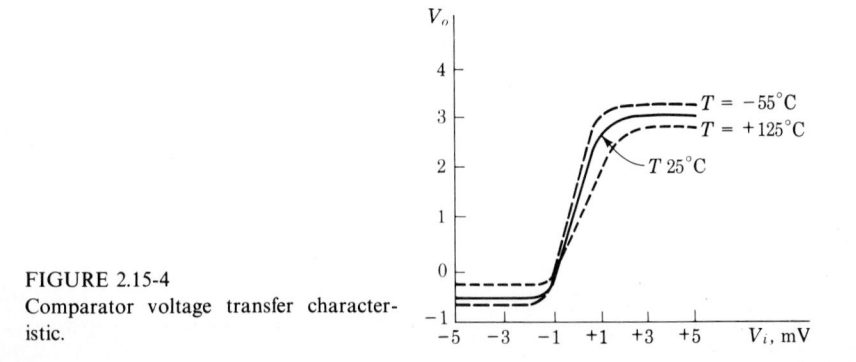

FIGURE 2.15-4
Comparator voltage transfer characteristic.

2.16   CHARACTERISTICS OF PHYSICAL COMPARATORS

Since comparators, like op-amps, are high-gain difference amplifiers, it is not surprising that some of the features specified by manufacturers for comparators are the same as those specified for op-amps.   Comparators have *input offset voltages* which are about the same as for op-amps and have *input offset currents* rather larger than for op-amps.   The offset current is of the order of 5 μA for a comparator as against about 50 nA for an op-amp.   (In comparators these offsets are measured with the output midway in its range.)

Both op-amps and comparators have common-mode rejections ratios in the range 60 to 100 dB.   In a comparator a high CMRR is necessary in order to assure that, when the reference voltage changes, the comparison point will change by precisely an equal amount.   But while op-amps have gains of the order of 100 dB, comparator gains are more usually about 60 dB.   High gain is of course of advantage in both cases.   In the op-amp the high open-loop gain allows us, through the use of negative feedback, to build a device which ends up with rather small amplifier gain but whose characteristic is very nearly independent of the amplifier itself and depends only on the external, generally passive, components in the feedback network.

Negative feedback is not employed with a comparator.   Thus the stability introduced by such feedback is not present.   With increasing gain the comparator amplifier displays increasing instability and inclination to oscillate.   If oscillations do occur, they will take place, of course, only while the comparator is in its high-gain region and not when the output is clamped at one of its limiting output levels.   Suppose that the comparator is swept through its transition from level to level in a time which is quite short in comparison with the oscillation period of the comparator.   In such a case, the instability, as with a

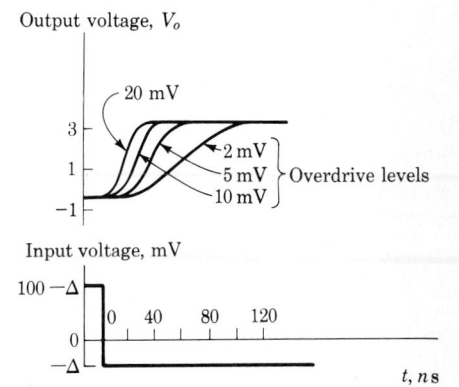

FIGURE 2.16-1
Response of comparator to a 100-mV
input signal for various overdrives.

Schmitt circuit (see below), may actually speed the transition and thereby serve a useful purpose. If, however, the input signal is a slowly varying waveform which keeps the comparator in its transition region for a long time, many oscillation cycles may occur and the effectiveness of the comparator will be impaired.

A matter of obvious interest is the *response time* of a comparator. To provide information on this point, manufacturers may supply plots like that in Fig. 2.16-1 (p. 77), which applies to the circuit of Fig. 2.14-1. Here the difference input of the comparator is initially set at 100 mV. This large input is so much greater than the ± 1 mV required to bring the comparator to one or another of its output levels that we can be sure that some of the comparator transistors will be driven well into saturation. Let us initially contemplate an input swing of 100 mV, from 100 mV to 0 V, which just brings the comparator to the midpoint of its output range. Such an input swing is characterized as bringing the input to the level of *zero* overdrive. Next the input driving waveform is lowered by $\Delta = 2$ mV so that at its lower level the input is carried to -2 mV. The comparator output now crosses from one level to the other and displays the waveform shown in Fig. 2.16-1 marked "2-mV overdrive level." Plots are also shown for higher overdrive levels. These waveforms provide information about the delay between the input drive and the initial comparator output response (due principally to the delay in bringing transistors out of saturation) and the time required for the comparator to cross the transition region.

2.17 THE SCHMITT TRIGGER CIRCUIT

We have seen that a comparator is a high-gain amplifier of a difference signal in which the limits to which the output is restricted due to overload are adjusted to accommodate to the device to be driven. We have noted the presence in the comparator of Fig. 2.14-1 of positive feedback, a useful feature since such feedback increases the amplifier gain. We consider now a type of comparator called a Schmitt trigger circuit, in which the positive feedback has been increased to the point where the comparator displays a range of instability.

Consider then the difference amplifier shown in Fig. 2.17-1 in which feedback has been introduced by virtue of the coupling from the output back to the *noninverting* terminal of the amplifier. The feedback is *positive*, i.e., *regenerative*, since the effect of the feedback is to increase the gain with feedback $A_f \equiv \Delta V_o / \Delta V_s$ of the amplifier. Let the gain of the amplifier itself (without the feedback) to a difference signal V_i be $-A \equiv \Delta V_o / \Delta V_i$, and let $\beta \equiv R_1/(R_1 + R_2)$ be the gain of the feedback path. Then we have

$$\Delta V_o = -A \Delta V_i = -A[\Delta V_s - \beta \Delta V_o] \tag{2.17-1}$$

so that the gain with feedback is

$$A_f \equiv \frac{\Delta V_o}{\Delta V_s} = \frac{-A}{1 - \beta A} \tag{2.17-2}$$

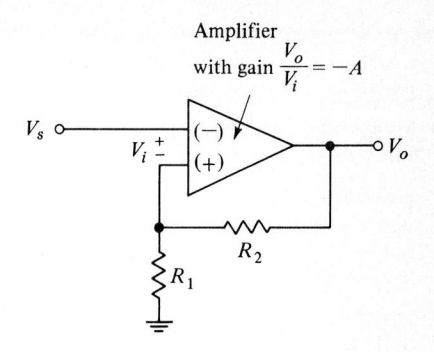

FIGURE 2.17-1
A basic Schmitt trigger.

The product $\beta A \equiv T$ is called the loop gain.[3] When $T = 0$, $A_f = -A$, and as T increases, the magnitude of A_f increases, becoming infinite at $T = 1$. Further increase in T reverses the sign of A_f.

Let us now assume that when $T = 0$, the amplifier has an input-output characteristic as in Fig. 2.17-2a. The amplifier limits at high and low voltage levels V_H and V_L. The excursion of the input $V_s(= V_i$ in this case) required

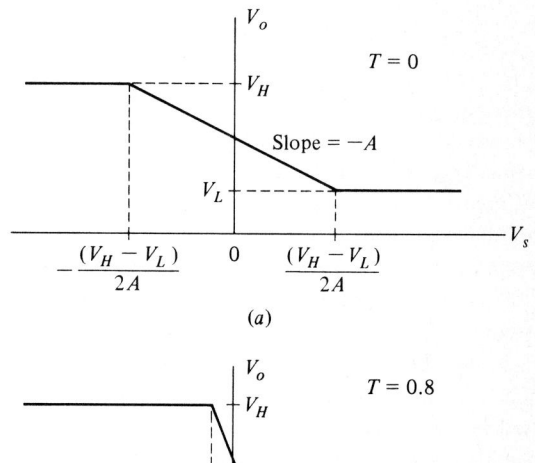

(a)

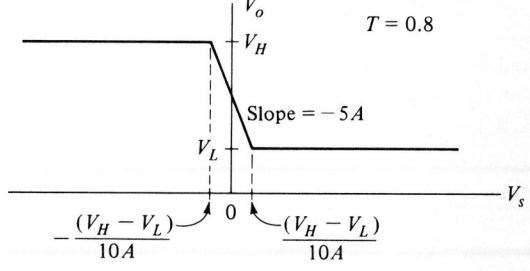

(b)

FIGURE 2.17-2
Transfer characteristic of the amplifier shown in Fig. 2.17-1 (a) when $T = 0$ and (b) when $T = 0.8$.

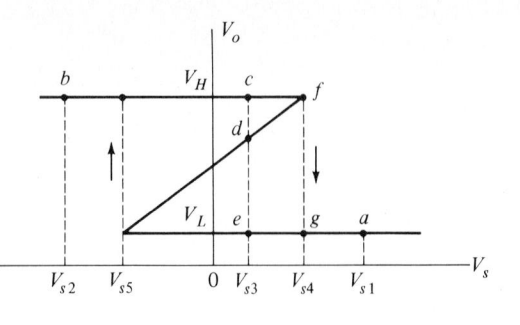

FIGURE 2.17-3
Transfer characteristic of the amplifier shown in Fig. 2.17-1 when $T > 1$. Here $A_f = A/(\beta A - 1) > 0$.

to swing the amplifier from limit to limit is symmetrical with respect to $V_s = 0$ extending over the range $\pm(V_H - V_L)/2A$. If T is increased to $T = 0.8$, Eq. (2.17-2) becomes $A_f = -5A$ and correspondingly the input-output plot appears as in Fig. 2.17-2b.

When T becomes larger than $T = 1$, the gain A_f becomes positive and the transfer characteristic displays the form shown in Fig. 2.17-3. When a comparator is operated in this manner, i.e., with $T > 1$, the comparator is called a *Schmitt trigger*. When V_s is V_{s1} or V_{s2}, the operating point of the Schmitt trigger is at a or at b, respectively. When $V = V_{s3}$, three equilibrium points c, d, or e are possible. The point d is an *unstable* equilibrium point. The Schmitt trigger may, in principle, reside indefinitely at d. However, if a slight disturbance or perturbation were to occur, the operating point would depart from d and eventually end up at c or e, depending on the direction of the perturbation. (The matter of the instability at d is explored in Prob. 2.17-1.) For the present, let it suffice to note that the gain equation (2.17-2) represents a steady-state relationship between V_o and V_s. (It does not take account of the time response of the amplifier to a change in input due to capacitance, etc.) With $V_s = V_{s3}$, two stable operating points are possible, c and e. At which of these points the Schmitt trigger finds itself depends, as we shall see, on past history.

Suppose then, that starting at b with $V_s = V_{s2}$, we increase V_s. Then the operating point will stay at the level $V_o = V_H$. When V_s passes $V_s = V_{s4}$, corresponding to operating point f, there is no longer an operating point at $V_o = V_H$ and the operating point must drop to g, where $V_o = V_L$. Similarly, with V_s initially at $V_s = V_{s1}$, if V_s is decreased, there will be a jump from V_L to V_H as V_s passes V_{s5}. We shall now show that in two important respects the Schmitt-circuit characteristic (Fig. 2.17-3) differs from the comparator characteristic (Fig. 2.17-2).

First we observe that in Fig. 2.17-3 the transitions from V_H to V_L and from V_L to V_H are made at different input voltages, i.e., at V_{s4} and V_{s5}. The Schmitt trigger thus exhibits *hysteresis*. The hysteresis lag $H \equiv V_{s4} - V_{s5}$, as can be verified (Prob. 2.17-2), is given by

$$H = \frac{T - 1}{A}(V_H - V_L) \qquad (2.17\text{-}3)$$

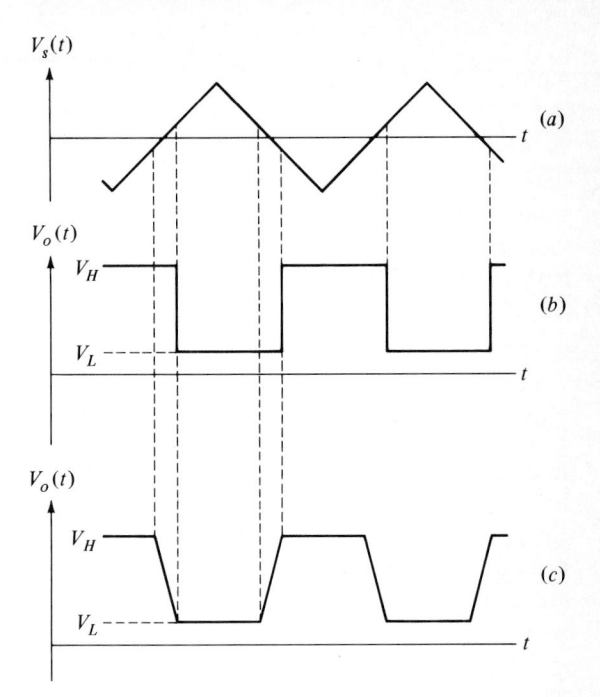

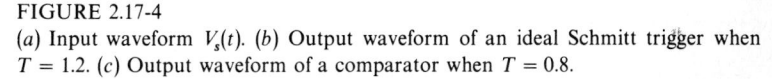

FIGURE 2.17-4
(a) Input waveform $V_s(t)$. (b) Output waveform of an ideal Schmitt trigger when $T = 1.2$. (c) Output waveform of a comparator when $T = 0.8$.

In a comparator application where hysteresis is not acceptable, the Schmitt trigger is simply not usable. In principle the hysteresis can be eliminated by setting T exactly equal to unity. In practice it is not possible to obtain a loop gain T which will maintain itself at unity over an extended period. As a consequence, comparators are operated with T comfortably set below unity while the Schmitt trigger is operated with T above unity.

Next it is to be noted that when a transition occurs in the output waveform V_o of a Schmitt trigger, the speed of the transition is determined by the response speed of the comparator amplifier itself. The time required to achieve the output transition is not determined, as in comparators, where $T < 1$, by the time required for the input signal to make an excursion through the input range of the amplifier. It is for this reason that the Schmitt trigger finds applicaton as a comparator where the rise and fall times of the input signal are slow and a very fast output rise and fall time is required. For example, there are devices which we shall want to have respond to input rising or falling waveforms which must be transmitted through small capacitors. In such a case the input waveforms will be severely attenuated if the speed of rise or fall is slow. In such cases the Schmitt circuit can be used very effectively to replace a slowly changing waveform by a fast waveform.

Figure 2.17-4 shows the response of a Schmitt trigger and a comparator to a slowly changing input. In each case the loop gain was adjusted so that $|A_f| = 5A$. Note, the output jump of the Schmitt trigger always occurs after the input waveform V_s crosses the origin. With regard to the transition time, we see that the transition time of the comparator is one-fifth of the transition time of the input V_s since the gain $A_f = 5$. However, the transition time of the Schmitt trigger is independent of the rise and fall times of V_s. Of course, there are delay, rise, and fall times associated with the comparator itself, which have not been shown in Fig. 2.17-4.

2.18   AN EXAMPLE OF A SCHMITT TRIGGER

A commercially available Schmitt trigger is shown in Fig. 2.18-1. Its output levels are appropriate for gates using emitter-coupled logic (ECL), discussed in Chap. 7. Transistors $T1$ and $T2$ form a difference amplifier while $T3$ and R_e constitute a constant current source. Two emitter-follower stages ($T4$ and $T5$) are provided and serve as buffers. Included in the integrated-circuit package is a regulated voltage supply (not shown in the figure), which provides the -3.9 V required at the base of $T3$. Resistors R_1 and R_2, through which positive feedback is furnished, are not part of the integrated-circuit package and are connected externally.

As before, we shall assume in the ensuing discussion that the base current of each transistor is small enough to be neglected, so that collector and emitter currents of the transistors are equal. Using $V_{BE3} = 0.7$ V, we find

$$I_{E3} = \frac{-3.9 - 0.7 + 5.2}{175} = 3.4 \text{ mA} \tag{2.18-1}$$

When the difference amplifier is at balance, the current $I_{E1} = I_{E2} = I_{E3}/2 = 1.7$ mA. Using Eq. (2.5-8), we then have that the gain of the difference amplifier, corresponding to $-A \equiv \Delta V_o/\Delta V_i$, where $V_i = V_{B1} - V_{B2}$, is

$$-A = \frac{-(I_{E3}/2)R_c}{2V_T} = -\frac{(1.7 \times 10^{-3})(300)}{2(25 \times 10^{-3})} = -10.2 \tag{2.18-2}$$

The output V_o is actually taken at the emitter of $T4$. We assume, however, that the emitter follower involving $T4$ has a gain of unity; hence the gain remains unaltered. (The output V'_o is made available as a convenience. When V_o is at V_H, V'_o is at V_L and vice versa.)

The high-level $V_o = V_H$ will occur when none of the current I_{E3} is flowing through the resistor $R_c = 300$ Ω in the collector of $T1$. The low-level $V_o = V_L$ will occur when all the current I_{E3} is flowing through R_c. Hence $V_H - V_L = R_c I_{E3}$. The hysteresis H is given by Eq. (2.17-3) as

$$H = \frac{T-1}{A}(V_H - V_L) = \frac{[R_1/(R_1 + R_2)]A - 1}{A} R_c I_{E3} \tag{2.18-3}$$

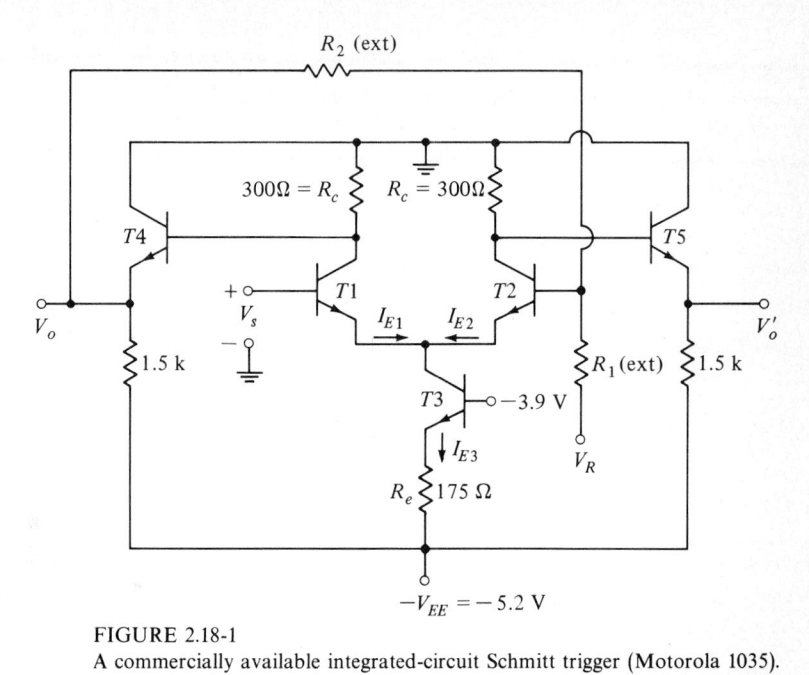

FIGURE 2.18-1
A commercially available integrated-circuit Schmitt trigger (Motorola 1035).

Using Eq. (2.18-2), we have

$$H = R_c I_{E3} \frac{R_1}{R_1 + R_2} - 4V_T \qquad (2.18\text{-}4)$$

If, typically, $R_1 = 200 \ \Omega$, $R_2 = 500 \ \Omega$, and with $I_{E3} = 3.4$ mA, $R_c = 300 \ \Omega$, and $V_T = 25 \times 10^{-3}$, we find $H = 190$ mV.

Adjusting the firing voltage of a Schmitt trigger We now turn our attention to the transfer function of the Schmitt trigger to determine the input voltage levels required to cause the Schmitt trigger to change state (these are called the *firing voltages*). We shall show that these levels are a function of the reference voltage V_R.

To determine the transfer function V_o versus V_s we refer to Fig. 2.18-1, from which we see that $V_i = V_{B1} - V_{B2}$ is

$$V_i = V_s - \left(\frac{R_1}{R_1 + R_2} V_o + \frac{R_2}{R_1 + R_2} V_R \right) \qquad (2.18\text{-}5)$$

In the linear region of operation [where $A_f = A/(T - 1)$] the input voltage $V_i = 0$ when $V_o = -I_{E3} R_c/2 - 0.7 \approx -1.2$ V. Using Eq. (2.18-5), we see that when $V_i = 0$ and $V_o = -1.2$ V, then

$$V_s = - \left[\frac{R_1}{R_1 + R_2} (1.2) - \frac{R_2}{R_1 + R_2} V_R \right] \qquad (2.18\text{-}6)$$

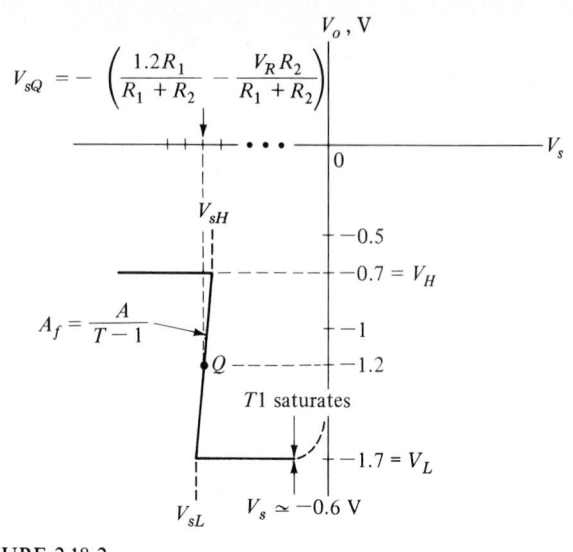

FIGURE 2.18-2
Transfer characteristic of the Schmitt trigger shown in Fig. 2.18-1.

Using Eqs. (2.18-6) and (2.17-2) enables us to determine the transfer function shown in Fig. 2.18-2. In this figure point Q represents the point $V_i = 0$, $V_o = -1.2$ V, and $V_s = V_{sQ}$ is given by Eq. (2.18-6). The slope of the transfer function at point Q is the gain $\Delta V_o / \Delta V_s = A_f = A/(T-1)$ [see Eq. (2.17-2)]. The maximum output voltage V_H occurs when all the current I_{E3} flows through $T2$ so that $T1$ is cut off. At this point $V_o = -0.7$ V. The minimum output voltage occurs when all the current I_{E3} flows through $T1$ so that

$$V_o = V_L = -0.7 - I_{E3} R_c \approx -0.7 - 1.02 \approx -1.7 \text{ V} \qquad (2.18\text{-}7)$$

It is interesting to note that if V_s increases above approximately -0.6 V, $T1$ begins to saturate and the output voltage rises. This result is readily obtained from Fig. 2.18-1, where we see that $V_{C1} = -1$ V when $V_o = -1.7$ V. To keep $T1$ comfortably out of saturation we require that $V_{CE1} \geq 0.3$ V. Since then $V_{BE1} \approx 0.7$ V, we find $V_s < -0.6$ V.

The firing voltages V_{sH} and V_{sL} shown in Fig. 2.18-2 are equal to $V_{sQ} \pm H/2$, where H is given by Eq. (2.18-4). For example if, as before, $R_1 = 200$ Ω, $R_2 = 500$ Ω, and if we require, say, that $V_{sQ} = -2$ V, V_R is found from Eq. (2.18-6) to be $V_R = -2.32$ V.

REFERENCES

1 Connelly, J. A. (ed.): "Analog Integrated Circuits," Wiley-Interscience, New York, 1975.
2 Wait, J. V., L. P. Huelsman, and G. A. Korn: "Introduction to Operational Amplifier Theory and Applications," McGraw-Hill, New York, 1975.
3 Schilling, D. L., and C. Belove: "Electronic Circuits: Discrete and Integrated," chap. 8, McGraw-Hill, New York, 1968.

LOGIC CIRCUITS

3.1 INTRODUCTION

In digital systems we encounter variables which are special in the sense that they are allowed to assume only two possible values. For example, in electrical digital systems, a typical variable voltage or current will have a waveform (ideally) like that indicated in Fig. 3.1-1. Here a voltage V makes abrupt transitions between the two voltage levels V_1 and V_2. Ideally, we consider that transitions, which occur at times $t = t_1$, t_2, etc., are so abrupt that we may say that at all times $V = V_1$ or $V = V_2$ and that V assumes no other value. Alternatively, if the transitions are not abrupt, we may take the attitude that we have no interest in the waveform except when $V = V_1$ or $V = V_2$. In Fig. 3.1-1 we have indicated that V_1 is positive while V_2 is negative. This feature is not essential. It may be that V_1 and V_2 are both positive or both negative; or it may even be that either V_1 or V_2 will have the value zero.

We may specify that value of the function $V(t)$ at any time by saying that $V = V_1$ or $V = V_2$; or, because of the very special character of V we may equally well say that V is "high" or "low," "up" or "down," etc. As a matter of fact, having selected any two words, we can arbitrarily assign one word

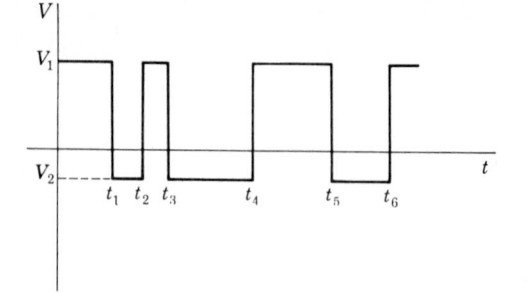

FIGURE 3.1-1
An idealized waveform encountered in
digital systems.

to one voltage level and the other to the second level. To suggest that V may
have one or the other level, at any particular time, but not both, a selection of
words implying opposite and mutually exclusive alternatives is most appropriate.
For this reason and for other reasons to be presented below we shall use the
words "true" (T) and "false" (F). It is necessary, in any given example, to
say whether true represents the level $V = V_1$ or $V = V_2$. Of course, then, false
represents the other level. If we decide arbitrarily that the more positive
voltage level represents true and the more negative represents false, we have
adopted the convention of *positive* logic. If we adopt the reverse convention, we
have *negative* logic.

In a particular digital system we may encounter many binary (two-valued)
variables such as the binary variable V shown in Fig. 3.1-1. We shall understand that in a given system, all such binary variables operate between the same
alternatives. For example, if in an electrical system, the binary variables are
voltages V_A, V_B, ..., then each voltage will operate between the same two
voltage levels, say $+6$ and -2 V.

3.2 FUNCTIONS OF A SINGLE BINARY VARIABLE

Let A be a binary variable, and let Z be a second binary variable which is a
function of A. Then

$$Z = f(A) \qquad (3.2\text{-}1)$$

What are the possible functional relationships between Z and A? Since A is
either true or false (T or F) and Z is either T or F, there are only two
possible functions. One possibility is that when A is true, Z is true and that
when A is not true, that is, A is false, Z also is not true (false). In this case
we would write

$$Z = A \qquad (3.2\text{-}2)$$

On the other hand, it might be that whatever the alternative assumed by
A, Z assumes the other alternative. In this case, when A is true, Z is false and
vice versa. Such a functional relationship is written

$$Z = \overline{A} \qquad (3.2\text{-}3)$$

and is read "Z equals '*not A*'" or "Z equals the '*complement* of A.'" Equation (3.2-3) says, for example, that when A is true, Z is not true, that is, Z is false. Equations (3.2-2) and (3.2-3) exhaust all the possible functions of a single binary variable.

It is to be noted that the complement, of the complement of a variable, is the variable itself, i.e.,

$$\bar{\bar{A}} = A \tag{3.2-4}$$

3.3 FUNCTIONS OF TWO BINARY VARIABLES

We consider now that the binary variable Z is a function of two binary variables, A and B, that is, $Z = f(A, B)$. A function is *defined* by providing a rule or other information through which Z, the dependent variable can be determined when the independent variables A and B are specified. Suppose the independent variables were continuous variables and therefore allowed an infinite range of values. In this case a function is usually defined by specifying the mathematical operation to be performed on the independent variables in order to determine the dependent variables. In the present case, however, since A and B may each assume only two values, there are only four possible combinations of the two variables. It is therefore feasible to define the function $Z = f(A, B)$ by specifically stating the values of Z for each of the combinations of A and B. Two different functions $f(A, B)$ and $g(A, B)$ are so specified in tabular form in Tables 3.3-1 and 3.3-2. In each table are listed each of the four possible combinations of A and B while in the right-hand column the corresponding value of Z is given. These tables and others like them are referred to as *truth tables* since we tabulate in them the conditions under which Z is true and false. The functions given in the truth tables of Tables 3.3-1 and 3.3-2 by way of example have been selected because they are of special interest and importance.

The AND operation The functions in Table 3.3-1 can be characterized in words by the statement "Z is true if and only if A is true *and* also B is true." Hence this function is referred to as the AND function, and the functional relationship is written $Z = A$ AND B. The AND operation (for reasons to be noted later) is also indicated in a more convenient notation, by placing a dot

Table 3.3-1

A	B	$Z = f(A, B)$
F	F	F
F	T	F
T	F	F
T	T	T

Table 3.3-2

A	B	$Z = g(A, B)$
F	F	F
F	T	T
T	F	T
T	T	T

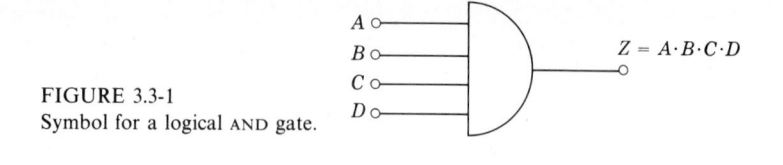

FIGURE 3.3-1
Symbol for a logical AND gate.

between the variables to suggest an operation of multiplication, although no actual operation of numerical multiplication is intended. Frequently, however, the dot is omitted so that the AND operation is written in any of the forms

$$Z = A \text{ AND } B = A \cdot B = AB \qquad (3.3\text{-}1)$$

As can be verified from the truth table, the AND operation is *commutative*, i.e.,

$$AB = BA \qquad (3.3\text{-}2)$$

The AND operation is also *associative*. Thus, suppose having formed the variable $Z = AB$, we then involve a third variable C and form the function $ZC = (AB)C$. Then this final function would be the same as would result if we had first formed BC and thereafter the function $A(BC)$. Thus

$$(AB)C = A(BC) \qquad (3.3\text{-}3)$$

If, then, many variables are connected by the AND operation, we need not indicate the order in which the variables were combined and can simply write $Z = ABCD \ldots$.

The symbol used to represent the connection of a number of variables through the AND operation is shown in Fig. 3.3-1. A device which effects such a connection between logical variables is called a *logical gate*. A logical AND gate which will accommodate four variables A, B, C, and D is indicated. Of interest to us is the situation in which the logical variables are represented by voltages. In such a case the intent of the symbol in Fig. 3.3-1 is to indicate that if (with respect to some reference point, "ground," not explicitly indicated) each of the input terminals A, B, C, and D is maintained at the voltage selected as representing true, then the output terminal Z will also be at this "true" voltage level. Otherwise if any inputs are maintained at the voltage representing false, the output will be at this "false" voltage level. Note that the AND gate singles out the situation of *simultaneity* or *coincidence*; i.e., all inputs must be simultaneously or coincidently true if the output is to be true. Accordingly an AND gate is sometimes called a *coincidence* gate.

The AND operation with switches We have noted that a voltage which can maintain only two allowable levels can be represented by a logical variable. A switch can also be represented by a logical variable. The switch can find itself in only two possible situations. It may be open, or it may be closed. Hence we may assign to it the logical variable A with the understanding that $A = \text{T}$ (true) means that the switch is closed while $A = \text{F}$ (false) means the switch

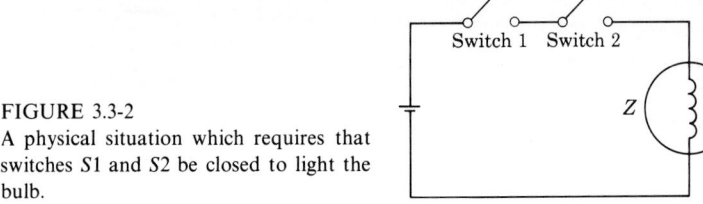

FIGURE 3.3-2
A physical situation which requires that switches $S1$ and $S2$ be closed to light the bulb.

is open. Or, if we please, we may reverse the assignment and agree that $A = T$ and $A = F$ represent that the switch is open and closed, respectively.

The AND operation performed by a circuit involving switches is indicated in Fig. 3.3-2. Let $A = T$ when $S1$ is closed, and $B = T$ when $S2$ is closed. Further let $Z = T$ when the light is lit. Then since the light will turn on when and only when the switches are closed simultaneously, the circuit operation is described by the logical equation $Z = AB$. If we should arbitrarily make the assignment that $A = F$ when $S1$ is closed, Z becomes $Z = \overline{A}B$.

3.4 THE OR FUNCTION

The truth table of Table 3.3-2 defines the OR operation. The appropriateness of the word "or" as a name for this function is to be seen from the fact that Z is true if A is true OR B is true OR if both A and B are true. The function $Z = A$ OR B is written

$$Z = A + B \qquad (3.4\text{-}1)$$

The notation of the plus sign to represent the OR function of course does not imply any operation of numerical addition. There is no sense in which A and B in Eq. (3.4-1) are being added. Nonetheless, as we shall see, the use of the plus sign (as well as the multiplication sign for the AND operation) when used in connection with other notation to be referred to shortly, will turn out to be a useful mnemonic device.

As can be verified from the truth table shown in Table 3.3-2, the OR operation, like the AND operation, is both commutative and associative. Thus

$$A + B = B + A \qquad (3.4\text{-}2)$$

and $$(A + B) + C = A + (B + C) \qquad (3.4\text{-}3)$$

The symbol for a logical OR gate is shown in Fig. 3.4-1.

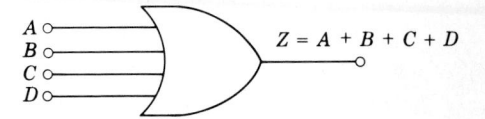

Figure 3.4-1
Symbol for a logical OR gate.

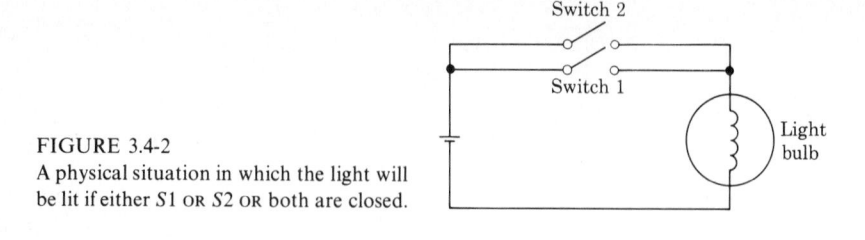

FIGURE 3.4-2
A physical situation in which the light will
be lit if either $S1$ OR $S2$ OR both are closed.

The OR operation accomplished with switches is shown in Fig. 3.4-2. If A represents the statement "$S1$ is closed," B represents "$S2$ is closed," and Z represents "the light is lit," then the operation of the circuit is described by $Z = A + B$. That is, the light is lit if $S1$ is closed OR if $S2$ is closed OR if both switches are closed.

3.5   THE NAND OPERATION AND THE NOR OPERATION

The NAND operation, i.e., $Z = A$ NAND B is represented symbolically by the logical equation

$$Z = A \uparrow B \tag{3.5-1}$$

and is defined in the third column of Table 3.5-1. In the fifth column we have tabulated $\overline{A \cdot B}$, that is, the complement of the AND operation applied to A and B. We note that the third and fifth columns are identical. Hence we can express the NAND operation as

$$Z = \overline{AB} \tag{3.5-2}$$

We shall invariably use Eq. (3.5-2) rather than Eq. (3.5-1) to represent the NAND operation. The name assigned to the operation is derived from Eq. (3.5-2), NAND being a contraction of NOT AND.

In the last column we have tabulated $\overline{A} + \overline{B}$ and note that

$$Z = \overline{AB} = \overline{A} + \overline{B} \tag{3.5-3}$$

Equation (3.5-3) is an example of *De Morgan's theorem*, considered again in Sec. 3.11.

Table 3.5-1

A	B	$Z = A \uparrow B$	$A \cdot B$	$\overline{A \cdot B}$	$\overline{A}$	$\overline{B}$	$\overline{A} + \overline{B}$
F	F	T	F	T	T	T	T
F	T	T	F	T	T	F	T
T	F	T	F	T	F	T	T
T	T	F	T	F	F	F	F

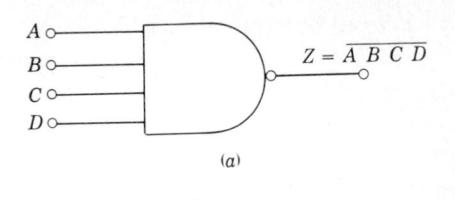

$$Z = \overline{A\ B\ C\ D}$$

(a)

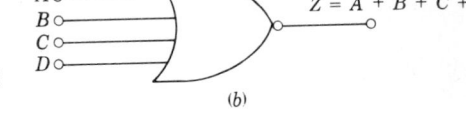

$$Z = \overline{A + B + C + D}$$

FIGURE 3.5-1
(a) Symbol for a NAND gate; (b) symbol for a NOR gate.

(b)

The NAND operation is commutative, i.e.,

$$\overline{AB} = \overline{BA} \tag{3.5-4}$$

However the NAND operation is *not* associative, i.e., as can be verified (Prob. 3.5-1),

$$(A \uparrow B) \uparrow C \neq A \uparrow (B \uparrow C) \tag{3.5-5a}$$

or, in the notation of Eq. (3.5-2),

$$\overline{(\overline{AB})C} \neq \overline{A(\overline{BC})} \tag{3.5-5b}$$

When the NAND operation is applied to more than two variables, it is not applied in the sense indicated by Eq. (3.5-5) but in the sense

$$Z = \overline{ABCD \cdots} \tag{3.5-6}$$

Equation (3.5-6) has an unambiguous meaning, while $Z = A \uparrow B \uparrow C \uparrow D \cdots$ is undefined until parentheses are added to indicate the order in which the variables are combined.

The symbol for the NAND gate is shown in Fig. 3.5-1a. It consists of an AND gate with a circle at the output. This circle represents the NOT operation.

The NOR operation The NOR operation, that is, A NOR B is represented symbolically by the equation

$$Z = A \downarrow B \tag{3.5-7}$$

and is defined in the third column of Table 3.5-2. As is seen in Table 3.5-2,

Table 3.5-2

A	B	$Z = A \downarrow B$	$A + B$	$\overline{A + B}$	$\overline{A}$	$\overline{B}$	$\overline{A}\,\overline{B}$
F	F	T	F	T	T	T	T
F	T	F	T	F	T	F	F
T	F	F	T	F	F	T	F
T	T	F	T	F	F	F	F

$Z = A \downarrow B$ can be expressed as

$$Z = \overline{A + B} \tag{3.5-8}$$

and we observe as well that

$$Z = \overline{A + B} = \overline{A}\,\overline{B} \tag{3.5-9}$$

which is again an example of *De Morgan's theorem*. We shall generally use the notation of Eq. (3.5-8) rather than (3.5-7). Equation (3.5-8) is also the source of the name assigned to the operation; i.e., NOR is a contraction of NOT OR.

Like the NAND operation, the NOR operation is commutative but not associative, i.e.,

$$\overline{\overline{A + B} + C} \neq \overline{A + \overline{B + C}} \tag{3.5-10}$$

When applied to more than two variables, the NOR operation is defined to mean

$$Z = \overline{A + B + C + D + \cdots} \tag{3.5-11}$$

The symbol for the NOR gate is given in Fig. 3.5-1b. It consists of an OR gate followed by a circle which represents the NOT operation.

As a matter of practice, most commercially available logical gates are NAND gates or NOR gates. However, the fact that these operations are not commutative makes it very inconvenient to design a logical system using such gates as building blocks. The procedure generally employed then involves a design using AND and OR gates and finally a conversion from this structure to one involving NAND and NOR gates. This matter is discussed further in Sec. 3.25.

3.6   THE EXCLUSIVE-OR OPERATION

The EXCLUSIVE-OR operation connecting two variables is represented symbolically by the equation

$$Z = A \oplus B \tag{3.6-1}$$

and is defined in the third column of Table 3.6-1. In words, Z is true if A is true OR if B is true provided that this condition of truth is *exclusive*, that is, A and B are not true simultaneously. From the last column of Table 3.6-1 we note that

$$Z = A \oplus B = A\,\overline{B} + \overline{A}\,B \tag{3.6-2}$$

Table 3.6-1

A	B	$Z = A \oplus B$	$\overline{A}$	$\overline{B}$	$\overline{A}\,B$	$A\,\overline{B}$	$A\,\overline{B} + \overline{A}\,B$
F	F	F	T	T	F	F	F
F	T	T	T	F	T	F	T
T	F	T	F	T	F	T	T
T	T	F	F	F	F	F	F

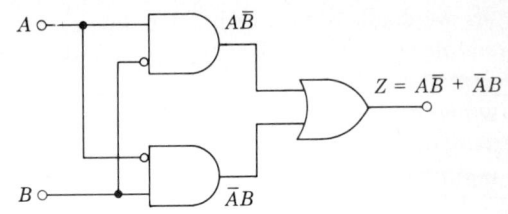

FIGURE 3.6-1
One realization of an EXCLUSIVE-OR gate.

This alternative form for Z expressed in words is "Z is true if A is true AND B is false OR if A is false AND B is true." One realization of $Z = A \oplus B$ based directly on Eq. (3.6-2) is shown in Fig. 3.6-1. The circuit uses two AND gates, one OR gate, and two NOT operations represented by the circles at the input to the AND gates. An implementation of the EXCLUSIVE-OR function with switches is considered in Prob. 3.6-1.

As can be verified, the EXCLUSIVE-OR operation is both commutative and associative. Hence when many variables are connected by the EXCLUSIVE-OR operation, we can write

$$Z = A \oplus B \oplus C \oplus D \cdots \tag{3.6-3}$$

without grouping the variables by parentheses. It is left for the student to show (Prob. 3.6-3) that where many variables are involved, as in Eq. (3.6-3), Z is true if an *odd* number of variables are true and Z is false if an even number of variables are true.

The symbol for an EXCLUSIVE-OR gate is shown in Fig. 3.6-2*a*. It turns out

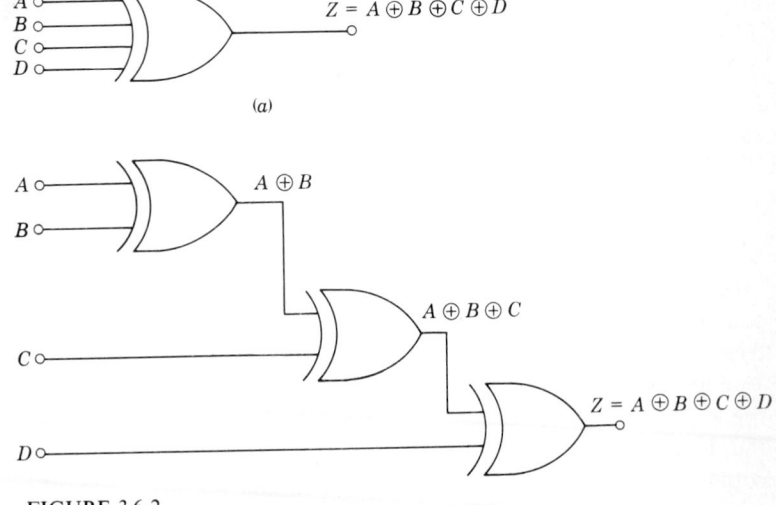

FIGURE 3.6-2
(*a*) Symbol for an EXCLUSIVE-OR gate. (*b*) An equivalent arrangement which accommodates four inputs using two input gates. Commercially available EXCLUSIVE-OR gates have only two inputs.

(as we shall discuss in great detail in succeeding chapters) to be a relatively simple matter to add additional inputs to an AND or NAND gate or to an OR or NOR gate. However, such is not the case with the EXCLUSIVE-OR gate. Hence while in Fig. 3.6-2a we have indicated a four-input gate, only two-input gates are available commercially. Thus, the arrangement of Fig. 3.6-2b which uses two-input gates is used to accommodate additional inputs.

3.7 OTHER FUNCTIONS

A function $Z = f(A, B)$ of two variables is defined by a truth table like Tables 3.3-1 and 3.3-2. Each different T and F entry in the Z column defines a different function. Since there are 4 entries in the column and each entry may be either T or F, there are 16 possible functions of 2 variables. We have considered five of the functions: AND, OR, NAND, NOR and EXCLUSIVE-OR. The remaining functions are considered in Prob. 3.7-1.

3.8 LOGICAL VARIABLES

The two-valued variables which we have been discussing are often called *logical* variables, while the operations such as the OR operation and the AND operation are referred to as *logical* operations. We shall now briefly discuss the relevance of such terminology, and in so doing we shall bring out the special aptness of the designations "true" and "false" to identify the possible values of a variable.

By way of example, suppose that you and two pilots are aloft in an airplane. You remain in the cabin, while the pilots, A and B, are in the cockpit. At some time, A joins you. This development causes you no concern. Suppose, however, that while you and A are in the cabin, you look up to find that B has also joined you. On the basis of your ability to reason *logically* you *deduce* that the plane is pilotless; and, presumably you sound an alarm so that one of the pilots will respond promptly to the urgency of the situation.

Alternatively, suppose that there had been attached to each pilot's seat an electronic device that provided an output voltage which is V_1 when the seat is occupied and V_2 when the seat is not occupied. Let us attach the designation "true" to the voltage level V_2 so that the level V_1 is "false." Let us further construct an electric circuit with two sets of input terminals and one set of output terminals. The circuit is to have the property that the output voltage will be V_2 if and only if both inputs, i.e., one input AND simultaneously the other, are at the level V_2. Otherwise the output is V_1. Finally let us connect the inputs to the devices on the chairs of pilots A and B and arrange that an alarm bell, connected to the output Z, respond when the output is V_2 ("true") and not otherwise. We have then constructed a circuit which performs the AND operation and is capable of making the *logical deduction* that the plane is unpiloted when, indeed, both pilots leave the cockpit.

To recapitulate, the situation is as follows: let the symbols A, B, and Z stand for the *propositions*

$A =$ It is true (T) that pilot A has left his seat.

$B =$ It is true (T) that pilot B has left his seat.

$Z =$ It is true (T) that the plane is pilotless and in danger.

Of course, $\overline{A}$, $\overline{B}$, and $\overline{Z}$ then represent the contrary propositions, respectively. For example, $\overline{A}$ represents the proposition that it is false (F) that pilot A has left the cockpit, etc.   The relationship among the propositions can now be written as

$$Z = AB \tag{3.8-1}$$

We have chosen to represent the *logical variables* A, B, and Z by electric voltages.   But it must be noted that actually Eq. (3.8-1) is a *relationship among propositions* and quite independent of the exact manner in which we choose to represent them or even whether we have any physical representation at all.   Equation (3.8-1) says that proposition Z is true if propositions A and B are both true and that otherwise proposition Z is false.

This algebra of propositions of which Eq. (3.8-1) is an example, is known as *Boolean algebra*.   Just as other algebras deal with variables which have a numerical significance, Boolean algebra deals with propositions and is an effective tool for analyzing the relationships between propositions which allow only two mutually exclusive alternatives.

EXAMPLE 3.8-1   A farmer named Jones has a large dog which is part wolf, a goat, and several heads of cabbage.   In addition, the farmer owns a north barn and a south barn. The farmer, dog, cabbages, and goat are all in the south barn.   The farmer has chores to perform in both barns.   However, if the dog is left with the goat when the farmer is absent, he will bite the goat, and if the goat is left with the cabbages, he will eat them.   To avoid either disaster, Jones asks us to build a small portable computer having four switches, representing the farmer, dog, goat, and cabbages.   If a switch is connected to a battery (V_1), the character represented by the switch is in the south barn; if the switch is connected to ground (V_2), the character is in the north barn.   The output of the computer goes to a lamp, which is to light whenever there is an impending disaster.   Thus the farmer can go about his chores, using the computer to tell him what he must take with him from one barn to another in order to avoid trouble.

How do we build this computer?

SOLUTION   To design the computer we must state precisely what it is we wish to do. We want the lamp to light if:

1   Farmer Jones is in the north barn AND the dog AND the goat are in the south barn, OR if

2   Farmer Jones is in the north barn AND the goat AND the cabbages are in the south barn, OR if

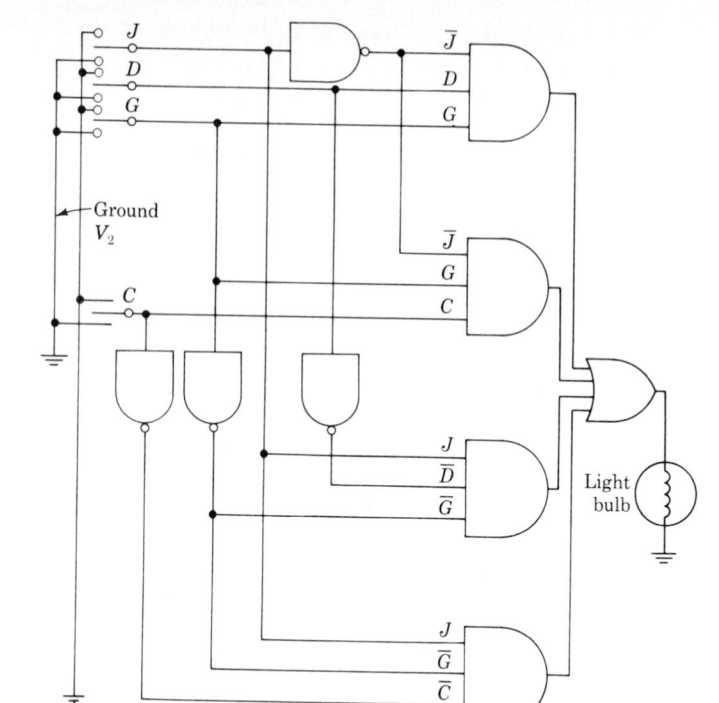

FIGURE 3.8-1
A special-purpose computer to solve the dog-goat-cabbage problem.

3 Farmer Jones is in the south barn AND the dog AND the goat are in the north barn, OR if
4 Farmer Jones is in the south barn AND the goat AND the cabbages are in the north barn.

We now use the logical variable J to represent the proposition "Jones is in the south barn." Correspondingly $\bar{J}$ represents the proposition "Jones is not in the south barn (he is in the north barn)." In this sense, similarly we have

D = dog in the south barn
$\bar{D}$ = dog in the north barn
G = goat in the south barn
$\bar{G}$ = goat in the north barn
C = cabbages in the south barn
$\bar{C}$ = cabbages in the north barn

We can now write a symbolic-logic statement which combines all the possibilities leading to a disaster:

$$L = \bar{J} \cdot D \cdot G + \bar{J} \cdot G \cdot C + J \cdot \bar{D} \cdot \bar{G} + J \cdot \bar{G} \cdot \bar{C}$$

where L, the logical variable representing the light, is $L = $ T when the light is ON.

To construct our special-purpose computer we need four switches (one for each of the characters J, D, G, and C), one lamp, one battery, and NOT, AND, and OR gates. The completed computer schematic is shown in Fig.3.8-1. Observe that to perform the NOT operation we have used a single-input NAND gate. As may be verified, a NAND gate with all its inputs joined so that the gate presents just one input terminal, will serve to perform the NOT operation. A single-input NOR gate would have served as well. The reader should study the circuit and verify that the lamp will light only if any of the four terms in the above expression for L is true.

3.9 THE 0, 1 NOTATION

We have previously referred to the two mutually exclusive possibilities for a proposition as true (T) or false (F). Then, depending on the situation, we would indicate that a proposition A is true or false by writing $A = T$ or $A = F$, respectively. We propose now, alternatively, to indicate that A is true by writing $A = \mathbf{1}$ and to indicate that A is false by writing $A = \mathbf{0}$. We use bold face for logical $\mathbf{0}$ and $\mathbf{1}$ to emphasize that $\mathbf{1}$ and $\mathbf{0}$ as used here are symbols, standing for T and F, and *not* numbers. These boldface symbols will not be used, however, in truth tables nor in Karnaugh maps (Sec. 3.18), since in such tables and maps there will be no ambiguity concerning the meaning of the symbol. With this notation, the truth tables for the AND and OR operations (Tables 3.3-1 and 3.3-2) appear in Tables 3.9-1 and 3.9-2.

If, further, we use "addition" to represent the OR operation and "multiplication" to represent the AND operation, the truth tables for OR and AND are represented equivalently by the following sets of equations:

$$
\begin{array}{cc}
\text{OR} & \text{AND} \\
A + B = Z & A \cdot B = Z \\
\hline
0 + 0 = 0 & 0 \cdot 0 = 0 \\
0 + 1 = 1 & 0 \cdot 1 = 0 \\
1 + 0 = 1 & 1 \cdot 0 = 0 \\
1 + 1 = 1 & 1 \cdot 1 = 1
\end{array}
\qquad (3.9\text{-}1)
$$

| Table 3.9-1 | TRUTH TABLE FOR AND USING 0, 1 NOTATION | | Table 3.9-2 | TRUTH TABLE FOR OR USING 0, 1 NOTATION | |

A	B	$Z = A$ AND B	A	B	$Z = A$ OR B
0	0	0	0	0	0
0	1	0	0	1	1
1	0	0	1	0	1
1	1	1	1	1	1

We then note that our notation constitutes an effective mnemonic device. For, if we pretend that addition and multiplication are really intended, and if we pretend hat the **0** and **1** are numbers, then with a simple exception $1 + 1 = 1$, all the above equations appear to conform to ordinary arithmetic.

3.10 NECESSARY AND SUFFICIENT OPERATIONS

We have noted that there are 16 possible functions of two variables. We have specifically discussed 5 of these and have observed that all can be expressed in terms of the AND, the OR, and the NOT operations. It turns out (Prob. 3.7-1) that *all* the possible functions can be expressed in terms of these three operations.

Pursuing the matter further, we find that actually we need not use all three operations because the AND operation itself can be expressed in terms of the OR and NOT operations, or, the other way around, the OR operation can be expressed in terms of the AND and NOT operations. The operator OR accomplished through the use of NOT gates and an AND gate is shown in Fig. 3.10-1a. The truth table to verify that the output Y is indeed $Y = A + B$ is given in Table 3.10-1. The column headed $Y = \overline{Z}$ in Table 3.10-1 is determined from the circuit of Fig. 3.10-1a. The last column headed $A + B$ gives $A + B$ by the definition of the OR operation. The complete equivalence of these two columns establishes that $Y = A + B$.

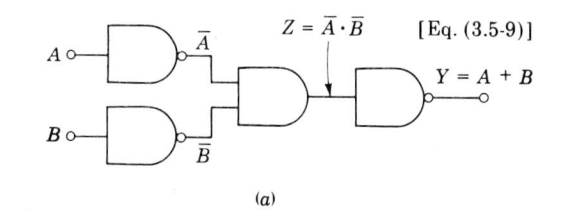

(a)

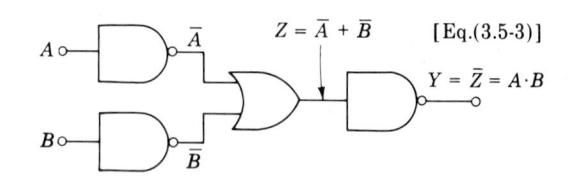

(b)

FIGURE 3.10-1

(a) Generation of the OR operation using NOT and AND gates. [A NAND (or a NOR) gate with all of its inputs joined serves as a NOT gate.] (b) Generation of the AND operation using NOT and OR gates.

NAND gate

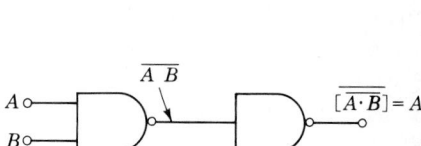

(a)

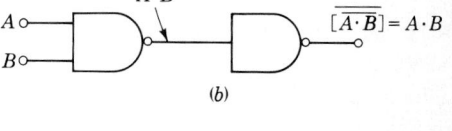

(b)

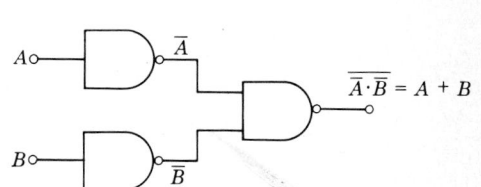

(c)

FIGURE 3.10-2
Showing that the NAND operation is sufficient. (a) the NOT operation, (b) the AND operation, (c) the OR operation.

NOR gate

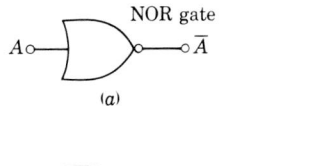

(a)

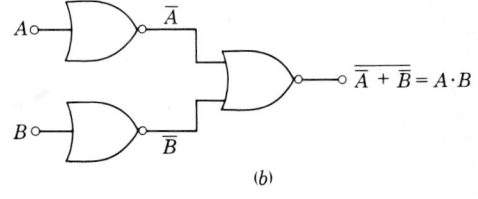

(b)

FIGURE 3.10-3
Showing that the NOR operation is sufficient. (a) the NOT operation, (b) the AND operation, (c) the OR operation.

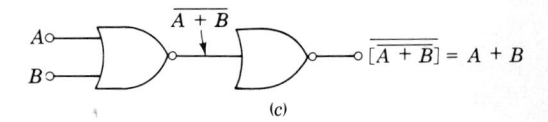

(c)

Table 3.10-1

A	B	$\overline{A}$	$\overline{B}$	$Z = \overline{A} \cdot \overline{B}$	$Y = \overline{Z}$	$A + B$
0	0	1	1	1	0	0
0	1	1	0	0	1	1
1	0	0	1	0	1	1
1	1	0	0	0	1	1

A circuit which performs the AND operation using only NOT gates and an OR gate is shown in Fig. 3.10-1*b*. A truth table like Table 3.10-1 establishes that in this case Y is indeed $Y = AB$.

While in principle the NOT operation together with either the AND operation or the OR operation is sufficient, in the analysis and design of logical systems, we shall use all three operations since it is by far more convenient to do so.

Finally, it is most instructive to note that the NAND operation above or the NOR operation also is sufficient to generate all possible functions. The use of the NAND gate to perform the NOT, AND, and OR operations is shown in Fig. 3.10-2. The use of the NOR gate to achieve the same purpose is shown in Fig. 3.10-3.

3.11 BOOLEAN ALGEBRAIC THEOREMS

We shall now develop a number of theorems which are useful in simplifying expressions of logical variables. We shall restrict our attention to theorems involving the OR and AND connectives and the NOT operation. We shall use the symbols for addition and multiplication to represent these connectives, and we shall employ the **0**, **1** notation to represent that a logical proposition is true or false, respectively.

We may note, initially, in connection with the NOT operation, that the following results follow rather self-evidently from the fact that a variable A can have only two values, $A = 1$ or $A = 0$. Thus,

$$\bar{\bar{A}} = A \qquad (3.11\text{-}1)$$

since if $A = 1$, then $\bar{A} = 0$ and $\bar{\bar{A}} = 1$ again. Similarly $\bar{0} = 1$ and $\bar{1} = 0$. Also

$A + 0 = A$	$(3.11\text{-}2a)$	$A \cdot 1 = A$	$(3.11\text{-}2b)$
$A + 1 = 1$	$(3.11\text{-}3a)$	$A \cdot 0 = 0$	$(3.11\text{-}3b)$
$A + A = A$	$(3.11\text{-}4a)$	$A \cdot A = A$	$(3.11\text{-}4b)$
$A + \bar{A} = 1$	$(3.11\text{-}5a)$	$A \cdot \bar{A} = 0$	$(3.11\text{-}5b)$

These eight theorems involve a single variable. The proof of any individual theorem is readily accomplished by taking advantage of the fact that the variable can have only two possible values. Thus, it is feasible to prove a theorem by actually considering every possible value of the variable and showing that the theorem holds in each case. Consider, for example, Eq. (3.11-5a). If $A = 1$, we have $1 + 0 = 1$, which is correct. If $A = 0$ we have $0 + 1 = 1$, which again is correct.

We may take note of an interesting relationship between the theorems listed in the left-hand column above and those in the right-hand column. Given an equation in one column, the corresponding equation in the other column can be written by (1) interchanging $+$ and $\cdot$ signs and (2) interchanging **0** and **1**. Theorems which are related to one another by this double interchange are called *duals.* That this duality should exist is not surprising in view of the duality apparent in the truth tables of the OR and AND operations. Examining truth tables 3.9-1 and 3.9-2 and Eq. (3.9-1), we find that the first equation in the OR table and the fourth equation in the AND table are duals. Similarly, the second OR equation and the third AND equation are duals, and so on for the remaining equations.

The *distributive law* applies to our algebra of logical variables. That is, given an expression involving a parenthesis like $A(B + C)$, we can remove the parenthesis by "multiplying" through as in ordinary algebra and we get $A(B + C) = AB + AC$. In its dual forms the distributive law appears as

$$A + BC = (A + B)(A + C) \tag{3.11-6a}$$

and

$$A(B + C) = AB + AC \tag{3.11-6b}$$

Note, again, in Eq. (3.11-6), there being no **1**s or **0**s in the equation, that the equations differ only in the interchange of additions and multiplications. Equation (3.11-6b) is rather intuitively acceptable because of its correspondence with ordinary algebra. Equation (3.11-6a) looks a little strange and needs getting used to. In any event, the student can verify Eq. (3.11-6) by substituting all possible combinations of variables in the equation, there being eight such combinations.

We now tabulate some other useful theorems, in each case in their dual forms.

$$A + AB = A \tag{3.11-7a}$$
$$A(A + B) = A \tag{3.11-7b}$$

$$A + \bar{A}B = A + B \tag{3.11-8a}$$
$$A(\bar{A} + B) = AB \tag{3.11-8b}$$

$$AB + A\bar{B} = A \tag{3.11-9a}$$
$$(A + B)(A + \bar{B}) = A \tag{3.11-9b}$$

$$AB + \bar{A}C = (A + C)(\bar{A} + B) \tag{3.11-10a}$$
$$(A + B)(\bar{A} + C) = AC + \bar{A}B \tag{3.11-10b}$$

$$AB + \bar{A}C + BC = AB + \bar{A}C \tag{3.11-11a}$$
$$(A + B)(\bar{A} + C)(B + C) = (A + B)(\bar{A} + C) \tag{3.11-11b}$$

EXAMPLE 3.11-1   Prove that Eq. (3.11-8a) is correct.

SOLUTION   The verification of Eq. (3.11-8a) follows directly from a comparison of the truth tables for $A + \bar{A}B$ and for $A + B$, shown below.

A	B	$Z_1 = A + \bar{A} \cdot B$	$Z_2 = A + B$
0	0	$0 + 0 = 0$	$0 + 0 = 0$
0	1	$0 + 1 = 1$	$0 + 1 = 1$
1	0	$1 + 0 = 1$	$1 + 0 = 1$
1	1	$1 + 0 = 1$	$1 + 1 = 1$

Thus $Z_1 = Z_2$, and Eq. (3.11-8a) is correct.

EXAMPLE 3.11-2 Prove that Eq. (3.11-11b) is correct.

SOLUTION The verification of (3.11-11b) follows directly from a comparison of the truth tables for $(A + B)(\bar{A} + C)(B + C)$ and for $(A + B)(\bar{A} + C)$, shown below. Note that there are eight different entries in the truth table. This results since A, B, and C can each, independently take on one of two possible values. Thus A, B, and C can have $2 \cdot 2 \cdot 2 = 2^3$ different values. The system of ordering employed in the truth table is explained in Sec. 3.13.

A	B	C	$Z_1 = (A + B)(\bar{A} + C)(B + C)$	$Z_2 = (A + B)(\bar{A} + C)$
0	0	0	$0 \cdot 1 \cdot 0 = 0$	$0 \cdot 0 = 0$
0	0	1	$0 \cdot 1 \cdot 1 = 0$	$0 \cdot 1 = 0$
0	1	0	$1 \cdot 1 \cdot 1 = 1$	$1 \cdot 1 = 1$
0	1	1	$1 \cdot 1 \cdot 1 = 1$	$1 \cdot 1 = 1$
1	0	0	$1 \cdot 0 \cdot 0 = 0$	$1 \cdot 0 = 0$
1	0	1	$1 \cdot 1 \cdot 1 = 1$	$1 \cdot 1 = 1$
1	1	0	$1 \cdot 0 \cdot 1 = 0$	$1 \cdot 0 = 0$
1	1	1	$1 \cdot 1 \cdot 1 = 1$	$1 \cdot 1 = 1$

Since $Z_1 = Z_2$ for all possible combinations of A, B, and C, the theorem is verified.

Finally, we take note of one last additional theorem, of sufficient importance to deserve being singled out. Known as *De Morgan's theorem*, it applies to an arbitrary number of variables and in its dual forms is given by

$$\overline{A \cdot B \cdot C \cdots} = \bar{A} + \bar{B} + \bar{C} + \cdots \qquad (3.11\text{-}12a)$$

$$\overline{A + B + C + \cdots} = \bar{A} \cdot \bar{B} \cdot \bar{C} \cdots \qquad (3.11\text{-}12b)$$

In words, these equations say that (1) the complement of a product of variables is equal to the sum of the complements of the individual variables and (2) the complement of a sum of variables is equal to the product of the complements of the individual variables.

The proof of De Morgan's theorem for two variables was presented in Sec. 3.5 in the discussion of the NAND and NOR operations [see Eqs. (3.5-3) and (3.5-9)]. The extension of the theorem to an arbitrary number of variables is left as a problem (Prob. 3.11-5).

3.12 AN EXAMPLE

As an illustration of the application of the theorems of the previous section, let us consider the following example.

Suppose that a student at a university consults the school bulletin to determine whether or not he is eligible to take a particular course in electronics. He finds that a student may take the course if and only if he satisfies the following conditions:

1   He has completed at least 60 credits and is an engineering student in good standing, or

2   He has completed at least 60 credits and is an engineering student and has departmental approval, or

3   He has completed fewer than 60 credits and is an engineering student on probation, or

4   He is in good standing and has departmental approval, or

5   He is an engineering student and does not have departmental approval.

Let us see how these rather involved specifications can be simplified. For this purpose we introduce logical variables A, B, C, D, and Z and make assignments as follows:

A = student has completed at least 60 credits

B = student is an engineering student

C = student is in good standing

D = student has departmental approval

Z = student may take the electronics course

Thus, by way of example, $C = 1$ represents the proposition that it is true that the student is in good standing. Correspondingly $C = 0$, or equivalently $\bar{C} = 1$, represents the proposition that the student is on probation, etc. Then we can write the entire specification in the logical algebraic equation

$$Z = ABC + ABD + \bar{A}B\bar{C} + CD + B\bar{D} \tag{3.12-1}$$

with the understanding that any combination of values of variables which yields $Z = 1$ results in satisfying the required conditions.

Combining the second and fifth terms in Eq. (3.12-1) and factoring out the B from these terms, as is allowed by Eq. (3.11-6b), we have

$$Z = ABC + \bar{A}B\bar{C} + CD + B(\bar{D} + DA) \tag{3.12-2}$$

From Eq. (3.11-8a) $\bar{D} + DA = \bar{D} + A$; making this simplification and then removing the parenthesis from Eq. (3.12-2), we have

$$Z = ABC + AB + \bar{A}B\bar{C} + CD + B\bar{D} \tag{3.12-3}$$

Factoring out the product AB from the first two terms of Eq. (3.12-3) and noting, as in Eq. (3.11-3a), that $C + 1 = 1$, we have

$$Z = AB + \bar{A}B\bar{C} + CD + B\bar{D} \tag{3.12-4}$$

Factoring out the B from the first two terms of Eq. (3.12-4) and recognizing, as in Eq. (3.11-8a), that $A + \bar{A}\bar{C} = A + \bar{C}$, we see that Eq. (3.12-4) becomes

$$Z = AB + B\bar{C} + CD + B\bar{D} \tag{3.12-5}$$

From Eq. (3.11-11a) we note that given an expression $CD + B\bar{D}$, which appears as the last two terms in Eq. (3.12-5), we can add the term BC without changing the truth value of the expression. Adding such a term and combining it with the second term in Eq. (2.13-5), we have

$$Z = AB + B(\bar{C} + C) + CD + B\bar{D} \qquad (3.12\text{-}6)$$

Since, as given in Eq. (3.11-5a), $\bar{C} + C = 1$ and $B \cdot 1 = B$, as in Eq. (3.11-2b), we then find that

$$Z = AB + B + CD + B\bar{D} \qquad (3.12\text{-}7)$$

We now combine the first, second, and fourth terms in Eq. (3.12-7) so that Z is

$$Z = B(1 + A + \bar{D}) + CD \qquad (3.12\text{-}8)$$

Using Eq. (3.11-3a), which states that $1 + X = 1$, and letting $X = A + \bar{D}$ in Eq. (3.12-8) yields

$$Z = B + CD \qquad (3.12\text{-}9)$$

Thus it appears that the rather involved statement above can be simplified to read "the student may take the course if he is an engineering student, OR if he is a student in good standing AND has departmental approval."

The method employed in this section to simplify the expression of Eq. (3.12-1) leaves something to be desired. It requires some ingenuity and foresight to know how best to combine terms, when to add terms with a view toward eventual simplification, and which theorems to use. Beginning in Sec. 3.16 we shall discuss an alternative simplification method which employs a much more routine and systematic procedure.

3.13 THE BINARY NUMBER SYSTEM

In the number system of everyday life, the *decimal* system, ten digits are employed, 0, 1, ..., 9. A number larger than 9 is represented through a convention which assigns a significance to the *place* or *position* occupied by a digit. For example, by virtue of the positions occupied by the individual digits in the number 6,903, this number has a numerical significance calculated as

$$6{,}903 = 6 \times 10^3 + 9 \times 10^2 + 0 \times 10^1 + 3 \times 10^0 \qquad (3.13\text{-}1)$$

We note, as in Eq. (3.13-1), that a number is expressed as the sum of *powers of ten* multiplied by appropriate coefficients. In the decimal system, ten is called the *radix* or *base* of the system.

It is, of course, entirely feasible, and often very useful, to work with a system which has a base other than ten. In digital systems, a number system with the base *two* is especially useful. Such a system is called a *binary* system and uses just two digits, 0 and 1. The advantage of the use of a binary system in connection with digital systems lies in the fact that we may arrange a one-to-one correspondence between the two digits (numbers) 0 and 1 and the two possible truth values (F or T) of a logical variable represented by the symbols (not numbers) **0** and **1**. As a matter of fact, the association becomes, on occasion, so intimate and so convenient that it is useful to lose sight of the distinction.

In the binary system, the individual digits represent the coefficients of powers of *two* rather than *ten*, as in the decimal system. For example, the decimal number 19 is written in the binary representation as 10011 since

$$10011 = 1 \times 2^4 + 0 \times 2^3 + 0 \times 2^2 + 1 \times 2^1 + 1 \times 2^0 \qquad (3.13\text{-}2a)$$

$$16 \quad + \quad 0 \quad + \quad 0 \quad + \quad 2 \quad + \quad 1 \quad = 19 \quad (3.13\text{-}2b)$$

A short list of equivalent numbers in decimal and binary notation is given in Table 3.13-1.

The conversion of a number from binary to decimal representation is illustrated in Eq. (3.13-2). The conversion from decimal to binary representation is just as readily accomplished by the procedure indicated in Table 3.13-2. The procedure is the following: the decimal number (in this case 19) is placed at the extreme right. We divide by 2 and place the *quotient* to the left and the *remainder* beneath the quotient. The process is repeated (for the next column we have 9/2 yields a quotient of 4 and a remainder of 1) until a quotient of 0 is obtained. The array of 1s and 0s in the second row is the binary representation of the number originally expressed in decimal form. In this example, we find that decimal 19 = 10011 binary.

Conversion of numbers less than unity A number which is not an integer can be expressed using *decimal-point* notation, for example, the number 1.8125 has the numerical significance

$$1.8125 = 1 \times 10^0 + 8 \times 10^{-1} + 1 \times 10^{-2} + 2 \times 10^{-3} + 5 \times 10^{-4} \qquad (3.13\text{-}3)$$

Table 3.13-1   EQUIVALENT NUMBERS IN DECIMAL AND BINARY NOTATION

Decimal notation	Binary notation	Decimal notation	Binary notation	Decimal notation	Binary notation	Decimal notation	Binary notation
0	00000	6	00110	12	01100	18	10010
1	00001	7	00111	13	01101	19	10011
2	00010	8	01000	14	01110	20	10100
3	00011	9	01001	15	01111	21	10101
4	00100	10	01010	16	10000	22	10110
5	00101	11	01011	17	10001	23	10111

Table 3.13-2   DECIMAL-TO-BINARY CONVERSION

Divide by 2	0	1	2	4	9	19 decimal
Remainder	1	0	0	1	1	Binary

Similarly, we can express a noninteger number using *binary-point* notation. Thus, the binary number

$$1.1101 = 1 \times 2^0 + 1 \times 2^{-1} + 1 \times 2^{-2} + 0 \times 2^{-3} + 1 \times 2^{-4} \quad (3.13\text{-}4)$$

Thus, *bits* to the right of the binary point multiply 2^{-n}, where n is the distance of the bit to the right of the binary point.

Conversion from a decimal number to a binary number when the number is *less* than unity proceeds as shown in Table 3.13-3. Here we see the conversion of the decimal number 0.8125 to the binary number 0.1101. The procedure is as follows. The decimal number (in this case 0.8125) is placed at the extreme *left*. We *multiply* by 2. If the resulting product is greater than or equal to 1, we subtract 1 from the product and place the difference to the right. We also put a 1 underneath. If the resulting product is less than unity, we put a 0 underneath. This procedure is repeated until the difference is zero. In this example, we find that $0.8125 = 0.1101$. Generally, unlike the situation which appears in the present case, even when the decimal number has a finite number of significant digits, the binary equivalent may not terminate.

3.14 THE GREY REFLECTED BINARY CODE

The binary number system discussed in Sec. 3.13 represents numbers by a sequence of 0s and 1s. The sequence is a "natural" one and generally understood since it follows the pattern of significance assigned to digit positions in the more familiar decimal representation. If we please, however, we can represent a number by a sequence of 0s and 1s which is entirely arbitrarily contrived. To avoid ambiguity, it would, of course, be necessary that each numerical value have assigned to it a distinctive and individual sequence.

Number representation systems other than the natural representation are called number *codes* since we must be given the "code" (the rules of assignment of sequence to number) in order to determine the numerical value represented by the sequence. Many binary codes, useful in a variety of different applications in digital systems have been devised. We shall consider here the *Grey reflected binary code*, which is often used in digital systems.

In the Grey code, the first two numbers, zero and one, are represented as in the natural binary representation by the digits 0 and 1. The next two numbers, two and three, are expressed in a representation arrived at as shown in Fig. 3.14-1a. A "mirror" represented by the dashed line is placed below the

Table 3.13-3   DECIMAL-TO-BINARY NUMBER CONVERSION FOR NUMBERS LESS THAN 1

Decimal 0.8125	$0.8125 \times 2 = 1.625$ $1.625 - 1 = 0.625$	$0.625 \times 2 = 1.25$ $1.25 - 1 = 0.25$	$0.25 \times 2 = 0.5$ 0.5	$0.5 \times 2 = 1.0$ $1 - 1 = 0$
Binary	1	1	0	1

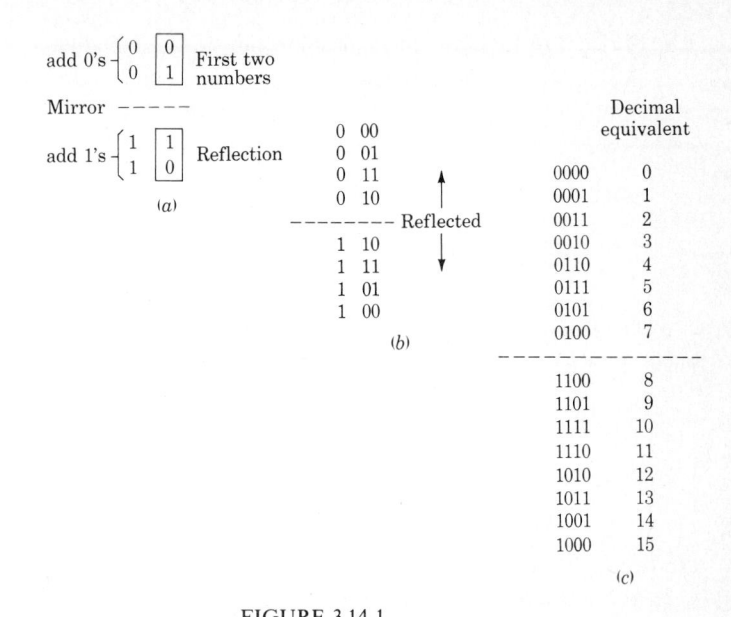

FIGURE 3.14-1
The development of the Grey binary reflected code.

first two numbers, giving rise to the reflection indicated. Thereafter the digit 0 is added above the mirror and the digit 1 added below the mirror. Thus the decimal numbers 0, 1, 2, 3 are represented respectively in the Grey code by 00, 01, 11, 10. Successive repetitions of the reflective process allow us to extend the Grey-code representation. The first eight Grey-code representations are given in Fig. 3.14-1b. The two right-hand rows of digits above and below the mirror are reflections of one another. The leftmost column is 0 above and 1 below the mirror. Figure 3.14-1c lists the first 16 Grey-code representations.

An especially useful feature of the Grey code is that representations of successive numbers differ from one another in only a single digit. By way of example, note that the number 7 (decimal) = 0100 (Grey) while the number 8 (decimal) = 1100 (Grey), a change in only the first digit. By contrast, note that 7 = 0111 (natural binary) while 8 = 1000 (natural binary). In this case all four digits have changed. This feature of the Grey code is of advantage in a number of situations.

3.15 STANDARD FORMS FOR LOGICAL FUNCTIONS: THE STANDARD SUM OF PRODUCTS

With a view toward developing a procedure for simplifying logical functions, we introduce at this point two *standard forms* in which logical functions may be expressed. The first of these, the *standard sum-of-products* form, is illustrated in the following examples.

EXAMPLE 3.15-1   Given the logical function of four variables

$$f(A, B, C, D) = (\bar{A} + BC)(B + CD) \tag{3.15-1}$$

express the function as a *sum of products*.

SOLUTION   Using the distributive law [Eq. (3.11-6b)], we find

$$f(A, B, C, D) = (\bar{A} + BC)B + (\bar{A} + BC)CD \tag{3.15-2a}$$
$$= \bar{A}B + BBC + \bar{A}CD + BCCD \tag{3.15-2b}$$
$$= \bar{A}B + BC + \bar{A}CD + BCD \tag{3.15-2c}$$

since, from Eq. (3.11-4b) $BB = B$, etc.

EXAMPLE 3.15-2   Given the logical function of five variables

$$f(A, B, C, D, E) = (A + \overline{BC})(\overline{D + BE}) \tag{3.15-3}$$

express the function as a *sum of products*.

SOLUTION   Using De Morgan's theorem and the distributive law, we find

$$f(A, B, C, D, E) = (A + \overline{BC})(\overline{D + BE}) \tag{3.15-4a}$$
$$= (A + \bar{B} + \bar{C})[\bar{D}(\overline{BE})] \tag{3.15-4b}$$
$$= (A + \bar{B} + \bar{C})[\bar{D}(\bar{B} + \bar{E})] \tag{3.15-4c}$$
$$= (A + \bar{B} + \bar{C})(\bar{B}\bar{D} + \bar{D}\bar{E}) \tag{3.15-4d}$$
$$= A\bar{B}\bar{D} + A\bar{D}\bar{E} + \bar{B}\bar{D} + \bar{B}\bar{D}\bar{E} + \bar{B}\bar{C}\bar{D} + \bar{C}\bar{D}\bar{E} \tag{3.15-4e}$$

These examples indicate how an arbitrary logical expression can be written as a *sum of products*. If only individual variables appear complemented, as in the first example, we need only the distributive law. If a complement sign appears over a combination of variables, as in the second example, we must first use De Morgan's theorem. In any event, it is always possible to write a logical expression as a simple sum of terms, each term being a product of some combination of variables, some complemented, some not. The same variable need never appear twice in a product. For if such a repetition of a variable or a complemented variable should develop in the course of multiplying out, we can eliminate the repetition through the use of the theorem $AA = A$ or $\bar{A}\bar{A} = \bar{A}$. If, on the other hand, we should find in a term a product $A\bar{A}$, the entire term can be dropped, since $A\bar{A} = \mathbf{0}$.

We notice that in the sum-of-products expressions of Eqs. (3.15-2c) and (3.15-4e) the individual terms do not all involve the same number of variables, and, as happens in these particular examples, no term involves all the variables. A further standardization leading to an expression in which *all* terms do involve *all* the variables (complemented or uncomplemented) can be achieved as indicated in the following illustration.

EXAMPLE 3.15-3   Given the logical function of three variables

$$f(A, B, C) = A + \bar{B}C \tag{3.15-5}$$

in which the individual terms do not involve all three variables, rewrite Eq. (3.15-5) so that all three variables do appear in each term.

SOLUTION   In the first term of Eq. (3.15-5) neither the variable B nor C appears. We therefore multiply this term by $(B + \bar{B})(C + \bar{C})$. This operation does not change the significance of the logical function since $B + \bar{B} = C + \bar{C} = \mathbf{1}$, as noted in Eq. (3.11-5a). Similarly, since the variable A does not appear in the second term of Eq. (3.15-5), we multiply that term by $A + \bar{A}$.   Equation (3.15-5) then becomes

$$f(A, B, C) = A(B + \bar{B})(C + \bar{C}) + (A + \bar{A})\bar{B}C \tag{3.15-6}$$

Using the distributive law, we have

$$f(A, B, C) = ABC + AB\bar{C} + A\bar{B}C + A\bar{B}\bar{C} + A\bar{B}C + \bar{A}\bar{B}C \tag{3.15-7}$$

Eliminating the duplication of the term $A\bar{B}C$, we find

$$f(A, B, C) = ABC + AB\bar{C} + A\bar{B}C + A\bar{B}\bar{C} + \bar{A}\bar{B}C \tag{3.15-8}$$

The form of Eq. (3.15-8), in which a sum of products appears, each product involving all the variables, is called the *standard sum-of-products* form. Each individual term is referred to as a *minterm*.   The relevance of this terminology will appear in Sec. 3.19.

The merit of the standard sum-of-products form as in Eq. (3.15-8) is that certain information about the logical function becomes immediately available upon inspection.   Since the form involves a logical sum of terms, the function f has the truth value $f = 1$ whenever any one (or more) of the terms has the logical value **1**.   Consider, then, the first term in Eq. (3.15-8).   This term, ABC, being a product, will have the logical value **1** only when $A = 1$ and $B = 1$ and $C = 1$.   Thus, this term tells us that $f = 1$ when $A = 1$, $B = 1$, and $C = 1$. Similarly the presence of the second term $(AB\bar{C})$ tells us that $f = 1$ when $A = 1$ and $B = 1$ and $\bar{C} = 1$ or equivalently when $A = 1$ and $B = 1$ and $C = 0$. Each minterm, then, specifies a combination of truth values of the individual variable for which the function has the truth value **1**.   And all the minterms, collectively, specify all the combinations of truth values of the variables for which the function has the truth value **1**.   With N variables there are 2^N minterms. If all minterms are present, $f = 1$ identically.

Each minterm corresponds to a row of the function truth table in which the function f has the value $f = 1$.   Thus, for example, suppose we examine a three-variable truth table and find in examining one row that $f = 1$ when $A = 0$, $B = 1$, and $C = 0$.   Then, if f is expressed as a sum of minterms we shall find that one of the minterms is $\bar{A}B\bar{C}$.   The number of minterms in an expression for f is the same as the number of rows of the truth table in which $f = 1$.   In the next section we consider a second standard form.

3.16   THE STANDARD PRODUCT OF SUMS

In the second standard form, the *standard product of sums*, a function is expressed as a *product of terms*, each term consisting of a sum, the sum involving *all* of the variables in either complemented or uncomplemented form.

In the previous section we arrived at a standard sum of products form by using the rule [Eq. (3.11-5a)] $A + \bar{A} = 1$ and using the distributive rule in the form of Eq. (3.11-6b). We can now establish an alternative standard form, the *standard product-of-sums* form, by using the duals of these two rules. Thus we use the distributive rule in the form $A + BC = (A + B)(A + C)$ [see Eq. (3.11-6a)], and we use the rule $A\bar{A} = 0$. To illustrate the development of the standard product of sums we use again the three-variable function of Eq. (3.15-5).

EXAMPLE 3.16-1   Given the logical function of three variables

$$f(A, B, C) = A + \bar{B}C \qquad (3.16\text{-}1)$$

express f in the standard product-of-sums form.

SOLUTION   Using the distributive law in the form of Eq. (3.11-6a), we have

$$f(A, B, C) = A + \bar{B}C \qquad (3.16\text{-}2a)$$
$$= (A + \bar{B})(A + C) \qquad (3.16\text{-}2b)$$

Since the first factor lacks the variable C and the second lacks the variable B, we add $C\bar{C} = 0$ to the first and $B\bar{B} = 0$ to the second. We then have

$$f(A, B, C) = (A + \bar{B} + C\bar{C})(A + B\bar{B} + C) \qquad (3.16\text{-}3a)$$
$$= (A + \bar{B} + C)(A + \bar{B} + \bar{C})(A + B + C)(A + \bar{B} + C) \qquad (3.16\text{-}3b)$$

To obtain this result we have noted that, using Eq. (3.11-6a),

$$(A + \bar{B}) + C \cdot \bar{C} = [(A + \bar{B}) + C][(A + \bar{B}) + \bar{C}] \qquad (3.16\text{-}4)$$

Since the first and fourth factor in Eq. (3.16-3b) are a duplication, and since $XX = X$, we eliminate the duplication and have

$$f(A, B, C) = (A + \bar{B} + C)(A + \bar{B} + \bar{C})(A + B + C) \qquad (3.16\text{-}5)$$

Each of the factors in Eq. (3.16-5) is referred to as a *maxterm*.

Just as the standard sum-of-products form makes apparent, by inspection, the combination of truth values of variables for which $f = 1$, so the standard product-of-sums form, as in Eq. (3.16-5), specifies the combination of truth values of the variables for which $f = 0$. For $f = 0$ whenever one (or more) maxterms have the truth value 0, and the maxterm will have the truth value 0 only when *all* terms in the sum have the truth value 0. Thus, from the first factor in Eq. (3.16-5) we observe that $f = 0$ when $A = 0$ and $\bar{B} = 0$ and $C = 0$ or, equivalently, when $A = 0$ and $B = 1$ and $C = 0$.

Each maxterm corresponds to a row of the function truth table in which the function f has the value $f = 0$. Thus, for example, suppose we examine a three-variable truth table and find, in examining one row, that $f = 0$ when $A = 0$, $B = 1$ and $C = 0$. Then, if f is expressed as a product of maxterms, we shall find that one of the maxterms is $(A + \bar{B} + C)$. The number of maxterms in an expression for f is the same as the number of rows of the truth table in which $f = 0$.

In summary, then, a logical function can be written in the standard sum-of-products form or in the standard product-of-sums forms. In the first case the function is expressed in terms of minterms, which specify the 1s of the function. In the second case the function is expressed in terms of maxterms, which specify the 0s of the function.

The same function [Eq. (3.15-5)] is expressed in terms of minterms in Eq. (3.15-8) and in terms of maxterms in Eq. (3.16-5). We can readily verify the consistency between these two forms. We note that since there are 3 variables involved, there are $2^3 = 8$ possible combinations of variables. It turns out that there are 5 combinations for which $f = 1$, and therefore there are 5 minterms in Eq. (3.15-8). We then anticipate that when f is expressed in terms of maxterms, there should be $8 - 5 = 3$ such terms, as indeed there are in Eq. (3.16-5). Next, and by way of example, we note that the maxterm $A + \bar{B} + C$ means that $f = 0$ if $A = 0$ and $B = 1$ and $C = 0$. If this combination of variables were supposed to yield $f = 1$, then in Eq. (3.16-1) we would have to have a term $\bar{A}B\bar{C}$.

3.17 MINTERM AND MAXTERM SPECIFICATIONS OF LOGICAL FUNCTIONS

The standardization of logical function forms introduced in the previous sections allows us to introduce a very convenient shorthand notation to express logical functions.

Let us, by way of example, consider a logical function involving three variables, A, B, and C. Let us agree that in writing a particular minterm we shall always write the minterm with the variables in that order. That is, a typical minterm would be written, say, $A\bar{B}C$ and never $\bar{B}AC$ or $AC\bar{B}$, etc. Next we assign the binary numerical digit 0 to each complemented variable and the binary digit 1 to each uncomplemented variable. On this basis the minterm $A\bar{B}C$ would be represented by the number 101 (binary), which is equal to 5 (decimal). We then refer to the minterm $A\bar{B}C$ by the symbol m_5. In this same way minterm $\bar{A}\bar{B}\bar{C}$ becomes m_0 and minterm ABC becomes m_7.

When we deal with maxterms, the order of assigning binary digits is reversed. A complemented variable is assigned the binary numerical digit 1 and an uncomplemented variable the digit 0. The maxterm $A + \bar{B} + C$ is then represented by the number 010 (binary), which is equal to 2 (decimal). We then refer to the maxterm $A + \bar{B} + C$ by the symbol M_2. In this same way, maxterm $\bar{A} + \bar{B} + \bar{C}$ becomes M_7, and maxterm $A + B + C$ becomes M_0. The assignment of symbols to minterms and to maxterms for the three-variable case is given in Table 3.17-1.

A logical function may be expressed most conveniently in terms of our new notation. Thus suppose that in a function of three variables we find that when it is expressed in minterm form, only minterms m_0, m_3, m_6, and m_7 are present. Then we would write this function $f(A, B, C)$ as

$$f(A, B, C) = m_0 + m_3 + m_6 + m_7 \qquad (3.17\text{-}1)$$

or, even more simply,

$$f(A, B, C) = \sum m(0, 3, 6, 7) \qquad (3.17\text{-}2)$$

Similarly, if we found that in some function $f(A, B, C)$, expressed in terms of maxterms, there appeared only the maxterms M_0, M_2, M_3, and M_7, we would write

$$f(A, B, C) = M_0 \cdot M_2 \cdot M_3 \cdot M_7 \qquad (3.17\text{-}3)$$

or, more simply

$$f(A, B, C) = \prod M(0, 2, 3, 7) \qquad (3.17\text{-}4)$$

As an additional example involving minterms we write below a function the binary designations of the minterms, the decimal designation, and finally the shorthand expression for the function:

$$f(A, B, C, D) = \bar{A}\bar{B}\bar{C}\bar{D} + \bar{A}BC\bar{D} + A\bar{B}C\bar{D} + ABCD \qquad (3.17\text{-}5)$$
$$\qquad\quad 0000 \qquad 0110 \qquad 1010 \qquad 1111$$
$$\qquad\qquad 0 \qquad\quad 6 \qquad\quad 10 \qquad\quad 15$$

$$f(A, B, C, D) = \sum m(0, 6, 10, 15) \qquad (3.17\text{-}6)$$

Table 3.17-1   DESIGNATION OF MINTERMS AND MAXTERMS

Variable complemented (C) or uncomplemented (U)			Minterm and designation	Maxterm and designation
A	B	C		
C	C	C	$\bar{A}\bar{B}\bar{C} = m_0$	$\bar{A} + \bar{B} + \bar{C} = M_7$
C	C	U	$\bar{A}\bar{B}C = m_1$	$\bar{A} + \bar{B} + C = M_6$
C	U	C	$\bar{A}B\bar{C} = m_2$	$\bar{A} + B + \bar{C} = M_5$
C	U	U	$\bar{A}BC = m_3$	$\bar{A} + B + C = M_4$
U	C	C	$A\bar{B}\bar{C} = m_4$	$A + \bar{B} + \bar{C} = M_3$
U	C	U	$A\bar{B}C = m_5$	$A + \bar{B} + C = M_2$
U	U	C	$AB\bar{C} = m_6$	$A + B + \bar{C} = M_1$
U	U	U	$ABC = m_7$	$A + B + C = M_0$

As an example using maxterms, consider

$$f(A, B, C, D)$$
$$= (A + B + C + D)(A + \bar{B} + C + \bar{D})(\bar{A} + B + \bar{C} + \bar{D})(\bar{A} + \bar{B} + \bar{C} + \bar{D})$$

0000	0101	1011	1111
0	5	11	15

$$\text{(3.17-7)}$$

$$f(A, B, C, D) = \prod M(0, 5, 11, 15) \qquad \text{(3.17-8)}$$

There is a complementary type of relationship between a function expressed in terms of minterms and the same function expressed in terms of maxterms. For suppose we are dealing with a function of, say, three variables. Let us say that when $A = 1$, $B = 0$, and $C = 1$, the function is $f = 0$. Then, when f is expressed as a sum of minterms, the minterm $A\bar{B}C$, that is, minterm m_5 (101) will not be present. Since, however, $f = 0$ for this combination of variables, the term $\bar{A} + B + \bar{C}$ will indeed appear as one of the maxterms. This maxterm is M_5. Thus, in general, if minterm m_k is *not present* in the minterm list, maxterm M_k *will be present* and vice versa. For example, if a function of 3 variables with 8 combinations of variables, numbered from 0 to 7, is $f = \prod M(0, 3, 6, 7)$ then this *same function* can be written as $f = \sum m(1, 2, 4, 5)$.

Similarly, suppose we write a function f in terms of the variable combinations which make $f = 1$, that is, in terms of minterms. Then suppose we write a second function g related to f in that each minterm has been replaced by an equal-numbered maxterm. Then the **0**s of g will be the **1**s of f and vice versa, so that $g = \bar{f}$. For example, for a three-variable case, if

$$f = \sum m(0, 3, 6, 7) \qquad \text{then} \qquad \bar{f} = \prod M(0, 3, 6, 7) \qquad \text{(3.17-9)}$$

These relationships between minterms and maxterms of the same and of complementary functions has an exact correspondence in a truth table. The fact that a maxterm M_k will appear in a function if minterm m_k is not present corresponds to the fact that rows where the function f is not $f = 1$ must be a row where $f = 0$. Similarly, a function f and its complement $\bar{f}$ have truth tables in which the **0**'s and **1**'s for f are interchanged.

3.18   THE KARNAUGH-MAP REPRESENTATION OF LOGICAL FUNCTIONS

A Karnaugh map (K map) is a diagram which provides an area to represent every row of a truth table. The usefulness of the K map rests in the fact that the particular manner of locating areas makes it possible to simplify a logical expression by visual inspection. In the following sections we shall see how the K map allows such simplification. In the present section we shall only explain the relationship between the K map and the truth table. We shall illustrate the relationship with the case of a function $f(A, B)$ of two variables.

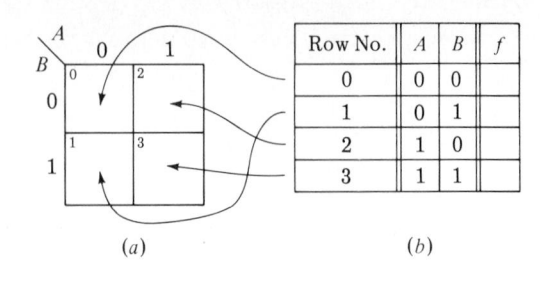

FIGURE 3.18-1
The relationship between the Karnaugh
map in (*a*) and the truth table in (*b*).   (*a*)   (*b*)

A K map for the two-variable case $f(A, B)$ is shown in Fig. 3.18-1*a*.   A two-variable truth table is shown in *b*.   The column for the value of the function $f(A, B)$ is left blank since we are interested, for the moment, in the table itself and not in use of the table to define a specific function *f*.   The rows of the table have been labeled decimally.   These row numbers have been arrived at by assigning a numerical significance to the **0**'s and **1**'s in the truth table. Thus the row with $AB = \mathbf{1\ 0}$ is read as row $1\ 0 = 2$ (decimal), etc.   The K map in *a* has four boxes corresponding, as shown by the arrows, one to one with the rows in the truth table.   The K map boxes have been numbered in two equivalent manners.   On the one hand, the decimal-row number has been written in the corner of the box.   On the other hand, the columns and rows have been labeled in binary fashion.   Thus, for example box 2 appears at the intersection of column $A = \mathbf{1}$ and row $B = \mathbf{0}$, i.e., at $AB = 10(=2)$ corresponding to row 2 of the truth table.

The K map may now be used as a replacement for the truth table.   That such is the case is to be seen in Fig. 3.18-2.   Here we have assumed a function $f(A, B)$ and the entries in each row have been transferred to the appropriate K-map box.   Thus the K map and the truth table convey the same information.

The function *f*, defined in Fig. 3.18-2, may be written in terms of its minterms.   We find, since **1**'s appear in rows 0 and 3 that

$$f(A, B) = \bar{A}\bar{B} + AB = m_0 + m_3 \tag{3.18-1}$$

Row No.	A	B	f
0	0	0	1
1	0	1	0
2	1	0	0
3	1	1	1

FIGURE 3.18-2
The K map presents the same information
as a truth table.   (*a*)   (*b*)

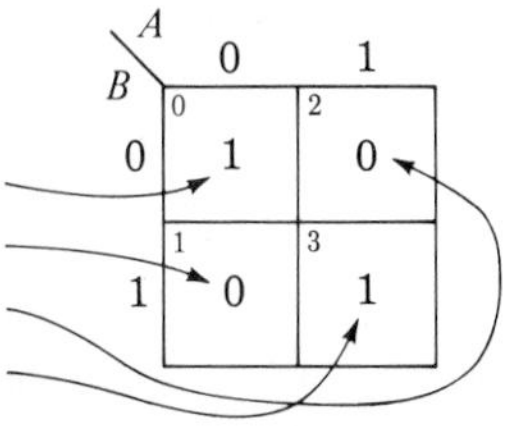

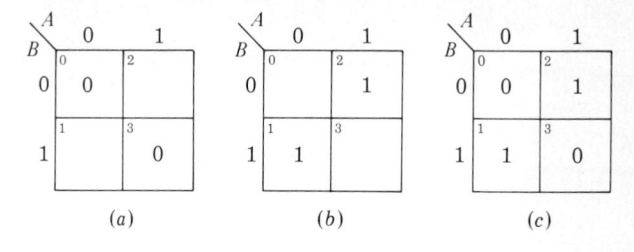

FIGURE 3.18-3

The function $f(A, B) = (A + B)(\bar{A} + \bar{B})$ is represented in (a) in terms of its maxterms and in (b) in terms of its minterms. The map in (c) displays, redundantly, both maxterms and minterms.

Similarly, the function may be written in terms of its maxterms. We find, since **0**'s appear in rows 1 and 2 that

$$f(A, B) = (A + \bar{B})(\bar{A} + B) = M_1 M_2 \qquad (3.18\text{-}2)$$

In Eqs. (3.18-1) and (3.18-2) we have expressed the function as a sum of minterms or as a product of maxterms by noting the location of **1**'s and **0**'s in the truth table or, equivalently, in the K map. The procedure may be reversed. Given the minterms of the function we may represent the function on the K map by locating **1**'s in the corresponding boxes. Or, given the maxterms of the function, we may represent the function on the K map by locating **0**'s in the corresponding boxes. To represent a function we may also enter **1**'s and **0**'s on the K map. However, for simplicity it is adequate to enter only **1**'s or only **0**'s. Where **1**'s are not entered **0**'s are understood and vice versa.

By way of example, let us find the K map representation of $f(A, B) = (A + B)(\bar{A} + \bar{B})$. We have

$$f(A, B) = (A + B)(\bar{A} + \bar{B}) = M_0 M_3 \qquad (3.18\text{-}3)$$

and we also have

$$f(A, B) = (A + B)(\bar{A} + \bar{B}) = A\bar{B} + \bar{A}B = m_2 + m_1 \qquad (3.18\text{-}4)$$

Equation (3.18-3) yields the K map representation shown in Fig. 3.18-3a, and Eq. (3.18-4) yields the equivalent K map in b. Figure 3.18-3c, in which entries of both **1**'s and **0**'s are made, is clearly equivalent to a and to b.

A point which is especially to be noted in connection with entering minterms and maxterms in a K map is the following. The box in which minterm m_i is to be entered is the *same* box as the box in which maxterm M_i is to be entered. Thus, if a function f has a maxterm, say, $\bar{A} + B = M_2$, then a **0** is entered in the box identified by 2 (decimal) with the row and column coordination $A = 1$ and $B = 0$. If a function g has a minterm $A\bar{B} = m_2$, a **1** is entered in the same box.

FIGURE 3.19-1
A two-variable K map.

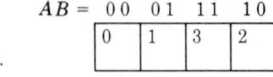

$AB = $

00	01	11	10
0	1	3	2

3.19 KARNAUGH-MAP REPRESENTATIONS FOR TWO, THREE, AND FOUR VARIABLES

An alternative K map for two variables is given in Fig. 3.19-1. Note especially the ordering of identification numbers for the boxes. Observe that the ordering is that of the Grey reflected binary code discussed in Sec. 3.14. The reason for this ordering will be discussed shortly. But because of this ordering, the minterm and maxterm designations proceed in the order 0, 1, 3, 2 rather than 0, 1, 2, 3.

We may note at this point the significance of the expressions minterm and maxterm. Suppose we have, by way of example, a function $f(A, B) = A\bar{B}$, that is, a function given by a single minterm. Then the corresponding K map appears as in Fig. 3.19-2a. Here we have chosen to include the 0s which are normally omitted. We now observe that a minterm fills with 1s the *minimum* possible area of the K map short of filling no area at all. Next suppose we have, as an example, a function $g(A, B) = A + \bar{B}$, that is, a function given by a single maxterm. Then the corresponding K map appears as in Fig. 3.19-2b, where we have chosen to include the usually omitted 1s. We note that a maxterm fills with 1s the *maximum* possible area of the K map, short of filling the entire area.

A K map for three variables, A, B, and C is shown in Fig. 3.19-3. The boxes have again been labeled in accordance with the minterm (or maxterm) they represent. The numbering has been arranged in such manner that the digit representing the truth value of the variable A is the most significant digit while C is represented by the least significant digit.

A K map for four variables is shown in Fig. 3.19-4. The numbering of the boxes again conforms to the convention that the variable A is represented by the most significant digit and D by the least significant. Note that in numbering vertically the third column is the last numbered and in numbering horizontally the third row is the last numbered.

K maps for larger numbers of variables can also be drawn. A 5-variable map has $2^5 = 32$ boxes, and a 6-variable map has $2^6 = 64$ boxes. We shall defer drawing such maps to Sec. 3.22 and instead turn our attention to the matter of how K maps can be used to simplify logical expressions.

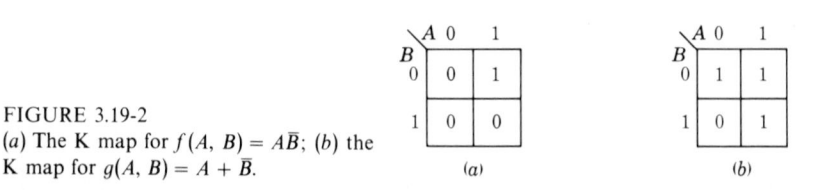

FIGURE 3.19-2
(a) The K map for $f(A, B) = A\bar{B}$; (b) the K map for $g(A, B) = A + \bar{B}$.

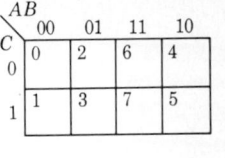

FIGURE 3.19-3
K map for three variables.

3.20 SIMPLIFICATION OF LOGICAL FUNCTIONS WITH KARNAUGH MAPS

The essential feature of the K map is that adjoining boxes, horizontally and vertically (but not diagonally) correspond to minterms, or maxterms, which differ in only a single variable, this variable appearing complemented in one term and uncomplemented in the other. It is precisely to achieve this end that the Grey reflected code is used to number the columns in Fig. 3.19-3 and both the rows and columns in Fig. 3.19-4. To see the benefit of this feature, consider by way of example minterms m_8 and m_{12}, which adjoin horizontally on the K map of Fig. 3.19-4. We have

$$m_8(8 = 1000) = A\bar{B}\bar{C}\bar{D} \tag{3.20-1}$$

$$m_{12}(12 = 1100) = AB\bar{C}\bar{D} \tag{3.20-2}$$

These two minterms differ only in that the variable B appears complemented in one and uncomplemented in the other. They can then be combined to yield

$$A\bar{B}\bar{C}\bar{D} + AB\bar{C}\bar{D} = A\bar{C}\bar{D}(\bar{B} + B) = A\bar{C}\bar{D} \tag{3.20-3}$$

Thus two terms, each involving four variables, have been replaced by a single term involving three variables. The variable which appeared complemented in one term and uncomplemented in the other has been eliminated. Now if the terms of Eq. (3.20-3) had appeared together with other terms in a logical function, we would eventually, by dint of comparing each term with every other, have noted that they might be combined. On the other hand, suppose that we had noted the presence of these two minterms by placing 1s in the appropriate boxes of a

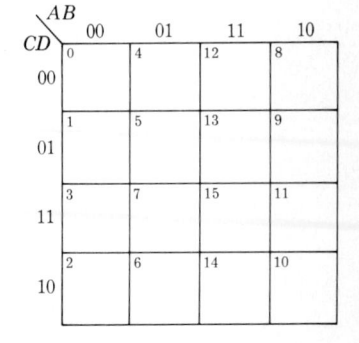

FIGURE 3.19-4
K map for four variables.

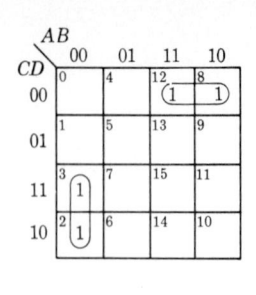

FIGURE 3.20-1

Minterms in adjacent boxes of a K map may be combined.

K map as in Fig. 3.20-1. Then we would have noted immediately that these minterms might be combined by the fact that the minterms correspond to *adjoining boxes*. The great merit of the K map is that it permits easy recognition through geometric visualization of combinations of minterms which can be combined into simpler expressions.

A general principle, then, which applies to a K map is that *any pair of adjoining minterms can be combined into a single term involving one variable fewer than the minterms themselves*. This combined term is deduced by starting with either of the minterms and striking out the variable which is complemented in one and uncomplemented in the other. Applying this rule to m_8 and m_{12} in Fig. 3.20-1, we note that for both these minterms $C = 0$, $D = 0$, $A = 1$ while m_8 has $B = 0$ and m_{12} has $B = 1$. Then two minterms combine to

$$m_8 + m_{12} = A\bar{C}\bar{D} \tag{3.20-4}$$

The combining of these minterms is illustrated on the K map by the encirclement of the 1s in the two adjoining boxes. In the same way we deduce from the K map of Fig. 3.20-1 that

$$m_2 + m_3 = \bar{A}\bar{B}C \tag{3.20-5}$$

To continue, it should now be pointed out that adjoining boxes or adjacent boxes refer not only to boxes which are geometrically touching but also to boxes which are not touching but which nevertheless represent minterms

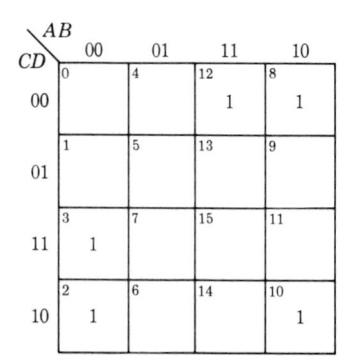

FIGURE 3.20-2

Adjacencies of peripheral rows and columns of a K map.

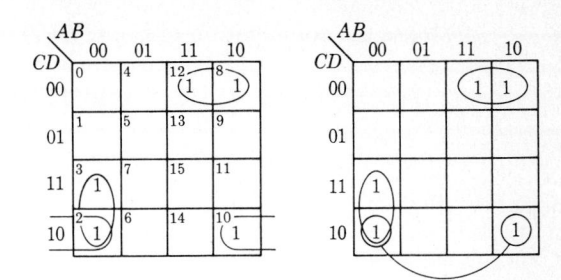

FIGURE 3.20-3
Alternative notations for indicating combinations of K-map boxes which are not geometrically adjacent.

or maxterms which differ in a single variable. On this basis each box in the leftmost column adjoins the box in the rightmost column on the same row. Thus, m_0 adjoins m_8, m_1 adjoins m_9, etc. Similarly, the topmost row adjoins the bottommost row, so that m_0 adjoins m_2, m_4 adjoins m_6, etc.

Consider now a K map with minterm entries as in Fig. 3.20-2. The entries are the same as in Fig. 3.20-1 except that m_{10} has been added. This m_{10} can be combined with either m_2 or m_8. In the first case we find

$$m_2 + m_{10} = \bar{B}C\bar{D} \qquad (3.20\text{-}6)$$

In the second case we find

$$m_8 + m_{10} = A\bar{B}\bar{D} \qquad (3.20\text{-}7)$$

If we use the combination in Eq. (3.20-6), the function represented in the K map of Fig. 3.20-2 is, from Eqs. (3.20-4) and (3.20-5),

$$f(A, B, C, D) = \sum m(2, 3, 8, 10, 12) = A\bar{C}\bar{D} + \bar{A}\bar{B}C + \bar{B}C\bar{D} \qquad (3.20\text{-}8)$$

The combinations leading to this result are indicated in two alternative notations in Fig. 3.20-3.

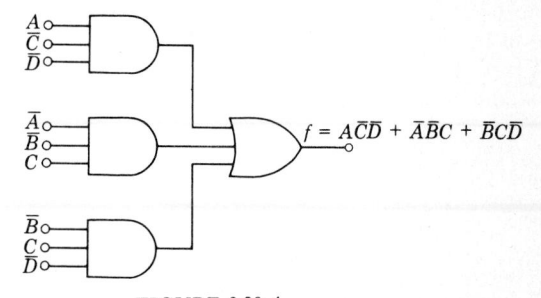

FIGURE 3.20-4
Implementation of function of Eq. (3.20-8).

If we use the combination of Eq. (3.20-7), we find

$$f(A, B, C, D) = \sum m(2, 3, 8, 10, 12) = A\bar{C}\bar{D} + \bar{A}\bar{B}C + A\bar{B}\bar{D} \quad (3.20\text{-}9)$$

Either of these results, Eq. (3.20-8) or (3.20-9), is equally acceptable, and they are equally economical.   They both require a single OR gate and three AND gates. In both cases each of the required AND gates requires three input terminals. The implementation of Eq. (3.20-8) using AND and OR gates is shown in Fig. 3.20-4.

A point that possibly needs a brief comment is the following.   In arriving at Eq. (3.20-8), by the combinations indicated in Fig. 3.20-3, we used minterm m_2 twice, one in combination with m_3 and once in combination with m_{10}.   Alternatively in arriving at Eq. (3.20-9) we would have used minterm m_8 twice.   This repetitive use of a minterm is allowable since in using, say, m_2 twice we have simply taken advantage of the theorem of Eq. (3.11-4a) which in the present case yields

$$m_2 = \bar{A}BC\bar{D} = \bar{A}BC\bar{D} + \bar{A}BC\bar{D} + \cdots$$

3.21   LARGER GROUPINGS ON A K MAP

We have seen that two K-map boxes which adjoin can be combined, yielding a term from which one variable has been eliminated.   In a similar way, whenever 2^n boxes adjoin, they can be combined and yield a single term from which n variables have been eliminated.   Typical groups of four boxes are indicated in Fig. 3.21-1.   In Fig. 3.21-1a the combinations $m_1 + m_5$ and $m_3 + m_7$ yield

$$m_1 + m_5 = \bar{A}\bar{C}D \quad (3.21\text{-}1)$$
$$m_3 + m_7 = \bar{A}CD \quad (3.21\text{-}2)$$

so that

$$(m_1 + m_5) + (m_3 + m_7) = \bar{A}\bar{C}D + \bar{A}CD = \bar{A}D(\bar{C} + C) = \bar{A}D \quad (3.21\text{-}3)$$

This result would, of course, have been the same if the grouping had been made in the order $(m_1 + m_3) + (m_5 + m_7)$.

Similarly we see that for Fig. 3.21-1b we have

$$f(A, B, C, D) = \sum m(1, 5, 9, 13) = \bar{C}D \quad (3.21\text{-}4)$$

Also, in Fig. 3.21-1c, the corner terms are considered adjacent and therefore

$$f(A, B, C, D) = (m_0 + m_2) + (m_8 + m_{10}) = \bar{A}\bar{B}\bar{D} + A\bar{B}\bar{D} = \bar{B}\bar{D} \quad (3.21\text{-}5)$$

a result obtained by inspection, by noting that the four terms do *not* differ in

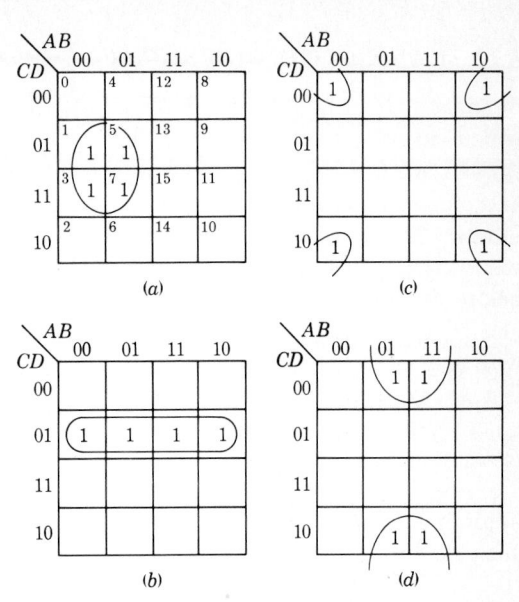

FIGURE 3.21-1
Representative adjacencies of four boxes.

$\bar{B}$ AND $\bar{D}$. As a final illustration of this type of combination we consider Fig. 3.21-1*d*, where it is easily shown that

$$f(A, B, C, D) = \sum m(4, 6, 12, 14) = B\bar{D} \qquad (3.21\text{-}6)$$

Typical groups of eight boxes are shown in Fig. 3.21-2. In these figures the 8 terms differ in all but one variable; in Fig. 3.21-2*a*, $f = \bar{A}$, while in Fig. 3.21-2*b*, $f = \bar{D}$. The next larger grouping would be sixteen boxes. In the four-variable case, sixteen 1s on a K-map would mean that the function is $f = 1$ independently of any variable. However, such groupings of sixteen are significant on a K-map for five variables (Sec. 3.22), which has thirty-two boxes etc.

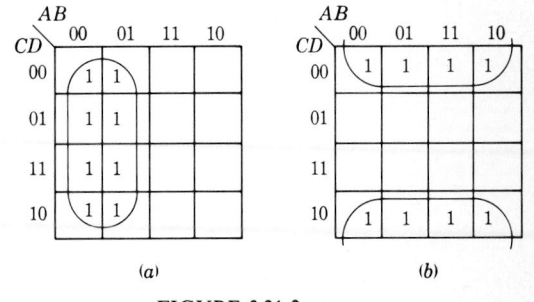

FIGURE 3.21-2
Representative adjacencies of eight boxes.

3.22 KARNAUGH MAPS FOR FIVE AND SIX VARIABLES

Suppose that in establishing a K map for five variables we follow the pattern which led us from the one-variable map to the four-variable map. We would then have a five-variable map, as in Fig. 3.22-1. Here we have added a variable and plotted two four-variable maps side by side. In ordering the rows we have followed the reflected binary code. This map preserves the features of the previous maps. Geometrically neighboring boxes continue to be adjoining, and, as before, the leftmost and rightmost columns, continue to adjoin, as do the top and bottom rows. But, now, as can be verified, boxes symmetrically located with respect to the vertical centerline (the dashed line in Fig. 3.22-1) also adjoin. For example m_7 adjoins m_{23}, m_{13} adjoins m_{29}, etc. Similarly, the previous adjacencies of the four-variable K map persist. Thus, m_1 and m_9 adjoin, as do m_2 and m_{10}, etc.

Now, we need, of course, to take maximum possible advantage of the adjoining of minterms, and the merit of the K map is precisely that it makes such adjoining minterms immediately apparent by visual inspection. Keeping these new adjoining boxes in the map of Fig. 3.22-1, it is entirely feasible to use this map as the five-variable K map. There is, however, an alternative arrangement which makes the visualization much easier. This alternative five-variable K map, and the one which is rather generally used, is shown in Fig. 3.22-2. Here, all the adjacent terms previously established for the four-variable K map continue to apply both for the left-hand four-variable section corresponding to $A = 0$ and for the right-hand four-variable section corresponding to $A = 1$. In addition, however, each box in the section $A = 0$ is adjacent to the corresponding box in the section $A = 1$. For example, m_5 is adjacent to m_{21}, m_{15} adjacent to m_{31}, etc. These adjacent terms between the two sections suggest that in our mind's eye we place one section on top of the other. It is then easy to keep in mind that boxes vertically above and below one another are adjacent.

Following the same considerations which led to the five-variable map, a six

ABC \ DE	000	001	011	010	110	111	101	100
00	0	4	12	8	24	28	20	16
01	1	5	13	9	25	29	21	17
11	3	7	15	11	27	31	23	19
10	2	6	14	10	26	30	22	18

FIGURE 3.22-1
A possible five-variable K map.

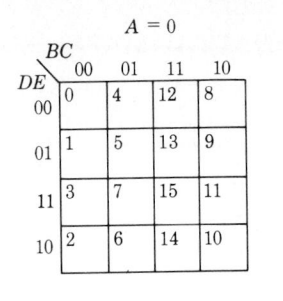

FIGURE 3.22-2
The five-variable K map.

variable K map is drawn in Fig. 3.22-3. The usual adjacent terms apply within each four-variable subsection of the map. In addition there are adjacent terms horizontally and vertically between corresponding boxes in the subsection. For example, m_5 is adjacent to m_{21} and also to m_{37}; m_{63} is adjacent to m_{31} and m_{47}; etc.

The Karnaugh map has the merit of allowing visualization of adjacent terms. When the number of variables becomes large, however, say seven or more, the K map becomes so expansive that its value as an aid to recognizing adjacent terms becomes questionable. Hence, for the many-variable case, tabular procedures are preferred.

In the following sections we illustrate several of the uses of the K map. To simplify the calculations required we shall restrict our study to four or fewer variables.

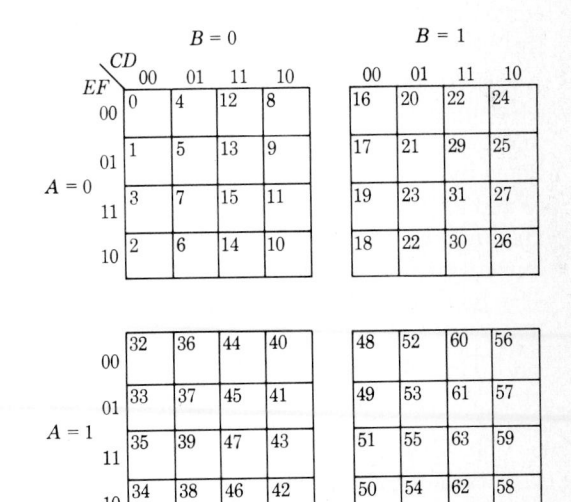

FIGURE 3.22-3
The six-variable K map.

3.23 USE OF KARNAUGH MAPS

When a logical function has been expressed in standard form in terms of its minterms, the K map may be used to simplify the function by applying the following principles:

> *1* The combinations of boxes (minterms) which are selected must be such that each box is included at least once. As noted, however, a particular box may be involved in a number of different combinations.
>
> *2* The individual combinations should be selected to encompass as many boxes as possible so that all boxes will be included in as few different combinations as possible.

The combinations are sometimes referred to as *products* and sometimes as *prime implicants*. It may indeed turn out that it is not necessary to use all possible prime implicants to include every box at least once. In the example of Fig. 3.20-3 we saw just such a case. Here the prime implicants were $p_1 = m_2 + m_3$, $p_2 = m_8 + m_{12}$, $p_3 = m_2 + m_{10}$, and $p_4 = m_8 + m_{10}$. The function in question could be expressed, however, either as $f = p_1 + p_2 + p_3$ or as $f = p_1 + p_2 + p_4$. In either case we had no alternative but to use p_1 since otherwise m_3 would not be accounted for. For this reason p_1 is called an *essential prime implicant*. Similarly, p_2 is essential since, without p_2, m_{12} would not be accounted for. On the other hand, since we are at liberty to select or not to select p_3, this prime implicant is not essential. A similar comment applies to p_4.

When we have expressed a function as a sum of prime implicants, for each such implicant we shall require an AND gate, as in the structure of Fig. 3.20-4. Further, the number of inputs to each such gate decreases as the number of boxes encompassed in the prime implicant increases. The economy of a gate structure is judged first of all by how few gates are involved. In different structures with equal numbers of gates economy is judged to be improved in the structure with the fewer total number of gate inputs.

Our preoccupation with finding prime implicants on a K map which encompass as many boxes as possible poses a potential hazard. It is illustrated in Fig. 3.23-1, where we might be tempted to combine $m_5 + m_7 + m_{13} + m_{15}$ as noted by the dashed encirclement. If we did so, we would still find it necessary to add four additional prime implicants to take account of the four remaining 1s. Having done so, we then find that all the boxes of the original combination of four have been accounted for and hence this original combination is superfluous. The function is written out under the K map. A second, similar example is illustrated in Fig. 3.23-1b.

The following algorithm applied to a K map will lead to a minimal expression for a logical function and avoid the hazard referred to above:

> *1* Encircle and accept as essential prime implicants any box or boxes that cannot be combined with any other.
>
> *2* Identify the boxes that can be combined with a single other box in *only*

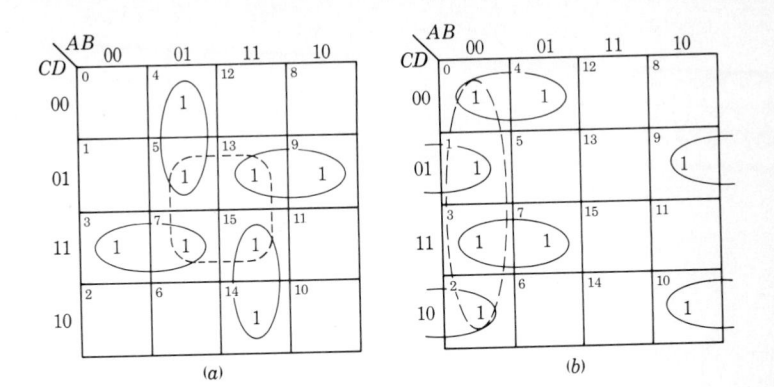

FIGURE 3.23-1
Two illustrations of a hazard associated with forming combinations on a K map.
(a) $f = (m_4 + m_5) + (m_3 + m_7) + (m_{14} + m_{15}) + (m_9 + m_{13})$
$\quad = \bar{A}B\bar{C} + \bar{A}CD + ABC + A\bar{C}D$
(b) $f = (m_0 + m_4) + (m_1 + m_9) + (m_3 + m_7) + (m_2 + m_{10})$
$\quad = \bar{A}C\bar{D} + \bar{B}\bar{C}D + \bar{A}CD + \bar{B}C\bar{D}$

one way. Encircle such two-box combinations. A box which can be combined into a two-grouping but can be so combined in *more than one way* is to be temporarily bypassed.

3 Identify the boxes that can be combined with three other boxes in *only one way.* If *not all* of the four boxes so involved are already covered in groupings of two, encircle these four boxes. Again, a box which can be encompassed in a group of four in *more than one way* is to be temporarily bypassed.

4 Repeat the preceding for groups of eight, etc.

5 After the above procedure, if there still remain some uncovered boxes, they may be combined with each other or with other already covered boxes in a rather arbitrary manner. Of course, however, we shall want to include these left-over boxes in as few groupings as possible.

This algorithm is illustrated in the following two examples. In the first example, the solution is completely determined by the algorithm. In the second example an easy exercise of judgment allows us to satisfy the requirements of step 5.

EXAMPLE 3.23-1   A four-variable function is given as

$$f(A, B, C, D) = \sum m(0, 1, 3, 5, 6, 9, 11, 12, 13, 15) \qquad (3.23\text{-}1)$$

Use a K map to minimize the function.

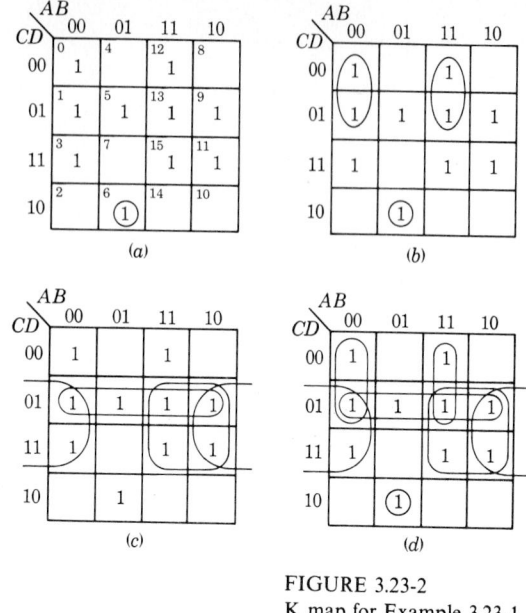

FIGURE 3.23-2
K map for Example 3.23-1.

SOLUTION   The K map for the function of Eq. (3.23-1) is drawn in Fig. 3.23-2a. We note that m_6 can be combined with no other box.   Hence, we encircle it and accept it as an essential prime implicant.   Next we note that m_0 and m_{12} can be combined in two-groups in only one way.   We therefore encircle each of these two-groups, as in Fig. 3.23-2b.   Other boxes which can combine in a two-group in more than one way are passed over.   We then observe that m_3, m_5, and m_{15} can be incorporated into four-groups each in only one way, and we note also that the four-groups so formed involve other boxes not all of which are incorporated in two-groups.   Hence, we encircle these three four-groups as is indicated in Fig. 3.23-2c.   Finally, in Fig. 3.23-2d, all encirclements have been combined, and we observe that all boxes have been accounted for.   Reading from this map, we find

$$f(A, B, C, D) = \bar{A}BC\bar{D} + \bar{A}\bar{B}\bar{C} + AB\bar{C} + \bar{C}D + \bar{B}D + AD \qquad (3.23\text{-}2)$$

EXAMPLE 3.23-2   A four-variable function is given as

$$f(A, B, C, D) = \sum m(0, 2, 3, 4, 5, 7, 8, 9, 13, 15) \qquad (3.23\text{-}3)$$

Use a K map to minimize the function.

SOLUTION   The K map for the function of Eq. (3.23-3) is drawn in Fig. 3.23-3a. Applying steps 1 and 2 of the algorithm does not result in any selection of prime implicants.   All four boxes m_5, m_7, m_{13}, and m_{15} satisfy the condition of step 3.   Applying the procedure of step 3 to any one of them leads to the encirclement shown in Fig. 3.23-3b.   Step 4 does not apply in the present case.   We do find that a number of boxes are not yet accounted for.   As required by step 5, we combine them arbitrarily.   It is

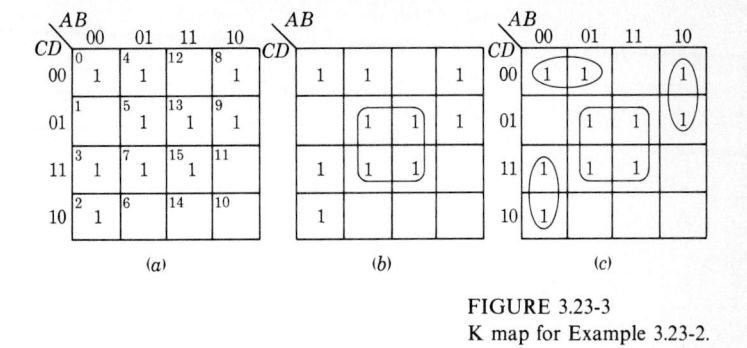

FIGURE 3.23-3
K map for Example 3.23-2.

rather obvious that the manner indicated in Fig. 3.23-3c leads to the fewest additional prime implicants. The solution, read directly from Fig. 3.23-3c, is

$$f(A, B, C, D) = \bar{A}\bar{C}\bar{D} + \bar{A}\bar{B}C + A\bar{B}\bar{C} + BD \qquad (3.23\text{-}4)$$

For variety we now consider a problem in which a function is specified in terms of its **0**s rather than its **1**s, that is, in terms of its *maxterms* rather than its *minterms*. In this case **0**s rather than **1**s are entered in the K map, and the solution appears in the form of a product of sums rather than a sum of products. The algorithm for combining minterms applies equally to maxterms; however, we may note a change in terminology. Corresponding to the term *prime implicant*, defined as a product term in a sum representing a function, we have instead *prime implicates*, defined as a sum term in a product.

EXAMPLE 3.23-3   A four-variable function is given as

$$f(A, B, C, D) = \prod M(0, 3, 4, 5, 6, 7, 11, 13, 14, 15) \qquad (3.23\text{-}5)$$

Use a K map to minimize the function.

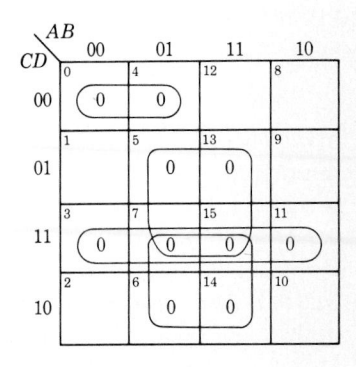

FIGURE 3.23-4
K map for Example 3.23-3.

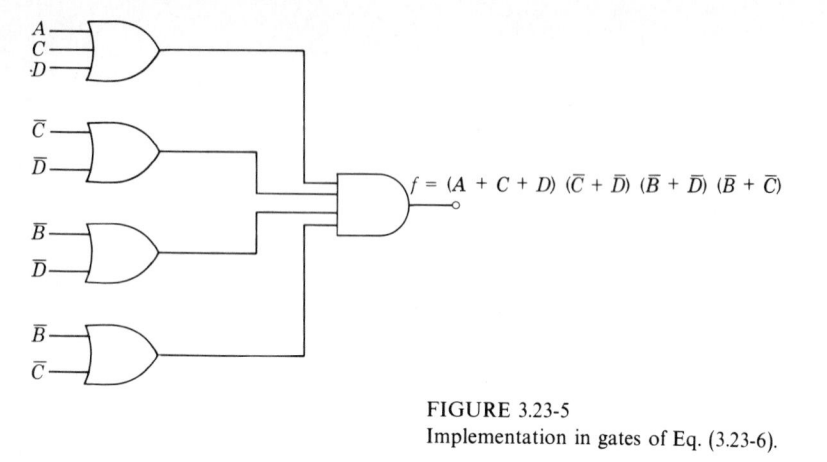

FIGURE 3.23-5
Implementation in gates of Eq. (3.23-6).

SOLUTION   The K map is shown in Fig. 3.23-4.   The algorithm leads uniquely to the groupings indicated.   We then find directly from the map that

$$f(A, B, C, D) = (A + C + D)(\bar{C} + \bar{D})(\bar{B} + \bar{D})(\bar{B} + \bar{C}) \tag{3.23-6}$$

The implementation of Eq. (3.23-6) using AND and OR gates is shown in Fig. 3.23-5. Observe that each term in Eq. (3.23-6) requires an OR gate and that all OR-gate outputs are combined in a single AND gate.   Figure 3.23-5 is to be compared with Fig. 3.20-4, where, to implement a sum-of-products function, the order of the OR and AND gates is reversed.

3.24   MAPPING WHEN FUNCTION IS NOT EXPRESSED IN MINTERMS

Our discussion of mapping suggests that if a function is to be entered on a K map, the function must first be expressed as a sum of minterms (or maxterms).   In principle, such is the case.   As a matter of practice, however, if the function is not so expressed, it is not necessary to expand the function algebraically into its minterms.   Instead the expansion into minterms can be accomplished in the process of entering the terms of the function on the K map.   To illustrate, consider that we propose to enter on a K map the function

$$f(A, B, C, D) = \bar{A}\bar{B}\bar{C}\bar{D} + B\bar{C}D + \bar{A}\bar{C} + A \tag{3.24-1}$$

in which only the first term is in the form of a minterm.   As in Fig. 3.24-1a, this first minterm may be entered directly on the K map.   The second term $B\bar{C}D$ corresponds on the K map to the location when $B = 1$, $C = 0$, and $D = 1$ and is *independent of A*.   Hence, this term is entered where indeed $B = 1$, $C = 0$, and $D = 1$ for *both values* of A, as in Fig. 3.24-1b.   Similarly the third term, as in Fig. 3.24-1c, is entered where $A = 0$ and $C = 0$ and for both values of B and for both values of D.   Finally, the term A is entered as in Fig. 3.24-1d where $A = 1$

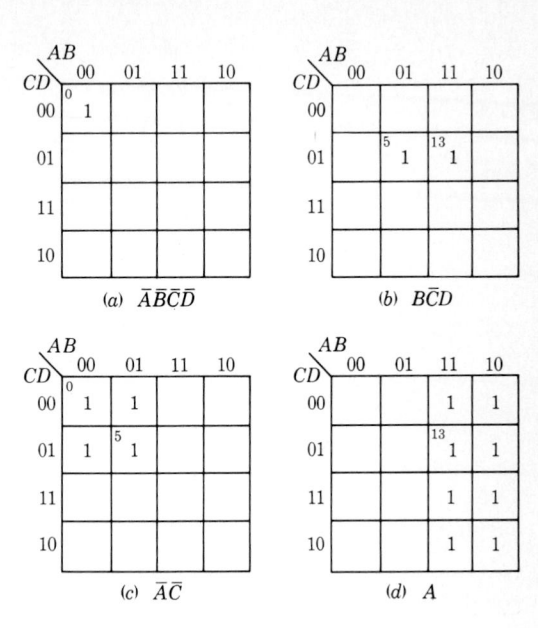

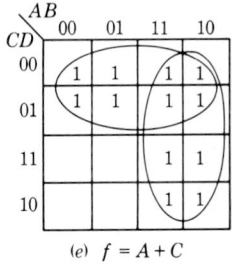

FIGURE 3.24-1
The representation of a K map of a function not explicitly given in terms of minterms.

and for both values of each of the other three variables. The complete K map, shown in Fig. 3.24-1*e*, results from combining the individual maps for the individual terms. We find that

$$f(A, B, C, D) = A + \bar{C} \tag{3.24-2}$$

We note that in several instances, in the cases of minterms m_0, m_5, and m_{13}, 1s appear in the maps of more than a single term. This situation causes no difficulty, for, as we have noted, a minterm added to itself still yields a single minterm.

EXAMPLE 3.24-1   Refer to Example 3.8-1. Write a truth table for all of the possibilities. Indicate a disaster by the symbol $L = 1$. Enter the $2^4 = 16$ entries onto a K map

Table 3.24-1   **TRUTH TABLE FOR EXAMPLE 3.8-1**

1 = south barn,   0 = north barn, L = light

J	D	G	C	L	J	D	G	C	L
0	0	0	0	0	1	0	0	0	1
0	0	0	1	0	1	0	0	1	1
0	0	1	0	0	1	0	1	0	0
0	0	1	1	1	1	0	1	1	0
0	1	0	0	0	1	1	0	0	1
0	1	0	1	0	1	1	0	1	0
0	1	1	0	1	1	1	1	0	0
0	1	1	1	1	1	1	1	1	0

and simplify. Show that the solution given for Example 3.8-1 is indeed correct and in simplest form.

SOLUTION   We first obtain the truth table shown in Table 3.24-1. Note that in row 1, $L = 0$ since no disaster occurs when the farmer, dog, goat, and cabbage are in the north barn. However, we observe from row 4 that a disaster occurs, $L = 1$, if the farmer is in the north barn while the goat AND cabbage are in the south barn. In a similar manner we determine L for each row.

Next we enter each row where $L = 1$ *directly* onto the K map as shown in Fig. 3.24-2. To see how this is done refer to row 4 of the truth table. Here, $J = 0$, $D = 0$, $G = 1$, $C = 1$. Thus, we choose the JD column which is **00** and its intersection with GC row where $GC = 11$. The other entries are obtained in a similar fashion.

The result for L in terms of minterms is

$$L = \bar{J}GC + \bar{J}DG + J\bar{G}\bar{C} + J\bar{D}\bar{G} \tag{3.24-3}$$

which is the same as the result obtained for Example 3.8-1.

3.25 SYNTHESIS USING NAND OR NOR GATES

If we synthesize a logical function in the sum-of-products form, its physical realization consists of a number of AND gates followed by a single OR gate, as in Fig. 3.20-4. Any input variable is transmitted first through an AND gate, which

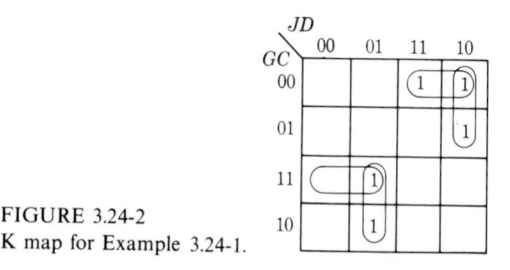

FIGURE 3.24-2
K map for Example 3.24-1.

is hence described as being at the first level, and thereafter through the OR gate, which is at the second level. The structure of Fig. 3.20-4 is therefore often referred to as a two-level AND-OR gate system. If we synthesize a logical function as a product of sums, we arrive at a two-level OR-AND structure, as in Fig. 3.23-5, where the input gates are OR gates and the second-level gate is an AND gate. We may note in passing that gates with more than two levels may sometimes be simpler than two-level gates. However, there is no easy procedure for designing multilevel gate systems as there is for two-level gate systems. Further the fewer the levels the smaller the propagation time through the structure.

We have noted that every logical operation can be achieved with NAND gates alone or with NOR gates alone. We have further noted that there are good practical reasons for using NOR or NAND gates. We shall now see how to find a physical realization of a logical function entirely with NAND gates or entirely with NOR gates. We shall illustrate the procedure by an example.

Consider a logical function Z whose K map appears as in Fig. 3.25-1a. As a sum of products we read

$$Z = \bar{A}B + A\bar{C} \tag{3.25-1}$$

As a product of sums we read

$$Z = (A + B)(\bar{A} + \bar{C}) \tag{3.25-2}$$

The two-level AND-OR gate for Z is shown in Fig. 3.25-1b, and the two-level OR-AND gate is given in Fig. 3.25-1c. Applying De Morgan's theorem twice to the sum of products in Eq. (3.25-1), we have

$$\bar{Z} = \overline{\bar{A}B + A\bar{C}} = (\overline{\bar{A}B})(\overline{A\bar{C}}) \tag{3.25-3a}$$

Hence
$$Z = \bar{\bar{Z}} = \overline{(\overline{\bar{A}B})(\overline{A\bar{C}})} \tag{3.25-3b}$$

which is realized as in Fig. 3.25-1d using only NAND gates. Observe that the circuit in Fig. 3.25-1d is *exactly* as in Fig. 3.25-1b except that every gate is replaced by a NAND gate. Applying De Morgan's theorem twice to the product of sums in Eq. (3.25-2), we have

$$\bar{Z} = \overline{(A + B)(\bar{A} + \bar{C})} = \overline{(A + B)} + \overline{(\bar{A} + \bar{C})} \tag{3.25-4a}$$

$$\bar{\bar{Z}} = Z = \overline{\overline{(A + B)} + \overline{(\bar{A} + \bar{C})}} \tag{3.25-4b}$$

which is realized as in Fig. 3.25-1e using only NOR gates. Observe that the circuit in Fig. 3.25-1e is exactly as in Fig. 3.25-1c except that every gate is replaced by a NOR gate.

In summary, to arrive at a NAND-gate circuit, start by expressing Z as a sum of products. Draw the corresponding two-level AND-OR configuration, and then change *all* gates to NAND gates. To arrive at a NOR-gate circuit, start by expressing Z as a product of sums. Draw the corresponding two-level OR-AND configuration and then change *all* gates to NOR gates.

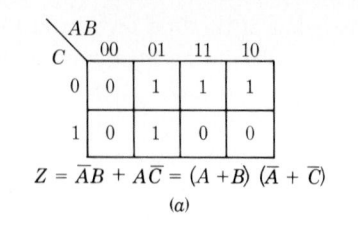

$$Z = \overline{A}B + A\overline{C} = (A + B)(\overline{A} + \overline{C})$$

(a)

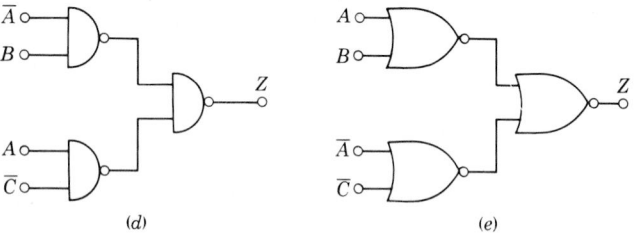

(b)   (c)

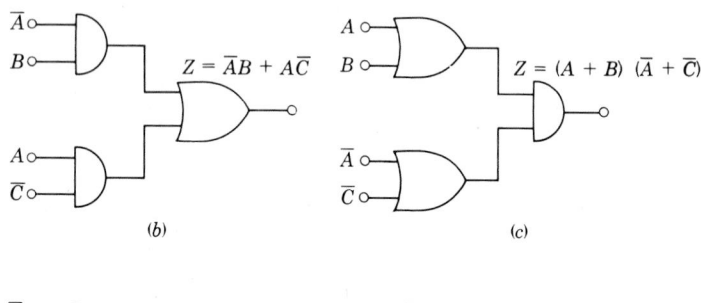

(d)   (e)

FIGURE 3.25-1

(a) K map for Z; (b) AND-OR circuit for Z; (c) OR-AND circuit; (d) circuit using only NAND gates; (e) circuit using only NOR gates.

3.26   INCOMPLETELY SPECIFIED FUNCTIONS

A logical function f is defined by specifying for each possible combination of variables whether the function has the value $f = 1$ or $f = 0$. Such a specification allows one immediately to enter the minterms or maxterms on a K map and thereafter to express the function in its simplest form.

Now suppose that we undertake to write in simplest form a function f which is specified for some (but not all) possible combinations of the variables. In such a case a number of different functions are possible, all of which satisfy the specifications. They will differ from one another in the values assumed by the function for combinations of the variables which are not specified. The question then arises how, from among the many allowable functions, to arrive directly at the simplest function.

Such incomplete specification arises, as a matter of practice, in two ways. Sometimes, it does indeed happen that we simply *do not care* what value is

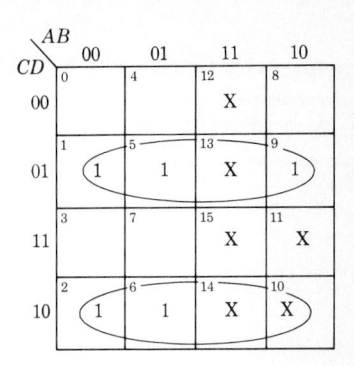

FIGURE 3.26-1
An incompletely specified function.

assumed by the function for certain combinations of variables. On other occasions, it may happen that we know that certain combinations of the variables will simply never occur. In this case, we may, as it were, pretend that we do not care, since the net effect is the same.

To illustrate the procedure, using K maps, to simplify an incompletely specified function, consider that a function is defined by

$$f(A, B, C, D) = \sum m(1, 2, 5, 6, 9) + d(10, 11, 12, 13, 14, 15) \qquad (3.26\text{-}1)$$

In this equation the d stands for "don't care," so that our function has the value $f = 1$ corresponding to minterms $1, 2, \ldots$, and is unspecified for combinations of variables corresponding to minterms $10, 11, \ldots$. On the K map, as in Fig. 3.26-1, we locate 1s where specified, and where a don't care is indicated we locate an X. The procedure thereafter is to interpret an X as a **1** if in so doing we effect a simplification and to ignore it otherwise. If the X's were all ignored, the map of Fig. 3.26-1 would yield

$$f = (m_1 + m_5) + (m_1 + m_9) + (m_2 + m_6) \qquad (3.26\text{-}2a)$$

$$= \bar{A}\bar{C}D + \bar{B}\bar{C}D + \bar{A}C\bar{D} \qquad (3.26\text{-}2b)$$

Let us, however, interpret as 1s the X's in m_{10}, m_{13}, and m_{14}. We then find that the function simplifies to

$$f = (m_1 + m_5 + m_9 + m_{13}) + (m_2 + m_6 + m_{10} + m_{14}) \qquad (3.26\text{-}3a)$$

$$= \bar{C}D + C\bar{D} \qquad (3.26\text{-}3b)$$

The remaining X's in m_{11}, m_{12}, and m_{15} cannot serve either to reduce the number of terms in the function or to reduce the number of variables in a term. Hence, these X's are simply ignored, i.e., judged to be **0**s.

4

RESISTOR-TRANSISTOR LOGIC (RTL) AND INTEGRATED-INJECTION LOGIC (IIL)

We begin our discussion of logic gates (Chaps. 4 to 8) by considering the *resistor-transistor-logic* (RTL) gate. This type of gate is no longer used in new design, but for a number of reasons the RTL gate is a useful starting point. On the one hand, this gate is elegantly simple and hence may be used conveniently to develop concepts useful in connection with all types of gates. Further, the RTL gate is historically the first gate to have been used extensively, and many installations employing this type of gate are still in operation. Finally, topologically at least, RTL is a forerunner of *integrated-injection-logic* (IIL) which, at the present writing, is one of the newest of the commercially available LSI logic families.

4.1 THE RESISTOR-TRANSISTOR-LOGIC (RTL) GATE

An N-input RTL gate (Fig. 4.1-1) consists of N transistors all of whose emitters are connected to a common ground and all of whose collectors are tied through a common collector resistor R_c to a supply voltage V_{CC}. Input voltages V_i, $(i = 1, 2, ..., N)$ representing logic levels, are applied to the bases through

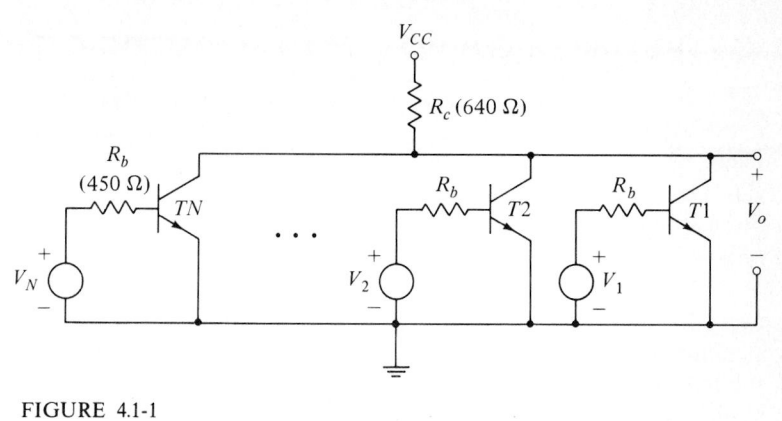

FIGURE 4.1-1

An N-input RTL gate. A medium-power commercial gate has the component values indicated in parentheses.

resistors R_b. If we adopt the convention of *positive logic*, then, as we can now verify, the gate performs NOR logic.

Let V_L and V_H be the voltages representing the two possible values of a logical variable. The subscripts L and H stand for "low" and "high." That is, $V_L < V_H$. In positive logic V_L corresponds to the case where the logical variable is "false," i.e., to logic level **0**. The voltage is V_H when the variable is "true," i.e., at logic level **1**. In the RTL gate V_L is required to be low enough for the corresponding transistor to be cut off when V_L is applied at a gate input. The voltage V_H is a voltage which, applied at an input, will drive the corresponding transistor to saturation. Note that V_L and V_H are not unique voltages but are characterized by the fact that V_L must be *less than* some unique voltage and V_H must be *greater than* some unique voltage. In the present case it is necessary that $V_L < V_\gamma$, the cut-in voltage, and V_H be equal to or higher than the voltage which when applied through R_b will bring the transistor to saturation.

The truth table for a two-input gate is given in Table 4.1-1. In one truth table we have used the logical values LOW and HIGH in the other **0** and **1**. Both types of tables are common.

Table 4.1-1

V_1	V_2	V_o	V_1	V_2	V_0
L	L	H	0	0	1
L	H	L	0	1	0
H	L	L	1	0	0
H	H	L	1	1	0

4.2 THE DIRECT-COUPLED TRANSISTOR-LOGIC (DCTL) GATE

The earliest gates of the type shown in Fig. 4.1-1 were more economically constructed by omitting the base resistors R_b. Because input connections are made directly to the base, such gates are called *direct-coupled transistor-logic* (DCTL) gates. DCTL gates, like RTL gates, perform NOR logic. We shall discuss these gates briefly calling attention to a fundamental deficiency on account of which they were never widely used.

As noted in Sec. 3.10, the NOR operation is functionally complete. That is, every logic operation can be performed using the NOR operation alone. In a switching system of any sophistication, then, it is likely that each of the inputs to the NOR gate can be derived from the outputs of other similar NOR gates, and, in turn, each NOR gate may be required to furnish input logic levels to other NOR gates. Hence the situation of the DCTL gate $G11$, shown in the dashed box of Fig. 4.2-1 is representative of the milieu in which a typical gate may have to function.

Let us focus our attention on gate $G11$. It receives input V_i from gate $G0$, which is also required to furnish a signal to one input of $N - 1$ other gates $G12$ through $G1N$. Similarly, gate $G11$ must furnish signals to one input of N other gates $G21$ through $G2N$. The other gate input terminals, not shown connected in Fig. 4.2-1, will derive their inputs from sources not specified here. If any of these gates has more inputs than required for its function, the unused gate input terminals are to be *grounded*. The corresponding transistor will then be at cutoff and will have no effect on the operation of that gate.

We now determine the voltage levels present at the input and output of the DCTL gates shown in Fig. 4.2-1. Suppose that all inputs of $G0$ are low enough to keep all transistors in $G0$ cut off. Then the common collector of $G0$ will rise, driving transistors $T1, T2, \ldots, TN$ to saturation. As noted earlier (Sec. 1.1), at room temperature the base-emitter voltage V_σ of a saturated transistor is 0.75 V. Thus the gate voltage level corresponding to the **1** state at the output of gate $G0$ is $V_\sigma = 0.75$ V. The current supplied to the bases of each of the transistors $T1, T2, \ldots, TN$ is supplied from the source V_{CC} through the resistor R_c. In this discussion we are assuming that R_c is small enough for adequate current to be available to drive all the transistors $T1, T2, \ldots, TN$ to saturation. That is, the current in R_c is equal to

$$\frac{V_{CC} - V_\sigma}{R_c} = \frac{V_{CC} - 0.75}{R_c}$$

Hence, each base current in $T1, \ldots, TN$ is $I_B = (V_{CC} - 0.75)/NR_c$ if $T1, \ldots, TN$ are identical. We are assuming that the current is adequate to saturate $T1, \ldots, TN$. Note that the collectors of gate $G0$ rise no higher than $V_\sigma = 0.75$ V.

With at least one input of $G11$ at the logical level **1**, the logic level at the output of $G11$ is **0**, since $V_{CE} = V_{CE}(\text{sat})$. This voltage level, $V_{CE}(\text{sat})$, is typically between 0.1 and 0.2 V, depending on the current flowing in resistor R_c

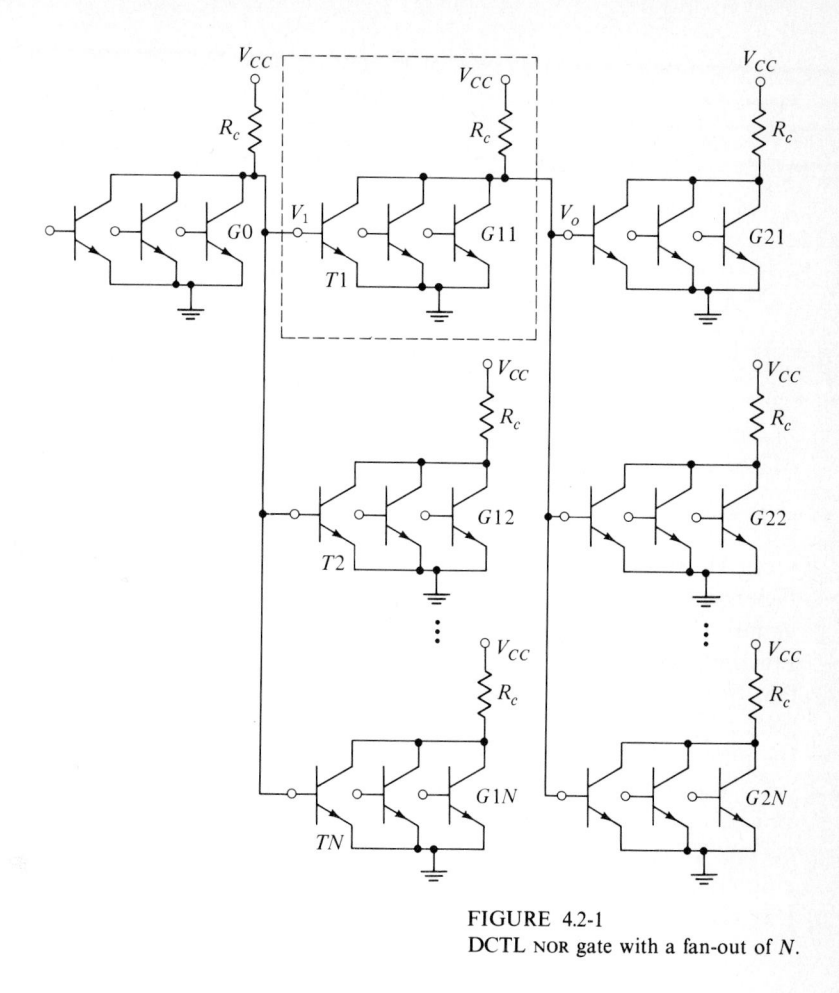

FIGURE 4.2-1
DCTL NOR gate with a fan-out of N.

of the gate. Since, as we have seen, the cut-in point of a silicon transistor occurs at a base-emitter voltage of about 0.65 V, the saturation voltage $V_{CE}(\text{sat}) = 0.1$ or 0.2 V, applied to an input of a succeding gate is low enough to keep the corresponding transistor cut off.

It is to be observed that the total voltage separation, referred to as the *logic swing* between the voltages corresponding to the logic levels **1** and **0**, is of the order of $0.75 - 0.1 = 0.65$ V.

Input-output characteristic of DCTL The above results are neatly summarized by the (idealized) *input-output characteristic* shown as the solid plot in Fig. 4.2-2. This figure plots the output V_o of gate $G11$ as a function of an input voltage V_i applied to one of the transistors in the gate. The other remaining transistors in the gate are assumed to be cut off; i.e., the other inputs are in the low state.

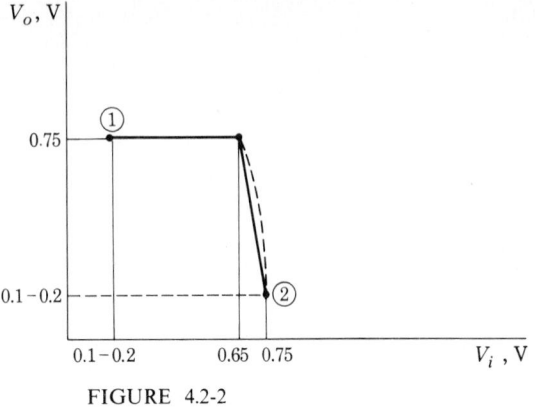

FIGURE 4.2-2
Input-output characteristic of a DCTL NOR gate.

At point 1, V_i is at its minimum voltage of $V_{CE}(\text{sat}) = 0.1$ to 0.2 V, depending on the saturation voltage of gate $G0$. At this input voltage, V_o is equal to 0.75 V due to the loading of gates $G21$, $G22$, etc. As V_i increases, $T1$ remains cut off until V_i reaches the cut-in voltage of the transistor, which is 0.65 V at room temperature. As V_i increases from 0.65 to 0.75 V, transistor $T1$ goes from cutoff to saturation, as shown. V_i is clamped at 0.75 V, and hence the plot is terminated at this voltage (point 2).

A more realistic input-output characteristic is shown by the dashed plot in Fig. 4.2-2. As we have noted (see Sec. 1.7), in the active region the collector current of a transistor I_C is nearly equal to the emitter current I_E, and the emitter current is related to the emitter-base voltage by the diode equation [Eq. (1.7-3)]. Hence, since $V_{BE} = V_i$, we have

$$I_C \approx I_E \approx I_{E0}\,\epsilon^{V_i/V_T} \tag{4.2-1}$$

For values of V_o less than 0.65 the base currents of the transistors being driven by $G11$ fall to zero and we may write

$$V_o = V_{CC} - R_c I_C = V_{CC} - R_c I_{E0}\,\epsilon^{V_i/V_T} \tag{4.2-2}$$

Hence the input-output characteristic is a falling exponential with ever increasing magnitude of slope. In the range $0.65 \le V_o \le 0.75$ V, Eq. (4.2-2) is less exact since it does not take account of the volt-ampere characteristic of the base-collector circuits of the driven transistors.

4.3 CURRENT HOGGING IN DCTL GATES

DCTL gates suffer from a difficulty referred to as *current hogging*. To appreciate this problem we need only recognize that while transistors of similar manufacture are generally quite similar in performance, they are not precisely identical. Thus, referring to Fig. 4.2-1, suppose that the output of gate $G0$ is at logic

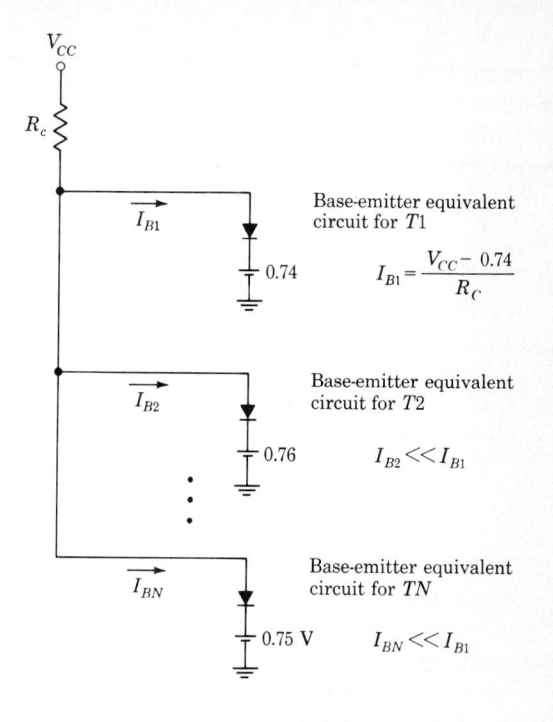

FIGURE 4.3-1
Equivalent circuit to illustrate current hogging in DCTL.

level **1**, so that $T1$, $T2$, etc., are all to be driven to saturation. Suppose, by way of example, that when $T1$ attains saturation, its base-emitter voltage is 0.74 V. But suppose that $T2$ requires a base-emitter voltage of 0.76 V for saturation while 0.74 V leaves it in the active region. This difference may be due not only to the fact that the transistors are different but also to the fact that $T1$ and $T2$ may be in different integrated-circuit packages and hence operating at different temperatures. The temperature difference, in turn, may be due to the different physical location of the two units and possibly a difference in power dissipation in the two integrated circuit packages.

In any event, as seen from Fig. 4.3-1, which shows the output circuit of $G0$ and the base-emitter equivalent circuits of $T1$, $T2$, ..., TN, $T1$ will hog much of the available current furnished by V_{CC} through R_c, and other transistors having base-emitter voltages greater than 0.74 V, such as $T2$, will be starved for base current and not be able to attain saturation as required. This current-hogging phenomenon, which is characteristic of DCTL, explains why this type of logic is not presently in wide use, and we shall not consider it further.

4.4   RESISTOR-TRANSISTOR LOGIC (RTL)

The current-hogging difficulty of the DCTL gate can be largely eliminated by introducing resistors in series with the base of each transistor as has been done in the RTL gate of Fig. 4.1-1 (see Prob. 4.4-1). In a medium-power commercial

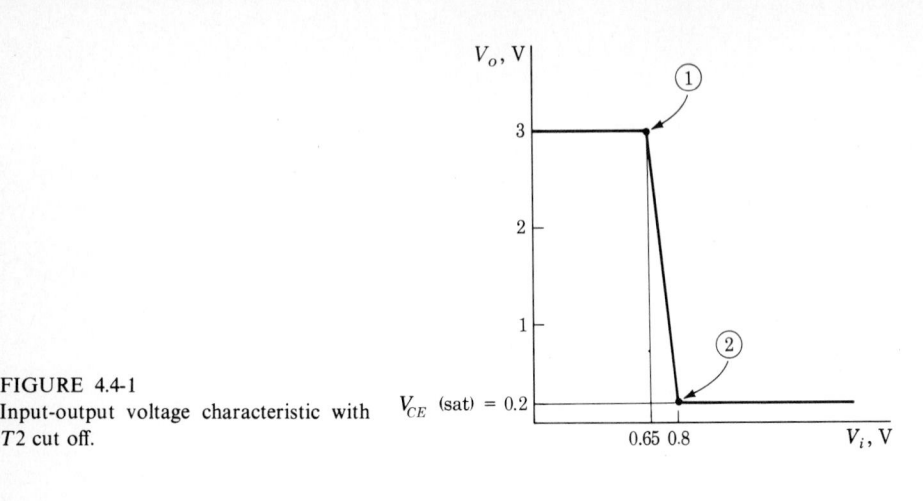

FIGURE 4.4-1
Input-output voltage characteristic with V_{CE} (sat) = 0.2
T2 cut off.

gate we find that $R_c = 640\ \Omega$ and $R_b = 450\ \Omega$, as noted in parentheses in Fig. 4.1-1. The supply voltage is typically $V_{CC} = 3$ V.

The input-output voltage characteristic of the RTL gate is shown in Fig. 4.4-1. Here we have plotted the output V_o as a function of one input V_i under the condition that all other input signals are at voltages which ensure that the corresponding transistor is cut off. Referring to Fig. 4.4-1, we note that if V_i is less than the cut-in voltage, which is 0.65 V at room temperature, T1 is cut off and $V_o = V_{CC} = 3$ V.

When V_i exceeds 0.65 V, T1 enters the active region and V_o decreases until T1 saturates. It is useful to calculate the minimum value of V_i needed to bring T1 into saturation. For the purpose of illustration let us assume that $V_{CE}(\text{sat}) = 0.2$ V, $h_{FE} = 50$, and $h_{FC} = 0.1$. Then from Fig. 1.10-1 we find that $\sigma \equiv I_C/50I_B = 0.85$. Using $R_c = 640\ \Omega$, the collector current of T1 is found from Fig. 4.1-1 to be

$$I_C = \frac{3 - 0.2}{640} = 4.4 \text{ mA} \qquad (4.4\text{-}1)$$

and hence

$$I_B = \frac{4.4 \times 10^{-3}}{50(0.85)} \approx 0.1 \text{ mA} \qquad (4.4\text{-}2)$$

Knowing I_B and assuming that $V_{BE}(T1) = V_\sigma = 0.75$ V since T1 is saturated, we can calculate V_i. Referring to Fig. 4.1-1, we have

$$V_i = 450I_B + V_{BE} = 450(0.1 \times 10^{-3}) + 0.75 \approx 0.8 \text{ V} \qquad (4.4\text{-}3)$$

Further increases in V_i produce only a small change in $V_o = V_{CE}$ since in saturation V_{CE} is typically between 0.1 and 0.2 V. Thus, as seen in Fig. 4.4-1, V_o has dropped to 0.2 V at $V_i = 0.8$ V, and we have idealized the plot by ignoring any further small change in V_o for $V_i > 0.8$ V. The plot is further idealized in that the plot from point 1 to point 2, which should be exponential in character, is drawn as a straight line.

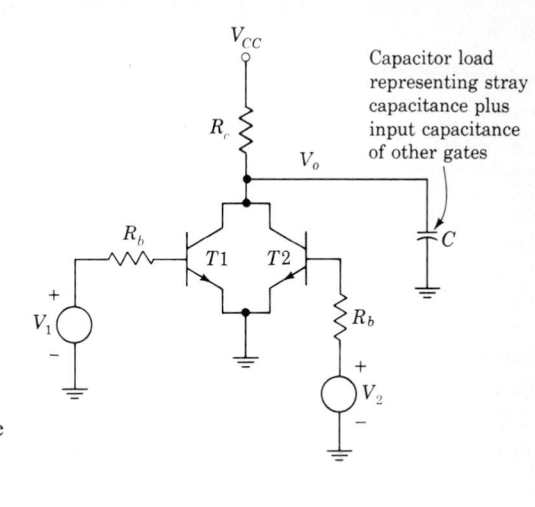

FIGURE 4.4-2
A two-input RTL gate with a capacitive
load.

Types of RTL NOR gates As noted, commercially available integrated-circuit RTL NOR gates typically have the collector resistor $R_c = 640 \ \Omega$ while the base resistors $R_b = 450 \ \Omega$. A low-power RTL NOR gate is also available having $R_c = 3.6 \ \mathrm{k}\Omega$ while $R_b = 1.5 \ \mathrm{k}\Omega$. A disadvantage of this low-power gate is that the larger resistors result in slower operation: i.e., the propagation delay time is longer. The reason for this longer delay is that all stray capacitors and capacitors inherent in the active devices must charge and discharge through these larger resistors.

Pull-up resistors The collector resistor R_c in the RTL gate is often called a *passive pull-up resistor*. The reason for this terminology is to be seen in connection with Fig. 4.4-2, where an RTL gate drives a capacitive load C. This capacitance is due in part to stray capacitance and in part to the capacitance associated with the base-emitter junctions of gates driven by the gate shown.

Assume, initially that V_1 is high and V_2 low, so that $T1$ is saturated, $T2$ cut off, and V_o in the **0** state. Now let V_1 fall to the **0** state. $T1$ now cuts off, and V_o rises toward V_{CC} with a time constant $\tau = R_c C$. We say that V_o is pulled up to V_{CC} by the *pull-up* resistor R_c. Since R_c is passive, we say that the gate shown employs *passive pull-up*. Active pull-up is also available for RTL and many of the other gates to be discussed subsequently. The operation of active pull-up is postponed until Sec. 4.7, where we discuss the RTL buffer.

4.5 FAN-OUT

It will generally be necessary, in a physical switching system, for one NOR gate to provide the input logic levels to a number of other such gates. The number of gates driven by a single gate is referred to as the *fan-out* of that driving gate.

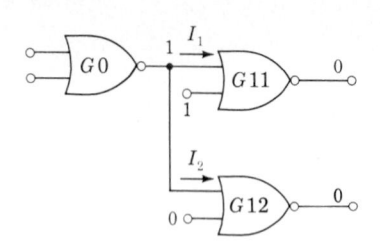

FIGURE 4.5-1
NOR gate G0 driving NOR gates G11 and
G12.

Fan-out is limited by the fact that when the output of the driving gate is at the logic 1 level, the transistors of the driven gate must all be furnished with enough current to saturate them.   The current required is supplied by the supply voltage V_{CC} through the collector resistor R_c of the driving gate and hence is limited by the simple constraints of Ohm's law.

It is of some interest to note that when a driving gate, with its limited available driving current, is fanned out to a number of gates, those driven gates which make the greatest drain on the available current are precisely the gates which do not make effective use of the current.   This point is made clearer by considering the situation indicated in Fig. 4.5-1.   Here a driving gate G0 has its output at logic level 1 and is driving two gates G11 and G12.   Gate G0 will provide a current I_1 to the base of a transistor in G11 and a current I_2 to the base of a transistor in G12.   In gate G11 the other input is at logic level 1; hence the output of G11 would be logic level 0 even if it were not furnished with current I_1.   In this sense the current I_1 is "wasted."   On the other hand, the second input of gate G12 is at logic level 0, and hence the current I_2 is essential to keep the output of G12 at level 0.   It is therefore somewhat disconcerting to note that the "wasted" current I_1 is larger than the current I_2.

To explore this point we need to consider the volt-ampere characteristic seen looking into the base of a transistor operating in the common-emitter configuration.   This volt-ampere characteristic is a plot of I_B, the base current, as a function the base-emitter voltage V_{BE}.   This current and voltage are indicated in Fig. 4.5-2.   From the Ebers-Moll equations (1.7-3) and (1.7-4) we find that if the transistor is in the active region, the base current $I_B = I_E - I_C$ is related to V_{BE} approximately by

$$I_B = \frac{I_{E0}(1 - \alpha_N)}{1 - \alpha_N \alpha_I} \epsilon^{V_{BE}/V_T} \tag{4.5-1}$$

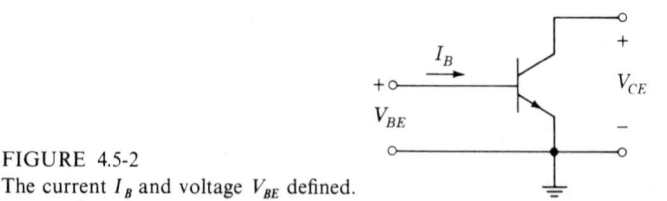

FIGURE 4.5-2
The current I_B and voltage V_{BE} defined.

In arriving at this equation we have taken account of the fact that $\epsilon^{V_{BE}/V_T} \gg 1$ and $\epsilon^{V_{BC}/V_T} \ll 1$. Since α_I is small ($\alpha_I \approx 0.1$ or even less), we may reasonably carry the approximation somewhat further and write

$$I_B = I_{E0}(1 - \alpha_N)\epsilon^{V_{BE}/V_T} \tag{4.5-2}$$

Finally, since $1 - \alpha_N = 1/(h_{FE} + 1) \approx 1/h_{FE}$, we have in the *active* region

$$I_B = \frac{I_{E0}}{h_{FE}} \epsilon^{V_{BE}/V_T} \tag{4.5-3}$$

Next, let us consider that the transistor has been driven to *saturation* to the maximum extent possible with $\sigma \equiv I_C/h_{FE}I_B = 0$. We then verify, using Eqs. (1.7-3) and (1.7-4), that

$$I_B = I_{E0} \epsilon^{V_{BE}/V_T} \tag{4.5-4}$$

Equations (4.5-3) and (4.5-4) appear eminently reasonable. When the transistor is driven to the point where $\sigma = 0$, the collector current may be ignored in comparison with the base current. If we ignore the collector current, we may equivalently consider that the collector terminal of the transistor is floating unconnected. In this case looking into the transistor between base and emitter, we see a simple diode, i.e., the base-emitter junction. The volt-ampere characteristic is then given straightforwardly by the diode equation (4.5-4). However, in the active region, the fraction $h_{FE}/(1 + h_{FE})$ of the emitter-junction current continues on to the collector junction, while the remaining fraction $1/(1 + h_{FE}) \approx 1/h_{FE}$ becomes base current. The base current I_B is again of the form of the diode equation except smaller by the factor h_{FE}, as given by Eq. (4.5-3).

We have referred to the cut-in voltage V_γ of a diode as the voltage at which the diode current just becomes large enough to be of significance. It is apparent that the cut-in point of the diode described by Eq. (4.5-3) is higher than the cut-in point of the diode described by Eq. (4.5-4), since in the first case the base current is always smaller by the factor h_{FE}. Since

$$\frac{1}{h_{FE}} = \exp\left(\ln \frac{1}{h_{FE}}\right) = \exp(-\ln h_{FE}) \tag{4.5-5}$$

we may write Eq. (4.5-3) in the form

$$I_B = I_{E0} \exp\left[\frac{V_{BE} - V_T \ln h_{FE}}{V_T}\right] \tag{4.5-6}$$

A plot of Eq. (4.5-6) would be identical to a plot of Eq. (4.5-4) except that the plot for Eq. (4.5-6) would be shifted in the positive V_{BE} direction by amount $V_T \ln h_{FE} = 0.1$ V for $V_T = 25$ mV and $h_{FE} = 55$. Thus, in a typical case the cut-in point of the active-transistor base input circuit will be about 0.1 V higher than for a transistor driven to the limit of saturation.

Now consider the situation represented in Fig. 4.5-3, where two transistors are paralleled like a two-input NOR gate. Suppose that $T2$ is cut off. Then

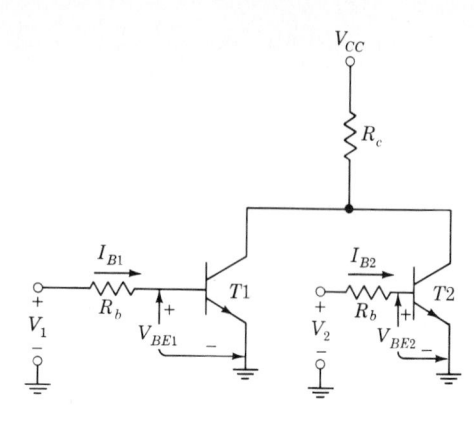

FIGURE 4.5-3
The input volt-ampere characteristic of
$T1$ depends on the operating condition
of $T2$.

the volt-ampere input characteristic at the base of $T1$ is given by Eq. (4.5-3), the presence of $T2$ having no effect. But suppose, on the other hand, that $T2$ is in saturation with V_{CE}, as usual, in the range 0.2 to 0.1 V. In this case when V_1 rises to bring $T1$ out of cutoff ($V_{BE1} = 0.65$ V), the collector junction will also find itself forward-biased. That is, with $V_{BE1} = 0.65$ V and, say, $V_{CE} = 0.2$ V, $V_{BC} = 0.65 - 0.2 = +0.45$ V. Hence, $T1$ is either cut off or in saturation and is *never in the active region*. Hence, when $T1$ is ON, the input volt-ampere characteristic at the base of $T1$ is given by Eq. (4.5-4).

Next, let us consider the worst-case situation represented in Fig. 4.5-4. Here a gate $G0$ drives one input of N different gates. In every case except one, the transistor paralleling the driven transistor is in saturation. The transistors $T12$, $T13$, ..., $T1N$ are in saturation if they are not cut off. The transistor $T11$ may operate in the active region. Allowing for this worst-case condition, we now estimate the allowable fan-out.

As we have seen, the base current I_{B1} required to drive $T11$ is given by Eq. (4.4-2) as 100 μA. We can make a rough estimate of the currents I_{B2}, ..., I_{BN} by considering that the base voltages of $T12$, $T13$, ..., $T1N$ are lower by 0.1 V then the base voltage of $T11$. Since V_o is common to all transistors, each current I_{B2}, ..., I_{BN} is larger than I_{B1} by $0.1/R_b = 0.1/450 = 220$ μA. Therefore I_{B2}, ... is

$$I_{B2} = 100 + 220 = 320 \ \mu A \tag{4.5-7}$$

We have also calculated, as given by Eq. (4.4-3), that at room temperature V_o [referred to as V_i in Eq. (4.4-3)] is 0.8 V. The maximum available current from the driver is I_o, given by

$$I_o = \frac{V_{CC} - V_o}{R_c} = \frac{3 - 0.8}{640} = 3.4 \text{ mA} \tag{4.5-8}$$

If the fan-out is to be N, we require that

$$I_o = I_{B1} + I_{B2} + \cdots + I_{BN} \tag{4.5-9a}$$

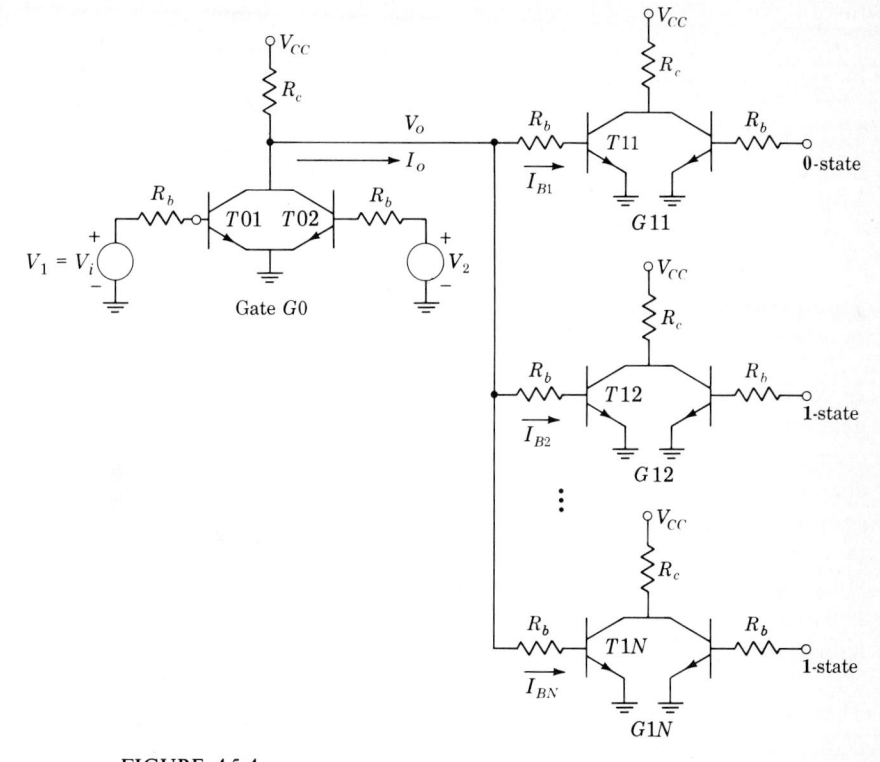

FIGURE 4.5-4
Gate $G0$ drives N two-input gates. In all cases except one, the transistor paralleling the driven gate is in saturation.

and
$$3,400 = 100 + (N - 1)320 \qquad (4.5\text{-}9b)$$

Thus we find that at room temperature

$$N = 11 \qquad (4.5\text{-}10)$$

At lower temperatures the allowable fan-out would be reduced. This situation results from the temperature sensitivity of the base-emitter voltage $V_{BE}(-2\,\text{mV}/°\text{C})$ and because h_{FE} decreases with decreasing temperature. At $-55°\text{C}$ we would find that h_{FE} has fallen to about half its value at room temperature. With h_{FE} reduced to so low a value, it is easy to verify that at $-55°\text{C}$ the fan-out must be reduced to about $N = 7$.

A further possibility may make it necessary to restrict the fan-out even slightly more than indicated by the preceding discussion. Referring to Fig. 4.5-4, we have assumed that when V_o is at logic **1**, transistors $T01$ and $T02$, being cut off, are drawing no current. As a matter of practice it turns out that when these transistors are cut off, each may, in a worst case, draw as much as 75 μA each of leakage current (see Sec. 4.9). Suppose then, for example, that gate $G0$ is a three-input gate. In this case the leakage current would be $3(75) = 225\ \mu A$. The unavailability of this leakage current as drive current for succeeding gates may well reduce the fan-out by one unit.

Altogether, on the basis of the preceding discussion and taking into account also the variability of transistor and resistor parameters to be anticipated from sample to sample of an integrated circuit, it is not surprising that manufacturers, with befitting conservatism, specify the fan-out of RTL gates at $N = 5$.

4.6 INPUT-OUTPUT VOLTAGE CHARACTERISTIC OF CASCADED RTL GATES

In connection with logic gates it is often useful to have available a plot of the gate output voltage V_o as a function of the gate input voltage V_i under the circumstances that the gate is fanning out into a number of loads. Such transfer characteristics incorporate in a single figure a great deal of information about operating points, logic swing (the change in voltage levels corresponding to the two logic levels **1** and **0**), and compatibility between gate input and output voltages. Such characteristics may also be used to determine a measure of the sensitivity of a gate to noise at any of its inputs and to display the effect of temperature and supply-voltage variation.

We shall plot the transfer characteristic for the RTL gate for a fan-out of 5. That is, referring to Fig. 4.5-4, we shall plot V_o as a function of V_i assuming $N = 5$ and that $T02$ is cut off. We shall also assume that in each of the driven gates all transistors, other than the transistors connected to $G0$, are cut off. Note that in the present situation the base of each right-hand transistor $G1I$–$G1N$ is to be in the **0**-state and not, as appears in the figure, in the **1**-state.

When the voltage $V_i = 0$ V, the output V_o of gate $G0$ is in the **1** state. All the driven gates, to which the driving-gate fans out, are in saturation. As we have discussed in Sec. 1.9, we shall assume that looking into the base, i.e., between base and emitter, a saturated transistor appears as a voltage source $V_\sigma = 0.75$ V (at room temperature) or more generally [see Eq. (1.2-1)] as a source

$$V_\sigma = 0.75 - (2 \times 10^{-3})(T - 25°C) \tag{4.6-1}$$

The equivalent circuit needed to calculate the collector voltage V_o of gate $G0$ is as given in Fig. 4.6-1. In this circuit we have implicitly assumed that all driven transistors are identical. Applying Kirchhoff's laws to the circuit of Fig. 4.6-1, we find that (since $450/5 = 90\ \Omega$)

$$V_o = \frac{3}{640 + 90}\,90 + \frac{V_\sigma}{640 + 90}\,640 \tag{4.6-2}$$

Transistor specifications are usually presented at three temperatures, $-55°C$, room temperature $T = 25°C$, and $+125°C$. The low and high temperatures represent the extreme operating temperatures, while room temperature represents

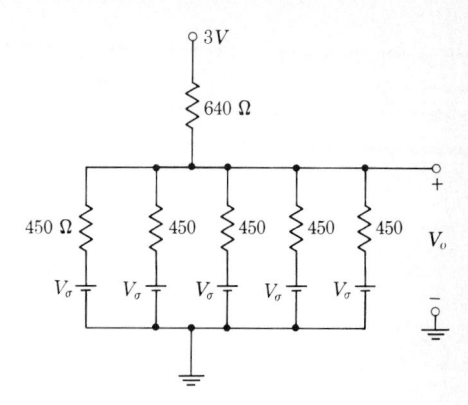

FIGURE 4.6-1
Equivalent circuit to calculate V_o when
$G0$ is OFF.

typical operating conditions. The output voltage V_o at these three temperatures is found, using Eqs. (4.6-1) and (4.6-2), to be

$$V_o = \begin{cases} 0.85\ \text{V} & T = 125°\text{C} \\ 1.03\ \text{V} & T = 25°\text{C} \\ 1.17\ \text{V} & T = -55°\text{C} \end{cases} \quad (4.6\text{-}3)$$

As V_i is increased from 0 V, a point will be reached where transistor $T01$ enters the active region. Again, as discussed in Sec. 1.2, we shall consider that this cut-in point occurs at

$$V_{BE} = V_\gamma = 0.65 - (2 \times 10^{-3})(T - 25°)$$

$$= \begin{cases} 0.45 & T = 125°\text{C} \\ 0.65 & T = 25°\text{C} \\ 0.81 & T = -55°\text{C} \end{cases} \quad (4.6\text{-}4)$$

Increasing V_i farther, we find that V_o decreases and eventually will drop to $V_{CE}(\text{sat})$, which we shall again assume to be in the range 0.2 to 0.1 V. The base-emitter voltage of the driving-gate transistors will again be $V_{BE} = V_\sigma$ at saturation.

The input voltage at the point where $T01$ in gate $G0$ reaches saturation has already been calculated in Eq. (4.4-3) to be $V_i = 0.8$ V at room temperature. Using Eq. (4.6-1), we can similarly find V_i at other temperatures:

$$V_i = \begin{cases} 0.6\ \text{V} & T = 125°\text{C} \\ 0.8\ \text{V} & T = 25°\text{C} \\ 0.96\ \text{V} & T = -55°\text{C} \end{cases} \quad (4.6\text{-}5)$$

Using the results given in Eqs. (4.6-3) to (4.6-5), we can construct the plots shown in Fig. 4.6-2. To see how these plots are obtained consider $T = 25°\text{C}$. Then for $V_i \leq 0.65$ V (cut-in), $V_o = 1.03$ [Eq. (4.6-3)]. The value of cut-in is given in Eq. (4.6-4). The value of V_i needed to have $V_o = 0.2$ V (saturation) is $V_i = 0.8$, which is obtained from Eq. (4.6-5). The maximum value of V_i is 1.03 V

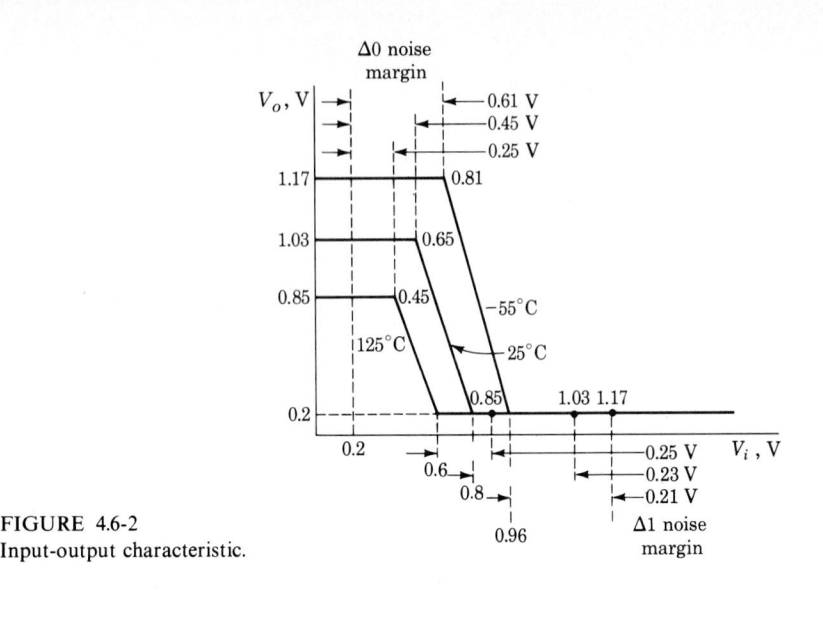

FIGURE 4.6-2
Input-output characteristic.

and is obtained by assuming that gate $G0$ is driven from another gate (not shown) having the same values of V_o that $G0$ has. Using a similar argument, we find that V_i cannot be less than 0.2 V, which occurs when the gate driving $G0$ saturates.

These plots are, of course, idealized and representative rather than real and exact. However, they are close enough to the actual plots that would be obtained in a representative real situation to be of great value. By way of comparison we compare the plots of Fig. 4.6-2 with the plots of Fig. 4.6-3 determined experimentally for typical gates. The real plots indicate some dependence of the temperature on the transistor saturation collector-emitter voltage. This, of course, should be expected, since, referring to Eq. (1.10-3), we see that $V_{CE}(\text{sat})$ is proportional to $V_T = kT/q$ and therefore $V_{CE}(\text{sat})$ *increases* with an increase in temperature. Our idealized plots have ignored this temperature dependence since the variation is small. We also note that while we have assumed that $V_o(\text{sat}) = 0.2$ V at room temperature, the characteristic shown in Fig. 4.6-3 has a value of 0.16 V. Starting with $V_{CE}(\text{sat}) = 0.16$ V at room temperature and since $0°\text{C} = 273°\text{K}$, we then have

$$V_{CE}(\text{sat}, 125°\text{C}) = \left[\frac{273 + 125}{273 + 25} \right] 0.16 = 0.21 \text{ V} \qquad (4.6\text{-}6)$$

and we correspondingly find $V_{CE}(\text{sat}, -55°\text{C}) = 0.12$ V. These results agree with Fig. 4.6-3.

Noise margin The plots of Fig. 4.6-2 are useful because they provide the input and output voltage ranges corresponding to the two logic levels **1** and **0**. We keep in mind that generally the input of one gate is the output of a preceding gate. Thus, when the logic level **0** appears at the output of a driving gate and

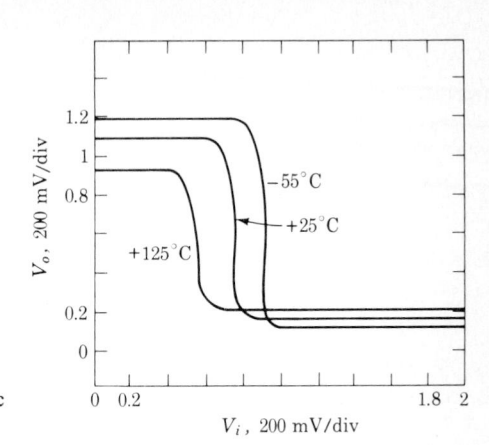

FIGURE 4.6-3
Typical RTL-gate transfer characteristic
for three temperatures at fan-out = 5.

hence at the input to a driven gate, this logic level will correspond to the voltage $V_{CE}(\text{sat}) \approx 0.2$ V. We now note from Fig. 4.6-2 that such a voltage, 0.2 V, is low enough to represent properly the logic level **0**. For over the entire range of temperature contemplated in Fig. 4.6-2, the driven gate transistor will be cut off and the output of the driven gate will therefore be at a high voltage. Further, we note that this voltage $V_{CE}(\text{sat}) = 0.2$ V provides a *margin of safety* in assuring cutoff. For even in the least assured case, which occurs at $T = 125°C$, there is a voltage difference $0.45 - 0.20 = 0.25$ V between $V_i = V_{CE}(\text{sat})$ and the minimum input voltages which would allow the driven transistor to enter its active region, $V_i = 0.45$ V. Having marked off the voltage $V_{CE}(\text{sat}) = 0.2$ V on the abscissa of Fig. 4.6-2, we can indicate the margins of safety for the three temperatures in the manner shown in the figure. This margin of safety is also referred to as the *noise margin for an input corresponding to logic level* **0** and is represented by $\Delta 0$. We have that

$$\Delta 0 = 0.65 - k(T - 25°C) - V_{CE}(\text{sat}) = 0.45 - k(T - 25°C) \qquad (4.6\text{-}7)$$

where k, as noted, is $k = 2$ mV/°C. We note the decrease of $\Delta 0$ with increasing temperature.

The sensitivity to temperature of $\Delta 0$ is actually somewhat larger than indicated in Eq. (4.6-7), for we see in Fig. 4.6-3 that $V_o = V_{CE}(\text{sat})$ increases somewhat with temperature. Thus, the single dashed line erected vertically from the abscissa at $V_i = V_{CE}(\text{sat}) = 0.2$ V in Fig. 4.6-2 should be replaced by three lines, corresponding to the three temperatures. The reduction in $\Delta 0$ below the values given in Fig. 4.6-2 would be greatest for $\Delta 0$ at 125°C, the case where there is already the smallest noise margin. We may, as a matter of fact, observe that this limited $\Delta 0$ encountered with increasing temperature constitutes a disadvantage of consequence in RTL circuitry.

Now let us turn our attention to the case in Fig. 4.5-4, where V_i has increased sufficiently for $T01$ to be driven into saturation. In this case, of course, the higher the voltage V_i the more assurance we have of saturating transistor $T01$.

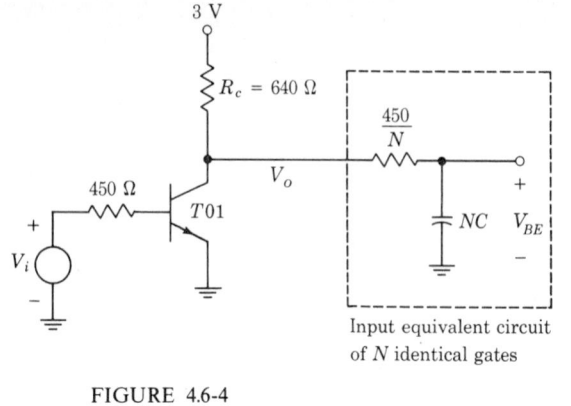

FIGURE 4.6-4
Equivalent circuit to determine the rise time of V_{BE}.

We note from Sec. 4.5 that the fan-out of the gate driving $G0$ limits the extent to which V_i can rise, just as the fan-out of $G0$ limits the maximum value of V_o. The greater the fan-out the lower the available voltage.   Hence, we must consider that the maximum available input voltage to drive a gate is not the unloaded output voltage of a driving gate but the output voltage which is loaded by a maximum fan-out.   We have drawn Fig. 4.6-2 for a fan-out of 5 precisely because this fan-out is nominally the maximum intended with RTL gates.   The output voltages (1.17, 1.03, and 0.85 V) for the fan-out of 5 are, of course, the input voltages available to drive succeeding transistors to saturation.

We can now mark off on the abscissa of Fig. 4.6-2 these output voltages corresponding to a fan-out of 5.   The noise margins $\Delta 1$ for an input corresponding to logic level **1** can be read off as indicated in the figure.   Thus, at $T = -55°$ the output is 1.17 V, and the minimum output which will keep a driven transistor in saturation is 0.96 V.   The noise margin is therefore $\Delta 1 = 1.17 - 0.96 = 0.21$ V.   We note from Fig. 4.6-2, that the $\Delta 1$ noise margins are not sensitively dependent on temperature.

Rise time   The rise time of gate $G0$ is affected by the number of gates it drives. To illustrate this consider that in Fig. 4.5-4 V_i is in the **1** state and V_o is in the **0** state.   Then $T11$ to $T1N$ are cut off, and the base-emitter junction of each of these transistors appears to be a capacitor.   Let us approximate this capacitor as a real constant capacitor C.   Then the equivalent circuit of $G0$ and the driven gates in the region where V_i changes from its **1** state to its **0** state (with V_2 remaining in the **0** state) is shown in Fig. 4.6-4.   Here we have assumed that each of the $N = 5$ driven gates is identical.

Now let V_i fall instantaneously from its **1** state to its **0** state. $T01$ cuts off, and the base-emitter voltage of $T11$ to $T1N$ rises with a time constant

$$\tau = \left(R_c + \frac{450}{N}\right) NC = (640N + 450)C \tag{4.6-8}$$

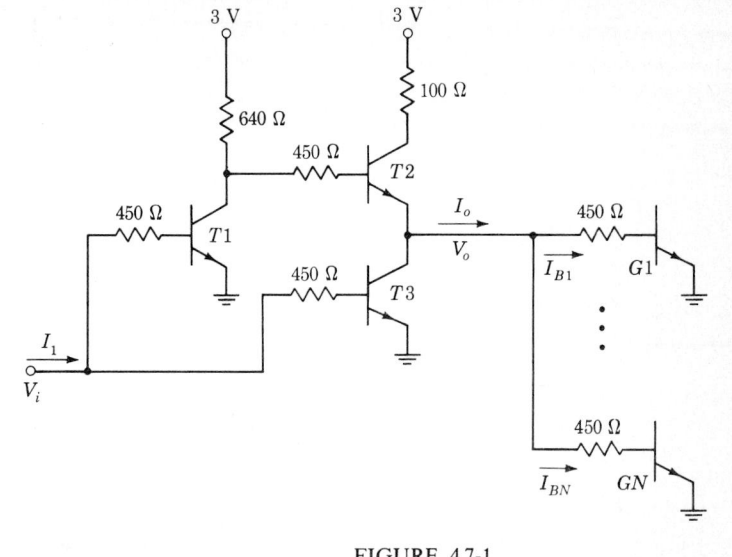

FIGURE 4.7-1
An RTL buffer inverter driving N gates.

The manufacturer generally specifies that $C \approx 5$ pF; thus with $N = 5$, $\tau = 18$ ns. In the next section we shall see that an RTL gate using an active pull-up can have the same value for τ and a significantly larger fan-out.

It should be kept in mind that the input capacitance to a transistor is not constant and is a function of the current in the base. Thus, the time constant indicated here is of only qualitative rather than quantitative value.

4.7 AN RTL BUFFER

The RTL gates discussed in the previous sections employed a *passive pull-up* R_c. This limited the output current, available to drive other gates, to a value less than V_{CC}/R_c. The *buffer* is an RTL gate using an *active pull-up* to achieve a very low output impedance. Thus, the output-current capability of the buffer is significantly greater than that of the ordinary gate. As a result, while the ordinary RTL gate has a fan-out of 5, the buffer has a fan-out of 25. In addition, since the buffer has a low output impedance, the *rise time* when driving a capacitive load is significantly less than that obtained when using a passive pull-up.

A typical RTL buffer is shown in Fig. 4.7-1. The transistor $T2$ is the active pull-up which replaces the passive pull-up resistor R_c in the ordinary NOR gate. The transistor $T1$ serves as a logic inverter. When V_i is at logic **0**, the output of $T1$ is at logic **1** and vice versa. Thus, the logic levels at the bases of $T2$ and $T3$ are always different, and when one of these transistors is conducting, the other is cut off. When $T2$ is OFF, its collector current is zero; $T3$ is saturated

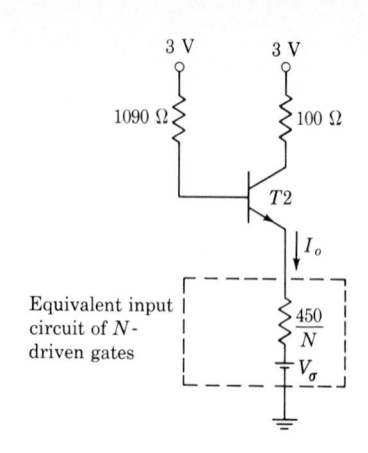

FIGURE 4.7-2
Equivalent circuit of the buffer when V_i
is in the **0** state.

in the extreme with $\sigma = I_C/h_{FE} I_B = 0$, and (as can be seen in Fig. 1.10-1) its
voltage $V_{CE}(\text{sat}) = 60$ mV.

When $T3$ is OFF and $T2$ is ON and driving N gates, its output current I_o
can be calculated from the circuit of Fig. 4.7-2. Assume first that $T2$ is in its
active region with $V_{BE}(T2) = 0.7$ V and assume also $h_{FE} = 50$. At room temperature, as usual, we take $V_\sigma = 0.75$ V. Applying Kirchhoff's voltage law to the
loop which includes the base-emitter junction of $T2$, we find

$$I_o = \frac{3 - 0.7 - 0.75}{1090/50 + 450/N} = \frac{7.1 \times 10^{-2}}{1 + 20.6/N} \tag{4.7-1}$$

If, instead, we assume that $T2$ is saturated, we have $V_{BE} = 0.75$ V and $V_{CE} = 0.2$ V.
In this case we find (Prob. 4.7-4) that

$$I_o(\text{sat}) = \frac{21.8 \times 10^{-3}}{1 + 4.9/N} \tag{4.7-2}$$

To tell whether or not $T2$ is saturated we need only compare the two currents.
For if $I_o(\text{sat}) > I_o$, the transistor is not saturated, while if $I_o(\text{sat}) < I_o$, the
transistor is saturated. It can be verified that the transistor is not saturated for
$N = 1$ and is saturated for $N \geq 2$.

The buffer is ordinarily employed to drive a large number of gates. We
shall therefore assume that $N \geq 2$ and that the transistor $T2$ is saturated. Since
$I_o(\text{sat})$ is supplied to all N gates, the current supplied to each gate is
$I_{B1} = I_{B2} = I_{BN} = I_B$, where

$$I_B = \frac{I_o(\text{sat})}{N} = \frac{21.8 \times 10^{-3}}{N + 4.9} \tag{4.7-3}$$

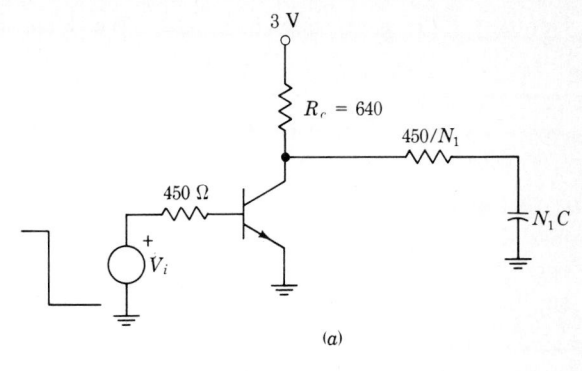

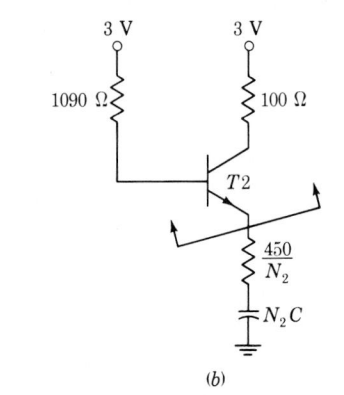

(b)

FIGURE 4.7-3
(a) RTL gate circuit to determine time constant τ_i. (b) Equivalent circuit of buffer just after $T1$ and $T3$ cut off.

For the purpose of comparing the fan-out of the buffer to the fan-out of the ordinary RTL gate we assume that $I_B = 320\ \mu A$, as in Eq. (4.5-7). Then

$$\frac{21.8 \times 10^{-3}}{N + 4.9} \geq 0.32 \times 10^{-3} \qquad (4.7\text{-}4a)$$

and
$$N \leq 65 \qquad (4.7\text{-}4b)$$

Note that in Eq. 4.5-10) we found $N = 11$. Thus, we have improved the fan-out by a factor of 6. Manufacturer's specifications, typically conservative, state that the maximum fan-out of a buffer is $N = 25$, which is a factor of 5 greater than the specification of fan-out for an RTL gate.

We compare now the rise times of a buffer with a passive pull-up NOR gate. The rise time of a gate is proportional to the RC time constant of the gate and load. In Fig. 4.7-3a we have the equivalent circuit of an RTL gate driving N_1

identical RTL gates (see Sec. 4.6). We have represented the base-emitter junction of each transistor by a capacitor C. The time constant τ_1 is

$$\tau_1 = \left(640 + \frac{450}{N_1}\right)N_1 C = (640N_1 + 450)C \qquad (4.7\text{-}5)$$

Figure 4.7-3b is the equivalent circuit of a buffer, driving N_2 RTL gates, just after the input voltage V_i dropped from its **1** to **0** state. $T1$ and $T3$ are cut off and $T2$ is saturated. Since $T2$ is saturated, looking back into the emitter, we see just the 100-Ω collector resistor. Hence, the time constant is

$$\tau_2 \approx \left(100 + \frac{450}{N_2}\right)N_2 C = (100N_2 + 450)C \qquad (4.7\text{-}6)$$

If the rise time of the RTL gate and buffer are to be the same, their time constants are the same. To obtain a relationship between N_1 and N_2 we therefore equate Eqs. (4.7-5) and (4.7-6), which yields

$$640N_1 = 100N_2 \qquad (4.7\text{-}7a)$$

Thus $\qquad\qquad\qquad\qquad N_2 = 6.4N_1 \qquad\qquad\qquad\qquad (4.7\text{-}7b)$

which is about the same ratio that we obtained from static considerations.

Current limiting Referring to Fig. 4.7-2, we note that with $T2$ saturated, the collector current is limited by the 100-Ω collector resistor. This resistor is designed to allow $T2$ to saturate and yet limit its power dissipation.

To illustrate, let $T2$ be saturated so that $V_{CE} \approx 0.2$ V. Then with $N = 25$, $I_o(\text{sat}) = 18$ mA. We can show (see Prob. 4.7-3) that $I_B = 1$ mA, so that $I_C = 17$ mA. Thus, the collector dissipation in $T2$ is $P_C = 3.4$ mW. However, if the 100-Ω resistor were short-circuited, $T2$ would not saturate, since $V_{CB} > 0$. Then from Eq. (4.7-1), I_o $(N = 25) = 39$ mA, and from Fig. 4.7-2,

$$V_{CE} = 3 - (450/25)(39 \times 10^{-3}) - V_\sigma = 1.55 \text{ V} \qquad (4.7\text{-}8)$$

The dissipation is now 60 mW, a considerable increase and a sizable dissipation for an integrated-circuit transistor.

4.8 AN RTL EXCLUSIVE-OR GATE

The EXCLUSIVE-OR gate was introduced in Sec. 3.10, where it was defined as a device which performs the operation

$$Z = \overline{A} \cdot B + A \cdot \overline{B} \equiv A \oplus B \qquad (4.8\text{-}1)$$

An RTL circuit capable of performing this logic operation is shown in Fig. 4.8-1a. The operation of this circuit is as follows. Consider that B is in the

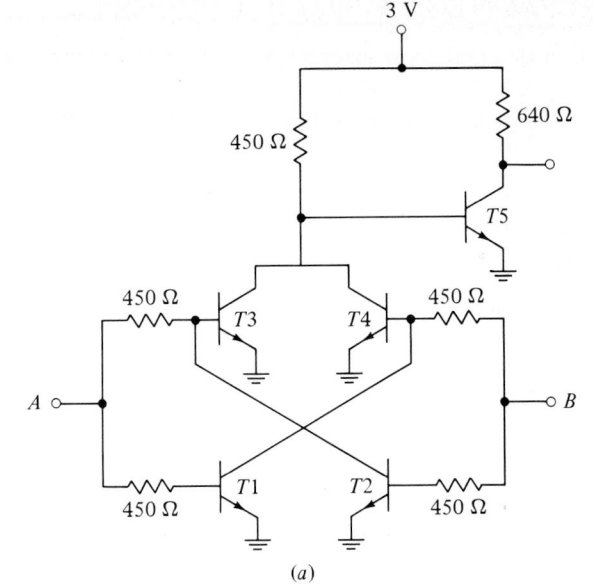

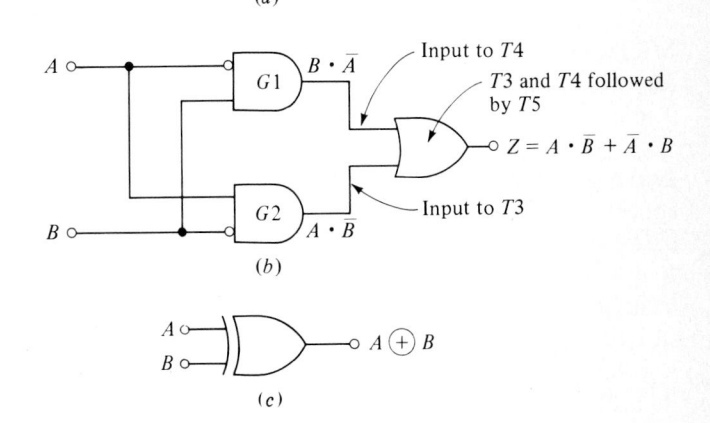

FIGURE 4.8-1

(a) An EXCLUSIVE-OR circuit using RTL. (b) Logic-gate representation and (c) symbol for EXCLUSIVE-OR circuit.

logic **0** state. Then $T2$ is OFF. Thus, the input to $T3$ is A. If, however, B is in the **1** state, $T2$ is saturated and $T3$ is cut off. Hence, the input to $T3$ is $A \cdot \overline{B}$. Similarly, the input to $T4$ can be shown to be $\overline{A} \cdot B$. These two operations are represented by $G2$ and $G1$, respectively, in Fig. 4.8-1b.

Referring to Fig. 4.8-1a, we see that transistors $T3$ and $T4$ represent a NOR operation. Transistor $T5$ inverts the NOR output, producing an OR gate. This is shown in Fig. 4.8-1b. Thus, the output of $T5$ is $Z = A \cdot \overline{B} + \overline{A} \cdot B$, as required.

4.9   MANUFACTURER'S SPECIFICATIONS

This section is devoted to a discussion of the manufacturer's specifications of RTL gates.

There are four basic types of RTL gates. They are the medium-power gate (MRTL), where $R_c = 640 \ \Omega$ and $R_b = 450 \ \Omega$; the low-power gate (LRTL), where $R_c = 3.6 \ \text{k}\Omega$ and $R_b = 1.5 \ \text{k}\Omega$; the buffer, which is available in either the medium- or low-power class; and the EXCLUSIVE-OR gate. Some of the more important characteristics of each of these gates are presented in Fig. 4.9-1 for room temperature, $T = 25°\text{C}$, and for a typical gate. Corresponding specifications at -55 and $+125°\text{C}$ are also available in manufacturer's literature.

Input current, I_{in} This is the maximum current that will be drawn by a gate input. We have estimated [Eq. (4.5-7)] that this current is about 320 μA. The manufacturer even more conservatively allows a worst case of 435 μA. This manufacturer's specification is about right to account for his specification that the allowable fan-out is 5.

Output current, I_{AN} This is the minimum current available to drive other gates under the circumstance of a fan-out of N. We note from Fig. 4.9-1 that for an MRTL gate, $I_{A5} = 2.54$ mA. Since $I_{in} = 435 \ \mu$A, we should expect that $I_{A5} = 5I_{in}$. We find that $5I_{in} = 5 \times 435 \ \mu\text{A} = 218$ mA. The difference $2.54 - 2.18 = 0.36$ mA provides an extra margin of safety and allows some current for leakage.

Leakage current When a transistor input is at logic **0**, we would expect the collector current to be zero. However, such is not always the case since the base-emitter voltage is always somewhat positive. For example, if the transistor is driven by a gate which is saturated, then the base-emitter voltage of the transistor is equal to the saturation voltage of the driving gate. In a worst-case situation this saturation voltage may be as high as 0.4 V. As can be seen in Fig. 4.9-1, the manufacturer guarantees that the leakage current I_L is always less than 218 μA for the MRTL gate and buffer and 100 μA for the LRTL gate and buffer.

Input and output loading factors To provide guidance concerning allowable fan-out of gates, buffers, etc., manufacturers often assign numbers to the input and output terminals. These numbers, called input and output *loading factors*, are proportional (to the nearest integer on the conservative side) to the current required by an input or available from an output. Thus, as appears in Fig. 4.9-1, the input loading factor of an MRTL gate is 1, and the output loading factor N_G is 5. These numbers are interpreted to mean that the available output current is at least 5 times (but less than 6 times) the required input current. Consequently, when a gate drives a gate, the allowable fan-out is 5. We note that a buffer has an input loading factor of 2. Hence, a gate may be fanned out to two buffers but not three. The allowable fan-out of an LRTL gate is 4, and hence the input and output loading factors are 1 and 4, respectively.

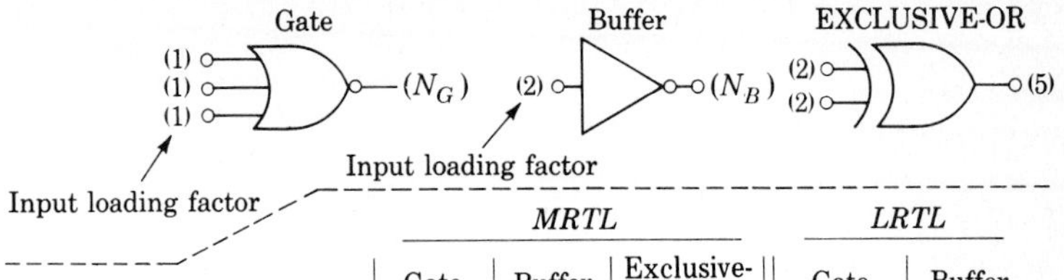

		MRTL			LRTL	
		Gate	Buffer	Exclusive-or	Gate	Buffer
Fan out $\begin{cases} N_G \\ N_B \end{cases}$	N_G	5		5	4	
	N_B		25			30
Propagation delay time, t_{pd}		12 ns	20 ns	12 ns	27 ns	57 ns
Power dissipation, P_d						
Inputs high		19 mW	16 mW	72 mW	4.8 mW	5.5 mW
Inputs low		5 mW	45 mW	72 mW	0.5 mW	16 mW
Input current, I_{in}		435 μA	870 μA	870 μA	130 μA	260 μA
Output current, I_{A5}		2.54 mA	12.7 mA	2.54 mA	815 μA	4 mA
Output leakage current, I_L		218 μA	218 μA	218 μA	100 μA	100 μA
Saturation voltage		210 mV	210 mV	210 mV	220 mV	220 mV

FIGURE 4.9-1
Manufacturer's specifications.

The loading factors are modified when MRTL and LRTL gate families are used together. The loading factors of the LRTL gate remains 1 and 4, but the loading factors of the MRTL gate become 3 and 16. These numbers indicate that an MRTL gate is able not only to drive five other MRTL gates but can simultaneously drive an LRTL gate, that is, $16 = 5 \times 3 + 1$. In general, in fanning gates out we need only observe the rule that the output loading factor of the driving gate must be equal to or less than the sum of the input loading factors of the driven gates.

4.10   PARALLELING RTL GATES

To increase the number of inputs available in an RTL gate we can operate these gates in parallel. We consider now how the input and the output loading factors are affected by such paralleling.

One way of paralleling is shown in Fig. 4.10-1, where we have connected the output terminals of the two gates A and B and in each case have connected the top end of the collector resistor R_c to the supply voltage V_{CC}. As a consequence, the composite gate operates with a collector resistor $R_c/2$. It will be recalled that the output loading allowed on an RTL gate is limited by the fact that the output current required to drive other gates must be furnished from V_{CC} through the collector resistor. Since the paralleling has reduced the collector

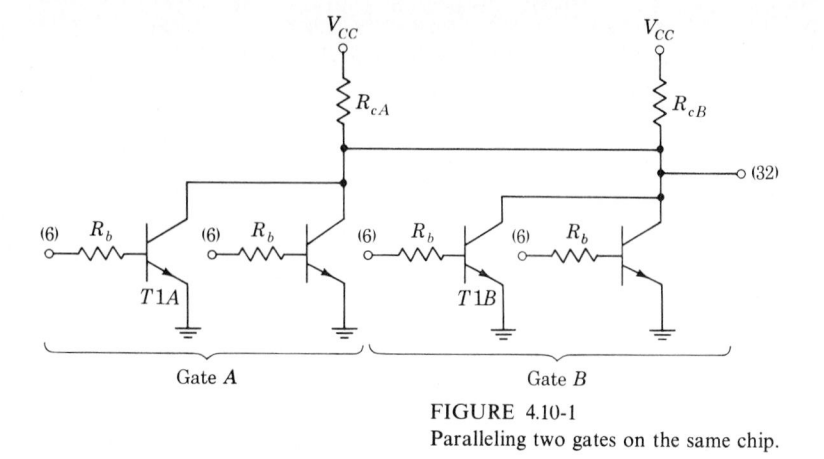

FIGURE 4.10-1
Paralleling two gates on the same chip.

resistor by a factor of 2, the effect of the paralleling is to increase the output loading factor by this same factor of 2. If MRTL gates are involved which individually have an output loading factor of 16, then, as indicated in Fig. 4.10-1, the paralleling has increased the output loading factor to 32. Altogether, then, the paralleling has doubled the output loading factor and has doubled as well the number of available input terminals to the gate.

The price to be paid for these advantages, as noted in Fig. 4.10-1, is that the input loading factor must also be increased by a factor of 2, increasing it to 6, while for a single gate the input loading factor is 3. This increased input load results from the fact that the collector resistance is $R_c/2$ rather than R_c. For with $R_c/2$ the collector saturation current increases by a factor of 2, and the base current required to drive a transistor to saturation must increase correspondingly by 2 (for the same σ and the same saturation voltage). Similarly if N gates are paralleled, the output loading factor becomes $16N$ and the input loading factor becomes $3N$.

These results for paralleling apply when the individual gates involved are part of the same integrated circuit, i.e., are on the same chip. When the individual gates are on different chips, the input loading factor must be increased beyond $3N$ for a reason now to be discussed.

A characteristic of integrated circuits is that there may be considerable variability in the values of resistors in any given circuit. Thus, from sample to sample, in an RTL gate, R_c and R_b may vary widely. On the other hand, there is substantially less variability in the ratio of resistors R_c/R_b. This greater uniformity of the resistor ratio serves to provide a measure of compensation in the operation of a gate operating individually. For suppose, in a particular gate, that R_b is considerably larger than average. Then in a typical operating situation the base current of a transistor in such a gate will be smaller than average. On the other hand, if R_b is larger, R_c will be also. Hence, the transistor saturation current will be smaller, and the smaller available base current will be adequate to drive the transistor to saturation. Similarly if R_c is smaller, the saturation current will be larger. But the correspondingly smaller R_b will serve to increase the base current, as required. Finally we may note that if the resistors in a gate depart from the average, all the resistors in all the gates *on that same chip* depart from the average in the same direction.

Now let us consider the situation when the paralleled gates are on *different* chips. In this case the resistor on one gate may be high and on the other gate low. Thus, for example, in Fig. 4.10-1, consider that in gate A the collector resistor (call it R_{cA}) is high and in gate B, R_{cB} is low. Suppose that the paralleled gate is to be driven to logic level **0** by driving transistor TIA to saturation. Since R_{cB} is smaller than R_{cA}, the parallel combination of R_{cA} and R_{cB} is less than $R_{cA}/2$. Therefore, when gate B is paralleled with gate A, the current required to drive TIA to saturation is *more than doubled*. Hence, the input loading factor at the input to TIA is also more than doubled by the paralleling. Thus, the input loading factor must be increased above 6. To put the matter another way, we say that to drive the extra current required in R_b of TIA we must assure a higher voltage at the input to TIA. This result can be accomplished by restricting somewhat the allowed loading on the driving gate. If we specify a higher input loading factor for gate A, we shall have accomplished precisely this end.

Referring again to Fig. 4.10-1 and assuming again that $R_{cB} < R_{cA}$, if the drive is to be applied to the base of transistor TIB instead of TIA, the loading factor at this input might be set at *less* than 6. On the other hand, when gates are paralleled, unless we take special pains to investigate, we shall not know which gate has the higher resistors and which the lower. Hence, the only thing we can do is to allow an extra margin of input loading factor at *every* gate input.

In the matter of the extra margin of input loading factor we must seek guidance from the manufacturer, who will base his recommendations on what he knows his manufacturing tolerances to be. Typically, we find that in MRTL, when a gate on a first chip is to be paralleled with a gate on a second chip, it is recommended that the input loading factor be increased by 0.75 load. Thus, in Fig. 4.10-1 we would set the input loading factors at 6.75. Finally consider that a total of N gates are to be paralleled with N_A of the gates on chip A and N_B of the gates on chip B $(N_A + N_B = N)$. Then, by an easy extension of the present discussion it would appear that the input loading factor on chip A is $3N + 0.75N_B$ and on chip B is $3N + 0.75N_A$.

4.11 SPECIFICATION OF OPERATING VOLTAGES

Figure 4.6-2 indicated the input-voltage–output-voltage variation of an RTL gate having a maximum fan-out of 5, a supply voltage $V_{CC} = 3$ V, and operating temperatures of -55, $+25$, and $+125°$C. However, the figure does *not* take account of the variability of transistor and resistor parameters due to manufacturing tolerances. To take account of such variability the manufacturer specifies

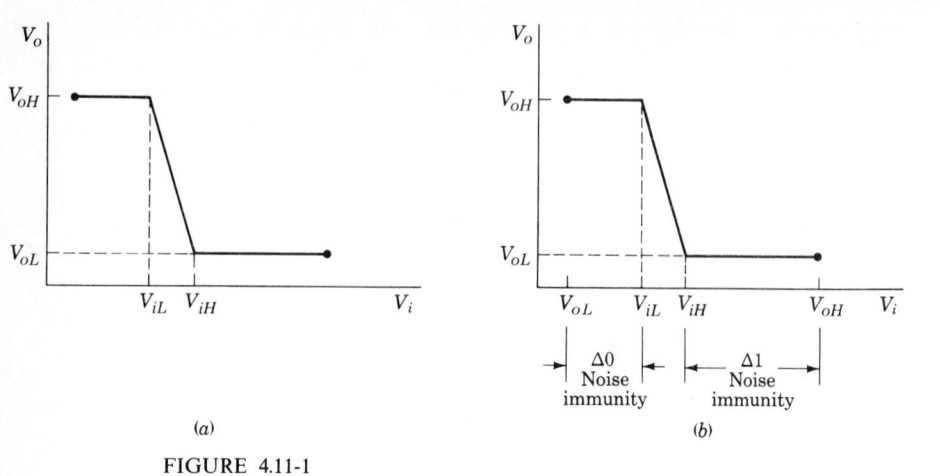

(a)   (b)

FIGURE 4.11-1
Input-output characteristic showing (a) worst-case parameters and (b) $\Delta 0$ and $\Delta 1$ noise immunities.

the worst-case parameters V_{oH}, V_{iH}, V_{oL}, V_{iL} shown in Fig. 4.11-1a. These parameters are specified at a given temperature, $V_{CC} = 3V \pm 10$ percent, maximum fan-out, and worst-case manufacturing tolerances. The definitions of these parameters are:

V_{oH} The *minimum* voltage which will be available at a gate output when the output is supposed to be at logic **1**

V_{iH} The *minimum* gate input voltage which will unambiguously be acknowledged by the gate as corresponding to logic **1**

V_{oL} The *maximum* voltage which will appear at a gate output when the output is supposed to be at logic **0**

V_{iL} The *maximum* gate input voltage which will unambiguously be acknowledged by the gate as corresponding to logic **0**

Thus, if V_i and V_o are gate input and output voltages, respectively, it is guaranteed that if $V_i \leq V_{iL}$, then $V_o \geq V_{oH}$ and if $V_i \geq V_{iH}$, then $V_o \leq V_{oL}$.

In Fig. 4.11-1b we have marked off the voltages V_{oL} and V_{oH} on the input voltage axis. It now appears that the noise immunities are

$$\Delta 0 = V_{iL} - V_{oL} \qquad \text{and} \qquad \Delta 1 = V_{oH} - V_{iH}$$

Since these four parameters V_{oH}, V_{iH}, V_{iL} and V_{oL} completely describe the worst-case input-output voltage characteristic and determine the $\Delta 0$ and $\Delta 1$ noise immunity, there is hardly any need for plotting the characteristic of Fig. 4.11-1. Instead, the same information is often presented in the manner shown in Fig. 4.11-2.

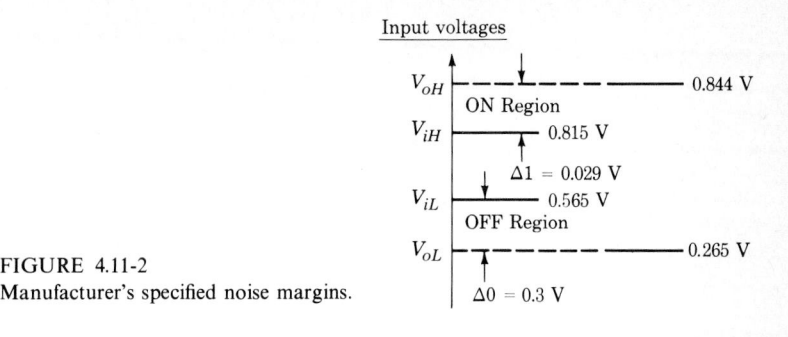

FIGURE 4.11-2
Manufacturer's specified noise margins.

EXAMPLE 4.11-1   For a given RTL gate operating at 25°C, the worst-case parameters as specified by the manufacturer are V_{oH} = 844 mV, V_{iH} = 815 mV, V_{iL} = 565 mV, and V_{oL} = 265 mV.   Find the $\Delta 0$ and $\Delta 1$ noise immunity.

SOLUTION   The worst-case parameter values are shown in Fig. 4.11-2.   Thus, the worst-case $\Delta 0$ noise immunity is 300 mV, while the worst-case $\Delta 1$ noise immunity is 29 mV.   These results are significantly poorer than those shown in Fig. 4.6-2, where the $\Delta 0$ noise margin is 450 mV and the $\Delta 1$ noise margin is 230 mV.   These results seem worse than they really are.   The manufacturer readily concedes that the results are extremely conservative and that larger noise margins actually exist.

4.12   PROPAGATION DELAY TIME

We have noted (Sec. 1.17) that there are propagation delays associated with logic gates.   In RTL gates these delays are generally specified in terms of the parameters defined in Fig. 4.12-1.   Here an input waveform (presumably the output of a preceding gate) is shown which makes a transition from logic **0** to logic **1** and thereafter a reverse transition.   The corresponding output waveform is also shown.   As is indicated, propagation delays are measured from the points on the waveform which are 0.5 V positive with respect to the logic **0** level.   Since logic **0** is at about 0.2 V, the points on the waveform are at 0.7 V, which is very nearly the midpoint of the voltage range at which the transistors of the gates are passing through their active region (see Fig. 4.4-1).   The time $t_{pd}(HL)$ is the propagation delay associated with a transition of the output waveform from its higher to its lower voltage while $t_{pd}(LH)$ is the corresponding delay as the output swings from its lower to its higher voltage level.   The two propagation delays are not ordinarily equal.

Propagation delays result, in part, from the fact that, as voltages change, capacitors must be charged and discharged. The total capacitance with which the gate must contend depends on the fan-out and also on the *fan-in* (i.e., the number of inputs on the gate).   Hence the propagation times are themselves

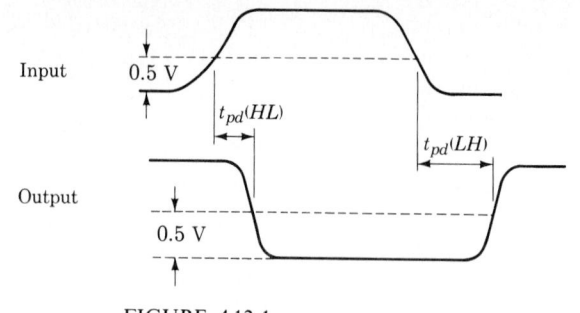

FIGURE 4.12-1
The propagation delay times $t_{pd}(HL)$ and $t_{pd}(LH)$.

functions of fan-out and fan-in.   Typically in RTL gates propagation times are of the order of 10 ns.   Such times compare favorably with propagation times associated with other types of gates yet to be considered.   Unfortunately, in spite of the favorable propagation times, RTL gates are not widely used because of their relatively poor noise margins and fan-out capabilities.

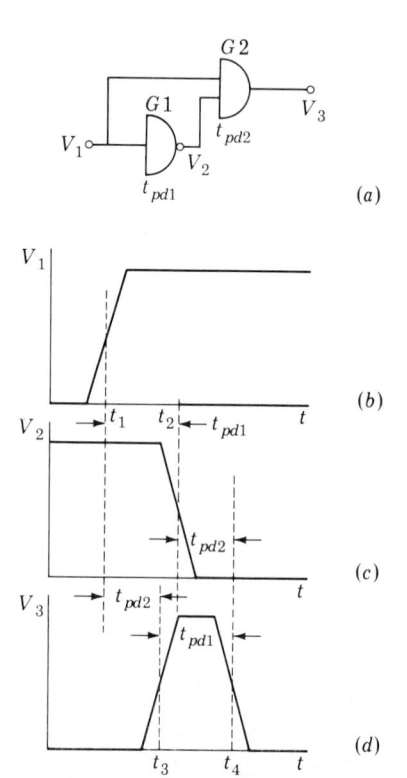

FIGURE 4.12-2
(a) A gate structure whose response is adversely affected by the propagation time delay t_{pd1}. (b) The input waveform. (c) The output of G1. (d) The output pulse from G2.

Propagation delay hazards As an example of the difficulty which may be caused by propagation delay in gates, consider the situation represented in Fig. 4.12-2. The NAND gate (used as an inverter) has a propagation delay t_{pd1} and the AND gate a delay t_{pd2}. It is intended that V_3 shall be in the **0** state independently of V_1, as is indeed consistent with the logic of the circuit. However, as we may see from the waveforms in the figure, a transition in V_1 from logic **0** to logic **1** will result in a positive pulse of duration t_{pd1} as appears in (d). V_1 makes its transition at t_1, taken, for simplicity, to be at the midpoint of the voltage excursion of the waveform V_1, the waveform V_1 having a finite rise time since it is itself the output of some preceding gate (not shown). V_2 then makes an excursion from **1** to **0** after a time t_{pd1} at time t_2. (We take the fall time of V_2 to be about the same as the rise time of V_2.) Since now V_1 rises to **1** before V_2 falls from **1**, there will be an interval when the AND gate output V_3 will be **1**. V_3 rises to **1** at $t = t_3$ delayed by an interval t_{pd2} from t_1. At $t = t_2$, after a delay t_{pd1}, V_2 falls back to **0** and finally, at $t = t_4$, after a delay again of t_{pd2}, V_3 will fall back to **0**. In summary, it then appears that the step in V_1 has given rise to a pulse in V_3. Note that the time of occurrence of the pulse is determined by t_{pd2} but that the width of the pulse is determined by t_{pd1}. Furthermore, the very *existence* of the pulse results from the delay t_{pd1}. Unintended results due to such gate propagation delays are referred to as *hazards*.

Charge compensation to reduce propagation time As noted, propagation delays result, in part, from the necessity to establish and remove base charge as transistors are turned ON and OFF. (See Sec. 1.17.) One way by which this process may be hastened is to provide for the flow of impulsive (very large albeit short-duration) currents into and out of the base. In RTL, such impulsive currents may be provided by bridging capacitors across the base resistors R_b, as shown in Fig. 4.12-3a.

Suppose $T1$ is in saturation because its input is at logic level **1**. When V_1 drops to logic **0**, $T1$ should cut off as the stored base charge is reduced to zero. In the absence of the capacitor C, this charge must dissipate in part through recombination in the base and in part by flowing out of the base through R_b. With the capacitor, however, an impulsive current can flow out of the base, removing the base charge much more rapidly. If the capacitor is adequately large it is possible, in principle at least, to transfer all the base charge to the capacitor instantaneously thereby turning the transistor OFF with limitless speed. The use of capacitors to draw charge abruptly out of a semiconductor is called *charge compensation*.

In integrated circuitry, conventional capacitors are used only infrequently because they take up a large amount of area on the *IC* chip. However, the junction capacitance of a transistor is sometimes employed when capacitors are required. In Fig. 4.12-3b such junction capacitors have been bridged across the resistors R_b through the addition of transistors $T3$ and $T4$. The capacitances of the base-emitter junctions of these added transistors bridge the resistors R_b.

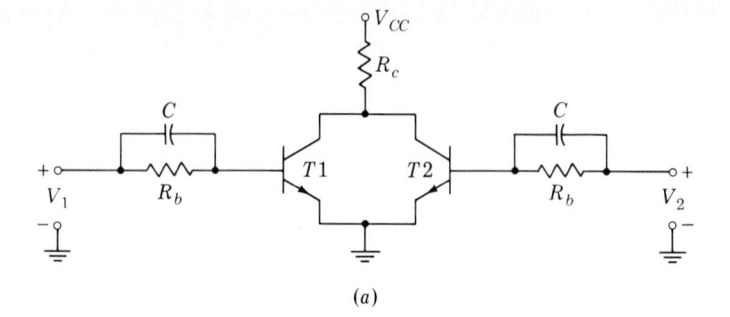

(a)

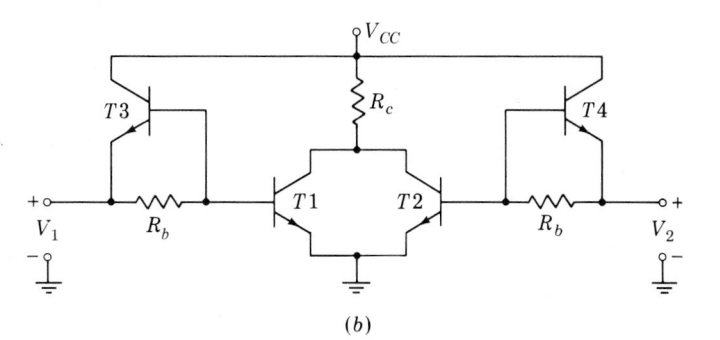

(b)

FIGURE 4.12-3
(a) An RTL gate with capacitors added to provide charge compensation.   (b) A practical application of charge compensation using the capacitances across the base-emitter junctions of the added transistors.

A family of RTL gates as in Fig. 4.12-3b with $V_{CC} = 4.0$ V, $R_c = 1.9$ kΩ, and $R_b = 1.2$ kΩ is manufactured by the Western Electric Co. for use by the Bell System.

4.13   INTEGRATED-INJECTION LOGIC (IIL)

We have see in Sec. 4.5 that when transistors are paralleled, as in RTL gates, the base current drawn by one transistor becomes inordinately large if one of its paralleling transistors goes to saturation.   This increase in base current becomes progressively more pronounced as more of the paralleling transistors are turned ON and may be particularly pronounced in high fan-in gates.   This feature of operation of paralleled transistors accounts, in large measure, for the relatively low fan-out of RTL gates.   In addition DCTL suffers from the difficulty of current hogging, discussed in Sec. 4.3.   For these reasons DCTL has never found any extensive applications.   And while RTL did enjoy a brief period of popularity, it has now fallen into disfavor and is not presently incorporated into new digital systems.

In succeeding chapters we shall explore a number of other families of logic which do not have the limitations of DCTL and RTL. Some of these involve bipolar transistors, others use field-effect devices. These other bipolar transistor gates, particularly, have the disadvantage that they are appreciably more complicated than DCTL or RTL. Hence each such gate occupies more real estate on the integrated-circuit silicon chip than, say, DCTL would. This greater area requirement is, of course, a disadvantage in LSI and even in MSI. As we shall see, FET devices are more economical of area but unfortunately are appreciably slower than bipolar transistor gates.

RTL and DCTL were the first logic families commercially developed. Thereafter a number of other logic families were introduced; these families are described and analyzed in succeeding chapters more or less in the order of their development. *Integrated-injection logic* (IIL or I^2L) is the most recent logic system to be introduced to commercial application. We discuss it at this point because it has a special relationship to DCTL. IIL has the elegant simplicity of DCTL. A typical gate uses very little real estate and consumes very little power. For these reasons IIL is eminently suited for medium- and large-scale integration applications.

When a logic family is to be used in medium- or large-scale integration, where gates are to be crowded as close together as possible, the power dissipation per gate is a matter of great concern. It is generally true, as we have already seen for RTL, that a trade-off can be made by sacrificing power dissipation to speed. Hence a figure of merit which is relevant in comparing one logic family with another is the *speed-power* product. Speed is measured by the propagation time and power by the power dissipation of a typical gate. It is impressive to compare the speed-power product of IIL with the corresponding product for other logic families. By way of example, consider the comparison between IIL and TTL (transistor-transistor logic, discussed in detail in Chap. 6). At the present time TTL is the most popular family of logic, certainly in small-scale integration, and it has extensive applications in medium-scale integration and some application in large-scale integration. We find that for TTL the speed-power product [dimensionally, (time × energy)/time = energy] is typically 100 pJ, while for IIL this product is in the range 0.1 to 0.7 pJ. And while TTL gates can be packed with a density of about 20 gates per square millimeter, IIL gates allow a packing density in the range from 120 to 200 gates per square millimeter.

At the present writing (summer 1976) IIL logic is not commercially available in small-scale integration. Packages containing one or several gates, as available in RTL and in other logic families, have not been marketed. On the other hand, medium- and large-scale-integration chips are available.

Basic configuration of IIL The DCTL gate shown in Fig. 4.2-1 and the RTL gate shown in Fig. 4.1-1 do indeed look like gates. Each exhibits multiple inputs and a single output, as expected of a gate. (The other families of logic discussed in succeeding chapters also look like gates.) On the other

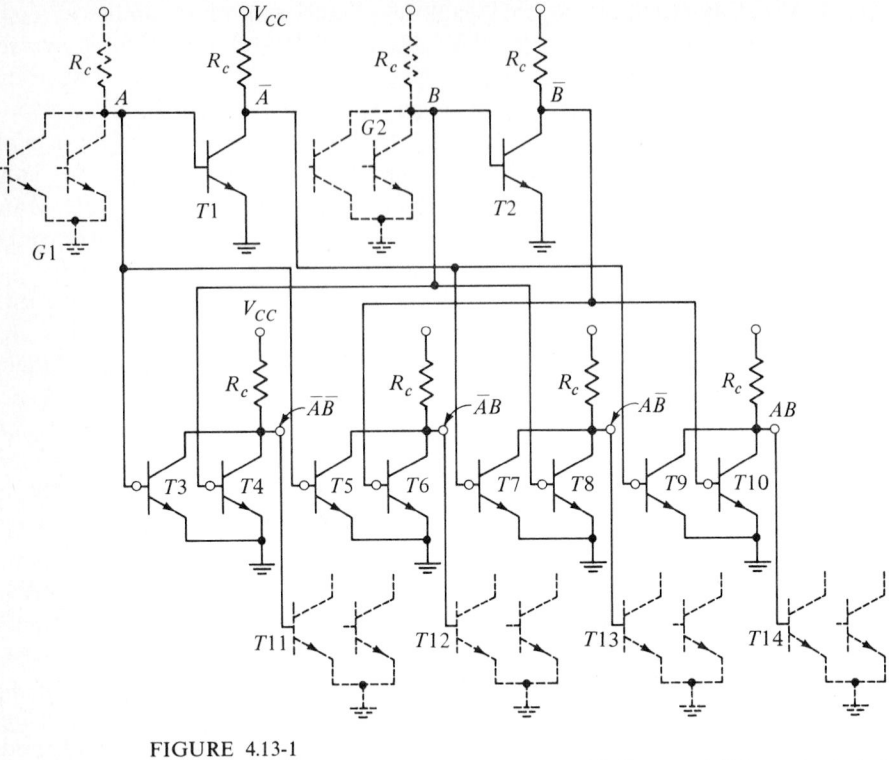

FIGURE 4.13-1
A DCTL gate structure which generates AB, $A\overline{B}$, $\overline{A}B$, and $\overline{A}\overline{B}$ from the logical variables A and B.

hand, the basic structure of an IIL exhibits a single input and multiple outputs, and some explanation is in order.

To pursue the matter, by way of illustration, let us consider that we have two logical variables to deal with, A and B, and that we need to generate the functions AB, $A\overline{B}$, $\overline{A}B$, and $\overline{A}\overline{B}$. We shall use DCTL logic. The variables A and B may themselves be functions of still other variables and hence will themselves be initially available as the outputs of other DCTL gates, $G1$ and $G2$, as shown in Fig. 4.13-1. Since these gates are external to the system we are to assemble, we have drawn them with dashed lines. (Two transistors are indicated in $G1$ and $G2$, but of course the number is arbitrary.)

We need $\overline{A}$ and $\overline{B}$, and these functions of A and B are generated by the single-input gates involving transistors $T1$ and $T2$. Finally, with A, B, $\overline{A}$, and $\overline{B}$ available, the functions required are generated by four two-input DCTL gates. The generated functions may themselves serve as inputs to other gates. To allow for this possibility we have connected the output $\overline{A}\overline{B}$ to the base of $T11$, the output $\overline{A}B$ to the base of $T12$, etc., these transistors, $T11$ through $T14$, providing one input of these other external gates.

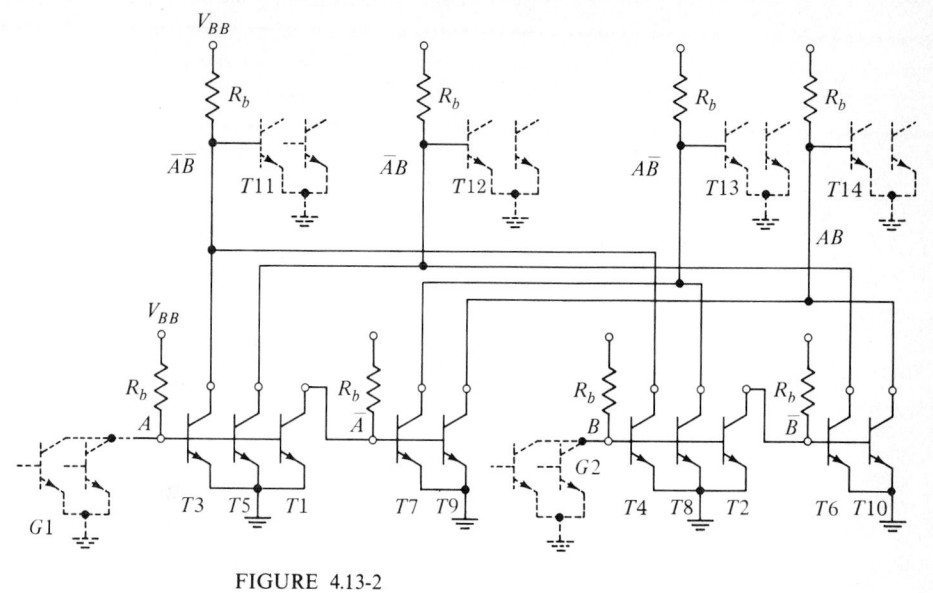

FIGURE 4.13-2
Figure 4.13-1 redrawn to group transistors with common base connections.

We have redrawn Fig. 4.13-1 to appear now as in Fig. 4.13-2. In redrawing that part of the figure of interest to us, i.e., the part drawn in solid lines, we have reorganized the drawing to group together transistors with common base connections. In Fig. 4.13-1 the transistors grouped together have common collector connections. Further, in going from Fig. 4.13-1 to Fig. 4.13-2 we have adopted a different interpretation of the function of the resistors. In Fig. 4.13-1 we view the resistors as common-collector resistors and accordingly label them R_c. In Fig. 4.13-2 the resistors are viewed as common-base resistors and are labeled R_b. Thus the resistor which was previously viewed as the common-collector resistor of the transistors of gate $G1$ is viewed instead as the common-base resistor of transistors $T1$, $T3$, and $T5$. Correspondingly, in one case the voltage to which the resistor has been returned is called V_{CC} and in the other it is called V_{BB}.

We have redrawn Fig. 4.13-2 to appear as in Fig. 4.13-3. Here groups of transistors with common emitters and bases (such as $T3$, $T5$, and $T1$) have been represented as a single multiple-collector transistor. The basic structure of the IIL gate is a transistor with multiple collectors and a single emitter together with a mechanism for supplying base current. The logic operation performed, as in DCTL, is the NOR operation. Thus, if a collector of a gate whose input is A is connected to a collector of a gate whose input is B and the joined collectors are in turn connected to the base of another gate, the logic that appears at the base is $\overline{A + B}$.

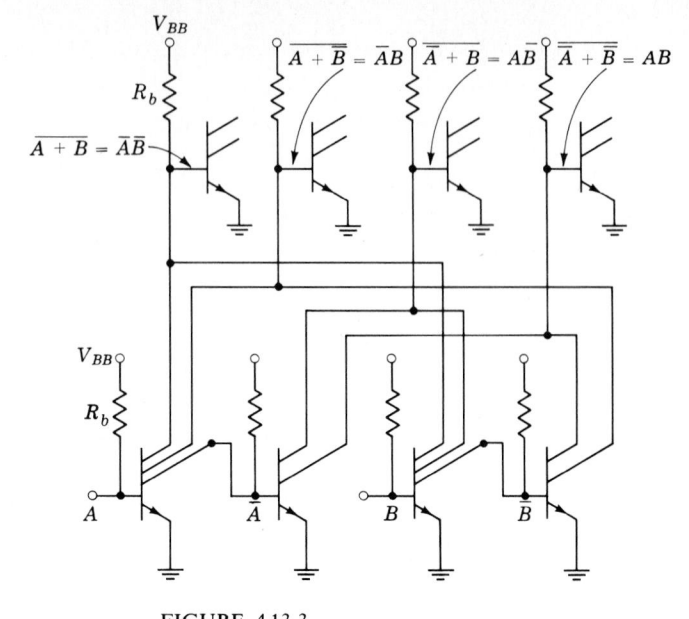

FIGURE 4.13-3
Figure 4.13-2 redrawn with multiple-collector transistors.

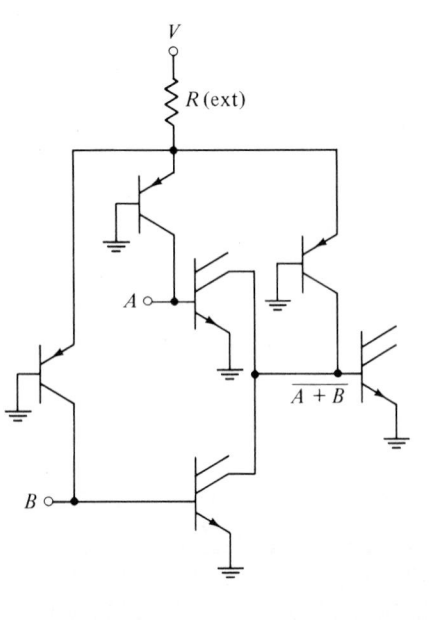

FIGURE 4.13-4
The basic configuration of IIL gates.

Up to this point we have made it appear that the base current of the multicollector transistors is supplied through a resistor. We now need to take note of the fact that actually this current is supplied through a transistor. Thus, finally, the basic configuration encountered in IIL appears as in Fig. 4.13-4. The transistor, it turns out, can be incorporated in the structure of the integrated circuit in a manner which uses much less chip area than would be required by a resistor. Observe that, for reasons to be discussed later, this added transistor is of the *pnp* type.

4.14 PHYSICAL LAYOUT OF IIL

In order to appreciate the features of IIL which account for its comparative merits, it is necessary to give some consideration to the physical construction of an IIL integrated circuit. We shall consider the matter in a simplified manner.

In integrated circuits, for reasons concerned with the technology of fabrication, transistors are of the *npn* type. The layout of an array of transistors is shown in Fig. 4.14-1a. (In Fig. 4.14-1 for simplicity we have deliberately omitted a number of features needed for the proper operation of the device but not essential to our discussion.) The substrate is *p*-type silicon, and the *p*-type material extends upward from the substrate to the integrated-circuit surface to provide isolation from one transistor to the next. The collector, base, and emitter are *n*-type, *p*-type, and again *n*-type, respectively, as shown. Isolation is effected by arranging that the substrate shall be always negative with respect to the collector. The various parts of the transistors are formed by diffusions of impurities from the surface of the integrated circuit. The collector is relatively lightly doped, the base doped more heavily, and the emitter doped quite heavily.

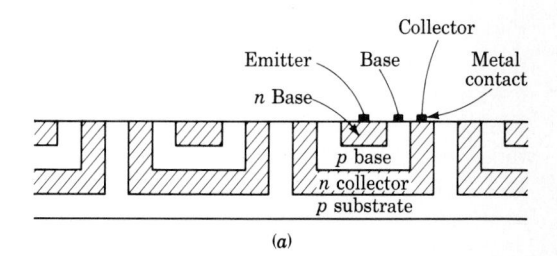

(a)

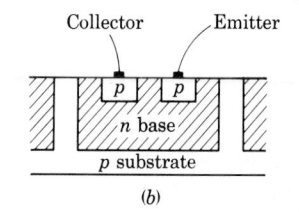

FIGURE 4.14-1

(a) The physical structure (simplified) of an integrated-circuit transistor. (b) The physical structure of a *pnp* lateral transistor.

(b)

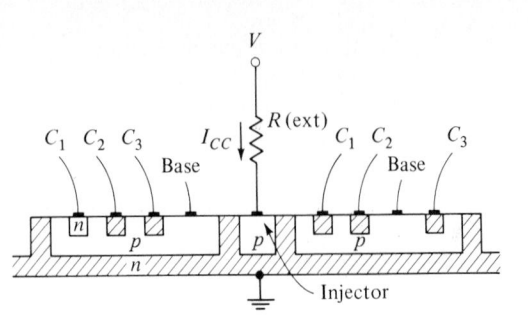

FIGURE 4.14-2
The physical structure of the basic gate configuration in IIL.

Both the geometry of the structure and the doping serve to ensure that the common-base gain α_N shall be close to unity and that $h_{FE} = \alpha_N/(1 - \alpha_N)$ shall be large. Since the emitter is doped much more heavily than the base, the carriers which cross the emitter junction are predominantly electrons. It is important that such be the case since holes crossing from base to emitter constitute an emitter current which will not be collected at the collector junction. Since the base is very thin, very few of the electrons injected into the base will be lost in the base as a result of recombination. Finally, we note that the emitter is almost surrounded by the collector, i.e., when viewed from the emitter, the collector subtends a large angle. As a consequence, an electron leaving the emitter, from no matter what point and in which direction, can hardly fail to reach the collector. For all these reasons we find $\alpha_N \approx 0.98$ or even higher. In the transistor configuration of Fig. 4.14-1a the axis of symmetry of current flow is in the vertical direction. Hence such a transistor is called a *vertical transistor*.

The transistor structure can be modified to allow a *pnp* transistor. Such a transistor is shown in Fig. 4.14-1b. The substrate is again *p* type. However the layer above the substrate becomes not the collector but the base. As shown, the emitter and collector are formed by *p*-type diffusions into the base. Observe that in the *pnp* case the angle subtended by the collector at the emitter is relatively small. For this reason the current gain h_{FE} of the *pnp* transistor is rather low, being in the range 0.5 to 5 compared with 50 to 150 for an *npn* transistor. Finally we note that the general direction of current flow is horizontal, i.e., to the side rather than vertical, hence the configuration of Fig. 4.14-1b, is called a *lateral transistor*.

Turning now to the IIL gate, we see the structure of such a gate in Fig. 4.14-2. (In this figure, we have again taken some liberties in the direction of simplification.) In this figure we display two three-collector gates side by side. The lower *n* layer, which in the conventional integrated-circuit transistor is the collector, is here the emitter. Since all the emitters in all the transistors operate at the same voltage (ground), a *p*-type isolation, necessary in Fig. 4.14-1a, is not required in the circuit of Fig. 4.14-2. In the conventional transistor the small *n* region, embedded in the *p*-type base, is the emitter. In the IIL transistor these small *n* regions are the multiple collectors. However, in the IIL transistor the relative doping is geometrically but not functionally the same as in the conventional transistor; i.e., while in the conventional transistor the emitter is most heavily doped and the collector least heavily doped, in the IIL transistor the situation is reversed. Hence the IIL transistor operates in a mode which we would describe as the *inverse* mode if we were dealing with a conventional transistor. Accordingly we expect that the current gain for the IIL transistor will be low. Such is indeed the case. Yet it turns out to be possible to get current gains as large as 5 and even higher. A current gain of this magnitude is entirely adequate; for, as can be seen in Fig. 4.13-1, each collector needs to provide for a fan-out of only 1.

As indicated in Fig. 4.14-2, the location of the base connection for the *npn* transistor is not the same from transistor to transistor. This feature is useful in that it allows base connections and collector regions to be located as required to accommodate the necessary interconnections.

Figure 4.14-2 also shows the *pnp* transistor which supplies base current to the *npn* transistor (see Fig. 4.13-4). The *p* region marked "injector" is the emitter of this transistor, its collector is the *p*-type base of the *npn* transistors, and its base is the *n*-type emitter of the *npn* transistor. Altogether the two transistors, one *npn* and one *pnp*, are formed with only four separate regions, the two transistors using two regions in common. For this reason IIL is also called *merged-transistor logic* (MTL).

Observe that the *pnp* transistor through which base current is *injected* into the *npn* transistor is pictured in Fig. 4.14-2 as serving two transistors on each side. Note also that the *pnp* transistors are *lateral* transistors. Current I_{CC} is supplied for the injector from a supply source *V* through an external resistor *R*. The voltage from the *p*-type injector region to ground is the voltage across a forward-biased junction and turns out in the present case to be about 0.85 V. Fortunately this injector-to-ground voltage tracks very well from injector to injector throughout the integrated-circuit chip. Hence it turns out to be possible to operate all the injectors in parallel so that all the injector current required by a chip can be supplied through a single external resistor.

Adjustability of speed   We have seen that in RTL gates it is possible to sacrifice power dissipation for the sake of an improvement in speed. This trade-off is effected by changing the sizes of the resistors. Small resistors allow larger currents, so that capacitances can charge more rapidly albeit at the expense of greater power dissipation. What is to be noted here is that different speeds are associated with *different* integrated-circuit chips. In IIL, as we shall now see, it turns out that the trade-off between speed and power can be effected on a single chip by the simple expedient of changing the current I_{CC} injected into the chip.

Since the voltage at the injector is constant (≈ 0.85 V), the input power is *proportional* to the average input current I_{CC}. (On a large-scale-integration chip we may reasonably expect that a nominally fixed fraction of the transistors will be conducting at any one time and hence that I_{CC} will be rather constant.)

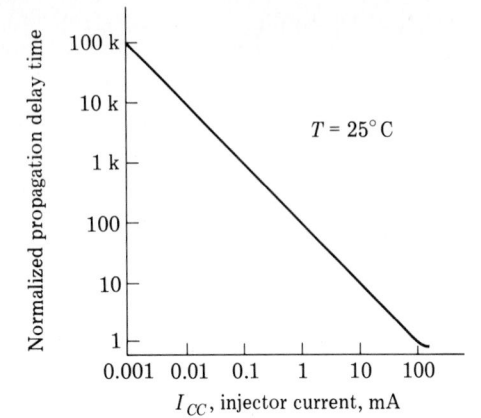

FIGURE 4.14-3
Normalized propagation delay time versus
injector current.

A fraction I_{CC}/n of this current will be available, on the average, to operate a gate, n being the average number of gates which are conducting. At *low current levels* the propagation-delay time is principally the time required to charge junction and parasitic capacitors. The time required to charge these capacitors is *inversely proportional* to the available charging current I_{CC}/n. Accordingly the speed-power product is constant independently of I_{CC}. Increasing I_{CC} (by increasing V or decreasing R in Fig. 4.14-2), we can decrease the propagation delay at the expense of a proportionate increase in dissipation.

At *medium current levels* the principal source of propagation delay is the need to establish and remove the excess minority-carrier base charge in the transistors (see Sec. 1.17). This charge is proportional to the transistor current available to establish or remove it, and hence the propagation delay is independent of the current. Thus, in this current range, an increase in I_{CC} will increase the dissipation without reducing the delay.

At *high current levels* the transistors are driven into saturation. As discussed in Sec. 1.18, in saturation the stored base charge increases more than in proportion to transistor current. Hence, in this current range an increase in I_{CC} will not only increase the dissipation but will also increase the delay.

Figure 4.14-3 shows a plot of the normalized propagation delay time as a function of I_{CC} for the Texas Instrument (SBPO400) IIL chip. In the range of I_{CC} from 0.001 to 100 mA we can trade off dissipation and speed. Beyond about 100 mA an increase in dissipation yields no continuing decrease in speed. While the plot is not carried to the high-current region, at such high currents the plot would no doubt have a positive slope.

4.15 AN IIL DECODER

In Fig. 4.14-2 we show an injector serving two transistors. Actually a single injector can serve many transistors. When this is intended, the injector is extended into a long strip parallel to the surface of the chip and is referred to as an *injector rail*.   The transistors are fabricated on both sides of the rail and extend perpendicular to the rail.

As an example of an alternative layout we have the 3-bit decoder shown in Fig. 4.15-1.   The function of decoders and their logic is discussed in Sec. 12.9. It will be sufficient here to note that the present decoder has as input the three logic variables A, B, and C and is intended to make available on eight separate output lines the eight minterms (see Sec. 3.17) of the input variables. Hence, at any time, only one output line will be at logic **1**; all others will be at

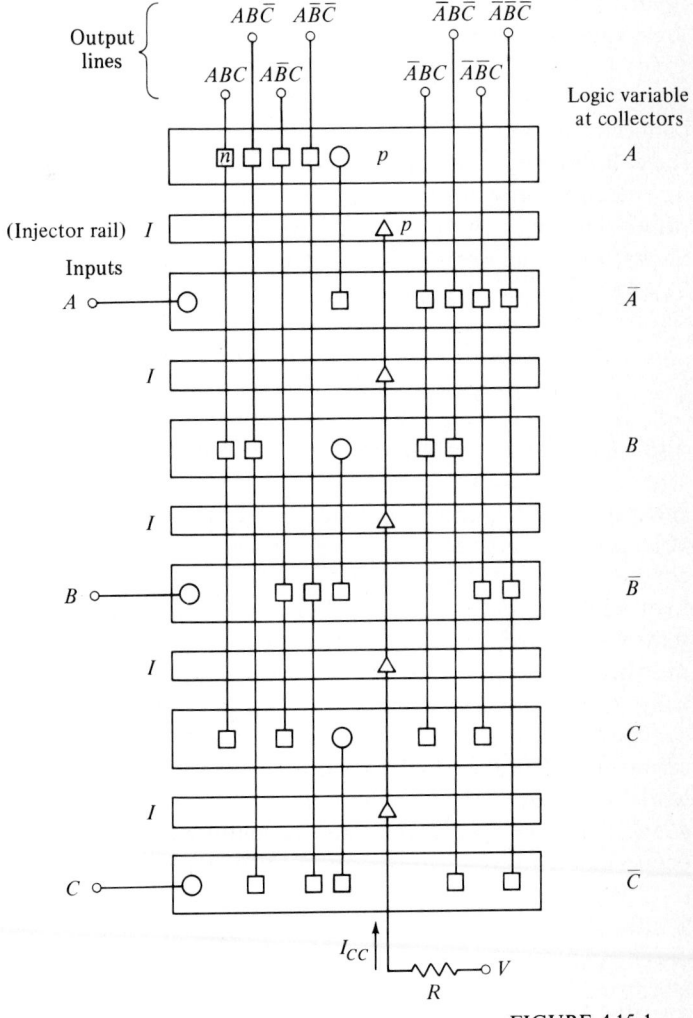

FIGURE 4.15-1
A 3-bit IIL decoder.

logic **0**. The line selected to be at logic **1** is determined by the logic levels of the inputs.

The view shown is seen looking down at the surface of the chip. The common *n*-type emitter of the *npn* transistors (see Fig. 4.14-2) (which is also the base of the injector transistor) is not explicitly shown. We do see the *p*-type injector rail and the *p*-type base of the *npn* transistors (which is also the base of the injector transistor). Connections for the injected current are shown as triangles on the injector rail. The base connections are shown as circles and the collectors as squares. To the right of the figure are specified the logic variables present at the collector on each multicollector transistor. Thus the transistor whose input is A has $\overline{A}$ at each collector. One of these collectors provides $\overline{A}$ at the input to a second transistor, and at the collectors of this second transistor the logic variable is A. As we have already noted, the simple joining of collectors yields the logic product of the variables present at the collectors. It is to be understood, even though not shown in the figure, that each output line must be connected to the input of another gate so that the injector of that gate can provide the collector currents required for the transistors shown in the figure.

In the present case the transistors have been arranged to be parallel rather than perpendicular to the injector rails. Observe that, as a consequence, the pattern of interconnections is especially simple, requiring no crossovers.

4.16 CURRENT AND VOLTAGE LEVELS

Internal to an IIL chip the voltage levels are about 0.7 and 0.1 V. The higher voltage is measured at a base when all collectors connected to that base are not conducting. The lower level is the voltage at a collector and any base coupled to it when the collector is part of a transistor that is conducting. Thus the total voltage swing between logic levels is 0.6 V. The currents carried by the transistor are rather small in comparison with RTL and other gate types to be considered in succeeding chapters. Typically currents are in the range 1 to 10 μA.

When the logic signals internal to the chip have undergone whatever logical operations are called for, a signal may be taken from the chip to the outside world from a collector of the last transistor. This last collector is not to be connected to a base and hence to develop a signal the collector must be connected to a supply voltage through an external resistor. Supply voltages up to 10 V are reasonable, and correspondingly the output-voltage swing will be only a little less than 10 V. There may well be a difficulty because the available output current is inadequate. In such a case additional transistors may be fabricated on the chip to provide for appropriate interfacing between the chip and its external load.

REFERENCES

1 Hart, C. M., A. Slob, and H. E. J. Wulms: Bipolar LSI Takes a New Direction with Integrated Injection Logic, *Electronics*, Oct. 3, 1974, pp. 111–118.
2 Berger, H. H., and S. K. Wiedmann: Merged-Transistor Logic (MTL): A Low-Cost Bipolar Logic Concept, *IEEE J. Solid-State Circuits*, October, 1972, pp. 340–346.
3 Hart, K., and A. Slob: Integrated Injection Logic: A New Approach to LSI, *IEEE J. Solid-State Circuits*, October 1972, pp. 346–351.
4 Alstein, J.: I^2L: Today's Versatile Vehicle for Tomorrow's Custom LSI, *EDN*, Feb. 20, 1975, pp. 34–38.
5 De Troge, N. C.: Integrated Injection Logic: Present and Future, *IEEE J. Solid-State Circuits*, October 1974, pp. 206–211.
6 Horton, R. L., J. Englade, and G. McGee: I^2L Takes Bipolar Integration a Significant Step Forward, *Electronics*, Feb. 6, 1975, pp. 83–90.

5

DIODE-TRANSISTOR LOGIC

In this chapter we consider a logic family, *diode-transistor logic* (DTL), whose circuitry is somewhat more involved than RTL. Although DTL has the advantage of greater fan-out and improved noise margins, it suffers from somewhat slower speed.

5.1 DIODE-TRANSISTOR-LOGIC (DTL) GATE

A diode-transistor-logic gate as realized with discrete components is shown in Fig. 5.1-1. Transistor $T0$ is the output transistor (corresponding to $T2$) of a preceding gate. Three inputs are indicated, but, of course, more can be added by adding more diodes to the array of diodes DA, DB, and DC.

The gate performs the NAND operation for positive logic. Consider that the driving gate $T0$ is at logic level **0**. Then $T0$ is in saturation, and the corresponding collector voltage is $V_i \approx 0.2$ V or lower. We can verify that when the input in Fig. 5.1-1 is $V_i = 0.2$ V, $T2$ is cut off and the output at the

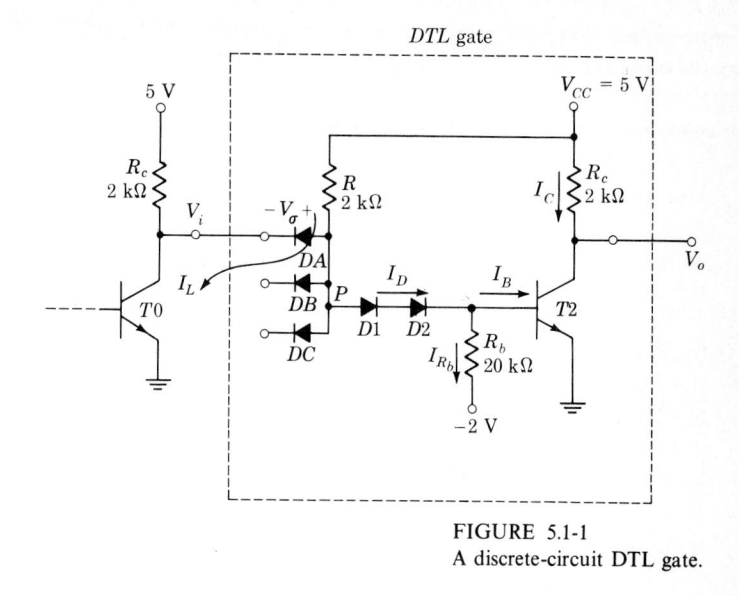

FIGURE 5.1-1
A discrete-circuit DTL gate.

collector is $V_{CC} = 5$ V, which corresponds to logic level **1**. For, with the input at 0.2 V, the voltage at point P is

$$V_P = V_{CE}(\text{sat}, T0) + V_{DA} = 0.2 + 0.75 = 0.95 \qquad \text{at } T = 25°C \qquad (5.1\text{-}1)$$

Here we have assumed that the current in diode DA, which is approximately 2 mA, is adequate to produce a voltage drop of 0.75 V across diode DA. The base-to-ground voltage of $T2$ is then

$$V_{B2} = V_P - V_{D1} - V_{D2} = 0.95 - 0.65 - 0.65 = -0.35 \text{ V} \qquad (5.1\text{-}2)$$

Here we have taken $V_{D1} = V_{D2} = 0.65$ since, as will appear, the currents in $D1$ and $D2$ are very small. This voltage V_{B2} is less than the cut-in voltage $V_\gamma = 0.65$ V of transistor $T2$, and therefore $T2$ is cut off. Thus, the current in diodes $D1$ and $D2$ continues through resistor R_b, and this current is

$$I_{R_b} = I_D = \frac{V_{B2} + 2}{R_b} = \frac{-0.35 + 2}{20 \times 10^3} = 0.08 \text{ mA} \qquad (5.1\text{-}3)$$

This result confirms our initial assumption that the diode current is indeed quite small. Finally, since $T2$ is cut off, the output voltage $V_o = 5$ V, corresponding to logic level **1**.

We have shown that if V_i is at logic level **0**, V_o is at logic level **1**. Similarly we can show that if all of the inputs are at logic level **0**, V_o remains at logic level **1**. For, if all the inputs are simultaneously at logic level **0**, the current through resistor R will divide among the diodes DA, DB, DC, ... With a smaller current through, say, DA, the drop across this diode will decrease slightly. It should be noted that as a result V_P decreases slightly, so that $T2$ is forced further into cutoff. Hence, V_o remains at the **1** level.

Current sinking   The current I_L drawn out of diode DA, as indicated in Fig. 5.1-1, flows into the collector of the driving saturated transistor $T0$. This current I_L has its *source* in the supply voltage V_{CC}. The process of returning this current to ground, which is also the negative return of the supply source, is commonly called *sinking*. Thus, transistor $T0$ *sinks* the current drawn out of the input of the driven gate.

Saturation of $T2$ If any one of the inputs is at logic level **0**, $T2$ is cut off and the gate output is at logic **1**. If, however, all inputs are at logic level **1** (5 V), the current through R will flow through $D1$ and $D2$ and into the base of $T2$. Transistor $T2$ will be driven into saturation, and the output will drop to logic level **0**, as required for NAND operation.

To see that transistor $T2$ does indeed become saturated let us calculate the base current $I_B = I_D - I_{R_b}$. In this case, $V_{B2} = 0.75$ V, and the current in R_b is

$$I_{R_b} \approx \frac{0.75 + 2}{20 \times 10^3} \approx 0.14 \text{ mA} \qquad (5.1\text{-}4)$$

The voltage at point P is now $V_P = V_{D1} + V_{D2} + V_{B2} \approx 2.25$ V, where we have assumed that I_D and I_B are sufficiently large to cause the diode and base-emitter voltage to be approximately 0.75 V. Thus, the diode current I_D, which is also the current in resistor R, is

$$I_D = \frac{V_{CC} - V_P}{R} = \frac{5 - 2.25}{2 \times 10^3} \approx 1.4 \text{ mA} \qquad (5.1\text{-}5)$$

The base current of transistor $T2$ is the difference between the currents I_D and I_{R_b} and is

$$I_B = I_D - I_{R_b} \approx 1.26 \text{ mA} \qquad (5.1\text{-}6)$$

which is clearly sufficient to saturate transistor $T2$, since with $V_{CE}(\text{sat}) \approx 0.2$ V, $I_C = 2.4$ mA. Assuming that $h_{FE} = 50$, this collector and base current corresponds to $\sigma = 0.04$. Referring to Fig. 1.10-1, we find that $V_{CE}(\text{sat})$ is more nearly 0.1 V than 0.2 V.

Base resistor R_b   If transistor $T2$ is in saturation, and if then one or more of the gate inputs returns to logic level **0**, point P falls to $V_P = 0.95$ V. The equivalent circuit of the gate *at this instant* is as shown in Fig. 5.1-2. Note that the voltage drop across both diodes $D1$ and $D2$ is 0.2 V and therefore these diodes are cut off. Hence, $I_D = 0$, and there is no current source to supply I_B.

If there were no charge storage in the system, I_B would instantly drop, causing the base-emitter voltage to fall below its cut-in voltage, and therefore $T2$ would immediately cut off. The steady-state base-emitter voltage would then be -0.35 V [see Eq. (5.1-2)]. However, there is charge stored in the transistor.

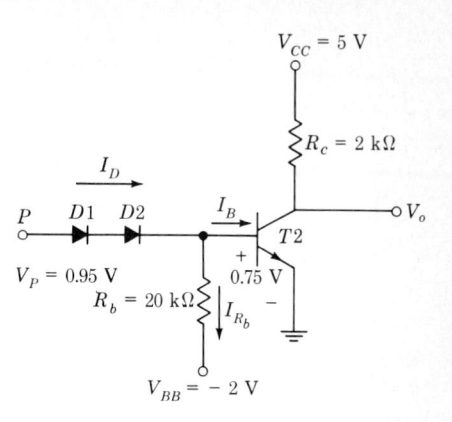

FIGURE 5.1-2
Equivalent circuit of the DTL gate at the
instant that V_P is decreased to 0.95 V.

When point P drops, momentarily cutting off diodes $D1$ and $D2$, this stored charge leaves through resistor R_b. Thus, resistance R_b provides a discharge path for the charge stored in the transistor. Resistor R_b is connected to the -2 V supply to increase the rate of discharge.

The selection of $R_b = 20$ kΩ and $V_{BB} = -2$ V (rather than, say, $R_b = 1$ kΩ and $V_{BB} = -4$ V) is the result of a compromise. Many values of resistance and supply voltage are possible, and most will result in satisfactory operation. To increase the rate of charge removal and therefore decrease the time needed to cut off $T2$, we would like R_b to be small and V_{BB} to be very negative. However, when $T2$ is cut off and we attempt to turn it on quickly, we would like all the diode current I_D to flow into the base of $T2$. In this case we would like R_b to be very large and V_{BB} to be positive. Typical values of R_b range from 5 to 30 kΩ, and values of V_{BB} vary from 0 to -5 V.

5.2 FAN-OUT

Figure 5.2-1 shows the DTL gate driving N other gates. When the gate output transistor $T0$ is sinking the current of a gate $G1$ that it is driving, it encounters its heaviest loading when all the other inputs of the driven gate are at logic level **1** (or equivalently when all other inputs are left floating). For, in this case all the current through R continues through DA and into the collector of $T0$. On the basis of the previous discussion [see Eq. (5.1-1)] this current is

$$I_L = \frac{5 - 0.95}{R} = \frac{4.05}{2 \text{ k}\Omega} \approx 2 \text{ mA} \tag{5.2-1}$$

In any event, it is clear that to minimize the input loading due to DTL gate $G1$ it is advantageous to make R a large resistance.

On the other hand, consider that all inputs of gate $G1$ are at logic level **1**. In this case current I_D flows from V_{CC}, through R and the diodes $D1$ and $D2$, and finally divides between R_b and the base of $T2$. Transistor $T2$ should now

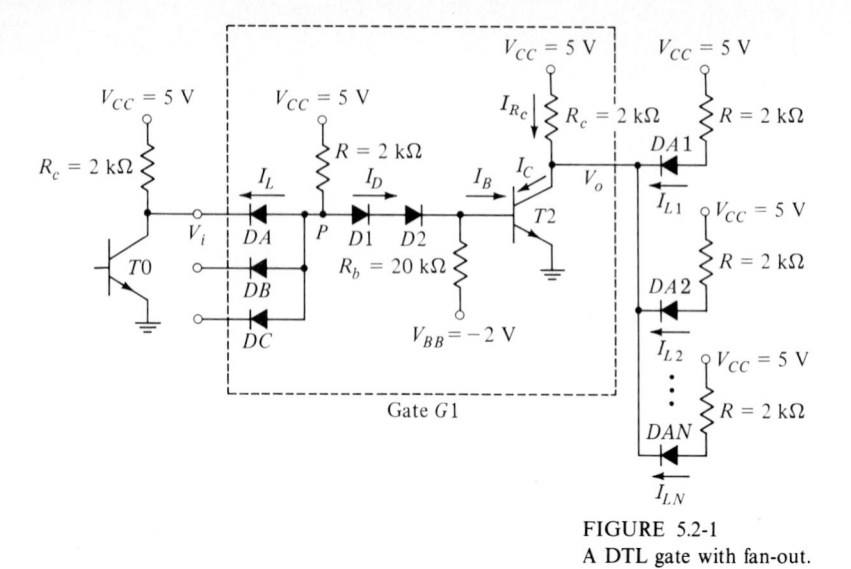

FIGURE 5.2-1
A DTL gate with fan-out.

be in saturation. Now, to see the relationship between the fan-out N of gate $G1$ and the resistor R, let us assume that the collector voltage of $T2$, when saturated, is $V_o = 0.2$ V. Then the current in each of the N loading gates is

$$I_{L1} = I_{L2} \cdots = I_{LN} \approx \frac{V_{CC} - V_{DA1} - V_o}{R} = \frac{5 - 0.75 - 0.2}{R} \approx \frac{4}{R} \qquad (5.2\text{-}2)$$

To arrive at this result we have assumed that all the current flowing in R, in each of the driven gates, flows into $T2$. We are therefore neglecting the current flow in diodes $D1$ and $D2$ of these driven gates. This neglected current is approximately 0.08 mA [see Eq. (5.1-3)] and is negligible.

The total collector current in $T2$ is then

$$I_C = I_{R_c} + NI_{L1} = \frac{V_{CC} - V_o}{R_c} + \frac{4N}{R} = 2.4 \text{ mA} + \frac{4N}{R} \qquad (5.2\text{-}3)$$

The base current in $T2$ is I_B and is found, as before, by noting that $V_P = 2.25$ V when $T2$ is saturated. Again, neglecting the current in R_b, we have

$$I_B \approx I_D = \frac{V_{CC} - V_P}{R} = \frac{2.75}{R} \qquad (5.2\text{-}4)$$

Since we require that $T2$ be driven to saturation, we also have

$$I_C = \sigma h_{FE} I_B \qquad (5.2\text{-}5)$$

in which the value of σ can be estimated from Fig. 1.10-1 after we have decided what value we want for $V_{CE}(\text{sat})$, that is, how deeply $T2$ is to be driven to saturation.

Combining Eqs. (5.2-3) to (5.2-5) and solving for N, the fan-out, we find

$$N \approx 0.7\sigma h_{FE} - 0.6R \times 10^{-3} \qquad (5.2-6)$$

Observe that to increase N we must increase σ; that is, we must restrict the extent to which $T2$ is driven into saturation. As we increase N, a point will be reached where $\sigma = 1$ and $T2$ is no longer saturated. This situation develops because each additional load on the driving gate requires $T2$ to sink an additional 2 mA without a compensating increase in base current in $T2$.

Assume $h_{FE} = 50$, $R = 2$ kΩ, and that $\sigma = 0.85$. This value of σ corresponds to a modest excursion into saturation and yields $V_{CE}(\text{sat}) = 0.2$ V (see Fig. 1.10-1). In this case the first term in Eq. (5.2-6) is $0.7\sigma h_{FE} = 30$, while the second term is $0.6(2) = 1.2$. Thus if we decide that the output transistor need be but moderately saturated, the fan-out $N = 29$. This result would not change appreciably if R were doubled or halved. Hence, we may neglect this second term and estimate that $N = 0.7\sigma h_{FE}$, depending only on h_{FE} and not on R. This relative independence of N from R is to be anticipated on the basis of the following consideration. If R changes, say decreases, the base current of $T2$ increases. However, the current which $T2$ must sink for each additional load increases correspondingly, so that there is no net increase in allowable fan-out. It should be noted that if we decided that $V_{CE}(\text{sat}) = 0.1$ V under maximum fan-out, then $\sigma = 0.1$ and $N = 2$. Now the fan-out depends on σh_{FE} and R.

In the next section, when we consider the DTL gate in integrated-circuit form, we shall see how a modification of the circuitry devises to increase the base drive of the output transistor without a corresponding increase in the current load imposed by each additional gate. As a result the fan-out is significantly increased.

5.3 INTEGRATED-CIRCUIT DTL GATES

The DTL gate in integrated form is shown in Fig. 5.3-1. It is commercially available with either $R_c = 2$ kΩ or $R_c = 6$ kΩ. In the former case the power dissipation is higher than in the latter case. However, all capacitance shunting the collector of $T2$ to ground must charge through R_c when $T2$ goes off. Hence, with $R_c = 2$ kΩ the propagation delay time is smaller than with $R_c = 6$ kΩ.

We observe that in the integrated circuit of Fig. 5.3-1 transistor $T1$, whose collector is connected to a tap on R, is used in lieu of diode $D1$ in the discrete circuit of Fig. 5.1-1. It is to be recalled, however, that in integrated circuitry a diode is normally a transistor with its collector tied to its base. Hence, the change involved is not so much the replacement of a diode by a transistor as simply a new connection for the collector. This new transistor $T1$ operates in its active region and provides current gain between its base and emitter. Hence, it makes available more current for the base of $T2$ without requiring a reduction in the resistance R.

We now explain the effect of this new transistor and the tap position. Let

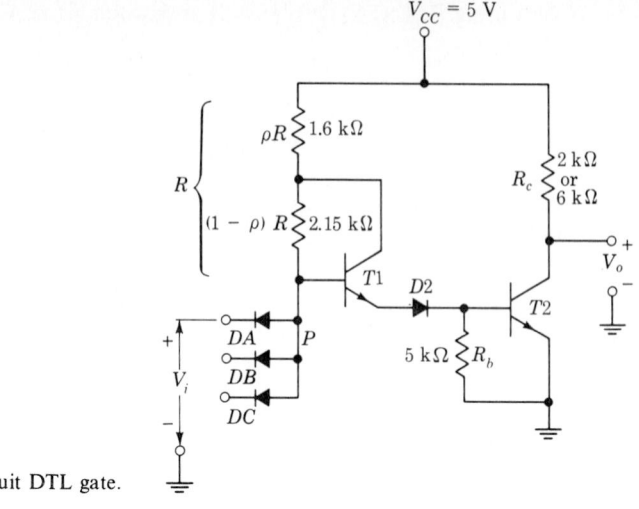

FIGURE 5.3-1
An integrated-circuit DTL gate.

us begin by calculating the fan-out of the gate shown in Fig. 5.3-2. Let us assume, as before, that when $T2$ is saturated, $V_o = 0.2$ V. Then, if the gate is fanned out to N loading gates, the load currents which must sink through $T2$ are given, as in Eq. (5.2-2), as

$$I_{L1} = I_{L2} = \cdots = I_{LN} \approx \frac{5 - 0.95}{R} \approx \frac{4}{R} \qquad (5.3\text{-}1)$$

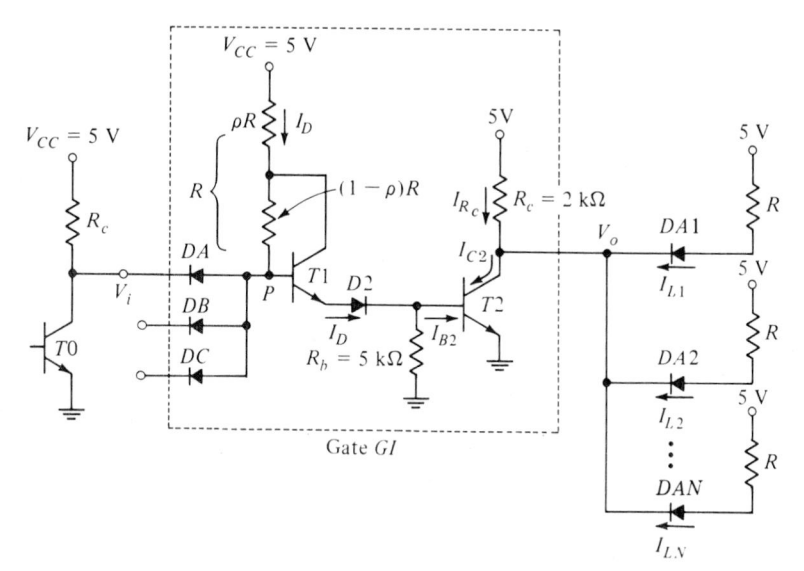

FIGURE 5.3-2
DTL gate with a fan-out of N.

Since the current in resistor R_c is $4.8/(2 \times 10^3) = 2.4$ mA, the collector current is

$$I_{C2} = 2.4 \times 10^{-3} + \frac{4N}{R} \qquad (5.3\text{-}2)$$

This result, of course, is identical to that obtained in Eq. (5.2-3).

We now calculate the base current I_{B2}, again assuming that the current in R_b is negligibly small. First, we note that the voltage at point P is

$$V_P = V_{BE1} + V_{D2} + V_{BE2} = 0.75 + 0.75 + 0.75 = 2.25 \text{ V} \qquad (5.3\text{-}3)$$

The calculation of $I_{B2} = I_D$ can now be easily performed with the help of Fig. 5.3-3. Let the base current in $T1$ be called I. Then $I_{B2} = (h_{FE} + 1)I$. Hence

$$V_{CC} - V_P = \rho R(h_{FE} + 1)I + (1 - \rho)RI \qquad (5.3\text{-}4)$$

Solving for $(h_{FE} + 1)I = I_{B2}$ yields

$$I_{B2} = (h_{FE} + 1)I = \frac{V_{CC} - V_P}{R[\rho + (1 - \rho)/(h_{FE} + 1)]} \qquad (5.3\text{-}5)$$

in which $V_{CC} - V_P = 2.75$ V.

Since $\sigma h_{FE} I_{B2} = I_{C2}$, we have, combining (5.3-2) and (5.3-5),

$$\frac{2.75\sigma h_{FE}}{R[\rho + (1 - \rho)/(h_{FE} + 1)]} = 2.4 \times 10^{-3} + \frac{4N}{R} \qquad (5.3\text{-}6)$$

Solving for the fan-out N yields

$$N \approx \frac{0.7\sigma h_{FE}}{\rho + (1 - \rho)/(h_{FE} + 1)} - 0.6R \times 10^{-3} \qquad (5.3\text{-}7)$$

When $\rho = 1$, Eq. (5.3-7), as required, reduces to Eq. (5.2-6). As ρ decreases, the fan-out increases. In typical commercial DTL integrated-circuit gates $1/\rho$ is in the range 2 to 10. With $h_{FE} \approx 50$ and assuming, as in the previous section,

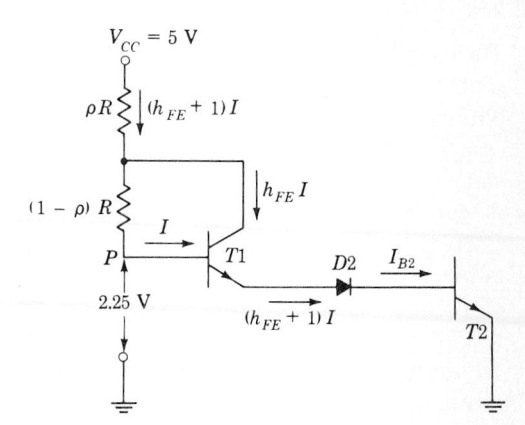

FIGURE 5.3-3
Circuit to show the effect of splitting the base resistor R.

that σ is in the range $\sigma \approx 0.85$, we have approximately that

$$N = \frac{0.7\sigma h_{FE}}{\rho} \qquad (5.3\text{-}8)$$

Thus, we conclude that the new circuit increases the fan-out by the factor $1/\rho$ or, equivalently, that if we keep N fixed, we may drive the output transistor further into saturation by decreasing σ by the factor $1/\rho$.

The mechanism of increase in fan-out is the following. When the gate input V_i in Fig. 5.3-2 is at logic **0**, $T1$ is OFF and the driving gate sees the same load as in the circuit of Fig. 5.2-1. When, however, V_i is at logic **1**, $T1$ is ON and provides current gain. Hence the base current available for the output transistor is larger (by the factor $1/\rho$) than it would be if all the base current had to be furnished directly through the entire resistor R.

Equation (5.3-8) suggests that it might be advisable in Fig. 5.3-1 to move the tap on R closer to V_{CC} (letting $\rho \to 0$), thereby increasing the base drive of $T2$ and the fan-out. The difficulty with such a procedure is that increasing the base current will result in an increase in the power dissipated in $T2$. Furthermore, if the gate is operated with a fan-out which is less than maximum, $T2$ will be heavily saturated, thereby increasing the propagation delay time required to remove $T2$ from saturation. The component values employed in the DTL gate represent a compromise, taking account of fan-out, propagation delay, and power dissipation.

A last feature to be noted in the gate on Fig. 5.3-1 is that the negative supply return of R_b has been eliminated. It is, of course, a great convenience to be able to operate the gate with a single supply voltage. We observe, however, that since we require that the stored charge in the base be drawn out of $T2$ through R_b, the resistor R_b is much smaller in the integrated-circuit gate than in the discrete circuit.

5.4 INPUT-OUTPUT CHARACTERISTIC

The input-output characteristic of the DTL gate of Fig. 5.4-1 has the form shown, somewhat idealized, in Fig. 5.4-2. We now show how this characteristic is obtained. When $T0$ is saturated, the input voltage V_i is $V_i \approx 0.2$ V. In this case $T2$ is cut off, and the output voltage $V_o = V_{CC} = 5$ V. We can estimate the value of V_i at which V_o will begin to fall as follows. When $T2$ is just at the edge of cutoff, its base-emitter voltage will be about $V_y = 0.65$ V (at $T = 25°C$). Correspondingly, at this cut-in point, the current $I_{B2} \approx 0$ and $I_D = V_y/R_b \approx 0.13$ mA. With this amount of current flowing in diode $D2$ and out of the emitter of $T1$ both voltage drops across $D2$ and the base-emitter junction of $T1$ will be ≈ 0.7 V. Hence, the voltage at point P will be $2(0.7) + 0.65 = 2.05$ V. With $T1$ drawing so little current most of the current through R will be flowing through diode DA. It can be verified (Prob. 5.4-1) that at this point the current

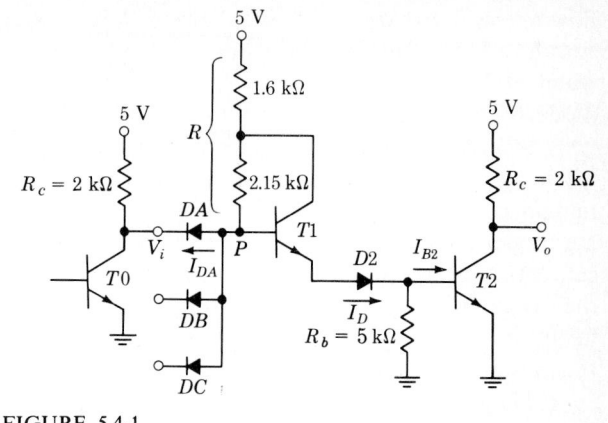

FIGURE 5.4-1
Circuit used to calculate the unloaded input-output characteristic.

$I_{DA} \approx 0.75$ mA. With this current, the drop across the input diode DA will be more nearly $V_{DA} \approx 0.75$ V. Hence (at $T = 25°C$) the input voltage is $V_i = 2.05 - 0.75 = 1.3$ V, as indicated in Fig. 5.4-2.

We now calculate the minimum value of V_i needed to saturate $T2$. When $T2$ switches from the edge of cutoff, $V_{BE2} = V_\gamma = 0.65$ V, to saturation, $V_{CE2} = 0.2$ V, the collector current I_{C2} changes by approximately $\Delta I_{C2} \approx 2.4$ mA and the base current changes by $\Delta I_{B2} = \Delta I_{C2}/\sigma h_{FE} \approx 60$ μA $(\sigma \approx 0.85)$. Furthermore, since V_{BE2} changes by 0.1 V, from V_γ to V_σ, the change in the current

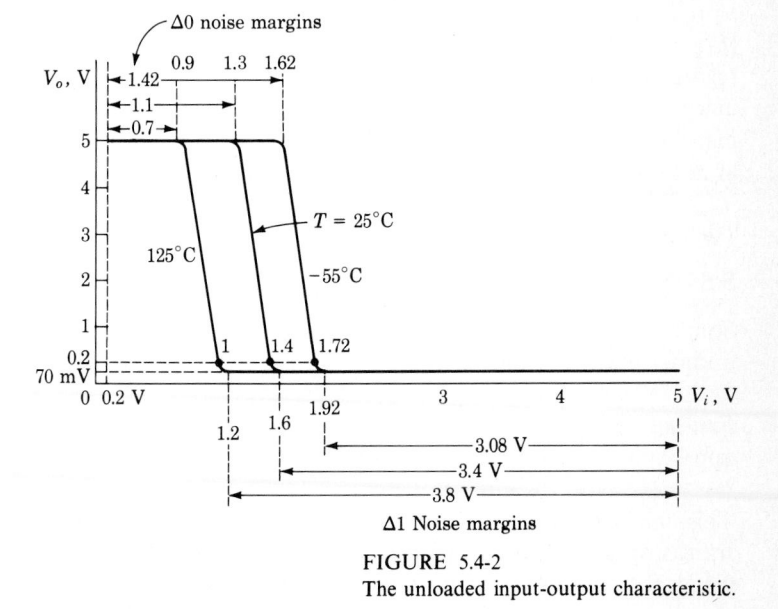

FIGURE 5.4-2
The unloaded input-output characteristic.

flowing in R_b is $\Delta I_{R_b} = 20$ μA.   Hence, $\Delta I_{E1} = 80$ μA and $\Delta I_{B1} \approx \Delta I_{E1}/h_{FE} = 1.6$ μA.   This increase in base current results in a decrease in diode current I_{DA} which is also approximately 1.6 μA.   Hence, a very small change in the current in diode DA is all that is necessary to change $T2$ from cutoff to saturation.

Since the current in $T1$ and diode $D2$ has changed by 80 μA, from 130 μA when $V_{BE2} = V_\gamma$ to 210 μA when $V_{BE2} = V_\sigma$, we can assume that V_{BE1} and V_{D2} remain essentially unchanged.   This assumption is certainly valid when considering diode DA since the current in this diode changes by only 1.6 μA.   Thus, we can conclude that when the input voltage increases by 0.1 V from 1.3 to 1.4 V, the output transistor $T2$ moves from cutoff to saturation.   When V_i increases further to 1.6 V, diode DA cuts off (since $V_p = 2.25$ V) and $I_{B2} \approx 1.5$ mA.   At this point σ decreases to $\sigma \approx 0.03$ and $V_{CE}(\text{sat}) \approx 70$ mV (see Fig. 1.10-1).

These results are shown in Fig. 5.4-2.   In the plots for $T = -55°$C and $T = +125°$C, also given in this figure, we have assumed, as usual, that the temperature sensitivity of a junction voltage is -2 mV/$°$C.   However, as before, we have ignored the temperature dependence of $V_{CE}(\text{sat})$.   The details of the calculations leading to the input-output characteristics are left as problems.

The input-output characteristic shown in Fig. 5.4-2 is somewhat idealized inasmuch as we performed the calculation for the case where $T2$ is unloaded, i.e., has a fan-out of 0.   A typical fan-out for a DTL gate is 8.   If we had taken the fan-out into account, as illustrated in Fig. 5.3-2, we would see that when $T2$ is cut off, V_o is still equal to 5 V since $DA1, DA2, \ldots, DAN$ are also cut off.   As discussed in Sec. 5.3, the effect of too large a fan-out is to increase the collector current in $T2$, thereby raising $T2$ out of saturation and increasing V_o above 0.2 V when $V_i = 5$ V.   For the maximum fan-out of 8 specified by the manufacturer this problem does not arise.

It is interesting to compare the input-output characteristic of the DTL gate (Fig. 5.4-2) with the input-output characteristic of the RTL gate (Fig. 4.6-2), with regard to the $\Delta 0$ and $\Delta 1$ noise margins.   Here we see that at 25$°$C the $\Delta 0$ noise margin for DTL is approximately 1.2 V while for RTL it is 0.45 V; the $\Delta 1$ noise margin for DTL is 3.4 V while for RTL it is 0.24 V.   This significant increase in noise margin is one of the major advantages of DTL over RTL.

5.5   MANUFACTURER'S SPECIFICATIONS OF DTL GATES

Some typical manufacturers' specifications for DTL are shown in Table 5.5-1 and discussed in this section.

Fan-out   We noted [see Eq. (5.3-7)] that the DTL gate has a fan-out which is approximately equal to $0.7\sigma h_{FE}/\rho$.   Taking $h_{FE}(T2) = 50$ and $\sigma = 0.85$, then with $1/\rho = 2.3$, as in the circuit of Fig. 5.4-1, we find the fan-out is $0.7(42)(2.3) \approx 68$. This fan-out number corresponds to a circumstance where the gate output transistor $T2$ is brought to $V_{CE}(\text{sat}) = 0.2$ V.   If we assume that $V_{CE}(\text{sat}, T2) = 0.1$ V, then $\sigma = 0.1$ and the maximum fan-out is reduced to 8.

The manufacturer specifies a typical fan-out of 8 for the DTL gate and 25 for the buffer. The buffer employs an active pull-up similar to that employed in RTL. A detailed analysis of the DTL buffer is left for Prob. 5.4-6. Furthermore, each gate loading the driving gate appears as a 10-pF capacitor. Thus, a load of eight gates acts like a 80-pF capacitor and therefore affects the propagation delay time of the loaded gate. Thus, the specified fan-out of 8 ensures that the propagation delay time requirement of 30 to 40 ns is satisfied and that the collector-emitter voltage of the output transistor is approximately 0.1 V.

Voltage levels The specified voltage levels V_{oH}, V_{iH}, V_{iL}, and V_{oL} are shown in Fig. 5.5-1 for $T = 25°C$. The voltage V_{oH} is the minimum output voltage of a gate corresponding to the logic **1** state. The value of the load current I_L when $V_o = V_{oH}$ is called I_{oH}. The value of 2.6 V shown in Table 5.5-1 is recommended by one manufacturer. Another manufacturer lists $V_{oH} = 4.3$ V. To estimate V_{oH} we refer to Fig. 5.4-1 and let $T2$ be cut off. Then assuming $V_{CC} = 5$ V $\pm$ 10 percent, $R_c = 2$ kΩ, a leakage current $I_L = 50$ μA, and $I_{oH} = -0.12$ mA, we find

$$V_o = V_{oH} = 4.5 - 2,000(50 + 120) \times 10^{-6} = 4.16 \text{ V} \qquad (5.5\text{-}1)$$

If V_{oH} is calculated for $I_{oH} = 0$, that is, *no load*, we find

$$V_{oH} = 4.4 \text{ V} \qquad (5.5\text{-}2)$$

This is similar to the value presented by one manufacturer.

Table 5.5-1   TYPICAL MANUFACTURERS SPECIFICATIONS AT 25°C FOR DTL NAND GATE AND BUFFER

	Gate	Buffer
Output loading factor (fan-out)	8	25
Input loading factor	1	1
Power dissipation, mW	11	42
Input voltage, V: V_{iL}	1.1	1.1
V_{iH}	2	2
Output voltage, V: V_{oL}	0.4	0.4
V_{oH}	2.6*	2.6†
Output leakage current I_L, μA	50	50
Reverse diode current I_R, μA	2	2
Forward current of input diodes I_F		
for a unit input load, mA	1.6	1.6
Propagation delay time $t_{pd}(HL)$	30	40
$t_{pd}(LH)$	80	80

* $I_{oH} = -0.12$ mA
† $I_{oH} = -2.5$ mA.

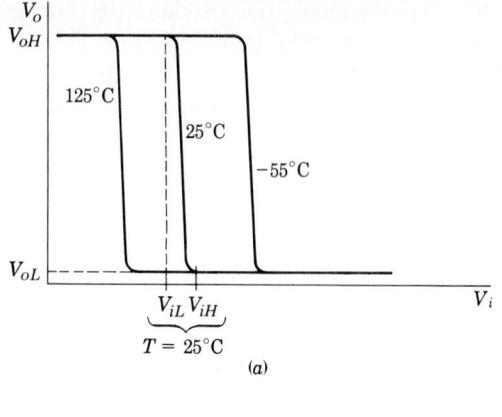

(a)

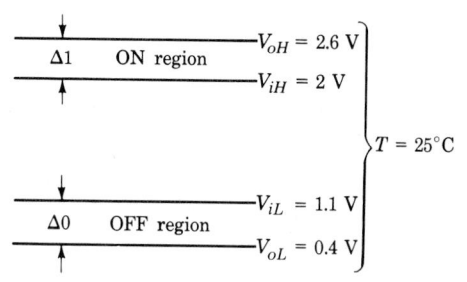

FIGURE 5.5-1
(a) Input-output characteristic defining
voltage levels. (b) Worst case voltage levels.

(b)

The maximum output voltage of the gate corresponding to the logic level
0 is V_{oL}. We have estimated that at a fan-out of 8, $V_{oL} = V_{CE}(\text{sat}) = 0.1$ V.
The manufacturer, in order to be extremely conservative, will only say that the
collector-emitter voltage V_{oL} will never exceed 0.4 V.

The maximum allowable input voltage which will reliably be recognized
as corresponding to logic level **0** is V_{iL}, and the minimum allowable input
voltage which will reliably be recognized as corresponding to logic level **1** is V_{iH}.
The temperature dependence of these parameters is brought out in Fig. 5.5-1.
Our calculations indicate that, at 25°C, $V_{iL}(\text{max})$ should be approximately 1.3 V
and $V_{iH}(\text{min})$ about 1.6 V. We find (see Table 5.5-2) that the manufacturer's

Table 5.5-2

	−55°C	25°C	125°C
V_{iH}, V	2.1	2	1.9
V_{iL}, V	1.4	1.1	0.8
Δ0	1.0	0.7	0.4
Δ1	0.5	0.6	0.7

specification on V_{iL} is 1.1 V, while the specification on V_{iH} is 2 V. Again, we note that these specified values represent worst-case conditions.

Noise immunity   The $\Delta 0$ and the $\Delta 1$ noise immunity are given by

$$\Delta 0 = V_{iL} - V_{oL} \qquad (5.5\text{-}3)$$

$$\Delta 1 = V_{oH} - V_{iH} \qquad (5.5\text{-}4)$$

The parameters V_{iH} and V_{iL} as specified by the manufacturer are summarized in Table 5.5-2.   Using $V_{oH} = 2.6$ V and $V_{oL} = 0.4$ V (since these values do not change significantly with temperature), we can calculate $\Delta 0$ and $\Delta 1$, which are also tabulated in Table 5.5-2.   The value $V_{oH} = 2.6$ V seems unreasonably conservative.   It may have been selected by the one manufacturer in order simply to arrange that the noise margins $\Delta 0$ and $\Delta 1$ be nearly equal to one another. Note the significant increase in the noise immunity of DTL compared with that of RTL.

Propagation delays   Propagation delays in DTL gates are of the order of 30 to 80 ns.   The delay associated with turning on the output transistor [the turn-on delay $t_{pd}(HL)$] is smaller than the delay associated with driving that transistor back to cutoff [the turn-off delay $t_{pd}(LH)$].   The turnoff delay is generally substantially larger than the turn-on delay, often by a factor of 2 or 3.   For, at turn-on, any capacitance shunting to ground the output of the gate can discharge rapidly through the low impedance of a transistor in saturation. At turnoff, however, this shunt capacitor must charge through the relatively large pull-up resistor R_c.   In addition, at turnoff, there is a storage-time delay in the output transistor $T2$ which is not encountered at turn-on.

5.6   THE WIRED-AND CONNECTION

A useful extension of the logic capability of DTL gates can be achieved by joining the outputs of two DTL gates.   Such a connection of outputs into a single common output is indicated in Fig. 5.6-1a.   In Fig. 5.6-1b is shown the output transistor of each of the two gates with their collectors tied together to a common output Y.   For the inputs given in Fig. 5.6-1a the logic variables which appear at the bases of these two transistors are $X_1 = ABC$ and $X_2 = DE$, as shown in Fig. 5.6-1b.   Additionally, it is readily verified (compare with the RTL gate of Fig. 4.4-1) that $Y = \overline{X_1 + X_2}$.   Hence, altogether we have

$$Y = \overline{X_1 + X_2} = \overline{ABC + DE} \qquad (5.6\text{-}1)$$

or, using De Morgan's theorem,

$$Y = \overline{X_1}\,\overline{X_2} = (\overline{ABC})(\overline{DE}) \qquad (5.6\text{-}2)$$

From Eq. (5.6-1) it appears as though tying the outputs together resulted in a NOR operation.   On the other hand, from Eq. (5.6-2) we can interpret the result

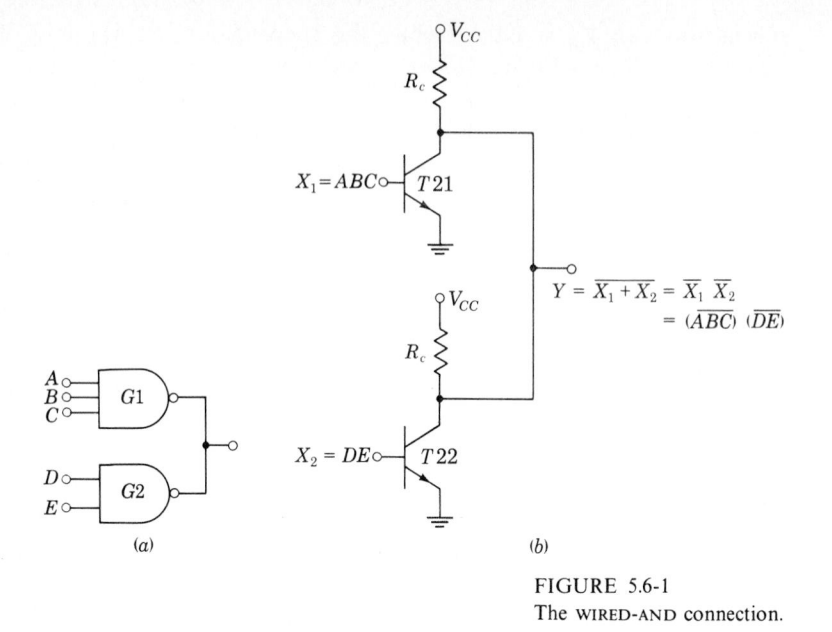

(a)

(b)

FIGURE 5.6-1
The WIRED-AND connection.

as what would be obtained if the outputs of the individual DTL gates, when not joined, were combined in an AND gate. This result applies as well if more than two gates are connected at their outputs. This consideration leads to the description of the arrangement of Fig. 5.6-1 as a WIRED-OR or a WIRED-AND connection of DTL gates. In this text we shall use the terminology WIRED-AND.

To appreciate the usefulness of the WIRED-AND connection, consider that we did indeed want to generate the logical function Y given in Eq. (5.6-1) or (5.6-2); and suppose that we were restricted to using the DTL NAND gates in the conventional manner. Then, as already noted, we should have to combine the DTL-gate outputs $\overline{ABC}$ and $\overline{DE}$ in an AND gate. An AND gate, when constructed from NAND gates, requires two such NAND gates. Hence, altogether, the equivalent of the gating circuit of Fig. 5.6-1a would appear as shown in Fig. 5.6-2 and require a total of four gates.

Loading rules In the WIRED-AND connection of a number of DTL gates, it may happen that only one output transistor is conducting while the others are

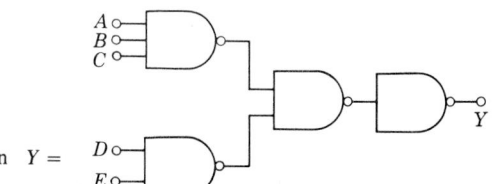

FIGURE 5.6-2
A NAND gate system to obtain $Y = \overline{(ABC)}\,\overline{(DE)}$.

all cut off. Then this transistor must not only sink the current of the loading gates and the current due to its own pull-up resistor but must also sink the current in the pull-up resistor of the other output transistors. To allow for this situation it is necessary to reduce the allowable fan-out of each gate in the WIRED-AND connection. This fan-out reduction is calculated in the following example.

EXAMPLE 5.6-1 A number K of DTL gates are connected in a WIRED-AND connection. Calculate the reduction required in the output loading factor as a function of K.

SOLUTION Refer to Fig. 5.6-3. Here we see the outputs $T1$, $T2$, ..., TK of K gates connected in a WIRED-AND configuration and driving N DTL gates. The driven gates are represented by their input diodes and series resistors R since the load affects the operation of the driving gate only when the driving gates are saturated, in which case all the current in R flows through its series diode.

Now assume that X_1 is in logic level **1** while X_2, ..., X_K are in logic level **0**. Then $T1$ is saturated. Let $V_{CE}(\text{sat}) \approx 0.2$ V. If the current in each diode, $D1$, $D2$, ..., DN, is called I_L, we have

$$I_L = \frac{V_{CC} - V_D - V_{CE}(\text{sat})}{R} = \frac{5 - 0.75 - 0.2}{3.75 \times 10^3} = 1.08 \text{ mA} \qquad (5.6\text{-}3)$$

The currents flowing in the collector resistors of the WIRED-AND output transistors are each equal to I_1, which is

$$I_1 = I_2 = \cdots = I_K = \frac{V_{CC} - V_{CE}(\text{sat})}{R_c} = \frac{5 - 0.2}{2 \times 10^3} = 2.4 \text{ mA} \qquad (5.6\text{-}4)$$

Thus, the collector current in $T1$ is

$$I_{C1} = KI_1 + NI_L = 2.4K + 1.08N \qquad \text{mA} \qquad (5.6\text{-}5)$$

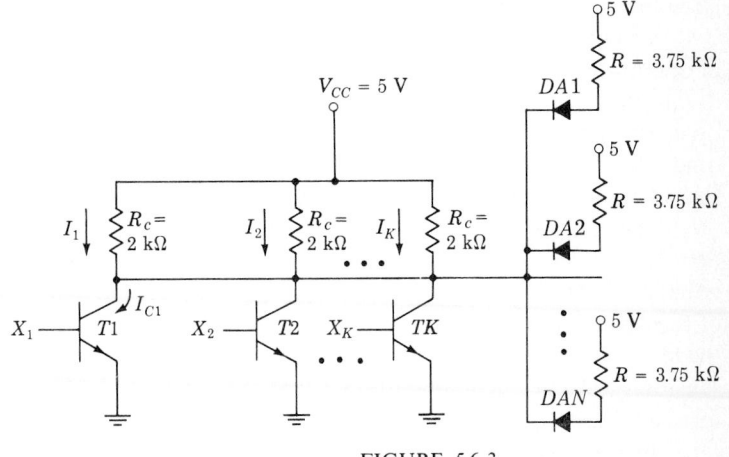

FIGURE 5.6-3
The WIRED-AND connection for K gates.

We see from this expression that increasing K by 1 gate is equivalent, as far as the increase in collector current is concerned, to increasing the fan-out by $\Delta N = 2.4/1.08 = 2.2$ gates.

Thus, for each gate connected in parallel with $T1$ to form a WIRED AND, we must reduce the maximum output loading factor by 2.2 gates. This results in leaving the collector current fixed, and therefore $T1$ remains saturated. The manufacturer specifies a load reduction of 2.5 gates.

It is of great practical importance to note that DTL gates using an active pull-up, as in the buffer shown in Fig. $P5.4$-6, should not be used in the WIRED-AND mode. The reason for this restriction is made clear in Sec. 6.14.

5.7 HIGH-THRESHOLD LOGIC (HTL)

There are circumstances where logic circuits must operate in environments which are very noisy electrically. For operation in such surroundings there is available a line of DTL logic circuits with thresholds, i.e., noise immunities $\Delta 0$ and $\Delta 1$, which are quite high in comparison with the thresholds of conventional DTL circuitry.

A high-threshold-logic (HTL) gate is shown in Fig. 5.7-1. Comparing Fig. 5.7-1 with the conventional gate of Fig. 5.4-1, we note that in the HTL gate the supply voltage has been raised from 5 to 15 V. This feature, as we shall see, accounts for the increased noise immunity. Because of the higher supply voltage, the diode $D2$ in Fig. 5.3-1 must sustain a higher voltage and has hence been replaced by a zener diode. We note, additionally, that in the HTL gate the resistance values are appreciably larger than in the conventional gate. This increase in the resistance values is necessary because of the increased supply voltage; for, otherwise, the increased supply voltage would result in a large increase in current and therefore in power dissipation in the HTL gate.

The use of these larger resistance values has an adverse affect on the speed of operation of the HTL gate. For now, when capacitances need to charge or discharge through these resistances, they do so in circuits having relatively larger time constants. Thus, while conventional DTL gates have propagation delay times which are, typically, some tens of nanoseconds, HTL gates have propagation times which may be as high as hundreds of nanoseconds.

The operating principles and characteristics of zener diodes are discussed in Sec. 1.4. The zener diode in the HTL gate of Fig. 5.7-1 has an operating voltage of about 6.9 V. Further, at this voltage the temperature sensitivity of the operating voltage of a zener diode is about comparable in magnitude to the temperature sensitivity of a forward-biased diode. However, the temperature sensitivity of the zener diode is positive, while the temperature sensitivity of a forward-biased junction is negative.

In the conventional DTL gate we noted, as in Fig. 5.4-2, that the input voltage at which the gate makes its transition between logic levels is dependent

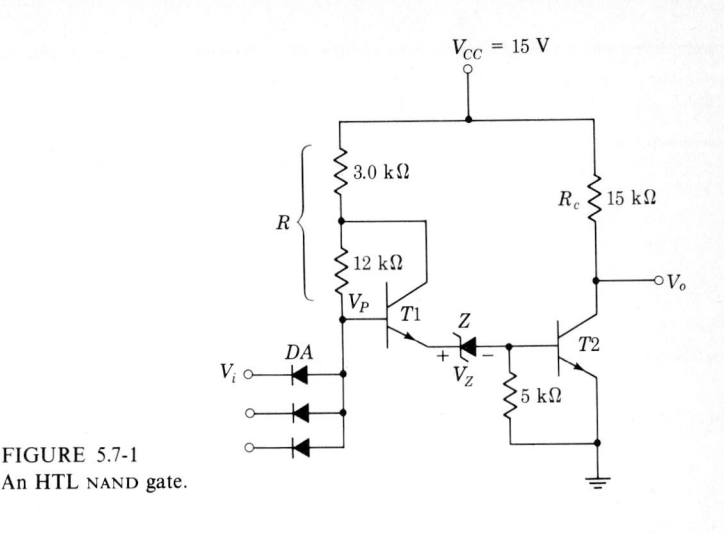

FIGURE 5.7-1
An HTL NAND gate.

on temperature. The temperature sensitivity of this voltage is equal to the temperature sensitivity of two diodes in series. For, in Fig. 5.4-1, since the input diode and the base-emitter junction of $T1$ are in polarity opposition, the temperature sensitivities of these two junctions cancel. We are then left with the combined temperature sensitivities of $D2$ and the base-emitter junction of $T2$. In the HTL gate of Fig. 5.7-1 the input diode and the junction of $T1$ cancel, as before. But now we also have a cancellation of the temperature sensitivities of the zener diode and the junction in $T2$ since the temperature sensitivities of these two are approximately equal and are also in opposite directions. The result is that the temperature sensitivity of the HTL gate is significantly less than that indicated in Fig. 5.4-2 for the DTL gate. We therefore ignore temperature effects.

5.8   INPUT-OUTPUT CHARACTERISTIC OF THE HTL GATE

The input voltage at which transistor $T2$ begins to come out of cutoff is (see Prob. 5.8-1)

$$V_i = V_{BE2} + V_Z + V_{BE1} - V_{DA}$$
$$= 0.65 + 6.9 + 0.7 - 0.75 = 7.5 \text{ V} \tag{5.8-1}$$

The input voltage at which transistor $T2$ is at the edge of saturation is (see Prob. 5.8-2)

$$V_i = V_{BE2} + V_Z + V_{BE1} - V_{DA}$$
$$= 0.75 + 6.9 + 0.7 - 0.75 = 7.6 \tag{5.8-2}$$

Using the results in Eqs. (5.8-1) and (5.8-2), we can draw the input-output characteristic shown in Fig. 5.8-1.

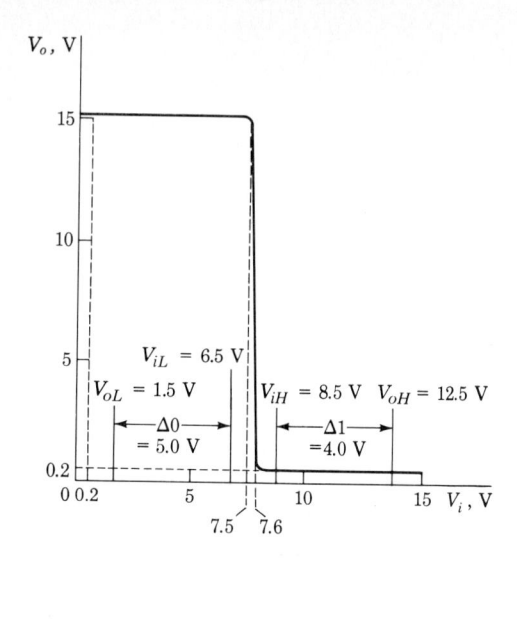

FIGURE 5.8-1
Input-output characteristic of HTL.

5.9   MANUFACTURER'S SPECIFICATIONS

Manufacturer's voltage specifications for HTL are typically

$$V_{iL} = 6.5 \text{ V} \qquad V_{oL} = 1.5 \text{ V} \qquad I_{oL} = 12 \text{ mA}$$
$$V_{iH} = 8.5 \text{ V} \qquad V_{oH} = 12.5 \text{ V} \qquad I_{oH} = -30 \text{ } \mu\text{A} \tag{5.9-1}$$

These voltages have been marked on the plot of Fig. 5.8-1. The voltage range V_{iL} to V_{iH} bridges the transition region rather symmetrically and leaves a reasonable margin of safety. The voltage $V_{oH} = 12.5$ V is also rather reasonable. For, with a 15-kΩ pull-up resistor, as in Fig. 5.7-1, a leakage current in the output transistor of 100 μA, and with $I_{oH} = -30$ μA, the output voltage of the gate is $V_o = V_{CC} - 15 \times 10^3 \times (130 \times 10^{-6})$. Letting $V_{CC} = 14$ V (the manufacturer assumes that V_{CC} can vary by ± 1 V from the nominal value of 15 V), we have $V_o = 12$ V, which is actually somewhat less than the value of V_{oH} given.

The voltage $V_{oL} = 1.5$ V seems artificially high. The manufacturer states that even if 12 mA enters the output transistor $T2$ from an outside source, $V_o \leq 1.5$ V. Presumably, if a gate is operating properly, the output transistor should be in saturation and the gate output should be 0.2 or at most, 0.4 V (see Fig. 1.10-1). This value of V_{oL} is specified, we may surmise, to make the noise immunities $\Delta 0$ and $\Delta 1$ more nearly comparable. For we find that the noise margins $\Delta 1$ and $\Delta 0$ are

$$\Delta 1 = V_{oH} - V_{iH} = 12.5 - 8.5 = 4.0 \text{ V} \tag{5.9-2}$$

and

$$\Delta 0 = V_{iL} - V_{oL} = 6.5 - 1.5 = 5.0 \text{ V} \tag{5.9-3}$$

Propagation delay time   The propagation delay time of HTL gates is typically

$$t_{pd}(LH) = 200 \text{ ns} \qquad t_{pd}(HL) = 100 \text{ ns}$$

which is approximately 3 times as large as the delay found when using DTL.

Fan-out   Manufacturers typically specify a fan-out of 10 for the HTL gate.   As we shall now see, this figure is, as usual, conservative.

The derivation of fan-out for the DTL gate which leads to Eq. (5.3-7) applies to the present case as well.   Using, in the derivation $V_{CC} = 15$ V, $R = 15$ kΩ and $V_Z = 6.9$ V, we find that Eq. (5.3-7) becomes

$$N = \frac{0.5\sigma h_{FE}}{\rho + (1 - \rho)/(h_{FE} + 1)} - 0.07R \times 10^{-3} \qquad (5.9\text{-}4)$$

In the HTL gate, $\rho = 0.2$, so that $N \approx 2.5\sigma h_{FE}$.   If we use $h_{FE} = 50$ and $\sigma = 0.85$, which brings the output transistor just slightly into saturation, where $V_{CE}(\text{sat}) = 0.2$ V, we find $N = 106$.   If we require that the output transistor be driven well into saturation [$\sigma = 0.1$, $V_{CE}(\text{sat}) = 0.1$ V], we find $N \approx 12$.   Thus, $N = 10$ seems conservative yet reasonable.

REFERENCES

1 Bohn, R., and R. Seeds: Collector Tap Improves Logic Gating, *Electronic Design*, August 3, 1964, pp. 51–55.

6

TRANSISTOR-TRANSISTOR LOGIC

6.1 TRANSISTOR-TRANSISTOR LOGIC (TTL)

The usefulness of a DTL gate is limited by its speed of operation. A principal source of this limitation can be appreciated by considering the circuit of Fig. 6.1-1, where a single-input gate is shown. With a view toward the TTL gate shortly to be introduced, we have explicitly shown the input diode as a diode-connected transistor, i.e., a transistor with collector and base connected.

Consider, now, that the input to the gate is at logic level **1**. Then current flows through $T2$ and diode D and into the base of $T3$. The transistor $T3$ is driven to saturation, and the gate output is at logic level **0**. Now change the input to logic level **0**. Then the output of the gate should go to logic level **1**. However, this transition to logic level **1** will not take place until transistor $T3$ comes out of saturation, passes through the active region, and eventually goes to cutoff. Cutoff, however, will not be reached until the stored base charge of $T3$ has been removed (see Sec. 1.20). During this removal of the stored base charge, transistor $T1$ is ON, but transistor $T2$ and diode D are cut off. Hence, there is no alternative: the base charge must leak off through the resistor R_b

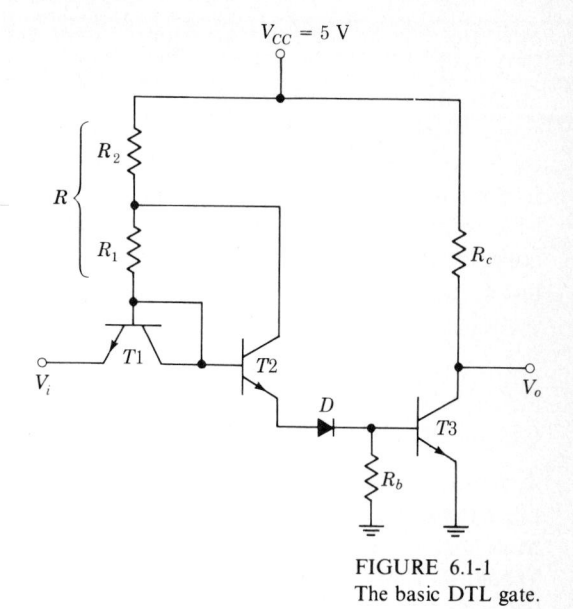

FIGURE 6.1-1
The basic DTL gate.

or dissipate by recombination. This relatively slow mechanism for the removal of stored charge establishes the essential speed limitation of the DTL gate.

The DTL speed limitation is overcome in the *transistor-transistor-logic gate* (TTL). In its simplest and most elemental form this gate appears as shown in Fig. 6.1-2. (For the moment, we picture a single input gate.) There is a certain similarity between the TTL gate of Fig. 6.1-2 and the DTL gate of Fig. 6.1-1. In the TTL gate the input transistor $T1$ is connected as a transistor, the collector-base connection being removed. Transistor $T2$ and diode D of the DTL gate are removed, and the collector of $T1$ is connected *directly* to the base of $T3$.

When, in the TTL gate, the input is high, the emitter-base junction of $T1$ will be back-biased and current will flow, through R and through the forward-biased base-collector junction of $T1$ into the base of $T3$. In this mode of

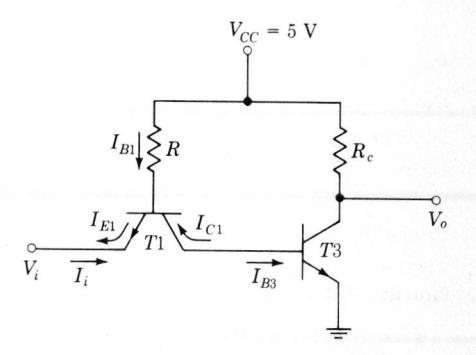

FIGURE 6.1-2
A basic TTL circuit.

operation the *collector* of transistor $T1$ operates as an *emitter* and the *emitter* as a *collector*. Transistor $T1$ is operating in the *inverse mode* (see Sec. 1.10). Transistor $T3$ will be driven into saturation, and the gate output will be low (logic level **0**). Now let the input drop to logic level **0**. The emitter junction of $T1$ will become forward-biased. We note that the base of $T3$ is connected to the collector terminal of $T1$. The stored charge in the base of $T3$ no longer leaks off through a resistor, as in the DTL gate, but flows out through the collector of transistor $T1$. Note that in the TTL gate, a base resistor for $T3$ has not even been included. A simple comparative calculation is made in the next section.

6.2 A COMPARISON BETWEEN TTL AND DTL

A simple comparison between the DTL gate and the TTL gate will illustrate the great advantage of the latter in the speed of removal of stored base charge from $T3$. When $T3$ is in saturation, its base-emitter voltage is $V_\sigma = 0.75$ V. When, in Fig. 6.1-1, diode D is cut off, the initial current out of the base of $T3$ is $V_\sigma/R_b = 0.75/2 \text{ k}\Omega \approx 0.38$ mA.

On the other hand, in the TTL gate, consider that the input is grounded. Then the base-emitter junction of $T1$ is forward-biased, and we shall assume that the voltage across the junction is 0.75 V. As we shall see, in a typical case the resistor R in Fig. 6.1-2 is $R = 4 \text{ k}\Omega$. Then the base current in $T1$ is $I_{B1} = (5 - 0.75)/4 \text{ k}\Omega \approx 1.1$ mA. Initially, transistor $T1$ is operating in its active region, since the base-to-ground voltage of $T3$, which is also the collector-to-ground voltage of $T1$, is initially $V_\sigma = 0.75$ V and therefore $V_{CE1} = 0.75$ V. We then observe that the initial collector current in $T1$, which is also the rate at which the stored base charge of $T3$ is being removed, is $h_{FE} I_{B1}$, where h_{FE} is the current gain of $T1$. Even if we allow an h_{FE} no larger than $h_{FE} = 20$, we find the discharge rate to be $h_{FE} I_{B1} = 20(1.1) = 22$ mA, which is to be compared with 0.38 mA in the DTL case.

This faster removal of the charge stored in $T3$ results in TTL gates which operate at propagation delay times that are one-tenth those of DTL gates.

6.3 THE INPUT TRANSISTOR

When the gate input is at logic **1** (say nominally at 5 V), transistor $T3$ is in saturation with a base-to-ground voltage $V_\sigma = 0.75$ V. It is then apparent that the base-emitter junction of $T1$ is reverse-biased. Transistor $T1$ is therefore operating in its inverse active region (see Sec. 1.10). In this inverse region the transistor operates with an inverse common-base current gain α_I, the corresponding common-collector current gain being $h_{FC} = \alpha_I/(1 - \alpha_I)$. Referring to Fig. 6.1-2, we see that if the base current of $T3$ is I_{B3}, the input current I_i, is $I_i = \alpha_I I_{B3}$. The remainder of the base current for $T3$ is supplied by the base

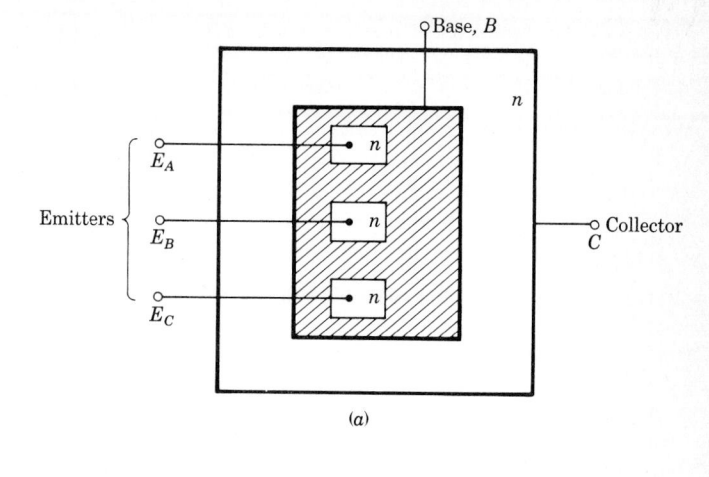

(a)

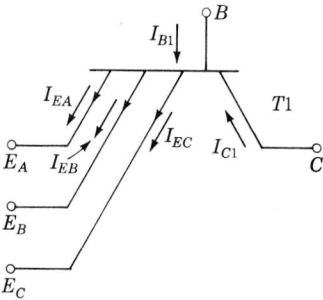

(b)

FIGURE 6.3-1
(a) Pictorial representation and (b) symbol of multiemitter transistor.

current I_{B1} of $T1$, so that $I_{B1} = (1 - \alpha_I)I_{B3}$. In a TTL gate, the input transistor $T1$ is deliberately designed to have a very low value of inverse current gain. Values in the range $\alpha_I \approx 0.02$ or even lower are typical of TTL gates. With such a value of α_I only 2 percent of the required base current of $T3$ need be supplied by the input driving source, while the other 98 percent is supplied through R from V_{CC}. Thus, the low value of α_I has the advantage of minimizing the loading on a driving source, at least when the input is at logic level **1**.

For the circuit of Fig. 6.1-2 to serve as a gate, additional inputs must be provided. These additional inputs may be made available by paralleling $T1$ with additional input transistors. All such input transistors would then have their collectors tied together and their bases similarly joined. In practice we do not parallel input transistors but construct a transistor with a single common collector, a single common base, and multiple emitters. The physical structure of such a *multiple-emitter transistor* is shown in Fig. 6.3-1a, and its circuit symbol is shown in Fig. 6.3-1b. A three-input TTL gate using such a multiple-emitter

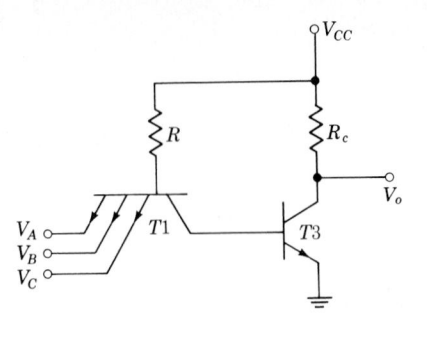

FIGURE 6.3-2
A basic TTL gate with three inputs.

transistor is shown in Fig. 6.3-2. Like the DTL gate, the TTL gate of Fig. 6.3-2 is a NAND gate. If any one of the inputs is at logic level **0**, transistor $T3$ is cut off and the gate output is at level **1**. If all the inputs are at logic **1**, transistor $T3$ is in saturation and the output is at logic **0**.

A discussion of the effect of the multiple-emitter transistor on the operation of the TTL gate is postponed until Sec. 6.7.

6.4 THE ACTIVE PULL-UP

Providing as it does a mechanism for the rapid removal of the output transistor base charge, the TTL gate has only one principal remaining limitation on the gate speed. This results from the effects of capacitance which appears across the output of the gate, from the collector of $T3$ to ground. This capacitance is composed of the capacitance of the output transistor itself, of the capacitance to ground of wires which connect the gate output to other gates, and of the input capacitances of these other gates or other devices which are being driven. When $T3$ is driven to cutoff, this output capacitance must charge from V_{CC} through the pull-up resistor R_c. If the output capacitance is C_o, the capacitance charges and the output rises from logic **0** to logic **1** with a time constant $R_c C_o$. This time constant can be reduced by reducing the resistance of R_c. Such a reduction, however, would increase the power dissipation in R_c and, of course, in transistor $T3$ while $T3$ is conducting. In addition, the reduction in R_c would make it more difficult to saturate $T3$.

The expedient used in TTL gates to hasten the charging of output capacitance substantially without introducing an unacceptable increase in power dissipation is shown in Fig. 6.4-1. Here, the pull-up resistor R_c in Fig. 6.3-2 is replaced with the active devices, transistor $T2$ and diode D. This circuit is recognized to be an *active pull-up* similar to the active pull-up used in RTL gates (Fig. 4.7-1) as well as in DTL gates. We now discuss, qualitatively, the operation of the circuit of Fig. 6.4-1.

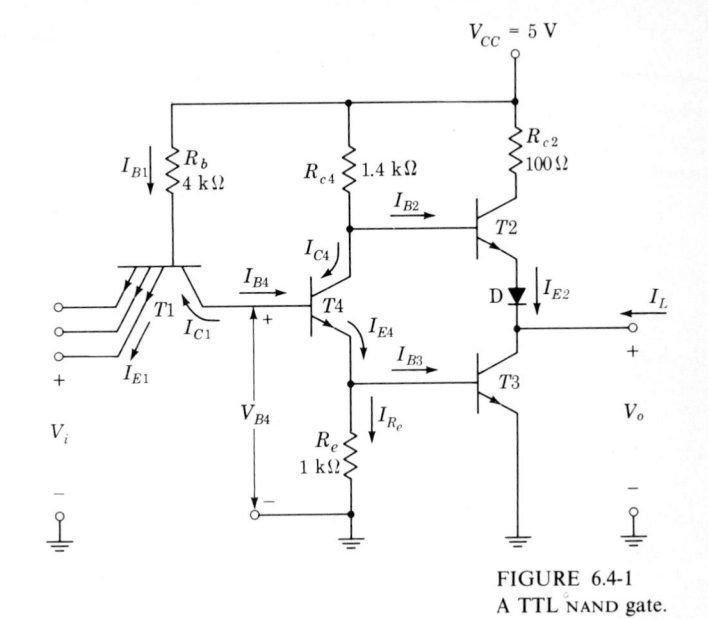

FIGURE 6.4-1
A TTL NAND gate.

It is intended, in the active pull-up circuit of Fig. 6.4-1, that when the output V_o is at logic **0**, transistor $T3$ will be in saturation and $T2$ cut off. Alternatively, when the output goes from logic **0** to logic **1**, $T3$ is to cut off, $T2$ is to go ON, and the output capacitance is to charge through the series combination of the emitter-follower transistor $T2$ and diode D. This switching of transistors $T3$ and $T2$ is accomplished by transistor $T4$, which is a *phase splitter* used to provide the bases of $T3$ and $T2$ with voltages which swing in opposite directions so that when one transistor of the $T3$-$T2$ *totem-pole* pair is being driven ON, the other is being driven OFF and vice versa.

To appreciate the need for the diode D located between the output transistor pair, consider the situation when the inputs are all at logic **1**, in which case the output should be at logic **0**. With all inputs at logic **1**, transistors $T4$ and $T3$ will both be in saturation (we shall verify later that such is indeed the case). Then the collector voltage of $T3$ is $V_{CE}(\text{sat}) \approx 0.2$ V while the collector voltage of $T4$ is $V_{CE4}(\text{sat}) + V_{BE3} \approx 0.2 + 0.75 = 0.95$. In this case, if diode D were not present, the base-emitter voltage of $T2$ would be $V_{C4} - V_{C3} = 0.95 - 0.2 = 0.75$ V, and $T2$ would also be in saturation. We require however that at the logic **0** output $T2$ be cut off. With the diode D present, the 0.75 V drop between the collectors of $T4$ and $T3$ must divide between diode D and the base-emitter junction of $T2$. In this case neither diode D nor transistor $T2$ will be forward-biased sufficiently to pass any appreciable current.

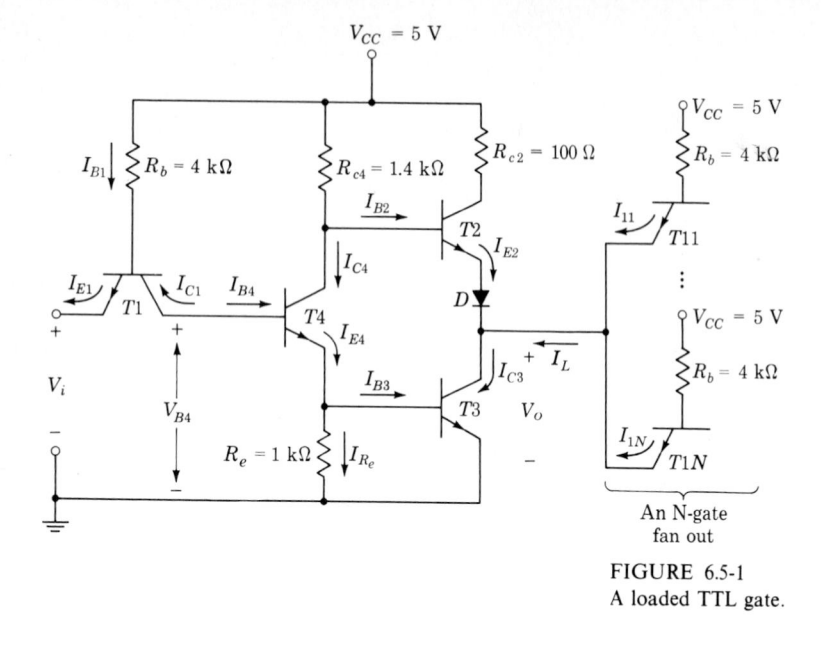

FIGURE 6.5-1
A loaded TTL gate.

6.5 INPUT-OUTPUT CHARACTERISTIC NEGLECTING THE INPUT TRANSISTOR

We shall now undertake to describe and make calculations concerning the operation of the TTL gate shown in Fig. 6.5-1. Since this gate is substantially more complicated than either the RTL gate or the DTL gate, we find it convenient to consider at the outset not the overall input-output characteristic but the characteristic relating the gate output voltage V_o to the base voltage V_{B4} of transistor $T4$. In Sec. 6.6 we shall consider the relation between V_{B4} and the input voltage V_i; and by combining these two characteristics, we shall deduce the overall characteristic.

Let us start with $V_{B4} = 0$ V. In this case $T4$ and $T3$ are cut off. Transistor $T2$ supplies a current I_{E2} to the N driven gates. The emitter current I_{E2} is equal to the current $-I_L$, I_L being the load current. Following common convention, the positive direction of I_L is taken *into* the gate.

The circuit from which to calculate the output voltage V_o is given in Fig. 6.5-2. The output voltage V_o is given by

$$V_o = V_{CC} - R_{c4}I_{B2} - V_{BE2} - V_D \tag{6.5-1}$$

where V_{CC} ($=5$ V) is the supply voltage, $R_{c4}I_{B2}$ is the drop across R_{c4}, V_{BE2} is the base-emitter voltage of $T2$, and V_D is the drop across the diode. Assuming

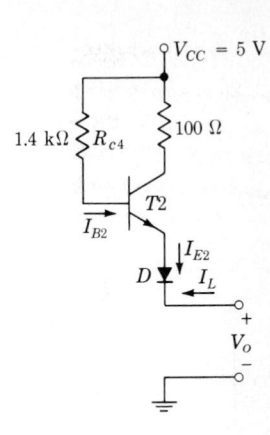

FIGURE 6.5-2
Output circuit used to calculate V_o when
$T3$ and $T4$ are cut off.

that $T2$ is operating in its active region, the base current is $I_{B2} = I_{E2}/(h_{FE} + 1) = -I_L/(h_{FE} + 1)$. Equation (6.5-1) becomes, since $h_{FE} \approx h_{FE} + 1$,

$$V_o = V_{CC} + \frac{R_{c4} I_L}{h_{FE}} - V_{BE2} - V_D \qquad (6.5-2)$$

When $T3$ is cut off, V_o is at its logic **1** level. To calculate the load current I_L we observe that the input transistors of the N driven gates in Fig. 6.5-1 are each operating in their inverse mode, just as for the input transistor in the basic circuit shown in Fig. 6.1-2. In each of the *driven* gates (only the input transistors of which are shown in Fig. 6.5-1) transistors $T3$ and $T4$ are in saturation. Hence, $V_{B4} = V_{BE3} + V_{BE4} = 0.75 + 0.75 = 1.5$ V. The base current of $T4$ is being supplied almost entirely through the collector junction of the input transistor. This input transistor is operating in its active (albeit inverse) mode, and so we shall allow a drop of 0.7 V across its base-collector junction. Altogether the voltage at the base of the input transistor is $1.5 + 0.7 = 2.2$ V. The drop across each base resistor R_b is therefore 5.0 V $- 2.2$ V $= 2.8$ V. The current through R_b is $2.8/(4 \text{ k}\Omega) = 0.70$ mA. Assuming that the input transistors $T11 - T1N$ each have $h_{FC} = 0.02$, the input current to each driven gate, is $0.70(0.02) = 14 \mu A$.

If the driving gate $T2$ were fanned out to a single gate, with the emitter junction and diode carrying so small a current (14 μA) we could reasonably take $V_{BE2} = V_D = V_\gamma = 0.65$ V. In this case also, the drop across R_{c4} [the second term in Eq. (6.5-2)] is negligible, and we have

$$V_o \approx V_{CC} - 2V_\gamma = 5.0 - 2(0.65) = 3.7 \text{ V} \qquad (6.5-3)$$

Suppose, on the other hand, we drive 10 gates (as we shall see, a fan-out of 10 is the conservative figure generally specified by manufacturers) and suppose we allow as a worst case that h_{FC} may be as large as 0.1. In this case the load current would be of the order of 1 mA. Even with this current the drop across

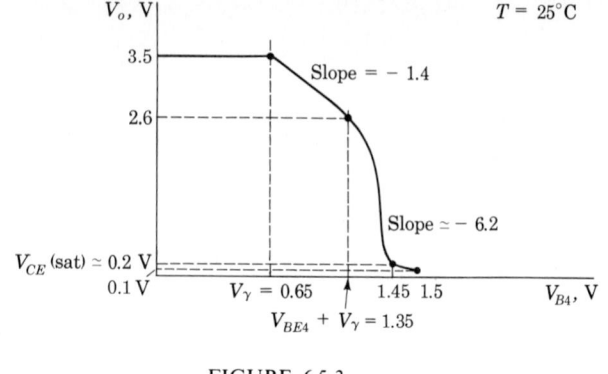

FIGURE 6.5-3
Input-output characteristic of the TTL gate.

R_{c4} would continue to be negligible, but it would be more reasonable to take $V_{BE2} = V_D = V_\sigma = 0.75$ V. The output voltage would then be

$$V_o = V_{CC} - 2V_\sigma = 5.0 - 2(0.75) = 3.5 \text{ V} \tag{6.5-4}$$

We therefore see that when V_o is in the **1** state, the value of V_o is only slightly dependent on the load current and hence on the fan-out. We shall accordingly simplify matters by assuming that when $T4$ (and hence $T3$) are cut off, the output voltage V_o is as given in Eq. (6.5-4). This value of the output voltage is given in the plot of the input-output characteristic shown in Fig. 6.5-3.

Phase splitter   Before continuing, it will be well to digress briefly to review some of the gain and impedance characteristics of a phase splitter. Such a stage is shown in Fig. 6.5-4a. The transistor is assumed to be biased into its active region, although the biasing arrangements are not explicitly shown. The stage has a collector resistor R_c and an emitter resistor R_e and is driven by a source of impedance r_i. We assume that, in the active region, the transistor can be represented by the equivalent circuit involving hybrid parameters, as shown in Fig. 6.5-4b. We then find that the impedances and the voltage gains are given by the equations in Fig. 6.5-4. Rather typically we might assume that $h_{FE} = 50$ and $h_{ie} = 1$ kΩ. If $h_{FE} R_e \gg h_{ie} + r_i$, these equations for impedances and gains can be simplified to the useful approximations

$$R_i \approx h_{FE} R_e \tag{6.5-5a}$$

$$A_E = \frac{V_E}{V_i} \approx 1 \tag{6.5-5b}$$

$$A_C = \frac{V_C}{V_i} = -\frac{R_c}{R_e} \tag{6.5-5c}$$

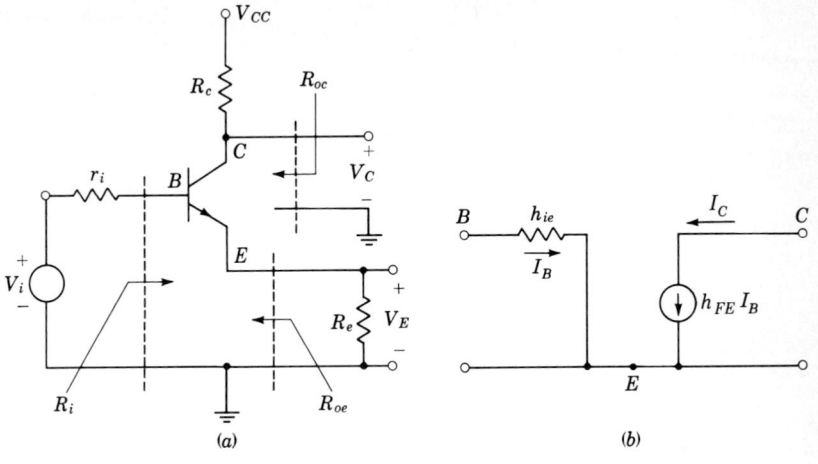

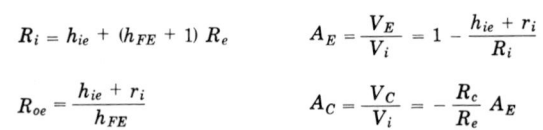

$$R_i = h_{ie} + (h_{FE} + 1) R_e \qquad A_E = \frac{V_E}{V_i} = 1 - \frac{h_{ie} + r_i}{R_i}$$

$$R_{oe} = \frac{h_{ie} + r_i}{h_{FE}} \qquad A_C = \frac{V_C}{V_i} = -\frac{R_c}{R_e} A_E$$

$$R_{oc} = R_c$$

FIGURE 6.5-4

(a) A phase splitter. (b) Equivalent circuit of a transistor.

Returning to the circuit of Fig. 6.5-1, let us allow V_{B4} to increase from zero. When V_{B4} attains a value of about $V_{B4} = V_\gamma = 0.65$ V, transistor $T4$ will begin to enter its active region. However, because of the drop across the base-emitter junction of $T4$, $T3$ will remain cut off. Hence, at this point, the circuit used to calculate the response of the output voltage V_o to V_{B4} is as shown in Fig. 6.5-5.

We now calculate the *slope* of the transfer characteristic in the region where $T3$ is cut off. The gain $A_2 \equiv \Delta V_o / \Delta V_{B2}$ from the base of $T2$ to the output is the gain of an emitter-follower and, hence, as in Eq. (6.5-5b) is $A_2 \approx 1$. The gain provided by $T4$ is [from Eq. (6.5-5c)] $A_4 = -R_{c4}/R_e = -1.4$. The overall gain $A = \Delta V_o / \Delta V_{B4}$ is $A_2 A_4 = -1.4$, as indicated in Fig. 6.5-3.

With further increase in V_{B4}, $T3$ will eventually reach the cut-in point. At this point $T4$ is in the active region, and we shall correspondingly assume a base-emitter voltage drop of $V_{BE4} = 0.7$ V. Assuming, as usual, that $T3$ comes out of cutoff when the voltage drop across its base-emitter junction is $V_{BE3} = V_\gamma = 0.65$ V, we estimate that $T3$ begins to turn on when $V_{B4} = V_{BE4} + V_{BE3} = 0.7 + 0.65 = 1.35$ V. At the point where $T3$ just turns ON, the current through R_e and, hence, very nearly, the current through R_{c4} is $V_\gamma / R_e = 0.65/1$ k$\Omega = 0.65$ mA. The corresponding drop through $R_{c4} = 1.4 \times 0.65 = 0.9$ V. Hence,

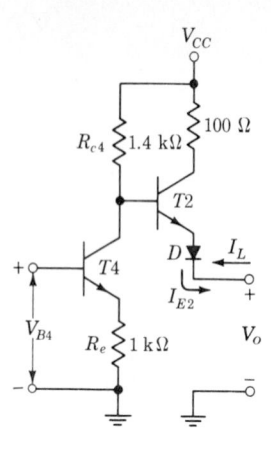

FIGURE 6.5-5
Circuit used to calculate V_o versus V_{B4}
when $T4$ is ON and $T3$ is still OFF.

as appears in Fig. 6.5-3, when $V_i = 1.35$ V, V_o is 0.9 V lower than the value in Eq. (6.5-4), that is $V_o = 3.5 - 0.9 = 2.6$ V.

When $T3$ turns ON, the incremental gain $A = \Delta V_o / \Delta V_{B4}$ increases and the output voltage V_o begins to drop more sharply with increasing V_{B4} than when $T3$ was OFF. This increase in gain has a twofold source. On the one hand V_o drops, simply because $T3$ begins to conduct. On the other, with $T3$ ON, the emitter resistor R_e is shunted by the impedance seen looking into the base of $T3$. The resistance in the emitter circuit of $T4$ is thereby decreased, and, as indicated in Eq. (6.5-5c), the gain of $T4$ increases. The following illustrative calculation will display these two sources of increased gain.

EXAMPLE 6.5-1　In the circuit of Fig. 6.5-1, let $T3$, $T2$, and $T4$ be assumed to be in the active region, each with parameters $h_{ie} = 1$ kΩ and $h_{FE} = 50$. Calculate the incremental gain $A = \Delta V_o / \Delta V_{B4}$.

SOLUTION　With $h_{ie} = 1$ kΩ shunting $R_e = 1$ kΩ, the equivalent resistor in the emitter of $T4$ is $R'_e = 500$ Ω. The condition $h_{FE} R'_e \gg h_{ie}$ continues to apply. Hence, the gain from base to emitter of $T4$, which is also the gain from the base of $T4$ to the base of $T3$, continues to have the value unity. In Fig. 6.5-6 we have redrawn the relevant portion of Fig. 6.5-1, with R_e replaced by R'_e, the connection from the emitter of $T4$ to the base of $T3$ removed, and with a generator $\Delta V_{B3} = \Delta V_{B4}$ applied at the base of $T3$. Let us now apply the principle of superposition and calculate separately the output voltage ΔV_o due individually to each of the generators ΔV_{B4} and ΔV_{B3}. The gain, referring to the generator ΔV_{B4}, is calculated as before, except that R_e is now replaced by R'_e. We then have

$$A_4 = \frac{\Delta V_{C4}}{\Delta V_{B4}} = -\frac{R_{c4}}{R'_e} = -\frac{1.4}{0.5} = -2.8 \tag{6.5-6}$$

Next we calculate the gain due to ΔV_{B3}. Referring to Fig. 6.5-4, we can readily verify

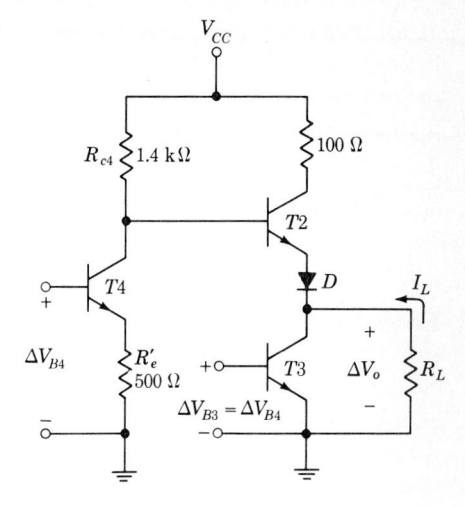

FIGURE 6.5-6
Circuit used to calculate incremental gain
when $T3$ is in the active region.

that a common-emitter amplifier stage operating *without* an emitter resistor and with a
collector resistor R_c has a gain

$$A_3 = \frac{\Delta V_o}{\Delta V_{B3}} = -\frac{h_{FE}}{h_{ie}} R_c \qquad (6.5\text{-}7)$$

The equivalent collector resistor R_c with which $T3$ operates is the incremental impedance
seen looking back into the diode D and the emitter of $T2$. The transistor impedance
$\Delta V_{BE}/\Delta I_B \equiv h_{ie}$. The impedance $\Delta V_{BE}/\Delta I_E = \Delta V_{BE}/h_{FE}\Delta I_B = h_{ie}/h_{FE}$. Since the same
current flows in the diode as in $T2$, we let the diode impedance equal the transistor
impedance h_{ie}/h_{FE}. To find R_c we add the diode impedance to the impedance seen looking
into the emitter of $T2$, as given by the equation for R_{oe} in Fig. 6.5-4. We have

$$R_c = \frac{h_{ie}}{h_{FE}} + \frac{h_{ie} + R_{c4}}{h_{FE}} = \frac{10^3}{50} + \frac{(1.4 + 1) \times 10^3}{50} = 68 \ \Omega \qquad (6.5\text{-}8)$$

a value small enough in comparison with any load the gate will encounter to permit
us to ignore the load. From Eqs. (6.5-7) and (6.5-8) we have

$$A_3 = \frac{\Delta V_o}{\Delta V_{B3}} \approx -\frac{h_{FE}}{h_{ie}} R_c = -\frac{1.4 + 2}{1} = -3.4 \qquad (6.5\text{-}9)$$

The total incremental output voltage ΔV_o is therefore $\Delta V_o = A_4 \Delta V_{B4} + A_3 \Delta V_{B3}$. However, $\Delta V_{B3} = \Delta V_{E4} \approx \Delta V_{B4}$ since $T4$ is an emitter follower. Hence, $\Delta V_o = (A_4 + A_3) \Delta V_{B4}$,
and the overall gain is

$$A = \frac{\Delta V_o}{\Delta V_{B4}} = A_4 + A_3 = -2.8 - 3.4 = -6.2 \qquad (6.5\text{-}10)$$

The calculation in Example 6.5-1 indicates an abrupt change in the
magnitude of the gain from 1.4 to 6.2 when $T3$ turns on. Actually we

assumed an abrupt change in the input impedance h_{ie} of $T3$ from h_{ie} infinite below cutoff to $h_{ie} = 1$ kΩ above cutoff. It is the finite value of h_{ie} that yields the nonzero gain A_3 as given in Eq. (6.5-7), and it is the finite value of h_{ie} that increases the gain A_4 above its initial value of 1.4. However, the parameter h_{ie} depends on the emitter current of $T3$ and the extent to which the transistor has been driven into saturation. It is readily established that

$$h_{ie} \approx (1 + \sigma h_{FE}) \frac{V_T}{I_E} \tag{6.5-11}$$

Hence, as $T3$ turns ON, h_{ie} changes gradually from an extremely large value to a value which gets progressively smaller as $T3$ draws more and more current and tends toward saturation. Thus, the magnitude of the incremental gain in the circuit of Fig. 6.5-6 starts at 1.4 when $T3$ is OFF, increases when $T3$ turns ON, has the value 6.2 when $h_{ie} = 1$ kΩ, and increases still more as the current in $T3$ increases further. These features appear in the plot of the input-output characteristic shown in Fig. 6.5-3.

Referring once more to Fig. 6.5-1, we see that with still further increases of V_{B4}, $T3$ is driven eventually to saturation. Assuming, as we have, that at saturation the base-to-emitter voltage of a transistor is $V_\sigma = 0.75$ V, we would have saturation $V_o \approx 0.2$ V when

$$V_{B4} = V_{BE4} + V_\sigma = 0.7 + 0.75 = 1.45 \text{ V} \tag{6.5-12}$$

This estimate has been included on the plot of Fig. 6.5-3.

When V_{B4} is increased beyond 1.45 V to 1.5 V both transistors $T4$ and $T3$ are saturated and $T2$ is cut off. In this case $T3$ is *deeply* saturated and $V_o \approx 0.1$ V. Then collector current $I_{C3} = I_L$, and the load current is now positive. This current is calculated in Sec. 6.9.

6.6 INPUT-OUTPUT CHARACTERISTIC OF THE INPUT TRANSISTOR

The plot of Fig. 6.5-3 relates the output voltage V_o to the base voltage V_{B4}. Of much greater interest is the input-output characteristic that relates V_o to the input voltage V_i, shown in Fig. 6.6-1. We now study the operation of transistor $T1$ to enable us to determine the relation between V_i and V_o.

As we have noted, when V_i is at a high voltage corresponding to logic **1**, the input transistor $T1$ operates in the active inverse mode. The collector current I_{C1} which is the base current I_{B4} of $T4$ is supplied principally by the base current I_{B1} of $T1$. The current gain h_{FC} in this inverse mode is extremely low by deliberate transistor design ($h_{FC} \approx 0.02$), so that the input source V_i supplies only about 2 percent of the base current of $T4$. Under these circumstances $T4$ and $T3$ are in saturation, $V_{B4} = 0.75 + 0.75 = 1.5$ V, and the gate output V_o is at logic **0**. This situation prevails so long as V_i is sufficiently positive with respect to V_{B4} ($2V_\sigma = 1.5$ V) to maintain $T1$ in the active inverse region.

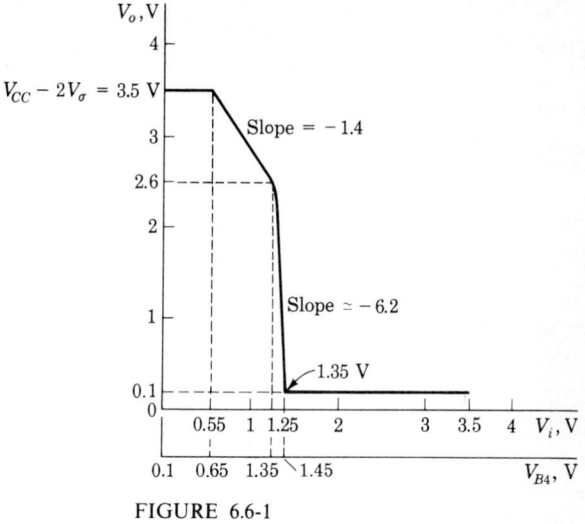

FIGURE 6.6-1
Input-output characteristic of a loaded TTL gate.

We shall now show that as V_i falls, then, at least up to the point where $V_i = 2V_\sigma$ ($=1.5$ V), the transistors $T4$ and $T3$ remain in saturation and consequently V_o remains at logic **0**. With $V_i = 2V_\sigma$, $V_{CE1} = 0$ V, and $T1$ is in saturation. From Eq. (1.10-11) we find that when $V_{CE1} = V_{CE}(\text{sat}) = 0$, $\sigma = -1/(h_{FE} + h_{FC}) \approx -1/h_{FE}$ since $h_{FE} \gg h_{FC}$. Applying the definition of σ to transistor $T1$, we have that $\sigma = I_{C1}/h_{FE} I_{B1}$. Substituting in this definition the value $\sigma = -1/h_{FE}$, we find that $-I_{C1} = I_{B1}$. Hence, the emitter current of $T1$ is now zero. However, the emitter current of $T1$ supplied only 2 percent of the base current of $T4$, and this small loss of base current will hardly affect $T4$ or $T3$. Hence, V_o remains at logic **0**.

As V_i falls to voltages lower than $2V_\sigma = 1.5$ V, the base current I_{B1} will be diverted away from the base-collector junction of $T1$ into its base-emitter junction. Eventually, the base current of $T4$ will be reduced to a value where the transistor $T4$ will be at the verge of coming out of saturation. We calculate now the voltage V_i corresponding to this situation. We have often noted that when a transistor is comfortably inside the saturation region, its collector-emitter voltage is $V_{CE}(\text{sat}) = 0.2$ V. At the very edge of saturation a more reasonable number is $V_{CE}(\text{sat}) = 0.3$ V. Referring to Fig. 6.5-1, since $T2$ is OFF when $T3$ is ON, we have

$$V_{CC} = R_{c4} I_{C4} + V_{CE4}(\text{sat}) + V_{BE3} \tag{6.6-1}$$

or

$$5 = 1.4 \times 10^3 I_{C4} + 0.3 + 0.75 \tag{6.6-2}$$

from which

$$I_{C4} = 2.8 \text{ mA} \tag{6.6-3}$$

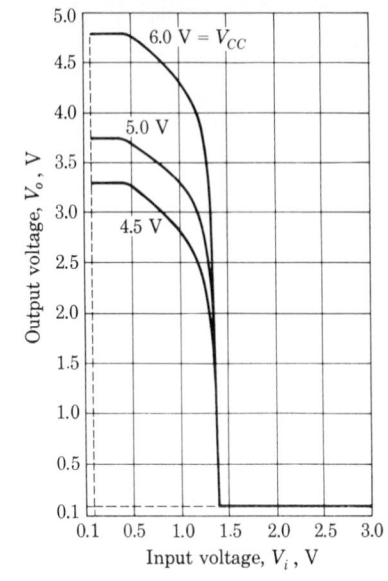

FIGURE 6.6-2
Medium-speed TTL transfer characteristics V_{CC} = 4.5, 5.0, and 6.0 V; T_A = 25°C.

Since $T4$ is at the edge of saturation, $I_{C4} = h_{FE} I_{B4}$; therefore using $h_{FE} = 50$, we have

$$-I_{C1} = I_{B4} = \frac{I_{C4}}{h_{FE}} = \frac{2.8 \text{ mA}}{50} = 56 \ \mu\text{A} \qquad (6.6\text{-}4)$$

The base voltage V_{B4} is 1.5 V. We allow a voltage drop of 0.7 V across the collector junction of $T1$ since this junction is carrying only 56 μA. The base-to-ground voltage of $T1$ is $1.5 + 0.7 = 2.2$ V. Consequently the base current of $T1$ is $I_{B1} = (5 - 2.2)/(4 \text{ k}\Omega) = 0.7$ mA. We can now calculate the parameter σ for transistor $T1$; we find

$$\sigma = \frac{I_{C1}}{h_{FE} I_{B1}} = \frac{-56 \times 10^{-6}}{50(0.7) \times 10^{-3}} = -1.6 \times 10^{-3} \qquad (6.6\text{-}5)$$

This value of σ is so close to $\sigma = 0$ that we shall make no error in assuming $\sigma = 0$. In this case $V_{CE}(\text{sat})$ is given by Eq. (1.10-4), and we have

$$V_{CE1} \approx V_T \ln \frac{1}{\alpha_I} \qquad (6.6\text{-}6)$$

For $\alpha_I = 0.02$, $V_{CE} \approx 100$ mV. Thus, the voltage V_i at which $T4$ comes out of saturation is about 0.1 V lower than the corresponding voltage V_{B4}. Hence, while in Fig. 6.5-3, the saturation region begins at $V_{B4} = 1.5$ V, in Fig. 6.6-1 it begins at $V_i = 1.4$ V.

As V_i continues to decrease, the current $I_{B4} = -I_{C1}$ continues to decrease, becoming zero when $T4$ cuts off and remaining zero thereafter. Hence, σ, which has the value given by Eq. (6.6-5) when $T4$ is in saturation, falls to $\sigma = 0$ when

$T4$ cuts off and remains at zero thereafter. Hence, again V_i is 0.1 V lower than V_{B4}.

It is now apparent that a plot of the overall input-output characteristic (V_o versus V_i) of the TTL gate is the same as the plot of V_o versus V_{B4} shifted along the voltage axes to the left by amount 0.1 V. Such an overall characteristic is shown in Fig. 6.6-1 and is to be compared with the characteristics shown in Fig. 6.6-2. These latter characteristics are typical measured characteristics supplied by manufacturers.

6.7 THE MULTIEMITTER TRANSISTOR

In the previous section we ignored the fact that the input transistor has several emitters. We shall now take this feature into account. For convenience, in Fig. 6.7-1 we have represented the *multiemitter transistor* as an array of individual transistors with collectors joined and bases joined. The inputs V_{iA}, V_{iB}, ..., V_{iN} are applied to the individual emitters.

Suppose now that all inputs are at the logic level **1**, corresponding nominally to $V_{iA} = V_{iB} = \cdots = V_{iN} = V_{CC} - 2V_\sigma = 3.5$ V. All input transistors $T1$ are now operating in the inverse active region. We assume tentatively that transistors $T3$ and $T4$ are in saturation. Thus, the voltage $V_{B4} = V_\sigma + V_\sigma = 1.5$ V while $V_{B1} = 0.7 + 1.5 = 2.2$ V. The current through R is then $I_{B1} = (5.0 - 2.2)/(4 \text{ k}\Omega) = 0.7$ mA. In this mode of operation the combined emitter current $I_E = -h_{FC}I_{B1}$, where h_{FC} is the inverse current gain corresponding to α_I. Now $h_{FC} = \alpha_I/(1 - \alpha_I) \approx \alpha_I$ since $\alpha_I \ll 1$. Using $\alpha_I = 0.02$, we have $I_{E1} = -0.02(0.7)\text{ mA} = -0.014\text{ mA}$. Thus, the total base current of $T4$ is $I_{B4} \approx 0.7$ mA. This is the same result calculated in Sec. 6.6 for a single-emitter input transistor $T1$.

We note again that the base current required to drive $T4$ comes almost entirely from V_{CC} through R_b. The total current required from the gate driving $T1A$, $T1B$, ..., $T1N$ is 2 percent of the base current I_{B4}. If there are N inputs, each driving gate need supply only $2/N$ percent of this current.

Next, let us consider that one of the input emitters, say the emitter of $T1A$, is at logic level **0**, which corresponds to about 0.2 V, the saturation voltage of the driving gate. Then, as we have seen in Sec. 6.6, $T1A$ will be in saturation, and $V_{B4} = V_{iA} + V_{CE}(T1A) \approx 0.2 + 0.1 = 0.3$ V. Transistors $T4$ and $T3$ are therefore cut off, as required. Now, while $T1A$ is operating in saturation with its emitter junction forward-biased and carrying an emitter current very large in comparison with its collector current, let us assume that the emitters of each of the other "transistors" $T1B$, ..., $T1N$ are in the logic **1** state, with $V_i = 3.5$ V. These transistors are operating in the inverse active mode with their base-collector junctions forward-biased and their emitters reverse-biased. We note further that since $V_{iA} = 0.2$ V and $T1A$ is saturated, $V_{B1} = V_{iA} + V_{BE}(T1A) = 0.2 + 0.75 = 0.95$ V. Also, the collector voltages of transistors $T1B$, ..., $T1N$ are equal to V_{B4}, where $V_{B4} = 0.3$ V. Hence, the voltage drops across the

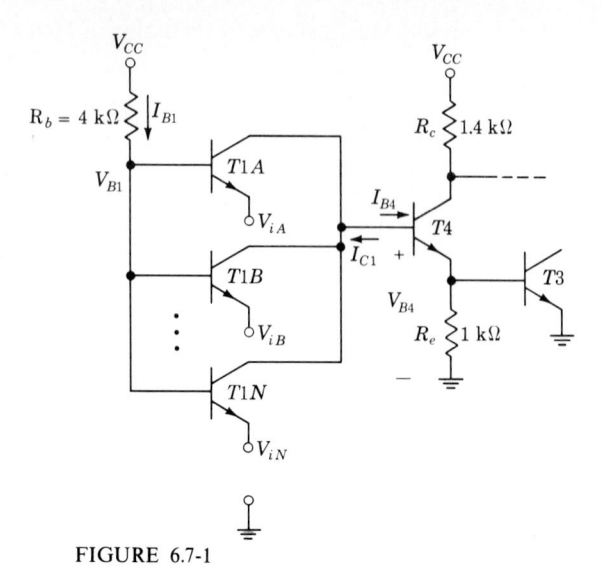

FIGURE 6.7-1
Circuit to show operation of the multiemitter transistor.

forward-biased base-collector junctions of $T1B$, ..., $T1N$ are 0.65 V, which is 0.1 V *less* than the forward bias across the base-emitter junction of $T1A$. On many occasions, we have noted that a change of 0.1 V in voltage will carry a transistor from saturation to the edge of cutoff. Hence, in Fig. 6.7-1, we may estimate that when one (or more) inputs are set at the logic level **0**, the other transistors are, to all intents and purposes, cut off.

One additional characteristic of the multiemitter transistor should be mentioned. Referring to Fig. 6.3-1, we see that transistors are formed not only between each of the three emitters and the collector but also between the emitters E_A and E_B, E_A and E_C, and E_B and E_C since they also are *npn* structures. However, because of the geometry of the "transistors," their current gains are very small (less than 0.01) and the presence of these transistors may be neglected.

6.8 INPUT VOLT-AMPERE CHARACTERISTIC OF THE TTL GATE

Because of the relative complexity of TTL logic, it is useful to consider the input and the output volt-ampere characteristic in addition to the transfer characteristic. We consider here the input characteristic. The output characteristic is discussed in Sec. 6.9.

In Fig. 6.8-1a we are interested in the input volt-ampere characteristic at one emitter input. Other emitter inputs (only one additional input is shown in the figure) are assumed to be at logic level **1**. For the reasons presented in Sec. 6.7 we neglect the effect of these other emitter terminals.

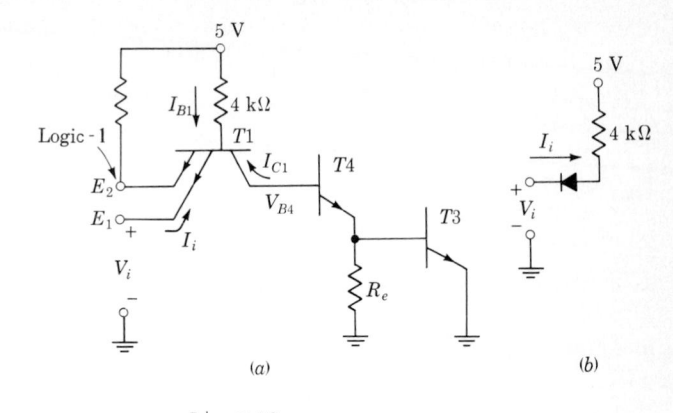

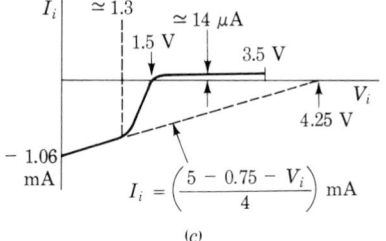

FIGURE 6.8-1
Input volt-ampere characteristic.

When $V_i = 0$, $T4$ is cut off and therefore the collector current $I_{C1} = 0$. In this case, neglecting the current contribution from E_2 (see Sec. 6.7), we can represent $T1$ as a diode. An equivalent circuit for calculating the current I_i is then shown in Fig. 6.8-1b. When we assume the usual 0.75 V junction voltage, $I_i = -(5 - 0.75)/(4 \text{ k}\Omega) = -1.06$ mA, as noted in the plot of Fig. 6.8-1c. As long as all the current through the 4-kΩ resistor flows through the E_1 junction, the volt-ampere characteristic is a straight line with current intercept -1.06 mA and voltage intercept $5 - 0.75 = 4.25$ V. This straight line is indicated (dashed) in the plot.

As V_i increases, the V_i-I_i plot departs from the straight line as $T4$ begins to turn on. When $I_i = 0$, all the current through the 4-kΩ resistor flows into the collector of $T1$ and into the base of $T4$. At this point both $T4$ and $T3$ will be in saturation, and the voltage V_{B4} will be 1.5 V. Since $I_i = 0$, the emitter current in $T1$ is zero and the collector-emitter voltage is given by Eq. (1.10-9):

$$V_{CE1} = -V_T \ln \frac{1}{\alpha_N} \qquad (6.8\text{-}1)$$

With $V_T = 25$ mV and $\alpha_N = 0.98$, we have $V_{CE1} \approx -0.5$ mV. Thus, the input voltage V_i is 0.5 mV less than $V_{B4} = 1.5$ V. Hence, we can approximate $V_i \approx 1.5$ V when $I_i = 0$. This point $V_i = 1.5$ V and $I_i = 0$ is noted in the plot of Fig. 6.8-1c.

We can also estimate the voltage at which the characteristic will begin to depart from the straight dashed line in Fig. 6.8-1c. This departure will begin when the base current of $T4$ becomes significant. We may judge that such will be the case when $T4$ and $T3$ are each in its active region. At this time $T1$ will be saturated, so that altogether we shall have $V_i = V_{BE3} + V_{BE4} - V_{CE1} \approx 0.7 + 0.7 - 0.1 = 1.3$ V.

With V_i greater than 1.5 V, the emitter current in $T1$ reverses direction and the transistor enters its inverse active region. The emitter E_1 then acts as a collector and $I_i = h_{FC} I_{B1}$. Using $h_{FC} \approx 0.02$ and $I_{B1} = 0.7$ mA, we have $I_i \approx 14$ μA. This current remains approximately constant as V_i increases to its maximum voltage.

6.9 OUTPUT VOLT-AMPERE CHARACTERISTIC OF THE TTL GATE

When the gate input in Fig. 6.5-1 is at logic **0**, transistors $T4$ and $T3$ are cut off. The gate output is now at logic **1**, and the gate is *sourcing* current, i.e., furnishing current to a load so that I_L is negative. When the gate input is at logic **1**, $T2$ is cut off, $T3$ is in saturation, and the gate will be *sinking* current, that is, I_L will be positive.

Consider first the case where the input is at logic level **0**, $T3$ is cut off, and the gate is sourcing current. Then the output characteristic of the gate can be deduced from Fig. 6.9-1, where $-I_L = I_S$ is the source current. Since, in this mode, transistor $T2$ is in the active region, the base current is $I_S/(h_{FE} + 1)$. Starting at the 5 V supply and taking account of the voltage drops across R_{c4}, across the base-emitter junction of $T2$ (0.65 V), and across the diode D (0.65 V), we find

$$V_o = 3.7 - \frac{R_{c4}}{h_{FE} + 1} I_S \qquad (6.9\text{-}1)$$

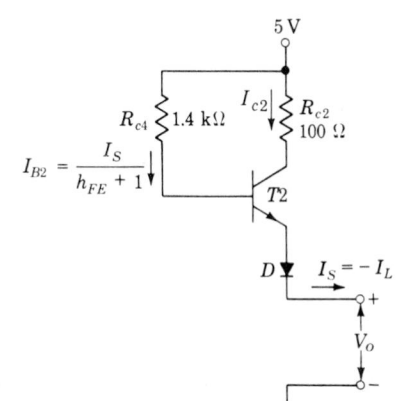

FIGURE 6.9-1
Equivalent circuit to find output volt-ampere characteristic when $T3$ is cut off.

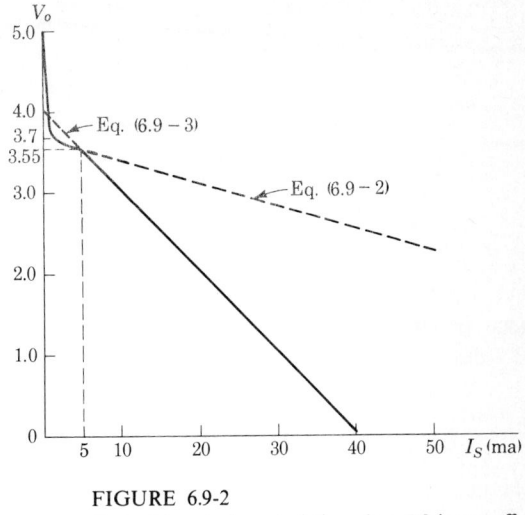

FIGURE 6.9-2

Output-input characteristics when $T3$ is cut off.

Thus, the output has a Thevenin representation which consists of a voltage of 3.7 V in series with an output impedance $R_{c4}/(h_{FE} + 1)$. With $R_{c4} = 1.4$ kΩ and using $h_{FE} + 1 = 50$, we find that the output impedance is 28 Ω, so that

$$V_o = 3.7 - 28 I_S \qquad (6.9\text{-}2)$$

When transistor $T2$ is in saturation, $V_{BE2} = 0.75$ and $V_{CE2} = 0.2$ V. Also $V_D = 0.75$ V. In this case $I_{B2} = (3.5 - V_o)/1,400$ and $I_{C2} = (4.05 - V_o)/100$. Since $I_S = I_{C2} + I_{B2}$ and $I_{B2} \ll I_{C2}$, we have

$$V_o \approx 4.05 - 100 I_S \qquad (6.9\text{-}3)$$

The straight-line plots for Eqs. (6.9-2) and (6.9-3) intersect at $I_S \approx 5$ mA. Hence, for $I_S > 5$ mA the transistor is in saturation, and Eq. (6.9-3) applies, while for $I_S < 5$ mA the transistor is in the active region, and Eq. (6.9-2) applies. The output characteristic is then given by the solid-line plot of Fig. 6.9-2. At very low currents that plot must depart from the straight line of Eq. (6.9-2). For at $I_S = 0$, there need be no drop across R_{c2} nor any drop across the base-emitter junction of $T2$ or the diode D, and, in principle, V_o should become equal to $V_{CC} = 5.0$ V.

When all the inputs to the gate of Fig. 6.5-1 are at logic level **1**, $T2$ is cut off while $T3$ and $T4$ are in saturation. The output looks back directly across the saturated transistor $T3$, as in Fig. 6.9-3, which now *sinks* a current I_L. The volt-ampere characteristic at these output terminals is now precisely the common-emitter collector characteristic of the transistor corresponding to the base current I_{B3}. This characteristic is similar to that shown in Fig. 1.10-1, the difference being that we now plot $V_o = V_{CE3}(\text{sat})$ as a function of $I_L = \sigma h_{FE} I_{B3}$ rather than as a function of σ. We calculate I_{B3} as follows. Referring to Fig. 6.5-1, we have

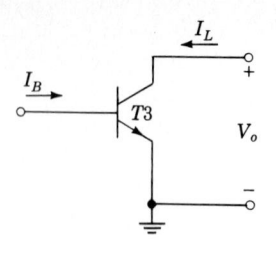

FIGURE 6.9-3
Equivalent circuit for calculating output
volt-ampere characteristic when $T3$ is
saturated ($T2$ is cut off).

(see Sec. 6.5) $I_{B1} \approx 0.7$ mA and therefore $I_{B4} \approx 0.7$ mA. With $T4$ saturated and $T2$ cut off,

$$I_{C4} = \frac{V_{CC} - V_{BE3} - V_{CE4}(\text{sat})}{R_{c4}} = \frac{5 - (0.75 + 0.2)}{1.4 \text{ k}\Omega} \approx 2.9 \text{ mA} \qquad (6.9\text{-}4)$$

Hence
$$I_{E4} = I_{B4} + I_{C4} = 0.7 + 2.9 = 3.6 \text{ mA} \qquad (6.9\text{-}5)$$

Since the current in resistor R_e is

$$I_{R_e} = \frac{0.75}{1 \text{ k}\Omega} = 0.75 \text{ mA} \qquad (6.9\text{-}6)$$

we have

$$I_{B3} = I_{E4} - I_{R_e} = 3.6 - 0.75 = 2.85 \text{ mA} \qquad (6.9\text{-}7)$$

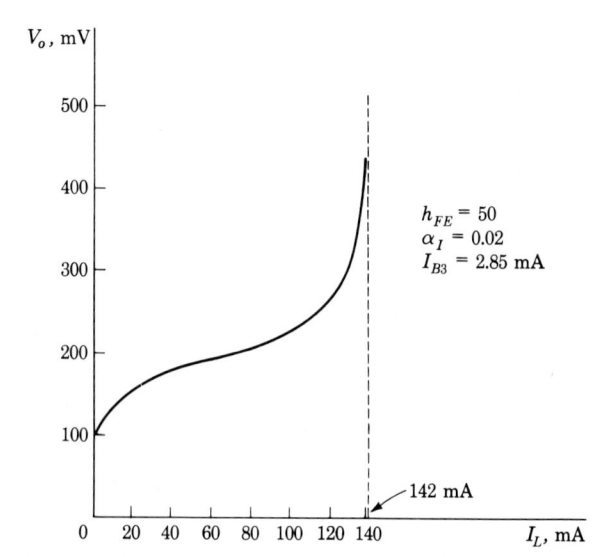

FIGURE 6.9-4
Output volt-ampere characteristic when all inputs to a TTL gate are in state **1**.

A typical TTL output characteristic is shown in Fig. 6.9-4. This characteristic corresponds to a base current $I_{B3} = 2.85$ mA as given by Eq. (6.9-7). At low currents the output is $V_{CE}(\text{sat}) \approx 0.1$ V, the value we have associated with an unloaded transistor in saturation. At higher currents the drop across the saturated transistor increases. Finally, at a current in the neighborhood of $I_L = h_{FE} I_{B3} = 50(2.85 \times 10^{-3}) = 142$ mA, transistor $T2$ comes out of saturation.

6.10 MANUFACTURER'S DATA AND SPECIFICATIONS; TEMPERATURE DEPENDENCE AND NOISE IMMUNITY

As with other gates, the characteristics of TTL gates are temperature-dependent. As before, the principal source of this dependence is the temperature dependence of the base-emitter and base-collector junction voltages. Because of the similarity with calculations already made, these calculations of temperature effects are left for the problems. Instead we shall take note of this temperature dependence as it is evidenced in typical average characteristics published by manufacturers.

Figures 6.10-1 to 6.10-4 all apply to the Texas Instrument type 54/74 gate. The type 74 gates are intended for the ambient temperature range 0 to 70°C and allow for a supply-voltage range from 4.75 to 5.25 V. The type 54 are held to tighter tolerances, are intended for a temperature range -55 to 125°C, and allow a supply range from -4.5 to 5.5 V. This gate is a "standard" type TTL NAND gate and represents a useful compromise between good speed and modest power dissipation. Figure 6.10-1 gives typical transfer characteristics for a variety

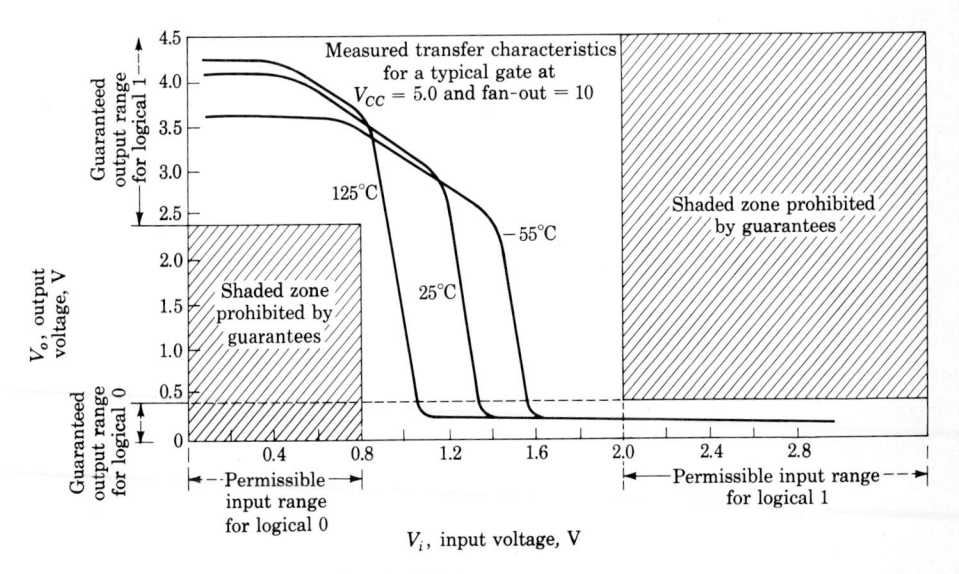

FIGURE 6.10-1
Voltage transfer characteristics for typical SN54/74 TTL NAND gate.

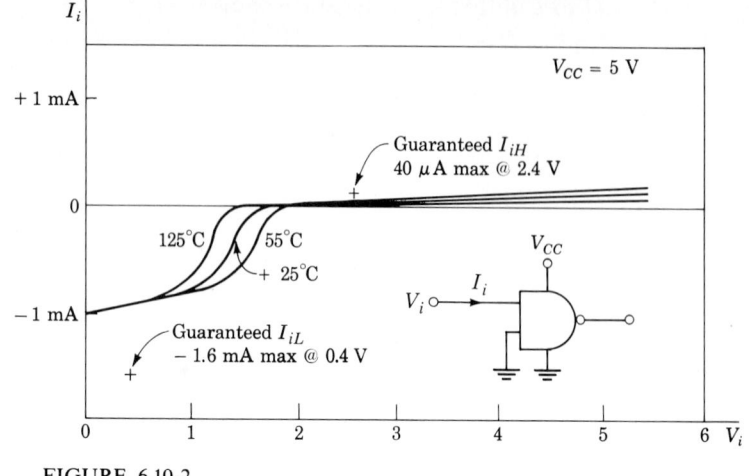

FIGURE 6.10-2
Input characteristics as a function of temperature for standard series SN54/75.

of temperatures and for a fan-out of 10.   Note that the manufacturer guarantees that any input voltage greater than 2.0 V will be acknowledged by the gate to correspond to the logic level **1**; that is, $V_{iH} = 2.0$.   Similarly it is guaranteed that the gate output at logic level **1** will never be less than 2.4 V; that is, $V_{oH} = 2.4$ V.   Thus, the **Δ1** noise margin is

$$\Delta 1 = V_{oH} - V_{iH} = 2.4 - 2.0 = 0.4 \text{ V} \tag{6.10-1}$$

We also note that $V_{iL} = 0.8$ V and that $V_{oL} = 0.4$ V, yielding

$$\Delta 0 = V_{iL} - V_{oL} = 0.8 - 0.4 = 0.4 \text{ V} \tag{6.10-2}$$

These noise margins are, as expected, extremely conservative.

Fan-out   The fan-out is limited by the amount of current $T3$ (Fig. 6.5-1) can sink when it is in saturation.   We find from Fig. 6.10-2 that when a driving-gate output is at logic level **0**, it must sink about 1.0 mA from each driven gate.   This result is verified in Fig. 6.5-1, where, with $T3$ saturated,

$$I_{11} = \frac{V_{CC} - V_{BE}(T11) - V_o}{R_b} = \frac{5 - 0.75 - 0.2}{4 \text{ k}\Omega} \approx 1 \text{ mA}$$

We note also from Fig. 6.10-3 that the ability of the gate to sink current while keeping $T3$ in saturation is most severely limited at the lowest temperature, $-55°C$.   Even at this temperature we note that $T3$ does not leave saturation except for currents in excess of 30 mA.   We might then estimate a fan-out capability of $30/1.0 = 30$.   We find however, that the manufacturer recommends a fan-out of only 10 to keep V_{oL} well below 0.4 V.   Further the fan-out affects propagation delay time as well as saturation.

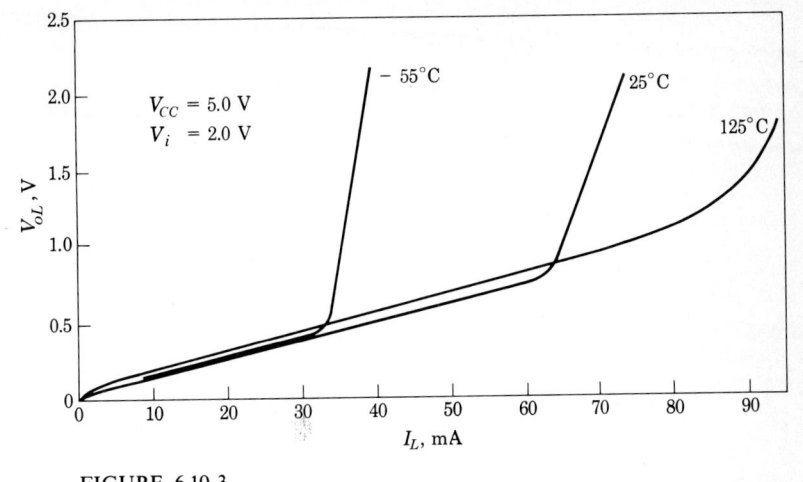

FIGURE 6.10-3
Output volt-ampere characteristic when all inputs are in **1**-state ($T3$ saturated).

Figure 6.10-4 is a plot of the output voltage V_{oH} when the gate is in the high state, as a function of the sourcing current, I_S. Note that if $V_{oH} > 2.5$ V $I_S < 11$ mA at room temperature.

Propagation delay time Propagation delays in TTL gates are defined and measured in nuch the same manner as with other gates. The times $t_{pd}(HL)$ and $t_{pd}(LH)$ are as defined in Fig. 6.10-5a. Here we see that the times are

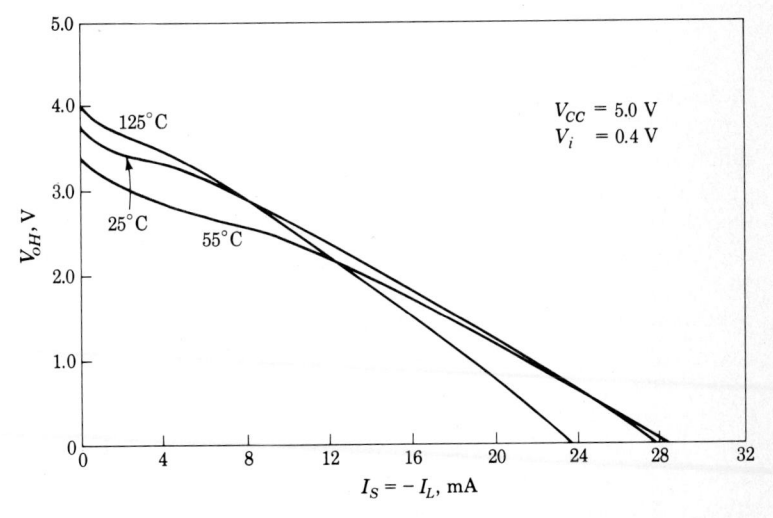

FIGURE 6.10-4
Output volt-ampere characteristic when V_i is in the **0** state ($T3$ cut off).

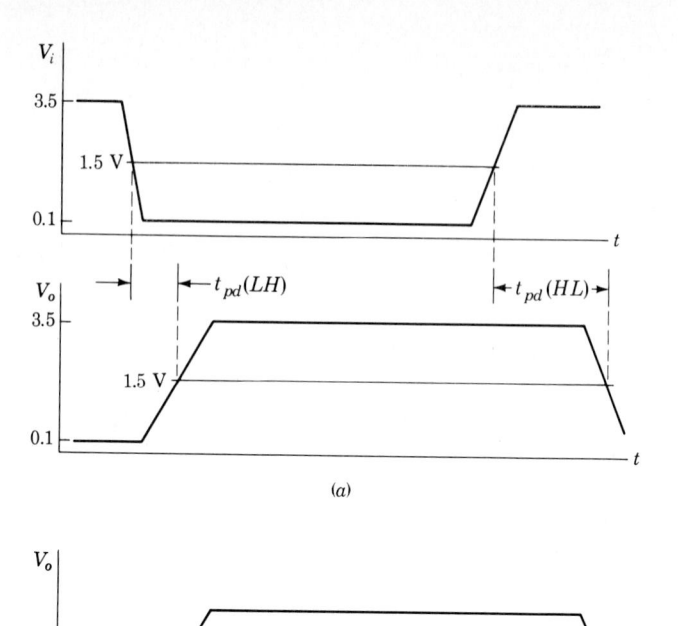

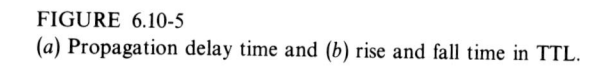

FIGURE 6.10-5
(a) Propagation delay time and (b) rise and fall time in TTL.

measured when the input and output voltages are each equal to 1.5 V. The voltage value is not standard, each manufacturer defining the delay time somewhat differently.

The propagation delay times are a function of capacitive loading and therefore of the fan-out. With a capacitive load of the order of 15 pF and a fan-out of 10, the average propagation delay time $t_{pd} \equiv \frac{1}{2}[t_{pd}(HL) + t_{pd}(LH)]$ is typically 10 ns. Of course, increased capacitive loading increases the propagation delay. Schottky TTL (see Sec. 6.13) gates have propagation delay times of 2 to 4 ns.

TTL gates are used at high speeds, where the *rise time* and the *fall time* are important. As noted in Fig. 6.10-5b, the rise and fall times of TTL gates are often measured between the 1- and 2-V points of the output voltage. Alternatively, sometimes the rise and fall times are defined as the time of transition between the 10 and 90 percent points of the output waveform. Typical values are $t_r = 8$ ns and $t_f = 5$ ns for a medium-speed TTL gate and 4 to 8 ns for a high-speed TTL gate. Schottky TTL gates have rise and fall times of about 3ns. ,

Because of its active pull-up, the TTL, like the RTL buffer (see Sec. 4.7), can accommodate a heavy capacitive load like that presented by 10 driven gates. When V_i drops to logic **0**, $T4$ cuts off and the base voltage of $T2$ rises abruptly. The capacitive load due to the fan-out keeps the output voltage V_o initially at 0.1 V. Hence, as is easily verified, $T2$ saturates and the current initially flowing out of the emitter of $T2$, $I_{E2} \approx 38.5$ mA. This large transient current (current spike) rapidly changes the load capacitance bringing V_o abruptly to the logic **1** level.

Unfortunately, it turns out that, when $T2$ turns ON and saturates, it will do so before $T3$ comes out of saturation. As a consequence a substantial portion of the current I_{E2} which should be available to charge the loading capacitance will instead be diverted into $T3$.

6.11 POWER-SUPPLY CURRENT DRAIN

The current drain from the power supply used with TTL gates is not the same in the two logic levels. The drain is larger for logic level **0** than for logic level **1**. In a typical case, say in the gate of Fig. 6.5-1, the steady-state drain is roughly 3 mA in one level and roughly 1 mA in the other (see Prob. 6.11-1). The difference depends principally on the fact that in one case $T4$ conducts and in the other is cut off. Of much more importance, however, is the fact that the current spike in I_{E2}, discussed in the preceding section, must be supplied by the power supply.

The exact waveform of the supply current depends both on the type of TTL gate involved and on the capacitive loading. Spikes range in amplitude up to about 40 mA, as noted, and have durations from several nanoseconds to many tens of nanoseconds in the case of heavy capacitive loading. A typical current waveform is shown in Fig. 6.11-1.

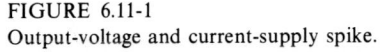

FIGURE 6.11-1
Output-voltage and current-supply spike.

These current spikes occur as frequently as the gate makes transitions, and hence the average current drawn is a function of the frequency at which the gate is being operated. It is not unusual for the average current drain to increase by a factor of 2 or 3 when the gate is being operated at some tens of megahertz.

If permitted to flow through the internal impedance of the power supply, these spikes of current would produce spikes of voltage, which would then be distributed throughout the system being serviced by the power supply. To avoid such a situation, it is necessary to provide capacitive bypassing to ground at the point where the power-supply lead is connected to the integrated-circuit chip.

6.12 TYPES OF TTL GATES

A number of types of TTL gates are available, differing principally in the compromise made between speed and power dissipation. Thus, the gate of Fig. 6.5-1 is considered a medium-speed gate and has an average power dissipation of about 10 mW and, as noted, a propagation delay of about 10 ns. The power dissipation can be decreased, at the expense of speed, by increasing the resistor values. Thus, we find that a commercially available gate with (see Fig. 6.5-1) $R_b = 40$ kΩ, $R_{c4} = 20$ kΩ, $R_e = 12$ kΩ, and $R_{c2} = 500$ Ω has a power dissipation of only 1 mW, but its average propagation time is 33 ns.

The speed of a gate can be increased by lowering resistor values. However, when speed is at a premium, additional changes are incorporated into the gate. A typical high-speed TTL gate is shown in Fig. 6.12-1. We note that R_b has been reduced to 2.4 kΩ, R_{c4} to 800 Ω, and R_{c2} to 60 Ω. We note three additional changes as well: (1) Diodes shunt each input to ground; (2) the emitter resistor of $T4$ has been replaced by a circuit consisting of an additional transistor $T5$ and two resistors; and (3) the connection from the collector of $T4$ to the base of $T2$ is no longer direct, as in Fig. 6.5-1, but is made instead through an emitter follower, forming a *Darlington amplifier*, composed of $T6$ and a 3.5-kΩ emitter resistor. We now consider each of these modifications in turn.

The input diodes Input diodes are common to all TTL gates, except the slowest gates. The diodes act as input "clamps" to suppress the ringing that results from the fast voltage transitions found in TTL systems. Consider, for example, that the output voltage of a TTL gate suddenly changes from the **1** level to the **0** level. The wire connecting this gate to a driven gate now carries this signal. If the wire, which acts like a transmission line, is not terminated properly, ringing results (see Appendix A), as illustrated in Fig. 6.12-2a.

The input diodes clamp the negative undershoot at approximately −0.75 V and absorb enough of the applied signal energy to prevent a large positive overshoot which might turn the gate ON again. This suppression of the ringing is illustrated in Fig. 6.12-2b.

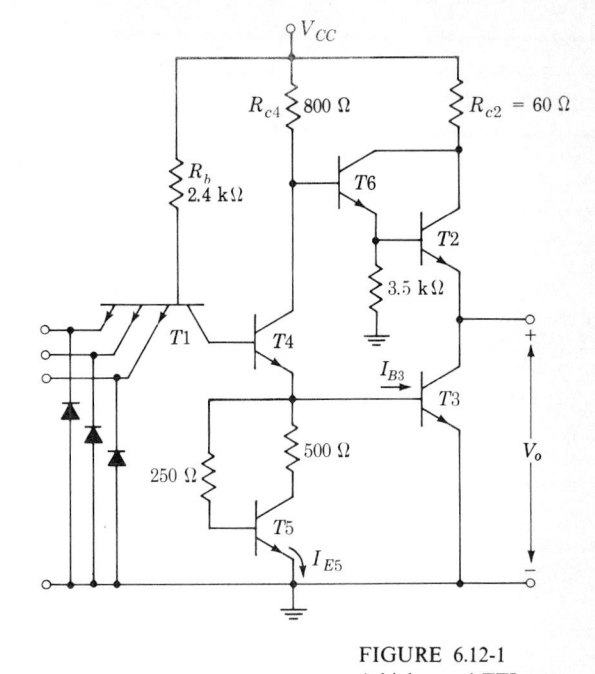

FIGURE 6.12-1
A high-speed TTL gate.

The Darlington circuit We consider now the effect of $T6$ in the circuit of Fig. 6.12-1. First, we may note, as a matter of incidental interest, that the diode D encountered in the circuit of Fig. 6.5-1 is not used in Fig. 6.12-1. It will be recalled that the diode was included to ensure that $T2$ will be cut off when $T4$ and $T3$ are saturated. In the configuration shown here the voltage drop across the base-emitter junction of $T6$ serves the same function provided by the diode voltage drop, and $T2$ remains cut off when $T3$ is ON.

Returning for a moment to the circuit of Fig. 6.5-1, we recall that when $T2$ is in its active region, the output resistance seen at the gate output terminal is approximately R_{c4}/h_{FE}, for which, in Sec. 6.9 [see Eq. (6.9-2)] we estimated the value 28 Ω. In Fig. 6.12-1, we would calculate the output resistance as follows. First the output resistance, seen looking back into the emitter of $T6$, is $R_{c3}/h_{FE} = 800/50 = 16$ Ω. Then, repeating the calculation for $T2$, we find that the gate output resistance is $16/h_{FE} = 16/50 \approx 0.3$ Ω. We have neglected the resistance of the transistor in these calculations. As a result, we find that in the gate of Fig. 6.12-1 the measured output resistance is more nearly 10 Ω. In any event, the important characteristic of the gate of Fig. 6.12-1, is that when $T6$ and $T2$ are conducting, the gate output resistance is substantially lower than in the circuit of Fig. 6.5-1. (We recognize, of course, that transistors $T6$ and $T2$ are in a Darlington configuration, one characteristic of which is precisely this reduced output resistance.) This lowered output impedance increases

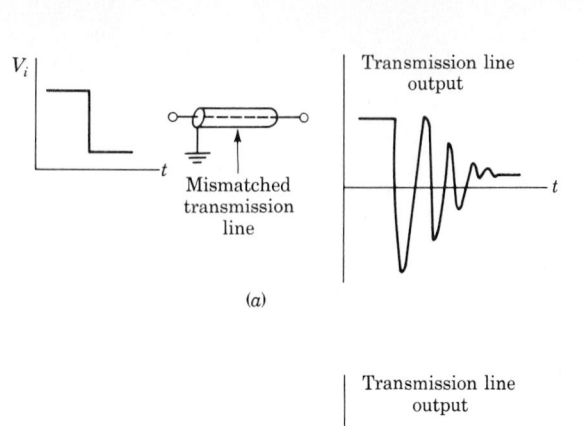

(a)

(b)

FIGURE 6.12-2
(a) Output of a transmission line to a negative-transition step, showing ringing.
(b) Output of the transmission line loaded by a diode.

the speed of operation of the gates. For with the lowered output resistance, any capacitance shunting the gate output will be able to charge more rapidly.

Of course, these considerations concerning output resistance do not apply when transistors $T6$ and $T2$ are cut off or in saturation. However, over a considerable range of the transition region from one logic level to the other, both transistors are in the active region, and the above discussion does apply. In this connection it is of interest to note that while in Fig. 6.5-1 transistor $T2$ goes to saturation when the gate output goes to logic level **1**, such is not the case in the circuit of Fig. 6.12-1. In a Darlington circuit only the input transistor ($T6$ in Fig. 6.12-1) and not the output transistor ($T2$) can be driven to saturation. For no matter whether $T6$ is saturated or in its active region, V_{CE6} is positive. Since $V_{CB2} = V_{CE6}$, the collector-base junction of $T2$ can never be forward-biased. Hence $T2$ can never saturate.

The active pull-down   The emitter resistor R_e in Fig. 6.5-1 provides the connection to ground for the base of the output transistor $T3$. Thus, R_e is the *pull-down* resistor which pulls the base of $T3$ down to ground when $T4$ cuts off. This terminology is analogous to use of the term *pull-up* resistor used in connection with collector resistors which are returned to the supply voltage. Extending the analogy, we shall refer to the circuit associated with $T5$ in Fig. 6.12-1 as an *active pull-down*. We consider now the advantages which accrue from the use of such an active pull-down.

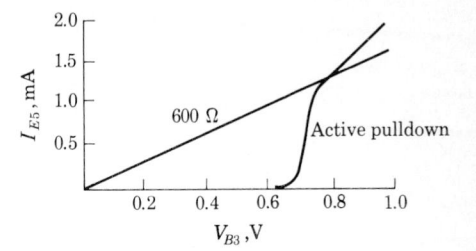

FIGURE 6.12-3
The effect of the active pull-down.

In Fig. 6.12-3, we compare the volt-ampere characteristics of a 600-Ω resistor and the active pull-down. Calculations to show that the curve representing the active volt-ampere pull-down characteristic is reasonable are left as a student exercise. In high-speed TTL gates which use the Darlington configuration ($T6$ and $T2$ in Fig. 6.12-1) but which use a passive pull-down, the pull-down resistor has the value 600 Ω; hence the selection of a 600-Ω resistor for comparison.

Now consider that transistor $T3$ is to be turned ON. It will go ON when its base voltage exceeds 0.65 V and will reach saturation when the base voltage is 0.75. We note from Figs. 6.12-1 and 6.12-3 that over this range of voltages the active pull-down diverts *less* current from the base of $T3$ that it would with a passive 600-Ω pull-down resistor. Hence, in this mode of operation $T5$ appears as a "resistor" which is larger than 600 Ω. As a result we have as a first advantage that $T3$ turns on faster.

To see a second advantage we note that when $T3$ is ON and the fan-out is small, the base current in $T3$ is substantially larger than is required to bring it to saturation. This excessive base current prolongs the storage time and delays the turnoff of the transistor. It also develops that $T3$ can be driven so far into saturation that its base-emitter voltage may well exceed even 0.8 V. In such a case, again to be noted in Fig. 6.12-3, the active pull-down diverts *more* current from the base than a passive pull-down would. Hence, in this mode the active pull-down appears as a smaller resistor than the 600-Ω resistance. Thus, excessive base current is somewhat suppressed.

In considering the *turnoff* of $T3$, we might imagine, however, that once the base voltage drops below 0.8 V, the active pull-down would be at a disadvantage, since it draws less current than a 600-Ω resistor out of the base of $T3$. Such, however, is not the case. We must keep in mind that the active pull-down characteristic plotted in Fig. 6.12-3 applies only in steady state. Actually, at turnoff, the active pull-down continues to draw the larger current for some few nanoseconds until the charge distribution within $T5$ itself has adjusted to the change.

The active pull-down also offers an important advantage in connection with the operation over a wide range of temperature. In a TTL gate without active pull-down, the turnoff time for $T3$ increases with increasing temperature because $T3$ is driven further into saturation. The reasons for the increases are twofold:

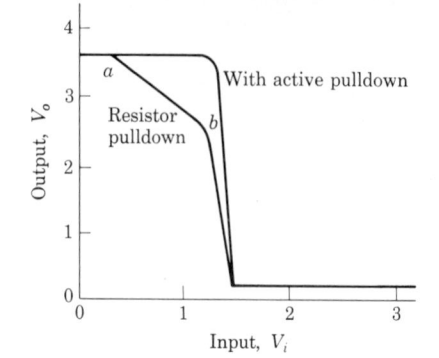

FIGURE 6.12-4
Comparison of transfer characteristics of TTL gates: resistor pull-down versus active pull-down.

(1) since junction voltages decrease with increasing temperature, with an increase in temperature more current will flow into the base of $T3$; (2) the current gain of a transistor increases with temperature. Hence, even a fixed base current will drive a transistor farther into saturation as the temperature increases. With higher temperature, however, the active pull-down shown in Fig. 6.12-1 actually draws more current, thereby keeping I_{B3} fairly constant. This compensation results since an increase in temperature decreases V_{BE5}, thereby increasing I_{B5} and I_{C5}. A calculation to determine the change in I_{B3} resulting from a change in temperature is left as a student exercise.

A final advantage of the active pull-down is to be seen in its effect on the input-output voltage characteristic of the TTL gate. Characteristics with active pull-down and resistor pull-down are compared in Fig. 6.12-4. With resistor pull-down the plot exhibits a region (between a and b), as discussed in Sec. 6.5, where the slope has a magnitude equal to the ratio of collector to emitter resistors of $T4$. In this region $T4$ is in its active region and provides gain. Point a marks the turn-on of $T4$, and point b marks the turn-on of transistor $T3$.

In the circuit of Fig. 6.12-1, until such time as $T3$ turns on, the active pull-down provides no path for the emitter current of $T4$. Thus, $T4$ and $T3$ go on simultaneously, and the region a-b is absent, with active pull-down. As a result the characteristic exhibits the much more abrupt change between levels, as indicated. The fact that the transition between logic levels is achieved with much smaller change in input voltage is, of course, of advantage in the matter of noise immunity.

6.13 SCHOTTKY TTL

In all TTL gates, transistors are driven into saturation and diodes are turned ON. A good part of the speed limitation of the gates results from the storage time delay associated with turning OFF and allowing transistors to recover from

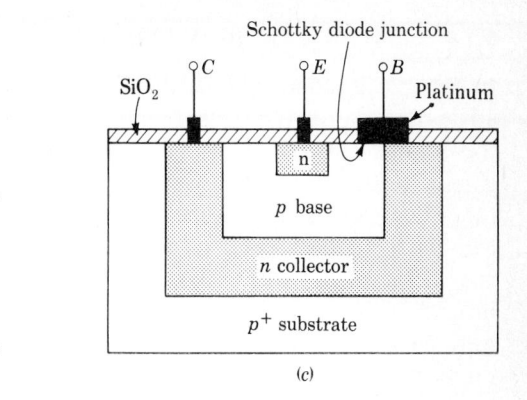

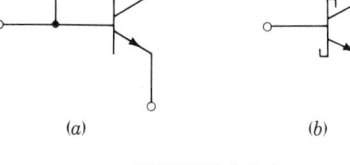

(a)　　　　　　　(b)　　　　　　　(c)

FIGURE 6.13-1

(a) Schottky diode connected from base to collector. (b) Schottky transistor. (c) Fabrication of the Schottky transistor.

saturation.　The use of Schottky diodes (Sec. 1.19) reduces the transistor turnoff time to negligible proportions.

In Sec. 1.19 we considered how a diode bridged between collector and base of a transistor can serve to restrain a transistor from entering saturation (see Fig. 1.19-1).　To review the matter, we note again that a transistor in saturation operates with a collector junction forward-biased to the extent of about 0.55 V or more.　The transistor may therefore be kept out of saturation by the addition of a *diode clamp*, which does not allow the collector junction to become adequately forward-biased.

A transistor with a Schottky-diode clamp is indicated in Fig. 6.13-1a. Such a diode-transistor combination is referred to as a *Schottky transistor* and generally represented by the figure shown in Fig. 6.13-1b.　The geometry (simplified and idealized) of an integrated-circuit Schottky transistor is shown in Fig. 6.13-1c.　Observe that the metal contact to the *n*-type collector has been allowed to overlap the *p*-type base, thereby creating a metal-semiconductor (Schottky) diode junction between the metal and the collector.　As discussed in Sec. 1.19, the base doping is graded so that the junction of the base and the metal is not rectifying.

There is available a family of high-speed TTL gates of which the schematic is as given in Fig. 6.12-1 but in which the input diodes are Schottky diodes and all transistors, except T2, are Schottky transistors.　An exception is made of T2 since, as we have noted, T2 does not saturate.　This Schottky family of gates has propagation times as low as 2 ns and rise and fall times in the range 2 to 3 ns. The power-supply current spikes are also reduced, being about 20 percent as large as those encountered in TTL gates that do not use Schottky transistors.　In these gates the improved speed results, in part, from the fact that Schottky transistors have areas which are roughly one-half the areas used in conventional TTL transistors.

6.14   OTHER LOGIC WITH TTL GATES

In Sec. 5.6 in connection with DTL NAND gates we noted the feasibility and advantages of the WIRED-AND connection. Such a connection, which directly ties the outputs of two or more gates together must be disallowed with TTL NAND gates. For consider that with the gates so joined, one of the gates is OFF while the other gate is ON. Then there would be a path from the supply voltage through a small collector resistor R_{c2} and through transistor $T2$ of the OFF gate into $T3$ of the ON gate. The current drawn would be approximately 40 mA, and instead of lasting a short time the current would last until the logic levels were changed. This would increase the drain on the power supply, but—most important—the dissipation in transistors $T2$ and $T3$ and the 100-Ω resistor would be excessive.

Still, the economy possible with the WIRED-AND connection is such that manufacturers find it advantageous to make available a modified TTL gate which allows such a connection. The modification consists in deleting the upper totem-pole transistor; i.e., these gates are intended to be used with an external passive pull-up, (resistor), and, of course, the advantages of an active pull-up are thereby relinquished.

While it is possible to perform all logic functions, it is often very convenient and economical to have available an AND gate. Of course, two NAND gates may be cascaded to achieve the AND function. However, gates are available in which a second inversion is incorporated within the gate itself and at low power level.

NOR gates and gates which perform the AND-OR-INVERT (AOI) function are also available in TTL logic. The circuitry of these gates is somewhat different from the circuity of NAND gates (see Probs. 6.14-1 to 6.14-4).

REFERENCE

1   Morris, R. L., and J. R. Miller, eds.: Designing with TTL Integrated Circuits, McGraw-Hill, New York, 1971.

EMITTER-COUPLED LOGIC

7.1 INTRODUCTION

All the other logic forms considered (RTL, DTL, TTL) suffer from a common and fundamental limitation on their speed of operation. This limitation occurs because in all these logic types transistors are driven into saturation, resulting in an increased propagation delay time. This consideration prompts us to inquire whether a logic form is possible in which transistors are switched from cutoff to an operating point in the active region. Certainly the forms so far considered cannot be operated in this way; for the range of base-emitter voltages in the active region extends over only some tens of millivolts, and there are so many variable factors (temperature variations, variability due to manufacture, etc.) to contend with. Suppose we arranged that at one logic level a transistor stage is operating in its active region. Then a small drift of the applied input voltage would be enough to drive the transistor either to cutoff or saturation with a correspondingly large change in the stage output voltage.

It is possible to establish a transistor in its active region, with stability, by introducing negative feedback through the simple expedient of using a large emitter resistor. This is precisely what is accomplished when using an emitter

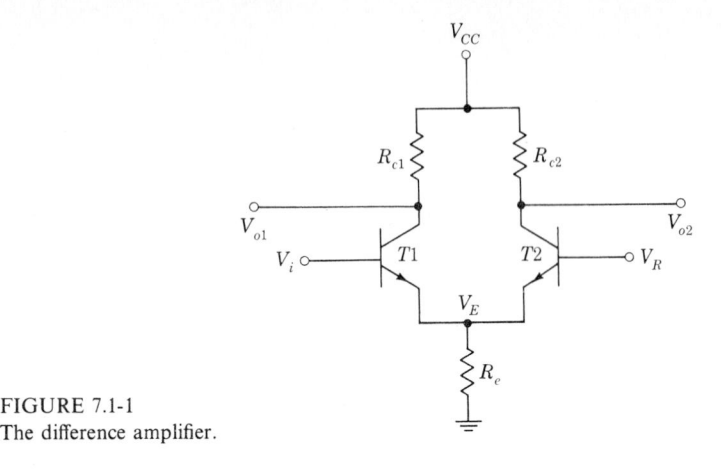

FIGURE 7.1-1
The difference amplifier.

follower or a phase splitter. The difficulty with this arrangement is that with a large emitter resistor, a large input-voltage swing at the base is required to carry the transistor from cutoff well into the saturation region. We can achieve both ends, however, i.e., active-region operation with stability and switching between this region and cutoff with a small input-voltage swing. These ends are accomplished by devising a circuit which will not turn a transistor current ON and OFF but will rather switch a current from one transistor to another.

The basic circuit configuration employed in the logic type under consideration, shown in Fig. 7.1-1, will, of course, be recognized as the *difference amplifier*, which was discussed in Sec. 2.1 in connection with its use in operational amplifiers and comparators. Since in this difference amplifier the emitters of the two transistors are connected, the logic family based on this circuit is referred to as *emitter-coupled logic*.

In the present application as a logic gate, the base of transistor $T2$ is held at a fixed reference voltage V_R while an input voltage V_i is applied to the base of transistor $T1$. When V_i is sufficiently lower than V_R, transistor $T1$ will be cut off and current will flow through $T2$. The reference voltage V_R and the resistors R_{c2} and R_e are selected to assure that $T2$ operates in its active region and is not saturated. When V_i rises to equal V_R, the currents in the two transistors will be nominally equal. Finally, as V_i continues to increase, the emitter voltage V_E increases, since $V_E = V_i - V_{BE1}$ and V_{BE1} is approximately constant, and eventually $T2$ will cut off. We now have $T1$ operating in its active region. In summary, then, it appears that a variation of V_i will switch the current from one transistor to the other. As a matter of fact, we shall show that as V_i changes from the point where $T1$ is cut off to the point where $T2$ is just cut off, the total emitter current through R_e changes less than 2 percent. Thus, the mechanism of operation consists in switching a nominally fixed emitter current from one transistor to the other.

7.2 THE ECL GATE

An ECL gate, incorporating the basic structure of Fig. 7.1-1, is shown in Fig. 7.2-1. The input transistor $T1$ in Fig. 7.2-1 is shown here paralled by a number of transistors to provide for multiple gate inputs.

Outputs are taken at the collectors through emitter followers. The emitter followers provide buffering and low impedance at the output terminals. An accurate, temperature-controlled reference voltage $V_R = -1.175$ V (at room temperature) is employed at the base of $T2$. The supply voltage employed is $-V_{EE} = -5.2$ V. Note that, contrary to the usual pattern, the collector resistors of the transistors are grounded and the emitter transistors are connected to a negative supply voltage. Thus, all the voltages encountered here will be negative. The reason for this reversal is discussed in Sec. 7.11.

As with other gates, ECL gates are commercially available in a number of types differing largely in the values of resistor components. Higher-resistance units dissipate less power and operate at a lower speed. Lower-resistance units dissipate more power but are faster. The gate with components and reference voltage in Fig. 7.2-1 is a Motorola unit of medium speed and power dissipation, designated as MECL II.

The ECL gate can operate as either a NOR gate or as an OR gate. It can be readily verified that if V_{o1} is taken as the output, a NOR gate is realized, while if V_{o2} is used, an OR gate results. In many an application either one or the other output will be used, while in some applications it is a great convenience to have available both outputs, which are, of course, complements of one another.

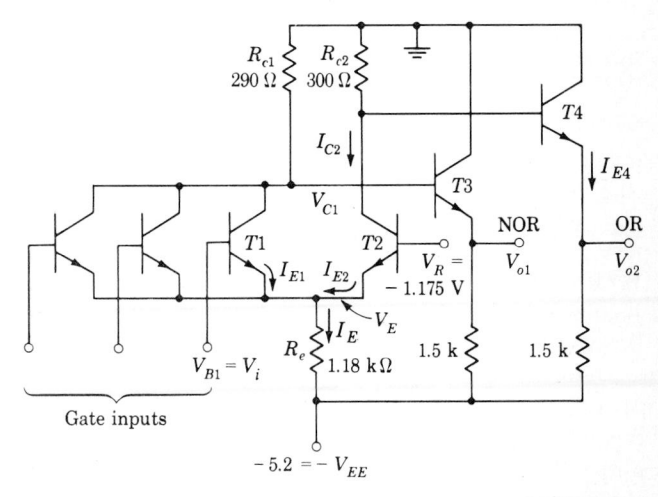

FIGURE 7.2-1
An ECL gate.

7.3   ECL-TRANSISTOR VOLTAGES

The saturation current of a transistor is $I_{sat} = [V_{CC} - V_{CE}(\text{sat})]/R$, in which V_{CC} is the supply voltage and R is the total resistance in series with the transistor. If, as usual, $V_{CE}(\text{sat}) \ll V_{CC}$, then $I_{sat} \approx V_{CC}/R$ rather independently of any transistor characteristic.   The base-emitter voltage $V_{BE} (= V_\sigma)$ required to cause this current I_{sat} to cross the emitter junction depends on the cross-sectional area of the emitter junction.   That is, at least at the onset of saturation.

$$I_{sat} = I_C \approx I_E = I_{E0} \, \epsilon^{V_\sigma/V_T} \tag{7.3-1}$$

Here I_C and I_E, are the collector and emitter currents, respectively, and I_{E0} is the emitter-junction reverse saturation current, which is proportional to the area of the emitter junction.   Hence, altogether V_σ depends on V_{CC}, R, and the emitter-junction area.

For the gates considered so far it has turned out that the supply voltages, resistor values, and the geometry of the integrated transistor have been such that using $V_\sigma \approx 0.75$ V gives reasonable agreement with measured voltages.   For ECL gates, however, a better value is $V_\sigma = 0.8$ V.   This higher voltage value is a result principally of the smaller physical dimensions of the ECL transistor, a feature which serves as well to reduce capacitance and hence improve speed.

Consistent with previous procedures we may again consider that cutoff is at a voltage about 100 mV lower than V_σ.   On this basis we have $V_\gamma = 0.70$ V. And, as has been our practice, we shall assume that when the transistor is in its *active* region, the base-emitter voltage, which we shall call V_{BEA}, is $V_{BEA} = 0.75$ V, that is, midway between V_γ and V_σ.   In summary, at room temperature, for the transistors in the ECL circuit of Fig. 7.2-1 we shall use

$$V_{BE} = \begin{cases} V_\gamma = 0.70 \text{ V} & \text{cut-in} \\ V_{BEA} = 0.75 \text{ V} & \text{active region} \\ V_\sigma = 0.80 \text{ V} & \text{saturation} \end{cases} \tag{7.3-2}$$

For other families of ECL gates corresponding voltages are slightly different. We shall neglect these differences.   In any event, as usual, we take the temperature sensitivity of these voltages to be -2 mV/°C.

7.4   TRANSFER CHARACTERISTIC: THE OR OUTPUT

With all except one input transistor cut off, the transfer characteristic between V_i and V_{o2} (the OR output) is as given in Fig. 7.4–1a.   We consider now how this characteristic comes about.

Referring to Fig. 7.2-1, we see that when V_i is sufficiently high, $T1$ will be ON and $T2$ will be OFF.   Let us initially neglect the voltage drop through R_{c2} due to the base current of $T4$.   Then $V_{C2} = V_{B4} = 0$ V, and allowing a voltage of 0.75 V from base to emitter of $T4$, we find $V_{o2} = -0.75$ V.   However, as we shall now see, the neglect of the drop through R_{c2} is not entirely justifiable, and a small correction is in order.

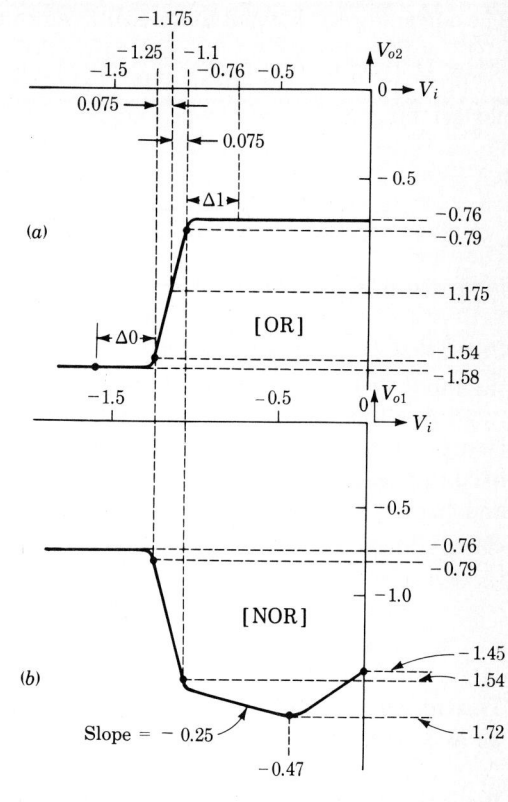

FIGURE 7.4-1
Transfer characteristic of an ECL gate.

The emitter current of $T4$ is

$$I_{E4} = \frac{-0.75 + 5.2}{1.5} \approx 3 \text{ mA} \qquad (7.4\text{-}1)$$

Transistors in ECL gates have current gains h_{FE} in the range from about 40 to 150. We use $h_{FE} = 100$ as a typical value. The base current of $T4$ is $I_{B4} = I_{E4}/(h_{FE} + 1) = 3/101 \approx 0.03$ mA. This base current flows through $R_{c2} = 300 \ \Omega$ and produces a voltage drop $300(-0.03 \times 10^{-3}) \approx 0.01$ V. Hence $V_{C2} = V_{B4} \approx -0.01$ V rather than 0 V, and correspondingly $V_{o2} \approx -0.76$, as indicated in Fig. 7.4-1a.

Now let V_i decrease so that $T1$ turns OFF and $T2$ enters its active region. The emitter-to-ground voltage V_E is then

$$V_E = V_R - V_{BEA}(T2) = -1.175 - 0.75 = -1.925 \text{ V} \qquad (7.4\text{-}2)$$

The emitter current is

$$I_E = \frac{V_E - (-V_{EE})}{R_e} = \frac{-1.925 + 5.2}{1.18} = 2.78 \text{ mA} \qquad (7.4\text{-}3)$$

The current in R_{c2} is equal to the emitter current I_E *diminished* by the base current of $T2$ and *augmented* by the base current of $T4$. Since these two base currents are both small in comparison with I_E and are nearly equal to each other, we can neglect the base currents when calculating V_{C2} in the logic **0** state. We find $V_{C2} = V_{B4} = -R_{c2} I_E = -300(2.78) = -0.83$ V. Finally the gate output voltage is

$$V_{o2} = V_{B4} - V_{BEA}(T4) = -0.83 - 0.75 = -1.58 \text{ V} \qquad (7.4\text{-}4)$$

The two logic levels $V_{o2} = -0.76$ V (logic **1**) and $V_{o2} = -1.58$ V (logic **0**) are indicated in Fig. 7.4-1a.

Proof that I_E changes by less than 2.0 percent   When transistor $T1$ goes from the point of just being cut off (all emitter current flowing through $T2$) to the point where it carries all the emitter current (no current through $T2$), the total change in emitter voltage is $\Delta V_E = 50$ mV. Since the base voltage of $T2$ is fixed, this change ΔV_E is all that is required to change V_{BE2} from 0.75 to 0.70 V and hence to carry $T2$ from its active region to cutoff [see (Eq. 7.3-2)]. The change in emitter current is $\Delta I_E = \Delta V_E/R_e = (50 \times 10^{-3})/(1.18 \text{ k}\Omega)$; from Eq. (7.4-2), $I_E = 2.78$ mA. Hence

$$\frac{\Delta I_E}{I_E} = \frac{(50 \times 10^{-3})/(1.18 \times 10^3)}{2.78 \times 10^{-3}} = 1.5\% \qquad (7.4\text{-}5)$$

This relative change is so small that we do not make a serious error when we assume that I_E is constant.

Transition width   As V_i increases, V_{o2} changes from -1.58 V (logic **0** assuming positive logic) to -0.76 V (logic **1**). This change in logic level occurs as the input V_i swings through a transition region, which, as we shall now show, is 150 mV in width. We have

$$I_{E1} + I_{E2} = I_E = I = \text{const} \qquad (7.4\text{-}6)$$

Over the transition region, $T1$ and $T2$ are both in their active regions; hence

$$I_{E1} \approx I_{E0}\,\epsilon^{V_{BE1}/V_T} = I_{E0}\,\epsilon^{(V_{B1} - V_E)/V_T} \qquad (7.4\text{-}7)$$

and
$$I_{E2} \approx I_{E0}\,\epsilon^{(V_R - V_E)/V_T} \qquad (7.4\text{-}8)$$

The ratio of emitter currents is

$$\frac{I_{E1}}{I_{E2}} = \epsilon^{(V_{B1} - V_R)/V_T} \qquad (7.4\text{-}9)$$

Combining Eq. (7.4-9) with Eq. (7.4-6), we have

$$I_{E1} = \frac{I}{1 + \epsilon^{(V_R - V_{B1})/V_T}} \qquad (7.4\text{-}10a)$$

and
$$I_{E2} = \frac{I}{1 + \epsilon^{(V_{B1} - V_R)/V_T}} \qquad (7.4\text{-}10b)$$

Suppose we define one edge of the transition region to correspond to the condition $I_{E1} = 0.05I$ and $I_{E2} = 0.95I$ while the other edge corresponds to $I_{E1} = 0.95I$ and $I_{E2} = 0.05I$. Then, as is easily verified, the total input voltage difference $\Delta V_{B1} = \Delta V_1$, corresponding to the total width of the transition region, is

$$\Delta V_1 = V_{B1}(I_{E1} = 0.95I) - V_{B1}(I_{E1} = 0.05I)$$
$$\approx 2V_T \ln 20 = 6V_T = 150 \text{ mV} \tag{7.4-11}$$

As appears in Fig. 7.4-1, this transition range is symmetrically disposed with respect to the reference voltage V_R. The limits of the transition region occur at $V_R + 0.075$ and $V_R - 0.075$ V at -1.1 and -1.25 V. The corresponding values of V_{o2} are -0.79 and -1.54 V.

Noise margin It is now of special interest to observe that the reference voltage and resistors in the ECL gate of Fig. 7.2-1 have been selected so that the gate output voltages symmetrically straddle the input-voltage transition region. We note that the mean of the output voltages is $\frac{1}{2}(-0.76 - 1.58) = -1.170$. This is very nearly equal to the reference voltage $V_R = -1.175$ V. As a result the noise margins are very nearly equal.

The output of a driving gate is -0.76 V at logic **1**. We note in Fig. 7.4-1 that in order for a driven gate to recognize an input as being at logic **1**, this input must not be less than -1.1 V. Hence the $\Delta\mathbf{1}$ noise margin is

$$\Delta\mathbf{1} = -0.76 - (-1.1) = 0.34 \text{ V} \tag{7.4-12}$$

Similarly the $\Delta\mathbf{0}$ noise margin is

$$\Delta\mathbf{0} = -1.25 - (-1.58) = 0.33 \text{ V} \tag{7.4-13}$$

It is to be noted that these noise margins are typical and not worst-case margins.

7.5 THE NOR OUTPUT

We have defined the edges of the transition region of the OR output as being the points where $I_{C2} \approx I_{E2} = 0.05I$ and $0.95I$, where I is the nominally constant current through R_e. Since the current $I = I_{E1} + I_{E2}$ is constant, $I_{C1} \approx I_{E1} = 0.05I$ when $I_{C2} \approx I_{E2} = 0.951I$ and vice versa. Hence, the transition points for the OR and NOR output occur for the same values of input V_i as shown in Fig. 7.4-1b. Further, if we neglect the small difference between R_{c1} and R_{c2}, the corresponding output voltages are also the same for OR and NOR outputs.

When V_i is low enough for $T1$ to be cut off, the output voltage is $V_{o1} = -V_{BEA}(T3) - I_{B3}R_{c1} = -0.75 - 0.01 = -0.76$ V, just as for the logic **1** level of the OR output. When, however, V_i has increased to the point where all the emitter current has transferred to $T1$, a further increase in V_i will result in a further increase in I_{C1} and the output V_{o1} will continue to fall. With $T2$ cut off, the gain A from input to collector of $T1$ is (see Sec. 6.5)

$$A = \frac{\Delta V_{C1}}{\Delta V_i} = -\frac{R_{c1}}{R_e} = -\frac{0.290}{1.18} = -0.25 \qquad (7.5\text{-}1)$$

Hence, as indicated in Fig. 7.4-1, V_{o1} falls with this negative slope, until transistor $T1$ begins to approach saturation.

We now estimate the input voltage V_i at which saturation begins to make itself felt. A transistor which is well into saturation has, as we have frequently noted, a collector-emitter voltage in the range 0.1 to 0.2 V. However, as shown in Fig. 1.10-1, when a transistor is just entering the region of saturation, the collector-emitter voltage is more nearly about 0.3 V. With 0.3 V between collector and emitter, the voltages V_{C1} and V_E are

$$V_{C1} = -(5.2 - 0.3)\frac{R_{c1}}{R_{c1} + R_e} = -\frac{4.9(.290)}{1.47} = -0.97 \text{ V} \qquad (7.5\text{-}2)$$

and
$$V_E = -0.97 - 0.3 = -1.27 \text{ V} \qquad (7.5\text{-}3)$$

The output voltage V_{o1} and its corresponding input voltage are then

$$V_{o1} = V_{C1} - V_{BE3} = -0.97 - 0.75 = -1.72 \text{ V} \qquad (7.5\text{-}4)$$

and
$$V_i = V_E + V_\sigma = -1.27 + 0.8 = -0.47 \text{ V} \qquad (7.5\text{-}5)$$

in which we have used 0.75 and 0.8 V, respectively, as the base-emitter drop for $T3$ (in the active region) and $T1$ (in saturation).

With a still further increase in V_i, the output V_{o1} begins to rise because additional emitter current in $T1$ is diverted to the base and away from the collector. When $V_i = 0$ V, $T1$ is deeply in saturation. Using 0.8 and 0.1 V, respectively, as the base-emitter voltage and the collector-emitter voltage of $T1$, we have $V_E = -0.8$ V and $V_{C1} = -0.8 + 0.1 = -0.7$ V. The output voltage is then

$$V_{o1} = -0.7 - 0.75 = -1.45 \text{ V} \qquad (7.5\text{-}6)$$

as indicated in Fig. 7.4-1b.

When in Fig. 7.2-1, $T1$ and some of its paralleling transistors conduct, the current in R_{c1} may be somewhat larger than the current in R_{c2} when $T2$ is ON. To reduce the corresponding difference in the logic **0** levels of OR and NOR outputs, R_{c1} is made smaller than R_{c2}.

7.6 MANUFACTURER'S SPECIFICATIONS: TRANSFER CHARACTERISTIC

The ECL circuit shown in Fig. 7.2-1 corresponds in all respects, component values, supply and reference voltages, etc., to the line of ECL logic manufactured by Motorola and designated MECL II. The plots of Fig. 7.6-1 are published by

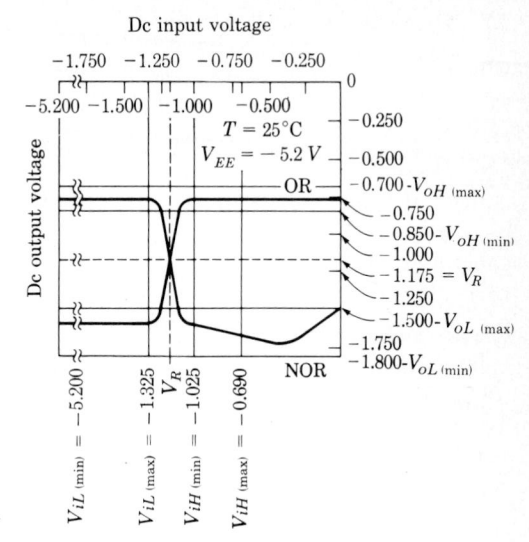

FIGURE 7.6-1
Typical ECL transfer characteristics.

Motorola as *typical, average* transfer characteristics for the gate. There is generally good agreement between the plots of Fig. 7.6-1 and the derived transfer characteristic given in Fig. 7.4-1.

The maximum range of variability to be anticipated in these gates is indicated in Fig. 7.6-1. Thus, while the available output voltage V_{oH} is nominally -0.76 V, the actual output may range from $V_{oH}(\text{max}) = -0.70$ V to $V_{oH}(\text{min}) = -0.85$ V. Similarly the output voltage V_{oL} ranges from $V_{oL}(\text{max}) = -1.5$ V to $V_{oL}(\text{min}) = -1.8$ V. The manufacturer specifies that an input voltage not more negative than $V_{iH}(\text{min}) = -1.025$ V is guaranteed to be to the right of the transition region and hence to be acknowledged by the gate as corresponding to a logic **1** input. The $\Delta 1$ margin, allowing for worst-case possibilities, is

$$\Delta 1 = V_{oH}(\text{min}) - V_{iH}(\text{min}) = -0.850 + 1.025 = 0.175 \text{ V} \qquad (7.6\text{-}1)$$

rather than the typical value 0.34 V given in Eq. (7.4-10). Similarly, the worst-case $\Delta 0$ margin is

$$\Delta 0 = V_{oL}(\text{max}) - V_{iL}(\text{max}) = 1.500 - 1.325 = 0.175 \text{ V} \qquad (7.6\text{-}2)$$

The $V_{iL}(\text{min}) = -5.2$ V is specified just to indicate that even if the most negative voltage available should happen to be impressed on the input, no disadvantage would result. The voltage $V_{iH}(\text{max}) = -0.690$ V is so specified to indicate that a voltage in excess of this value will cause transistor $T1$ to carry an unnecessarily large current which will eventually produce saturation.

Notice again that it is required (to assure equal noise margins $\Delta 0 = \Delta 1$) that the midpoints of the transition regions of the OR and NOR outputs occur at the coordinates $V_i = V_R$ and $V_{o2} = V_{o1} = V_R$.

7.7 FAN-OUT

When the output of a gate is at logic **0**, it need furnish no input current to the driven gate. Input current is required when the logic level is logic **1**, and the question of allowable fan-out arises.

We have seen that at logic **1** V_{o2} (or V_{o1}) is at -0.76 V when *no current* is being drawn. We have also noted that with $V_i = -0.76$ the $\Delta 1$ noise margin is $\Delta 1 = 0.34$ V [see Eq. (7.4-12)]. If, then, we fanned out to a number of gates, the output voltage V_{o2} would fall below -0.76 V and there would be a corresponding reduction in the noise margin. Thus, in the present ECL case (as for the RTL gates as well) the allowable fan-out is a continuous function of what we decide is an acceptable noise margin. This situation is different from that which prevails in the DTL and TTL gates, where the outputs were taken across a saturated transistor. In those cases, as long as the allowable fan-out is not exceeded, the output voltage of the output transistor remains essentially constant and so does the noise margin.

A fan-out calculation is given in the following illustrative example.

EXAMPLE 7.7-1 The output V_{o2} of the gate of Fig. 7.2-1 is to be fanned out to N similar gates, as shown in Fig. 7.7-1. Find N at room temperature if the $\Delta 1$ noise margin is to be 0.3 V. Assume the following worst-case conditions. The resistors of the driving stage are 20 percent higher than typical, $R_{c2} = 300(1.2) = 360\ \Omega$, the emitter resistor $= 1.5(1.2) = 1.8$ kΩ; the resistors of the driven stages are 20 percent lower than typical, $R_e = 1.18(0.8) = 940\ \Omega$. The supply voltage is 10 percent high, $V_{EE} = 5.2(1.1) = 5.7$ V. The transistors have current gains $h_{FE} = 40$. (These departures from typical values are all in the direction to reduce the fan-out.)

SOLUTION Refer to Fig. 7.7-1. At the edge of the transition region $V_i = V_{o2} = -1.1$ V, as indicated in Fig. 7.4-1. This voltage value is not affected by the change in the value of R_e. If the noise margin is to be 0.3 V, we require that $V_i = V_{o2} = -1.1 + 0.3 = -0.8$ V. Assuming, as usual, that $V_{BE\,A} = 0.75$, we have $V_E = -0.8 - 0.75 = -1.55$ V, $I_E = [-1.55 - (-5.7)]/940 = 4.4$ mA and

$$I_i = \frac{I_E}{h_{FE} + 1} = \frac{4.4}{41} = 107\ \mu A$$

Turning now to the driving stage, we have $V_{B4} = V_{o2} + 0.75 = -0.8 + 0.75 = -0.05$ V, $I_{B4} = 0.05/360 = 139\ \mu A$, $I_{E4} = (h_{FE} + 1)I_{B4} = 41(139) = 5.7$ mA. $I_4 = [-0.8 - (-5.7)]/1.8 = 2.7$ mA, so that

$$I_o = I_{E4} - I_4 = 5.7 - 2.7 = 3\ \text{mA}$$

The fan-out is

$$N = \frac{I_o}{I_i} = \frac{3,000}{107} = 28$$

The number $N = 28$ calculated in Example 7.7-1 is to be compared with $N = 25$ given by the manufacturers as a "dc fan-out." On the other hand, if

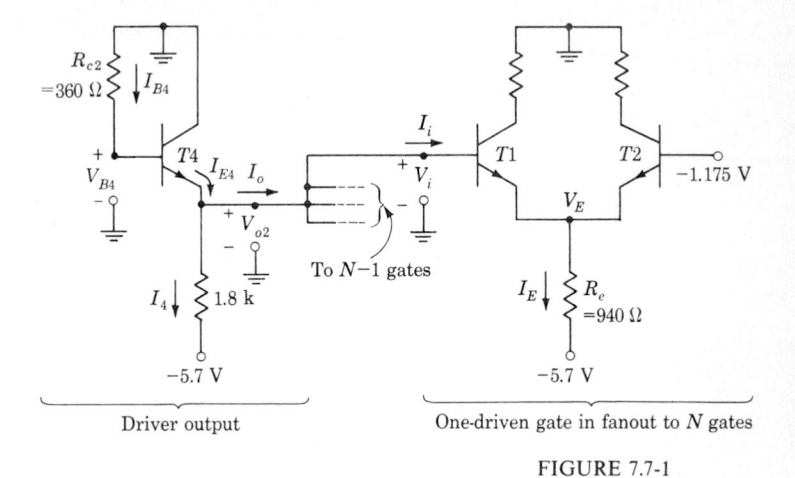

FIGURE 7.7-1
A calculation of fan-out.

we allow $\Delta 1$ to fall to 0.1 V, then, as can be verified (Prob. 7.7-1), N becomes $N \approx 250$. In any event, it is apparent that fan-out in ECL gates is no problem if our only concern is the availability of enough driving current. However, the principal merit of ECL logic is its high speed. This speed is adversely affected by increasing the fan-out since each additional loading gate increases the loading capacitance correspondingly. Hence, the manufacturer specifies as well an "ac fan-out." This fan-out, usually about 15, is the fan-out recommended to keep rise and fall times and propagation times below specified limits.

7.8 SPEED OF OPERATION

Of all the gates considered, ECL is the fastest. When unloaded, a MECL II gate will exhibit a propagation delay time of the order of 4 ns. An even faster gate (MECL III) has a typical propagation delay time of about 1 to 2 ns. However, the speed of an ECL gate is adversely affected by capacitive loading. This situation is a result of the characteristic of the emitter followers used at the gate outputs. For, when the base voltage of an emitter follower changes sharply in the direction to increase emitter current, the emitter follows. Any capacitance hanging across the output charges rapidly through the low output impedance ($\approx 6\,\Omega$ in MECL II) of the emitter follower. When, however, the input changes in the reverse direction, the emitter voltage remains fixed momentarily due to the coupled capacitor. Since the base voltage has dropped, the emitter follower cuts off and the capacitance must discharge through the relatively large emitter resistance ($-1.5\,\text{k}\Omega$ in Fig. 7.2-1). However, as we can now see, in the present ECL case, if the capacitive loading is moderate, the discharge time need not be excessively large because the logic levels are separated by a voltage difference which is small in comparison with the separation between the logic levels and the supply voltage.

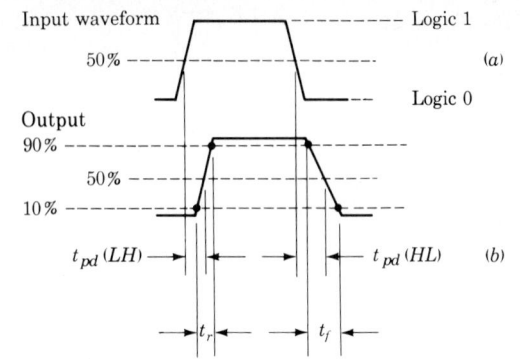

FIGURE 7.8-1
Propagation delay and rise and fall times.

For consider the OR output of the ECL gate. Initially the output is at logic level **1** at $V_{o2}(1) = -0.76$ V. When the emitter follower ($T4$ in Fig. 7.2-1) cuts off, the output starts to fall, heading asymptotically toward $-V_{EE} = -5.2$ V. Logic level **0** is reached when the output voltage becomes $V_{o2}(0) = -1.58$ V. The time T required to make the transition between logic levels can be shown to be

$$T = RC \ln\left[\frac{V_{o2}(1) - (-V_{EE})}{V_{o2}(0) - (-V_{EE})}\right] \tag{7.8-1}$$

Using the values given above, we find that

$$T \approx 0.2RC \tag{7.8-2}$$

Assuming a capacitive load $C = 5$ pF, we find $T = 1.5$ ns.

Consider then a gate input as in Fig. 7.8-1a. With a substantial capacitive load on the gate output, the OR output would exhibit the characteristic to be noted in Fig. 7.8-1b, where the output falls more slowly than it rises. Propagation times in ECL gates are measured at a voltage midway between the two logic levels, the 50 percent point noted in the figures. We observe that the propagation time for a negative-going output $t_{pd}(HL)$ is longer than $t_{pd}(LH)$, the propagation time for a positive-going output.

In MECL II, it turns out that the input capacitance of a gate averages about 3.3 pF. Allowing for the capacitance associated with wired interconnections, we reasonably consider that each added fan-out contributes 5 pF. At no fan-out, we find that $t_{pd}(HL)$ and $t_{pd}(LH)$ are about equal, each being approximately 3.5 ns. However, at a fan-out of 20, corresponding to a capacitive load of $5(20) = 100$ pF, the manufacturer informs us that $t_{pd}(LH)$ has increased to only about 5 ns while $t_{pd}(HL)$ has increased to about 18 ns. Similarly, we find that at no fan-out, the rise and fall times t_r and t_f (see Fig. 7.8-1) are 6 and 4 ns, respectively. At a fan-out of 20, t_r increases to about 8 ns while t_f increases to about 30 ns. The times $t_{pd}(HL)$ and t_f can be reduced, of course, by shunting the emitter resistor of the output emitter follower by an external resistor, at the expense, however, of increased power dissipation. In any event, it is apparent that if we wish to preserve the high speed inherent in ECL gates we must restrict the fan-out.

7.9 TEMPERATURE-COMPENSATED BIAS SUPPLY

In all of the preceding discussion we have assumed operation at a single temperature, 25°C. Of course, the transfer and other characteristics of ECL gates are temperature-dependent, as with other gates. And, as with the other gates, the principal source of the dependence is the temperature variation of the voltage drop across the forward-biased base-emitter junctions of the transistors.

There is one special point to be made in connection with the temperature variation of an ECL gate, which we shall now consider. We noted earlier that the ECL gate was designed so that the $\Delta 1$ and $\Delta 0$ noise margins should be approximately equal. If the reference voltage V_R is kept fixed, this condition can apply at only a single temperature. It is possible, however, to supply a reference voltage which is itself temperature-dependent so as to ensure that the symmetry of the noise margins is maintained over a wide range of temperatures. The bias supply circuit which accomplishes this for MECL II gates, and which is often incorporated directly on the integrated-circuit chip, is shown in Fig. 7.9-1. The operation of the circuit is analyzed below.

Assuming that diodes $D1$ and $D2$ operate at a forward bias of 0.70 V, at a temperature $T = 25°C$, we find (neglecting the base current in $T5$) that the base voltage of $T5$ is -0.425 V. Assuming also a drop of 0.75 V from the base to emitter of $T5$, we find that $V_R = -0.425 - 0.75 = -1.175$ V, as expected.

Let us now assume a temperature change ΔT. In this case each forward-biased junction voltage changes by the amount $\delta = -k \, \Delta T \ (k = 2 \text{ mV/°C})$. We calculate now the effect of such a change on the reference voltage V_R and on the output voltage levels $V_o(1)$ and $V_o(0)$ corresponding to logic **1** and logic **0**

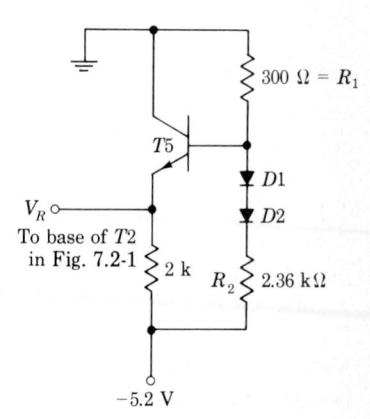

FIGURE 7.9-1
Reference supply circuit for ECL gates.

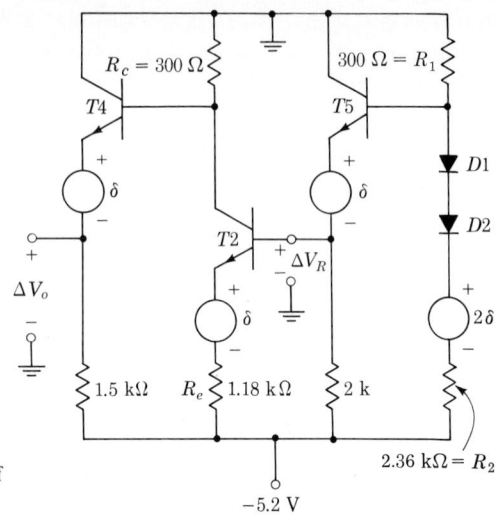

FIGURE 7.9-2
Equivalent circuit for calculating effect of
temperature variation.

levels of the OR output (a calculation for the NOR output is left for the
problems). The circuit of Fig. 7.9-2 includes the bias supply of Fig. 7.9-1 and
that part of the gate circuit of Fig. 7.2-1 which generates the OR output. For
ease of identification, the transistors in Fig. 7.9-2 have been given the same
designations as in Figs. 7.2-1 and 7.9-1. Generators (δ and 2δ), representing the
voltage increments introduced into the circuit due to a temperature change,
have also been included.

Assuming that the gain through the emitter follower ($T5$) is unity, we
calculate the change in reference voltage V_R to be

$$\Delta V_R = \frac{2\delta R_1}{R_1 + R_2} - \delta = \delta\left(-1 + \frac{2}{1 + R_2/R_1}\right)$$

$$= \delta\left(-1 + \frac{2}{1 + 2.36/0.3}\right) \approx -0.77\delta \qquad (7.9\text{-}1)$$

When $T2$ conducts, the output is at logic level **0**. The increment in this
level is, using Eq. (7.9-1),

$$\Delta V_o(0) = -\Delta V_R \frac{R_c}{R_e} + \delta \frac{R_c}{R_e} - \delta$$

$$= (0.77\delta + \delta)\frac{R_c}{R_e} - \delta$$

$$= \left(1.77 \frac{0.3}{1.18} - 1\right)\delta = -0.55\delta \qquad (7.9\text{-}2)$$

Note, that in Eqs. (7.9-1) and (7.9-2) only resistor ratios appear. This feature
is important because ratios of resistors can be held to much tighter tolerances

(≈ 2 percent) than the absolute values of resistors (≈ 20 percent). Finally when $T2$ is cut off and the output is at logic level **1**, the corresponding increment is

$$\Delta V_o(1) = -\delta \qquad (7.9\text{-}3)$$

From Eqs. (7.9-2) and (7.9-3) we can now calculate that the *average* increment of the two logic levels is

$$\frac{\Delta V_o(1) + \Delta V_o(0)}{3} = \frac{-\delta - 0.55\delta}{3} \approx -0.77\delta \qquad (7.9\text{-}4)$$

which is equal to the increment ΔV_R given by Eq. (7.9-1). Thus, as the temperature changes, the midpoint of the range from logic **0** to logic **1** changes by the same amount as the reference voltage V_R.

Since the reference voltage lies midway between the two output-voltage levels, independently of the temperature variation, the $\Delta \mathbf{1}$ and $\Delta \mathbf{0}$ noise immunities are the same. This therefore maximizes the noise immunity. However, the noise immunity does change with temperature (Prob. 7.9-2). Further, the above analysis neglected the effect of fan-out. It can be shown (see Prob. 7.9-3) that the result is relatively independent of fan-out.

The bias supply of Fig. 7.9-1 also provides a measure of compensation for variations in the supply voltage $-V_{EE}$. However, the compensation is not as exact as for temperature variations (see Prob. 7.9-5).

7.10 LOGIC VERSATILITY OF ECL GATES

As with DTL logic, it is possible to extend the logic capabilities of ECL gates simply by connecting the gate outputs together. For example, it can readily be verified that two or more emitter followers, operating into a common load,

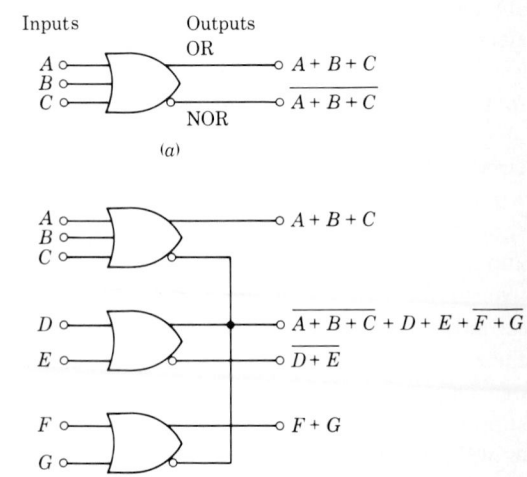

FIGURE 7.10-1
The WIRED-OR connection.

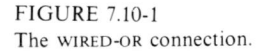

constitute an OR gate for positive logic.   Hence, if the outputs of ECL gates are joined, this connection provides the logical sum (OR operation) of the outputs that would otherwise appear at the individual gates.   Such a connection is therefore referred to as the WIRED-OR connection.

An example of the WIRED-OR connection is shown in Fig. 7.10-1.   In Fig. 7.10-1*a* is shown the symbol for the ECL gate.   Two outputs are shown. However, as a matter of convenience and to allow greater fan-out, some commercially available gates are provided with multiple OR and multiple NOR outputs.   On the other hand, some gates are provided with only one OR and one NOR output.   Figure 7.10-1*b* shows a WIRED-OR connection involving both OR and NOR outputs.

7.11   THE NEGATIVE SUPPLY VOLTAGE

In RTL, DTL, and TTL gates, the negative end of the power-supply voltage is grounded.   In ECL gates, as we have noted, it is common practice to ground the positive end of the supply.   We consider now the advantages of such an arrangement with ECL gates.

Let us initially put aside the question of grounding entirely and address ourselves to another matter.   In Fig. 7.11-1 we have drawn an ECL gate.   (To simplify the drawing we have omitted the part of the gate that provides the NOR output.   This simplification will have no bearing on our discussion.)   A supply voltage V_{CC} has been bridged across the gate.   It has been our practice to consider that the output voltage of the gate is the voltage V_o taken between the emitter of $T4$ and the positive side of the power supply.   But it readily appears that no fundamental change would be involved if we had chosen instead to take as the output voltage the voltage V_o' between emitter and negative side of the supply.   As a matter of fact, as far as the signal, i.e., voltage changes, is concerned, the positive and negative sides of the supply are the same electrical point.

Next, we must recognize that closed-circuit loops (one of which is indicated in Fig. 7.11-1*a*) are formed by the connection to V_{CC}.   Through these loops magnetic flux can thread, the flux being produced by the currents in the circuit shown or by currents in neighboring circuits.   This flux, when changing, will produce an electromagnetic field in the loops in which V_{CC} is included.   We should then really include in the power-supply loop (distributed) self-inductance and mutual inductance.   Equivalently, we have chosen, in Fig. 7.11-1*b* to include instead, in series with the supply, a "noise" source V_n which is to represent all induced voltages.   The characterization of V_n as noise is appropriate since these induced voltages will be random and unpredictable.   A particular source of noise which V_n may represent is the following.   Suppose that the power supply in Fig. 7.11-1 supplies not only the gate shown but also other circuits as well, which are paralleled across the supply.   As these other circuits respond to the changing logic levels of their input signal, the current they draw from the

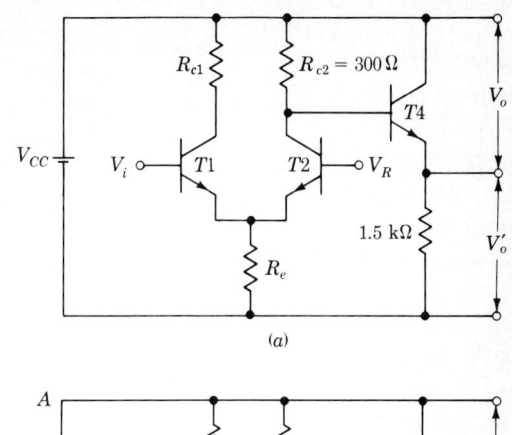

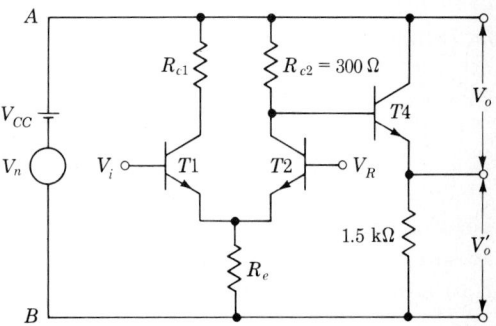

FIGURE 7.11-1
A negative supply configuration minimizes
external-noise transfer.

power supply will change either transiently or more permanently. In any
event, these current changes, flowing through the inductances or even through
the finite impedance of the power supply itself, will generate voltages which may
be represented by V_n.

It is now clear that the two sides of the power supply A and B are no
longer equivalent and V_o and V'_o similarly are no longer equivalent. The voltages
V_o and V'_o will reflect the noise to different extents. By way of example let us
assume that $T2$ is cut off. Then the circuit to calculate V_o and V'_o is as appears
in Fig. 7.11-1c. The impedance between collector and emitter of $T4$ is
$R_{c2}/(h_{FE} + 1) = 300/100 = 3 \ \Omega$ for $h_{FE} = 99$. Hence, $V_o = (3/1,500)V_n = 0.002V_n$

while $V'_o = (1,500/1,503)V_n \approx V_n$. [A corresponding calculation for the case where $T2$ is conducting is left as a problem (Prob. 7.11-2).] It is clear that the advantage lies with using V_o rather than V'_o.

It is well known that when a circuit is to operate in the proximity of a large mass of metal, it is advantageous electrically to connect this mass to some very common node in the circuit. The metal is then referred to as ground and provides a measure of shielding even through it need not physically intrude between the elements to be shielded from each other. This ground is often the chassis on the relay rack on which the circuit is built and will include the cabinet, if any, which houses the unit. It is also very common practice (though not necessarily universal) to arrange for the output terminals of the signal sources and the input terminals of signal-measuring devices (such as cathode-ray oscilloscopes, etc.) to use the ground as one signal terminal. This arrangement has the advantage that when units are interconnected, the chassis, cabinets, etc., can all be joined electrically.

Finally, returning to the ECL gate, it appears that we are initially at liberty to ground either the positive or negative side of the supply. Either would provide equivalent shielding. However, we find an advantage in using the positive side of the supply as one of the output terminals and a further advantage in using ground as one such terminal. Hence, altogether the practice is to ground the positive side of V_{CC}.

A second, less sophisticated reason for preferring the grounding arrangement normally employed with ECL is the possibility of an accidental short circuit developing between the output of a gate and ground. With the positive end of the supply grounded, as in Fig. 7.11-1, such a short places the entire 5.2-V supply across the 1.5-kΩ emitter resistor of the output emitter follower. The gate is able to tolerate such a short. On the other hand, with the negative end of the power supply grounded (Fig. 7.11-1), a short to ground from the gate output would place the entire supply voltage across the output transistor and at the same time apply the entire supply voltage to the transistor base through R_{c1} or R_{c2} (290 or 300 Ω). Such a short would promptly overheat and burn out transistor $T4$.

7.12 LEVEL TRANSLATION

It is often necessary to interconnect two different logic systems such as ECL and TTL (or DTL). One such application is the time-division multiplexing of M digital signals to form a single digital signal. Although the bit rate of each of the M signals may be handled using TTL, the bit rate of the composite signal is M times faster and may require ECL to process it.

Saturated logic to ECL translation The circuit of a commercial unit used to translate from saturated logic (TTL or DTL) to ECL is shown in Fig. 7.12-1. The circuit involving $T1$, $T2$, and $T4$ is recognized as the ECL gate in which

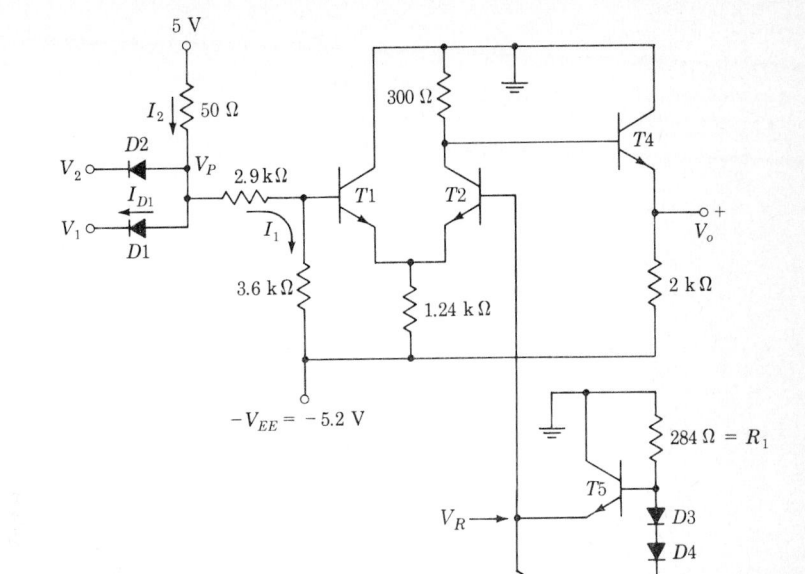

FIGURE 7.12-1
Saturated logic-to-ECL translator.

the transistor used to provide the NOR output ($T3$ in Fig. 7.2-1) has been omitted. The component values are slightly different from those which appear in the circuit of Fig. 7.2-1. The collector resistor of $T1$ has been omitted since no signal is taken from the collector. This omission has the advantage of assuring that $T1$ will never saturate. The circuitry associated with $T5$ provides the temperature-compensated reference voltage. When a single input variable is to be handled, one of the input diodes is returned to 5 V and is thereby cut off. If two logical variables are applied, then, as is readily verified, in addition to level translation the circuit performs the AND operation. The saturated logic levels of V_1 and V_2 are $V_{oL} \approx 0.2$ V and $V_{oH} \approx 3.5$ V (for TTL; DTL is generally higher). It is left as a student exercise (Prob. 7.12-1) to verify that these voltages, corresponding to logic **0** and logic **1**, respectively, will produce, at the base of $T1$, voltages, which correspond to logic **1** and logic **0** in ECL. Actually, the translation is accomplished by the input diodes and the three resistors bridged between $+5$ and -5.2 V. The rest of the circuitry provides temperature compensation and buffering to allow large fan-out.

The resulting transfer characteristic of the translator shown in Fig. 7.12-1 is given in Fig. 7.12-2.

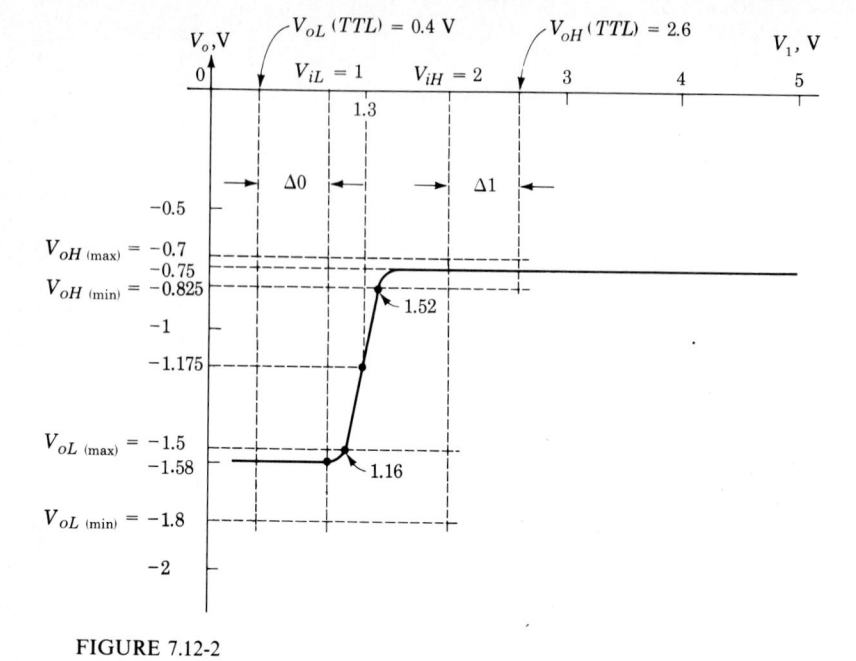

FIGURE 7.12-2
Transfer characteristic of translator shown in Fig. 7.12-1. $V_{iL} = 1$ V and $V_{iH} = 2$ V are manufacturer's specifications for the translator.

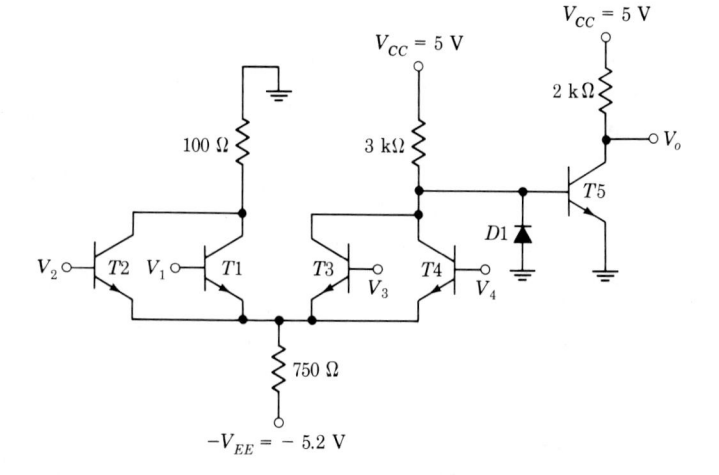

FIGURE 7.12-3
ECL-to-saturated-logic translator.

ECL-to-saturated-logic translation An ECL-to-saturated-logic translator is shown in Fig. 7.12-3. The circuit shown is capable of operating in either the OR or NOR logic mode. To provide the NOR operation V_4 is connected to the reference voltage V_R, and V_3 is returned to logic **0**. Then $V_0 = V_1$ NOR V_2. Similarly, if V_2 is connected to the reference voltage and V_1 is returned to logic **0**, $V_o = V_3$ OR V_4.

The operation of this translator is rather straightforward, the translation being performed in the collector of transistors $T3$ and $T4$ since the collector resistor is returned to 5 V rather than ground. Diode $D1$ is inserted, as in TTL, to damp out any ringing produced in the circuit. The details of these calculations and the determination of the transfer function are left for the problems.

7.13   ECL-GATE INTERCONNECTIONS

A pair of wires, a length of coaxial cable, etc., used to make interconnections between terminal pairs must, in principal at least, be viewed as a length of transmission line. The propagation of signals on transmission lines is considered in Appendix A. The transmission-line character of the interconnection makes itself especially apparent when the waveforms encountered make transitions between levels in times which are comparable to the time of propagation along the line. When the transition times are long in comparison to the time of propagation along the line, the line can be approximated by lumped-circuit elements.

Consider, then, the connection of the output of a driving ECL gate to the input of a driven gate. The emitter-follower driver has a low output impedance ($< 10\ \Omega$) while the input impedance of the driven gate may well be of the order of many thousands of ohms. In Appendix A, we consider the characteristic impedance of interconnecting wires of geometries such as might reasonably be found in electronic circuitry. We estimate that such impedance would be in the range of some tens to some hundred of ohms. Hence, as indicated in Fig. 7.13-1a, the interconnection between gates may reasonably be represented by a line of one-way delay t_d and characteristic impedance R_0. The sending-end termination is $R_s \ll R_0$, and the receiving-end termination is $R \gg R_0$.

If the input to the line V_i makes an abrupt transition between voltage levels, as indicated by the dashed waveform in Fig. 7.13-1b, the output V_o has the damped oscillatory waveforms shown by the solid plot. It is thus apparent that a transition at the driver-gate output from logic **0** to logic **1** might be interpreted at the driven gate as a sequence of several such transitions. If, on the other hand, the input waveform makes its transition in a time which is rather long in comparison with t_d, then, as shown in Fig. 7.13-1c, the output follows the input more closely; the oscillations of V_o about V_i are not nearly so pronounced.

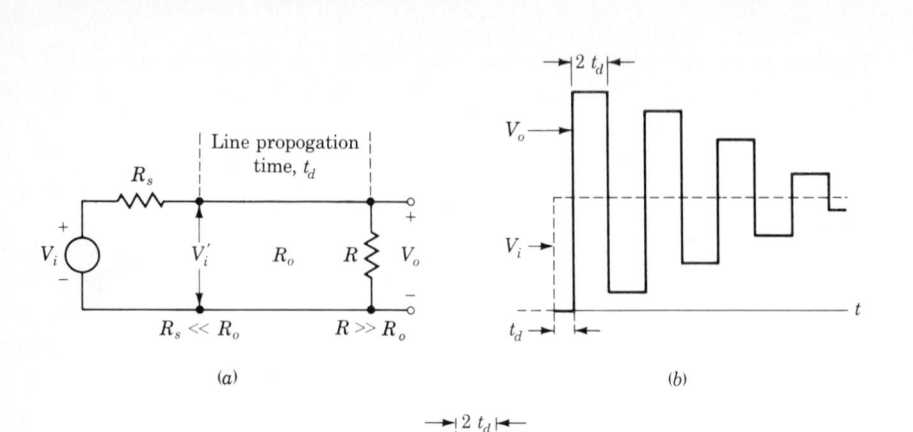

FIGURE 7.13-1
(a) A line of delay time t_d and characteristic impedance R_0. It is terminated at its sending end in $R_s \ll R_0$ and at its receiving end in $R \gg R_0$. (b) The response at the output for a step input. (c) The response for an input ramp which rises in a time long in comparison with t_d.

We have already noted oscillations of the type indicated in Fig. 7.13-1b in connection with TTL gates (see Sec. 6.12). In ECL the problem is even more urgent because of its higher speed. Transition times between logic levels of the order of 1 ns may occur in ECL. Depending on the dielectric constant of the insulation used to support the lines, 1-ns delays may occur in lines only 4 to 6 in long.

The oscillations can be suppressed by terminating the line at its receiving end in its characteristic impedance. In the absence of such termination, for an input logic swing of fixed rise time, the oscillation becomes more pronounced as the line gets longer. For this reason manufacturers generally provide guidance concerning the maximum allowable unterminated line lengths. This allowable length depends on the rise time, the fan-out, the characteristic impedance of the line, and the propagation time delay per unit length of line. By way of example, for Motorola type MECL 10,000, which has a rise time of 3.5 ns, using a *microstrip* line with $R = 50\ \Omega$, and with a fan-out of 1, the allowable length is 8.3 in. For Motorola type MECL III, which has a rise time of 1 ns, using again a microstrip line except with $R_o = 100\ \Omega$ and with a fan-out of 8, the allowable length is only 0.1 in.

A transmission-line interconnection between gates is indicated in Fig. 7.13-2.

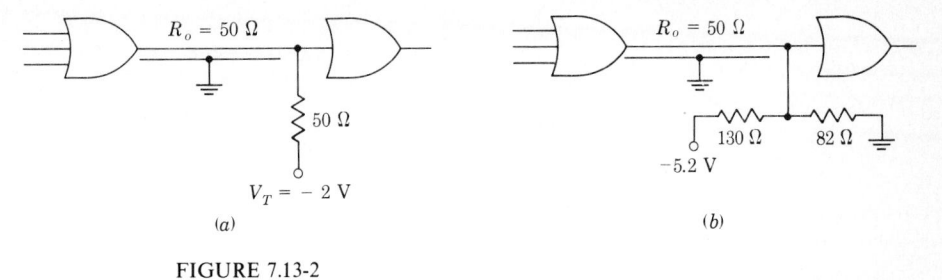

FIGURE 7.13-2
(*a*) A matched transmission-line interconnection between gates using an auxiliary supply voltage V_T for the termination. (*b*) An arrangement which avoids the auxiliary voltage.

Here a 50-Ω line and matching termination have been used. As far as dc operation is concerned, one end of this terminating resistor is connected to the emitter of the output emitter follower of the driving gate. If this terminating resistor were larger, it would be allowable to return the other end to the -5.2-V supply. In such a case the terminating resistor, would parallel the emitter resistor of the output emitter follower of the driving gate. In the present case, however, because of the small resistance of the termination, such a connection would result in excessive current through the output emitter follower, with consequent excessive dissipation in both transistor and termination. To circumvent this difficulty the terminating resistor is returned instead to an auxiliary terminating voltage V_T, which is commonly -2.0 V, as indicated in Fig. 7.13-2*a*. When the use to be made of the -2.0-V return is not adequate to justify a separate supply, the arrangement in Fig. 7.13-2*b* may be used. As can be verified, the -5.2-V supply and the 130- and 82-Ω resistors have a Thevenin equivalent replacement consisting of -2.0 V in series with 50 Ω. This latter arrangement limits the current in the output emitter follower but at the expense of considerable dissipation in the added resistors. Finally, we may note that in fast ECL gates, where it is virtually certain that a terminating arrangement as in Fig. 7.13-2 will be used, the manufacturer will often omit using an emitter resistor in the driving gate. In any event the output current diverted into the termination, and hence not available to drive gates, must be taken into account in estimating the allowable fan-out.

In Fig. 7.13-3 are shown a number of ways in which we can arrange the geometry of fan-out. In Fig. 7.13-3*a* all the driven gates are physically mounted very close to one another and are paralleled at the receiving end of a matched transmission line. In Fig. 7.13-3*b* the driving gate drives three transmission lines, two unmatched and one matched. The unmatched lines must be restricted in length in accordance with manufacturer's recommendations. In a representative case in an arrangement as in Fig. 7.13-3*b* we find with Motorola MECL III logic, using 50-Ω microstrip lines, the unterminated line with a fan-out of 1 may be no longer than 1.6 in. The other unterminated line, with a fan-out of 4 and

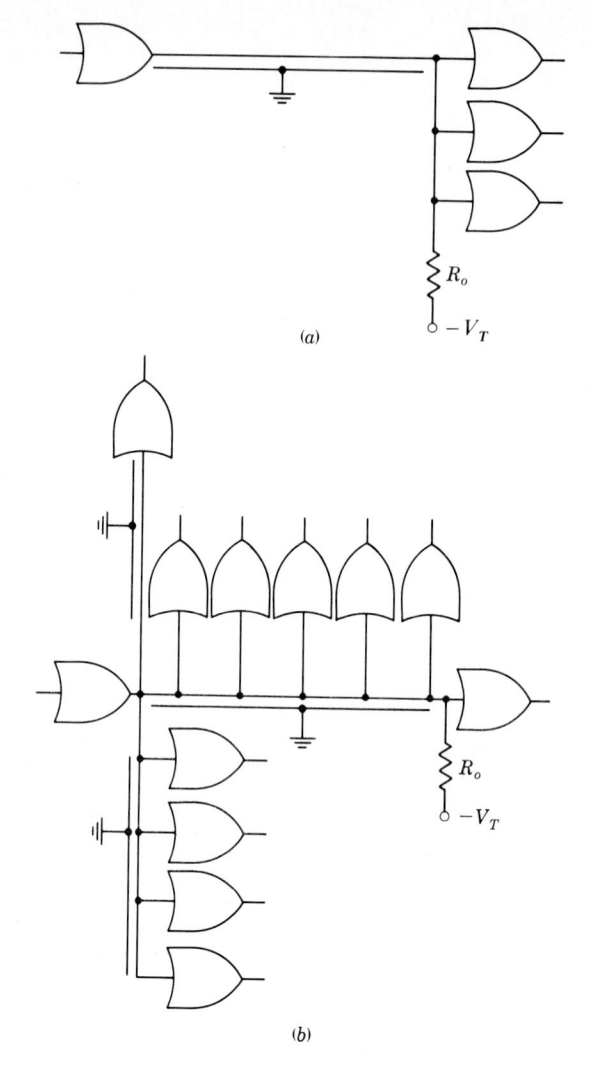

FIGURE 7.13-3
Fan-out arrangements for ECL gates: (*a*) the driven gates lumped at the end of a matched line; (*b*) gates distributed along matched and unmatched lines.

the load gates distributed nominally uniformly along the length of the line, the total length of the line may be no longer than 0.7 in.

In Fig. 7.13-3*b* the loads supplied by the matched line are distributed along the length of the line. A special consideration to be taken into account in connection with such an arrangement is discussed in the following illustrative example.

EXAMPLE 7.13-1 The matched line in Fig. 7.13-3*b* is 9 in long and drives 6 gates spaced at multiples of 1.5 in from the input end of the line. The line has capacitance per unit length and inductance per unit length $C = 2$ pF/in and $L = 0.02$ μH/in. The input capacitance of a gate is 5 pF. Estimate the value of the resistive termination required for the line.

SOLUTION Using Eq. (A.1-4), we find that the characteristic impedance of the line is

$$R_0 = \sqrt{\frac{L}{C}} = \sqrt{\frac{0.02 \times 10^{-6}}{2 \times 10^{-12}}} = 100 \ \Omega$$

The six gates have a total input capacitance of $6(5) = 30$ pF. Since the length of the line is 9 in, the gates add a capacitance per unit length $C' = 30/9 = 3.3$ μF/in. If this added capacitance were uniformly distributed along the line, the line impedance would be given exactly by Eq. (A.1-4) with C replaced by $C + C'$. While such is not the case, we may well expect that this replacement will nonetheless yield a good approximation. We then find

$$R_0 = \sqrt{\frac{0.02 \times 10^{-6}}{5.3 \times 10^{-12}}} = 61 \ \Omega$$

The propagation delay per unit length [the inverse of the velocity in Eq. (A.1-5)] is

$$t_{pd} = \sqrt{L(C + C')} = \sqrt{(0.02 \times 10^{-6})(5.3 \times 10^{-12})} \approx 0.3 \ \text{ns/in}$$

Series termination A line terminated at its receiving end is referred to as a *parallel-terminated line* since, as can be seen in Fig. 7.13-1, the matching resistor R is bridged across the line. A line may alternatively be matched at its input end by selecting $R_s = R_0$ in Fig. 7.13-1, in which case the line is referred to as a *series-terminated* line. It is feasible to provide such input matching, for the output impedance of a gate (see Fig. 7.2-1) is approximately $R_{c2}/h_{FE} = \frac{300}{50} = 6 \ \Omega$. We have noted that the line impedances are in the range 50 Ω to several hundred ohms. Hence, matching allows for the insertion of an additional resistor R_s in series with the output of the driving gate, as indicated in Fig. 7.13-4. Observe that series matching has the advantage that no auxiliary supply voltage is required. On the other hand, the series resistor limits the available output current and thereby restricts the fan-out to about 10.

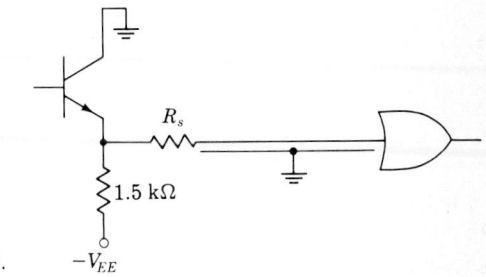

FIGURE 7.13-4
A line matched at its input end.

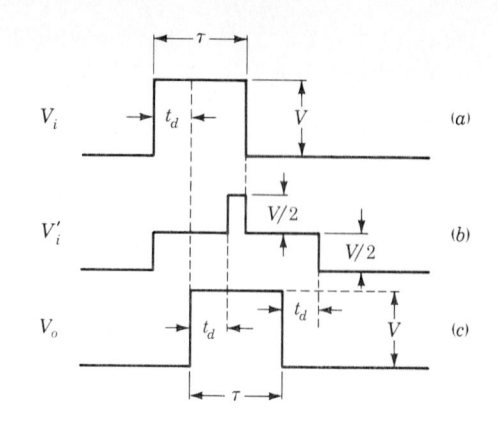

FIGURE 7.13-5

(a) A pulse of duration τ is applied to a line matched at its input end and open at its receiving end. Also shown is the waveform (b) at the input end and (c) at the receiving end of the line.

When driven gates are strung out on a parallel-terminated line, some difficulty may arise because signals from the driver arrive at different gates at different times. In series-terminated gates, of course, a similar situation prevails. However, an additional difficulty arises when employing the series-terminated line which does not occur in the parallel-terminated gate. Consider that in Fig. 7.13-1a, $R_s = R_0$, $R = \infty$, and V_i has the form indicated in Fig. 7.13-5a; that is, the driving gate changes its logic level, but the change persists only for a time τ. Suppose further that one of the driven gates is located at the input side of the

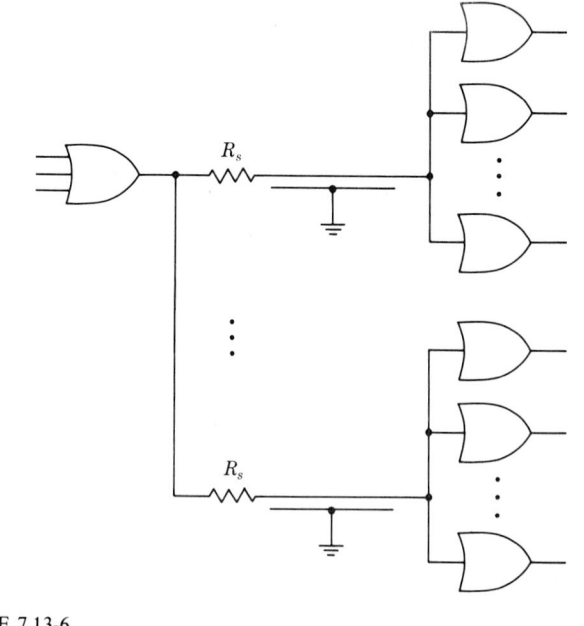

FIGURE 7.13-6

A number of series-matched lines used to accommodate a large fan-out.

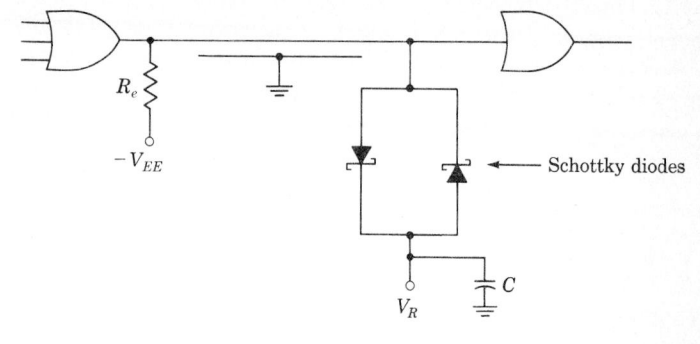

FIGURE 7.13-7
Diode termination of a line.

line, where the voltage is V'_i. It is left as a problem (Prob. 7.13-3) to verify that the input waveform and the waveform at the end of the line are given as in Fig. 7.13-5. We have drawn the waveforms for $\tau = 2.5t_d$, t_d being the one-way delay of the line. Note that the change in level in V_o persists for the same time τ as in V_i although there is a relative delay t_d. On the other hand, in V'_i, the full change persists only for a time $\tau - 2t_d$ and develops only after a time delay $2t_d$. If then it should happen that $\tau \leq 2t_d$, the full change in level would never appear in V'_i. The difficulty can be relieved by arranging the spacing between driven gates to be small in comparison with the spacing of the gates from the input side of the line. When many driven gates must be accommodated, the configuration of Fig. 7.13-6 is effective. Here a number of parallel lines are used, each with its own series termination.

Diode termination   An additional method of suppressing oscillations is indicated in Fig. 7.13-7. This method is a rather natural extension of the use of input diodes, encountered earlier in connection with TTL gates. Of course, Schottky diodes are called for. We have noted that the ECL levels symmetrically straddle the reference voltage (see Fig. 7.6-1). Hence, the diodes are returned to this reference voltage V_R. Schottky diodes which have a cut-in voltage $V_\gamma \approx 0.3$ V would allow the line voltage to swing freely through the range $V_R \pm V_\gamma$, while oscillations outside this range would be sharply damped.

Twisted pair lines   A second difficulty associated with the transmission of ECL waveforms is *crosstalk*, or unintended coupling of signals between circuits. Because of the speed of the signals, a large signal may be coupled from one signal path to another by a small stray capacitance or mutual inductance. Crosstalk can, of course, be minimized by using coaxial cables, but such cables are bulky and certainly do not readily allow a distribution of taps for driven gates.

   A feature of ECL gates which is of great use in suppressing crosstalk is the fact that gates have two outputs (OR and NOR) which are of opposite polarity.

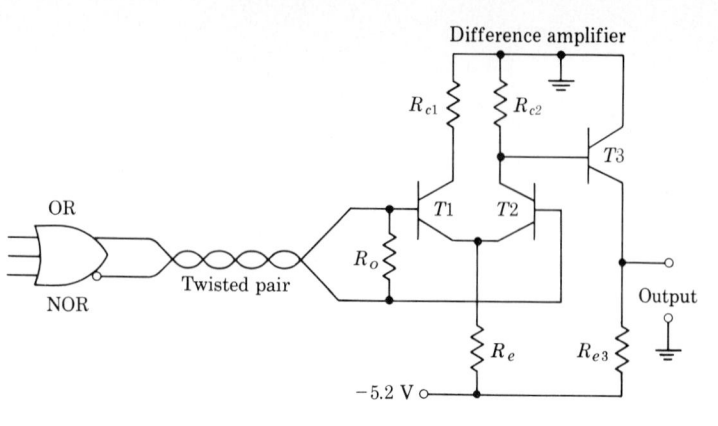

FIGURE 7.13-8
Twisted-pair transmission in ECL.

Whatever the change in voltage at one output, the change at the other output is equal and opposite. We take advantage of this feature as indicated in Fig. 7.13-8. The *difference* in output between OR and NOR output is transmitted over a twisted pair of wires to a *difference* amplifier. This difference voltage is nominally twice as large in amplitude as the signal available from the OR or the NOR output separately. The difference amplifier is normally made available by manufacturers for the present purpose and is generally referred to as a *receiver*. The twisting of the transmission wires keeps the wires together and also regularly reverses their relative positions. Hence, any signal path which might have induced in it a signal from one of the wires in the pair may well be expected to have an equal and opposite signal induced by the other wire. Hence, crosstalk from the twisted pair to other signal paths may be expected to be minimal. Similarly, crosstalk of other signals to the twisted pair will be introduced into the difference amplifier as a *common-mode* signal and hence will largely be restrained from appearing at the single-ended output. Of course, as appears, the twisted pair must be matched at its receiving end. A twisted pair may be used for transmission over many feet and may be used to distribute commonly used signals (such as clock waveforms) to many points. Generally, at each point when the signal is to be picked off the pair, a receiver will be required.

REFERENCE

1 Blood, W. R., Jr.: "MECL System Design Handbook," Motorola Semiconductor Products, Inc, Phoenix, Ariz., October 1971.

MOS GATES

MOS devices, as well as bipolar junction transistors (BJT), find application in logic gates. In this chapter we discuss the operation of PMOS (p-channel), NMOS (n-channel), and CMOS (complementary-symmetry) gates. CMOS is rapidly becoming the most favored because of its lower power dissipation, shorter propagation delay, and shorter rise and fall times.

8.1 ANALYTIC EQUATIONS FOR MOSFETS

Within a MOSFET, by definition, the charge carriers move away from the *source* and towards the *drain*. Therefore, in an n-channel device, where the carriers are negative, the conventional direction of current flow within the device is from drain to source. Thus the drain is positive with respect to the source, i.e., V_{DS} is positive as is also the current I_{DS}. Typical characteristics of n-channel MOSFETS are shown in Fig. 8.1-1. In Fig. 8.1-1a and b the MOSFET characteristics refer to an *enhancement* device. In such a device there is no channel between source and drain at $V_{GS} = 0$ V. No drain-to-source current I_{DS} flows until the gate-to-source voltage exceeds a threshold voltage V_T. This threshold

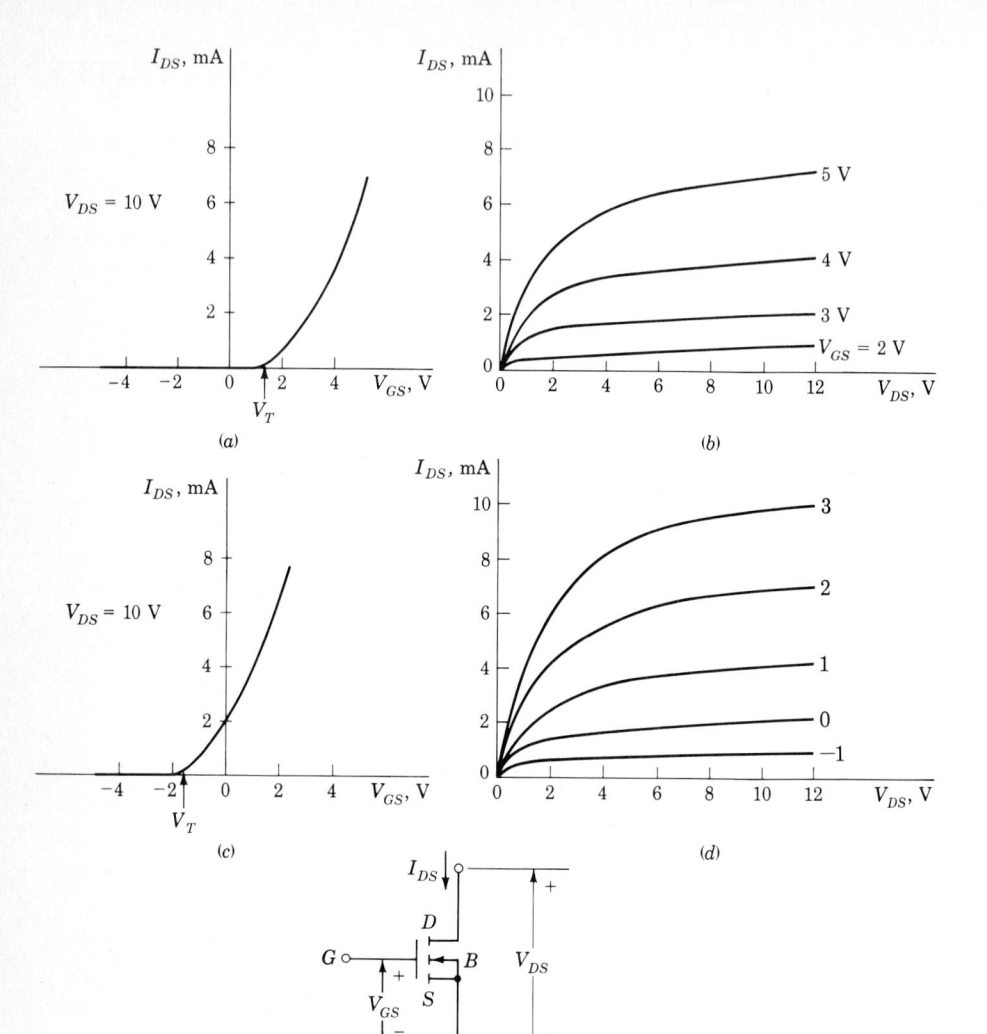

FIGURE 8.1-1

MOSFET characteristics. (*a* and *b*) Enhancement type. (*c* and *d*) Depletion type. (*e*) Defining voltages and currents. The device symbol represents an *n*-channel enhancement-type transistor.

voltage is of polarity which is the same as the polarity normally applied to the drain. Thus in an *n*-channel device, where V_{DS} is positive, so also is V_T, and a channel forms when $V_{GS} > V_T$.

In a *p*-channel device, where the carriers are positive, it is V_{SD} and I_{SD} (rather than V_{DS} and I_{DS}) which are positive. Furthermore, a channel forms to allow current I_{SD} to flow, when the source-to-gate voltage V_{SG} (rather than V_{GS}) exceeds a positive threshold voltage V_T, that is, when $V_{SG} > V_T$. We are using the symbol V_T with two meanings. In an n-channel device V_T is defined as a particular value of V_{GS}, while in a p-channel device V_T represents a particular value of V_{SG}. Where any confusion may result we shall use instead the symbols $V_T(n)$ and $V_T(p)$. The symbolism we are employing avoids inconvenient negative signs and absolute-value signs.

In Fig. 8.1-1c and d typical characteristics are shown (again for an n-channel device) for a *depletion* transistor. Here a channel exists when $V_{GS} = 0$, and the threshold voltage V_T is negative. Strictly, such an n-channel transistor operates in the depletion mode when V_{GS} is negative and in the enhancement mode when V_{GS} is positive. It is customary nonetheless to refer to such a device simply as a depletion MOSFET. Both enhancement and depletion transistors are used in logic gates.

Either transistor type (enhancement or depletion) may operate in the non-saturation region (also referred to as the *triode* region in fond memory of the days of vacuum tubes) or in the saturation region. In the triode region there is a continuous channel between source and drain, and I_{DS} varies "linearly" with V_{DS} for fixed V_{GS}. At the source, the channel depth is nominally proportional to the extent to which the gate-to-source voltage V_{GS} exceeds the threshold voltage V_T and is thus proportional to $V_{GS} - V_T$. For fixed V_{GS} the channel depth is fixed. At the drain, the channel depth is proportional to the extent to which the gate-to-drain voltage V_{GD} exceeds V_T. Hence at the drain the channel depth is proportional to $V_{GD} - V_T = V_{GS} - V_{DS} - V_T$. The channel is pinched off at the drain when $V_{GD} - V_T \leq 0$ or when

$$V_{DS} \geq V_{GS} - V_T \qquad (8.1\text{-}1)$$

When $V_{DS} \geq V_{GS} - V_T$, the transistor is in *saturation*. That is, because of the channel pinch-off, the current I_{DS} remains nearly constant, increasing only very slightly with increasing V_{DS}.

Just as we found it convenient to have analytic expressions for bipolar transistors (Ebers-Moll equations), so too is it useful to have analytic expressions for the MOSFET. In the triode region it is found that for an n-channel device

$$I_{DS} = k[2(V_{GS} - V_T)V_{DS} - V_{DS}^2] \qquad 0 \leq V_{DS} \leq V_{GS} - V_T \qquad (8.1\text{-}2)$$

In the saturation region

$$I_{DS} = k(V_{GS} - V_T)^2 \qquad 0 \leq V_{GS} - V_T \leq V_{DS} \qquad (8.1\text{-}3)$$

The constant k is given by

$$k = \frac{\mu \varepsilon}{2t} \frac{W}{L} \qquad (8.1\text{-}4)$$

where μ = mobility of carriers in channel (electrons in n-channel devices)
  ε = dielectric constant of oxide insulating layer
  t = thickness of oxide under gate
  W = channel width
  L = channel length

Typically, for n-channel devices $\mu\varepsilon/2t \approx 12 \ \mu\text{A/V}^2$ and for p-channel devices is smaller by about a factor of 3. The width-to-length ratio W/L may range from 0.1 for a load transistor to as high as 20 or 40 for a driver device. (See Sec. 1.13)

In a p-channel transistor, operating in the triode region the equations for the device current are more conveniently written in the form

$$I_{SD} = k[2(V_{SG} - V_T)V_{SD} - V_{SD}^2] \qquad 0 \le V_{SD} \le V_{SG} - V_T \qquad (8.1\text{-}5)$$

In the saturation region

$$I_{SD} = k(V_{SG} - V_T)^2 \qquad 0 \le V_{SG} - V_T \le V_{SD} \qquad (8.1\text{-}6)$$

These equations, like Eqs. (8.1-2) and (8.1-3), are approximations and do not include all effects which have an influence on device current; however, they are entirely adequate for our purposes of exploring the operation of FET logic gates.

8.2   TEMPERATURE EFFECTS

Equations (8.1-2), (8.1-3), (8.1-5), and (8.1-6) for the current I_{DS} (and I_{SD}) are affected by the temperature because both V_T, the threshold voltage, and the parameter k are temperature-sensitive. The temperature dependence of V_T is given approximately by

$$\frac{dV_T}{dT} \approx -2.5 \ \text{mV/}^\circ\text{C} \qquad (8.2\text{-}1)$$

The temperature sensitivity of k results almost entirely from the temperature sensitivity of μ [see Eq. (8.1-4)], the carrier mobility. The mobility decreases approximately inversely with the absolute temperature and hence so also does k. When there is a temperature increase, I_{DS} (or I_{SD}) increases because of the lowering of the magnitude of V_T and decreases because of the decreased carrier mobility. In a typical case we find that the effect of μ may be fivefold greater than the effect of V_T. The overall result is that generally the overall effect of a temperature increase is a decrease of current. In this respect the MOSFET differs from the bipolar transistor, where an increase in temperature increases the current both because the current gain h_{FE} increases and because the junction voltages decrease.

8.3   THE MOS INVERTER

As discussed in Chap. 1, the basic MOS switching-circuit configuration is an inverter which consists of a MOSFET switch driver driving a load which is itself a MOSFET device rather than a passive resistor. The driver is invariably an enhancement device since it is a great convenience that the driver be OFF when the gate voltage is at or near ground. When the driver is turned ON, it

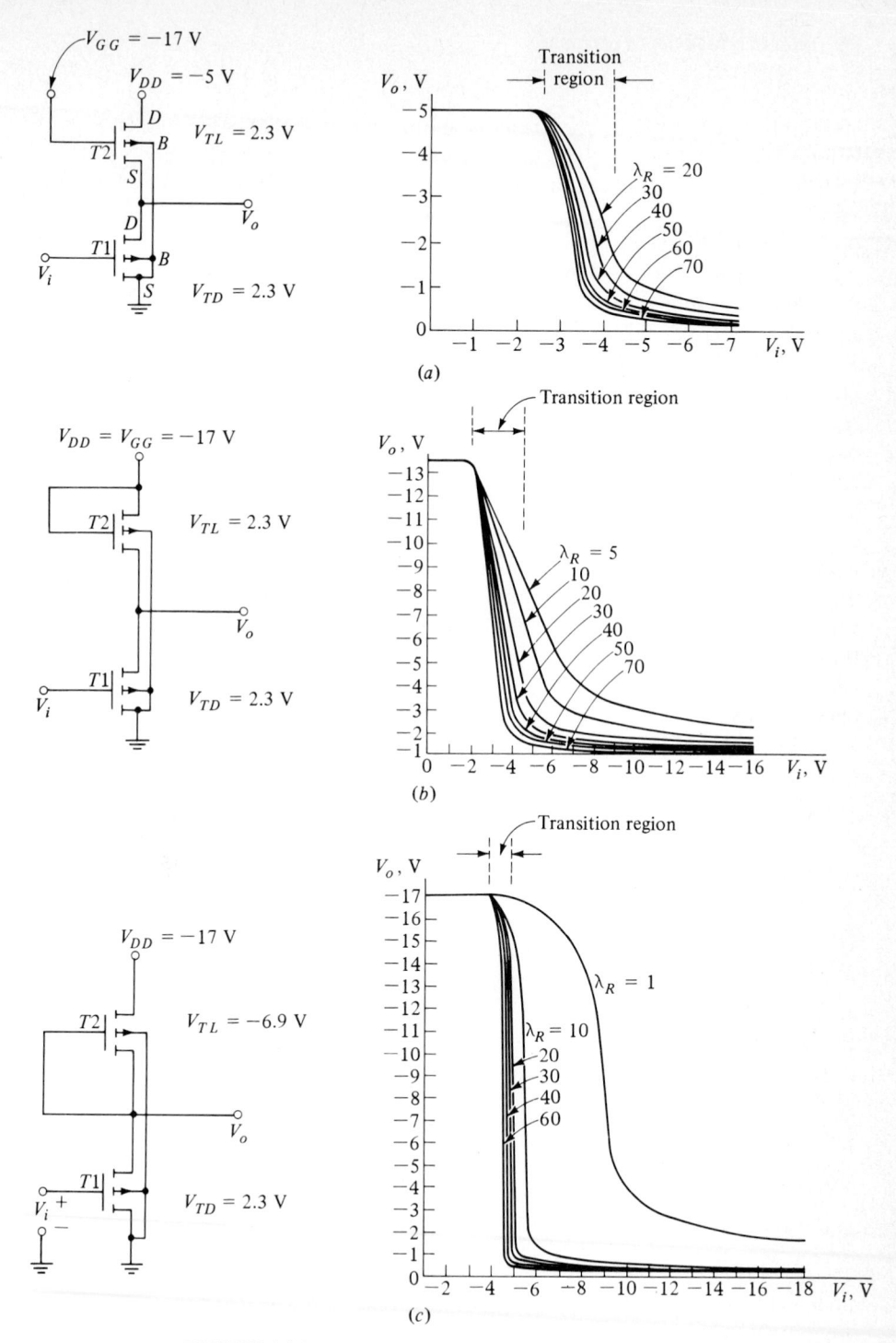

FIGURE 8.3-1
Input-output characteristics. (From "MOS/LSI Design and Applications," W. N. Carr, and J. P. Mize, McGraw-Hill, chap. 4, 1972).

261

invariably finds itself in the triode region. Such is the case since the gate voltage (furnished by another driving gate) will be at or near the supply voltage and the drain-to-source voltage will be at minimum magnitude. The load, on the other hand, may be an enhancement device or a depletion device and may operate in the triode or saturation region.

In Fig. 8.3-1 we display calculated input-output characteristics of PMOS inverters for three typical cases. In Fig. 8.3-1a both transistors are of the enhancement type, and both load and driver transistors have a threshold voltage $V_T = 2.3$ V. Since the transistors are p-channel devices, we apply the criterion given in Eq. (8.1-6) to determine whether we are in the triode or saturation region. Thus, to be in the triode region we require that $V_{SD} \leq V_{SG} - V_T$. This condition can be written as

$$V_{DG} \geq V_T \qquad (8.3\text{-}1)$$

Since $V_{DG} = V_D - V_G = -5 - (-17) = 12$ V is greater than $V_T = 2.3$ V, the load transistor is biased to operate in the triode region. In Fig. 8.3-1b the driven transistor remains as in Fig. 8-3-1a, but in this case the load transistor operates in the *saturation* region (the proof of this statement is left to the problems). In Fig. 8.3-1c the load transistor is a depletion device with a negative threshold voltage $V_T = -6.9$ V. The biasing of the load places it in the triode region when $V_o \leq -10.1$ V and in saturation when $V_o > -10.1$ V.

As we have discussed in Sec. 1.14 (see Fig. 1.14-1), we should expect the form of the input-output characteristic to depend principally on the parameter λ_R, defined by

$$\lambda_R \equiv \frac{\lambda_D}{\lambda_L} \equiv \frac{(W/L)_D}{(W/L)_L} \qquad (8.3\text{-}2)$$

where $(W/L)_D$ = width-to-length ratio of channel in driver transistor
    $(W/L)_L$ = width-to-length ratio for load
We noted that as λ_R increases, the transition of the output between its high and low levels becomes sharper. These expectations are confirmed in Fig. 8.3-1. In the calculations leading to the plots in Fig. 8.3-1 the effect of substrate bias is taken into account. Note also that the inverter using a depletion-mode MOSFET load has the steepest transition region.

8.4 THE CMOS INVERTER

The CMOS inverter is shown in Fig. 8.4-1a. The drains of a p-channel and an n-channel transistor are joined, and a supply voltage V_{SS} is applied from source to source. The output is taken at the common drain. The input V_i swings nominally through the range of V_{SS}. In the CMOS inverter shown, since we have grounded the source of the n-channel device, V_{SS} must be a positive voltage and V_i swings between ground and V_{SS}.

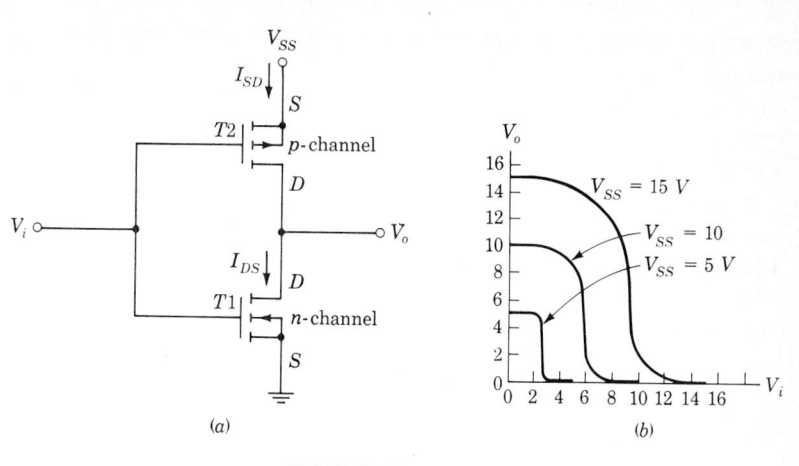

FIGURE 8.4-1
(a) A CMOS inverter and (b) its transfer characteristic.

Because of the complete symmetry of the circuit it seems intuitively clear that we shall want the two transistors to be reasonably alike. Therefore customarily it is arranged that the parameter k in Eqs. (8.1-2), (8.1-3), (8.1-5), and (8.1-6) are the same for the two transistors. The mobility of carriers in the p-channel device is smaller than the mobility in the n-channel device by a factor of 2 or 3. Hence to make the k's equal, the ratio W/L for the p-channel must be correspondingly larger by a factor of 2 or 3 than W/L for the n-channel device [see Eq. (8.1-4)]. However, even with such an adjustment of the W/L ratio the CMOS inverter is not necessarily entirely symmetrical since the threshold voltages of the p-channel and n-channel devices generally turn out to be somewhat different.

Usually CMOS gates are designed to operate with supply voltages in the range 5 to 15 V. Typical transfer characteristics are shown in Fig. 8.4-1b. For the device to which Fig. 8.4-1b applies, the magnitude of the threshold voltage is about 2 V for each of the transistors. Observe the abruptness of the transition and that the total swing in voltage is equal to V_{SS}. In the MOS inverters such a situation prevails only when we arrange that the *ratio* λ_R of the λ's be very large. In the present CMOS case, however, this situation prevails rather independently of the value of λ_R. Hence the CMOS inverter is often referred to as a *ratioless* inverter.

8.5 CALCULATION OF CMOS-INVERTER TRANSFER CHARACTERISTIC

It is instructive to use the device current equations to calculate the transfer characteristic of a CMOS inverter in a typical case. Referring to Fig. 8.4-1, we have that for $V_i \le V_T(n)$, $T1$ is OFF, $T2$ is ON, and $V_o = V_{SS}$. Similarly for

$V_i \geq V_{SS} - V_T(p)$, $T2$ is OFF, $T1$ is ON, and $V_o = 0$ V.   Further, $T1$ is saturated when $V_{DS1} \geq V_{GS1} - V_T$, i.e.,

$$V_T(n) \leq V_i \leq V_o + V_T(n) \tag{8.5-1}$$

while $T2$ is saturated when $V_{SD2} \geq V_{SG2} - V_T$, i.e.,

$$V_o - V_T(p) \leq V_i \leq V_{SS} - V_T(p) \tag{8.5-2}$$

Thus, if, say, $V_T(n) = V_T(p) = 2$ V and $V_{SS} = 10$ V, we would have $T1$ saturated when $2 \leq V_i \leq V_o + 2$ and $T2$ saturated when $V_o - 2 \leq V_i \leq 8$.   Or, to put the matter otherwise, $T1$ would be saturated when

$$V_o \geq V_i - 2 \tag{8.5-3}$$

and $T2$ would be saturated when

$$V_o \leq V_i + 2 \tag{8.5-4}$$

The currents I_{SD} and I_{DS} indicated in Fig. 8.4-1a are always equal. Accordingly, when $T1$ is in saturation and $T2$ is not, we have, using Eqs. (8.1-3) and (8.1-5),

$$k_n[V_i - V_T(n)]^2 = k_p\{2[V_{SS} - V_i - V_T(p)](V_{SS} - V_o) - (V_{SS} - V_o)^2\} \tag{8.5-5}$$

Here we have taken account of the fact that for the p-channel transistor $V_{SG} = V_{SS} - V_i$ and $V_{SD} = V_{SS} - V_o$.   Similarly we find that when $T2$ is in saturation and $T1$ is not, we have

$$k_p[V_{SS} - V_i - V_T(p)]^2 = k_n\{2[V_i - V_T(n)]V_o - V_o^2\} \tag{8.5-6}$$

Finally, when both transistors are in saturation, we find that

$$k_n[V_i - V_T(n)]^2 = k_p[V_{SS} - V_i - V_T(p)]^2 \tag{8.5-7}$$

Using Eqs. (8.5-5) to (8.5-7), we have plotted in Fig. 8.5-1 the input-output characteristic of a CMOS inverter for $V_T(n) = V_T(p) = 2$ V, $V_{SS} = 10$ V, and $k_p/k_n = 1$.   Above the line $V_o = V_i - 2$ [Eq. (8.5-3)] $T1$ is in saturation.   Below the line $V_o = V_i + 2$ [Eq. (8.5-4)] $T2$ is in saturation.   In the region between the two lines both transistors are in saturation.

Note that the simultaneous saturation of both transistors defines a unique voltage $V_i(\text{sat})$, calculated from Eq. (8.5-7) to be

$$V_i(\text{sat}) = \frac{\sqrt{k_p/k_n}[V_{SS} - V_T(p)] + V_T(n)}{1 + \sqrt{k_p/k_n}} \tag{8.5-8}$$

In Fig. 8.5-1 with $k_p/k_n = 1$, $V_T(p) = V_T(n)$, $V_i(\text{sat}) = 5$ V.   This voltage, at which there occurs an abrupt transition in output voltage is midway between 0 and V_{SS} because we have selected $k_p = k_n$.   If k_p were not equal to k_n, then even if $V_T(n)$ were equal to $V_T(p)$, complete symmetry would not prevail.   In any event we find from Eqs. (8.5-1) and (8.5-2) that the magnitude of the abrupt transition is given by

$$\Delta V_o = V_T(n) + V_T(p) \tag{8.5-9}$$

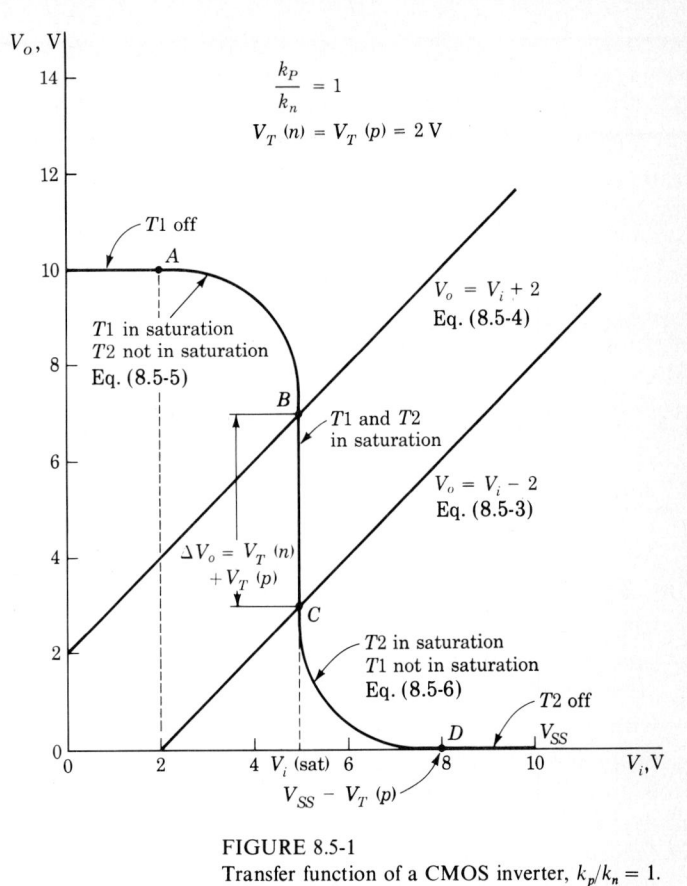

FIGURE 8.5-1
Transfer function of a CMOS inverter, $k_p/k_n = 1$.

The infinite slope displayed in Fig. 8.5-1 results from our assumption that in the saturation region the device current is absolutely independent of drain-to-source voltage, i.e., that the device is a constant current source. Such, of course, is not precisely so, and hence the transition from B to C in a physical situation would be sharp but not absolutely abrupt.

8.6 MOS GATES

Assuming that positive logic is intended, the NMOS circuit of Fig. 8.6-1 is a two-input NAND gate. The supply voltages V_{DD} and V_{GG} and the threshold voltage V_T are all positive. Logic **0** is represented by a voltage less than the threshold voltage and logic **1** by a voltage above the threshold voltage. The truth table given in Fig. 8.6-1 is readily verified. When either V_1 or V_2 or both are below threshold, only TL conducts. In this case V_o is $V_{GG} - V_T$ or V_{DD}, whichever is lower. When both V_1 and V_2 are above threshold, both $T1$ and $T2$ conduct.

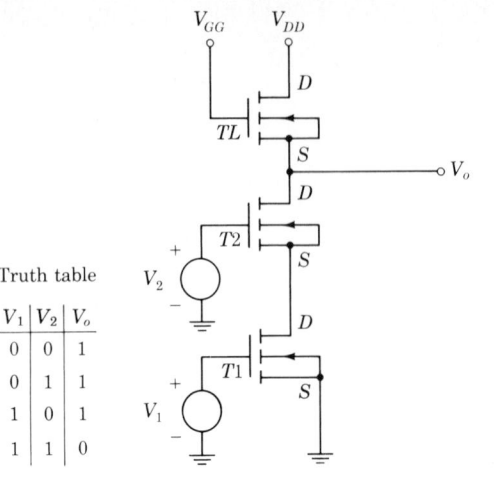

Truth table

V_1	V_2	V_o
0	0	1
0	1	1
1	0	1
1	1	0

FIGURE 8.6-1
The effect of stacking.

The output voltage V_o is the sum of the voltages across $T1$ and $T2$ and is presumably well below threshold.

Additional inputs can be provided by including additional devices in series with $T1$ and $T2$. The output V_o, when at logic **0**, is the sum of the voltage drops across all the series devices. It is, of course, necessary to keep V_o at logic **0** and therefore comfortably below the threshold V_T. Hence as more series devices are included, the voltage drop across each one individually must be reduced. This is accomplished by increasing the width-to-length ratio W/L of the devices in order to reduce the resistance of the channel. Thus, suppose that starting with a design for a two-input gate, we wanted to modify the gate to accommodate three inputs. Then the two initial driver gates would be replaced by three driver gates, each with a W/L ratio three-halves the W/L ratio of the original devices.

Since the same input voltage (with respect to ground) is on each gate input terminal, the gate-to-source voltage on each driven FET is not the same. The gate-to-source voltage is a maximum for the driver at the bottom of the stack and decreases as we go up the stack (see Fig. 8.6-1). As noted, driver transistors operate in the triode region where the device resistance is a function of gate voltage. Thus, we can compensate for this effect of stacking by making W/L progressively larger for transistors higher up the stack.

If the NMOS driver transistors are placed in parallel, as in Fig. 8.6-2, then, as can be verified, a NOR gate results.

If PMOS devices are used, the supply voltages V_{DD} and V_{GG} must be negative with respect to ground. It is left as a student exercise to verify (again for positive logic) that with drivers in series we have a NOR gate and with drivers in parallel we have a NAND gate. The feasibility of stacking MOS in series [which is not readily permitted with bipolar devices (see Prob. 8.6-3)] allows some

Truth table

V_1	V_2	V_o
0	0	1
0	1	0
1	1	0
1	0	0

FIGURE 8.6-2
An NMOS NOR gate.

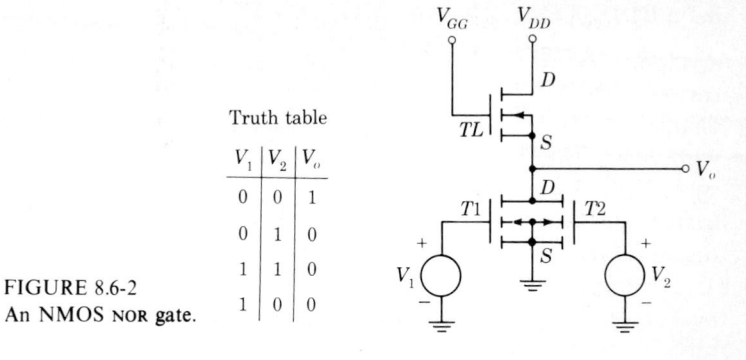

interesting gate configurations like those shown in Fig. 8.6-3. Here we have used PMOS devices. The circuit in Fig. 8.6-3a accomplishes the logic indicated in Fig. 8.6-3b, and the circuit in Fig. 8.6-3c performs the logic shown in Fig. 8.6-3d.

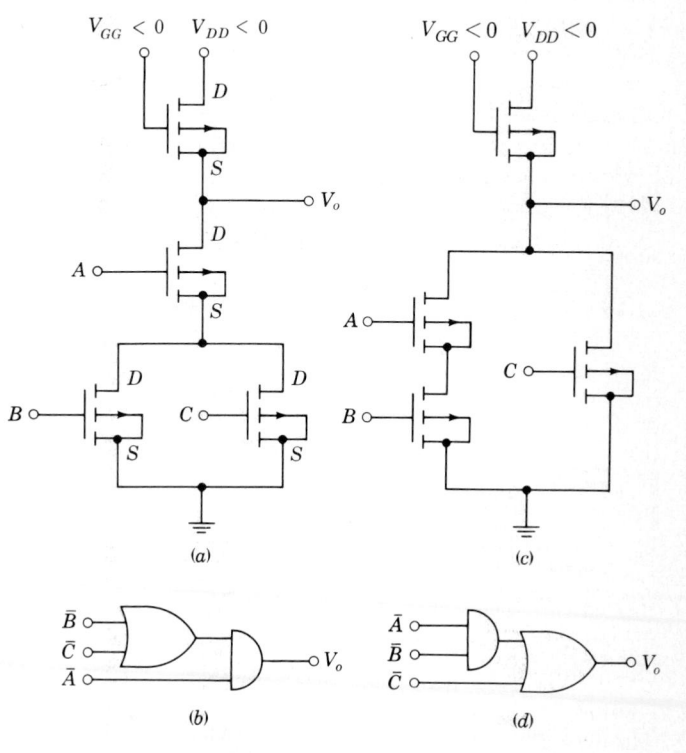

FIGURE 8.6-3
Illustrating the feasibility of stacking MOSFETs.

8.7 RISE TIME IN AN MOS GATE

A bipolar transistor is turned ON by establishing a distribution of minority-carrier charge in the transistor base and turned OFF by removing this minority charge. Additionally, capacitors must be charged and discharged. The speed with which bipolar transistor gates can be operated is therefore limited by the speed with which these operations can be performed. In MOS devices minority-carrier charge is not involved, and the speed of operation is determined only by the speed with which capacitors can be charged.

Consider then the situation represented in Fig. 8.7-1. This basic inverter becomes a multiple input gate if additional transistors are added in parallel or series with $T1$. The capacitance C_L represents the capacitance load on the gate and may well be the input capacitance of a succeding gate. We inquire now into the rise time of the output voltage V_o $(= V_C$, the capacitor voltage) as $T1$ is turned OFF. We assume that V_C starts from 0 V. We consider first the case where the load transistor TL operates in the saturation region, as would be the case if $V_{GG} = V_{DD}$.

In the saturation region we have, as in Eq. (8.1-3),

$$I_L = I_C = k_L(V_{GS} - V_T)^2 \tag{8.7-1}$$

The gate-to-source voltage of the load transistor with parameter $k = k_L$ is $V_{GS} = V_{DD} - V_C$. We therefore have that the rate of rise of V_C, which is $dV_C/dt = I_C/C_L$, is given by

$$\frac{dV_C}{dt} = \frac{I_C}{C_L} = \frac{k_L}{C_L}(V_{DD} - V_T - V_C)^2 \tag{8.7-2}$$

It is convenient to introduce the voltage $V_f \equiv V_{DD} - V_T$. This voltage V_f is the (final) voltage to which V_C will rise asymptotically. The solution of Eq. (8.7-2) subject to the initial condition that $V_C = 0$ at $t = 0$ is found to be (Prob. 8.7-1)

$$V_C = \frac{(k_L t/C_L)V_f^2}{1 + (k_L t/C_L)V_f} \tag{8.7-3}$$

If we define the rise time t_r to be the time required for V_C to rise from 0 V to $0.9V_f$, we find from Eq. (8.7-3) that

$$t_r = \frac{9C_L}{k_L V_f} \tag{8.7-4}$$

For example, if $C_L = 5$ pF (a typical input capacitance to a single gate) and $k_L = 20$ μA/V^2 (a typical value for a load transistor), and if $V_f = 10$ V, we find $t_r \approx 0.2$ μs. This rise time is extremely long in comparison with the rise time in BJT gates. The reason for this slow response is basically the necessity for making the channel in the load transistor long and narrow. And this long narrow channel is required, as we have noted, in order to allow the input-output characteristic of the gate to exhibit an abrupt transition between logic level. The long narrow channel limits the amount of current available to charge the capacitor C_L.

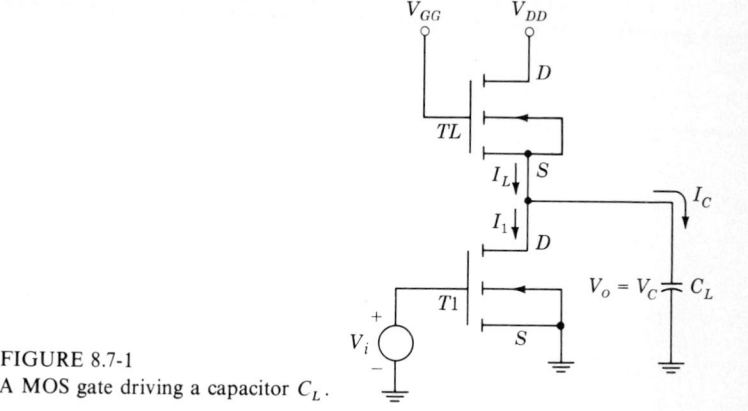

FIGURE 8.7-1
A MOS gate driving a capacitor C_L.

If, starting with a saturated load transistor, we increased the gate supply voltage V_{GG} to the point where $V_{GG} - V_T > V_{DD}$, the transistor would operate in the triode region. The increase in V_{GG} would increase the current through TL and hence the current available to charge the load capacitor. However, a modest increase in V_{GG} would result in only a small increase in current, and no very substantial improvement in rise time would result. The rise time would still be limited basically by the restricted current available through the long narrow channel of the load. [The equation for $V_C(t)$ in the triode case is given in Prob. 8.7-2.]

8.8 THE FALL TIME

When, in the inverter of Fig. 8.7-1, the driver transistor $T1$ is turned ON, the capacitor will discharge through $T1$ and eventually fall nominally to ground. While $T1$ is discharging, C_L, the current through TL, continues in the direction to charge C_L. But, as we have seen, the W/L ratio of the driver is very much larger than the corresponding ratio for the load. Hence the discharge current through $T1$ will be much larger than the charging current through TL, and we shall therefore neglect the charging current.

Suppose then that $T1$, which is initially OFF, is turned ON by the application to its gate of a gating voltage $V_{GS} > V_T$. The volt-ampere characteristic of the transistor for this gating voltage is indicated in Fig. 8.8-1. Originally the transistor is OFF and operating at P_1, where $V_{DS} = V_{CM}$, the maximum voltage drop across the capacitor. When $T1$ is turned ON, the operating point moves abruptly to P_2 since the capacitor voltage cannot change instantaneously. The capacitor voltage $V_C = V_{DS}$ then decreases, and the transistor makes an excursion through the region of saturation as shown in Fig. 8.8-1. [We have idealized the saturation region to correspond exactly to Eq. (8.1-3), which assumes I_{DS} precisely constant and independent of V_{DS}.] At P_3 the triode region begins, and the

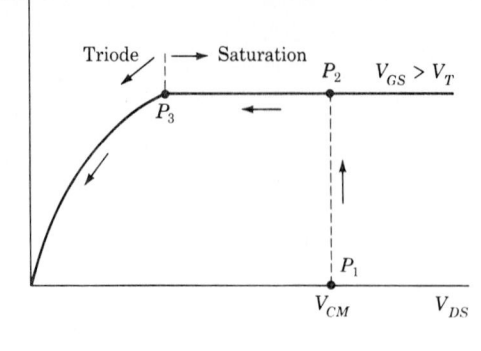

FIGURE 8.8-1
The operating path of $T1$ as the capacitor C_L discharges.

transistor and capacitor follow the triode characteristic, eventually to zero voltage.

Using Eq. (8.1-3), which applies in the saturation region, we find

$$V_C(t) = V_{CM} - \frac{I_{DS}}{C_L} t = V_{CM} - \frac{k_D}{C_L}(V_{GS} - V_T)^2 t \qquad (8.8\text{-}1)$$

The point P_3 is reached when $V_{DS} = V_{GS} - V_T$. The time $t \equiv t_{\text{sat}}$ at which $V_C(t) = V_{GS} - V_T$ is calculated from Eq. (8.8-1) to be

$$t_{\text{sat}} = \frac{C_L}{k_D}\left[\frac{V_{CM} - V_{GS} + V_T}{(V_{GS} - V_T)^2}\right] \qquad (8.8\text{-}2)$$

In the triode region, using Eq. (8.1-2), we find

$$\frac{dV_C}{dt} = -\frac{I_{DS}}{C_L} = -\frac{k_D}{C_L}[2(V_{GS} - V_T)V_C - V_C^2] \qquad (8.8\text{-}3)$$

The time in the triode region required for $V_C(t)$ to fall from V_{CM} to $0.1V_{CM}$ is

$$t_{\text{triode}} = -\int_{V_{CM}}^{0.1V_{CM}} \frac{dV_C}{(k_D/C_L)[2(V_{GS} - V_T)V_C - V_C^2]} = \frac{1.15C_L}{k_D(V_{GS} - V_T)} \qquad (8.8\text{-}4)$$

Let us take $k_D = 1$ mA/V². Since we have assumed $k_L = 0.5$ μA/V², we have $k_D/k_L = 50$, which is quite reasonable for an MOS gate. Assume also that $V_{CM} = 10$ V, $V_{GS} - V_T = 5$ V, and $C_L = 5$ pF. Then we calculate from Eqs. (8.8-2) and (8.8-4) that $t_{\text{sat}} = 0.004$ μs and $t_{\text{triode}} = 0.016$ μs. The total fall time is then $t_f = 20$ ns. The fall time is thus very appreciably smaller than the rise time previously calculated to be 0.2 μs. The principal reason for the large difference is the fact that $k_D \gg k_L$ because the channel of the driver is much wider and shorter than the channel of the load transistor.

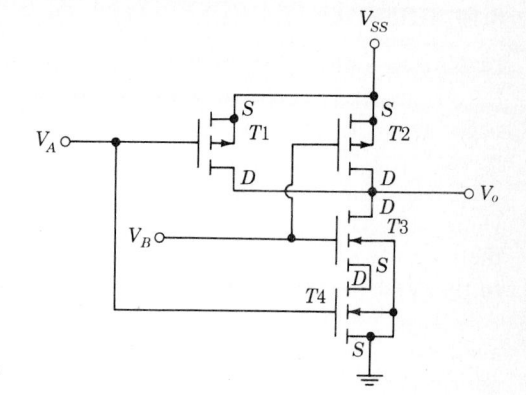

FIGURE 8.9-1
A CMOS NAND gate.

8.9 THE CMOS GATE

A two-input CMOS NAND gate is shown in Fig. 8.9-1. Note that the driver transistors are series-connected while the load transistors are paralleled. The individual input is applied simultaneously to a pair of transistors, one driver and one load. Assuming positive logic and taking logic **0** to be nominally ground voltage and logic **1** to be nominally V_{SS}, we can easily verify that the circuit is indeed a NAND gate. The output V_o will be at logic **0** (ground) only when both NMOS driver transistors are ON, in which case both PMOS load transistors will be OFF. The circuit in Fig. 8.9-2, in which the load devices are in series and the driver transistors in parallel, is readily verified to be a NOR gate.

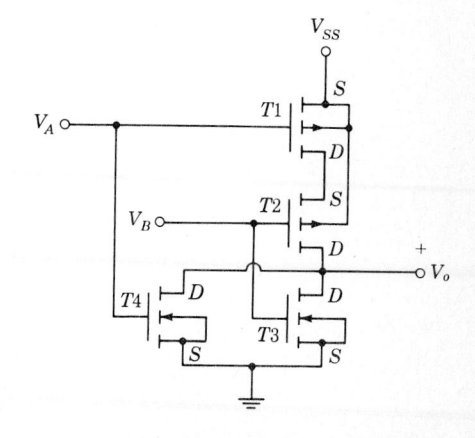

FIGURE 8.9-2
A CMOS NOR gate.

8.10 RISE AND FALL TIMES IN CMOS GATES

In NMOS and in PMOS the channel in the load devices must be very much longer and narrower than in the driver; i.e., we require $k_L \ll k_D$. This design is necessary to assure that when the driver is ON, the voltage drop across the driver transistor will be a very small fraction of the supply voltage. In CMOS, on the other hand, when the driver goes ON, the load is simultaneously driven OFF. Hence in CMOS, as already noted, $k_L \approx k_D$. Since the driver is n-channel and the load is p-channel, the W/L ratios are designed to be inversely proportional to the ratio of their mobility.

One of the major advantages of CMOS, then, is that there is always available a low-resistance channel path to charge and discharge a capacitive load across the gate output. Such a capacitive load at the output of a CMOS inverter is shown in Fig. 8.10-1. In an NMOS or in a PMOS gate the charging times of the load capacitor are widely different because of the difference in driver and load channel geometries, that is, k_D and k_L. In CMOS on the other hand, the charge and discharge times are quite comparable.

The capacitor-charging calculations given above for the NMOS inverter can be applied directly in the present case. Referring to Fig. 8.10-1, suppose that TL has been ON and TD OFF, so that $V_C = V_{SS}$. The capacitor C_L will now discharge when the input gate voltage (from the output of a preceding gate) goes to $V_{GS} = V_{SS}$, thereby turning TL OFF and TD ON. The fall time t_f is $t_f = t_{\text{sat}} + t_{\text{triode}}$, as given by Eqs. (8.8-2) and (8.8-4). In these equations $V_C(\text{max}) = V_{GS} = V_{SS}$. We may note that actually these equations apply somewhat more exactly in the present case than in the case (of NMOS or PMOS) for which they were derived; for it will be recalled that in that derivation we neglected the small charging current through the load transistor while the capacitor was discharging. In the present case there is no current through the load device.

The parameter k for both the load and driver transistors will be comparable to the value of k ordinarily designed into the driver transistor of a gate without complementary symmetry. Hence in CMOS the fall time will be comparable to the fall time calculated in Sec. 8.7. However, because of the symmetry of the CMOS structure of Fig. 8.10-1 the rise time will be the same as the fall time if

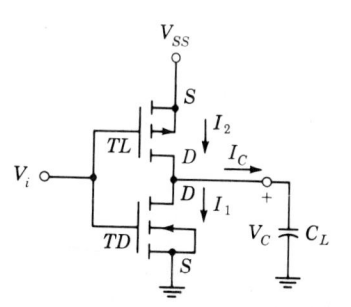

FIGURE 8.10-1
A CMOS gate driving a capacitor C_L.

$k_L = k_D$. As can be verified, except for the direction in which $V_L(t)$ is changing, the equations which described the charging of C_L are identical to the equations which describe its discharge.

8.11 MANUFACTURER'S SPECIFICATIONS

The specifications provided by manufacturers for CMOS devices are similar to those provided for BJT gates. These specifications deal with input and output currents and voltages, propagation delays, rise and fall times, etc. The specifications for the Motorola 4012 low-power NAND gate are given in Fig. 8.11-1. (Specifications for the type 4001 NOR gate are identical.) These specifications apply under the circumstances that a supply voltage $V_{SS} = 5$ V is employed. The definitions of the parameters V_{iH}, V_{iL}, V_{oH}, V_{oL} are given in Sec. 4.11. The current I_{iH} stands for the minimum current which must be supplied by a driving source if the CMOS-gate input is to be held at a voltage high enough, i.e., at V_{iH} or higher, for the gate to acknowledge that its input is at logic level **1**. The other current symbols have similar meanings. Observe that the input currents, being only of the order of 10 pA, may generally be ignored. Referring to the output specifications, we note that the gate output is able to sink a maximum of 0.4 mA and still stay low enough in voltage to remain in the logic **0** region, that is, $V_o \leq V_{oL}$. At the other end, the gate is able to serve as a source of at most 0.5 mA and still remain in the logic **1** region.

Figure 8.11-1*b* shows the worst-case transfer characteristics of the gate. The transfer characteristic of a typical gate will lie somewhere between the *worst-case* limiting plots shown. With a 5-V supply the manufacturer specifies that

$$\Delta 0 = V_{iL} - V_{oL} \approx 1.5 \text{ V} \qquad (8.11\text{-}1a)$$

$$\Delta 1 = V_{oH} - V_{iH} \approx 1.5 \text{ V} \qquad (8.11\text{-}1b)$$

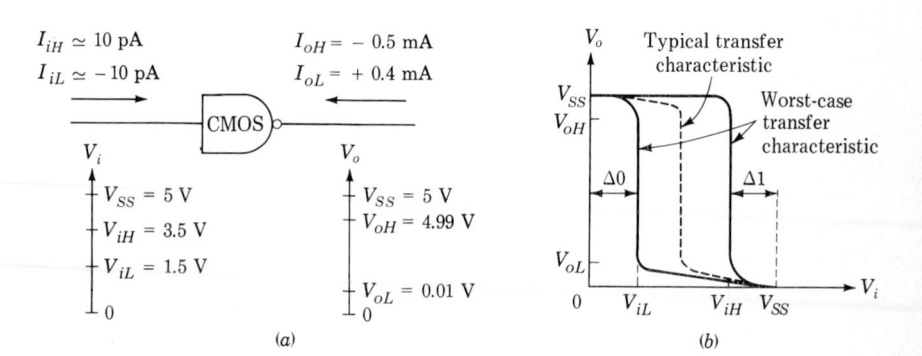

FIGURE 8.11-1
(*a*) Some manufacturer's specifications. (*b*) Typical worst-case transfer characteristic.

Table 8.11-1

	Time, ns
$t_{pd}(LH)$	30
$t_{pd}(HL)$	30
t_r	60
t_f	60

If we inquire about typical rather than worst-case noise immunities, we find that with $V_{SS} = 5$ V, these are about 2.25 V. Thus the noise immunities are larger than those encountered in BJT gates with comparable supply voltages.

The propagation delay and transition times for the Motorola gates are given in Table 8.11-1, assuming a load consisting of a capacitor $C_L = 15$ pF in parallel with a resistor $R_L = 200$ kΩ. We assume that the input impedance of a CMOS gate is a 5-pF capacitance, so that this capacitive load represents a fan-out of 3. Increasing the fan-out increases the delay and transition times linearly, since the circuit time constant is directly proportional to the total load capacitance.

The CMOS gate referred to in Table 8.11-1 is relatively slow. However, high-speed CMOS gates are available having propagation delay times of the order of 20 ns. Recent advances in fabrication techniques, in which the substrate material employed is *sapphire*, have resulted in fast, low-power CMOS gates. These gates, called SOS/CMOS (silicon-on-sapphire) have rise and fall times

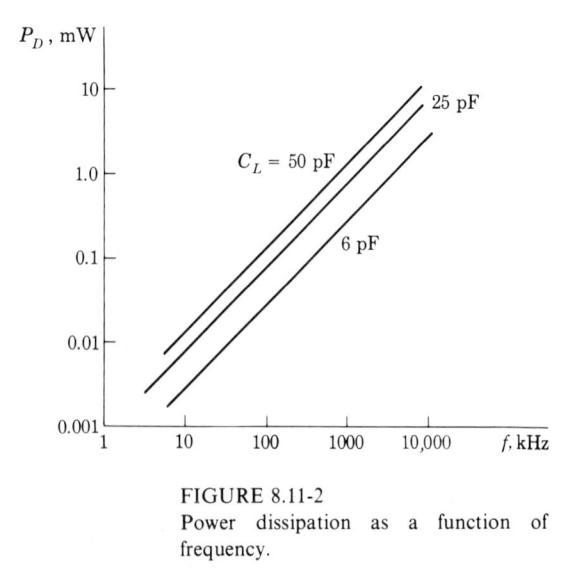

FIGURE 8.11-2
Power dissipation as a function of frequency.

which are less than 20 ns. High-speed CMOS gates are today comparable to DTL gates, and a few claim to be as fast as the slower versions of TTL.

Typical quiescent power disipation of these low-power CMOS units is 50 nW. However, at 100-KHz operation, the power dissipated is approximately 30 μW. This dissipation increases at the rate of 20 dB/decade and is also a function of the capacitive load. Plots of the dissipation of the Motorola 4012 are shown in Fig. 8.11-2. (See also Sec. 1.15.)

Buffers The CMOS gate is a low-current gate designed to drive other CMOS gates. When a CMOS gate is to be used to drive a TTL or DTL gate, a CMOS *buffer* is often employed. An RCA 4009A buffer is capable of sinking a load current of 4 mA (I_{oL}) when it is in the low state and can source 1.75 mA when the output voltage is 2.5 V. This 2.5-V value is specified since it represents V_{iH} for many TTL gates.

The circuit configuration of the buffer is the same as the standard gate configuration. However, to obtain extra current capability, the dimensions of the CMOS devices used are increased.

8.12 INTERFACING BJT AND CMOS GATES

MOS gates and (at the present writing) CMOS gates are slower than BJT gates. On the other hand, the MOS gates can be fabricated with a component density on a silicon die which exceeds that possible with BJT devices. There is a merit to conserving "real estate" on the silicon die. For as the area involved on the die increases, so does the likelihood that a crystal imperfection will render the device defective, i.e., a reject. As a result there is an advantage in using both BJT and MOS devices in combination. The BJT devices are used where speed is required, and the MOS devices are used where slower operation is allowed. Since BJT and MOS devices generally operate at different voltage and current levels, some consideration must be given to proper interfacing. In many cases, interfacing requires interposing between the two types of devices circuits involving discrete bipolar transistors. Such would be the case, for example, if we needed to interface ECL logic operating between 0 and -5.2 V with MOS logic operating between 0 and $+14$ V. In simpler cases, interfacing may require a relatively minor accommodation in the BJT or MOS gate.

To illustrate some of the measures which may serve to effect a required accommodation we consider the interfacing between TTL logic and CMOS logic. This is the most common interfacing employed since TTL gates are the most often used logic types. TTL operates from a positive 5-V supply. We shall consider the case in which the CMOS as well operates from $+5$ V. For this purpose we have listed in Fig. 8.12-1 some of the relevant specifications for the TTL gate.

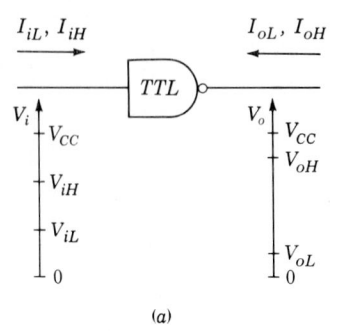

	TTL	
V_{iH}	2.8 V	
I_{iH}	100 μA	
V_{iL}	0.4 V	
I_{iL}	− 1.3 mA	
V_{oH}	2.4 V	3 V
I_{oH}	−0.7 mA	0
V_{oL}	0.4 V	
I_{oL}	10 mA	

(a)   (b)

FIGURE 8.12-1
(a) Input and output characteristics.   (b) Table of typical values.

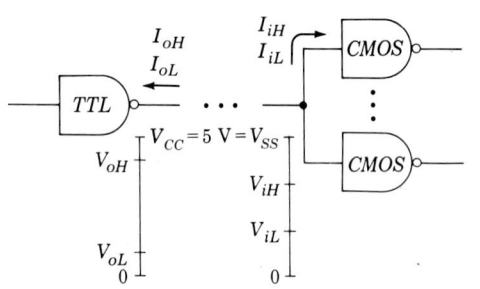

FIGURE 8.12-2
A TTL gate driving N CMOS gates.

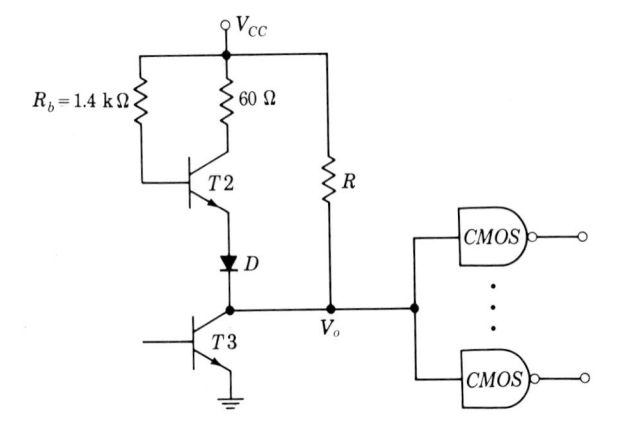

FIGURE 8.12-3
TTL with a passive pull-up.

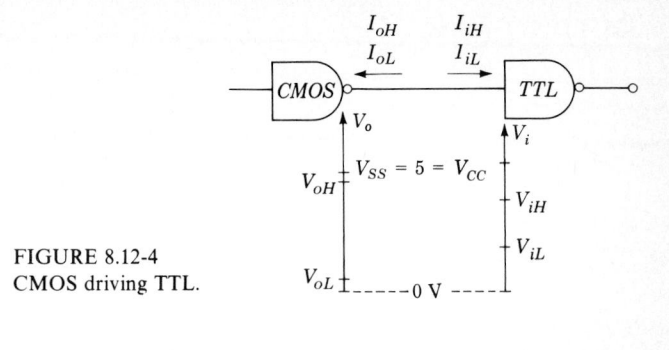

FIGURE 8.12-4
CMOS driving TTL.

We consider first the case of a TTL gate driving N CMOS gates, as in Fig. 8.12-2. Thus, the fan-out is N. For such an arrangement to operate successfully it is required that

$$-I_{oH}(\text{TTL}) \geq NI_{iH}(\text{CMOS}) \tag{8.12-1a}$$

$$I_{oL}(\text{TTL}) \geq -NI_{iL}(\text{CMOS}) \tag{8.12-1b}$$

$$V_{oL}(\text{TTL}) \leq V_{iL}(\text{CMOS}) \tag{8.12-1c}$$

$$V_{oH}(\text{TTL}) \geq V_{iH}(\text{CMOS}) \tag{8.12-1d}$$

As is readily verified from the data in Fig. 8.11-1 and in Fig. 8.12-1, Eqs. (8.12-1a) and (8.12-1b) are satisfied for any reasonable fan-out N. In addition, Eq. (8.12-1c) is also satisfied. However we find that Eq. (8.12-1d) is not satisfied since, even at "no load" $V_{oH}(TTL) = 3$ V while $V_{iH}(\text{CMOS}) = 3.5$ V. A frequently employed circuit modification used to raise $V_{oH}(TTL)$ above 3.5 V is shown in Fig. 8.12-3 where an external resistor R has been bridged between V_{CC} and the output. Typically R is in the range 2 to 6 kΩ.

When we consider a CMOS gate driving a TTL gate (see Fig. 8.12-4), we find that the condition $I_{oL}(\text{CMOS}) \geq -NI_{iL}(\text{TTL})$ is not satisfied even for $N = 1$ since $I_{oL}(\text{CMOS}) = 0.4$ mA while $-I_{iL}(\text{TTL}) = 1.34$ mA. There are available, however, a number of CMOS buffers having adequate available output current (up to 6 mA).

REFERENCES

1 Carr, W. N., and J. P. Mize: "MOS/LSI Design and Applications," McGraw-Hill, chap. 4, 1972.
2 RCA Solid State Databook: SSD-203B, COS/MOS Digital Integrated Circuits.

9

FLIP-FLOPS

9.1 INTRODUCTION

The circuit of Fig. 9.1-1, showing a pair of cross-coupled NOR gates, is called a *flip-flop*. It has a pair of input terminals S and R, standing for "set" and "reset," respectively. We shall use these symbols S and R not only to designate the terminals but also to specify the logical level at the terminals. Thus, $S = 1$ indicates that a voltage corresponding to logic level **1** is normally present at terminal S. Similarly, the output terminals and the corresponding output logic levels are Q and $\overline{Q}$. In this notation we have explicitly taken account of the fact that in normal operation, as we shall see, the logic levels at the outputs are complementary.

The fundamental, most important characteristic of a flip-flop is that it has a "memory." That is, given that the present logic levels at S and R, are **0** and **0**, it is possible, from an examination of the output, to determine what the logic levels were at S and R immediately before they attained these present levels.

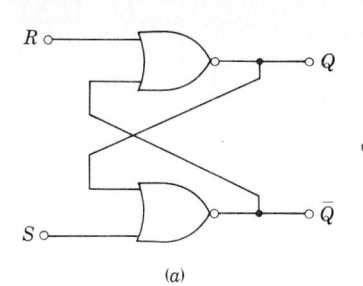

S	R	Q_{n+1}	$\bar{Q}_{n+1}$
0	0	Q_n	$\bar{Q}_n$
0	1	0	1
1	0	1	0
1	1	Not used	

(a)   (b)

FIGURE 9.1-1
(a) An SR flip-flop and (b) truth table.

9.2 TERMINOLOGY

In connection with the discussion to follow it is convenient to introduce some useful terminology, and it will be helpful to be aware of an attitude generally prevalent among logic-system designers.

In NAND and NOR gates (as well as an AND and OR gates), when it serves our purpose to do so, we can arbitrarily select one input terminal and view it as an *enable-disable input*. Thus, consider a NOR or an OR gate. If the one selected input is at logic **1**, the output of the gate is independent of all other inputs. This one selected input takes control of the gate and the gate is disabled with respect to any other input. (The term "inhibit" is used synonymously with "disable.") Alternatively, if the selected input is at logic **0**, it does not take control, and the gate is enabled to respond to other inputs. In a NAND or AND gate, a selected input takes over control and disables the gate when this input goes to logic **0**. For with one input at logic **0**, the gate output cannot respond to the other inputs. The difference between the NOR or OR gate, on the one hand, and the NAND or AND gate, on the other, is to be noted. In the first case, the control input achieves its control when it goes to logic **1**; in the second case, it achieves its control when it goes to logic **0**.

There is a generally prevailing attitude in digital systems to view logic **0** as a basic, undisturbed, unperturbed, quiescent state and to view the logic **1** state as the excited, active, effective state, i.e., the state arrived at "after something has happened." Thus, when an effect has been produced, the inclination is to define the resultant state as one corresponding to which some logical variable has gone to logic **1**. The logical variable is at logic **0** when "nothing has happened." Similarly, if an effect is to be produced by a change in a logical variable, it is preferred that the logical variable so involved be defined in such manner that the effect is achieved when the logic variable goes to logic **1**. We shall see examples of this attitude in our discussion of flip-flops.

9.3   THE FLIP-FLOP AS A MEMORY ELEMENT

In the flip-flop circuit of Fig. 9.1-1 S and R are control inputs.   Let us assume that $S = R = 0$.   In this case since NOR gates are involved, the control inputs do not exercise control and we may ignore them.   The gate outputs depend only on the logic levels present at the two other inputs.   Then let us assume that $\overline{Q} = 1$.   In this case the output of the upper NOR gate is at logic 0.   Hence, as anticipated, the outputs are indeed complementary, and we have $Q = 0$.   We assumed that $\overline{Q} = 1$, and we must now verify that this assumption is consistent.   Such is indeed the case, for the two inputs to the lower gate are $S = 0$ and $Q = 0$; hence the gate output is $\overline{Q} = 1$, as assumed.

With $S = R = 0$, if we had assumed instead that $\overline{Q} = 0$, we would have found that $Q = 1$; this assumption, like the previous assumption, is self-consistent.   Altogether, then, it appears that with $S = R = 0$, the flip-flop may persist in either of two situations (often referred to as *stable states*) one with $Q = 0$ and $\overline{Q} = 1$, the other with $Q = 1$ and $\overline{Q} = 0$.   We shall refer to the first of these states as the *reset state* and to the second as the *set state*.   The reset state is often called the *clear* state and is defined as the state in which $Q = 0$.   These terms are used in analogy with their use, say, with a desk calculator.   When one pushes the clear button, all registers return to zero, i.e., to the initial state where nothing has happened, and the machine is ready for another calculation.   Hence, correspondingly, this state in a flip-flop is identified by $Q = 0$.

If we arrange that $S = 0$ while $R = 1$, then, as can be verified, the flip-flop can establish itself in only one possible state, i.e., the reset state with $Q = 0$.   Similarly with $S = 1$ and $R = 0$ the flip-flop will find itself in the set state with $Q = 1$.

Next, let us suppose that the flip-flop is in the reset state with $S = 0$, $R = 1$ so that $Q = 0$, and then at some time say $t = T$, the input at R goes to $R = 0$.   As can be verified from Fig. 9.1-1 this change in R will not change the state of the flip-flop, for the change is taking place at a NOR-gate input whose other input is $\overline{Q} = 1$.   Thus, we now find, with $S = R = 0$, that the flip-flop remains in the reset state, and from this fact we can deduce that at a time earlier than $t = T$ the input levels were $S = 0$ and $R = 1$.   Thus, the flip-flop has *remembered* what the input levels were before $t = T$.   In other words, with $S = R = 0$, the state of the flip-flop depends on the immediately prior history of the input levels.   Similarly, if initially $S = 1$ and $R = 0$, the flip-flop will be in the set state with $Q = 1$; and if thereafter S should change so that $S = R = 0$, the flip-flop will remain in the set state.

Finally, consider that both S and R are set to $S = R = 1$.   In this case both outputs will be at logic 0.   (In this case the implication, in Fig. 9.1-1, that the outputs are complementary, is incorrect.)   If now it should happen that both inputs change simultaneously to $S = R = 0$, the resultant state of the flip-flop will not be predictable.   For with $S = R = 0$ two states are possible, and the state attained will be no more predictable than the outcome of tossing a coin.   Because of this ambiguity the case $S = R = 1$ is not used.

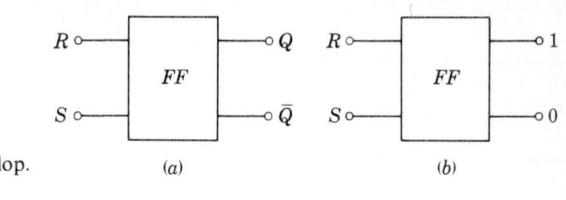

FIGURE 9.3-1
Typical representation of an SR flip-flop.

The results of this discussion of flip-flop operation are summarized in the truth table shown in Fig. 9.1-1b. Here we contemplate that the inputs S and R are changing their logic levels in an arbitrary fashion. There will be intervals in which $S = 0$ and $R = 0$, intervals when $S = 0$ and $R = 1$, etc. Let us number the intervals 1, 2, ..., n, $n + 1$, ... Then in interval $n + 1$ when $S = 0$ and $R = 1$ we shall find $Q_{n+1} = 0$ ($\bar{Q}_{n+1} = 1$) no matter what has been the past history of S and R in preceding intervals. Similarly, with $S = 1$ and $R = 0$, we shall have $Q_{n+1} = 1$. However, if $S = R = 0$ in interval $n + 1$, then the state of the flip-flop will be the same as it was in interval n.

Suppose that we should have $S = R = 1$ in interval n, while in interval $n + 1$, $S = R = 0$. Here we imagine a simultaneous change in S and R. Actually a *precisely* simultaneous change is not physically realizable. If actually S should change first so that we go from $S = R = 1$ to $S = 0$, $R = 1$ and then to $S = 0$, $R = 0$, we will find $Q_{n+1} = 0$. If R should change first we shall have $Q_{n+1} = 1$.

The flip-flop circuit element representation is shown in Fig. 9.3-1a. In Fig. 9.3-1b the outputs are marked by logic levels **1** and **0**, corresponding to Q and $\bar{Q}$, respectively. Of course, we do not intend to imply that there are fixed logic levels at these output terminals. The intention is to convey that the terminal marked 1 is at logic level **1** (the set situation) when the set terminal is excited, that is, $S = 1$ ($R = 0$).

9.4 FLIP-FLOP USING NAND GATES

A flip-flop using NAND gates is shown in Fig. 9.4-1a. We can readily verify that the truth table for this flip-flop is as given in Fig. 9.4-1b. In this case $\bar{S} = \bar{R} = 0$ makes the logic levels at the two gate outputs the same (logic level **1**) and hence is not used. In this truth table we have deleted the $\bar{Q}$ column, which is redundant when the Q column is available. Since NAND gates are employed, the inputs exercise control at logic level **0**. That is, the flip-flop changes state when either input is put to logic **0**. Since this situation is inconsistent with the preference discussed earlier, we have labeled the control inputs $\bar{S}$ and $\bar{R}$ rather than S and R.

The uncomplemented variables S and R do not appear in the circuit of Fig. 9.4-1a. Nevertheless, we may imagine two variables S and R of which $\bar{S}$ and $\bar{R}$ are the complements and then consider further that $S = 1$, $R = 0$ and $S = 0$, $R = 1$, respectively, set or reset the flip-flop. Using these two virtual, or

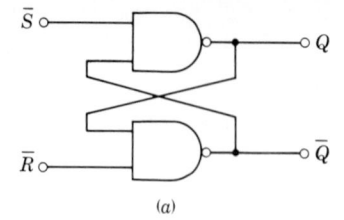

(a)

$\bar{S}$	$\bar{R}$	Q_{n+1}
0	0	Not used
0	1	1
1	0	0
1	1	Q_n

(b)

S	R	Q_{n+1}
0	0	Q_n
0	1	0
1	0	1
1	1	Not used

(c)

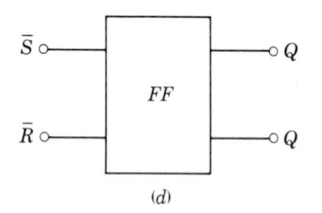

(d)

FIGURE 9.4-1
(a) An SR flip-flop using NAND gates, (b) truth table of the SR flip-flop in terms of $\bar{S}$ and $\bar{R}$, (c) truth table in terms of S and R, (d) symbol of the RS flip-flop shown in (a).

fictitious, variables and using the information in the truth table of Fig. 9.4-1b, we can construct the truth table of Fig. 9.4-1c, now seen to be identical with the table of Fig. 9.1-1b, which applied directly when NOR gates were employed. Thus, in terms of real or virtual variables, the truth table of Fig. 9.1-1b becomes a "universal" table which applies to a flip-flop independently of the type of gate used. A circuit representation of the flip-flop is shown in Fig. 9.3-1d.

9.5 THE CHATTERLESS SWITCH

As a simple, yet useful, application of the SR flip-flop, consider the situation represented in Fig. 9.5-1. In Fig. 9.5-1a a switch is used to make available an output voltage $V_o = 0$ or $V_o = V_S$. These voltages may be the voltages corresponding to the logic levels in a digital system. The circuit is intended to permit us manually to present one or the other logic level to the input of some digital system. However, a characteristic of a mechanical switch is that when the arm is

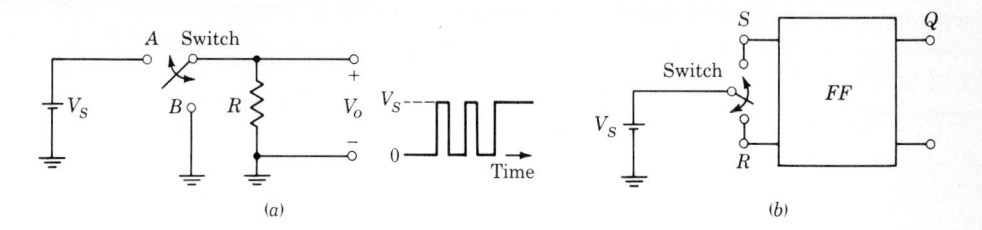

FIGURE 9.5-1
(*a*) A SPDT switch and (*b*) a chatterless switch.

thrown from one position to the other, this moving contact arm of the switch *bounces* or *chatters* several times before finally coming to rest in the position of contact. For example, suppose that we throw the switch with the intention that V_o should go from logic level **0** to logic level **1**. Then, quite typically, the switch will come in contact with position *A*, bounce off *A* (but *not* return to *B*), and make several transitions between levels before coming to rest. Note that when the switch is at position *A*, the output voltage is V_S, while if the switch is not touching *A*, the output voltage is 0 V since no current flows through the resistor *R*.

A *chatterless switch* is shown in Fig. 9.5-1*b*. Here we assume that the flip-flop operates in accordance with the truth table of Fig. 9.1-1. Such would be the case if the flip-flop were constructed, of, say, RTL NOR gates. In this case, a gate input terminal having no external connection (left "floating") can be considered to be at logic **0**. The switch shown in Fig. 9.5-1*b* can be thrown one way or the other to establish a logic level **1** at either *S* or *R*. It is now an easy matter to verify that if the switch is, say, at *R*, then $Q = \mathbf{0}$ or if the switch is then thrown to *S*, *Q* will change to $Q = \mathbf{1}$. Either change of state will occur at the first contact and not be affected by contact bouncing, since the switch rarely bounces all the way back to the other input. Hence, during the switch bounce, $S = R = \mathbf{0}$.

It is common practice to employ an *SR* flip-flop after a mechanical switch to avoid chatter. Either NOR or NAND gate flip-flops can be employed for this purpose.

9.6 CLOCKED FLIP-FLOP

Let us now examine the situation depicted in Fig. 9.6-1 in order to appreciate a problem which arises regularly in connection with the operation of flip-flops. We assume an *SR* flip-flop that operates according to the truth table of Fig. 9.1-1*b*. The circuit shown should be thought of as part of a larger digital system and the inputs to the flip-flop come from various parts of this system. We assume at the moment of interest that $R = \mathbf{0}$ and that $S = \mathbf{0}$ as well, being the combinatorial result of two logic signals *A* and *B* which happen to be at $A = \mathbf{1}$ and

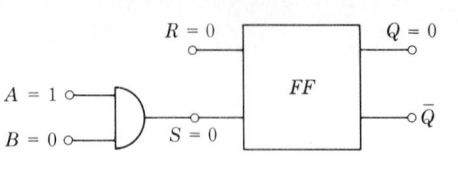

FIGURE 9.6-1
The need for a clocked flip-flop illustrated.

$B = 0$. With $S = R = 0$ both output states are possible, and we assume for the purpose of illustration that $Q = 0$.

Now suppose that both A and B change at about the same time and that this simultaneous change is not intended to change the state of the flip-flop. Toward this end, it is required that A change *before* B, thereby keeping $S = 0$. But in the design of the system it may be difficult to assure such a time sequence. If it should happen that A changes *after* B, the flip-flop will be unintentionally set at the change in A and will *remain set* in spite of the subsequent change in B. In a large digital system propagation delays through gates and other (not easily predictable) delays often cause difficulties of the type discussed here. See Fig. 4.12-3.

These delays cause some difficulties even in ordinary combinatorial circuits, i.e., circuits without memory, but the difficulty is exaggerated when memory is involved. For example, in Fig. 9.6-1, it is intended that S remain at $S = 0$. If B changes at a time Δt before A changes, then S will be in error for only this time Δt. On the other hand, the flip-flop may remain in the wrong state indefinitely.

It is therefore of great advantage to be able to arrange a flip-flop in which the S and R inputs determine the eventual state of the flip-flop but the exact moment of the response of the flip-flop to these inputs is determined by an auxiliary signal. State transitions can then be deferred until all input logic levels which are to influence a flip-flop have indeed established themselves.

A flip-flop which circumvents the timing difficulties discussed above is shown in Fig. 9.6-2. Two additional gates have been added to the basic circuit, and NAND gates are employed rather than NOR gates. The set and reset inputs to the basic flip-flop are here called $\bar{S}'$ and $\bar{R}'$ to reserve the symbols S and R for the set and reset terminals presented to the external world.

Also available as external terminals are $\bar{S}_d$ and $\bar{R}_d$, which are *direct inputs* to the basic flip-flop gates themselves. These inputs are not essential to the operation, and we shall briefly defer consideration of their purpose. We assume, at present, that $\bar{S}_d = \bar{R}_d = 1$, so that the output gates are *enabled*; i.e., when $\bar{S}_d = \bar{R}_d = 1$, neither $\bar{S}_d$ nor $\bar{R}_d$ affects the operation of the flip-flop.

The auxiliary logic input, designated by the letter C, is usually referred to as a *clock* input. When the clock signal applied to this input is at logic level 0, $\bar{S}' = \bar{R}' = 1$ independently of S and R. Either flip-flop state is possible, and the state in effect depends on the past history of S and R. Similarly when $S = R = 0$, the clock cannot influence the state of the flip-flop and again the state in effect depends on the past history.

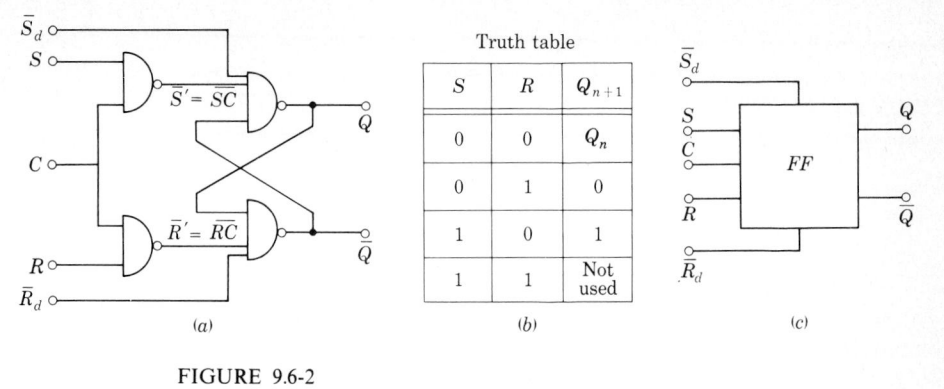

FIGURE 9.6-2

(a) A NAND-gate clocked flip-flop, (b) truth table for part (a), (c) circuit representation of a clocked flip-flop with direct inputs.

Now suppose that while the clock is at logic **0** ($C = 0$), we put $S = 1$ and $R = 0$. Then, at the time when C becomes $C = 1$, the flip-flop will go to the set state (if not already in that state). Thus, by placing $S = 1$ and $R = 0$ we have determined that the flip-flop shall go to the set state, but the moment at which the transition is to take place is determined by the clock, more specifically by the time of occurrence of the *positive-going transition of the clock*. Similarly, assume that $S = 0$ and $R = 1$; then at the positive-going transition of the clock the flip-flop will reset (or remain reset). In either case, as is readily verified, a subsequent return of the clock to $C = 0$ will leave the flip-flop state unaltered. Inputs $S = R = 1$ are not used; for, in that case, at $C = 1$ we would have $\overline{S}' = \overline{R}' = 0$ and when C returned to $C = 0$, the state to which the flip-flop would revert would be unpredictable.

In order to preserve this feature whereby state changes occur only at the time of the positive-going transitions of the clock, it is required generally that S and R shall not change except when the clock is low ($C = 0$). It may therefore be a matter of convenience to arrange that the clock waveform remain at $C = 1$ for no longer than is required to assure that the flip-flop make its state change. In this case the clock waveform would appear as a sequence of relatively narrow positive pulses; i.e., the interval between pulses is long in comparison with the duration of the individual pulses. In this case, the clock waveform is described as being a sequence of *clock pulses*.

In a digital system using clocked flip-flops, as in Fig. 9.6-2, each clock pulse advances the digital processing by one step. The rate at which the processing proceeds is then determined by the *rate* at which these pulses occur and hence the name clock is quite appropriate. In many a system the clock pulses occur at a regular rate, reminiscent of the ticking of a clock, and the term is even more appropriate. Since the entire digital system operates in synchronism with the clock, the system is called a *synchronous system*.

The operation of the flip-flop is summarized in the truth table shown in

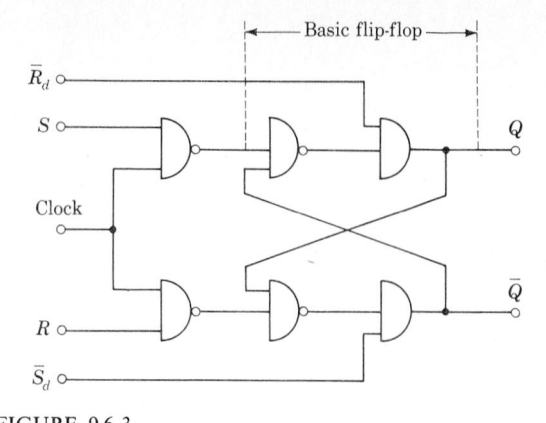

FIGURE 9.6-3
A direct input flip-flop where $\overline{R}_D$ and $\overline{S}_D$ *override* the clock.

Fig. 9.6-2*b*. The symbol Q_n stands for the flip-flop state before the positive nth clock transition and Q_{n+1} for the state after the nth transition. Thus, when $S = R = 0$, the state does not change, and so $Q_{n+1} = Q_n$.

Alternatively, let us consider that a cycle of the clocking waveform ends at the transition of the clock waveform which induces a state change if indeed a state change is called for. (In the present case, the cycle ends when the clock returns to logic **1**.) The symbol Q_n then stands for the state of the flip-flop during the nth clock cycle and Q_{n+1} is the state during the next, the $(n + 1)$st cycle.

Direct inputs The direct inputs $\overline{S}_d$ and $\overline{R}_d$, shown in Fig. 9.6-2*a*, are used to establish an initial state for the flip-flop or to hold the flip-flop in a particular state independently of the data present at the data input terminals S and R. Thus, in the absence of a clock pulse, the state of the flip-flop is determined entirely by the direct inputs.

The direct inputs can be used to override the data inputs. For we see that if $\overline{R}_d = \overline{S}_d = 1$, the data inputs take control. However, if we establish that $\overline{S}_d = 1$ and $\overline{R}_d = 0$ or $\overline{S}_d = 0$ and $\overline{R}_d = 1$, the flip-flop will find itself in the reset or set state, respectively, at the *end* of the clock pulse, i.e., after C returns to logic **0**, independently of the logic levels at S and R. Since S and R are intended to operate in connection with the clock pulse while S_d and R_d are not so intended, S and R are often called *synchronous* inputs and $\overline{S}_d$ and $\overline{R}_d$ are called *asynchronous* inputs. The circuit symbol for the circuit is shown in Fig. 9.6-2*c*.

The direct-input arrangement of Fig. 9.6-2*a* has a feature which is sometimes unacceptable. Consider, say, that $\overline{S}_d = 1$ and $\overline{R}_d = 0$ before, during, and after a clock pulse. Then, before the clock pulse, the flip-flop is reset with $Q = 0$ and $\overline{Q} = 1$ independently of S and R. In addition, we shall find that after the pulse ends (C again is at logic **0**) the flip-flop will be in the reset state. However, if we let $S = 1$ and $R = 0$, then for the duration of the clock pulse $(C = 1)$ while $\overline{Q}$ remains at $\overline{Q} = 1$, Q will change from 0 to 1. For when the clock pulse is at logic 1, $\overline{S'}$ becomes $\overline{S'} = 0$ and Q must go to $Q = 1$.

A flip-flop in which such a feature is absent and the output is completely unresponsive to the clock pulse when the direct inputs are applied is shown in Fig. 9.6-3. It is left to the problems for the student to verify that if $\overline{S}_d = 0$ and $\overline{R}_d = 1$, then $Q = 1$, or if $\overline{S}_d = 1$ and $\overline{R}_d = 0$, then $Q = 0$, each independent of the states of C, S, or R. Further, the student can show that only if $\overline{R}_d = \overline{S}_d = 1$ do the C, S, and R inputs control the flip-flop operation. Note also that because of the new points of application of $\overline{S}_d$ and $\overline{R}_d$, the ordering of these input terminals is different in Figs. 9.6-3 and 9.6-2.

9.7 INTERCONNECTION OF FLIP-FLOPS: THE MASTER-SLAVE FLIP-FLOP

In a clocked flip-flop there is one level of the clock waveform at which the input gates are disabled and the flip-flop is thereby isolated from the data present at the input terminals. (In the circuit of Fig. 9.6-2 this clock level is $C = 0$, but in other flip-flops it may as readily be $C = 1$.) In the circuit of Fig. 9.6-2 the change in output at Q and at $\overline{Q}$ takes place when the clock makes a transition from the level where the input gates are disabled to the level where the gates are enabled. We shall call this transition the *triggering* transition; i.e., the flip-flop responds to a clock transition *from disable to enable*. Such operation is often unacceptable. Much to be preferred is an operation in which the flip-flop responds to a clock transition *from enable to disable*. We shall now explain the reason for preferring the second alternative, and we shall describe one flip-flop circuit which operates in this manner. Other flip-flop circuits which satisfy this requirement will be discussed in later sections.

Two features are quite general in digital systems: (1) a common clocking waveform is used for all the flip-flops in the system, and (2) the data inputs of flip-flops may be derived entirely or in part from the outputs of other flip-flops. Where these two features prevail, a flip-flop that responds when the clock goes from disable to enable is beset by a difficulty. The problem is that the flip-flop responds not only to the data present at its input before the clocking transition but also to the *new data* presented as a result of the fact that other flip-flops have also changed output.

To clarify this point in a simple example we consider the situation represented in Fig. 9.7-1, where a cascade of two clocked flip-flops is shown. The flip-flops are of the type shown in Fig. 9.6-2, which responds when the clock goes from $C = 0$ (input gates disabled) to $C = 1$ (gates enabled). We assume that the direct inputs (not shown) have been used to place each flip-flop in the reset state initially so that $Q_1 = 0$ and $Q_2 = 0$. We intend the circuit to operate in the following manner. When the clock waveform is applied, during the first cycle of the clock, Q_1 should go to $Q_1 = 1$ while Q_2 should

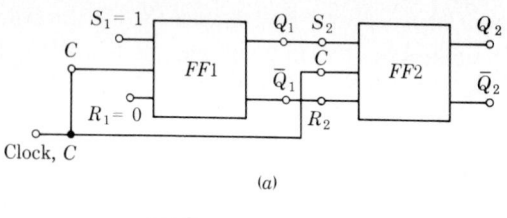

(a)

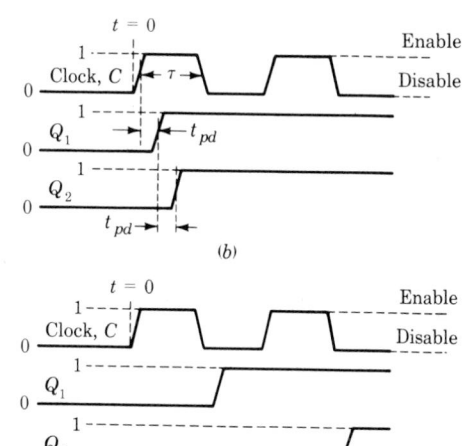

(b)

(c)

FIGURE 9.7-1

(a) A cascade of two flip-flops, (b) waveforms for the cascade shown in part (a), (c) desired waveforms.

remain at $Q_2 = 0$. During the second clock cycle Q_1 should remain at $Q_1 = 1$, but Q_2 should become $Q_2 = 1$. That is, the input data $(S1 = 1, R1 = 0)$ are to be transferred to $FF1$ during the first clock cycle and transferred as well to $FF2$ during the second clock cycle. (Note that if this two-flip-flop cascade worked, we could add additional flip-flops to the cascade and then have a circuit which, clock cycle by clock cycle, would shift input data down the cascade. Such a cascade is called a *shift register* and is discussed further in Sec. 10.1.

We may, however, see, as shown in the waveform of Fig. 9.7-1b, that this anticipated sequence will not take place. In this figure, finite rise and fall times and propagation delays have been indicated. The clock waveform is applied at $t = 0$, and at the first transition of C (from $C = 0$ to $C = 1$) Q_1 goes to $Q_1 = 1$ since $S_1 = 1$ and $R_1 = 0$. Initially Q_2 does not change since $S_2 = Q_1 = 0$ and $R_2 = \bar{Q}_1 = 1$. However, after the propagation delay through $FF1$, S_2 becomes $S_2 = 1$ and $R_2 = 0$. Since C is *still at* $C = 1$, the enabling level, $FF2$ makes a transition to $Q_2 = 1$. Hence, we find that $FF2$ responds and Q_2 goes to $Q_2 = 1$ during the first clock cycle rather than during the second clock cycle as intended. (If there were additional flip-flops in the cascade, they also would respond after each successive flip-flop propagation delay so long as C remained at $C = 1$.)

One way of relieving the difficulty is to arrange that the clock stay at the enabling level for a very short time. In such a case the clock waveform in Fig. 9.7-1 would have the appearance of a train of positive pulses, the pulse duration being very short in comparison with the interval between pulses. In this case, if the clock-pulse duration were shorter than the propagation delay through a flip-flop, the circuit would operate as required; for in such a situation we would have C returning to the disabling level $C = 0$ before the response of $FF1$ to the clock pulse became apparent at $Q_1 = S_2$ and $\overline{Q}_1 = R_2$. This solution, however, requires the generation and transmission throughout a digital system of very narrow pulses. As a matter of fact, in ECL, where propagation delays of only 2 ns are not uncommon, generating and handling appropriately narrow pulses might well represent state-of-the-art circuitry. In many situations, it may well be found that pulses narrow enough to avoid the timing difficulty described above may well be too narrow to trigger the flip-flop reliably.

A better solution is found through the use of flip-flops designed so that the triggering transition is the transition *from the enabling to the disabling level.* Corresponding waveforms are shown in Fig. 9.7-1c. Here $FF1$ responds when the clock goes to the disabling level. This response of $FF1$ changes Q_1 from $Q_1 = 0$ to $Q_1 = 1$ and changes $\overline{Q}_1$ from $\overline{Q}_1 = 1$ to $\overline{Q}_1 = 0$. However, these changes in $S_2 = Q_1$ and $R_2 = \overline{Q}_1$ do not occur until *after* the triggering transition, and after the triggering transition the input gates of $FF2$ are disabled and hence $FF2$ cannot respond. However, as shown, at the time of the second triggering transition $FF1$ has had time to establish itself in the set state, fixing S_2 at $S_2 = 1$ and R_2 at $R_2 = 0$, so that this second triggering transition produces a response in $FF2$.

We have seen the usefulness of arranging that triggering be done by the clock transition from the enable to disable level in connection with the coupling of flip-flops. We shall find this principle equally useful when connections from a flip-flop output are made back to the input of the very same flip-flop. Such is the situation in the case of the JK flip-flops to be discussed in Sec. 9.12.

One type of flip-flop that responds when the clock makes a transition to the disable level is a *master-slave* flip-flop. The master-slave flip-flop, shown in Fig. 9.7-2, employs two individual clocked flip-flops, each of the type shown in Fig. 9.6-2. One flip-flop is called the *master* and the other the *slave.* The clock C is applied to the input gates of the master, but the clock complement $\overline{C}$ is applied to the input gates of the slave. When the clock is high $(C = 1)$, the clock complement is low $(\overline{C} = 0)$. Hence, when gates $1A$ and $1B$ are enabled $(C = 1)$, gates $3A$ and $3B$ are disabled, and vice versa.

When the clock is high, the data at S $(= S_M)$ and R $(= R_M)$ are transferred to the master flip-flop. For example, if $S = 1$ and $R = 0$, then when the clock goes high, Q_M will become $Q_M = S_S = 1$ and $\overline{Q}_M$ will become $\overline{Q}_M = R_S = 0$. These levels Q_M and $\overline{Q}_M$ cannot affect the slave since when the clock is high, the slave gates $3A$ and $3B$ are disabled $(\overline{C} = 0)$. But Q_M and $\overline{Q}_M$ are the data inputs

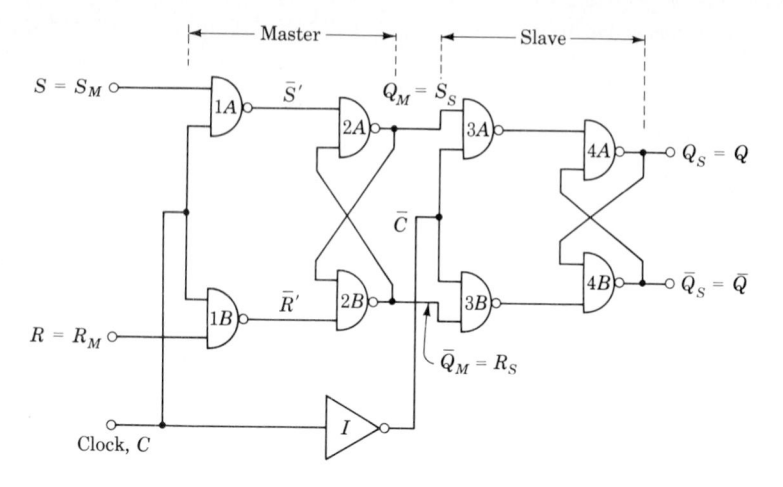

FIGURE 9.7-2
A master-slave flip-flop.

S_S and R_S of the slave flip-flop. Hence, when the clock goes low, so that $\overline{C}$ goes high, the data at Q_M and $\overline{Q}_M$ will be transferred to the outputs of the slave. Thus, to summarize, when C goes high, the data at the data input terminals S and R are registered in the master but restrained from passing on to the slave. When the clock pulse reverts to logic 0, at which time the input gates are disabled, the data in the master (Q_M) are transferred to the slave flip-flop and appear at the output Q_S and $\overline{Q}_S$.

The truth table for the master-slave flip-flop of Fig. 9.7-2 is identical to the truth table in Fig. 9.6-2. The distinctive feature of the master-slave circuit, however, is that with this arrangement, if changes in the output at Q and $\overline{Q}$ are to take place, they do so when the clock moves to the 0 state. Essential to the operation of the circuit is the requirement that as C changes from $C = 0$ to $C = 1$, gates $3A$ and $3B$ become disabled *before* the data entering gates $1A$ and $1B$ can be transmitted to gates $3A$ and $3B$. It is apparent from Fig. 9.7-2 that if the propagation delay time of each gate in the flip-flop is t_{pd}, then gates $3A$ and $3B$ are disabled a time t_{pd} before the state of Q_M can change. Hence the circuit always operates properly.

Direct inputs When direct (asynchronous) data inputs are to be made available on a master-slave flip-flop, they must be connected to *both* the master and slave, as shown in Fig. 9.7-3, in order to *override* the clock. It can be verified, for this circuit, that when the two $\overline{S}_d$ inputs are at $\overline{S}_d = 1$ and both $\overline{R}_d$ inputs are at $\overline{R}_d = 0$, then $Q = Q_M = 0$ and $\overline{Q} = \overline{Q}_M = 1$. Further, in this situation, both flip-flops, master and slave, are independent of the logic level at S and R and are entirely unresponsive to the clock pulse. Similarly with $\overline{S}_d = 0$ and $\overline{R}_d = 1$, $Q = Q_M = 1$ and $\overline{Q} = \overline{Q}_M = 0$. The flip-flop is released from direct control by establishing $\overline{S}_d = \overline{R}_d = 1$. Direct inputs with levels $\overline{S}_d = \overline{R}_d = 0$ are not used.

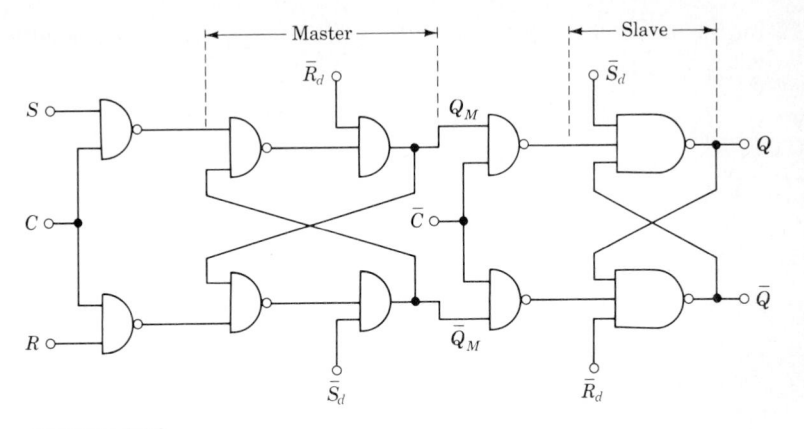

FIGURE 9.7-3
A master-slave flip-flop with direct inputs which override the synchronous inputs.

Some commercially available master-slave flip-flops provide direct-control terminals only for the slave flip-flop. In such a case the situation in connection with the slave is entirely similar to the situation depicted in Fig. 9.6-2. Here, as described in Sec. 9.6, the direct inputs do not completely override the level transitions of the clock pulse.

The circuit symbol for the flip-flop of Fig. 9.7-3 is shown in Fig. 9.7-4. The direct set and reset inputs exercise control at logic **0**, and this feature is indicated both by the complement bar over the words "set" and "reset" and by the complement circles on the set and reset leads. There is, of course, redundancy in this symbolism. Similarly, the fact that output changes occur at the negative-going transition of the gate is often indicated by both the complement bar over the clock and by the complement circle. In the matter of the clock, sometimes the redundancy is removed and only one symbolism or the other is employed. Sometimes, also, neither of these symbolisms is used, and instead the symbol which appears in the flip-flop box shown in the figure is employed.

The master-slave flip-flop has a flip-flop (the slave) which provides the output signal. It also has a second flip-flop (the master) which provides storage

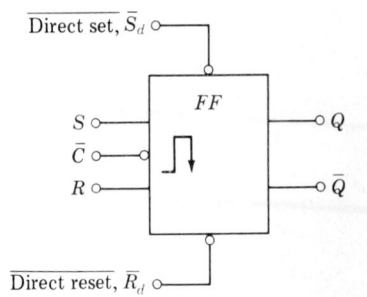

FIGURE 9.7-4
Circuit symbol for a clocked flip-flop.

for the input data.   Because of this storage, input data are available to the slave when the clock is at the level at which the input gates are disabled, and the flip-flop is thereby disconnected from the external world.   In succeeding sections we shall consider other flip-flops in which only a single flip-flop is employed (corresponding to the slave) and in which the disable mechanism is supplied by other means.

9.8   THE AC-COUPLED EDGE-TRIGGERED FLIP-FLOP

In this section we study the operation of the asynchronous ac-coupled flip-flop shown in Fig. 9.8-1.   This circuit represents the basic concept underlying the operation of the high-speed ECL flip-flop.   A study of the actual ECL circuitry is postponed until Sec. 9.18.   In Sec. 9.9 we consider the clocked ac-coupled flip-flop which serves the same function as a master-slave flip-flop, but uses only a single flip-flop.

An *ac-coupled* flip-flop is one in which the input terminals are capacitively coupled to an *RS* flip-flop, as shown in the figure.   As a result, the *RS* flip-flop changes state due to the *transition* of the *R* or *S* input rather than its *level*. Thus, the ac-coupled flip-flop is often called an *edge-triggered* flip-flop.   In Fig. 9.8-1a we have used NOR gates since the circuit is usually constructed using ECL; however, NAND gates can also be employed (see Fig. 9.8-3).

Because of the capacitive coupling of the data, fixed levels at the input terminals will have no effect on the flip-flop.   However, *changes* in logic levels may have an effect.   To indicate that the response, if any, is to *changes*, the data inputs are here labeled (following common usage) S_D and R_D, the subscripts having the significance *dynamic*.   This use should not be confused with the *direct* coupled inputs S_d and R_d discussed earlier.

We noted earlier that in cases where the *level* of a logical signal is important, the custom is to assign logic symbols to terminals as complemented or uncomplemented variables in such manner that logic **1** corresponds to the situation in which something may happen.   In the case where the *change* from level to level is important rather than the level itself, there appears to be no generally accepted convention.   We shall find it more generally consistent with practice, however, to assign dynamic symbols in such manner that where an *uncomplemented* dynamic symbol appears, it indicates that the change will be effective when the change is *from* logic level **1** *to* logic level **0**.   Correspondingly a change from logic level **0** to logic level **1** will be ineffective.   Whether the effective transition (or the ineffective transition) is positive-going or negative-going depends on the type of gates employed.

Finally, combining the conventions we have adopted, we note the following by way of example.   Suppose a terminal is marked R_D and suppose it is specified that $R_D = 1$.   We shall then interpret the statement to mean that there was an *effective* transition at the terminal, i.e., from **1** to **0**.   If, on the other hand it is specified that $R_D = 0$, we understand that *either* there was an *ineffective* transition

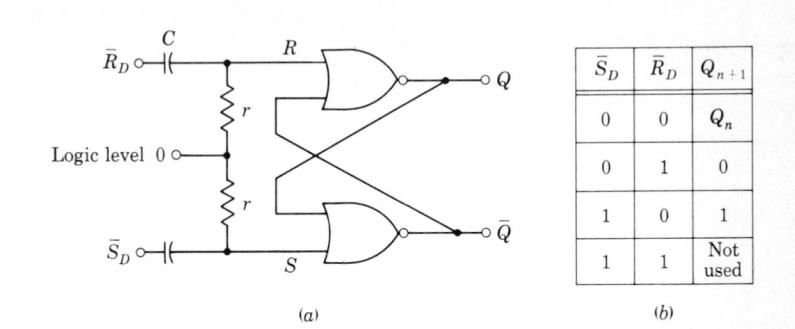

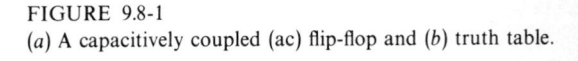

(a)   (b)

FIGURE 9.8-1
(a) A capacitively coupled (ac) flip-flop and (b) truth table.

from **0** to **1** or there was *no* transition. An *effective* transition at a terminal marked $\bar{S}_D$ would go from **0** to **1** and would be represented by $\bar{S}_D = 1$.

Let us assume that the NOR gates of Fig. 9.8-1 operates with positive logic. Suppose, then, that the voltage at $\bar{S}_D$ or at $\bar{R}_D$ executes a negative-going transition from logic **1** to logic **0**. Then a negative pulse-type waveform will appear at S or at R. Since logic **0** is not a fixed voltage level but is represented by any voltage less than some threshold voltage, the negative pulse will simply leave R or S at logic **0**. Hence a negative-going transition at $\bar{S}_D$ or $\bar{R}_D$ will have no effect on the flip-flop. Next, consider that there is a positive-going transition at, say, $\bar{S}_D$ alone ($\bar{S}_D = 1$, $\bar{R}_D = 0$). Then a positive pulse will appear at S. If this pulse is of adequate amplitude, S will be carried to the logic **1** level and the flip-flop will go to the set state or will be left in the set state if already there. The subsequent decay of the pulse which will return S to logic **0** will not affect the flip-flop. Similarly a positive-going transition at $\bar{R}_D$ alone ($\bar{R}_D = 1$, $\bar{S}_D = 0$) will reset the flip-flop or leave it in the reset state if it is already there. The truth table for the flip-flop is given in Fig. 9.8-1b.

Since, in Fig. 9.8-1a, $\bar{R}_D$ and $\bar{S}_D$ are applied to the RS flip-flop through capacitors, it is appropriate to consider the situation represented in Fig. 9.8-2a. Let the input voltage $V_i(t)$ be a positive pulse having a positive-going transition that appears to be a voltage *ramp*. Then if the pulse changes in voltage by an amount ΔV in a time interval T, the input voltage can be written as $V_i(t) = (\Delta V/T)t$ during this transition interval.

Figure 9.8-2b shows the output voltage $V_o(t)$ when the rise time T greatly exceeds the filter time constant $\tau = rC$. In this case the filter acts like a differentiator, and the output is a pulse having an amplitude $\tau \Delta V/T$ and a duration approximately equal to T. Figure 9.8-2c shows the output voltage $V_o(t)$ when T is decreased so that $T \ll \tau$. In this case the output voltage follows the input voltage rather closely and therefore attains a peak voltage approximately equal to ΔV.

We see that as the rise time T of the input pulse increases, the pulse amplitude presented to R or S decreases. Since the RS flip-flop requires some

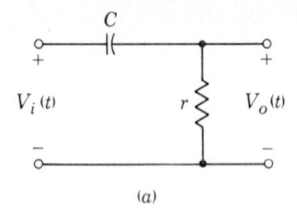

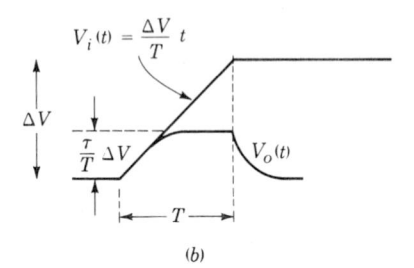

FIGURE 9.8-2
(a) The rC input circuit, (b) $T \gg \tau$,
(c) $T \ll \tau$.

minimum input voltage in order to change state, the rise time of the input signal T must be less than some maximum value. For example, consider that a level change at S or R of 0.75 V is needed to change the state of the flip-flop; then if the input voltage changes by 3 V and $\tau = 2$ ns, the rise time T must be less than 8 ns.

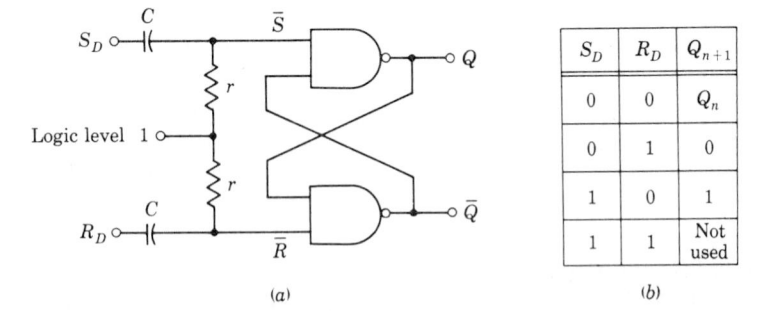

S_D	R_D	Q_{n+1}
0	0	Q_n
0	1	0
1	0	1
1	1	Not used

(a)   (b)

FIGURE 9.8-3
(a) Edge-triggered flip-flop using NAND gates and (b) truth table.

A capacitively coupled flip-flop using NAND gates is shown in Fig. 9.8-3a. Note that the resistors are returned to logic **1**. In this case transitions of S_D and R_D from logic **1** to logic **0** will be effective (negative-going transitions if the gates use positive logic). Hence the dynamic variables appear uncomplemented. The truth table is given in Fig. 9.8-3b.

9.9 THE CLOCKED AC-COUPLED FLIP-FLOP

A clocked ac-coupled flip-flop is shown in Fig. 9.9-1a. The circuit is a development of the ac flip-flop of Fig. 9.8-1, the modification consisting of the addition of the two input gates 1A and 1B. This clocked flip-flop has the basic configuration of the ac-coupled ECL flip-flop to be discussed in Sec. 9.18. The circuit may be controlled by its direct set and direct reset terminals, or it may be used as a clocked flip-flop or as an asynchronous ac-coupled flip-flop. Since there are three modes of use, three truth tables are required to describe its operation.

The truth table for the direct set and direct reset is given in Fig. 9.9-1b. This table assumes that R' and S' are kept at $R' = S' = 0$. One way this condition can be assured is by keeping the clock $\overline{C}$ at logic **1** so that both input gates are disabled.

The truth table for clocked operation is given in Fig. 9.9-1c. To verify that the table is correct consider, for example, that $\overline{S} = 0$ and $\overline{R} = 1$. Then gate 1A is disabled, and R' will not be affected by the clock, remaining always at $R' = 0$. Assume then that the clock $\overline{C}$ makes a transition from logic **0** to logic **1**. Then S' will go briefly from **0** to **1**. With $R' = 0$ and $S' = 1$, Q must go to $Q = 1$, the set state. The other entries in the truth table can be verified similarly.

If the clock logic level is held at $\overline{C} = 0$, both input gates are enabled and the $\overline{R}$ and $\overline{S}$ data inputs can be used just as in the ac flip-flop of Fig. 9.8-1. When this type of operation is intended, it is appropriate, as indicated, to relabel the $\overline{R}$ and $\overline{S}$ inputs as $\overline{R}_D$ and $\overline{S}_D$, respectively. The truth table for this asynchronous edge-triggered operation is the same as in Fig. 9.8-1b, and is duplicated in Fig. 9.9-1d.

Earlier, in Sec. 9.7, we noted that in synchronous digital systems the situation that generally prevails is one in which the data inputs to a flip-flop are derived entirely or in part from the outputs of other flip-flops. We noted further that to assure proper operation in such circumstances the flip-flops in the system must be designed in such a manner that the clock-triggering transition is the clock transition which carries the input gates from the *enable to the disable level*. We can now verify that the ac-coupled flip-flop does indeed satisfy this condition.

Refer to Fig. 9.9-1a. If $\overline{C}$ goes from $\overline{C} = 1$ to $\overline{C} = 0$, then depending on the levels at $\overline{S}$ and $\overline{R}$, the outputs of the gates 1A and 1B will either not respond at all or respond with a negative-going transition. In either case, the flip-flop gates 2A and 2B will not be affected. Next suppose that $\overline{S} = 0$ and

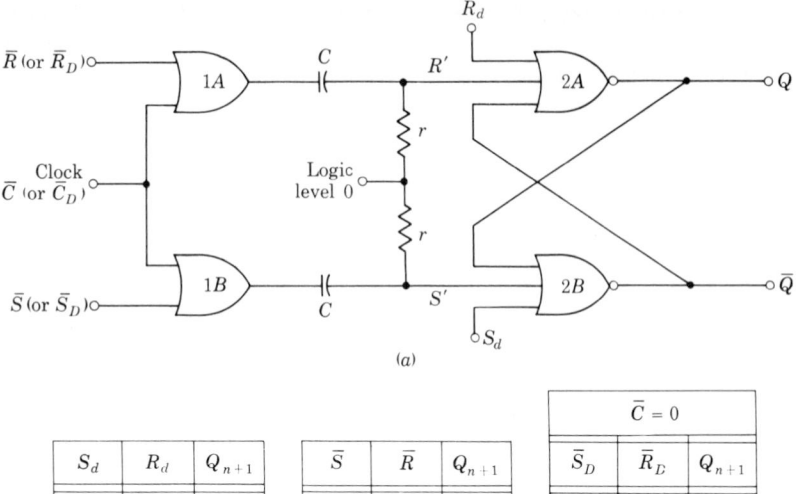

(a)

S_d	R_d	Q_{n+1}
0	0	Q_n
0	1	0
1	0	1
1	1	Not used

(b)

$\bar{S}$	$\bar{R}$	Q_{n+1}
0	0	Not used
0	1	1
1	0	0
1	1	Q_n

(c)

$\bar{C} = 0$		
$\bar{S}_D$	$\bar{R}_D$	Q_{n+1}
0	0	Q_n
0	1	0
1	0	1
1	1	Not used

(d)

FIGURE 9.9-1

(a) Clocked ac-coupled flip-flop, (b) asynchronous operation using direct inputs S_d and R_d, (c) clocked operation, (d) asynchronous ac operation; $\bar{C} = 0$ when this mode of operation is desired.

$\bar{R} = 1$, and let $\bar{C}$ make a positive-going transition from $\bar{C} = 0$ to $\bar{C} = 1$. Then this transition will have no effect at the output of gate $1A$, but the output of gate $1B$ will respond with a positive-going transition which will appear at the input of gate $2B$ as a positive pulse. As a result, the flip-flop will go to the set state $Q = 1$, as required by the truth table of Fig. 9.9-1c. Similarly for $\bar{S} = 1$ and $\bar{R} = 0$ the flip-flop will go to the reset state with $Q = 0$ when $\bar{C}$ goes from 0 to 1. In any event, and most importantly, allowing for the propagation time through the flip-flop, *by the time the flip-flop output has assumed its new state, the input gates will have settled into a disabled condition as required.*

We have already pointed out one difference between the master-slave flip-flop and the ac-coupled flip-flop, namely, that in the ac-coupled flip-flop there is a minimum transition time required of the clock. We consider, now, a second difference. In the master-slave, the input data may be changed at any time. The data may be changed when the input gates are disabled or when they are enabled. No matter when these changes are made, no change will appear at the output except at the triggering transition of the clock. At that time the data to which the flip-flop will respond are the most recent data present just before the triggering transition. Now consider the ac-coupled flip-flop of Fig. 9.9-1; when $\overline{C} = 1$, so that the input gates are disabled, we are again free to update the input data. But suppose $\overline{C} = 0$; then a positive-going change in $\overline{S}_D$ or $\overline{R}_D$ may possibly affect the flip-flop gates $2A$ or $2B$, depending on the state in which the flip-flop happens to find itself. Hence, in an ac-coupled flip-flop the general precaution needs to be observed that data changes are to be made only when the input gates are disabled, that is, $\overline{C} = 1$.

9.10 A CAPACITIVE STORAGE FLIP-FLOP

A second method by which capacitors are used in a flip-flop to achieve edge triggering and master-slave operation is shown in Fig. 9.10-1. This technique is often employed with TTL circuits.

The disable level of the clock is $C = 0$ $(\overline{C} = 1)$ since at this clock level gates $1A$ and $1B$ are disabled. With $C = 0$, S' and $R' = 0$. Now suppose, by way of example, that when $C = 0$, $S = 1$ and $R = 0$. Then when the clock goes to $C = 1$, we shall charge the capacitors so that $S' = 1$ and $R' = 0$. These levels S' and R' can have no effect on the output of the flip-flop since as the clock goes to $C = 1$, $\overline{C}$ becomes $\overline{C} = 0$ and gates $2A$ and $2B$ are disabled.

Now let the clock go back to its disable level, $C = 0$. Then if the capacitors were not present, S' and R' would immediately revert to $S' = R' = 0$, disabling gates $2A$ and $2B$. The capacitors, however, will hold the levels $S' = 1$ and $R' = 0$ for a short time. Hence as the clock reverts to $C = 0$, $\overline{C}$ goes to the **1** state, the output of gate $2A$ will go to logic level **0**, and the output of gate $2B$ will remain at logic level **1**. Hence, if not already in that state, the flip-flop will go to the set state, $Q = 1$.

Thus, we observe that the capacitors here provide the same data storage as that provided by the master in the master-slave flip-flop.

This capacitance storage circuit has two limitations. First a restriction is placed on the maximum allowable fall time of the clock waveform C, for the

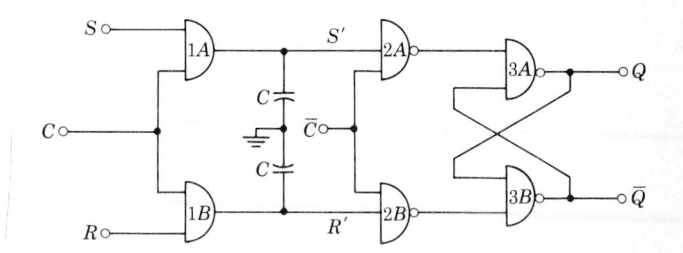

FIGURE 9.10-1
A capacitive storage flip-flop.

fall time must be short enough to ensure that in that interval the capacitors do not have the opportunity to discharge appreciably. A second restriction is that the clock pulse width must exceed a specified minimum value needed to ensure that the capacitors charge to a sufficiently high voltage. Typical values are that the clock fall time $t_f \leq 150$ ns and the clock pulse width ≥ 20 ns. For this reason very high-speed TTL circuits which operate in the 50-MHz range and above do not employ capacitive charge storage.

9.11 PROPAGATION-DELAY FLIP-FLOPS

We discuss now an additional method of constructing a flip-flop with edge triggering. This technique is often employed with RTL gates. Consider initially the gate circuit of Fig. 9.11-1a. If $B = 1$, then $C = 0$ and $D = 0$ independently of A. Hence, $B = 1$ *disables* the gate configuration. Next, let $B = 0$. In this case the two inputs A and C to gate 2 are always complementary. Hence, it would again appear that D is always $D = 0$. This is not the case. To illustrate this point assume $B = 0$, and let there be a level transition in A from $A = 0$ to $A = 1$, as in Fig. 9.11-1b. Then (assuming positive logic) there will correspondingly be, as shown, a negative-going transition in C. However, because of the *propagation delay* through gate 1, the transition in C will be somewhat delayed. Hence, there will be an interval t_{pd1}, as shown, when both A and C will be $A = C = 1$, and for such an interval (after some further delay through gate 2) we shall have $D = 1$.

It can further be verified that with $B = 0$, a *negative*-going transition in A will have no effect at D and that with $B = 1$ neither a positive- nor a negative-going transition will have an effect at D. This situation is similar to that described in connection with Fig. 4.12-3.

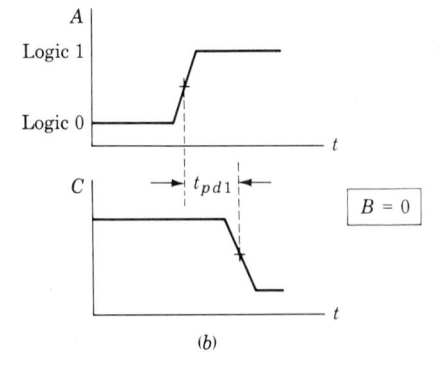

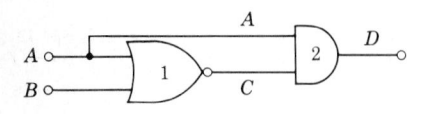

(a)

(b)

FIGURE 9.11-1

(a) Two gates connected to yield an output of **1** for a time approximately equal to the propagation delay time of gate 1; (b) waveforms for A and C, when $B = \mathbf{0}$, to show that $D = \mathbf{1}$ for a time interval approximately equal to t_{pd1}.

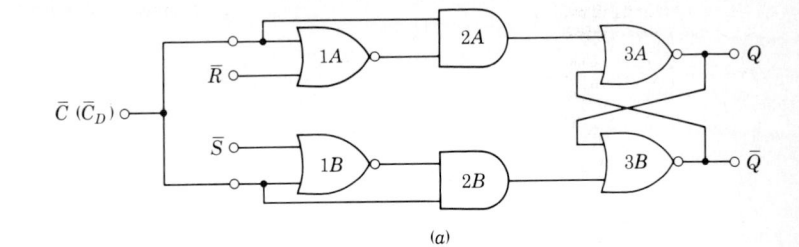

(a)

$\bar{S}$	$\bar{R}$	Q_{n+1}
0	0	Not used
0	1	1
1	0	0
1	1	Q_n

(b)

FIGURE 9.11-2
(a) An edge-triggered flip-flop using the principle of propagation delay; (b) truth table.

The two-gate structure shown in Fig. 9.11-1a is incorporated in the flip-flop of Fig. 9.11-2a. The clock level $\bar{C} = 1$ is the disable level. For with $\bar{C} = 1$ gates 1A and 1B are disabled and $\bar{S}$ and $\bar{R}$ can be changed without affecting the output state of the flip-flop. As we have seen, a negative-going transition in $\bar{C}$, from $\bar{C} = 1$ to $\bar{C} = 0$, will have no effect on the outputs of gates 2A and 2B no matter what the logic levels of $\bar{S}$ and $\bar{R}$ happen to be. Now let us assume, for example, that $\bar{S} = 0$ and $\bar{R} = 1$. Then when $\bar{C}$ makes a positive transition from $\bar{C} = 0$ to $\bar{C} = 1$, there will be an interval when the output of gate 2B goes to logic level 1. As a consequence the flip-flop will go to its set state with $Q = 1$ ($\bar{Q} = 0$).

As with the flip-flops involving capacitors, the propagation-delay flip-flops place a restriction on the maximum allowable rise time of the clock. The truth table describing the system operation is shown in Fig. 9.11-2b.

9.12 THE JK FLIP-FLOP

Each of the clocked flip-flops we have discussed has a pair of input gates, the inputs to which are the clock waveform together with data inputs. For example, in Fig. 9.6-2a the input gates are NAND gates, one with inputs S and C, the other with inputs R and C. A similar situation applies in Fig. 9.7-2, where the gates are labelled 1A and 1B. In Fig. 9.9-1, NOR gates are used, and the inputs are $\bar{C}$ and $\bar{S}$ and $\bar{C}$ and $\bar{R}$.

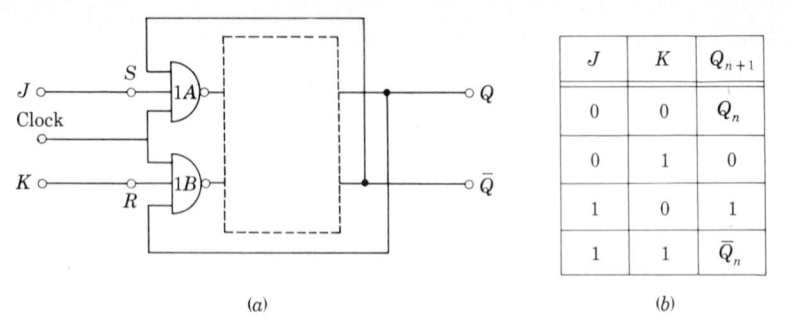

J	K	Q_{n+1}
0	0	Q_n
0	1	0
1	0	1
1	1	$\bar{Q}_n$

(a)   (b)

FIGURE 9.12-1
(a) The JK flip-flop and (b) truth table.

In Fig. 9.12-1 we show a clocked flip-flop with a modification. The dashed rectangle represents all of the flip-flop with the exception of the two input gates, which are shown explicitly. NAND gates are assumed here, but other types would serve just as well. The modification consists in providing an additional input terminal at each gate, and there is a connection, as shown, from the flip-flop outputs to these inputs. The data terminal previously identified as S is now called J, and data terminal R is now called K.

In the absence of this modification, the logic levels at S and R "steered" the clock waveform. That is, depending on S and R, one or the other of the gates $1A$ and $1B$ was enabled, and the clock set or reset the flip-flop. The point of the modification is to arrange that this clock steering be determined not alone by S and R but by the state of the flip-flop as well.

The operation of the JK flip-flop can be determined from the truth table given in Fig. 9.12-1b. First let $J = K = 0$. Then both input gates are disabled, and the clock will not change the flip-flop state. Hence, $Q_{n+1} = Q_n$.

Next, assume that $J = 0$ and $K = 1$ and that the flip-flop is in the reset state with $Q = 0$. Then gate $1A$ is disabled because $J = 0$, and $1B$ is disabled because $Q = 0$. Hence, the clock will not move the flip-flop out of the reset state. But suppose that while, as before, $J = 0$ and $K = 1$, the flip-flop is instead in the set state with $Q = 1$. Then the reset gate $1B$ is enabled, and the clock will cause a transfer of the flip-flop to the reset state with $Q = 0$. Similarly, with $J = 1$ and $K = 0$, the clock will set the flip-flop if it is not already in the set state.

Finally let $J = K = 1$. Then, which one of the gates $1A$ or $2A$ is enabled depends entirely on Q_n and $\bar{Q}_n$, that is, on the state of the flip-flop. If $Q_n = 0$, then $1A$ is enabled. The clock will set the flip-flop to $Q_{n+1} = 1$. If $Q_n = 1$, the clock will reset the flip-flop to $Q_{n+1} = 0$. Thus, each cycle of the clock waveform will change the state of the flip-flop. The word *toggle* is used to describe this change of state induced by each clock cycle. Note that in the JK truth table, $J = K = 1$ specifies $Q_{n+1} = \bar{Q}_n$. Note also that in the JK truth table, unlike the earlier SR truth tables, there is no combination of data inputs for which we find the term "not used."

We now take a closer look at the toggle mode of operation of the JK flip-flops. In so doing we shall come to appreciate again the need for the master-slave type of operation, discussed in Secs. 9.7 to 9.11, in which the output state of the flip-flop changes when the clock C goes to the *disable* level.

In the flip-flop of Fig. 9.12-1, since NAND gates are used at the input, the logic **0** level of the clock is the disable level and logic **1** is the enable level. Suppose then that $J = K = 1$, $Q = 0$, and $\overline{Q} = 1$. Now assume that the clock rises from logic **0** to logic **1** so that gate $1A$ is enabled. If the flip-flop responded at this transition, as would be the case for the flip-flop shown in Fig. 9.6-2, the flip-flop would set and we would have $Q = 1$ and $\overline{Q} = 0$. This in turn would enable gate $1B$, and the flip-flop would respond again, going back to $Q = 0$. With $Q = 0$, we start all over again. The result is that the flip-flop would oscillate between states until the clock returned to logic **0**. Aside from the disadvantage associated with this oscillation, the final state in which the flip-flop would end up would be unpredictable.

On the other hand, if the output levels of the flip-flop change only after the clock reverts to the disable level, these difficulties will be circumvented. For by the time Q and $\overline{Q}$ have changed, a change which will interchange the enabled-disabled condition of gates $1A$ and $1B$, the clock will have fallen and both input gates will thereby be disabled. As a matter of fact, a JK flip-flop would not be feasible except with a flip-flop that has the feature of triggering as the clock goes to the disable level, as in the master-slave configuration.

9.13 THE TYPE-D FLIP-FLOP

It is often required to delay a digital signal in time by the duration of one clock cycle. This operation can be performed by using an SR or a JK flip-flop, connected as shown in Fig. 9.13-1a. The logical signal, i.e., the data, is applied to the S terminal of an SR flip-flop or the J terminal of a JK flip-flop. The complement of the data is then applied to the R or K terminal. If at some time the datum D is $D = 0$, then, after the triggering transition of the next clock cycle, we shall have $Q = D = 0$. For with $S = 0$, $R = 1$ (or with $J = 0$, $K = 1$) the flip-flop goes to the reset state with $Q = 0$. Similarly, with $D = 1$, the next triggering transition will set the flip-flop to $Q = 1$.

The appropriate waveforms are shown in Fig. 9.13-1c. Here it is assumed that the negative-going transition of the clock is the triggering transition. It is hence the case that the data present when the clock is in the **1** state are transferred to the output of the flip-flop during the negative transition of the clock (in an edge-triggered flip-flop) or when the clock returns to the **0** level (in a master-slave flip-flop). With these assumptions, it is readily verified that, with the data sequence given, the Q output is as shown, i.e., identical to the data except delayed by one clock cycle.

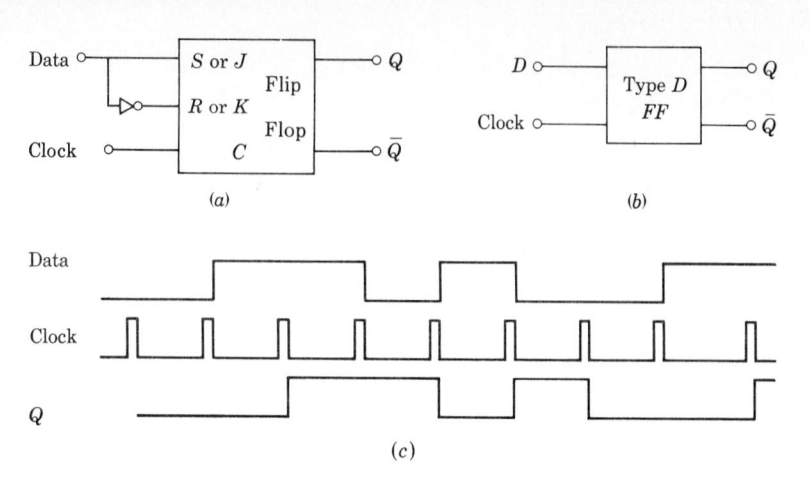

FIGURE 9.13-1

(a) The type-D flip-flop, (b) symbol for type-D flip-flop, and (c) waveforms.

In Fig. 9.13-1c we have made it appear as though the triggering edges of the clocking waveform occur precisely at the time at which the data were changing. If such were indeed the case, the response of the flip-flop would be ambiguous. As a matter of practice, the changes in data would occur slightly *after* the triggering transition.   The delay time would be the order of the propagation delay of a gate or of some other flip-flop.   Such would be the delay encountered if the delay itself were taken from the output of another flip-flop driven by the same clocking waveform.   Additionally there will be a delay (not indicated in the figure) between the output Q and the triggering edge of the clock waveform.

This delay operation does not require of the flip-flop the full versatility available in the SR or JK master-slave type of flip-flop.   Hence, in the design of a flip-flop intended only for the present application, some economies can be effected.   Such a special flip-flop is called a type D (for delay) flip-flop and represented by the circuit symbol shown in Fig. 9.13-1b.

A diagram of a type-D flip-flop is shown in Fig. 9.13-2, and we now consider its operation.   The output flip-flop that provides Q and $\bar{Q}$ consists of NAND gates 3A, 3B while gates 1A, 1B and 2A, 2B are two interconnected steering flip-flops.   It can now be verified (Prob. 9.13-1) that the following statements are valid:

1   When $C = 0$, $\bar{S} = \bar{R} = 1$ independently of D.   Hence gates 3A, 3B are enabled, and either state of the flip-flop is allowed.   This result does not depend on the sequence by which we arrive at the condition $C = 0$.   With either state allowed, the output Q therefore depends on the past history, and $Q_{n+1} = Q_n$.

2   Starting with $D = 0$ and $C = 0$, let C become $C = 1$.   Then $\bar{S}$ and $\bar{R}$ become $\bar{S} = 1$ and $\bar{R} = 0$, so that Q goes to $Q = 0$.   If now, while $C = 1$,

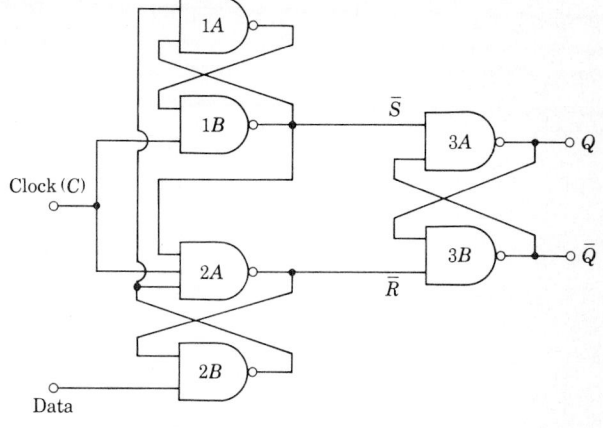

FIGURE 9.13-2
A type-D flip-flop.

there should be any subsequent changes in D, $\bar{S}$ and $\bar{R}$ will remain unaltered, as will Q. When C returns to $C = 0$, $\bar{S}$ remains $\bar{S} = 1$ and $\bar{R}$
becomes $\bar{R} = 1$, so that the output flip-flop remembers its previous input
data and Q remains $Q = 0$.

3 Starting with $D = 1$ and $C = 0$, let C become $C = 1$. Then $\bar{S}$ and $\bar{R}$
become $\bar{S} = 0$ and $\bar{R} = 1$ so that Q goes to $Q = 1$. If now, while $C = 1$,
there should be any subsequent changes in D, $\bar{S}$ and $\bar{R}$ will remain unaltered,
as will Q. When C returns to $C = 0$, $\bar{R}$ remains $\bar{R} = 1$ and $\bar{S}$ becomes
$\bar{S} = 1$ so that the output flip-flop remembers its previous input data and
Q remains at $Q = 1$.

In summary, then, when C makes a positive-going transition from $C = 0$
to $C = 1$, Q becomes $Q = 0$ if D was $D = 0$ and Q becomes $Q = 1$ if D was
$D = 1$. Q can be changed in no other way. Changes in D when $C = 1$ do not
affect Q. Hence, except for the fact that, in the present case, triggering takes place
on the positive-going clock transition, the behavior is precisely as discussed in
connection with Fig. 9.13-1.

As we have noted, when a flip-flop responds to the triggering transition of
the clocking waveform, the flip-flop must then find itself in a situation in which
it can no longer respond to a further change in input data. In the flip-flops
discussed earlier, this requirement is met by arranging that the triggering transition disable the input gates to which the input data is applied. In the present
case of the type-D flip-flop, this requirement is also met since as we have seen
the flip-flop responds when C goes to 1, but that with $C = 1$ the flip-flop no
longer responds to changes in D. Hence, altogether, the type-D flip-flop, like
the others, is suitable for use in a synchronous system in which the D input is
derived in part or in whole from the output of other flip-flops.

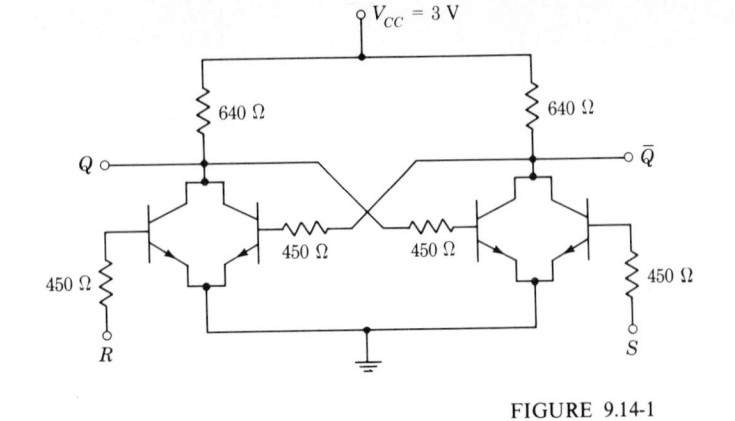

FIGURE 9.14-1
An RTL *RS* flip-flop.

The type-*D* flip-flop shown in Fig. 9.13-2 is constructed using NAND gates (TTL or DTL) and is sensitive to the level of the input clock. Type-*D* flip-flops are also available which employ NOR gates (ECL or RTL). In addition, edge-triggering varieties can be obtained using either NAND or NOR logic.

9.14 AN RTL *SR* FLIP-FLOP

In this and the following sections we shall look at the circuitry of some representative commercially available integrated-circuit flip-flops. Flip-flops, as we have seen, are collections of gates. A gate intended for use in unforseeable situations will have a circuit which is of one of the forms we have previously considered in some detail. When, however, a gate is used internally to a flip-flop and hence in a quite specific and circumscribed situation, economies are sometimes possible.

An unclocked *SR* flip-flop using RTL NOR gates is shown in Fig. 9.14-1. Two-input RTL NOR gates are employed, and the circuit is a direct realization of the flip-flop of Fig. 9.1-1.

9.15 A DTL FLIP-FLOP

As an example of a DTL flip-flop we shall consider the Motorola type 931, a clocked flip-flop which can be used in the *JK* mode. Before examining the circuit of this flip-flop, it will be well to consider several preliminary matters.

First, let us consider using as a gate the single-transistor circuit of Fig. 9.15-1*a*. Voltages representing logical variables *A* and *B* are applied as shown between base and ground and between emitter and ground. The gate output is *Z*, taken between collector and ground. Let the voltage level representing logic **0** be the

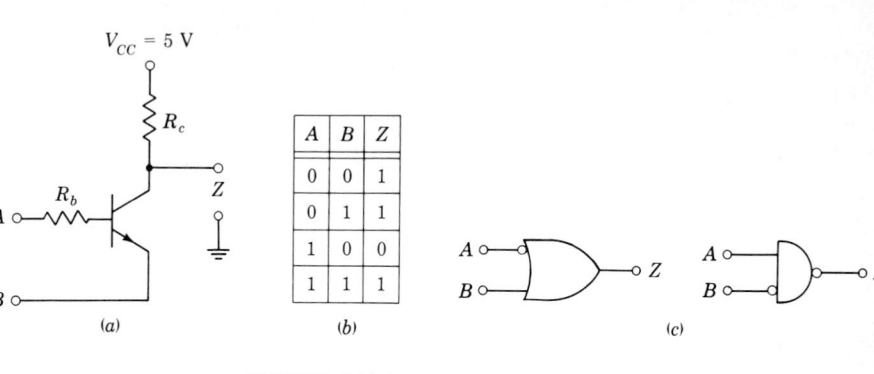

FIGURE 9.15-1
(a) A one-transistor gate, (b) truth table, and (c) circuit representation.

voltage across a saturated transistor, i.e., some tenths of volts. Let the voltage level representing logic **1** be a voltage adequate (when applied at A when B is at logic **0**) to saturate the transistor. Then, as can be verified, the truth table for the gate is as given in Fig. 9.15-1b, and the gate function is therefore $Z = \overline{A} + B = \overline{A \cdot B}$. This function is represented by the gate symbols shown in Fig. 9.15-1c.

Next, consider the gate shown in Fig. 9.15-2a. With voltage levels as specified above, Z will be at logic **0** if either A or B is at logic **1**. Hence, as shown in Fig. 9.15-2b, this circuit is a NOR gate.

Finally, consider the circuit of Fig. 9.15-3a. Our interest is in the volt-ampere characteristics seen looking into its terminals. To that end we first assume that V is sufficiently small for the base-emitter drop to be less than the cut-in voltage $V_\gamma \approx 0.65$ V. Then the transistor is cut off, and the volt-ampere characteristic can be represented by a 4.65-kΩ resistance, as shown in Fig. 9.15-3b. The reader can readily verify that for $V_{BE} \leq V_\gamma$, $V \leq 1$ V.

Now consider that V is sufficiently large for the transistor to be in its active region. Then $V_{BE} = 0.7$ V. In this mode of operation the current $I = I_C + I_2$,

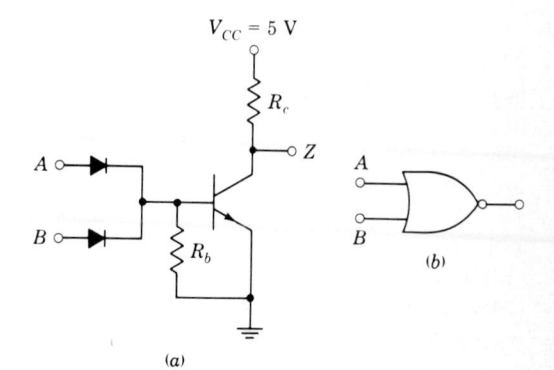

FIGURE 9.15-2
(a) A DTL NOR gate and (b) circuit representation.

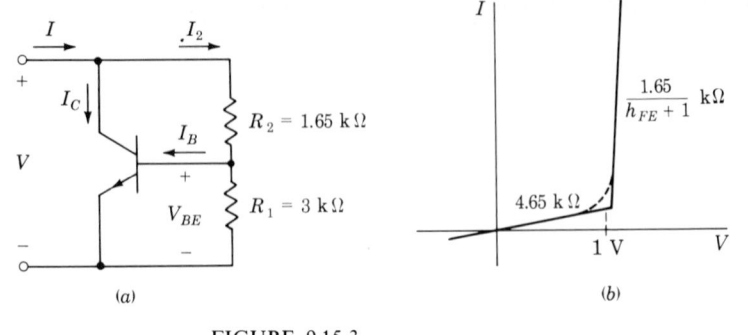

FIGURE 9.15-3
(a) A diode-type circuit and (b) volt-ampere characteristic.

where $I_C = h_{FE} I_B$, $I_2 = (V - 0.7)/1.65$ mA, and $I_B = I_2 - 0.7/3$ mA. Hence

$$I = \frac{h_{FE} + 1}{1,650} V - \left[\frac{0.7}{1,650} (h_{FE} + 1) + \frac{0.7}{3,000} h_{FE} \right] \qquad (9.15\text{-}1)$$

The slope of the volt-ampere characteristic in this region is therefore a piecewise linear resistor with a resistance of $1,650/(h_{FE} + 1)$. If $h_{FE} \approx 40$, this resistance is approximately 40 Ω.

The reader should note that the volt-ampere characteristic shown in Fig. 9.15-3b is similar to that of a diode. However, the break occurs at 1 V rather than $V_\gamma = 0.65$ V. Moreover, this circuit is more versatile than a diode since its two modes of operation result in two fixed-value resistors while a diode resistance varies continually. Furthermore, the two resistor values and the break voltage are easily controlled.

Turning now to the flip-flop which is shown in Fig. 9.15-4, we find that this unit is a master-slave flip-flop which uses gates in the pattern of Fig. 9.15-5. This circuit is different from the one considered in Fig. 9.7-2. However, as can be verified (Prob. 9.15-1), this present pattern will function as a master-slave flip-flop as effectively as the circuit of Fig. 9.7-2.

Let us now investigate the schematic diagram shown in Fig. 9.15-4 to show that the representation of Fig. 9.15-5 is correct. The various sections of the circuit which perform gating functions have been labeled to correspond with the labeling of the gates in Fig. 9.15-5. It will be recalled that the DTL gate consists of a diode-resistor AND gate followed by a transistor inverter. In the type 931, the AND gate portion of the DTL is shown as gates 1A and 1B while the inverter portions of the gates are cross-coupled to construct the master flip-flop. In Fig. 9.15-4, diodes D1A, D2A, and D3A, together with R_{1A}, constitute the AND gate 1A. The transistor T1A and its associated circuitry involving D4A and the transistor T2A (with its resistors) are the NOR gate 2A. That such is the case can be seen by comparison with Figs. 9.15-2 and 9.15-3. Transistor T3A is the gate 3A, as is seen by comparison with Fig. 9.15-1. The output gate

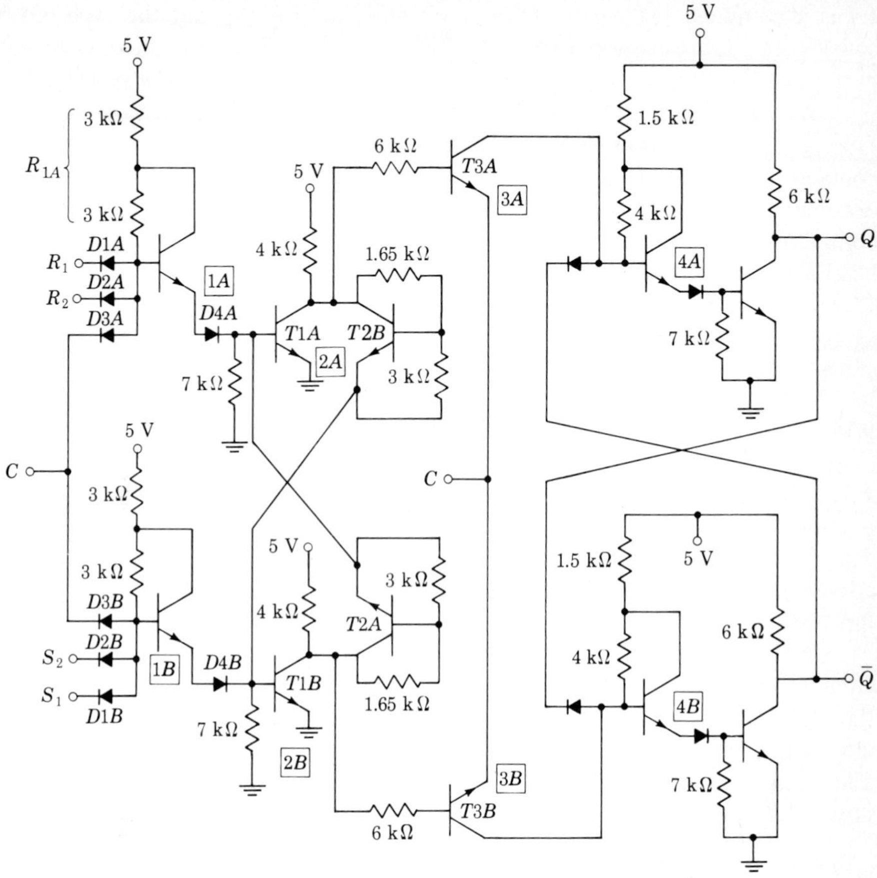

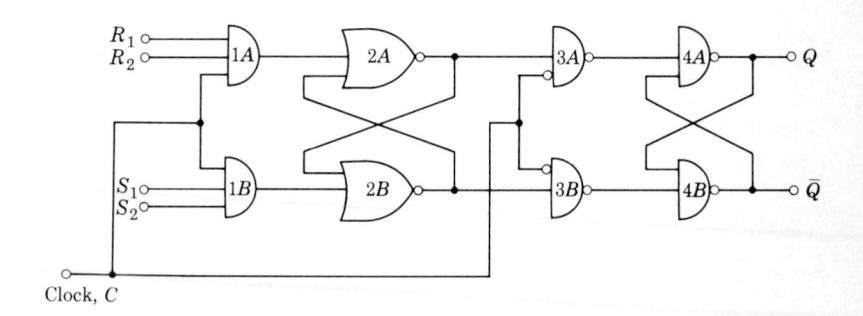

FIGURE 9.15-4
A master-slave *RS* flip-flop (motorola 931).

FIGURE 9.15-5
A gate representation of DTL flip-flop.

$4A$ is a complete DTL gate. We can readily appreciate that the type 931 is considerably more economical of components than a circuit, say, having the gate pattern of Fig. 9.7-2 in which every NAND gate is a full and complete DTL gate.

As the circuit in Fig. 9.15-4 appears, it is an SR master-slave flip-flop. Observe however, that two set and two reset inputs are provided. If we coupled the $\overline{Q}$ output back to, say, S_1 and Q back to R_1, the circuit would become a JK flip-flop with S_2 becoming J and R_2 becoming K. In the commercial circuit, direct set and clear inputs are also provided. This feature has been omitted from the schematic.

9.16 AN RTL PROPAGATION-DELAY FLIP-FLOP

A commercially available RTL propagation-delay flip-flop (Motorola type 926) is shown in Fig. 9.16-1. The pattern of gates is identical to the pattern of Fig. 9.11-2. However, in the present circuit, each input NOR gate has an additional terminal which is connected to Q and $\overline{Q}$ in order to construct a JK flip-flop.

Transistor $T0$ is simply a buffer stage intended to reduce loading of the clock source. It is not essential to the operation. However, because it inverts the clock waveform, the flip-flop triggers on a negative-going transition of the clock rather than on the positive-going transition. Transistors $T1A$, $T2A$, $T3A$ and transistors $T1B$, $T2B$, $T3B$ are the transistors which form the two three-input NOR gates.

Transistors $T4A$ and $T4B$ serve as AND gates. To see that such is the case consider the situation represented in Fig. 9.16-2a. Here, as with $T4A$ and $T4B$, input logic levels are applied to the base and to the collector of a transistor. The gate output Z is taken across the emitter resistor. With logic levels as encountered in RTL, we can verify that the truth table for this gate is as given in Fig. 9.16-2b and hence that the circuit constitutes an AND gate.

Transistors $T5A$ and $T5B$ are used to couple the input NOR gate ($T1$, $T2$, and $T3$) to the AND gate $T4$. The operation of this coupling circuit has previously been discussed in Sec. 9.15.

Transistors $T6A$, $T7A$, $T8A$ and $T6B$, $T7B$, $T8B$, together with their associated resistors, are the cross-coupled NOR gates which constitute the output flip-flop. Observe that direct set and reset facility is provided. In the commercial type 926 two additional transistor stages are included (not shown in Fig. 9.16-1) as output buffers for Q and $\overline{Q}$.

Transistors $T1A$ and $T1B$ provide inputs to allow feedback, as is required in a JK flip-flop. When $\overline{J} = \overline{K} = 1$ the input gates are disabled and the flip-flop does not respond to the clock. Toggling will take place when $\overline{J} = \overline{K} = 0$. The input data terminals, as indicated, are therefore appropriately labeled $\overline{J}$ and $\overline{K}$ rather than J and K.

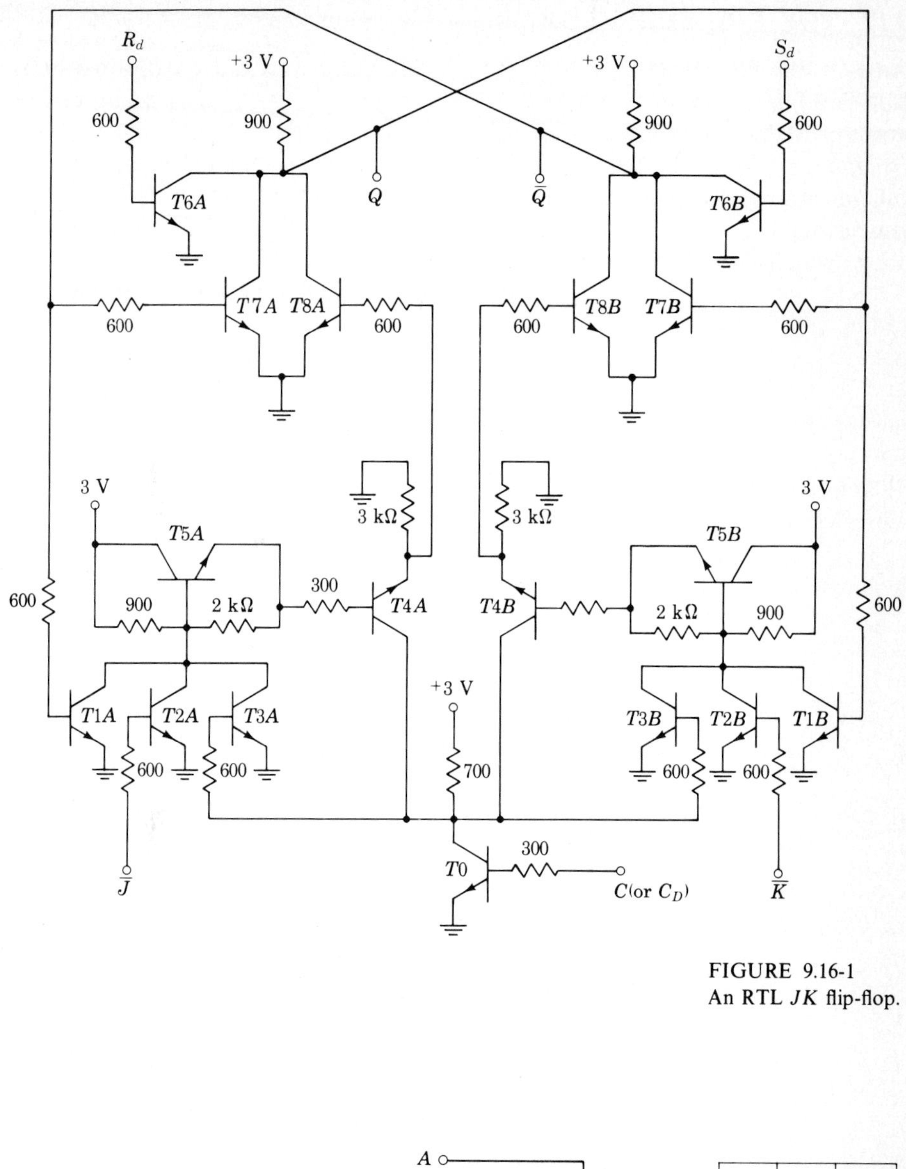

FIGURE 9.16-1
An RTL JK flip-flop.

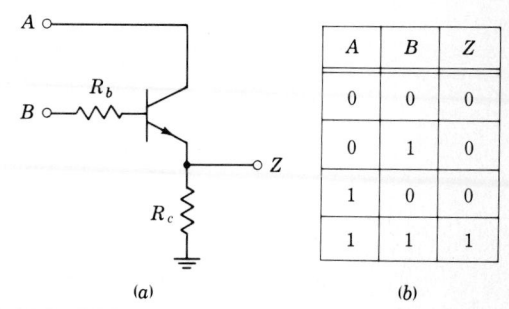

FIGURE 9.16-2
(a) A one-transistor AND gate and (b)
truth table.

A	B	Z
0	0	0
0	1	0
1	0	0
1	1	1

(a)

(b)

9.17   THE ECL FLIP-FLOP

An ECL flip-flop, assembled in an entirely straightforward way, using two ECL NOR gates, is shown in Fig. 9.17-1. Some economy can be achieved by taking account of the following considerations. When transistors $T1A$ or $T2A$ are conducting, transistor $T0A$ is cut off. Transistor $T0A$ is required in an ordinary difference amplifier to provide an alternate path for the current flowing in R_e when transistors $T1A$ and $T2A$ are both off. In addition, $T0A$ sets the reference level above which R must rise to reset Q to $Q = 0$ if Q was originally in the **1** state.

In the present special application we note that with $S = R = 0$ and, for example, with $Q = \mathbf{1}\,(-0.76\text{ V})$ and $\overline{Q} = \mathbf{0}\,(-1.58\text{ V})$, the emitters of $T1A$ and $T2A$ are at $V_R - V_{BE}(T0A) \approx -1.9$ V (since $V_R = -1.175$ V) while the emitters of $T1B$ and $T2B$ are at $V_Q - V_{BE}(T1B) = -1.5$ V. Since the emitter voltages are close, let us consider the possibility of removing the reference transistors $T0A$ and $T0B$ and connecting the emitters of $T1A$, $T1B$, $T2A$, and $T2B$. If this be done, a current path would always be available (as $T1A$ or $T1B$ would always be ON) and, in addition, a reference voltage would also always be available.

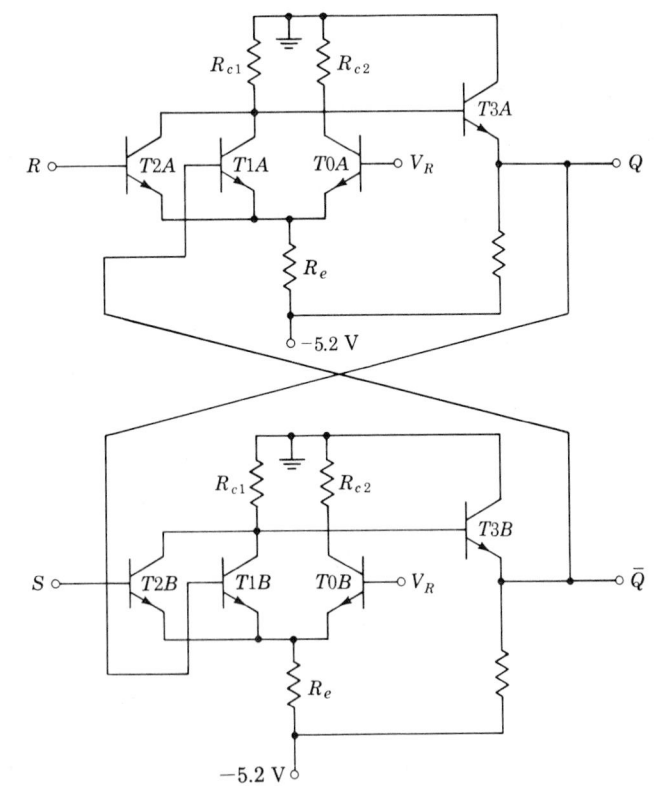

FIGURE 9.17-1
A gate-connected ECL flip-flop.

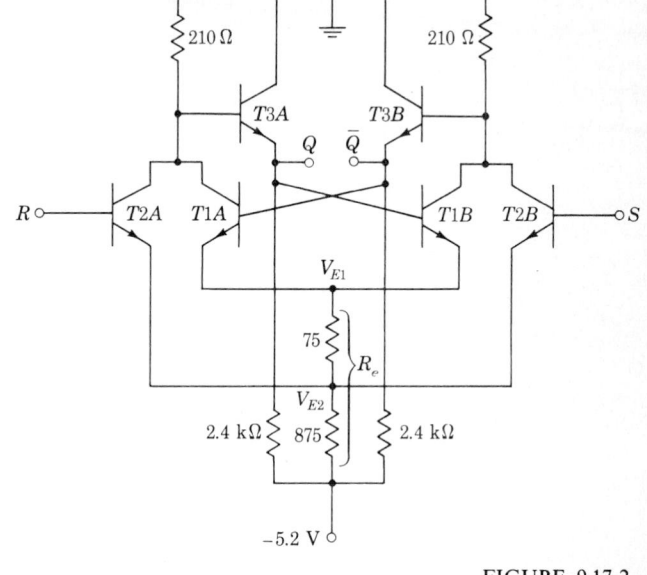

FIGURE 9.17-2
An ECL flip-flop.

An ECL flip-flop, which takes advantage of the economy obtained by eliminating transistors $T0A$ and $T0B$, is shown in Fig. 9.17-2. Comparing Figs. 9.17-1 and 9.17-2, we see that transistors $T0A$ and $T0B$ have been removed and the emitter resistors tied together. Actually the circuit shows that the emitters of $T2A$ and $T2B$ are tied together at a slightly negative potential with respect to the emitters of $T1A$ and $T1B$. The reason for this is as follows. Consider that $Q = 1$ and $\bar{Q} = 0$ and that $R = S = 0$. Then $T1B$ is ON, and $T2A$, $T1A$, and $T2B$ are OFF. Then, the voltage level of $Q = -0.76$ V, and the voltage $V_{E1} = -1.5$ V. Hence, the voltage $V_{E2} = V_{E1} - (V_{E1} + 5.2)(75/(75 + 875)) \approx -1.8$ V. Since R is at logic 0, the voltage of $R = -1.58$ V, and while $T2A$ is cut off, the base-emitter voltage of $T2A$ is $-1.58 - (-1.8) = 0.22$ V. This slight forward bias decreases the propagation delay time. Note that if the emitters of $T1$ and $T2$ were tied together, the base-emitter voltage on the transistors which are cut off would be $-1.58 - (-1.5) \approx -0.08$ V. We note that the voltage V_{E2} sets the reference level for the set and reset inputs. Thus, the reference level in Fig. 9.17-1, which is set by $T0$, is approximately -1.9 V, while the reference level in Fig. 9.17-2 is set by Q or $\bar{Q}$, whichever one is in the 1 state, at -1.8 V, a not-much-different value.

9.18 A *JK* AC-COUPLED ECL FLIP-FLOP

A *JK* ECL flip-flop is shown in Fig. 9.18-1. Component values are characteristic of the Motorola type 308. Transistors $T1$, $T2$, and $T3$ serve the same function in the present circuit as in the circuit of Fig. 9.17-2. The reader should verify that

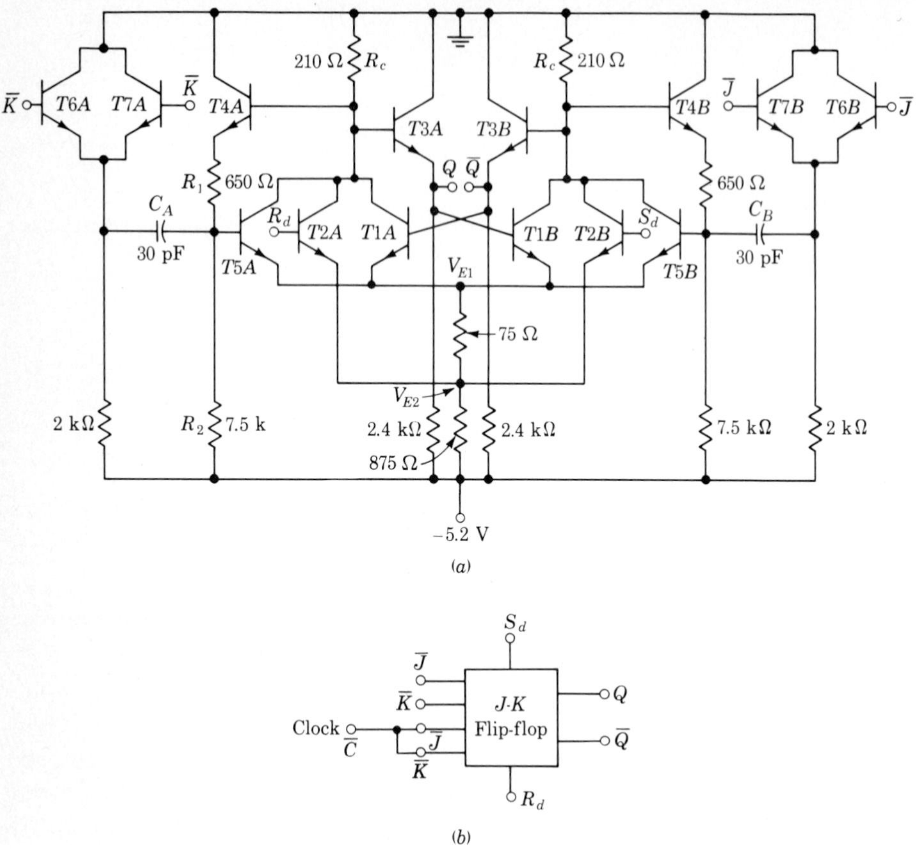

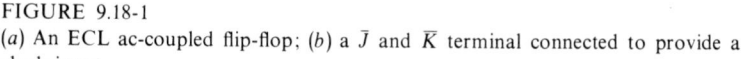

FIGURE 9.18-1
(a) An ECL ac-coupled flip-flop; (b) a $\bar{J}$ and $\bar{K}$ terminal connected to provide a clock input.

this flip-flop is constructed in accordance with the pattern of Fig. 9.9-1. Here transistors $T6$ and $T7$ and their associated 2-kΩ emitter resistors constitute the input OR gates. However, in this case we call the inputs $\bar{J}$ and $\bar{K}$ rather than $\bar{S}$ and $\bar{R}$. The clock terminal is obtained by connecting a $\bar{J}$ and a $\bar{K}$ input terminal (see Fig. 9.9-1) as is indicated in Fig. 9.18-1b.

Suppose that we wanted to modify the flip-flop of Fig. 9.9-1 to make a flip-flop which would be able to toggle. Then we would provide feedback connections from the outputs Q and $\bar{Q}$ to additional inputs on the input gates in the manner of Fig. 9.12-2. In the present case, however, the feedback is returned instead to the input side of gates $2A$ and $2B$. In Fig. 9.18-1 the bases of $T5A$ and $T5B$ are respectively the inputs R' and S' of gates $2A$ and $2B$. The single resistors r returned to logic $\mathbf{0}$ in Fig. 9.9-1 is replaced in Fig. 9.18-1 by transistors $T4A$ and $T4B$ and their associated 650-Ω and 7.5-kΩ emitter resistors. As is

$\bar{J}_D$	$\bar{K}_D$	Q_{n+1}
0	0	Q_n
0	1	0
1	0	1
1	1	$\bar{Q}_n$

(a)

$\bar{J}$	$\bar{K}$	Q_{n+1}
0	0	$\bar{Q}_n$
0	1	1
1	0	0
1	1	Q_n

(b)

FIGURE 9.18-2
(a) Truth table for asynchronous ac operation of the flip-flop of Fig. 9.18-1. $\bar{J}_D = 1$ means that J goes from 0 to 1. (b) Truth table for synchronous clocked operation.

verified below, with the $\bar{J}$ and $\bar{K}$ inputs quiescent, both $T5A$ and $T5B$ are cut off. However, one of the transistors is further into cutoff than the other. If, say, $Q = 0$ $(\bar{Q} = 1)$, the base of $T5A$ is at a lower voltage than the base of $T5B$ and, $T5A$ is the more deeply cut off. If now both $\bar{K}$ and $\bar{J}$ make a transition (i.e., if there is a clock waveform transition) from logic 0 to 1, a signal will be transmitted through $T5B$ but not through $T5A$. The flip-flop will change state, the situations of $T5A$ and $T5B$ will interchange and the flip-flop will again respond to a clock transition.

Let us now consider that $Q = 0$ $(\bar{Q} = 1)$ and let us calculate the base-to-emitter voltages of $T5A$ and $T5B$. The portion of the present circuit involving $T1A$, $T2A$, $T1B$, and $T2B$ is identical to the circuit of Fig. 9.17-2 for which we have calculated that $V_{E1} = -1.5$ V. As is to be expected, and as may be verified, the output logic levels are very nearly -0.76 V and -1.58 V. If $Q = 0$ $(V_Q = -1.58$ V), then the voltage at the emitter of $T4A$ is also -1.58 V, since the bases of $T4A$ and $T3A$ are connected. We now calculate that the voltage at the base of $T5A$, as determined by the voltage divider (650 Ω and 7.5 kΩ) returned to -5.2 V, is V_B $(T5A) \approx -1.9$ V. Hence V_{BE} $(T5A) = (-1.9) - (-1.5) = -0.4$ V. Similarly we find that V_B $(T5B) = -1.1$ V so that $V_{BE}(T5B) = (-1.1) - (-1.5) = +0.4$ V.

The total logic swing available to drive the flip-flop is $-0.76 - (-1.58) \approx 0.8$ V. This logic swing, presented at $\bar{K}$ and $\bar{J}$ by the clock waveform, will be transmitted through the input emitter-follower gates ($T6A$, $T7A$, $T6B$, and $T7B$) with nominally no attenuation. The 0.8-V logic swing applied through the capacitors will be adequate to turn ON $T5B$ but will not turn ON $T5A$.

The truth tables of the flip-flop is given in Fig. 9.18-2. Figure 9.18-2a describes the response when logic transitions are applied independently to a $\bar{J}$ or to a $\bar{K}$ input. (All unused inputs are held at logic 0). Thus the first row says that, if no logic transitions are applied or if the transitions are negative going, the flip-flop will not change state. The second row says that if a positive transition is applied to $\bar{K}$ only, the flip-flop will reset or remain in the reset state, etc. This feature of the ECL capacitor-coupled flip-flop, i.e., its ability to operate asynchronously (without the use of the clock) is very useful and is not shared by all the other types of flip-flops. Figure 9.18-2b describes the flip-flop response when it is driven by a clock (in the manner shown in Fig. 9.18-1b).

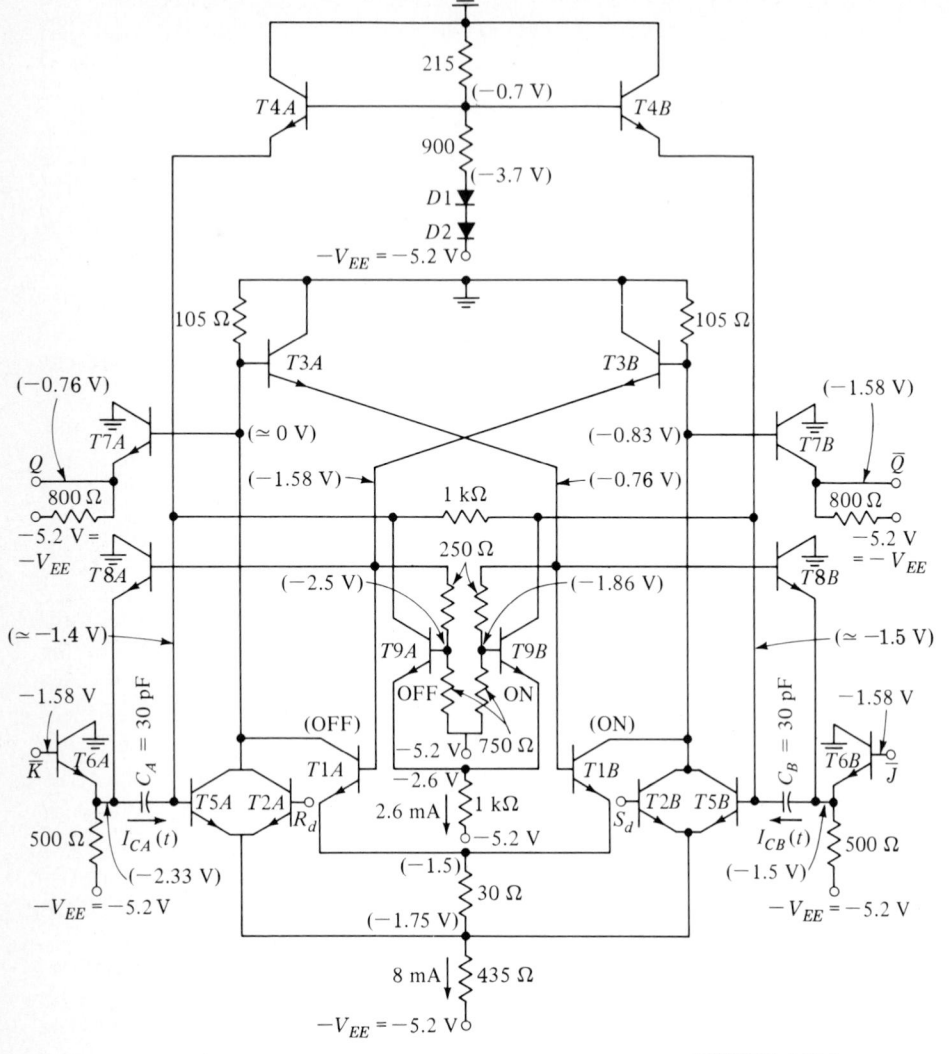

FIGURE 9.18-3
A high-speed ECL flip-flop.

The flip-flop of Fig. 9.18-1, discussed above, is a relatively slow device, having a maximum toggling rate of 15 MHz. A principle speed limitation results from the presence of the capacitors C_A and C_B. In either state of the flip-flop, set or reset, the voltage across C_A is different from the voltage across C_B. At each state change of the flip-flop these capacitor voltages must interchange. AC coupled ECL flip-flops intended for high speed operation incorporate additional components to provide a mechanism for rapid charge and discharge of these capacitors. In addition, resistor values are reduced in order to allow more rapid charging of stray shunt capacitance. The circuit diagram of the MECL II 1013 flip-flop which can toggle at 120 MHz is shown in Fig. 9.18-3. Voltages indicated correspond to the case in which the flip-flop is set ($Q = 1$). Some details of the operation of this circuit are explored in the problems.

9.19 MANUFACTURER'S SPECIFICATIONS

Typical manufacturer's specifications for the MC 10135 high-speed ECL master-slave *J-K* flip-flop are listed below.

> *Output voltage*   logic **1**: -0.81 to -0.96 V; logic **0**: -1.65 to -1.85 V.
> *Toggle frequency*   140 MHz. This is the maximum clock frequency at which the flip-flop will toggle reliably.
> *Propagation delay time* $t_{pd} = 3$ ns.
> *Rise time* $t_r = 6.5$ ns.
> *Fall time* $t_f = 8.5$ ns.
> *Set-up and hold times*   When the data input to a flip-flop changes, there is a transient period during which the gates change state. If the flip-flop is to respond properly to a new input, the new input must precede the clock triggering transition by some minimum time called the *set-up* (or settling) time. Similarly, the input must remain fixed for a given time after the clock transition, called the *hold* time. The set-up and hold times are often given by the manufacturer. For the MC 10135 high-speed ECL flip-flop the set-up time and the hold time is 1 ns.
> *Relative speed*   Of the various types of flip-flops discussed, the ECL flip-flop is the fastest, followed by the TTL flip-flop. The toggling rates of ECL flip-flops exceed 500 MHz while TTL flip-flops have toggling frequencies greater than 10 MHz.

9.20 TTL *JK* FLIP-FLOP

A TTL flip-flop which employs the charge storage principle is shown in gate form in Fig. 9.20-1. A detailed circuit diagram appears in Fig. 9.20-2. The similarity between Fig. 9.20-1 and 9.10-1 is apparent. There are some differences, however, one point of difference being the feedback connection in the present circuit from the outputs Q and $\bar{Q}$ to the input gates 1*A* and 1*B*. These connections are required to convert the *SR* flip-flop to a *JK* type. Secondly, the present circuit includes two additional gates 4*A* and 4*B*. These gates, as we shall see, are required to provide for rapid discharge of capacitors C_A and C_B as is required when the flip-flop is being called upon to toggle rapidly.

The level of the input clock C which disables the input gates (1*A* and 1*B*) is $C = 0$. When C rises to $C = 1$, the input data at J and K are stored on the capacitors, and when C returns to $C = 0$ the data at J and K are transferred to the output. To appreciate the operation of gates 4*A* and 4*B*, let us set the

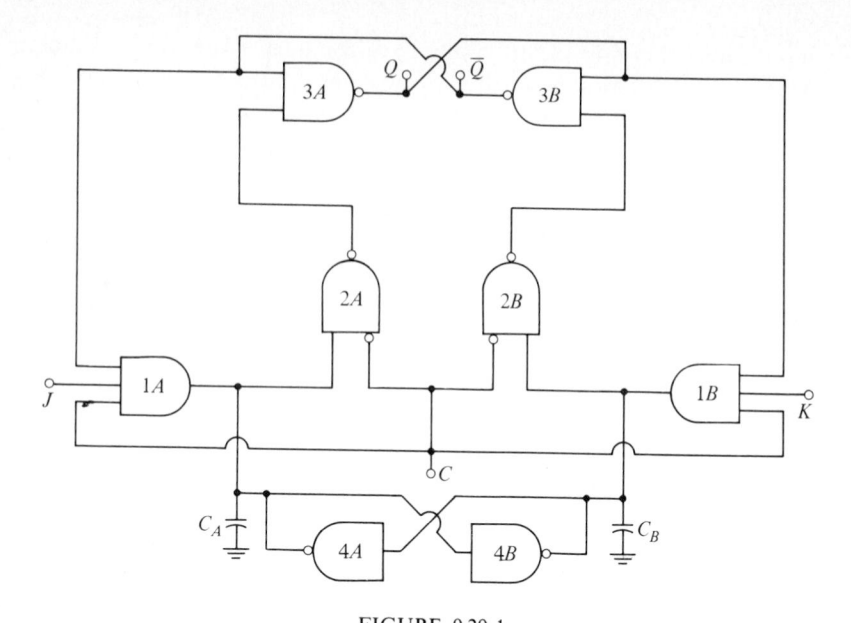

FIGURE 9.20-1
A TTL JK flip-flop which employs charge storage.

flip-flop to toggle ($J = K = 1$) and let us follow the operation as C goes through two successive clocking transitions closely spaced in time.

Let us assume initially then that $J = K = 1$, that $Q = 0$ ($\overline{Q} = 1$), and that C is $C = 0$ and has been $C = 0$ for a relatively long time. Then the output of both gates $1A$ and $1B$ are at logic 0 and both capacitors C_A and C_B are discharged, i.e., are also at logic 0. (As is to be seen in Fig. 9.20-2 there is a resistive path to ground across the capacitors so that the capacitors may discharge completely.) Now let C go to $C = 1$. Then the output of $1B$ remains at 0 while the output of $1A$ rises to 1 and rapidly charges C_A to 1. Capacitor C_A now holds the information that the flip-flop is in the reset state. When C returns to $C = 0$ a signal is transmitted through gate $2A$ and the flip-flop toggles. With C back at 0, we expect the output of gate $1A$ to return to 0 discharging C_A. As we shall see, in Fig. 9.20-2 the output of gate $1A$ is able to *source* current but not to *sink* it. Hence C_A will discharge only rather slowly as C_A provides some small current to the input of gate $2A$. This relatively slow discharge is of some advantage since it makes sure that the information stored on C_A is not lost prematurely.

Now suppose that while C_A is still charged, there is a second transition of C from 0 to 1. Then, in this case, C_B will charge to logic 1 as required, and, most inportantly, the logic 1 at the input of gate $4A$ will drive the output of $4A$ to logic 0 and C_A will discharge abruptly. As is to be seen in Fig. 9.20-2, when the input to gate $4A$ is 1, the gate provides a saturated transistor directly across the capacitor C_A.

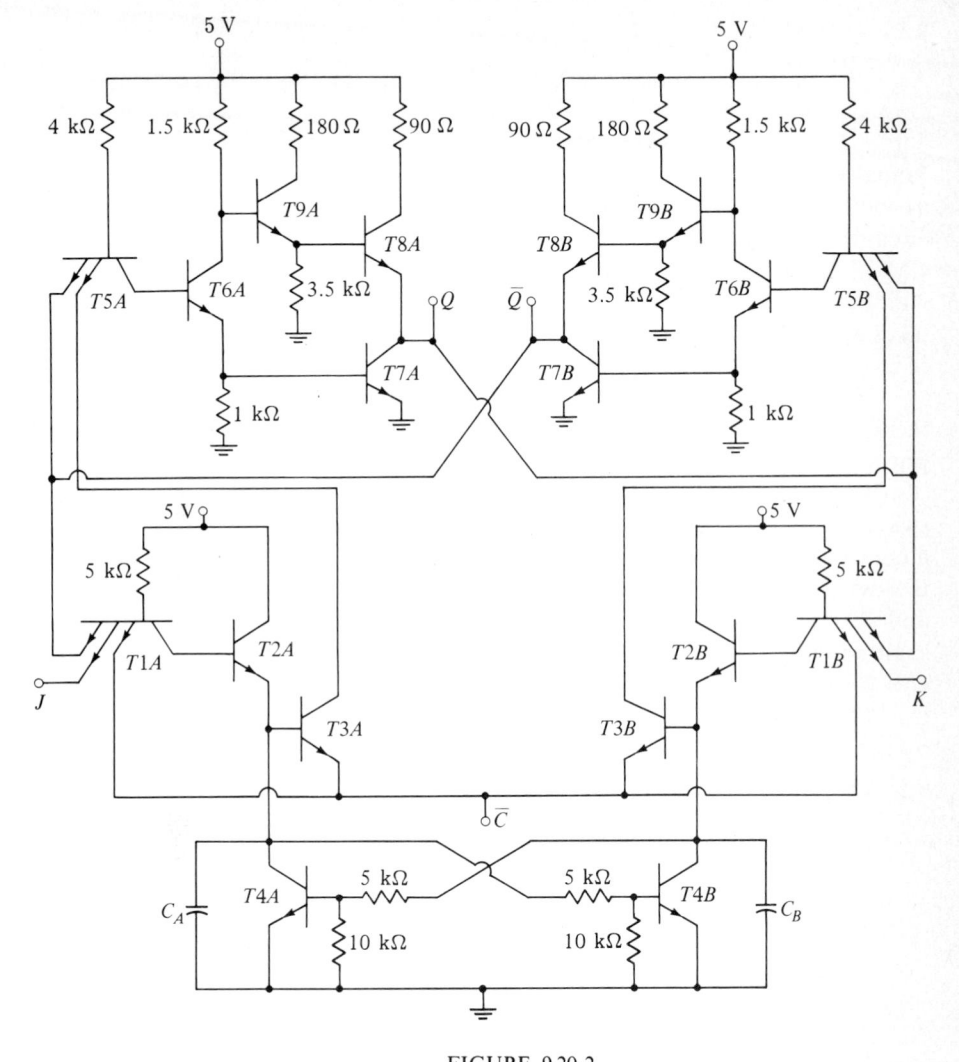

FIGURE 9.20-2
Schematic diagram of a charge-storage TTL flip-flop.

Turning now to Fig. 9.20-2 we note that gate $1A$ is comprised of the multiemitter transistor $T1A$ and the emitter follower $T2A$. Transistor $T3A$ operates in the manner described in connection with Fig. 9.15-1, and is a NAND gate with inversion on the emitter (clock) input. Observe, as suggested above, that, since current cannot flow back into the emitter follower ($T2A$), gate $1A$ can only source current to charge C_A and cannot sink current to discharge C_A.

Finally, we note, that while in ECL flip-flops thin-film capacitors are used, in the present TTL circuit capacitors C_A and C_B are the capacitors formed across reverse-biased diode junctions.

9.21 MOS FLIP-FLOPS

As is the case with bipolar transistors, MOS flip-flops may be constructed by cross-coupling NOR or NAND MOS gates. Figure 9.21-1 shows SR MOS flip-flops. CMOS devices are used in (a) and MOS devices are used in (b). Actually NMOS or PMOS flip-flops are generally not commercially available as individual units but they are widely used as components in MSI and LSI circuitry. On the other hand, CMOS flip-flops are available. When however, CMOS flip-flops are fabricated they are generally not constructed in accordance with the pattern of Fig. 9.21-1a. Instead they are designed in a manner which takes advantage of the fact that CMOS devices lend themselves to the realization of *transmission gates.*

The subject of transmission gates is discussed in Chap. 13. In particular, the CMOS transmission gate is considered in Sec. 13.10. For our present purpose it will be adequate to describe a transmission gate very briefly and somewhat superficially. A transmission gate, unlike a logical gate, is simply a switch as represented in Fig. 9.21-2a whose opening and closing is controlled by a control logic waveform C (C may be and often is a clock waveform). When the switch S is closed, signals (digital or analog) may be *transmitted* between A and B. When the switch is open, A and B are isolated from one another. We shall adopt the convention that when $C = 0$ the switch is open and that when $C = 1$ the switch is closed. If we intend to convey that the switch is open at $C = 1$ and closed at $C = 0$ we shall indicate that the switch is being controlled by $\overline{C}$.

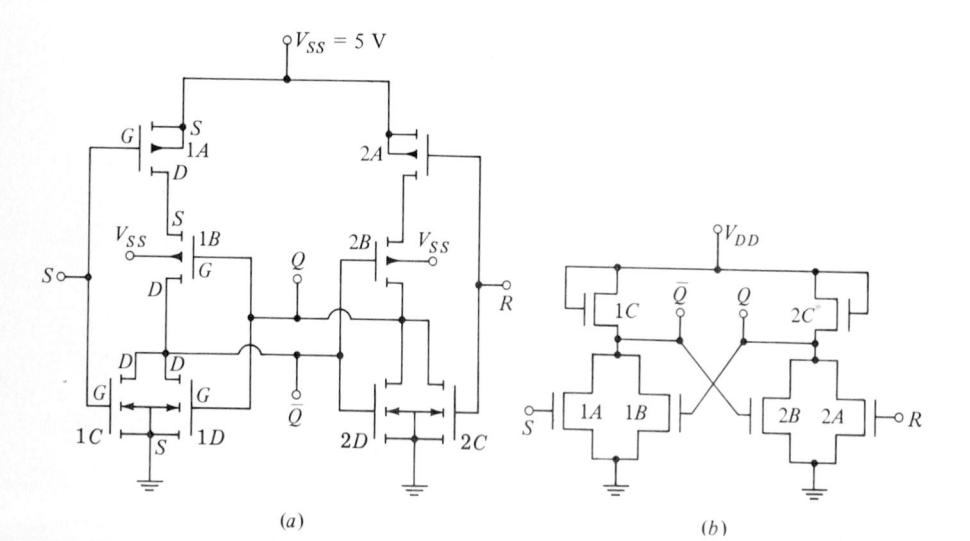

(a) (b)

FIGURE 9.21-1

(a) Circuit of a CMOS SR flip-flop and (b) Circuit of a MOS SR flip-flop.

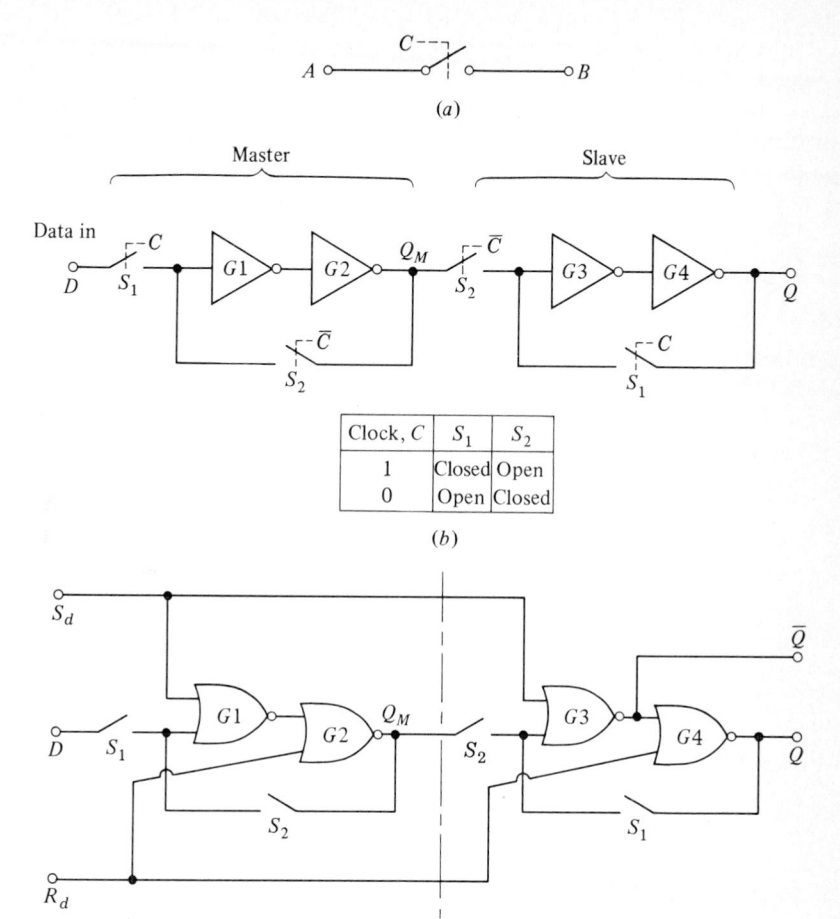

FIGURE 9.21-2
(a) Control switch (b) Master-slave type-D flip-flop and (c) type-D flip-flop with direct inputs.

Type-D flip-flop A type-D flip-flop is shown in Fig. 9.21-2b. It is useful at the outset to recognize that a cascade of two NOT gates (such as $G1$ and $G2$) with output returned to input (as when S_2 is closed) is a rudimentary flip-flop. It will hold either logic level indefinitely.

The clock waveform C drives the switches S_1 and S_2. Let us start with $C = 0$ so that switches S_1 are open and switches S_2 are closed. Then the master flip-flop is storing a bit Q_M whose logic value depends on past history. Thus the output $Q = Q_M$. Now let $C = 1$. Switches S_1 close and switches S_2 open. We require that S_2 open before S_1 closes so that the input D shall not be transmitted to the output Q. The closing of S_1 bridging $G3$ and $G4$ completes the flip-flop

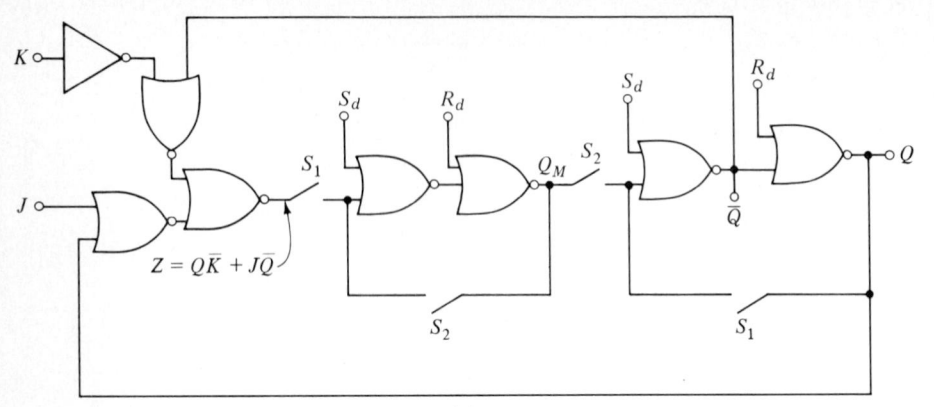

(a)

J	K	Q_{n+1}
0	0	Q_n
0	1	0
1	0	1
1	1	$\overline{Q}_n$

(b)

FIGURE 9.21-3
(a) A JK flip-flop and (b) truth table.

feedback and the original bit at Q is held in the slave flip-flop. During the brief interval between the opening of S_2 and the closing of S_1 the bit Q is held because of the stray capacitance to ground inevitably present in the gates. This transition of C to $C = 1$ establishes Q_M at the new logic level at D. Now let C return to $C = 0$. Then the new bit is stored in the master flip-flop and also transmitted to the output Q. Observe, again, as is required, that the new bit does not appear at Q until after the input switch S_1, connected to D, opens.

Direct inputs When direct (asynchronous) inputs are employed, the inverters shown in Fig. 9.21-2b are replaced by NOR gates, as shown in Fig. 9.21-2c. Thus, for example, if $S_d = 1$, then $Q_M = 1$ and $Q = 1$, independently of the clock. Hence, the direct inputs *override* the clock.

The JK flip-flop The JK flip-flop shown in Fig. 9.21-3a is seen to be identical to the D flip-flop of Fig. 9.21-2b except for the input circuit, where the input is (see Prob. 9.21-1)

$$Z = Q\overline{K} + J\overline{Q} \qquad (9.21\text{-}1)$$

Since Z will eventually determine the output state of the flip-flop, we can let $Z \equiv Q_{n+1}$ and $Q \equiv Q_n$. When this is done, Eq. (9.21-1) becomes

$$Q_{n+1} = Q_n \overline{K} + J\overline{Q}_n \qquad (9.21\text{-}2)$$

It is left as a problem (Prob. 9.21-3) to show that this results in the truth table for a JK flip-flop, which is given in Fig. 9.21-3b.

Typical Manufacturer's Specifications of a JK CMOS Flip-Flop

Quiescent power dissipation   Typically 50 nW.

Dynamic power dissipation   Typically 1 mW at 1 MHz. The dynamic power dissipation is directly proportional to the frequency of operation and the fan-out. The dissipation also increases with the supply voltage.

Output drive current -1.3 mA/$+2.5$ mA. Output drive current is considered positive if the current *enters* the flip-flop and negative when it leaves the flip-flop. Thus, the -1.3-mA current signifies current flow from the V_{DD} supply through the p-channel FET to the output.

Noise immunity   4.5 V in either state when $V_{SS} = 10$ V and 2.25 V in either state when $V_{SS} = 5$ V.

Output impedance   Approximately 800 Ω in either state.

Input impedance   5 pF.

Propagation delay   70 ns. Due to symmetry $t_{pd}(HL) = t_{pd}(LH)$. This delay is a direct function of the fan-out since each flip-flop contributes 5 pF. The propagation delay decreases with increasing supply voltage.

Maximum toggle frequency   Up to 10 MHz.

10

REGISTERS AND COUNTERS

A logical **1** or **0** is called a *bit* (*binary digit*). As we have seen, a flip-flop can store or remember or register a single bit. A flip-flop is therefore referred to as a *one-bit register*. If we require that N bits be remembered or registered, N flip-flops are required. When an array of flip-flops has a number of bits in *storage*, it becomes necessary on occasion to *shift* bits from one flip-flop to another in a manner now to be described. An array of flip-flops which permits this shifting is called a *shift register*.

10.1 THE SHIFT REGISTER

A 4-bit shift register constructed with type-D flip-flops is shown in Fig. 10.1-1a. More or fewer bits can be accommodated by adding or deleting flip-flops. Interconnections between flip-flops are such that (except for $FF0$) the logic level at a data input terminal is determined by the state of the preceding flip-flop. Thus, D_k is **0** if the preceding flip-flop is in the reset state with $Q_{k-1} = \mathbf{0}$, and $D_k = \mathbf{1}$ if $Q_{k-1} = \mathbf{1}$. The datum D_0 at $FF0$ is determined by an external source. In the shift register shown in Fig. 10.1-1b, type JK flip-flops (SR would serve as well)

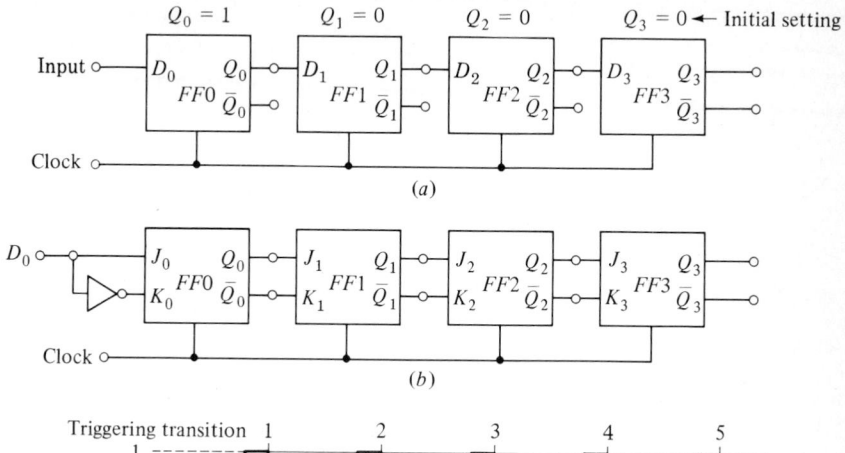

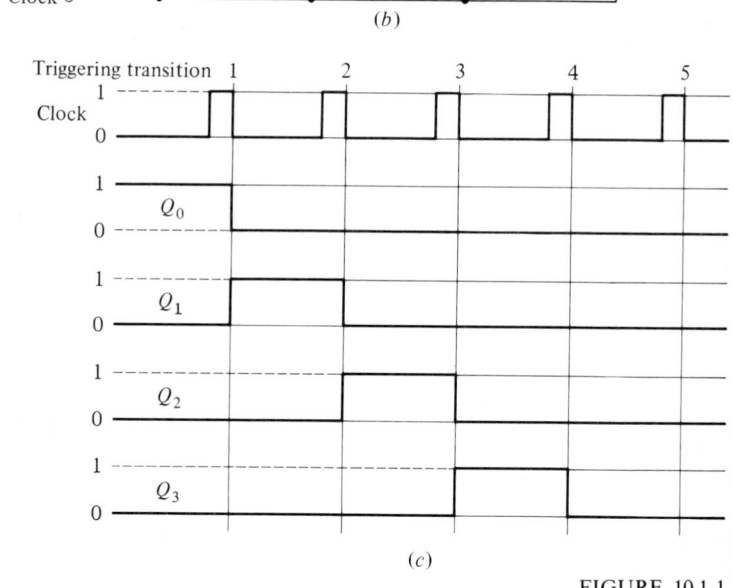

FIGURE 10.1-1
A four-bit shift register.

are being used in lieu of type-D flip-flops. As in Fig. 9.13-1, an inverter is used at the input to K_0. However, in the remaining flip-flops we can dispense with the inverters, since the complement $\overline{Q}$ is already available.

We recall that the characteristic of a type-D flip-flop is that immediately after the triggering transition of the clock, the flip-flop output Q goes to the state present at its input D just before this clock transition. Hence, whatever pattern of bits, **1**s and **0**s, is registered in the flip-flops shown, at each clock transition this pattern is shifted one flip-flop to the right. The bit registered in the last flip-flop ($FF3$ in Fig. 10.1-1) is lost, while the first flip-flop ($FF0$) goes to the state determined by its datum input D_0. This operation is exhibited in Fig. 10.1-1c. Here we have $D_0 = \mathbf{0}$, and we have initially set $Q_0 = \mathbf{1}$ and

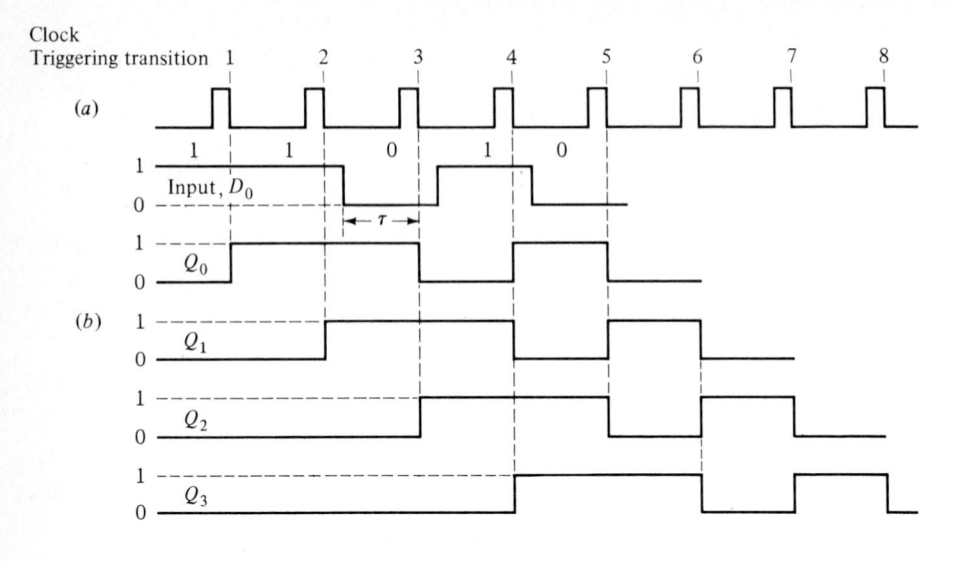

FIGURE 10.1-2
(a) Clock waveform, (b) waveforms for shift register of Fig. 10.1-1.

$Q_2 = Q_3 = Q_4 = 0$. We have here assumed arbitrarily that the flip-flop triggers on the negative-going transition of the clock waveform.

To understand the operation of the shift register better, let us observe its response to a sequence of bits presented at the input D_0, as in Fig. 10.1-2. The clock waveform is shown in Fig. 10.1-2a. An input sequence of bits is shown in Fig. 10.1-2b. The input sequence assumed is **11010**, and it is synchronous with the clock; i.e., when changes occur in logic level at the input, they occur at a fixed point in the cycle of the clock waveform. We assume also that initially, as shown, all flip-flops are in the reset state. This initial situation can be established by applying appropriate control logic to the direct inputs of the

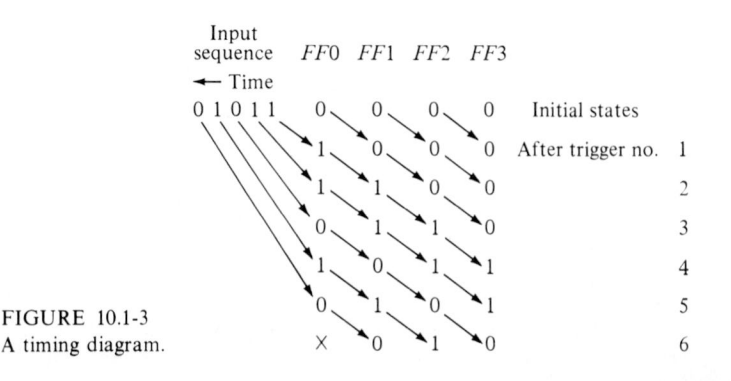

FIGURE 10.1-3
A timing diagram.

flip-flops. (These direct inputs are not shown in Fig. 10.1-1.) We may now observe, in Fig. 10.1-2, how at each clock pulse, the input sequence is shifted, bit by bit, into the shift register and down the register. The waveform of Q_0 is the waveform of the input sequence delayed by a time τ as shown in the figure, τ depending on where in the clock cycle the input-logic level changes occur. However, Q_1 is delayed behind Q_0, Q_2 behind Q_1, and Q_3 behind Q_2 by exactly the duration of the clock cycle. Since we have specified the input only for the duration of 5 bits, the waveform Q_0 is specified only up to the time before the sixth triggering transition of the clock. Q_1 is determined for one additional clock interval, while Q_2 and Q_3 are determined for two and three additional clock intervals, respectively.

In Fig. 10.1-3 we have repeated the information displayed in Fig. 10.1-2 in tabular form. For convenience in visualizing the sequential entering of the input data, we have run the time scale from right to left so that the input sequence appears backward.

10.2 CLOCKING

We must now emphasize that the shift register described in the previous section will operate as described only if the triggering clock transition is a transition which disables a flip-flop so that it cannot respond to a change in input data. After the clock transition the state of the kth flip-flop should be determined by the state of the $(k-1)$st flip-flop *before* the clock transition. The $(k-1)$st flip-flop may itself change as a result of the triggering transition. However, the kth flip-flop must be isolated from this new state of the $(k-1)$st flip-flop until the next clock cycle.

It was this precise situation that was explored in Sec. 9.7 and in Fig. 9.7-1. The circuit of Fig. 9.7-1 is now recognized as a 2-bit shift register. The situation analyzed at that point was precisely the response of this 2-bit shift register to an input sequence **11**.

10.3 SERIAL-PARALLEL DATA TRANSFER

Digital data may be presented in serial form or in parallel form. In serial form the data appear on a single line, one bit at a time, often synchronously with a clock. In parallel form, the data appear at one instant of time on as many lines as there are bits. A shift register may serve as a convenient means of transferring serial data to parallel data and vice versa.

To transfer N bits of data in serial form to parallel form, we allow the N serial bits to be the input sequence to an N-bit shift register. When the N bits are in the register, the data are present in parallel form at the N flip-flop outputs $Q_0, Q_1, ..., Q_{N-1}$.

Suppose, next, that the data are in parallel form in a shift register. The data may originally have been in serial form and have been entered into the register one bit at a time. Or the data may originally have been in parallel form and have been entered directly into the register flip-flop through the use of the direct set and direct reset controls. In any event, if we allow the clock to run, then at the output of the last flip-flop the data will appear, in serial form, bit-by-bit, in synchronism with the clock.

10.4 END-AROUND CARRY

In the shift register of Fig. 10.1-1, a bit shifted out of the last flip-flop is lost. When it is necessary to preserve the bits stored in the register, this can be achieved by coupling the output of the last flip-flop back to the data input of the first flip-flop. In such a register, the bits, as many as there are flip-flops, will circulate around the register, shifting by one flip-flop at each clocking. A register so connected is called an *end-around-carry* shift register or ring counter.

In an end-around-carry register, the bits to be circulated may originally have entered into the register in serial form, as in the circuits of Fig. 10.1-1. If such is the case, when the register is full, it is necessary to disconnect the register from the serial input and to make the end-around connection which will permit the bits to circulate. Gates and a control signal must be added to accomplish this purpose (see Fig. 12.2-1). Alternatively, the bits may be introduced into the register asynchronously through the use of the direct set and reset inputs.

10.5 SHIFT-RIGHT SHIFT-LEFT REGISTER

There are applications in which it is useful to have a register with the capabilities of shifting data to the right and to the left. Such a register, using type-D flip-flops, is shown in Fig. 10.5-1. In this circuit, D_R is the input for introducing serial data which are to be shifted to the right, and D_L is the input for data to

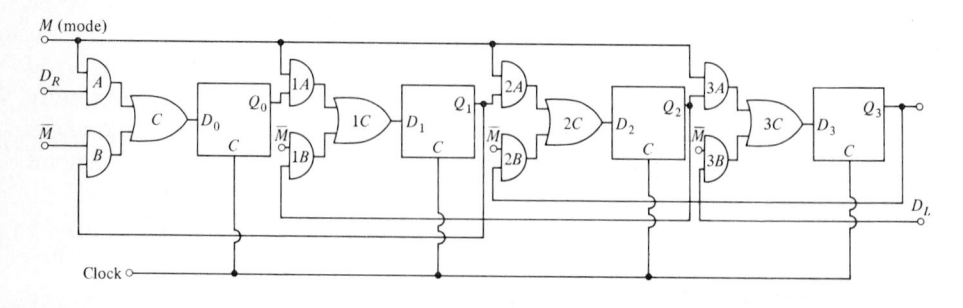

FIGURE 10.5-1
A shift-right, shift-left register.

be shifted to the left.   The direction of shift is determined by the logic level at the mode (M) input.   With $M = 1$ $(\overline{M} = 0)$ shifting is to the right, and with $M = 0$ shifting is to the left.

When $M = 1$, all gates A are enabled and gates B are disabled.   In this case, the register operates like the register of Fig. 10.1-1.   That is, the input is applied to D_0, Q_0 is connected to D_1, etc.; these connections are made through an AND and an OR gate.   When $M = 0$, gates A are disabled and gates B are enabled, and the connections between flip-flops (again through an AND and an OR gate) are reversed.   That is, Q_3 is connected to D_2, Q_2 connected to D_1, etc.   It is left for the problems to show that M should be changed only when $C = 0$. Otherwise the values stored in the register may be altered.

10.6 RIPPLE COUNTERS

A flip-flop has two states.   Correspondingly an *array* of N flip-flops has 2^N states, a *state of the array* being defined as a particular combination of states of the individual flip-flops.   A flip-flop counter is an array in which the flip-flops are interconnected in such manner that the array advances from state to state with each cycle of an input waveform.   If, starting from some initial state, the counter returns to that initial state after k cycles of the input waveform, the counter is called a *modulo-k* counter or a *base-k* counter.   If a counter has N flip-flops, and if the counter advances through every possible state (this mode of operation is not always either necessary or desirable), the counter is of modulo 2^N.

A mod-16 ripple counter is shown in Fig. 10.6-1a.   Four JK flip-flops are connected for the *toggling* mode with $J = K = 1$.   We observe that the interconnections between flip-flops consist only in Q_0 of the first flip-flop $FF0$ being connected to the clock input C_1 of $FF1$, Q_1 to the clock input of $FF2$, etc. The input signal whose cycles are to be counted is applied to the clock of $FF0$. We assume that the flip-flops toggle on the negative-going transition of the waveform at the clock input.   Then, as can be verified, the waveforms at the input and at the flip-flop outputs Q_0, Q_1, Q_2, and Q_3 appear as shown in Fig. 10.6-1b.   Note that each flip-flop changes state when and only when the preceding flip-flop or input waveform makes a change from logic 1 to logic 0. We have arbitrarily started at a point at which $Q_3 = Q_2 = Q_1 = Q_0 = 0$, which we shall henceforth write as $Q_3 Q_2 Q_1 Q_0 = 0000$.   The sequence of input pulses takes the counter through all its possible $2^4 = 16$ states, so that after the sixteenth clock pulse the counter has returned to its initial state.   Considering the count K to be $K = 0$ when $Q_3 Q_2 Q_1 Q_0 = 0000$, it can be verified that the count of input pulses is given by

$$K = 2^3 Q_3 + 2^2 Q_2 + 2^1 Q_1 + 2^0 Q_0 \qquad (10.6\text{-}1)$$

as indicated in the figure in binary and decimal notation.

The count in the counter can be "read" with the aid of a decoder like that shown in Fig. 10.6-2.   Here we have restricted the number of flip-flops to

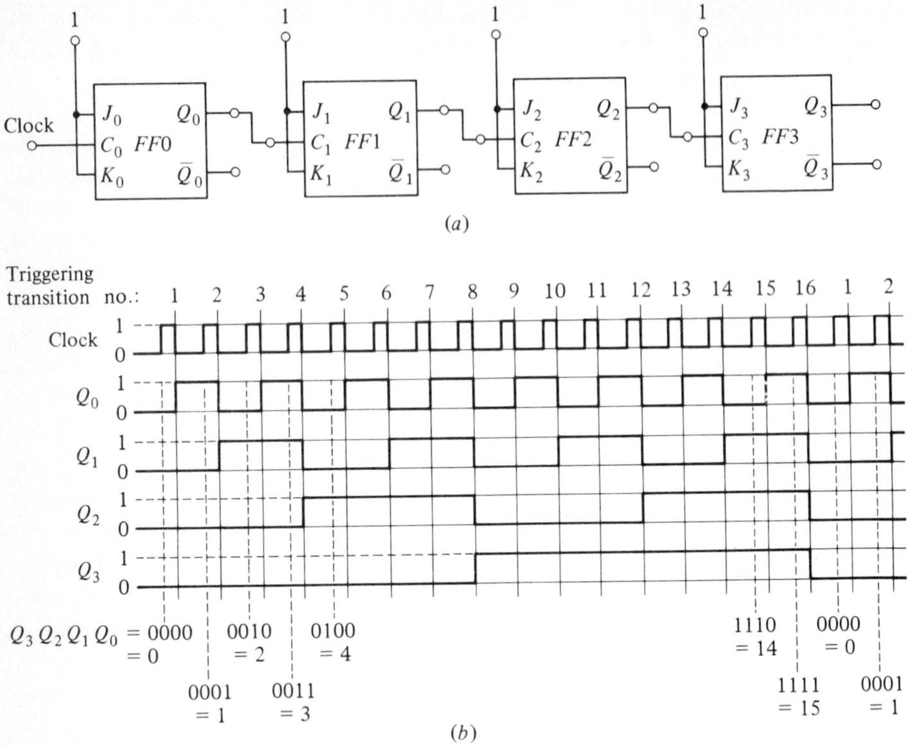

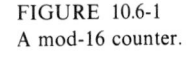

(a)

(b)

FIGURE 10.6-1
A mod-16 counter.

three, to make it easier to follow the wiring connections. The counter, then, is mod 8. Let us ignore initially the strobe input. More specifically, let us assume that the strobe input is at logic **1** so that the strobe input leaves all the AND gates enabled. As can be verified, when the counter is in the state $Q_2 Q_1 Q_0 =$ **000**, K_0 will be $K_0 = 1$ while $K_2, \ldots, K_7 = 0$. When the counter is at $Q_2 Q_1 Q_0 = $ **001**, K_1 will be $K_1 = 1$ while all other AND gate outputs will be at logic **0**, etc. Thus, the count of the counter can be read simply by noting which of the outputs $K_0, \ldots, K_7$ is at logic **1**.

To see the purpose of the strobe input to the AND gates in Fig. 10.6-2 we need to recognize that the waveforms in Fig. 10.6-1b are unrealistically drawn in that we have ignored the propagation delay time through the flip-flops. Accordingly, the waveforms (for a mod-8 counter) have been redrawn in Fig. 10.6-3 taking the propagation time t_{pd} into account. We have also drawn in Fig. 10.6-3 the waveform at K_0. Also shown in the figure is the *ideal* waveform K_0, that is, the waveform as it would appear if there were no propagation delay. Ideally, K_0 should be $K_0 = 1$ only from the moment the counter recycles at the time of

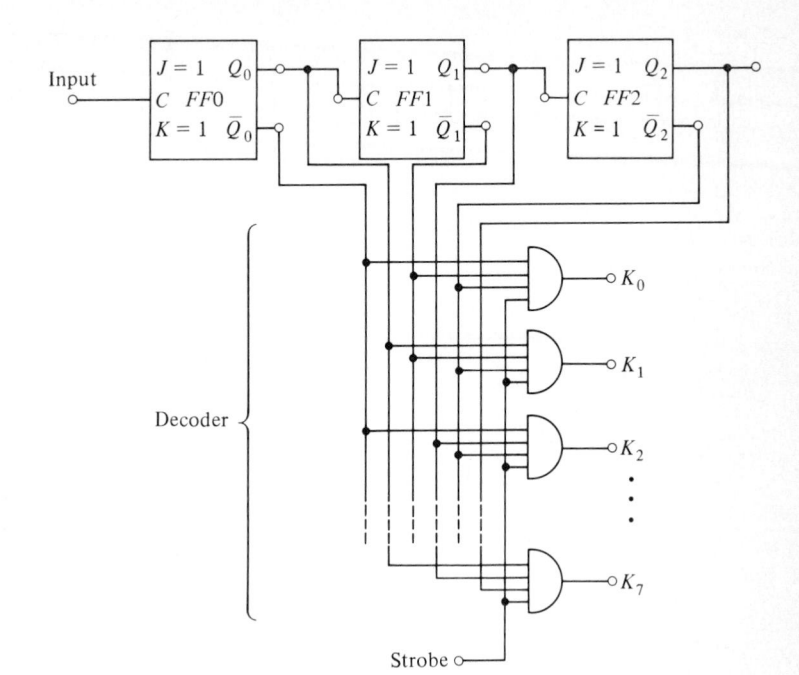

FIGURE 10.6-2
A binary-to-decimal converter.

the eighth triggering transition to the time of the next triggering transition, and otherwise we should have $K_0 = 0$. We observe, however, that because of the propagation delays there are *decoding errors*. Decoding errors appear as well in the other decoder outputs in Fig. 10.6-2 (see Prob. 10.6-1). To circumvent these errors we employ a strobe input to the AND gates. The strobe input is normally at logic **0**, so that all the decoder outputs are at **0**. When we want to read the decoder, we strobe it, i.e., take a quick look, by allowing the strobe line to go to logic level **1**. The strobe is allowed to go to **1** only after a time has elapsed after a clocking transition which is long enough to avoid the errors. The strobe level returns to logic **0** before the next clocking transition.

The counter of Fig. 10.6-1a is known as a *ripple counter*. The origin of this terminology can be seen from the following consideration. Suppose we have a counter with N flip-flops, and suppose all the flip-flops are in the set state, so that $Q_0 = Q_1 = \cdots = Q_{N-1} = 1$. At the next clocking transition Q_0 will go to $Q_0 = 0$. The transition in Q_0 will then cause Q_1 to go to **0**, which, in turn, will make Q_2 go to **0**, and so on. Thus, the input clocking transition will ripple through the counter from Q_0 to Q_{N-1}. When the rippling is completed, the counter will be at $Q_0 = Q_1 = \cdots = Q_{N-1} = 0$. The counter is also referred to as an *asynchronous counter* because there is no clocking signal applied *in common* to each flip-flop to induce transitions in each flip-flop simultaneously.

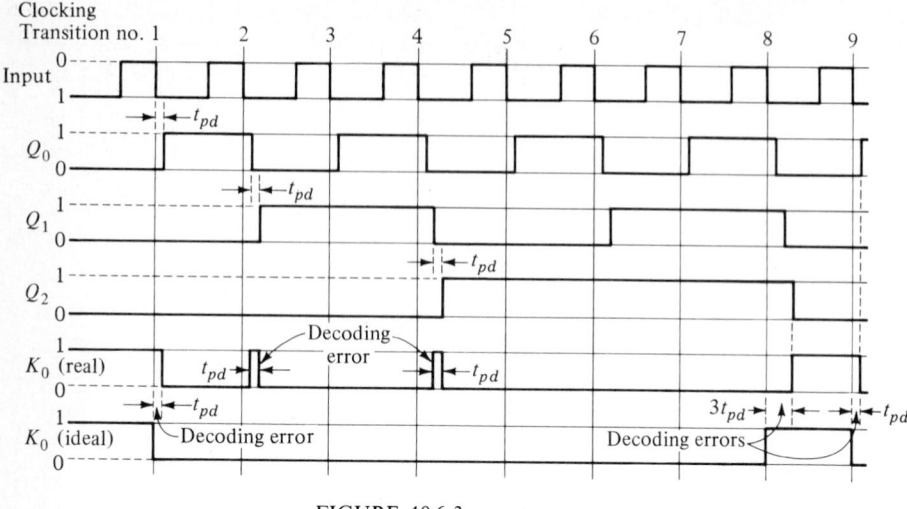

FIGURE 10.6-3
Waveforms for a mod-8 counter including propagation delay time.

The asynchronous character of the counter establishes an upper limit to the allowable frequency of the input waveform. To be specific, suppose we have a counter with $N = 4$ and that we are at the count where $Q_0 = Q_1 = Q_2 = Q_3 = 1$. Suppose, further, that the propagation time of a flip-flop is 50 ns. Then at the next input clocking transition, a ripple will be initiated which will not be completed for $4(50) = 200$ ns. When the ripple is completed, we should have $Q_0 = Q_1 = Q_2 = Q_3 = 0$ and we shall be able to read the counter and note that the counter has been reset. But suppose that the next input transition occurs after an interval of time which is less than 200 ns. Then the first flip-flop will start to go to $Q_0 = 1$ before the last flip-flop has established itself at $Q_3 = 0$. There will then be no interval when all four flip-flops are reset, and the decoder will not be able to tell us that the count has arrived at a point when all the flip-flops are reset. Put another way, the decoding error in the K_0 decoder output will have a duration as long as the interval between input clocking transitions. At a minimum, then, the interval T between clocking transitions must be Nt_{pd}, N being the number of flip-flops and t_{pd} the propagation delay in each. If, in addition, we want to have available an interval T_s for strobing, we require that

$$T \geq Nt_{pd} + T_s \qquad (10.6\text{-}2a)$$

and the maximum frequency of the counter is

$$f \leq \frac{1}{Nt_{pd} + T_s} \qquad (10.6\text{-}2b)$$

Often the circuit of Fig. 10.6-1a is not used for the purpose of counting the number of input pulses but to make available at its output a waveform which goes through a number of pulses or cycles smaller than, but integrally related to, the number of input pulses. Thus, we observe in Fig. 10.6-1a that Q_0 executes one-half the number of cycles executed by the input. Q_1 executes one-quarter the number of cycles executed by the input, etc. In general, at the output of an N-stage counter the waveform will execute fewer cycles than are executed at the input by the factor 2^N. When used in this application, the circuit is referred to as a *frequency divider* or *frequency scaler*, and a divider with N flip-flops is often called a 2^N *scaler*.

A flip-flop cascade can operate at a higher input frequency when used as a scaler than when used as a counter. For, when it is a scaler, we are interested only in the output waveform and have no interest in reading the count. When it is a scaler, the maximum input frequency is determined by the *toggling rate* of the first flip-flop $FF0$, that is, by the maximum rate at which this first flip-flop can toggle back and forth in response to an input waveform. The second flip-flop in line $FF1$ need toggle at only half that rate, and so on.

Finally we may note that whether the flip-flop cascade is used as a counter or as a scaler, the input waveform need not be periodic. In Figs. 10.6-1 and 10.6-2 we have assumed a periodic input waveform simply for convenience in constructing the waveform.

10.7 METHODS TO IMPROVE COUNTER SPEED

A sequential digital system is described as operating *synchronously* if all flip-flops are driven by the same clock waveform. In such a case all state changes in the flip-flops take place at the same time, i.e., in synchronism with the clock waveform. The ripple counter of Fig. 10.6-1 is not a synchronous system.

A synchronous serial counter The speed of counters can be increased by operating them synchronously. One such method of synchronous operation is shown in Fig. 10.7-1. Note that this counter is synchronous since the input clock waveform is applied to all flip-flop clock inputs simultaneously.

At a triggering clock transition a flip-flop changes state only if its J and K terminals (shown tied together in the circuit of Fig. 10.7-1) are at $J = K = 1$. Hence, $FF1$ responds only if $Q_0 = 1$. Similarly $FF2$ responds only if $Q_0 = Q_1 = 1$, and $FF3$ responds only if $Q_0 = Q_1 = Q_2 = 1$, etc. On this basis it can be verified that the sequence of states assumed by the flip-flop is exactly the same as for the ripple counter of Fig. 10.6-1a, and the waveform chart for this synchronous counter is precisely as given in Fig. 10.6-1b. We note that information concerning the state of the preceding flip-flops is carried to succeeding flip-flops through a series of AND gates. Hence, the present counter is referred to as a *serial-carry synchronous counter*. Or since, as we shall see, the synchronous feature increases the allowable speed, the counter is also called a *serial-carry fast counter*.

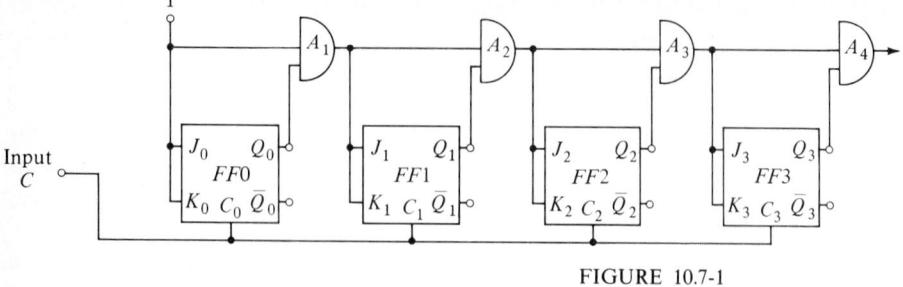

FIGURE 10.7-1
A serial-carry synchronous counter.

Since each flip-flop is clocked simultaneously, at any particular count every flip-flop which is to change state does so simultaneously. Hence, we might imagine that the present counter would not generate decoding errors. Such is not precisely the case, however, for the state changes in a flip-flop do not take place instantaneously. There is a finite rise and fall time at the output of a flip-flop. Further, in a particular flip-flop, rise and fall times are different, and there are differences between flip-flops both in rise and fall time and in propagation time. Thus, in the present synchronous counter there are decoding errors, and generally strobing is required, as in the ripple counter. However, in the synchronous counter the decoding error persists only for an interval which is of the order of the time required for a flip-flop to change state plus the maximum time difference in propagation time between flip-flops. But, most importantly, the duration of this decoding error interval *does not increase* with the number of flip-flops in the counter. On the other hand, in the ripple counter each additional flip-flop increases the interval of the decoding error by the propagation time of that flip-flop. Hence, insofar as speed is limited by the need to read correctly the count between input cycles the serial-carry counter has a marked advantage over the ripple counter.

The synchronous serial counter, however, has a speed limitation of its own, resulting from the need to propagate logic levels through the series of AND gates. Consider for example, that the counter is in the state $Q_3Q_2Q_1Q_0 = \mathbf{1110}$. Then in each flip-flop, except $FF0$, we have $J = K = \mathbf{0}$. At the next clock pulse (call it the ith) only $FF0$ responds, and we have $Q_3Q_2Q_1Q_0 = \mathbf{1111}$ after the change in Q_0 from $Q_0 = \mathbf{0}$ to $Q_0 = \mathbf{1}$. After the propagation delay of gate $A1$ we shall have $J_1 = K_1 = \mathbf{1}$, and after an additional gate propagation delay we shall have $J_2 = K_2 = \mathbf{1}$, etc. Thus, the effect of the change in Q_0 *ripples* through the AND gates. Suppose we allow adequate time for this change to make itself felt at the last flip-flop. Then in each flip-flop we shall have $J = K = \mathbf{1}$, and at the $(i + 1)$st clock pulse, each flip-flop will make a state change; we shall therefore have $Q_0Q_1Q_2Q_3 = \mathbf{0000}$, as required. If, however, we do not allow adequate time between the ith and $(i + 1)$st input for this serial transmission through the AND gates, gates at the end of the chain will not change state and the counter will be in error. Hence, an N-stage counter uses $N - 1$ AND gates and will count

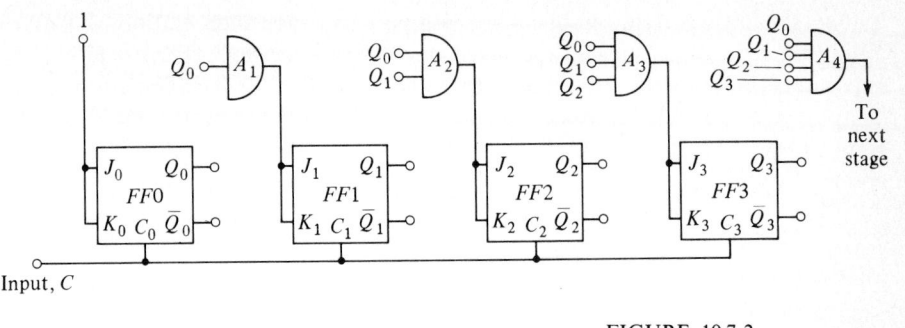

FIGURE 10.7-2
A synchronous parallel counter.

correctly as long as the duration between clock pulses T is greater than the propagation delay time $(N - 1)t_{pd}$ of all the AND gates and $FF0$. The duration of the strobe pulse also serves to limit the counting rate. The maximum clock frequency is then

$$f \leq \frac{1}{(N - 1)t_{pd}(\text{AND}) + t_{pd}(FF0) + T_s} \qquad (10.7\text{-}1)$$

In practice one usually omits gate $A1$ since Q_0 can be connected directly to the JK inputs of $FF1$ and gate $A2$. In this case the maximum clock frequency is somewhat higher.

In comparing the ripple counter with the synchronous serial counter, we note that the ripple counter has no limitation on the counting speed, other than the inherent speed of an individual flip-flop, but the speed must nonetheless be restricted to allow correct reading of the count. The synchronous serial counter, on the other hand, *counts incorrectly* at speeds where no problem is encountered in correct decoding. Still, it turns out that the synchronous serial counter has a net speed advantage over the ripple counter because the propagation delay through the simple AND gates is less than the propagation delay in a flip-flop. The price paid for this speed increase is the extra complexity of the AND gates, which are not required in the ripple counter.

A synchronous parallel counter A further improvement in counter speed is achieved in the synchronous parallel counter of Fig. 10.7-2. Here we have avoided the need for rippling logic levels through AND gates. Instead, each logic level required at an AND gate has been brought directly to that gate. As can be verified, the sequences of states assumed by the flip-flops and the waveform chart are again identical to those of Fig. 10.6-1, which applies as well to the straightforward ripple counter and to the synchronous serial counter.

While the synchronous parallel counter is the fastest of the three types considered, it is not without disadvantages. We observe that each succeeding AND gate requires one additional input. Further, each additional gate increases, by one, the fan-out required of each flip-flop. And each such additional load on a flip-flop generally decreases the speed of the flip-flop itself. Thus, $FF0$ has the maximum fan-out and the longest propagation delay time. The propagation delay as a function of fan-out F, $t_{pd}(F)$, is usually given by the manufacturer. With this parameter, the maximum clock frequency for the synchronous parallel counter is given by

$$f_{max} = \frac{1}{t_{pd}(F) + t_{pd}(\text{AND}) + T_s} \tag{10.7-2}$$

10.8 NONBINARY COUNTERS

All the counters considered up to this point have a modulo which is a power of 2. Thus, a counter with N flip-flops is a mod-2^N counter. Starting at some initial state, the system cycles through all possible 2^N states of the counter before returning to its initial state. We consider now how a counter can be modified to an arbitrary modulus, i.e., not a power of 2. For example, a *decade* (mod-10) counter is of special interest.

   Suppose we want a mod-k counter. Then we start with N flip-flops such that N is the smallest number for which $2^N > k$. Thus, for $k = 5$ or 6 or 7 we take $N = 3$, since $2^N = 8$. For $k = 9$ through 15 we take $N = 4$, etc. To construct a mod-k counter we start with a mod-2^N counter and arrange to omit some of its possible states. For a mod-k counter we shall have to omit $2^N - k$ states. We are at liberty to select which states are to be omitted. Hence, there are usually many ways in which the mod-k counter can be designed. A counter may operate a count display or be used to control the sequence of other digital operations. In any event, some sort of decoder will have to be interposed between the counter and whatever it is intended to operate. Often the decisions with respect to which states to omit are made with a view toward simplifying the decoder.

10.9 MOD-3 COUNTERS

There are too many ways in which a nonbinary counter can be designed to make it practical or even worthwhile to develop general procedures for designing nonbinary counters. Nonetheless, it is instructive, and certainly of interest, to explore some of the considerations leading to such design. Let us then begin by designing a mod-3 counter.

   A mod-3 counter requires *two* flip-flops. Let us arbitrarily specify that the counter is to cycle through its three states in the manner indicated in Table 10.9-1. Thus, there is a counter state $S = 0$ corresponding to $Q_1Q_0 = 00$. The next input cycle is to advance the counter to its second state $S = 1$, in which $Q_1Q_0 = 01$. The cycle following is to advance the counter to $S = 2$, in which

$Q_1Q_0 = 10$.   Finally, one additional input cycle should reset the counter back to $S = 0$.

We can see that a ripple counter is not possible in the present case.   In a ripple counter one of the flip-flops would receive the clock input from the gate output of the other flip-flop.   We observe that when S goes from $S = 2$ to $S = 0$, Q_1 changes state and hence requires a clocking transition.   However, Q_0 does not change state and hence cannot provide such a clocking transition. The situation is not improved if we interchange the states of Q_1 and Q_0.   We begin our design by assuming synchronous operation with the external clocking signal applied to both flip-flops.   We shall assume that we have JK flip-flops, and we must still determine how to arrange for the J and K inputs always to be at such logic levels that the two flip-flops make the proper transition at the proper time.

With this end in view it is convenient to present the truth table of a JK flip-flop in the form of Table 10.9-2, which specifies the required logic at J and K if a flip-flop is to go from a state Q_n (before the clocking transition) to a state Q_{n+1} (after the clocking transition).   The entry X means "don't care."

By reference to Tables 10.9-1 and 10.9-2 we can now fill in the K maps for J_0, K_0, J_1, and K_1 for the two flip-flops $FF0$ and $FF1$ which form our mod-3 counter.   The result is given in Fig. 10.9-1.   To see how this figure is obtained note that when the count is 0, so that the outputs of both flip-flops are $Q_1Q_0 = 00$, the next state of the counter is $Q_1Q_0 = 01$.   That is, Q_1 must remain at $Q_1 = 0$ so that we require that $J_1 = 0$ and $K_1 = X$ while Q_0 must go to 1 so

Table 10.9-1

Counter state S	Q_1	Q_0
0	0	0
1	0	1
2	1	0
0	0	0
⋮	⋮	⋮

Table 10.9-2

$Q_n \longrightarrow Q_{n+1}$		J	K
0	0	0	X
0	1	1	X
1	0	X	1
1	1	X	0

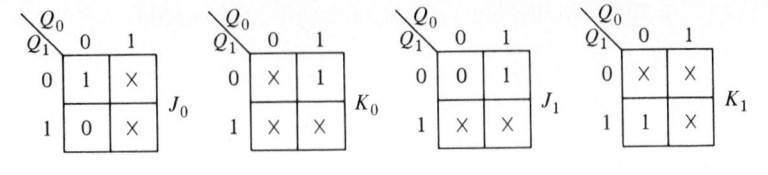

FIGURE 10.9-1
K maps for a mod-3 counter.

that we require $J_0 = 1$ and $K_0 = X$. The case $Q_1 Q_0 = 11$ never arises, so that the corresponding K-map boxes are all marked X. The remaining boxes for J_0, K_0, J_1, and K_1 when $Q_0 Q_1 = 01$ and 10 are left for Prob. 10.9-2. The K maps of Fig. 10.9-1 when put in simplest form (see Sec. 3.0) yield

$$J_0 = \bar{Q}_1 \qquad K_0 = 1 \qquad \text{and} \qquad J_1 = Q_0 \qquad K_1 = 1 \qquad (10.9\text{-}1)$$

The resulting counter circuit is shown in Fig. 10.9-2a. The specification of the logic values for Q_1 and Q_0 corresponding to each counter state, as in Table 10.9-1, is called a *state assignment*. Different state assignments will lead to different counter configurations, although in some cases the differences may be superficial (see Prob. 10.9-3). The waveform chart for the counter, which can be deduced from the circuit or simply read from the state assignment table, is shown in Fig. 10.9-2b.

10.10   MOD-5 COUNTERS

As a second example of counter design we consider a mod-5 counter to operate in accordance with the state assignment of Table 10.10-1. Three flip-flops are now required. Assuming synchronous operation, we can, as before, construct

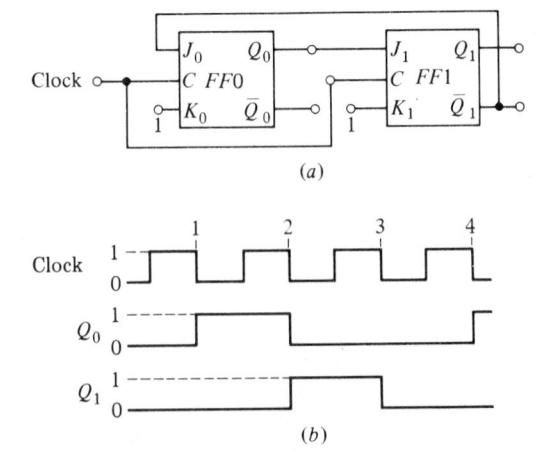

(a)

(b)

FIGURE 10.9-2
A mod-3 counter.

Table 10.10-1

Counter state S	Q_2	Q_1	Q_0
0	0	0	0
1	0	0	1
2	0	1	0
3	0	1	1
4	1	0	0
0	0	0	0

K maps for the J and K terminals through reference to Tables 10.9-2 and 10.10-1. These maps are given in Fig. 10.10-1. To review the construction of these maps, consider that the count is $S = 2$ so that $Q_2 Q_1 Q_0 = $ **010**. At the next clock pulse Q_0 will become **1** while Q_2 and Q_1 will not change state. Thus, we set $J_0 = 1, K_0 = X, J_1 = X, K_1 = 0, J_2 = 0$, and $K_2 = X$. The other portions of the K maps are completed in similar fashion (see Prob. 10.10-1). From these maps we read, after simplification,

$$J_0 = \overline{Q}_2 \qquad K_0 = 1$$
$$J_1 = K_1 = Q_0 \qquad (10.10\text{-}1)$$
$$J_2 = Q_1 \cdot Q_0 \qquad K_2 = 1$$

The circuit diagram and waveforms are given in Fig. 10.10-2.

In a particular case, the external AND gate in Fig. 10.10-2a may not be required. For we note, referring to Fig. 9.20-1, that the J input of a flip-flop may be one input terminal of an AND gate. It might be feasible to add an

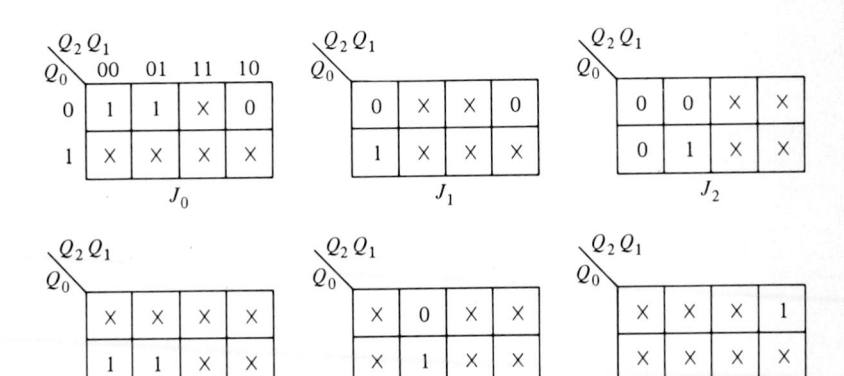

FIGURE 10.10-1
K maps for a mod-5 counter.

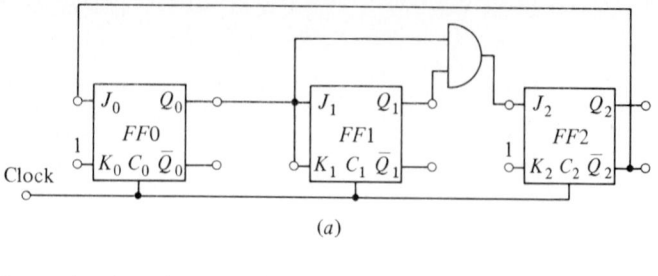

(a)

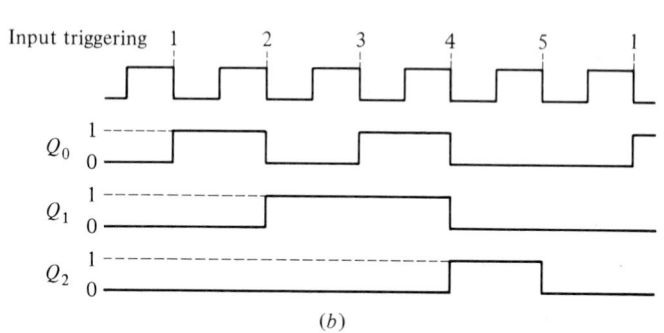

(b)

FIGURE 10.10-2

(a) A mod-5 counter. (b) Waveforms.

additional input to this input flip-flop gate. In such a case, the two J inputs of the flip-flop can be used to accommodate Q_1 and Q_0, and an external gate would not be required.

We note, from the state-assignment table (Table 10.10-1) or from the waveform chart of Fig. 10.10-2b, that $FF1$ changes state when and only when Q_0 goes from $Q_0 = 1$ to $Q_0 = 0$. This response of $FF1$ is precisely the response that would result if C_1 were connected to Q_0 as in a ripple counter. In Fig. 10.10-3 we have modified the mod-5 counter, incorporating this ripple connection and indicating the AND gate as being internal to $FF2$. In this last counter, then, the clocking of the flip-flops is in part *synchronous* and in part *ripple*.

To see the effect of the state assignment on a counter, we consider a second example of a mod-5 counter, this time with a state assignment as given in Table 10.10-2. From Tables 10.10-2 and 10.9-2 we can construct, as before,

Table 10.10-2

Counter state S	Q_2	Q_1	Q_0
0	0	0	0
1	0	1	1
2	1	1	1
3	1	1	0
4	1	0	1

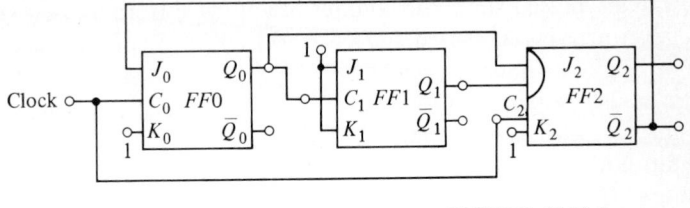

FIGURE 10.10-3
A modified mod-5 counter.

K maps for the J and K inputs of the three flip-flops. We find (Prob. 10.10-3) in this case, that

$$
\begin{array}{ll}
J_0 = 1 & K_0 = Q_2 \\
J_1 = \overline{Q}_0 & K_1 = \overline{Q}_0 \\
J_2 = Q_0 & K_2 = \overline{Q}_1
\end{array}
\qquad (10.10\text{-}2)
$$

The counter is shown in Fig. 10.10-4.

We may now note, from Table 10.10-2, that Q_1 changes when and only when Q_0 changes from $Q_0 = 0$ to $Q_0 = 1$. Hence, if we please, the synchronous drive applied to the clock input of $FF1$ can be replaced by a ripple connection to $FF0$. As in Fig. 10.10-3, we would set $J_1 = K_1 = 1$. However, in the present case we would connect C_1 to $\overline{Q}_0$ rather than to Q_0.

10.11 LOCKOUT

In the counter specified by Table 10.10-1, logic states $Q_2 Q_1 Q_0 = \mathbf{101}, \mathbf{110},$ and $\mathbf{111}$ are not used. Suppose that, by chance, the counter happened at some point to find itself in any one of these unused states. Such a situation might develop as a result of external noise affecting the state of a flip-flop. For each of the states listed in Table 10.10-1 we know the state to which the counter goes at the next input clock pulse. But we do not know the next state for an unused state. Hence, there is the further possibility that when a flip-flop is "thrown"

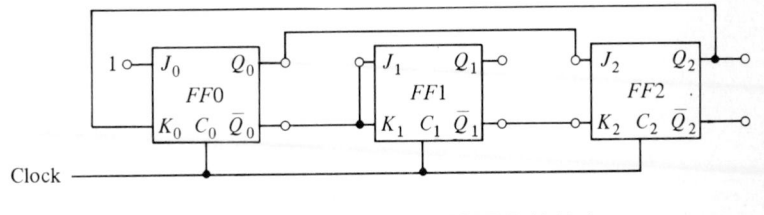

FIGURE 10.10-4
An alternate mod-5 counter design.

into an unused state, the counter might cycle from unused state to unused state, never arriving at a used state. In this case, of course, the counter is useless for its originally intended purpose. A counter whose unused states have this feature is said to suffer from *lockout*. To insure that at "start-up" the counter is in its initial state, external logic circuitry is provided which properly presets each flip-flop.

If, starting from an unused state, a counter eventually arrives at a used state either directly or after passing through other unused states, the operation is often acceptable. We might consider the time required for a used state to be reached as a warm-up time for the counter or as the counting error due to noise. In any event, in any counter design it is necessary to check *each* unused state to determine whether it leads to a *lockout* condition. It can be verified that lockout will not be encountered with any of the unused states of the counter of Fig. 10.10-2 (Prob. 10.11-1).

To ensure that lockout does not occur we design the counter so that the J's and K's for an unused state, which previously received the value X, now take on prescribed values which transfer the unused state to a used state. For example, if, in the mod-5 counter shown in Fig. 10.10-2, noise caused a change of state so that $Q_2 Q_1 Q_0 = 101$ (an unused state), the next clock pulse will change the counter state to $Q_2 Q_1 Q_0 = 010$ (a used state). To ensure that this occurs by design (rather than by chance) we require when $Q_2 Q_1 Q_0 = 101$ that $J_0 = X$, $K_0 = 1$, $J_1 = 1$, $K_1 = X$, $J_2 = X$, $K_2 = 1$. These values should be inserted in Fig. 10.10-1. However, referring to Eq. (10.10-1) and Fig. 10.10-1, we note that these are the values entered in the K maps or the values assigned to "don't cares" by the manner in which we have read the maps.

EXAMPLE 10.11-1   It is required to redesign the mod-5 counter shown in Fig. 10.10-1 so that if the unused states $Q_2 Q_1 Q_0 = 101$, 110, or 111 occur, the next clock pulse will reset the counter to $Q_2 Q_1 Q_0 = 000$.

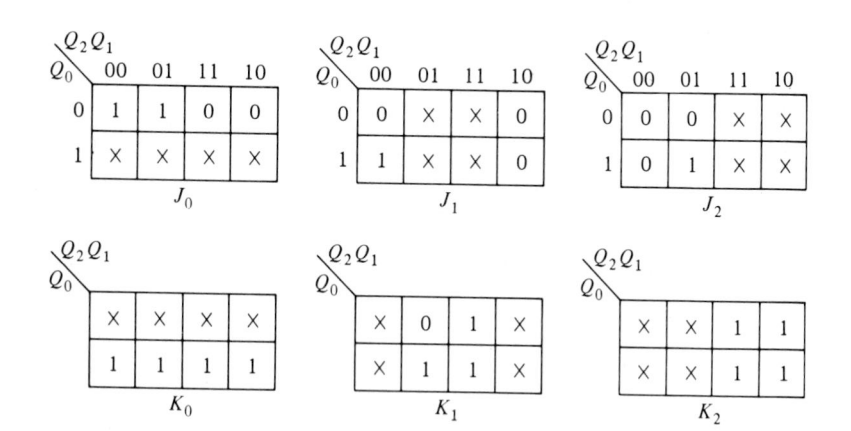

FIGURE 10.11-1
K maps to design a mod-5 counter having the feature that if an unused state appears, the counter will reset to 000 at the next clock pulse.

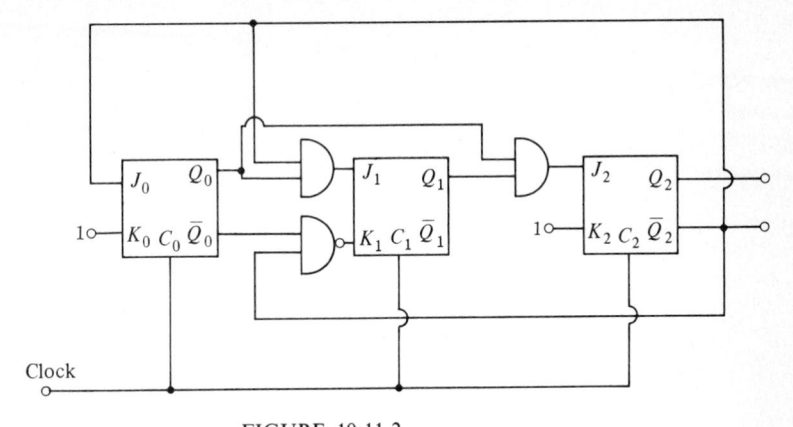

FIGURE 10-11-2
A mod-5 counter which resets if an unused state is reached.

SOLUTION The K maps needed to reset the counter are shown in Fig. 10.11-1. Here we see that the J's and K's have been altered so that an unused state, when clocked, produces $Q_2 Q_1 Q_0 = 000$. For example, to go from **101** to **000** requires $J_0 = X$, $K_0 = 1$; $J_1 = 0$, $K_1 = X$; and $J_2 = X$, $K_2 = 1$. To go from **110** to **000** requires that $J_0 = 0$, $K_0 = X$; $J_1 = X$, $K_1 = 1$; $J_2 = X$, and $K_2 = 1$. And to go from **111** to **000** requires that $J_0 = J_1 = J_2 = X$ and $K_0 = K_1 = K_2 = 1$.

The resulting simplified equations for the J's and K's, obtained from Fig. 10.11-1, are

$$J_0 = \overline{Q}_2 \qquad K_0 = 1$$
$$J_1 = Q_0 \overline{Q}_2 \qquad K_1 = Q_0 + Q_2 = \overline{\overline{Q}_0 \overline{Q}_2}$$
$$J_2 = Q_0 Q_1 \qquad K_2 = 1$$

The resulting mod-5 counter is shown in Fig. 10.11-2.

10.12 COMBINATIONS OF MODULO COUNTERS

A single flip-flop is a mod-2 counter. A chain of flip-flops, in ripple connection or synchronously operated, will make a counter with a base that is a power of 2. Now, however, that we have constructed a mod-3 and a mod-5 counter, we can combine these structures with mod-2 counters to make other counters which operate with a modulo which is not a power of 2. Thus, a mod-2 and mod-3 counter in cascade constitute a mod-6 counter. Two mod-3 counters make a mod-9 counter. A mod-2 counter and a mod-5 counter can be combined to make a *decade counter*, i.e., a mod-10 counter, etc. The connection between the individual counters may be a ripple connection, or the counters may be operated in synchronism with one another, independently of whether the individual counters are ripple or synchronous counters. Further, we are at liberty to permute the order of the individual counters in a chain of counters. Such permutation will

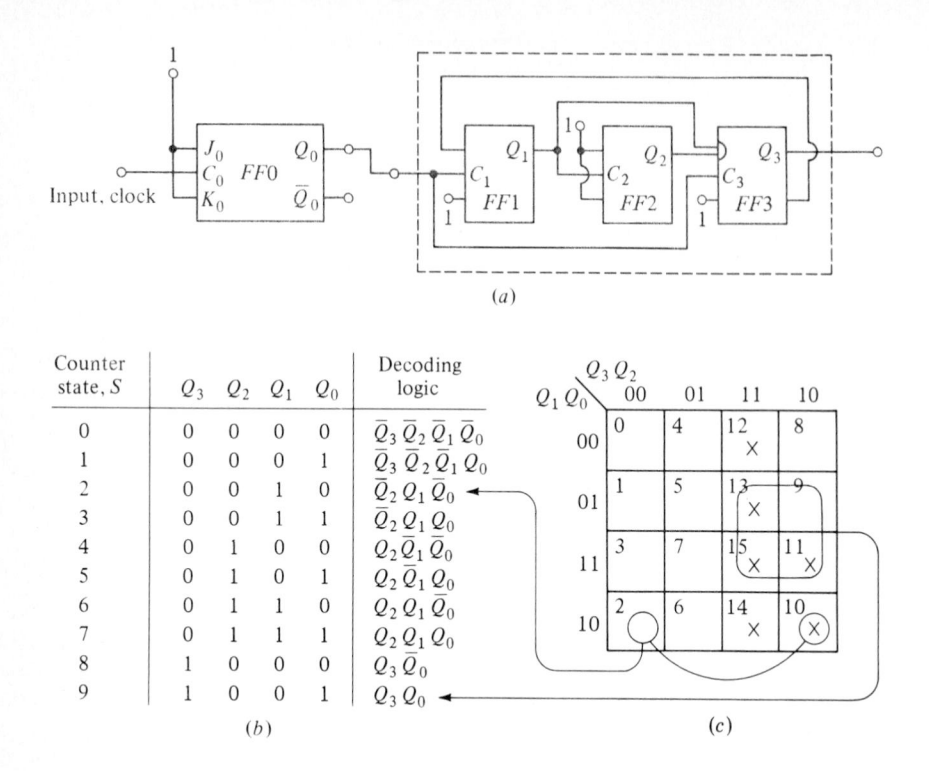

Counter state, S	Q_3	Q_2	Q_1	Q_0	Decoding logic
0	0	0	0	0	$\bar{Q}_3\bar{Q}_2\bar{Q}_1\bar{Q}_0$
1	0	0	0	1	$\bar{Q}_3\bar{Q}_2\bar{Q}_1 Q_0$
2	0	0	1	0	$\bar{Q}_2 Q_1\bar{Q}_0$
3	0	0	1	1	$\bar{Q}_2 Q_1 Q_0$
4	0	1	0	0	$Q_2\bar{Q}_1\bar{Q}_0$
5	0	1	0	1	$Q_2\bar{Q}_1 Q_0$
6	0	1	1	0	$Q_2 Q_1\bar{Q}_0$
7	0	1	1	1	$Q_2 Q_1 Q_0$
8	1	0	0	0	$Q_3\bar{Q}_0$
9	1	0	0	1	$Q_3 Q_0$

(b) *(c)*

FIGURE 10.12-1
A decade counter constructed from a mod-5 and a mod-2 counter.

not change the modulus of the composite counter but may well make a substantive difference in the code in which the counter state is to be read.

As an example, in Fig. 10.12-1a we have cascaded a mod-2 counter with the mod-5 counter of Fig. 10.10-3 to form a decade counter. Since we have placed the mod-2 flip-flop ahead of the mod-5 counter, we call this flip-flop $FF0$. Hence, the flip-flops in Fig. 10.10-3 advance in order so that $FF0$ becomes $FF1$, etc. Defining the counter state $S = 0$ as the state in which $Q_3Q_2Q_1Q_0 = \mathbf{0000}$, the state-assignment table, as can be verified, appears as in Fig. 10.12-1b. Observe that as the counter steps through its sequence, the logic levels at $Q_3Q_2Q_1Q_0$ read the count in the 8-4-2-1 (binary coded decimal (BCD) code).

Also given in Fig. 10.12-1b is the decoding logic for this counter. We have in mind a decoder consisting of AND gates, as in Fig. 10.6-2, except that in the present case we require 10 such gates with outputs $K_0, \ldots, K_9$. The interpretation of the entries in the decoding logic table is that, by way of example, the AND gate with output K_4 should have three inputs Q_2, $\bar{Q}_1$, and $\bar{Q}_0$.

The decoding logic entries can be read directly from the K map shown in Fig. 10.12-1c. In this map we have placed don't care signs (X) in all the boxes corresponding to combinations of $Q_3Q_2Q_1Q_0$ which do not occur in the state assignment table. By way of example, to read the code corresponding to $K = 9$, we locate the K-map box corresponding to $Q_3 Q_2 Q_1 Q_0 = \mathbf{1001}$. Because we are dealing with the 8-4-2-1 BCD code this box happens also to be the box for minterm 9. If we now place a logic 1 in this minterm 9 box, we find that it can be combined with the three surrounding don't cares to yield $Q_3 Q_0$. As a second example we have shown how the code for $K = 2$ comes to be $\overline{Q}_2 Q_1 \overline{Q}_0$.

In Fig. 10.12-1a if we permute the mod-2 and the mod-5 counters, we still have a mod-10 counter. However, the state-assignment tables and the decoding logic will be different (see Prob. 10.12-1).

10.13 OTHER COUNTER DESIGNS

Counters with bases which are a power of 2 may consist of a chain of flip-flops coupled in ripple fashion. The ripple connection avoids the need for gates external to the flip-flops. When the base is other than a power of 2, we can straightforwardly design a counter, even with a specified state assignment, provided we design for synchronous operation.

It is of considerable interest to consider an alternative scheme of counter design which allows ripple coupling throughout and at the same time permits an arbitrary modulo. The new feature to be added here involves the use of asynchronous flip-flop inputs (such as direct set), which were not involved in the counters considered heretofore.

A decade counter is shown in Fig. 10.13-1. The flip-flops are again of the JK type (J, K inputs not shown) with $J = K = 1$, so that the flip-flops are in the toggle mode. Note that the interconnections between successive flip-flops are all ripple connections. Since we shall use the direct-set terminals, these are explicitly shown. We assume that the direct-set terminal exercises control over the flip-flop when it is at logic level $\mathbf{0}$, and hence we have labeled these inputs $\overline{S}_d$. The direct-set inputs of $FF0$ and $FF3$ are not involved in the counting operation, and these inputs are left at logic $\mathbf{1}$.

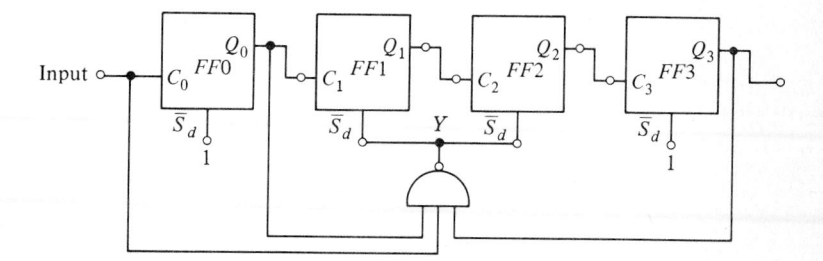

FIGURE 10.13-1

A decade counter using the direct input, $\overline{S}_d$.

Waveform charts, ignoring propagation delays for the flip-flop outputs and for the $\bar{S}_d$ inputs Y of $FF1$ and $FF2$, are given in Fig. 10.13-2. We note that starting with the initial state $Q_0 = Q_1 = Q_2 = Q_3 = 0$ and through the count of 9, the operation is precisely as in the mod-16 counter, since over this counting range we have $Y = 1$. Also, since the counter follows the normal ripple sequence, we can read the count directly from the four outputs Q_0, Q_1, Q_2, and Q_3 in the 8-4-2-1 BCD code.

At the positive-going input transition immediately following the ninth flip-flop triggering transition, each of the inputs to the NAND gate is at logic level **1**, and Y becomes $Y = 0$. As a result, $FF1$ and $FF2$ go to the set state, and we now have $Q_0 = Q_1 = Q_2 = Q_3 = 1$. The chain of flip-flops is now in the state corresponding to the state for count $15 = \mathbf{1111}$ in a straightforward ripple counter. Thus, the effect has been to advance the count from $\mathbf{1001} = 9$ to $\mathbf{1111} = 15$ at one step, when the input clock pulse is at the **1** level, skipping all the intermediate counts. When the input clock pulse returns to the **0** level, producing a negative-going transition, a ripple is started through the whole cascade of flip-flops, returning each flip-flop to the reset state so that the counter has returned to its initial state.

At the tenth negative-going input transition Y will return to $Y = 1$, releasing $FF1$ and $FF2$ from control. Also Q_0 will go to $Q_0 = 0$, inducing a state change in $FF1$ from $Q_1 = 1$ to $Q_1 = 0$. If such is indeed to be the case, it is necessary that Y go to $Y = 1$ *before* Q_0 goes to $Q_0 = 0$. This requirement should cause no difficulty, however, since ordinarily we expect that the propagation delay through a gate will be smaller than the delay through a flip-flop and that Y will return to **1** when the clock pulse becomes **0**.

In the 8-4-2-1 BCD code, decimal 9 is **1001**, and in a decade counter, this count should be followed by **0000**. We note, however, in the waveform

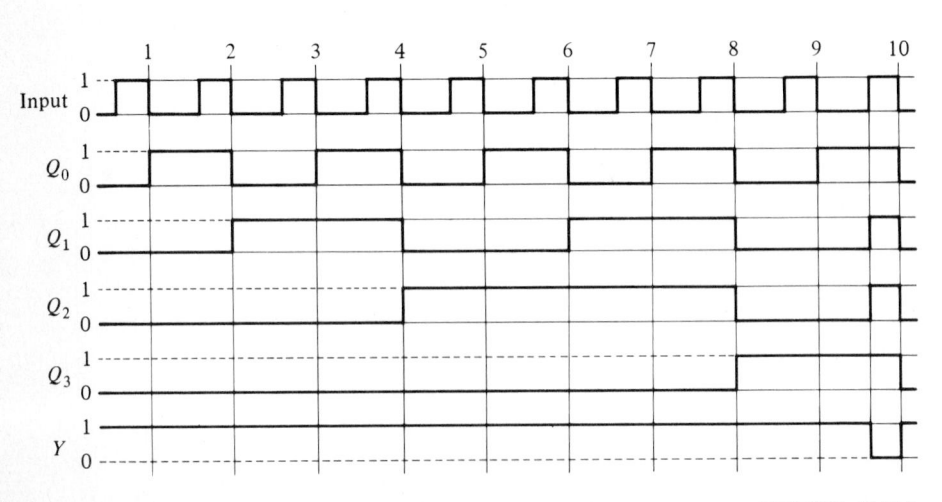

FIGURE 10.13-2
Waveforms.

charts of Fig. 10.13-2, that **1001** is followed by **1111** for the interval while the input clock is at logic **1**, before going to **0000**. To avoid this undesired counter state in reading the counter, it will be necessary, of course, to provide appropriate strobing, just as it is generally necessary to provide strobing to avoid errors in count readings due to propagation delays (as discussed in connection with Fig. 10.6-2).

Finally we may note that the scheme employed in Fig. 10.13-1 to construct a decade counter can be extended to construct a counter of arbitrary mod k. The design steps are the following:

1 The number of flip-flops N is determined by the condition $2^{N-1} < k < 2^N$.
2 The NAND gate has as inputs the counter input and the Q output of each flip-flop which is in the set state ($Q = 1$) at the count $k - 1$.
3 The NAND-gate output is connected to the direct-set inputs $\bar{S}_d$ of each flip-flop which is in the reset state ($Q = 0$) at the count $k - 1$.

Other synchronous ripple-type counters appear in Probs. 10.13-4 and 10.13-5.

10.14 THE UP-DOWN RIPPLE COUNTER

If we read the count in accordance with Eq. (10.6-1), the mod-16 ripple counter of Fig. 10.6-1*a* counts in the sequence 0, 1, 2, ..., 15, 0, 1, etc. Such a counter is called an *up* or *forward* counter. A *down* counter or a *backward* counter counts in the sequence 0, 15, 14, ..., 1, 0, 15, etc. It is left as a student exercise (Prob. 10.14-1) to verify that a *ripple-up* counter can be converted into a *ripple-down* counter simply by connecting the clock input of each flip-flop (other than the first, *FF*0) to the $\bar{Q}$ output of the preceding flip-flop, rather than to the Q output.

A counter which will count either up or down depending on a logic control signal is shown in Fig. 10.14-1. Gates have been interposed between flip-flops so that the logic level at M can determine whether the clock input of a succeeding flip-flop is connected to the Q or $\bar{Q}$ output of the preceding flip-flop. With the mode $M = 1$ ($\bar{M} = 0$) each clock input C is connected to a Q output, and the

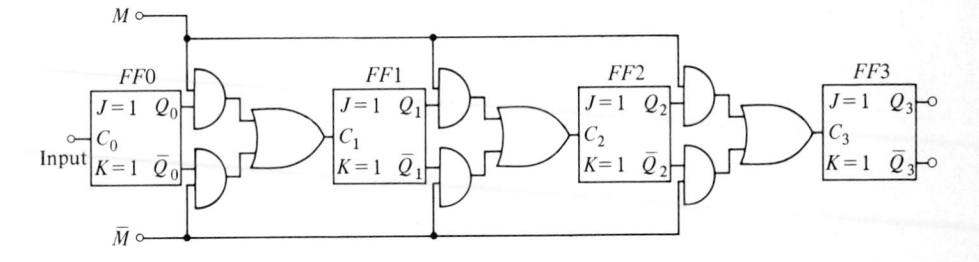

FIGURE 10.14-1
A forward-backward ripple counter.

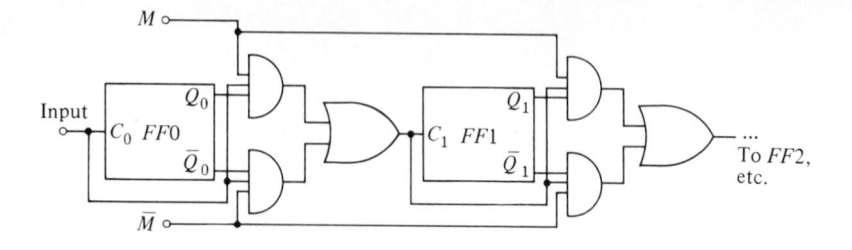

FIGURE 10.14-2
An up-down counter designed to permit a change in the direction of counting.

counting proceeds in the *up* direction. Alternately with $M = 0$ the counting proceeds in the *down* direction.

A precaution must be observed in connection with the use of the *up-down* counter of Fig. 10.14-1. We are not at liberty to change the count direction, with an arbitrary count registered in the counter, except at the risk of having the count change. For, suppose the counter is counting in the up direction ($M = 1$, $\overline{M} = 0$) and is at a point where one of the flip-flops, say $FF0$, is at $Q_0 = 1$. Then since $M = 1$ and $\overline{M} = 0$, we shall have $C_1 = 1$. Now suppose that we undertake to change direction by switching M to $M = 0$ and $\overline{M}$ to $\overline{M} = 1$. Then as a result of this change alone and in spite of the fact that $FF0$ has not changed state, the logic level at C_1 will go from 1 to 0 and $FF1$ will change state (we assume that the flip-flop output responds to the negative-going clock transition). Similarly, any flip-flop whose preceding flip-flop is in the set state will change state, and thereby the count registered in the counter will be changed. If we want to avoid this count change, we must change the direction from up to down only when each flip-flop (except for the last) is in the reset state. Alternatively, a change from the down direction to the up direction must be made only when every flip-flop is in the set state.

A modification of the up-down counter which allows changing direction at any point without loss of count is shown in Fig. 10.14-2. In this circuit, an additional input is provided to each AND gate. In the case of the AND gate which couples $FF0$ and $FF1$, this additional input is connected to C_0, in the next set of AND gates the connection is to C_1, etc. It is left as a problem (Prob. 10.14-3) to verify that the modified counter can be reversed at any count.

10.15 THE UP-DOWN SYNCHRONOUS COUNTER

To design a synchronous up-down counter we begin by designing two separate counters. The first (call it the *up counter*) is designed to step through a sequence of counter states in whatever manner is specified. The second (the *down counter*) is then designed to step through the same sequence in the reverse manner. Thereafter, external steering gates are added, as required, to change the connections

Table 10.15-1

Counter state S	Q_1	Q_0
0	0	0
1	1	0
2	0	1
0	0	0

from those appropriate for one direction of counting to those appropriate for the other.

As an example, let us design a mod-3 *up-down* synchronous counter. Let the *up* sequence be specified as in Table 10.9-1. For this case we have already found that the JK excitation equations for the two flip-flops are as given in Eq. (10.9-1). These equations are

$$J_0 = \overline{Q}_1 \qquad K_0 = 1$$
$$J_1 = Q_0 \qquad K_1 = 1 \qquad (10.15\text{-}1)$$

The state-assignment table for the down counter must then be as given in Table 10.15-1. It can be verified (Prob. 10.15-1) that for the down counter one possible set of JK excitation equations is

$$J_0 = Q_1 \qquad K_0 = 1$$
$$J_1 = \overline{Q}_0 \qquad K_1 = 1 \qquad (10.15\text{-}2)$$

It thus appears, in this especially simple case, that switching between *up* and *down* simply involves switching J_0 between $\overline{Q}_1$ and Q_1 and switching J_1 between Q_0 and $\overline{Q}_0$. The circuit diagram of the counter is given in Fig. 10.15-1. The mode control is M. When $M = 1$ ($\overline{M} = 0$), the count proceeds upward, and when $M = 0$, the count is downward.

It can also be verified that the direction of the count can be switched, as long as the clock input is **0**, without altering the count stored in the counter.

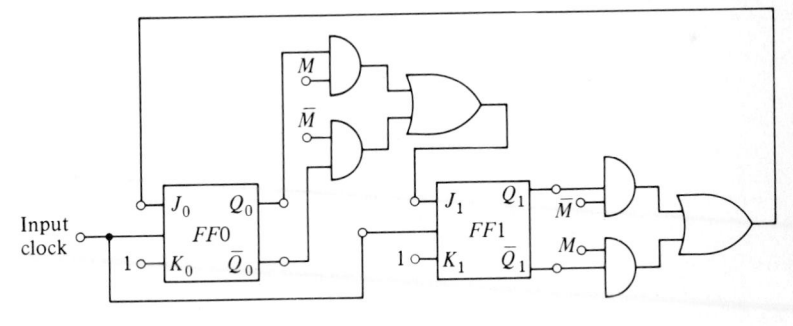

FIGURE 10.15-1
An up-down synchronous counter.

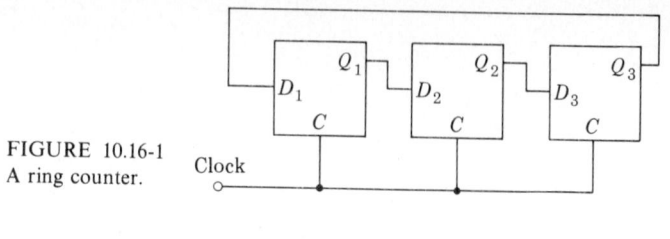

FIGURE 10.16-1
A ring counter.

10.16 RING COUNTERS

The end-around-carry shift register described earlier can be used as a counter. In such a counter the flip-flops are coupled as in a shift register, with the last flip-flop coupled back to the first so that the array of flip-flops is arranged in a ring. Such counters are consequently referred to as *ring counters*.

Suppose that the ring has k flip-flops and that we place one flip-flop in the set state while all others are in the reset state. Then with each input clock pulse the set condition will advance one flip-flop around the ring, returning to the initial flip-flop after k input cycles. Such a circuit shown in Fig. 10.16-1 constitutes a mod-k counter.

The ring counter has the advantage of requiring no decoder, since we can read the count by simply noting which flip-flop is set. Since it is entirely

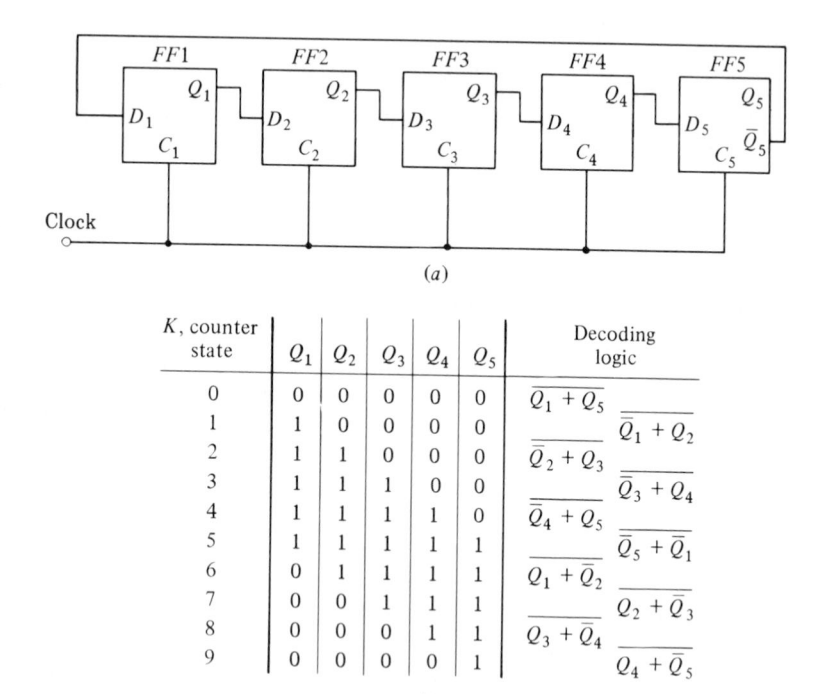

(a)

K, counter state	Q_1	Q_2	Q_3	Q_4	Q_5	Decoding logic
0	0	0	0	0	0	$\overline{Q_1 + Q_5}$
1	1	0	0	0	0	$\overline{\bar{Q}_1 + Q_2}$
2	1	1	0	0	0	$\overline{\bar{Q}_2 + Q_3}$
3	1	1	1	0	0	$\overline{\bar{Q}_3 + Q_4}$
4	1	1	1	1	0	$\overline{\bar{Q}_4 + Q_5}$
5	1	1	1	1	1	$\overline{Q_5 + \bar{Q}_1}$
6	0	1	1	1	1	$\overline{Q_1 + \bar{Q}_2}$
7	0	0	1	1	1	$\overline{Q_2 + \bar{Q}_3}$
8	0	0	0	1	1	$\overline{Q_3 + \bar{Q}_4}$
9	0	0	0	0	1	$\overline{Q_4 + \bar{Q}_5}$

(b)

FIGURE 10.16-2
(a) Switch-tail ring counter. (b) Truth table.

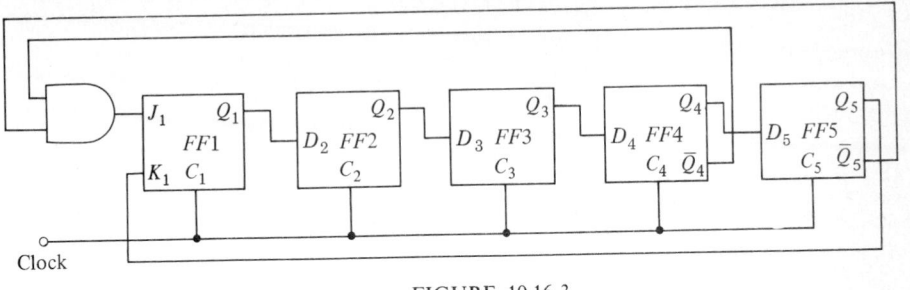

FIGURE 10.16-3
A switch-tail ring-counter designed to prevent *lockout*.

synchronous in operation and requires no gates external to the flip-flops, it has the further advantage of being extremely fast. It has the disadvantage of being uneconomical of flip-flops. Thus, a mod-16 ring counter requires 16 flip-flops, while a mod-16 ripple counter requires only 4 flip-flops.

Economy in the number of flip-flops can be improved by a modification of the simple ring counter into a *switch-tail* ring counter. Such a counter requires only half the number of flip-flops. A decade switch-tail counter, shown in Fig. 10.16-2, uses only five flip-flops. Except in one case, each flip-flop is coupled to the next flip-flop as in a shift register. In one case, however, the coupling of $FF5$ back to $FF1$, the connections have been switched, that is, output $\bar{Q}_5$ goes to D_1, rather than the other way around. It makes no difference where this single switching is effected, but it has become customary to draw the circuit diagram with the switching at the end, i.e., at the tail of the chain of flip-flops; hence the name. The state table and decoding logic are given in Fig. 10.16-2b. The verification of this table and logic are left as a student exercise (Prob. 10.16-1).

The ring counters, whether of the simple or switch-tail type, have the following disadvantage: if the counter should find itself in an unused state, it will persist in moving from one unused state to another and never find its way to a used state. This difficulty can be corrected by adding a simple AND gate, as in Fig. 10.16-3. With this addition, if the counter should initially find itself in an unused state, then after a number of clock cycles, depending on the state, the counter will find its way to a used state and thereafter follow the cycling sequence of the state assignments in Fig. 10.16-2b. In Prob. 10.16-3 the student is guided through an analysis of the error-correcting operation effected by the addition of the AND gate.

10.17 SEQUENCE GENERATORS

A *sequence generator* is a system which generates, in synchronism with a clock, a prescribed sequence of logic bits. Such generators may be used as a counter, a frequency divider for timing purposes, in code generation, etc. We shall consider in this section how shift registers can be adapted to serve as sequence generators.

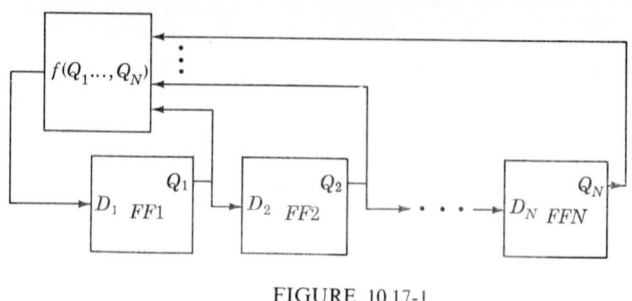

FIGURE 10.17-1
Basic structure of a sequence generator.

The *length* of a sequence is the number of successive bits in the sequence before the pattern of bits repeats. The sequence length is analogous to the period of a periodic analog waveform. For example, the sequence $\cdots$ **11101110111011** $\cdots$ has a *length* of 4 bits. Since the generator repeats the pattern over and over again, we can read the sequence in a number of ways, depending on our starting point. Thus, the sequence given may be described as a repetition of the bit pattern **1110** or of the patterns **1101** or **1011** or **0111**. Further, as we shall see, the bit sequences will become available at the outputs of flip-flops. If we use the sequence available at the complementary output of the flip-flop, the sequence will be read as **0001** or **0010** or **0100** or **1000**. Hence, a repetition of any of the above eight patterns of four bits constitutes the same sequence. It is of interest to note that in many applications the length of the sequence is more important than the exact pattern of repeated bits. It can be verified that for a length of 1, 2, or 3 bits there is only one possible sequence in each case. For a length of 4 bits there are two sequences.

The basic structure of a sequence generator is shown in Fig. 10.17-1. Here we have N flip-flops cascaded in the usual manner of a shift register. Type-D flip-flops have been used. A clock waveform, not explicitly shown, is applied to all flip-flops. The system has a feature reminiscent of a shift-register ring counter. In a ring counter the output Q_N of the last flip-flop becomes the data input D_1 of the first flip-flop. In the present case, however, the logic level of D_1 is determined not alone by Q_N but also by the output of other flip-flops in the cascade. That is,

$$D_1 = f(Q_1, Q_2, \ldots, Q_N) \qquad (10.17\text{-}1)$$

The prescribed sequence will appear at the output of each of the flip-flops. Of course, as we go from $FF1$ to $FF2$, etc., we shall find successive delays, by one clock interval, in the appearance of the sequence.

In order to build a sequence generator capable of generating a sequence of length S it is necessary to use, at a minimum, a number N of flip-flops, where N satisfies the condition

$$S \leq 2^N - 1 \qquad (10.17\text{-}2)$$

Clock interval k	Q_1	Q_2	Q_3	Q_4	$D_1 = f(Q_1 \ldots, Q_N)$
1	1	1	1	1	0
2	0	1	1	1	0
3	0	0	1	1	0
4	0	0	0	1	1
5	1	0	0	0	0
6	0	1	0	0	0
7	0	0	1	0	1
8	1	0	0	1	1
9	1	1	0	0	0
10	0	1	1	0	1
11	1	0	1	1	0
12	0	1	0	1	1
13	1	0	1	0	1
14	1	1	0	1	1
15	1	1	1	0	1
1	1	1	1	1	0
2	0	1	1	1	0
3	0	0	1	1	0
⋮	⋮	⋮	⋮	⋮	⋮

FIGURE 10.17-2

Truth table for sequence generator of sequence given by Q_1.

If the order of the **1**s and **0**s in the sequence is prescribed, generally it will not be possible to generate a length S in which the minimum number of flip-flops is used. On the other hand, for any N, there is at least one sequence for which the sequence length S is a maximum, that is, $S = 2^N - 1$.

To illustrate the design of a sequence generator let us consider, by way of example, the generation of the sequence $S = \cdots \mathbf{100010011010111} \cdots$, a sequence of length $S = 15$ and therefore requiring at least four flip-flops. We have no assurance at the outset that we shall be able to realize this prescribed sequence with only four flip-flops. However, as we shall see, it will be possible to do so in the present case.

In Fig. 10.17-2 corresponding to the 15 clock intervals of the sequence we have listed the prescribed sequence under Q_1. Correspondingly, then, the sequences listed under Q_2, Q_3, and Q_4 are this same sequence respectively delayed by one, two, and three clock intervals. The entries under D_1 are written by inspection of the Q_1 column. Thus, when $k = 2$, it is required that $Q_1 = 0$. Hence, when $k = 1$, it is required that $D_1 = 0$. From this truth table in Fig. 10.17-2 we can readily verify by using a K map (or in the present case simply by inspection) that D_1 is required to be

$$D_1 = Q_3 \oplus Q_4 \qquad (10.17\text{-}3)$$

Hence the sequence generator appears as in Fig. 10.17-3.

When applied to a sequential circuit as in Fig. 10.17-1, the word "state" refers to the collective states of the individual flip-flops. Thus in the first clock interval the system is in the state **1111** ($Q_1 Q_2 Q_3 Q_4 = \mathbf{1111}$), while in the second interval it is in the state **0111**, and so on. A sequential circuit with N flip-flops

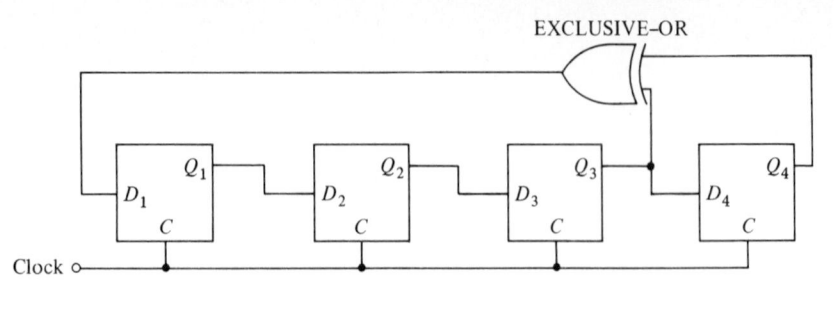

FIGURE 10.17-3
A sequence generator for the sequence given by $Q_1(k)$ in Fig. 10.17-2.

is capable in principle of being in 2^N states. Suppose, then, that we make a sequencing table, as in Fig. 10.17-2, for a maximum-length generator of N flip-flops. Then it is necessary that during the generation of the complete sequence, no state be repeated. Such is the case because a present state uniquely determines the future development of the sequence. Hence, the recurrence of a particular state implies a recurrence of the succeeding sequence. It can be verified that in Fig. 10.17-2 every possible state does indeed occur just once, except for the state **0000**. This all-zero state must be excluded since in this state $D_1 = \mathbf{0}$. Hence the next state, in the next clock interval, would again be **0000**. Thus, if this state ever occurred, the generator would stop the sequencing.

It is now apparent that a sequence generator, built on the pattern of Fig. 10.17-1 and in a manner which requires eliminating the all-zero state, has available $2^N - 1$ states and can therefore generate a sequence of maximum length. The trick involved is finding how to generate the data input D_1 to the first flip-flop in order to ensure that every allowable state is reached before any state recurs. When the required D_1 has been found, the sequence generator can be started after initially registering in the flip-flops any state other than the all-zero state. We now state, without proof, that for N up to $N = 15$, the logic required for D_1 to obtain a maximal-length sequence is as given in Table 10.17-1. It should be noted here that other maximal-length sequences are possible (providing of course, that different logic is employed for D_1). For example, for $N = 14$, 756 different maximal-length sequences are possible.

We have noted that it is necessary to avoid the situation in which a **0** is registered in all the flip-flops. Such a circumstance might develop by chance when the generator is turned on or by the effects of spurious signals (noise). It is left as a problem to show that if in the sequence generator shown in Fig. 10.17-4, all flip-flops are in the **0** state, the first clock pulse will take the system to the state $\mathbf{100 \cdots 0}$.

The randomness of maximal length sequences Suppose we watch the bits of a sequence as they appear in turn. Such an observation might be compared to taking note of the results of tossing a coin. Each bit is a logic **1** or a logic **0**.

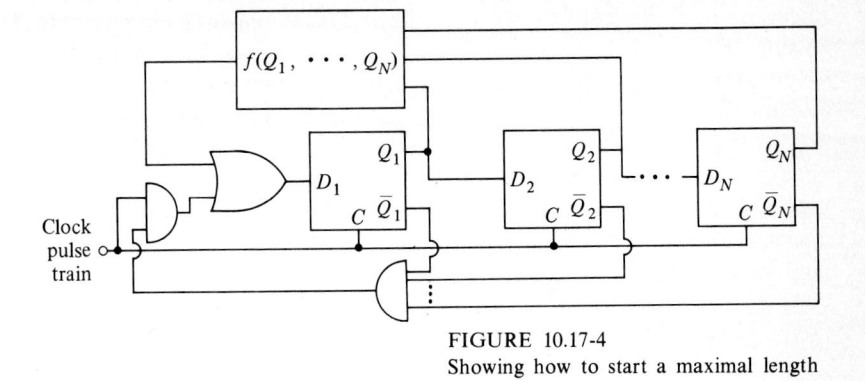

FIGURE 10.17-4
Showing how to start a maximal length
sequence generator.

Correspondingly each toss yields a head or a tail. With a fair coin, at each
individual toss we have no way of knowing beforehand whether the toss will
yield a head or tail. Further, we would find that over a large number of tosses,
heads and tails would appear with equal frequency. Finally, we would discover
that no amount of examination of the past history of results of tosses would
yield any information which would help us predict the result of the next toss.

Table 10.17-1   **LOGIC DESIGN FOR
MAXIMAL-LENGTH
SHIFT-REGISTER
SEQUENCES**

N	D_1, Logic design for $S = 2^N - 1$
1	Q_1
2	$Q_1 \oplus Q_2$
3	$Q_2 \oplus Q_3$
4	$Q_3 \oplus Q_4$
5	$Q_3 \oplus Q_5$
6	$Q_5 \oplus Q_6$
7	$Q_6 \oplus Q_7$
8	$Q_2 \oplus Q_3 \oplus Q_4 \oplus Q_8$
9	$Q_5 \oplus Q_9$
10	$Q_7 \oplus Q_{10}$
11	$Q_9 \oplus Q_{11}$
12	$Q_2 \oplus Q_{10} \oplus Q_{11} \oplus Q_{12}$
13	$Q_1 \oplus Q_{11} \oplus Q_{12} \oplus Q_{13}$
14	$Q_2 \oplus Q_{12} \oplus Q_{13} \oplus Q_{14}$
15	$Q_{14} \oplus Q_{15}$

Because of these features, we describe the outcome of a coin toss as a *random event.*

It turns out, most interestingly, that the maximal-length sequences generated as above also have a randomness about them. It is found that the past history of a sequence is of almost no help in predicting the next bit unless the examination of the past history extends back far enough to detect the fact that the sequence actually does repeat itself eventually. Once we had discovered the periodicity, we would be able to predict future bits by comparison with corresponding bits in the previous period. Because of this periodicity, and because the sequence is actually generated in a deterministic manner, the maximal length sequence is referred to as a *pseudo-random* sequence.

On the other hand, suppose we wanted to make it difficult for someone to guess the next bit in a sequence even if he knew that the sequence was generated by a sequence generator and hence not really random but only pseudo-random. Such might be the case in developing codes for the purpose of maintaining secrecy. We might then use a very large number of flip-flops (large N). In MOS large-scale integration (LSI) it is reasonable to have a 2,000-flip-flop register on a single chip. In such a case the sequence length would be $2^{2,000} - 1$. The prospective guesser of the next bit would be hard pressed indeed to keep in mind so large a number of bits in order to determine the length of the sequence.

A way of arriving at a numerical measure of the randomness of a sequence is the following. *Suppose we view the sequence as being represented by a waveform which is at $+1$ V for logic **1** and -1 V for logic **0**. Let us now multiply the voltage of the sequence by itself in each interval and add the products. The sum of the products would be $2^N - 1$. Now let us perform the same arithmetic using the sequence and the sequence shifted by one or more intervals. Suppose the sequence is really random, with no correlation between a particular bit and a bit obtained from its past history. We should then as likely find that in any interval the voltages are of opposite sign as that they are of the same sign. Hence when we multiply bits in an interval and add the products we would expect the sum to be zero. This sum of products is called the *autocorrelation* function of the sequence.[1]

Let $Q(k)$ represent the sequence, k being the number of the clock interval. Thus $Q(k) = +1$ or -1 V. Then $Q(k + i)$ is the same sequence except shifted

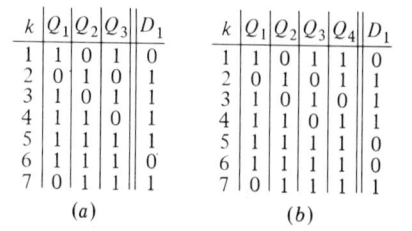

FIGURE 10.17-5

(a) Truth table for a "troublesome" sequence given by $Q_1(k)$. (b) Truth table with a fourth flip-flop.

k	Q_1	Q_2	Q_3	D_1
1	1	0	1	0
2	0	1	0	1
3	1	0	1	1
4	1	1	0	1
5	1	1	1	1
6	1	1	1	0
7	0	1	1	1

(a)

k	Q_1	Q_2	Q_3	Q_4	D_1
1	1	0	1	1	0
2	0	1	0	1	1
3	1	0	1	0	1
4	1	1	0	1	1
5	1	1	1	1	0
6	1	1	1	1	0
7	0	1	1	1	1

(b)

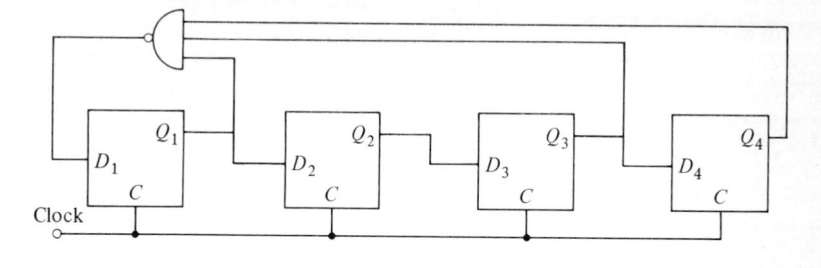

FIGURE 10.17-6
Sequence generator for sequence: 1 0 1 1 1 1 0.

(delayed or advanced) by i clock intervals. It can be shown that the auto-correlation function R is

$$R(i) = \sum_{k=1}^{2^N-1} Q(k)Q(k+i) = \begin{cases} 2^N - 1 & i = 0, 2^N - 1, 2(2^N - 1), \cdots \\ -1 & \text{otherwise} \end{cases} \quad (10.17\text{-}4)$$

If the sequence were truly random, we should have $R(i) = 2^N - 1$ only for $i = 0$ and $R(i) = 0$ for every other case.

Prescribed sequences   In the discussion above we were concerned principally with the length of a sequence.   Suppose, however, that we want to generate a pre-scribed sequence.   Then, following the procedure indicated in the truth table of Fig. 10.17-2 may lead to a difficulty.   Consider, for example, that we want to generate the sequence $S = \cdots \mathbf{1011110} \cdots$.   Since there are seven intervals, we may try three flip-flops since $2^3 - 1 = 7$.   We now find, as in Fig. 10.17-5a, that in the intervals $k = 1$ and $k = 3$ the state of the system is the same.   Yet we are required to generate $D_1 = 1$ in one case and $D_1 = 0$ in the other case.   Since D_1 must be some fixed function of the state of the system, two values for D_1 are not possible.   The same problem is seen to exist in states $k = 5$ and 6.

The problem can be resolved by increasing the number of flip-flops until each state in the truth table becomes unique.   In the present case a single addi-tional flip-flop is adequate, as can be seen in the table of Fig. 10.17-5b.   In this table there appears only 7 out of 16 possible states.   However, no state appears more than once.   It can be verified from this table that the required data input D_1 is

$$D_1 = \overline{Q_1 Q_3 Q_4}$$

The circuit of the sequence generator is given in Fig. 10.17-6.

REFERENCE

1   Taub, H., and D. L. Schilling: "Principles of Communications Systems," chaps. 1 and 2, McGraw-Hill, New York, 1971.

11
ARITHMETIC OPERATIONS

In this chapter we shall consider techniques and circuits for digitally performing the arithmetic operations addition, subtraction, multiplication, and division. Where mod-2 arithmetic is involved, we shall let the logic values **1** and **0** respectively represent the numerical digits 1 and 0. We shall not make a careful distinction between these *logic* values and *numerical* values since no confusion will result.

11.1 ADDITION OF TWO BINARY NUMBERS

Two binary digits, an *addend* and an *augend*, are added in the manner of Fig. 11.1-1, to generate a sum digit (S) and a carry digit (C). In truth-table form, the rules of Fig. 11.1-1 appear as in Fig. 11.1-2.

The carry digit is related to the addend and augend by AND logic

$$C = A \cdot B \tag{11.1-1}$$

while the sum is given by the EXCLUSIVE-OR operation (see Fig. 11.1-3a); i.e.,

$$S = A \oplus B = A \cdot \overline{B} + \overline{A} \cdot B \tag{1.1-2}$$

$$
\begin{array}{cccc}
0 & 0 & 1 & 1 \\
+0 & +1 & +0 & 1 \\
\hline
00 & 01 & 01 & 10
\end{array}
$$

Addend (A)
Augend (B)

Sum (S)
Carry (C)

FIGURE 11.1-1
Addition of two binary digits.

FIGURE 11.1-2
Truth table for mod-2 addition.

A	B	S	C
0	0	0	0
0	1	1	0
1	0	1	0
1	1	0	1

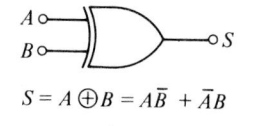

$$S = A \oplus B = A\bar{B} + \bar{A}B$$

(a)

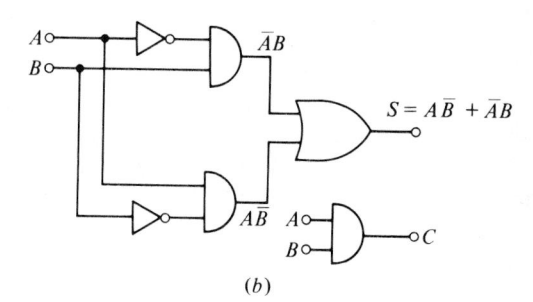

(b)

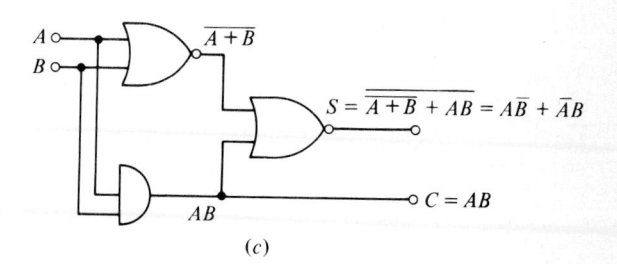

(c)

FIGURE 11.1-3
(a) An EXCLUSIVE-OR gate generates the sum bit, (b) the carry bit is generated by a separate gate, (c) an alternative method of generating the sum and carry bits.

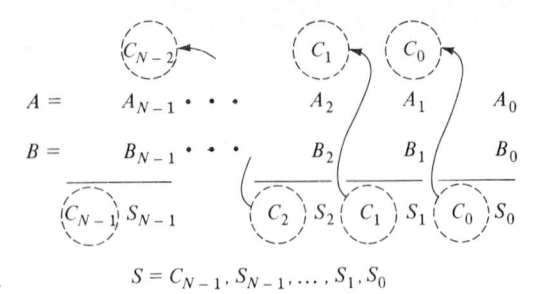

FIGURE 11.1-4
Addition of two N-bit numbers.

$$S = C_{N-1}.S_{N-1}, \dots, S_1.S_0$$

One realization of the EXCLUSIVE-OR gate is shown in Fig. 1.1-3 (an RTL EXCLUSIVE-OR circuit is discussed in Sec. 4.8). If the complements A and B happen to be available, then, the inverters are not required. A second EXCLUSIVE-OR realization is given in Fig. 11.1-3c. Here we note that $C = AB$ is available, so that a separate gate to generate C is not required.

Two mod-2 numbers of N digits, $A = A_{N-1}A_{N-2} \cdots A_0$ and $B = B_{N-1}B_{N-2} \cdots B_0$ are added, as shown in Fig. 11.1-4. We proceed as follows. The least significant bits (LSBs) A_0 and B_0 are added to yield a sum S_0 and a carry C_0. (This operation can be carried out with the gates of Fig. 11.1-3b or c, but the operation indicated in the second and other columns of Fig. 11.1-4 involves the addition of three digits and cannot be done with the EXCLUSIVE-OR gate.) In the second column we add C_0, A_1, and B_1 to form the sum S_1 and the carry C_1. This procedure continues until each pair of bits has been added. The result is the sum $S = C_{N-1}S_{N-1} \cdots S_0$.

A gate structure which performs such an addition of three digits is called a *full adder*. It turns out, as we shall see, that one way in which a full adder can be constructed is by using two of the gate structures of Fig. 11.1-3b or c. For this reason the structure shown in Fig. 11.1-3b or c is referred to as a *half adder*. A block-diagram representation of a half adder (HA) is shown in Fig. 11.1-5.

11.2 THE FULL ADDER

A full adder satisfies the truth table of Fig. 11.2-1. K maps and two-level gate implementations for the sum bit S_i and the carry bit C_i are given in Fig. 11.2-2. There are various other gate structures besides those given in Fig. 11.2-2b by

FIGURE 11.1-5
Block diagram to represent the half adder.

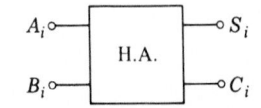

Inputs			Outputs	
Carry in C_{i-1}	Addend A_i	Augend B_i	Sum S_i	Carry out C_i
0	0	0	0	0
0	0	1	1	0
0	1	0	1	0
0	1	1	0	1
1	0	0	1	0
1	0	1	0	1
1	1	0	0	1
1	1	1	1	1

FIGURE 11.2-1
Truth table for a full adder.

which the full adder can be realized. For example, another way in which S_i and C_i can be generated is

$$S_i = C_{i-1} \oplus A_i \oplus B_i \tag{11.2-1}$$

and

$$C_i = A_i B_i + (A_i \oplus B_i)C_{i-1} \tag{11.2-2}$$

C_{i-1} \ A_iB_i	00	01	11	10
0		1		1
1	1		1	

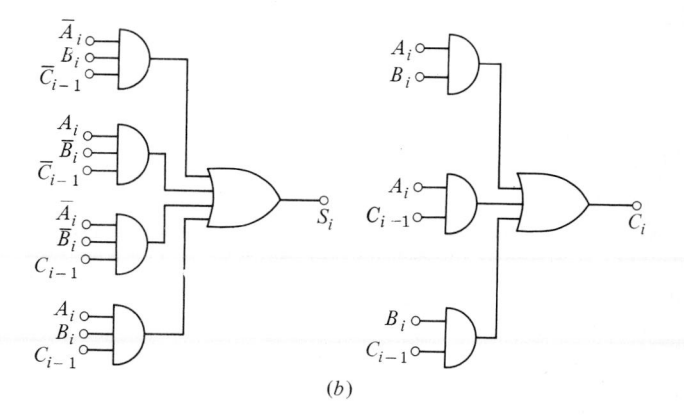

$S_i = \bar{A}_iB_i\bar{C}_{i-1} + A_i\bar{B}_i\bar{C}_{i-1} + \bar{A}_i\bar{B}_iC_{i-1} + A_iB_iC_{i-1}$

C_{i-1} \ A_iB_i	00	01	11	10
0			1	
1		1	1	1

$C_i = A_iB_i + A_iC_{i-1} + B_iC_{i-1}$

(a)

(b)

FIGURE 11.2-2
(a) Karnaugh maps for the sum and carry bits of a full adder; (b) two-level gate realizations.

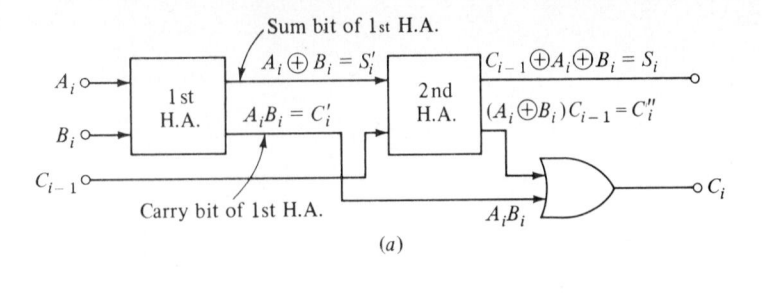

Sum bit of 1st H.A.

$A_i \oplus B_i = S'_i$

$A_i B_i = C'_i$

$C_{i-1} \oplus A_i \oplus B_i = S_i$

$(A_i \oplus B_i)C_{i-1} = C''_i$

Carry bit of 1st H.A.

$A_i B_i$

(a)

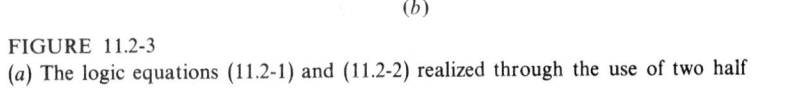

(b)

FIGURE 11.2-3

(a) The logic equations (11.2-1) and (11.2-2) realized through the use of two half adders; (b) symbol for a full adder.

It can readily be verified that these equations satisfy the truth table of Fig. 11.2-1. And, as is now to be seen in Fig. 11.2-3a, Eqs. (11.2-1) and (11.2-2) can be implemented by the use of two half adders. One half adder is used to combine two of the input digits and yields a sum digit S'_i and a carry C'_i. The second half adder combines the third digit with the first two. The circuit symbol for the full adder is shown in Fig. 11.2-3b.

11.3   A SERIAL ADDER

A serial adder for two N-bit numbers $A = A_{N-1}A_{N-2} \cdots A_0$ and $B = B_{N-1}B_{N-2} \cdots B_0$ is shown in Fig. 11.3-1. Here the columns, as in Fig. 11.1.-3, are added one at a time, starting with the least significant digit. The circuit involves three shift registers, a single full adder, and a single type-D flip-flop.

To see how the *serial adder* operates let us assume that, with the clock waveform quiescent, we have initially loaded the addend and augend into the two N-bit shift registers. We assume that the shift registers shift to the *right*, and we have therefore loaded the addend and augend into the registers with the least significant bit (LSB) at the right and the most significant bit (MSB) at the left. We assume also that initially the flip-flop is in the reset state with $Q = 0$. The sum will eventually appear in the $(N + 1)$st-bit sum register. During the sequence of operations which shift the sum into the sum register, any bits which are initially in the sum register will be shifted out and lost. Hence the initial states of the flip-flops in this register are of no concern and need not be initially reset.

Before the first clock pulse, the full-adder inputs are $A_i = A_0$, $B_i = B_0$, and $C_{i-1} = 0$. The outputs are $S_i = S_0 = A_0 \oplus B_0$ and $C_i = C_0 = A_0 \cdot B_0$. The first clock cycle then:

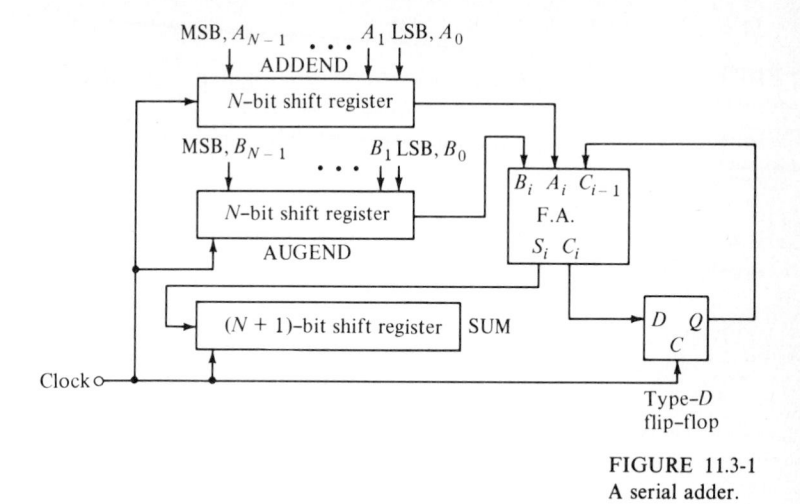

FIGURE 11.3-1
A serial adder.

1 Enters S_0 into the leftmost flip-flop of the sum register
2 Shifts the contents of the addend and augend registers one bit to the right so that A_i and B_i become $A_i = A_1$ and $B_i = B_1$
3 Transfers the data input $D = C_0$ to the output Q so that C_{i-1} becomes C_0

After this first clock cycle we have $S_i = S_1$ and $C_i = C_1$, as in the second column from the right in Fig. 11.1-4. The second clock pulse then enters S_1 into the sum register, shifting S_0 in the process one flip-flop ahead. At the same time this second clock cycle transfers $C_i = C_1$ to the input terminal C_{i-1}. In this way, one additional digit in the sum is determined and transferred to the sum register with each clock pulse. After N clock pulses, the addend and augend registers are entirely clear, and $A_i = B_i = 0$. With $A_i = B_i = 0$, the last carry will appear at the S output. The $(N + 1)$st clock pulse will transfer this last bit into the sum register, and the addition process is complete.

This discussion of serial addition leaves many peripheral questions un-answered. We might well be inclined to inquire where the augend and addend were kept until they were transferred to their respective registers, how these registers were loaded, how the adder is kept quiescent until the loading is complete, how the adder knows when the addition operation is complete, how the flip-flop is cleared as required at the start of the operation, etc. The answers to these questions would raise other questions until finally we would have worked our way back to the input-output (I/O) equipment requirements (key-boards, displays, printouts, etc) through which we physically enter a problem into a digital system and finally read the answer. All these matters are properly the concern of the designer of digital systems. Our present concern, on the other hand, is simply the basic process by which a *sum* can be evaluated. We shall therefore not digress at this point to answer these important and interesting questions.

11.4 PARALLEL ADDITION

In serial addition the addition resembles the action of a human calculator; i.e., digits of like significance are added one column at a time and then the digits of next higher significance. This method is economical of hardware but is relatively slow, since each pair of bits added requires one clock cycle. *Parallel addition,* in which all orders of augend and addend are added simultaneously is faster than serial addition but less economical of hardware, since a separate adder is required for each pair of bits. A parallel adder is shown in Fig. 11.4-1.

The *addend register* is a set of flip-flops, as many as there are digits in the addend, in which the addend can be stored locally for use by the adders. Typically each flip-flop may be a type-D flip-flop. The bits in the addend $A = A_{N-1} \cdots A_0$ are presented to the data-input terminals of the flip-flops, and after the next clock pulse each bit will be registered at the output of its corresponding flip-flop. The addend might have been in storage in some memory device. These bits are directed to the appropriate flip-flop by some gating structure (not shown in the figure). The advantage of the local addend register is that after the addend has been transferred to the register, the memory and the gates used to effect the transfer can be made available for other purposes.

Similarly the augend has an augend register. Note that the registers are not *shift registers* but *storage registers*. Since no serial sequence is to be followed, there is no need to shift bits along the length of the register. Hence, the

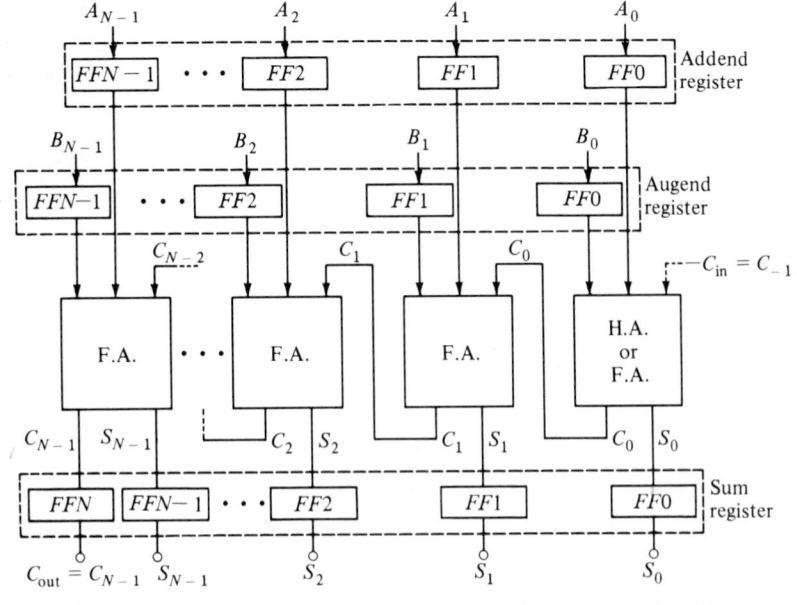

FIGURE 11.4-1
A parallel adder.

flip-flops in the registers of Fig. 11.4-1 are entirely independent of one another, except that they share a common clock waveform. (To avoid encumbering the figure, the clock line, which goes to every flip-flop is not shown.) When the sum $S = C_{N-1}S_{N-1} \cdots S_0$ has been formed, it can be registered in the sum register by a clock pulse.

In the lowest order, A_0 and B_0 can be combined in a half adder because these digits do not have to be combined with a carry. In all other positions, full adders are required. Nonetheless in a commercial parallel adder we might well find that a full adder is used even for the least significant bits. For in such a case we might use the adder as an extension of another adder to increase the number of bits which can be accommodated. If the full adder is used, we shall have a carry input to this first full adder (marked C_{in} in the figure). If the full adder is not being used, we set $C_{in} = \mathbf{0}$.

The parallel adder of Fig. 11.4-1 suffers a speed limitation because a carry may have to ripple through each one of the adders. For example, consider that $A = 1111$ while $B = 0001$ so that the sum is $S = 10000$. After a time equal to the propagation delay time of the lowest-order adder, a carry $C_0 = 1$ will appear at the output of this first adder. When this carry appears at the input to the next-order adder, it will, in turn, generate a carry C_1 which will appear after a second propagation delay. The last carry, which constitutes the digit 1 in the sum $S = 10000$, will not appear until after a total delay which is equal to the sum of the propagation delays in each full adder. The similarity to the comparable speed limitation in ripple counters is apparent. Fast adders, which partially circumvent this difficulty, are discussed in Sec. 11.6.

11.5 ADDITION OF MORE THAN TWO NUMBERS

The evaluation of the sum of M N-bit numbers $S = X(1) + X(2) + X(3) + \cdots + X(k) + \cdots$ can be done in a number of ways. One method is illustrated in Fig. 11.5-1. Here the N-bit full adder is simply a chain of N full adders, as in Fig. 11.4-1. The carry connections are not explicitly shown. Each individual input bit $X_i(k)$, that is, the ith bit of the kth word, is shown for identification, as is each individual sum output $S_i(k)$. The register to which are connected the adder sum outputs $S_0(k), S_1(k), \ldots, S_{N-1}(k)$ serves in the present case both as the addend register and as the sum register. Since the progressively increasing sum will appear in this register, it is referred to as an *accumulator* register.

Assume that initially all flip-flops in both registers are in the reset state and that the first number $X(1)$ has been presented at the data inputs of the augend register. Assume further that synchronously with the clock waveform (not explicitly shown but common to all flip-flops) the successive numbers $X(2)$, $X(3)$, etc., will be presented at the augend-register data inputs. The first triggering clock transition will register $X(1)$ in the augend register, making $B_0 = X_0(1)$, $B_1 = X_1(1)$, etc. After this clock transition we shall still have the

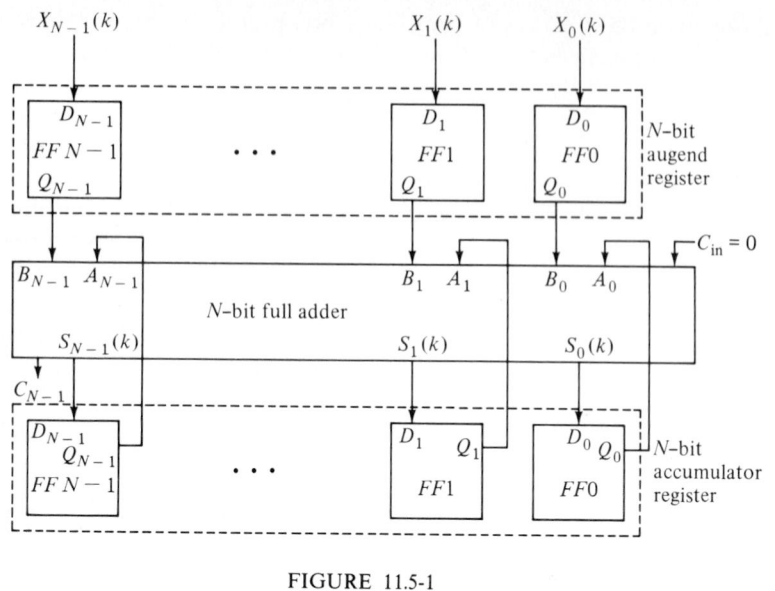

FIGURE 11.5-1
System to obtain the sum of more than two numbers.

number $00 \cdots 0$ registered in the accumulator register so that $A_0 = A_1 = \cdots A_{N-1} = 0$. The adder will add $X(1)$ to 0 giving $S_0(1) = X_0(1)$, $S_1(1) = X_1(1)$, etc. The second triggering clock transition will then register $X(1)$ in the accumulator register and register $X(2)$ in the augend register. We now have $B = X(2)$ and $A = X(1)$, so that the sum is $S(2) = X(2) + X(1)$. The third triggering clock transition will register this sum, $X(2) + X(1)$, in the accumulator and register $X(3)$ in the augend register, and so on.

The process of Fig. 11.5-1 for summing a sequence of numbers is a rather intuitive extension of the process of addition based on the properties of the full adder. It is interesting to note that in the adder of Fig. 11.4-1, intended for combining two numbers, the sum register is not essential to the addition process. On the other hand, in the scheme of Fig. 11.5-1 the accumulator register is essential, but there is no addend register. Thus, in Figs. 11.4-1 and 11.5-1 only two registers are essential.

The procedure for adding more than two numbers, is often referred to as *accumulation* rather than addition. An alternative *accumulator* which accommodates 3 input bits and can accumulate a 4-bit sum $S = S_3 S_2 S_1 S_0$ is shown in Fig. 11.5-2a. In this circuit the intention is to add an augend $X(k) = X_2(k)X_1(k)X_0(k)$ to an addend, the addend being the already accumulated sum $S(k-1) = S_2(k-1)S_1(k-1)S_0(k-1)$ registered in the 4-bit register.

The box marked *CG* is a *carry generator* and is shown in detail in Fig. 11.5-2b. This carry generator is identical to the gate structure of Fig. 11.2-2b and generates a **1** whenever two or all three of the inputs are **1**.

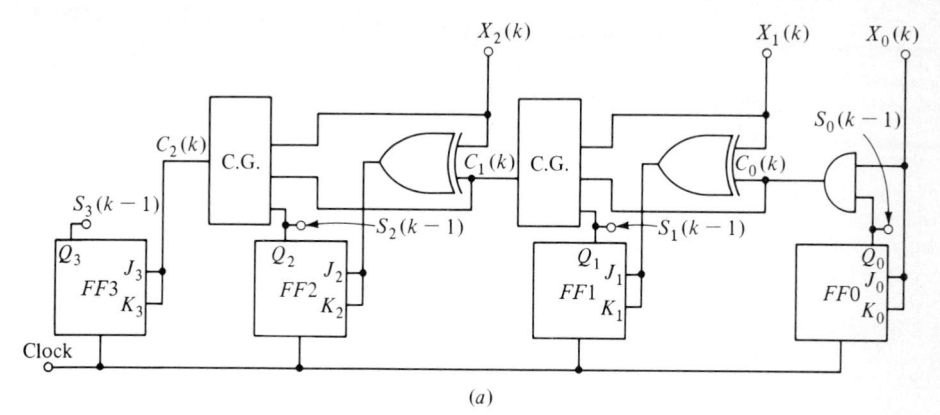

(a)

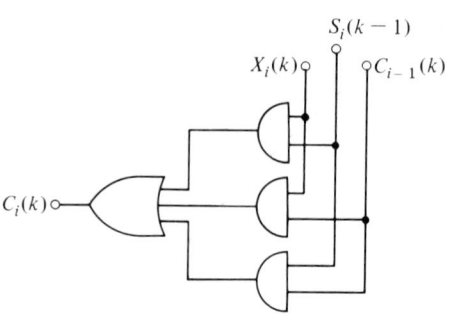

(b)

FIGURE 11.5-2
A 3-bit accumulator.

To appreciate how the sum digit is generated, consider again Eq. (11.2-1). Here we note that the sum digit is generated by combining the 3 bits (augend, addend, and carry) through two EXCLUSIVE-OR operations. (The EXCLUSIVE-OR operation is associative and commutative, so that the order of variables and the order of combination are of no consequence.) One way, then, to generate S_i is to start by combining two of the inputs, say the augend bit $X_i(k)$ and the carry bit $C_{i-1}(k)$, in an EXCLUSIVE-OR gate. Thereafter the output of this first EXCLUSIVE-OR gate can be combined with $S_i(k-1)$ in a second EXCLUSIVE-OR gate. Having generated the new sum bit $S_i(k)$ in this manner, we would then transfer this bit to a flip-flop in the register which is to be used to hold the sum. However, as we may now see, it is possible to use a JK flip-flop to perform an EXCLUSIVE-OR operation. Thus, the flip-flop may serve not only as a register for a sum bit but may also replace the second EXCLUSIVE-OR gate used to combine $S_i(k-1)$ with $X_i(k) \oplus C_{i-1}(k)$.

Let us then arrange, as in Fig. 11.5-2, that in the JK flip-flops $J = K$ and

$$J = K = X_i(k) \oplus C_{i-1}(k) \qquad (11.5\text{-}1)$$

$Q_i(k-1)$		$Q_i(k)$
Registration in ith flip–flop before the kth clock pulse, $S_i(k-1)$	$J = K = X_i(k) \ \oplus \ C_{i-1}(k)$	Registration in ith flip–flop after kth clock pulse, $S_i(k)$
0	0	0
0	1	1
1	0	1
1	1	0

FIGURE 11.5-3
Truth table describing operation of the JK flip-flop in the accumulator circuit of FIGURE 11.5-2.

When the flip-flop is clocked, it will toggle if $J = K = 1$ and remain in its same state if $J = K = 0$. It can now be verified that the behavior of the flip-flop is described by the truth table of Fig. 11.5-3 and that $S_i(k)$ is indeed given, as required, by

$$S_i(k) = S_i(k-1) \oplus X_i(k) \oplus C_{i-1}(k) \qquad (11.5\text{-}2)$$

In the present accumulator circuit, the flip-flops serve not only as the accumulator register but perform, in addition, an EXCLUSIVE-OR operation. For this reason the present circuit is somewhat more economical of hardware than the accumulator circuit of Fig. 11.5-1.

11.6 FAST ADDERS; LOOK-AHEAD CARRY

As mentioned earlier, the adder of Fig. 11.5-1 and the accumulator of Fig. 11.5-2a suffer from a speed limitation which results from the need for the carry to ripple from order to order. For example, in Fig. 11.5-1, consider the addition of $A = A_3 A_2 A_1 A_0 = 1111$ and $B = B_3 B_2 B_1 B_0 = 0001$. In the lowest order the two 1s generate a carry. This carry together with A_1 in the next order generates a carry again, and so on. Eventually when the carry has propagated through all orders, we shall have $A + B = C_4 S_3 S_2 S_1 S_0 = 10000$.

There is, at least in principle, a straightforward way of avoiding these propagation delays. We simply use gates to generate the sum digit in each order *directly* from all the input digits and without recourse to the carry signals. The first sum digit is simply $S_0 = A_0 + B_0$. The second sum digit S_1 however, will, involve gates which provide inputs for A_1, A_0, B_1, B_0 and their complements (see Prob. 11.6-1). The Nth-order sum digit will involve gates with $2N$ inputs, etc. It seems clear that with this approach the number of gates involved will rapidly get out of hand.

The parallel carry adder (where all sum digits are generated directly from input digits) is fast but uneconomical of hardware. The serial carry adder

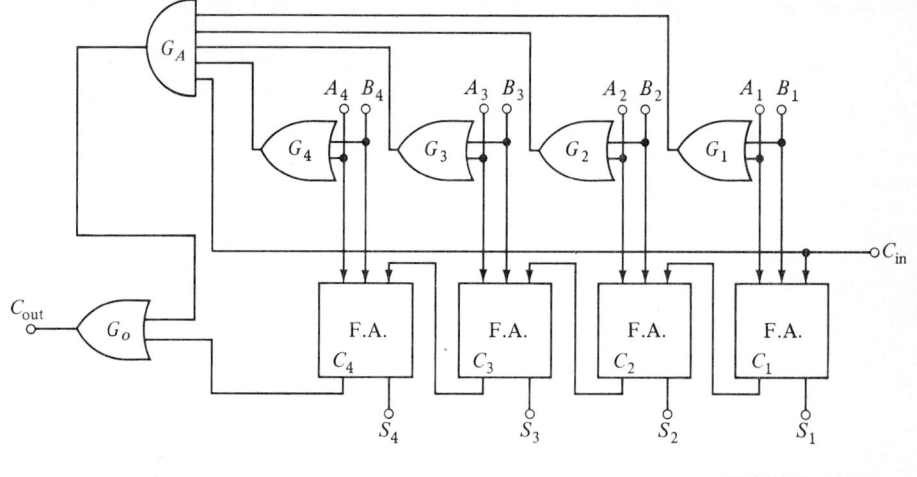

FIGURE 11.6-1
A high-speed adder.

(involving a serial propagation of carries) is slow but economical. All manner of compromise can be made between these two extremes. One possible compromise is to bypass carries around some adders. The bypass scheme is shown in Fig. 11.6-1. Here a 4-bit adder uses four full adders connected in the conventional manner, which involves the ripple of carries. For generality, we assume that A_1 and B_1 are not the lowest order. The lower orders are being added in some preceding adder (not shown), thereby generating a carry C_{in} to the adder shown in the figure.

We can now see that if there is an input carry C_{in}, and if the digits A_i and B_i are such that this input carry is to be propagated to the output, this output carry will be generated by virtue of the gate structure which has been added in Fig. 11.6-1 and hence need not ripple through the cascade of full adders. For assume that $C_{in} = 1$. Then if A_1 or B_1 or both are **1**, this carry will propagate through the first full adder. This carry will continue its propagation through the second full adder if A_2 or B_2 or both are **1**, and so on. Hence if C_{in} is indeed to be propagated through to the output, all the inputs of the AND gate will go to logic **1** and we shall have $C_{out} = 1$ as required. In Fig. 11.6-1 a bypass is shown bridging 4 bits. It is of course clear that the bypass is adaptable to any number of full-adder stages.

Of course, the carry bypass circuitry itself produces some propagation delay. This delay is the delay through three gates: the delay through the output OR gate G_{o1}, the delay through the AND gate G_A, and, assuming gates G_1 through G_4 to be identical, the delay of one of these OR gates. We note (see Fig. 11.2-2b) that the propagation delay in the generation of the carry is the delay of two gates. Hence we can estimate that the delay associated with a carry bypass is about a factor of 1.5 greater than the delay associated with a single full adder.

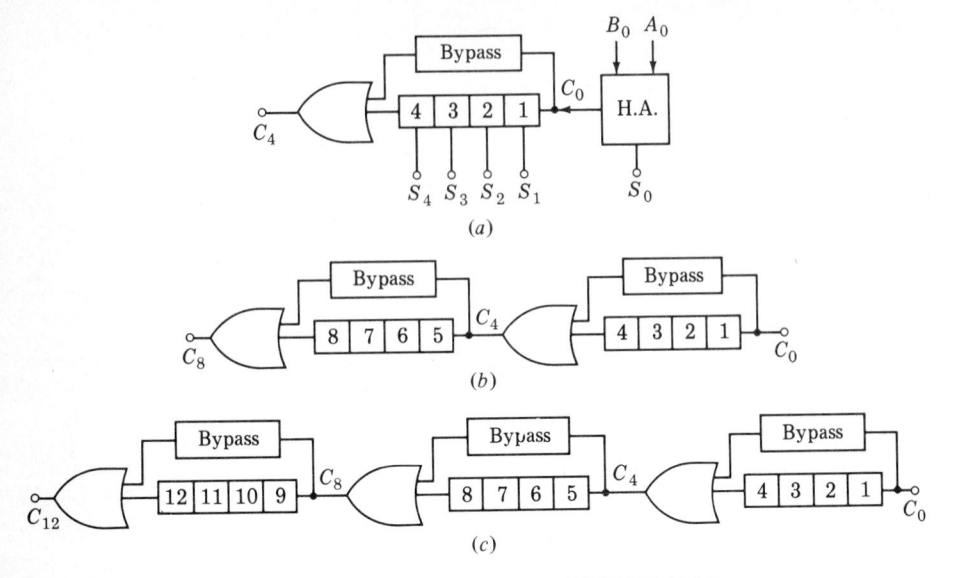

FIGURE 11.6-2
Adder stages shunted by carry bypasses.

To illustrate the applications of the bypass let us examine the effectiveness of a four-stage carry bypass. Such four-stage bypass units are made available by many manufacturers. In Fig. 11.6-2a is shown a four-stage bypass bridging the last four stages of a five-stage adder. The first stage has as inputs only the least significant bit and has no input carry which might need to propagate. This first stage, shown as a half adder, does, however, generate a carry C_0, and hence the bypass starts at this carry. The carry bypass receives, as inputs, the carry C_0 as well as the input digits A_1, B_1, A_2, B_2, etc. These inputs are not indicated in the figure. The carry-output bit may turn out to be the input carry C_0 propagated through the four full adders. In this case the carry will arrive first at the output of the bypass and then arrive again later as the carry output of the fourth full adder. On the other hand, there may be no input carry C_0, or the input carry may not propagate through to the output of the four-stage adder. Then again, it may happen that there is an output carry which is not due to C_0 but is due instead to a carry generated within the four-stage adder, and propagated to the output. Whatever the source of the output carry, if any, it will propagate through the output OR gate to the input of the next adder stage.

Let us first allow adequate time for C_0 to be established. Then we ask how much *additional* time we must allow to pass to be confident that all carries have propagated as far as they are going, so that we can read the adder output. We shall assume, as suggested earlier, that the propagation time of the bypass, is 1.5 times the propagation time d of one stage of a full adder. Without the bypass, the waiting time of a four-stage adder is always $T = 4d$. If C_0 is propagated to the output C_0, the propagation time through the bypass is

$T = 1.5d$; however, we must allow for the possibility that the input carry will only propagate as far as the output of full adder 3. We shall then have to allow a time $T = 3d$. Alternatively it may happen that $C_0 = 0$ but that a carry is generated at the output of full adder 1 and is then propagated to C_4. Again we shall have $T = 3d$. In both these cases the bypass cannot be effective, and the waiting time is $T = 3d$ as against $T = 4d$ without a bypass.

Next consider the situation represented in Fig. 11.6-2b, where an additional bypassed group of four adder stages now extends the adder-stage cascade. In the absence of the bypasses the maximum waiting time will be $T = 8d$. Now let us look for a situation which will thwart the potential effectiveness of the bypasses. We first assume, as a worst case, that a carry can be generated at the output of full adder 1 and propagate only as far as the output of adder stage 7. In this case the waiting time is $T = 6d$, again not a great saving.

Finally we consider the situation in Fig. 11.6-2c. Here, as can be verified, the longest waiting time corresponds to the case in which a carry is generated at the output of adder stage 1, propagates through the middle bypass, and ends at the output of adder stage 11. In this case the waiting time is $T = 3d + 1.5d + 3d = 7.5d$, compared with $T = 12d$ in the absence of the bypasses. In general, with n groups of four adder stages, each bypassed, the waiting time is

$$T = 6d + 1.5(n - 2)d \qquad (11.6\text{-}1)$$

as compared with $T = 4nd$ without bypasses. Hence as n increases, the bypasses become increasingly effective. In the limit the ratio of propagation time with bypasses to the propagation time without bypasses approaches 0.375, a significant reduction in time delay.

11.7 SUBTRACTION

The rules for subtraction of digits in any given order are specified by the truth table of Fig. 11.7-1. Here A is the *minuend*, B the *subtrahend*, $D = A - B$ the *difference*, and C the *borrow*. That is, when $A = 0$ and $B = 1$, it is necessary to borrow 1 from the minuend of the next higher order to allow the subtraction to be effected. When there is a borrow transferred to a next higher order, the operation to be performed in that order is first a subtraction of the subtrahend

A	B	D	C
0	0	0	0
0	1	1	1
1	0	1	0
1	1	0	0

FIGURE 11.7-1
Truth table for subtraction.

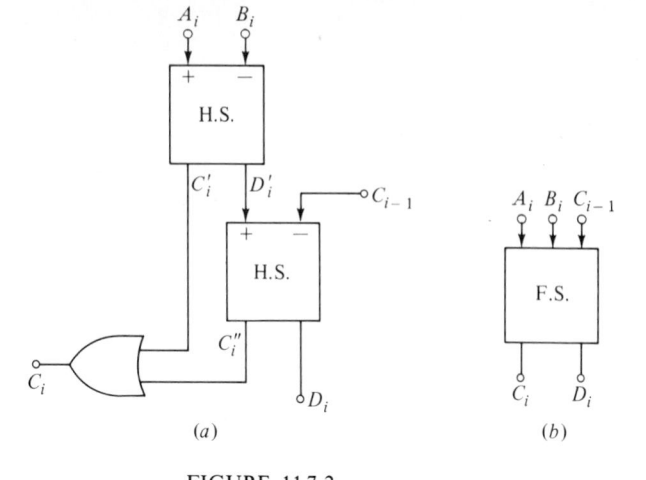

FIGURE 11.7-2
(a) A full subtractor; (b) symbol of a subtractor.

digit from the minuend digit and then a further subtraction of 1 from the difference (actually, the order in which the subtractions are performed is of no consequence).

A gate structure which accepts A and B as inputs and yields D and C as outputs is called a half subtractor (HS). We can verify that D and C are given by

$$D = A \oplus B \quad \text{and} \quad C = \overline{A} \cdot B \tag{11.7-1}$$

Thus D can be generated by the same gate structure as that used to produce the sum S in the half adder; however, the borrow carry is different from the addition carry.

A full subtractor (FS) accepts as inputs the minuend A, the subtrahend B, and a borrow from a previous order. It performs the two subtraction operations referred to in Eq. (11.7-1) and yields as outputs a final difference and a borrow carry for the next higher order. One way in which a full subtractor can be constructed is to combine two half subtractors, as in Fig. 11.7-2a. One half subtractor yields $A_i - B_i$ while the second half subtractor forms $(A_i - B_i) - C_{i-1}$. A borrow must be transferred to the next higher order if either or both of the subtractions results in a borrow. Hence, as shown, the two individual borrow outputs are transferred to the next order through an OR gate. The symbol for a full subtractor is shown in Fig. 11.7-2b.

Following the pattern of Fig. 11.4-1, an array of full subtractors can be combined as in Fig. 11.7-3 to effect the subtraction of $B = B_{N-1}B_{N-2} \cdots B_0$ from $A = A_{N-1}A_{N-2} \cdots A_0$. If there are no digits lower in order than A_0 and B_0, the subtraction in this first order can be accomplished in a half subtractor since

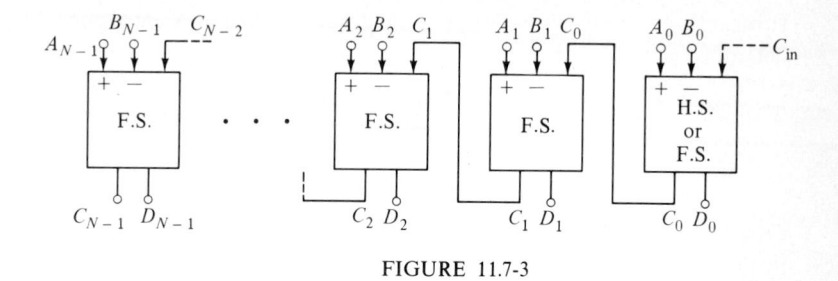

FIGURE 11.7-3
A subtractor for two N-bit numbers A and B.

$C_{\text{In}} = 0$. It should be noted that if $A > B$, the last borrow carry C_{N-1} will be $C_{N-1} = 0$, while if $A < B$, the last borrow carry will be $C_{N-1} = 1$.

We have not explicitly indicated minuend, subtrahend, and difference registers but in any physical situation such registers will be involved.

11.8 COMPLEMENTARY NUMBERS

In the arithmetic section of a computer or other digital processor we shall require facility for subtraction as well as for addition. As we shall see, subtraction can be performed by a process that involves addition and the use of complementary numbers. When subtraction is accomplished in this manner, an explicit subtractor is not required and often there is a significant saving in hardware. In this section we discuss *complementation*.

Twos complement Suppose we have a register which has four flip-flops. Then, with respect to that register, a number N has a complement N^* which can be defined in the following manner. We initially set the register to **0000** and then run the register *backwards* by the amount of the number N. For example if $N = \mathbf{0001}$, then $N^* = \mathbf{1111}$, if $N = \mathbf{0010}$, $N^* = \mathbf{1110}$, if $N = \mathbf{0011}$, $N^* = \mathbf{1101}$, etc.

A complement can be unambiguously determined only when the number of flip-flops in the register is specified. The number of such flip-flops can be specified in terms of the modulus of the register. As with the counter, so also in the register the modulus refers to the total number of different numbers which can be stored. Thus a four flip-flop register can register numbers from **0000** to **1111**, that is, from decimal 0 to 15, and has a modulus of $2^4 = 16$. Hence, the complements specified above are more precisely referred to as *complements* (*mod 2^4*) or *complements with a four-place modulus*. For example, the number 1 (**0001**) has the complement 15 (**1111**), since $15 + 1 = 16$. Note that $16 - 1 = 15$, that is, the modulo less the number is the complement.

The complement of a number can, of course, be evaluated by straightforward subtraction.   Thus we find that in the binary system the complements of $N = 0000$, 0001, 0010, and 0011, with four-place modulus, are

$$
\begin{array}{cccc}
10000 & 10000 & 10000 & 10000 \\
N = \quad 0000 & 0001 & 0010 & 0011 \\
\hline
N* = 1)0000 & 1111 & 1110 & 1101
\end{array}
$$

Observe that to allow the subtraction it was necessary to use a five-digit number in the minuend.   In all cases except the complement of 0 the result of the subtraction is a four-place number.   For $N = 0$, the subtraction yields a five-place number, and the leftmost digit must be discarded since the register has a four-place modulus.

Summarizing, we write the complement N_k^* with a k-place modulus as

$$N_k^* = (2^k - N)_k \tag{11.8-1}$$

Strictly, the subscript k on the right-hand member of Eq. (11.8-1) is necessary to remind us, as noted above, that when $N = 0$, $2^k - N$ has $k + 1$ digits and the leftmost digit must be discarded.   Still as a matter of convenience, when no confusion will result thereby, we shall simply write $N_k^* = 2^k - N$.

The evaluation of the complement by subtraction involves a series of subtractions, digit by digit, from the binary number 10.   Consider, for example, the subtraction involved in determining, as above, the complement of 0001.   We start in the least significant order and subtract 1 from 10, yielding 1 and transferring a borrow to the next order.   To account for this borrow in this next order we add 1 to the digit 0 in the subtrahend.   We now find that we are again subtracting 1 from 10 and so on for the rest of the subtraction.   In general, in evaluating any complement, after arriving at the first 1 in the subtrahend, note that each subtraction is a subtraction from binary 10 (decimal 2).   It is for this reason that the complement, as defined above, is customarily referred to as a *twos complement*.   Note further, that in evaluating a twos complement, once we generate a borrow, this borrow propagates to the very end of the computation. To distinguish the twos complement from a second type of complement defined below, we shall heretofore use the notation N_2^* for the twos complement.

A useful algorithm for determining a twos complement which does not formally involve a process of subtraction is the following:

1   Write the number using the number of bits appropriate to the modulus.
2   Leave unaltered the 1 bit in the position of least numerical significance and leave unaltered as well all 0 bits in places of even less numerical significance.
3   Complement all other bits.

As an example let us find the twos complement of the number $+20$ assuming a seven-place modulus.   We rewrite the number as 0010100.   The 1 in the position of least significance and the 0s to the right of that 1 are left unaltered. Complementing all other bits, we have 1101100.

Ones complement We now define a second type of complement N_1^* defined precisely like the twos complement, except that the register is initially set so that each flip-flop is at 1. For example, the complement, with a four-place modulus, of $N = 0001$ is $N_1^* = 1110$; if $N = 0010$, $N_1^* = 1101$; if $N = 0011$, $N_1^* = 1100$.

These complements can be evaluated by straightforward subtraction. We find

$$
\begin{array}{cccc}
1111 & 1111 & 1111 & 1111 \\
N = 0000 & 0001 & 0010 & 0011 \\
\hline
N_1^* = 1111 & 1110 & 1101 & 1100
\end{array}
$$

Observe that in evaluating this complement by subtraction, in each order we have a subtraction from 1. Hence this complement N_1^* is generally referred to as the *ones complement*. Note also that the evaluation of the ones complement is simpler than the evaluation of the twos complement since the subtraction involved in the ones complement generates no borrowing. Further, it is to be observed, as well, that in a sense, the ones complement can be determined without performing the process of subtraction at all. For the ones complement N_1^* is related to the number N in that *each digit in N_1^* is the "complement" of the corresponding digit in N*, that is, each 1 in N becomes 0 in N_1^* and each 0 in N becomes 1 in N_1^*.

It may now be noted that given N_1, its ones complement N_1^* can be made available in a number of simple ways, none requiring subtraction. Suppose, for example, that the number N is stored in a register with the number available at the Q outputs of the individual flip-flops. Then the complement N_1^* is immediately available at the $\overline{Q}$ outputs. Alternatively, if the flip-flops are JK flip-flops with $J = K = 1$, then the flip-flops are in the toggle mode. In this case a clock pulse applied to every flip-flop in the register will toggle each flip-flop, and N_1^* will appear at the Q outputs. A third method involves the use of EXCLUSIVE-OR gates: For a k-bit number let there be available k EXCLUSIVE-OR gates. The k digits of the number $A = A_{k-1} \cdots A_0$ are applied to one input of each of the k EXCLUSIVE-OR gates, as shown in Fig. 11.8-1. When the control signal is $C = 0$, the number at the gate outputs $X = X_{k-1} \cdots X_0$ is $X = A$. When $C = 1$, however, $X = A_1^*$.

The ones complement with a k-place modulus is

$$(N_1^*)_k = (2^k - 1) - N \tag{11.8-2}$$

The relationship between the ones and twos complements is therefore

$$(N_2^*)_k = (N_1^* + 1)_k \tag{11.8-3}$$

Strictly, the subscripts k in Eq. (11.8-3) are necessary. The equation $N_2^* = N_1^* + 1$ is not always correct. For consider, say, that $N = \mathbf{0000}$; then $N_1^* = \mathbf{1111}$, and $N_1^* + 1 = \mathbf{10000}$ while N_2^* is $\mathbf{0000}$. Again, however, as before, when the intent is clear, the subscript may be omitted.

Finally, in summary, we have defined two types of complements, and we have noted that the ones complement can be determined without subtraction

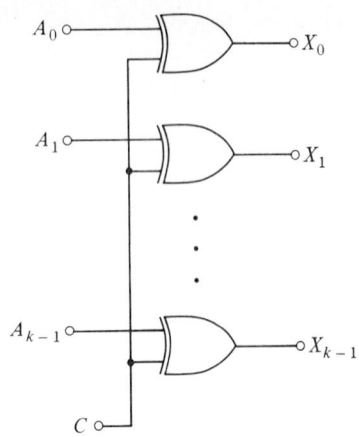

FIGURE 11.8-1
A method of generating the ones complement
of a number.

and, once the ones complement is available, the twos complement can be determined by adding 1 to the result, as noted in Eq. (11.8-3).

We have confined our interest to the binary system, but, of course, comparable complements can be defined in a number system with any base. In the decimal system, corresponding to the binary twos complement, we would have a *tens* complement and corresponding to the binary ones complement we would have a *nines* complement.

11.9 REPRESENTATION OF SIGNED NUMBERS IN REGISTERS

A four-place binary register has states **0000, 0001, ..., 1111**. It is rather natural to identify these states, in order, with the (decimal) numbers 0, 1, 2, ..., 15. Such an identification is indicated in the register of Fig. 11.9-1a. However, such an identification is not essential. We are at liberty to make any association we please between register position and the number represented. Thus, suppose we need to use a register to indicate negative numbers as well as positive numbers. The register has no manifest mechanism for indicating sign. Hence, it is clear that if negative numbers are to be represented, we have no alternative but *arbitrarily* to assign negative significance to some of the register positions. There are clearly a very large number of ways in which such assignments may be made. By way of example, we might decide that **0000** = 0 and that thereafter the register positions represent positive and negative numbers, in the manner **0001** = 1, **1001** = −1, **0010** = 2, **1010** = −2, etc. Of the various ways in which register positions may be identified with positive and negative numbers, three are of special interest and will now be considered.

The identification scheme shown in Fig. 11.9-1b is described as the *sign-magnitude* representation of negative numbers. The name derives from the fact that here the leftmost bit of the register indication serves to indicate sign.

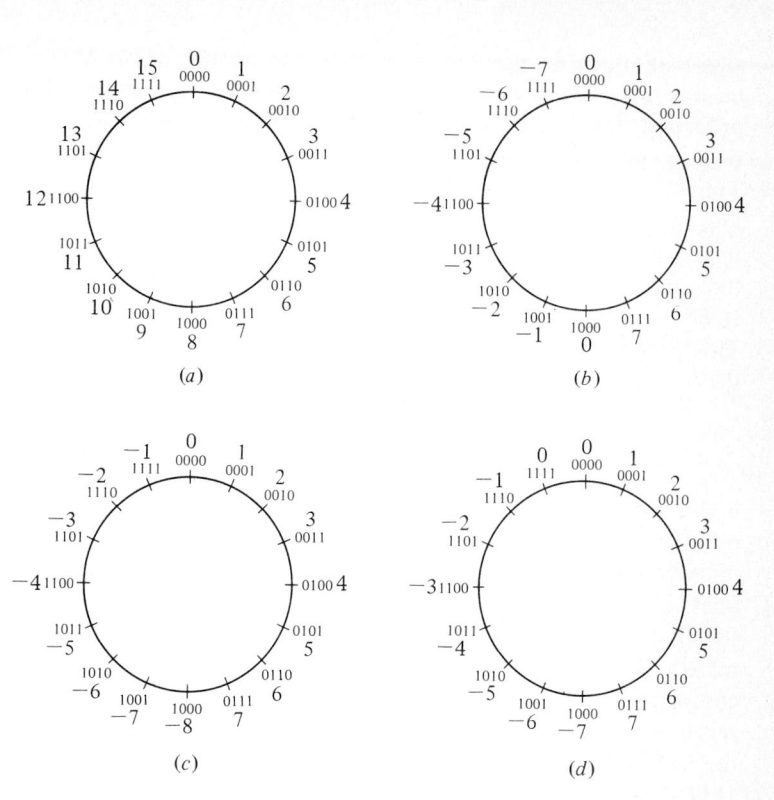

FIGURE 11.9-1
Possible assignments of numerical significance to register positions: (a) natural,
(b) sign magnitude, (c) twos complement, (d) ones complement.

A **0** bit represents a plus sign, while a **1** bit represents a minus sign. The remaining bits indicate the magnitude of the number. Thus, **0011** = +3, while **1011** = −3, etc. Observe that there are two zeros, **0000** (which might be read as "plus zero") and **1000** (which might be read as "minus zero"). An equal number of positive and negative numbers are represented. The four-stage register has no provision for registering a number larger than +7 or smaller than −7. This sign-magnitude register is of no great interest to us. We include it simply because it seems to represent such a "natural" way of dealing with negative numbers; i.e., since the register has no sign indicator, we simply use one of the register bits to indicate sign.

The association of register position with numbers shown in Fig. 11.9-1c is a *twos complement* association. That is, the register position assigned to a negative number is determined as the twos complement of the corresponding positive number. For example +3 = **0011**, while to locate −3 on the register we take the twos complement of 3, that is, **10000 − 0011 = 1101**, so that register position **1101** corresponds to −3. This twos-complement assignment differs from the sign-magnitude assignment in that the former has only one zero, has provision for one more negative number than positive number, and has, of course, a different ordering. The two assignments, however, have one feature in common. In both cases, when the leftmost digit is **1**, the number represented is negative. Thus, by way of example, suppose in the twos-complement representation, we find the representation **1011** and are asked for its numerical significance. Since the leftmost bit is **1**, we know that a negative number is represented. To find the magnitude of the number we determine its twos complement as **10000 − 1011 = 0101** = +5. Hence, the magnitude is 5 and **1011** represents −5. This procedure for evaluating the magnitude is valid since, given a number N, the twos complement of the twos complement is the number N itself, i.e.,

$$2^k - (2^k - N) = N \qquad (11.9\text{-}1)$$

The association of numbers and register positions indicated in Fig. 11.9-1d is the *ones-complement* association. That is, the register assigned to a negative number is determined as the ones complement of the corresponding positive number. For example +3 = **0011**, while to locate −3 on the register we take the ones complement of 3. that is, **1111 − 0011 = 1100**. Observe that in this register there are two zeros side by side. We shall see below that this feature is a matter of some consequence. Note further that in this register, as in the twos-complement and in the sign-magnitude register, when the leftmost digit in the register indication is **1**, the number represented is negative. Thus, for example, the representation **1011** indicates a negative number, and its magnitude is 4 since **1111 − 1011 = 0100** = +4. That is, **1011** represents −4.

11.10 SUBTRACTION THROUGH COMPLEMENTATION AND ADDITION

In Fig. 11.10-1 we have redrawn the twos-complement register, this time including a pointer which can be used to mark the register indication. Let us consider now how this register can be used to add and subtract. To add +2 (0010) to +3 (0011), we start with the pointer at zero and we advance the pointer, i.e., rotate clockwise, two positions and then three more positions. We find ourselves at +5 (0101). Now let us subtract +5 from +2. Then we advance the pointer two positions clockwise and then move the pointer counterclockwise by five steps. We end up at −3 (1101), as required. Thus, in general, by moving the pointer clockwise or counterclockwise to effect addition and subtraction, respectively, we can combine numbers. We need, of course, to assure ourselves beforehand that the result will not be larger than +7 or smaller than −8. Observe that this simple procedure for combining numbers will *not work* with the sign-magnitude register of Fig. 11.9-1b.

Now let us return to the matter of subtracting 5 from 2. Having advanced the pointer clockwise to a 0010, we need to move the pointer counterclockwise by 5. However, since there are exactly 16 positions in the register, we shall arrive at the same end point by *advancing* the pointer clockwise 16 − 5 = 11

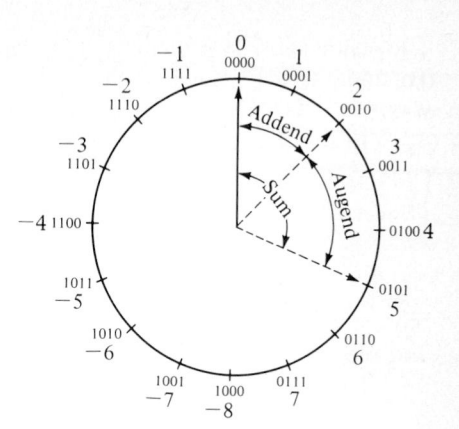

FIGURE 11.10-1

A twos-complement register used for addition and subtraction.

positions rather than counterclockwise by 5. Thereby we shall have performed a subtraction of 5 from 2 by a procedure that actually involves only addition. To put the matter more formally, suppose that we want to evaluate $D = A - B$ without actually performing an operation of subtraction. Then we write $D = A + (-B)$. The number $-B$ is represented, as in the twos-complement register, as $2^k - B$. Then starting at the register position corresponding to A, we *advance the register clockwise* (*add*) by a number of positions equal to $2^k - B$. We then end up at the register position $A + (2^k - B) = 2^k + (A - B)$. However, on a k-digit register an advance of 2^k positions corresponds to a complete rotation. Hence $2^k + (A - B) = A - B$.

11.11 TWOS-COMPLEMENT ADDITION AND SUBTRACTION

As examples of twos-complement operations let our operands be $A = 5$ and $B = 6$, and let us consider the four cases $(+5) + (+6), (+5) + (-6), (-5) + (+6)$, and $(-5) + (-6)$. At the outset we have to recognize that $5 + 6 = 11$ and $-5 - 6 = -11$. These numbers $+11$ and -11 are out of the range of a four-place register (see Fig. 11.9-1c), but it is easily seen that a five-place register will be adequate. In the first case we have

$$
\begin{array}{ll}
00101 & (+5) \\
+\ 00110 & (+6) \\
\hline
01011 & (+11)
\end{array}
$$

which requires no comment. In the second case we represent -6 as $100000 - 00110 = 11010$. We then have

$$
\begin{array}{ll}
00101 & (+5) \\
+\ 11010 & (-6) \\
\hline
11111 & (-1)
\end{array}
$$

The result 11111 is recognized as a negative number because the leftmost bit is 1. To find the magnitude of the number we take its twos complement, which is $00001 = +1$. Thus, 11111 represents -1.

In the third case we have

$$
\begin{array}{r}
11011 \quad (-5) \\
+ \ \ 00110 \quad (+6) \\
\hline
(1) \ 00001 \quad (+1)
\end{array}
$$

Since the register has only five places, the leftmost 1 will actually not appear and the register will read correctly $00001 = +1$.

In the last case we find

$$
\begin{array}{r}
11011 \quad (-5) \\
11010 \quad (-6) \\
\hline
(1) \ 10101 \quad (-11)
\end{array}
$$

Again the leftmost bit does not appear, and the register reads 10101. The number is negative, and its magnitude is 11 since $100000 - 10101 = 01011 = +11$.

11.12   ONES-COMPLEMENT ADDITION AND SUBTRACTION

Let us now turn to the ones-complement register of Fig. 11.9-1d. Here, to perform the operation of subtraction we can use an addition process similar to that used in connection with the register in Fig. 11.9-1c. Thus, to find $A - B$, we write $A - B = A + (-B)$. The augend $-B$ is represented by its ones complement $(2^k - 1) - B$, and, starting at A, we advance the register clockwise by $(2^k - 1) - B$ places. This number of positions of clockwise advance is one less than that used with the twos-complement register. However, comparing the two registers (Fig. 11.9-1c and d), we observe that in Fig. 11.9-1d the negative numbers are retarded by one position in comparison with their position in Fig. 11.9-1c. Hence, we shall end up at the right place.

There is a special situation in connection with the ones-complement register which we now examine. Consider that we want to perform the sum $7 + (-4)$. Then using the four-place register of Fig. 11.9-1d, we represent -4 as 1011 $(1111 - 0100)$ and then add $7 + (-4)$ with the result

$$
\begin{array}{r}
0111 \quad (7) \\
1011 \quad (-4) \\
\hline
(1) \ 0010 \quad (+2)
\end{array}
$$

The leftmost digit does not appear on a four-place register. Hence, we read $0010 = 2$. The result is in error, and the reason for the error is readily apparent. Having initially advanced the register to 0111 (seven), we then advanced the register an additional 1011 (eleven) positions. The total advance is

$7 + 11 = 18$ positions.   Since the register has only 16 positions, an advance of 18 positions carried the register through more than a complete revolution. *It is seen in the register shown in Fig. 11.9-1d that any such complete revolution carries the register through two positions which are marked zero. Hence after any such complete rotation, the register indication will be one unit less than the proper indication.*   The matter, then, can be remedied by including in our procedure a method of determining whether such a complete rotation has occurred and, if it has, advancing the register by one more position.

Let us now repeat the arithmetic of the previous section using the ones-complement representation of negative numbers.   We have

$$
\begin{array}{ll}
00101 & (+5) \\
00110 & (+6) \\
\hline
01011 & (11)
\end{array}
$$

Next we find

$$
\begin{array}{ll}
00101 & (+5) \\
11001 & (-6) \\
\hline
11110 & (-1)
\end{array}
$$

Again, as before, the 1 in the leftmost position indicates a negative number. The magnitude of this number is 1 since $11111 - 11110 = 00001 \; (= +1)$. Continuing, we have

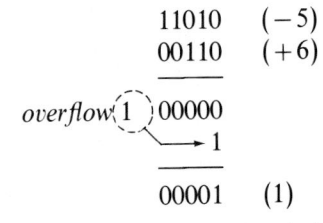

$$
\begin{array}{ll}
11010 & (-5) \\
00110 & (+6) \\
\hline
00001 & (1)
\end{array}
$$

In this case, the overflow into the sixth position indicates that the register has completed an entire revolution or more.   Hence, as indicated, we add 1 to the sum and keep only the five places corresponding to the five-position register. That is, the overflow 1 is disregarded in its overflow position but is carried around to the least significant position.   Finally, we have

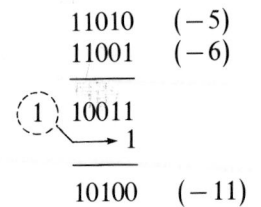

$$
\begin{array}{ll}
11010 & (-5) \\
11001 & (-6) \\
\hline
10100 & (-11)
\end{array}
$$

Figure 11.12-1 shows the simple modification required of an $(N + 1)$-bit adder to permit it to accomplish the overflow-carry-around operation required

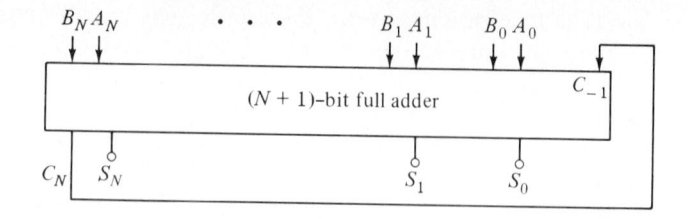

FIGURE 11.12-1
A basic adder-subtractor (ones-complement arithmetic).

in ones-complement arithmetic. The signed numbers are $A = A_N \cdots A_0$ and $B = B_N \cdots B_0$. The digits A_N and B_N carry the sign information. If the adder is constructed of N individual adders, even the adder in the least significant position must be a full adder. The output of the adder (sum or difference depending on the inputs) is read as $S_N \cdots S_0$. The output carry C_N is not part of the answer but is carried around to the carry output C_{-1}.

11.13   ADDITION AND SUBTRACTION OF A SEQUENCE OF SIGNED NUMBERS

Let us assume a ones-complement representation of negative numbers. Then the system of Fig. 11.5-1 can be modified to allow addition and subtraction of a sequence of signed numbers. If addition alone is required, we need only make provision for the carry around by connecting the C_{N-1} carry out back to the carry input C_{in}. It is, of course, understood that in the augend and accumulator registers the highest-order bit is a sign bit. In twos-complement arithmetic the *carry around* is not needed.

When the number in the augend register is to be subtracted rather than added, we must change each bit in the augend register to its complement. This we can do, among other ways, by interposing an EXCLUSIVE-OR gate in each of the connections from the augend outputs to the full-adder inputs, as shown in Fig. 11.8-1. A common input to each gate is a control signal which is at logic level **0** when addition is intended (this procedure must be modified for twos-complement arithmetic by also letting $C_{in} = 1$ during subtraction). The other input to each of the N gates is the corresponding Q output of the augend register.

In the course of combining a sequence of signed numbers, the algebraic sum may sometime exceed the capacity of the accumulator either in the positive or in the negative direction. Nonetheless, it is of interest to note that if the final algebraic sum is within the range of the accumulator, the final accumulator reading will be correct even if the accumulator overflowed in the course of arriving at this reading.

In the simplest case, it is easy to see that such is indeed so. For consider that we are adding a sequence of signed numbers and that the sum is being accumulated in a k-digit register. Let us imagine that the register was augmented by as many extra digit positions, $k + 1$, $k + 2$, etc., as are required to avoid overflow. Suppose that after overflowing into these extra digit positions, an accumulated sum is finally contained within the k digits of the register. Then the extra digit positions will all read zero, and it will have made no difference that these extra digit positions are not actually available. Similarly, in other cases, as when a carry around is involved, it may be shown generally, and by example, that the principle continues to apply (see Prob. 11.13-2).

11.14   A SATURATING ADDER

Suppose that we build an adder with a four-place register. If we use twos-complement arithmetic in which the leftmost place indicates sign, the largest sum which the adder can accommodate is $+7 = 0111$ and the most negative sum is $-8 = 1000$. Suppose now that we add $+5 = 0101$ and $+3 = 0011$. The sum is 8, which exceeds by only 1 the capacity of the adder. Yet the result $1000 (= 0101 + 0011)$ would be interpreted as -8 so that the error would be 16. A similar reversal of sign and exaggeration of error occurs when the capacity of the register is exceeded in the negative direction.

There are many situations where an error in the sum is simply not acceptable. Such would certainly be the case in keeping financial accounts in a bank. On the other hand, suppose it were possible to arrange that when a sum register overflowed, the register indication could be kept at its extreme value. That is, in the example above, suppose that the sum were held at $0111 = +7$ whenever the true sum was $+7$ or greater. In this way the error due to overflow would be kept at a minimum. Such a *saturating* adder might be acceptable if our interest were not specifically in the numerical value of the sum but in using the sum output to control some process. Thus, suppose the sum output was used to control the rate at which a process was proceeding, the rate being proportional to the sum. Then with a saturating adder, when overflow occurs, at least the process will be kept going at the fastest rate consistent with the capacity of the adder. We consider now how such a saturating adder can be constructed.

Figure 11.14-1a shows an $(N + 1)$-bit adder with an extra full-adder stage. As indicated, the sign bits A_N and B_N are applied both to this extra *overflow-detector stage* and to the last stage of the adder. The largest positive output of the basic adder is $2^N - 1$, and if there is no overflow beyond this extreme, the output of the adder will be $S_N DS_{N-1} \cdots S_0 = 00XX \cdots X$; that is, we shall have $S_N = D = 0$. Similarly if the output is negative but still within the capacity of the adder, we shall have $S_N = D = 1$.

If now an overflow occurs in the positive direction, then, as can be verified, we shall have $S_N = 0$ but $D = 1$. While if an overflow occurs in the negative direction, we shall have $S_N = 1$ but $D = 0$. We assume here that the overflow is such that the correct sum is not larger than $2^{N+1} - 1$. That is, we assume an overflow into only one bit position.

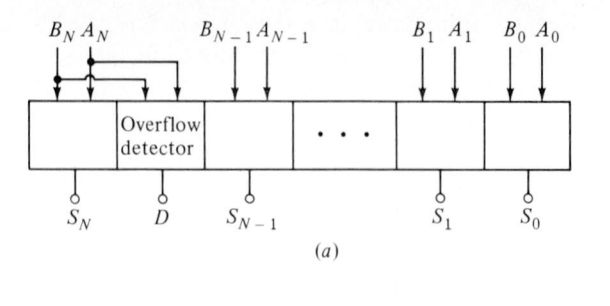

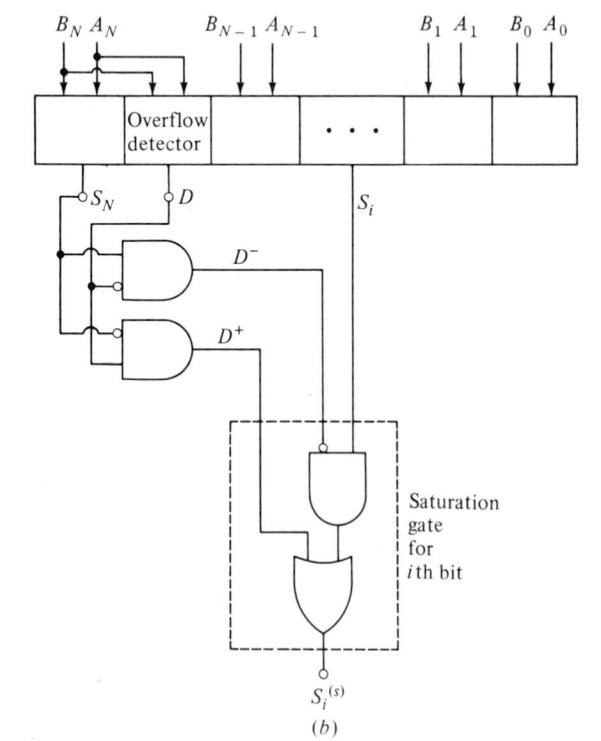

FIGURE 11.14-1
(a) An $(N + 1)$-bit adder with an additional 1-bit adder to detect overflow; (b) overflow-detection and saturation logic coding.

A saturating adder is shown in Fig. 11.14-1b. At a positive overflow $S_N = 0$ and $D = 1$, so that D^+ will be $D^+ = 1$. Otherwise D^+ will be $D^+ = 0$. Similarly at a negative overflow, and only then, we shall have $D^- = 1$. Each of the adder outputs from $S_{N-1}, \ldots, S_0$ (but not S_N and not D) is brought out through a separate *saturation* circuit (shown for S_i only in the dashed box) to make available $S_i^{(s)}$. It is readily verified that if overflow does not occur, $S_i^{(s)} = S_i$, the true sum bit and the output sum is $S_N S_{N-1} \cdots S_0$ (D is not brought to the output). If a positive overflow occurs, the output is $S_N S_{N-1}^{(s)} \cdots S_0^{(s)} = 01 \cdots 1$. If a negative overflow occurs, the output is $S_N S_{N-1}^{(s)} \cdots S_0^{(s)} = 10 \cdots 0$.

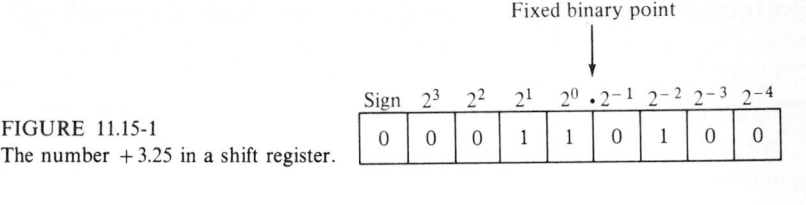

FIGURE 11.15-1
The number $+3.25$ in a shift register.

11.15 SCALING

The multiplication of a variable by a constant is called *scaling*. Of special interest is scaling in which the constant is a power of 2. Such scaling can be carried out conveniently by shift-register operations.

A binary number is multiplied by 2^n by moving the binary point n places to the right and is divided by 2^n by moving the binary point n places to the left. Let us start with the binary number $011.01 = 3.25$ (the leftmost zero in the binary number is the sign bit). Then $0110.1 = +2(3.25)$ and $01.101 = +\frac{1}{2}(3.25)$.

Suppose, now, that it turns out to be inconvenient to shift the binary point and we propose instead to leave the binary point in a fixed position and shift the digits instead. Then a left digit shift of n places would multiply by 2^n and a right shift would divide. In such a shifting operation the question arises what to do with the vacated rightmost position at each left shift and with the vacated leftmost position at each right shift. This question is answered by recognizing that a positive number is not changed by adding additional zeros to the right and to the left of the number. Thus,

$$011.01 \equiv \cdots 000011.01000 \cdots$$

and it is clear that zeros are to be placed into these vacated end positions.

In Fig. 11.15-1 the number $+3.25 = 011.01$ has been placed in a shift register. Two additional register positions with zeros have been included after the least significant bit, and two additional register positions with zeros have been added between the sign bit and the most significant bit of the number. These places are added to allow shifting without loss of significant bits. The number in the register can be divided by 2 or by 4 by shifting the bits one or two places to the right, respectively. If we attempt to divide by 8, which is a three-position shift, the least significant bit (the rightmost 1) will be shifted out of the register and lost. The error due to such an overflow beyond the rightmost register position is equal to one-half the numerical significance of the least significant position of the register. If the number of bits is large, such an error, called a *quantization error* may be acceptable.

In the register of Fig. 11.15-1 one or two left shifts are allowed to multiply the number by 2 or by 4. Three shifts would put a 1 in the sign-bit position and therefore are not allowed. If the leftmost position in Fig. 11.15-1 represented 2^4 rather than the sign bit, a third left shift would be permissible. In this case, however, a fourth left shift would lose the most significant bit, resulting generally in an unacceptable error called an *overflow error*. In any event, it is to be kept in mind that when shifting, we always shift zeros into the vacated end position of the register providing that the number is positive.

Negative numbers When negative numbers are represented by their twos complement, then just as with positive numbers, multiplication and division are accomplished by moving the binary point. Alternatively the binary point can be kept fixed and the digits moved. The number -3.25 in the twos-complement representation written into a register with the minimum number of register positions is calculated as

$$
\begin{array}{r}
1000.00 \\
-\ \ 011.01 \\
\hline
100.11
\end{array}
$$

If there were two additional register positions available on each side of the binary point, we would write instead

$$
\begin{array}{r}
100000.0000 \\
-\ \ 00011.0100 \\
\hline
11100.1100
\end{array}
$$

Comparing 100.11 with 11100.1100, both of which represent -3.25, it is clear that appending an arbitrary number of 1s at the left or an arbitrary number of 0s at the right does not change the value of the number. Therefore, in the case of the negative number, when shifting to the right we always shift 1s into the vacated leftmost register position. In shifting to the left, zeros are shifted into the vacated right most positions.

Shifting to the right will produce no error until the least significant bit (the rightmost 1) is shifted out of the rightmost register position. Shifting to the left will produce no error until the most significant bit (the leftmost 0) is shifted into the leftmost register position (the sign position).

Ones complement The number -3.25 in the ones-complement representation when written into a register with the minimum number of register positions is calculated as

$$
\begin{array}{r}
111.11 \\
011.01 \\
\hline
100.10
\end{array}
$$

If there were two additional register positions available on each side of the binary point, we would write instead

$$
\begin{array}{r}
11111.1111 \\
00011.0100 \\
\hline
11100.1011
\end{array}
$$

When we compare these results, it appears that, in the ones-complement case, appending an arbitrary number of 1s to either the left or to the right does not change the value of the number. Hence in this case 1s are to be shifted into the vacated end positions when shifting in either direction. Shifting will produce

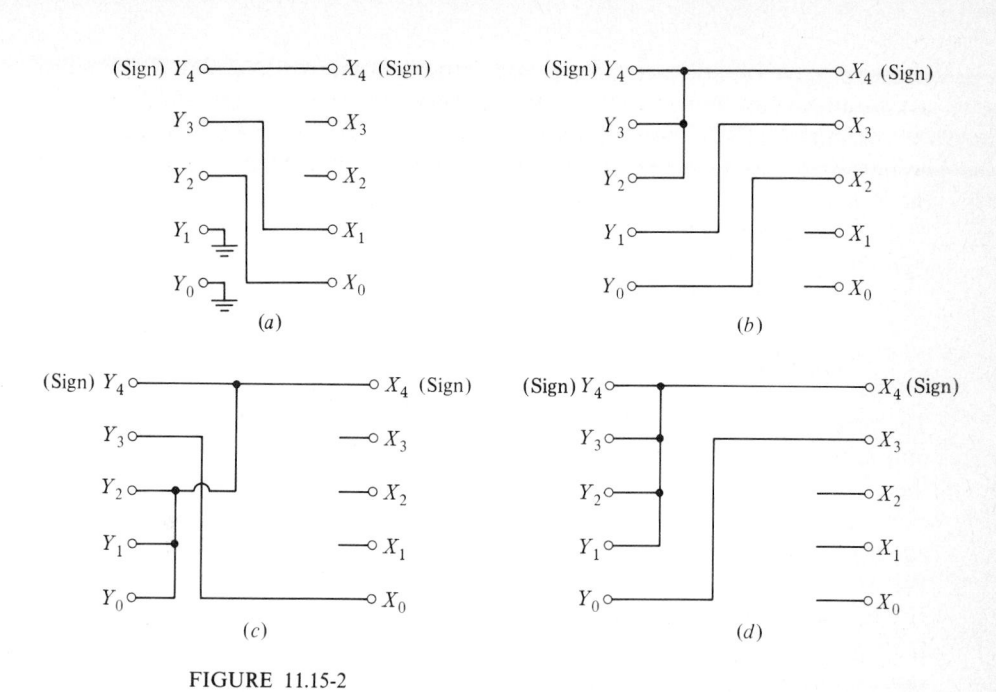

FIGURE 11.15-2

(a) Scaling by $a = 2^2$ when X and Y are in twos complement, $Y = 2^2 X$.
(b) Scaling by $a = 2^{-2}$ when X and Y are in twos complement, $Y = 2^{-2}X$.
(c) Scaling with $a = 2^3$ using ones-complement arithmetic, $Y = 2^3 X$. (d) Scaling
with $a = 2^{-3}$ using ones-complement arithmetic, $Y = 2^{-3}X$. (The ground symbol
represents logic **0**.)

no error until a right shift moves a zero out of the rightmost position or a left
shift shifts a zero into the position of the sign bit.

The procedure described above is often employed when performing
computations on a digital computer. However, the procedure is time-consuming
as each shift requires an additional clock pulse. In many digital systems we
would like to "hard-wire" the scaling circuit so that the variable $Y = 2^n X$ is
determined at the *same* time that X is given.

Figure 11.15-2a to d illustrates the hard-wire circuits employed in the
design of digital filters and other digital processing systems. For simplicity
5-bit numbers are shown. Figure 11.15-2a shows the wiring for $Y = 2^2 X$ using
twos complement. Note that $Y_1 = Y_0 = 0$ since 0s are shifted into the rightmost
bits. Note also that the sign bit is never shifted in practice since scaling
cannot cause a sign change. Further, X_3 and X_2 are lost as a result of scaling.
Thus, if either X_3 or X_2 is 1, an overflow error results. Saturation logic can
be designed to minimize overflow error using a procedure similar to the one
shown in Fig. 11.14-1.

Figure 11.15-2b shows the wiring circuit when $Y = 2^{-2}X$ using twos
complement. Note that the vacant bits Y_3 and Y_2 are connected to the sign
bit (Y_4). This is done because if X is positive, $Y_4 = Y_3 = Y_2 = 0$, while if X is

negative $Y_4 = Y_3 = Y_2 = 1$. Note also that the values of X_1 and X_0 are lost as a result of scaling.

Figure 11.15-2c shows the result of scaling X to form $Y = 2^3 X$ using ones complement. Here, since Y_0, Y_1, and Y_2 take on the same value as the sign bit Y_4, we connect them together. Figure 11.15-2d shows the circuit for $Y = 2^{-3} X$. Again the bits to be inserted, Y_1, Y_2, and Y_3, are connected to the sign bit Y_4.

11.16 MULTIPLICATION

The process of multiplication is illustrated in Fig. 11.16-1. The multiplicand is multiplied in turn by each digit of the multiplier. These partial products are then added with due consideration for the differing numerical significance of each digit of the multiplier. Each partial product is either identically zero or equal to the multiplicand, depending on whether the multiplier digit is 0 or 1. We note that the product has more digits than the multiplicand or multiplier. If these terms each have N digits, the product may have as many as $2N$ digits. If then the registers and other storage facilities in a digital system all have the same number of digit positions, some of the digits in the product may overflow.

A basic multiplier capable of multiplying two 3-bit words is shown in Fig. 11.16-2. It involves a 5-bit multiplicand shift register, a 3-bit multiplier shift register, and a 6-bit accumulator (we have chosen to use type-D flip-flops). Initially the multiplicand is loaded into the three right-hand flip-flops of the multiplicand register, the least significant bit into $FF0$, the most significant into $FF2$. The multiplier is loaded into the multiplier shift register, the least significant bit into $FF0$ the most significant into $FF2$. The accumulator register is assumed cleared.

Before the first clock pulse occurs, the first partial product has been formed. The multiplication is accomplished by the AND gates, and the partial product appears as $B_2 B_1 B_0$ at the adder inputs and also as $S_2 S_1 S_0$ at the adder outputs. At the first clock pulse, the partial product $S_2 S_1 S_0$ is registered in the three right-hand flip-flops of the accumulator registers and then appears as well as $A_2 A_1 A_0$ at the adder input. This same first clock pulse has move the next multiplier bit into $FF0$ of the multiplier shift register and has moved the three-

```
                        1  0  1  1        (Multiplicand)
                        1  1  0  1        (Multiplier)
                    ─────────────────
                        1  0  1  1    ⎫
                     0  0  0  0        ⎪
                  1  0  1  1            ⎬  (Partial products)
               1  0  1  1               ⎭
               ─────────────────
            1  0  0  0  1  1  1  1        (Product)
```

FIGURE 11.16-1
The process of multiplication.

Clock

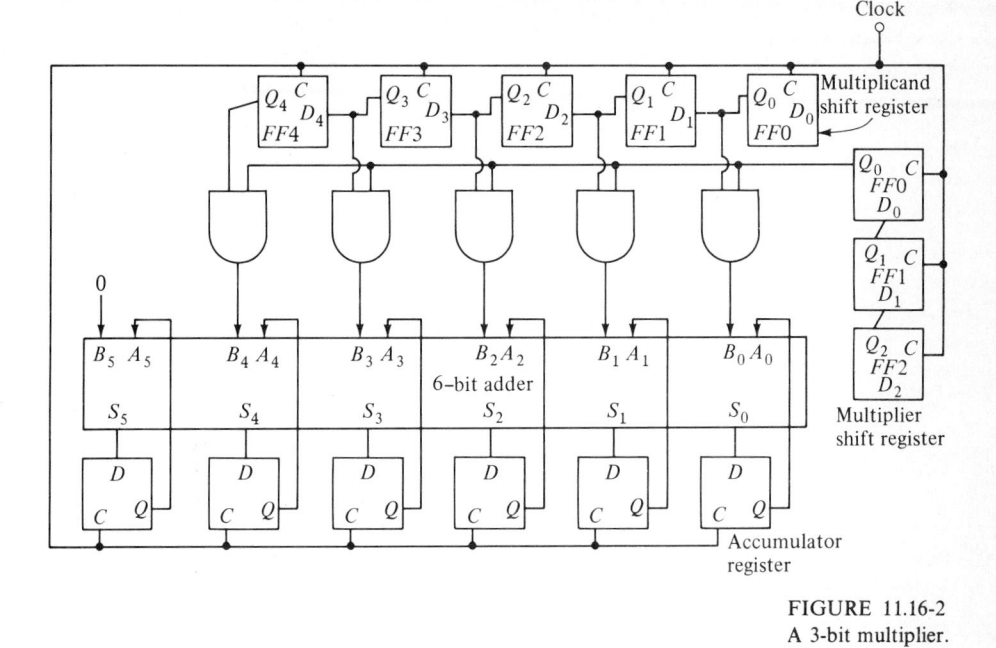

FIGURE 11.16-2
A 3-bit multiplier.

digit multiplicand one step to the left in the multiplicand shift register. Hence, the second partial product appears as $B_3 B_2 B_1$, and the sum of this partial product and the first partial product appears at the adder output as $S_4 S_3 S_2 S_1 S_0$. (The sum of the first two 3-digit partial products may yield a 5-digit word.) After the second clock pulse the sum of all three partial products, which is the required product, will appear at the adder output as $S_5 S_4 S_3 S_2 S_1 S_0$. The third pulse will register the sum of the three partial products in the accumulator register and will also clear the multiplier shift register. Hence the multiplication process is complete.

The serial multiplier is relatively slow as it requires additions to be performed one after the other. Parallel multipliers are presently available which do the additions in parallel, thereby significantly decreasing the duration of the multiplication.

11.17   DIVISION

Division can be performed by repeated subtraction. For example, to divide decimal 22 by 3 we subtract 3 from 22 until the remainder is less than the divisor. This subtraction can be performed 7 times, leaving a remainder of 1. Next we subtract 0.3 from 1 as many times as possible. This subtraction can be performed 3 times, leaving a remainder of 0.1. Next we subtract 0.03, etc., leading to the quotient 7.33 $\cdots$. As a simple example of this division procedure

Table 11.17-1

1101	(13)	1001	(9)
0110	(ones complement of 9)	0010	(ones complement of 13)
10011		1011	
└→1			
0100	(4)	− 0100	(−4)

we shall consider a circuit to accomplish the first order of this successive subtraction.

We shall deal only with positive numbers, and hence we shall find it convenient to omit the sign bit in the number representation, thereby saving one bit position in the registers. In the absence of a sign bit the procedure for subtracting when the minuend and subtrahend are positive is as shown by the example in Table 11.17-1, in which we subtract decimal 9 (1001) from decimal 13 (1101) and then subtract 13 from 9. When the difference is positive, there is an overflow 1 which is to be carried around and added. When the difference is negative, there is no overflow. In this case the correct answer is the ones complement of the sum with a minus sign attached. It is left as a student exercise to verify that such is the case generally. Of importance in our present application is the fact that when the difference is positive, there is an overflow 1 and when the difference is negative, there is no such overflow.

A scheme for dividing, to first order, a four-digit number $X = X_3 X_2 X_1 X_0$ by a divisor $Y = Y_3 Y_2 Y_1 Y_0$ by repeated subtraction is shown in Fig. 11.17-1.

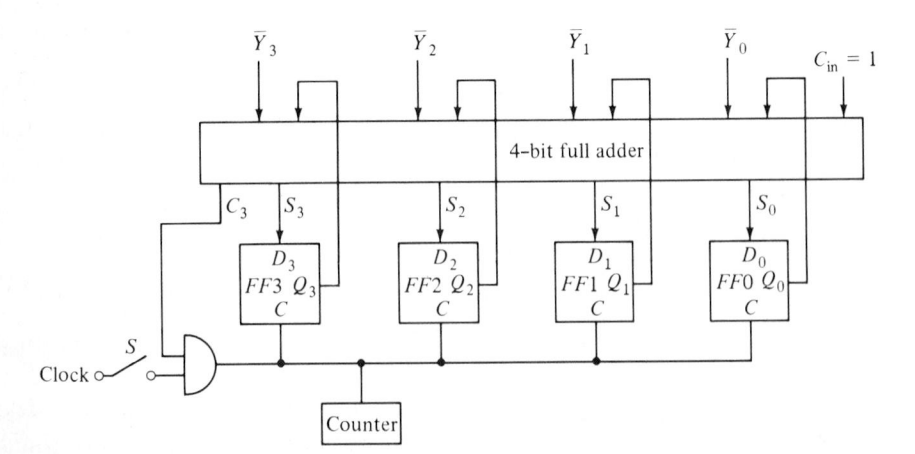

FIGURE 11.17-1
A divider by repeated subtraction.

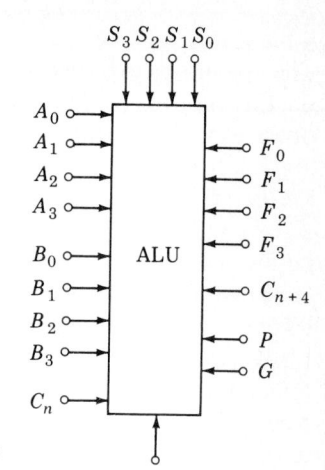

FIGURE 11.18-1
Representation of an arithmetic-logic unit
(ALU).

The dividend X is loaded into the register shown, consisting here of four type-D flip-flops. The register outputs are connected as one input to the adder while the other adder input is the ones complement of Y, given by $\overline{Y}_3 \overline{Y}_2 \overline{Y}_1 \overline{Y}_0$. In principle the carry output C_3 should be connected back to the carry input C_{in}. However, we are interested in this connection only so long as $C_3 = 1$. Hence, we have used the simpler arrangement of fixing C_{in} at $C_{in} = 1$. The counter is initially set at zero.

Assuming that $X > Y$, we shall have initially $C_3 = 1$ and hence the AND gate is enabled. The difference $X - Y$ appears at $S_3 S_2 S_1 S_0$. Now let us close the clock switch. The first clock pulse will load the difference $X - Y$ into the register and will also advance the counter by one step. We now have $(X - Y) - Y = X - 2Y$ at $S_3 S_2 S_1 S_0$. If C_3 is still 1, the AND gate will be enabled and the second clock pulse will load $X - 2Y$ into the register and advance the counter a second step, etc. If after n clock pulses the remainder becomes less than Y, we shall have $C_3 = 0$. The AND gate will now be disabled and the counter will stop. Hence the quotient is read from the counter.

11.18 THE ARITHMETIC LOGIC UNIT (ALU)

In ECL, TTL, and CMOS there are available integrated-circuit packages which are referred to as arithmetic-logic units (ALU). The logic circuitry in these units is entirely combinatorial (i.e., consists of gates with no feedback and no flip-flops). The ALU is an extremely versatile and useful device since it makes available, in a single IC package, facility for performing many different logical and arithmetic operations.

Functionally, the operation of a typical ALU such as the ECL type MC10181 unit is represented in Fig. 11.18-1. The unit accepts as inputs two

4-bit words $A = A_3 A_2 A_1 A_0$ and $B = B_3 B_2 B_1 B_0$ and, a carry input labeled C_n. The operations to be performed on these inputs are determined by the logic levels on the input $S = S_3 S_2 S_1 S_0$ and on the input M (mode). When $M = 1$, the operations are logical, while a combination of logical and arithmetic operations are selected when $M = 0$. The outputs are available at $F = F_3 F_2 F_1 F_0$ and the carry output C_{n+4}. In a typical case, as indicated in Table 11.18-1, if $M = 0$ and $S = 1\,0\,0\,1$ then the output F will be the arithmetic sum A plus B, taking account of any carry from a previous stage which may be present at C_n. Here, as in the table we use the word "plus" to indicate arithmetic addition in distinction to "$+$" which is used to represent the logical OR. When the unit is used for addition, and speed is important, we may want to bridge across the unit an external carry bypass, such as the type 10179.[1,2] With this contingency in mind the ALU provides terminals P and G. An input carry which is to *propagate* through the ALU will appear at terminal P. An output carry which is *generated* in the ALU unit itself will appear at terminal G.

We may wonder about the usefulness of some of the functions listed in Table 11.18-1. Consider, then, by way of example, that we need to compare two positive numbers A and B to determine which is larger. Such an operation (called *comparison*) is required in one algorithm for performing division. If we set $M = 0$ and $S = 0\,1\,1\,0$ then we find that $F = A$ minus B minus 1. If now $A > B$, the sign bit of F which is F_3 will be $F_3 = 0$. In the ALU it is

Table 11.18-1   **TABLE OF FUNCTIONS**

Function select				Logic functions ($M = 1$)	Arithmetic and logic ($M = 0$)
S_3	S_2	S_1	S_0	F	F
0	0	0	0	A	A minus 1
0	0	0	1	$A \cdot B$	$(A \cdot B)$ minus 1
0	0	1	0	$A \cdot \bar{B}$	$(A \cdot \bar{B})$ minus 1
0	0	1	1	0	minus 1 (in two's complement)
0	1	0	0	$A + B$	A plus $(A + \bar{B})$
0	1	0	1	B	$(A + \bar{B})$ plus $(A \cdot B)$
0	1	1	0	$A \oplus B$	A minus B minus 1
0	1	1	1	$\bar{A} \cdot B$	$A + \bar{B}$
1	0	0	0	$A + \bar{B}$	A plus $(A + B)$
1	0	0	1	$\overline{A \oplus B}$	A plus B
1	0	1	0	$\bar{B}$	$(A + B)$ plus $(A \cdot \bar{B})$
1	0	1	1	$\bar{A} \cdot \bar{B}$	$A + B$
1	1	0	0	1	A times 2
1	1	0	1	$\bar{A} + B$	A plus $(A \cdot B)$
1	1	1	0	$\bar{A} + \bar{B}$	A plus $(A \cdot \bar{B})$
1	1	1	1	$\bar{A}$	A

intended that negative numbers be presented in twos complement form. Hence, if $A \leq B$ then F will be negative and $F_3 = 1$.

The MC10181 ECL ALU is a very fast unit.   It adds two 4-bit numbers in approximately 6.5 ns, and typically an output carry is generated in less than 5 ns.

The ALU is the heart of every microprocessor unit (MPU).   In a microprocessor the ALU is combined with shift registers, ROMs and RAMs (see Chap. 12) so that a complex sequence of arithmetic and logic operations can be performed.

REFERENCES

1   MECL Integrated Circuits DATA BOOK: Motorola Semiconductor Products, Inc., 1973.
2   Motorola Application Note AN-488 (prepared by J. M. DeLaune): Motorola Semiconductor Products, Inc., 1971.

12

SEMICONDUCTOR MEMORIES

A digital processor generally requires a facility for storing information. The information so stored may consist of the numbers to be used in a computation, intermediate computational results, instructions which will direct a computation, or all three. Where no computation is involved, a storage facility may be called upon simply to store data. For example, a machine designed to address envelopes for mailing will need a storage facility for names and addresses. That part of a digital processor which provides this storage facility is called the *memory*. In this chapter we shall discuss the organization and the electronics of semiconductor memory devices.

12.1 TYPES OF MEMORIES

There are three types of memories: the sequentially accessed memory, the random-access memory (RAM), and the read-only memory (ROM).

A simple and straightforward example of a sequentially accessed memory is the punched paper tape shown in Fig. 12.1-1. The tape is a strip of paper of extended length across whose width at regular intervals are reserved spaces used

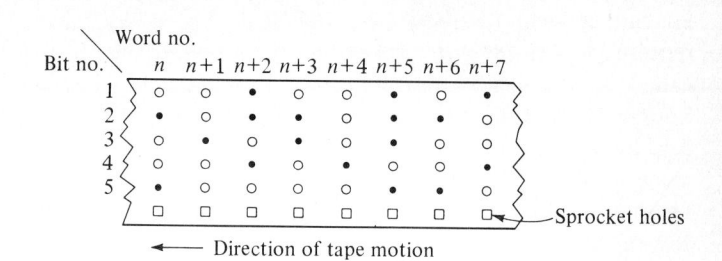

FIGURE 12.1-1
A punched paper tape.   An example of a sequentially accessed memory.

to represent logic **1** or logic **0**.   A logic **1** may be represented by punching a hole in the tape while a logic **0** is represented when no hole is punched in the reserved space.   (Of course, if we so choose, the punched hole may represent logic **0**, etc.)   The bits so recorded across the width of the tape comprise a word, and successive words appear in sequence along the length of the tape.   In Fig. 12.1-1 we have made provision for a 5-bit word.   The circles represent punchings through the paper and hence are read as logic **1**.   The dots represent no punching and hence are read as logic **0**.   Thus, the nth word is **10110**, the $(n + 1)$st word **11011**, etc.   The square holes at the bottom of the tape are sprocket holes used to advance the tape in synchronism with the system of which the tape memory is a part.   Words are *written* onto the tape by mechanically punching holes in the tape as the tape is advanced step by step under the *punching head*. Words are *read* in sequence as the tape advances step by step under a *reading head*.   The reading head may consist of five small light sources and an equal number of photosensitive devices.

The essential feature of the memory of Fig. 12.1-1 is that words are written and read *in sequence*.   If, say, the nth word happens to be under the reading head, then the $(n + k)$th word is not available for reading until after the tape has been advanced k steps.   Herein lies a principal limitation of a sequentially accessed memory.   If data needed for some purpose are located at random places in a sequential-access memory, a considerable time may be required to assemble the data in one location for processing.   On the other hand sequential-access memories have the merit of being relatively inexpensive and are very effective when it is possible to write data into the memory in the order in which they are later to be recalled.

In another type of sequential-access memory, the tape has a coating of magnetic material, and bits are registered on the tape by applying a magnetic field in one direction or the other.   As with the punched paper tape, words are written or read as the tape moves under a head.   Still other magnetic-coating memories use rotating disks or rotating drums.

In a random-access memory (RAM), *words* are again stored in *locations*. Provision is made for singling out a particular location, i.e., *addressing* the location, for *writing* data (a sequence of bits) into the addressed location, and for calling forth the data, i.e., *reading* from the addressed location. The time required to complete the operation of writing a word into a memory location is called the *write-access* time (and correspondingly for reading the *read-access* time). Suppose now that we have read a word from, or written a word into, some first location in a memory. Suppose further that we now turn our attention to some second location, chosen arbitrarily, i.e., at random, and access that location for reading or writing. In a random-access memory the access time to this second location is the same for all locations. This situation is different from that prevailing in a sequential memory system, where the access time to a second location depends on its location with respect to the first accessed location.

The read-only memory (ROM) is a type of random-access memory. The ROM differs from the RAM in that in a ROM no provision is made to write words into the memory while the system is operating, i.e., in real time. The content of the memory is usually established by the manufacturer or user and thereafter generally cannot be altered by the user.

12.2 SHIFT-REGISTER SEQUENTIAL MEMORIES

The sequential memories of the previous section, involving mechanically moving parts as they do, are limited in speed. A shift-register sequential memory is shown in Fig. 12.2-1. This memory uses N shift registers, each of K stages, and will provide storage for K words, each of N bits. Each register holds one of the N bits of each of the K words. With each clock cycle (the clock input

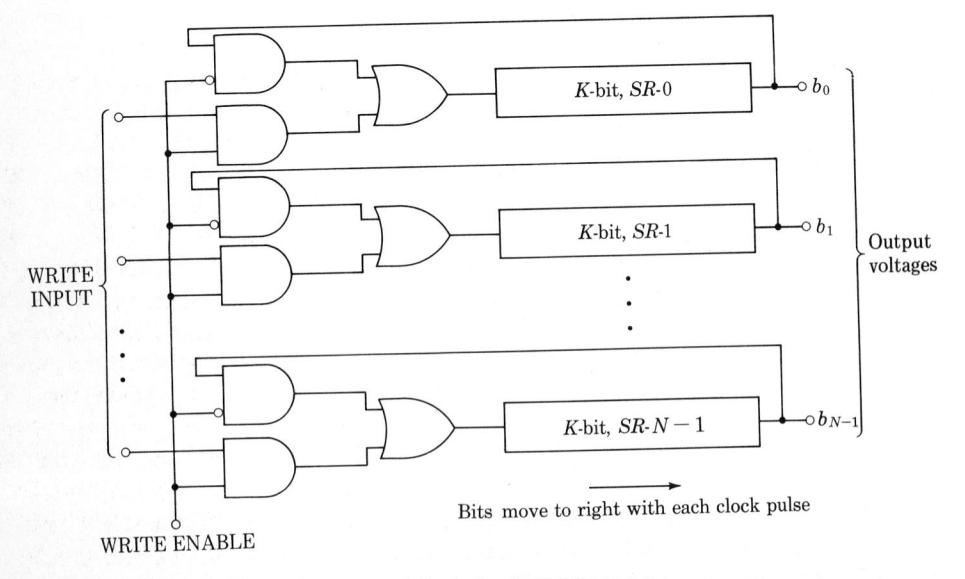

Bits move to right with each clock pulse

FIGURE 12.2-1
A sequential-access memory using shift registers.

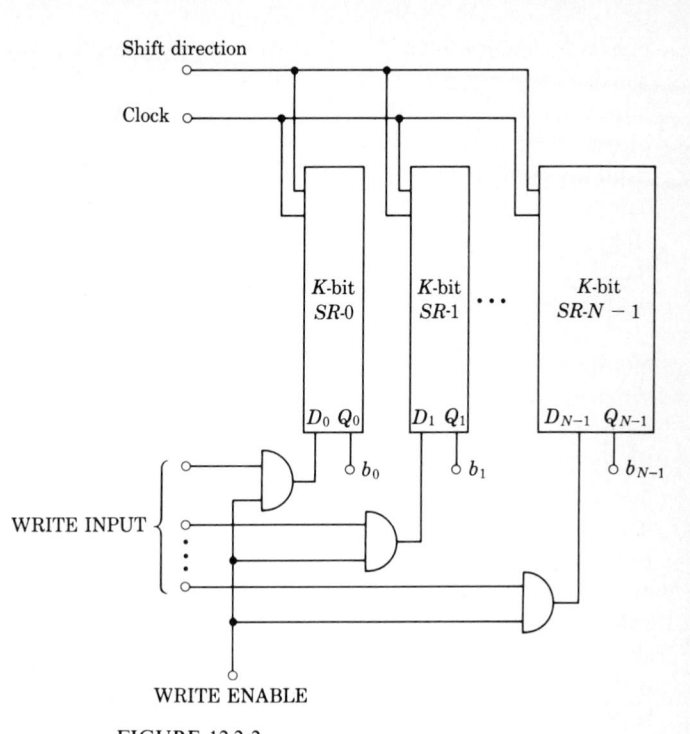

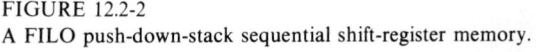

FIGURE 12.2-2
A FILO push-down-stack sequential shift-register memory.

is not explicitly indicated in Fig. 12.2-1) the bits in each register will advance one step, and the stored words will appear *sequentially* at the outputs b_0, b_1, ..., b_{N-1}. In this sequential memory the word being read, which is the ensemble of bits registered in the last register position, may be transferred back to the leftmost register position in the manner of a recirculating shift register. Thus, after the entire sequence of words has appeared at the memory output, the same sequence may be repeated endlessly as long as the clock continues to drive the registers. In this sense the present register corresponds to a tape memory in which the beginning and end of the tape have been joined so that the tape forms a closed loop.

If the logic level on the write-enable line is set at logic **1**, the recirculation path around the registers will be broken. In this case, the bits applied at the write-input lines, i.e., applied in synchronism with the clock, will successively fill register positions in the shift registers. Thus, with the write control at logic **1** we can *erase* and replace the contents of the memory.

Starting with the memory initially cleared, the first word written into the memory will be the first word that will appear at the memory output when the content of the memory is recalled. Hence, the present memory is described as a first-in first-out (FIFO) system.

A different mode of organization of a shift-register sequential memory is shown in Fig. 12.2-2. Here again N shift registers each of K bits are used to provide a memory for N-bit words, the memory having a capacity of K such words. The present memory differs from the memory of Fig. 12.2-1 in three important respects. The circulation feature of the previous circuit is not employed here, and the data input and data output are applied to and taken from the same register stage. Finally the shift registers have the feature that they can be shifted in either direction (see Sec. 10.5). If we think of a typical register stage in the shift register as a type-D flip-flop, then the data input is applied to the D (data) input of the bottom flip-flop in each shift register and the word is recalled by reading the bits of the Q outputs of these same flip-flops.

Let the *write enable* be at logic **1** and the *shift-direction* line set to shift bits upward, and let a sequence of words be presented at the *write-input* lines in synchronism with the clock. Then a total of K words can be stored in the memory. Each word enters at the bottom and moves up one stage to make room for the next word. Figuratively, the words are stacked one on top of another, the first input word being at the top of the stack. To read the remembered words in sequence we set the write enable at logic **0** and reverse the direction of shifting to the downward direction. With each clock cycle a remembered word appears at the memory output in the order opposite to that in which the words were entered. The present memory is hence referred to as a first-in last-out (FILO) system. Also, we may fancifully picture the operation as one in which a stack of words is pushed down from the top, thereby causing word after word to be squeezed out from the bottom. Such considerations account for the description of the present memory as a *push-down stack*.

12.3 MOS REGISTER STAGES

As noted, MOS technology is of special advantage in large-scale integration (LSI) because an MOS device generally occupies much less real estate on a silicon chip than a comparable bipolar device. Hence, MOS devices are widely used in the large-capacity sequential memories which employ shift registers. We shall consider now the characteristics and operating features of the MOS register stages which are cascaded into shift registers.

We have seen that shift registers are cascades of flip-flops and that these flip-flops must have the operating feature characteristic of the master-slave flip-flop or other equivalent flip-flops, as discussed in Chap. 9. That is to say, at the time in the clock cycle when the output of the flip-flop is responding to input data already present, any change in input data must not affect the output. In considering MOS shift-register stages, we may be hard-pressed to know whether they are to be called master-slave flip-flops, but we shall see that, within limitations, they accomplish the same function.

Consider the MOS inverter circuit of Fig. 12.3-1. In discussing this and subsequent circuits we shall arbitrarily introduce some symbols which will simplify the circuit diagrams and avoid excessive words. First we use the simplified symbol for an FET which omits the substrate and does not indicate

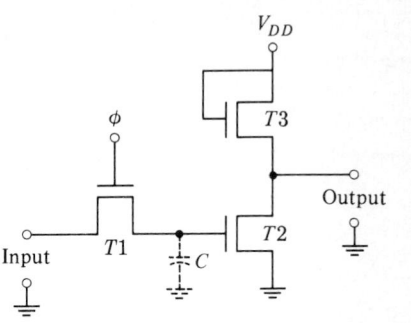

FIGURE 12.3-1
An MOS circuit which can store a bit
temporarily.

whether the device is a PMOS or an NMOS unit; however, we shall assume
throughout this chapter that NMOS is being employed. In this case the supply
voltage V_{DD} is positive, and an FET will be conducting when its gate voltage
is positive and will be cut off when the gate voltage is zero (at ground). We
assume, further, positive logic, so that an FET is conducting when its gate is
at logic **1** and not conducting when its gate is at logic **0**, which we shall take to
be nominally ground.

Returning now to the circuit of Fig. 12.3-1, let the voltage ϕ on the gate of
$T1$ go to logic **1** briefly. (We assume here, as in our subsequent discussions,
that logic **1** is a high enough voltage to ensure that when a gate is at this level,
the FET is ON even if its source is not at ground.) Then the capacitor C will
charge to the logic level on the input data line during the interval when $\phi = \mathbf{1}$.
(We assume the data line holds fixed during this interval.) The transistor $T1$ is
serving as a switch, called a *transmission gate* (see Sec. 13.5) to connect C to the
input data line.

When $T1$ turns OFF, the inverter formed by $T2$ and $T3$ will *remember* the
sampled data because of the charge stored on the capacitor. The capacitor is
shown dashed to indicate that it represents stray, incidental capacitance present
at the gate of a transistor and not a capacitor deliberately introduced.
Typically C may be in the range of 0.5 pF. Since the present circuit *stores*
or remembers a bit of data, it serves the same function as a flip-flop. However,
in comparison with the flip-flop the present circuit has the limitation that its
memory is short-lived. The charge on C will eventually leak off. This loss
of charge is due in some small extent to leakage through the insulation which
supports the gate on the transistor $T2$. To a much larger extent it is due to the
leakage through the reverse-biased junction formed between the substrate and the
drain of $T1$. Typically, it turns out that the capacitor is able to hold adequate
charge for an interval which is of the order of magnitude of about 1 ms. If bipolar devices were used in the present circuit, the storage time would be many
orders of magnitude smaller because of the relatively large base current required to
drive such a bipolar transistor. With bipolar transistors the storage time is so
small that the present circuit is generally entirely ineffective.

A shift-register stage can be constructed, as shown in Fig. 12.3-2, from

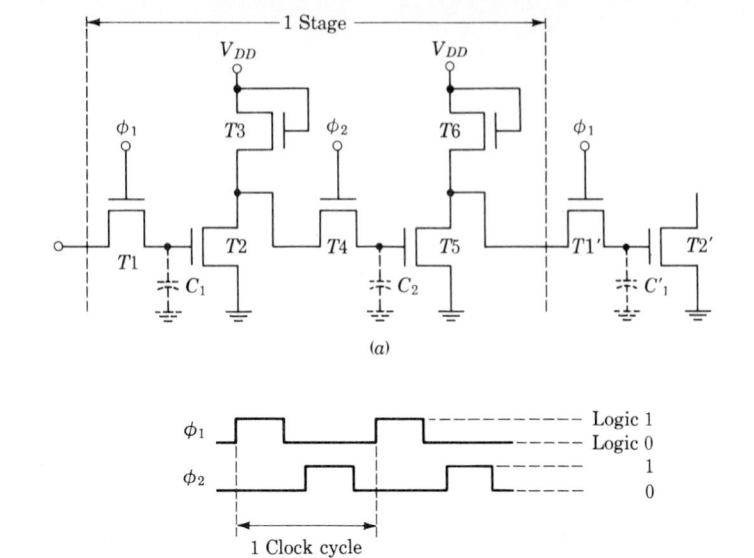

(a)

(b)

FIGURE 12.3-2
An MOS dynamic shift-register stage.

two inverter stages as in Fig. 12.3-1. Two gating voltages ϕ_1 and ϕ_2 for transmission gates $T1$ and $T4$ are required. These gating waveforms, as shown in Fig. 12.3-2b, are phased such that $T1$ and $T4$ will not be ON simultaneously. When $\phi = 1$, the input data bit to the stage is transferred to C_1. At the same time the complement of the bit stored on C_2 is transferred to C_1', the input storage capacitor of the next register stage. While $\phi_1 = 1$, $\phi_2 = 0$; hence, while the new input bit is being stored on C_1 and possibly changing the output of the input inverter ($T2$ and $T3$), this new input bit will not affect the output stage or the bit being transferred to the next stage. When ϕ_2 becomes $\phi_2 = 1$, the complement of the bit on C_1 will be transferred to C_2. The similarity between the operation of the present circuit and the operation of a master-slave flip-flop is apparent. The first inverter stage with its storage capacitor serves as the master flip-flop, and the second inverter stage with its storage capacitor serves as the slave flip-flop. The fact that two phases of gating voltage are required is to be compared with the fact that in the master-slave flip-flop the clock waveforms for master and slave are complementary waveforms.

It must be remembered, however, that the circuit shown in Fig. 12.3-2 is not really a flip-flop. For, if the time between ϕ_1 going to the **0** state and then to the **1** state is too great, the charge stored on capacitor C_1 will have leaked off. To avoid this possibility the shift register is operated "dynamically," and the maximum period of ϕ_1 (and ϕ_2) is typically less than 1 ms. Thus, this type of shift register is referred to as a *dynamic shift register*.

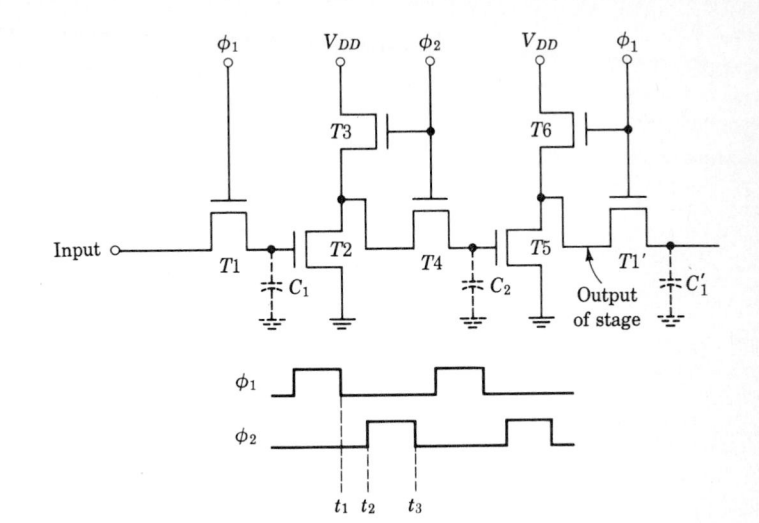

FIGURE 12.3-3
An MOS dynamic shift-register stage with load-transistor clocking.

In the circuit of Fig. 12.3-2, if the bit stored on C_1 is **1**, then during an entire bit interval $T2$ and $T3$ will be conducting, dissipating power, and drawing current from the supply V_{DD}. If the bit is a **0**, then $T3$ and $T4$ will not conduct but there will then be an interval equal to the duration of a clock cycle during which C_2 will be storing a **1** and during which $T5$ and $T6$ will conduct and dissipate power. Power dissipation can be significantly reduced through the circuit modification shown in Fig. 12.3-3. Here the gates of the inverter are not held at V_{DD} but are clocked so that $T3$ conducts only when $T4$ conducts and $T6$ conducts only when $T1'$ conducts. If the bit stored on C_1 or C_2 is a **0**, then the corresponding inverter will not conduct. If, however, a **1** is present on C_1 or C_2, then the corresponding inverter will conduct, but only during the time interval when either ϕ_1 or ϕ_2 is at logic **1**. The waveform ϕ_2 (or ϕ_1), however, need be held at logic **1** only long enough to allow C_2 (or C_1) to charge from V_{DD} through $T3$ and $T4$ (or through two other corresponding transistors). It turns out that this charging time is generally appreciably smaller than a full clock cycle (especially at low clock rates), and hence, the *clocked-load* arrangement of Fig. 12.3-3 can effect a good reduction in power dissipation. With the clocked load, however, the capacitive load on the clock driver is increased, and in the fabrication of the integrated-circuit chip there is the increased topological complexity occasioned by the need to distribute the external clock to all the load gates. As a result of the large current requirement, it often turns out to be necessary that the clock driver be a TTL circuit.

A variation of the clocked-load register stage is shown in Fig. 12.3-4. Here the clock for the transmission gates ($T1$, $T4$, etc.) is applied to the load gate of the succeeding inverter rather than, as in Fig. 12.3-3, to the load gate of the

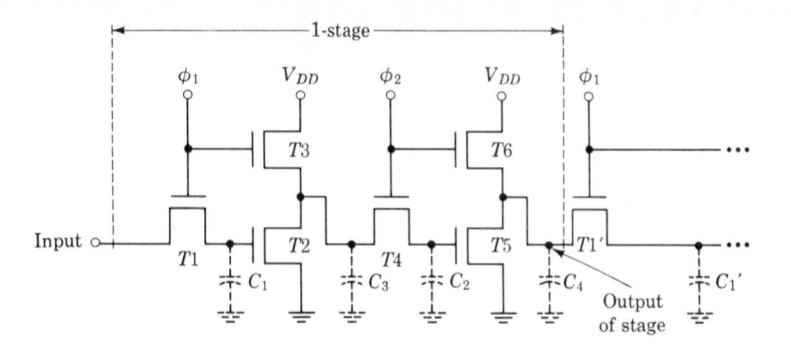

FIGURE 12.3-4
An alternative MOS dynamic shift-register stage with load-transistor clocking.

preceding inverter. In the present case the circuit operation requires the presence of an additional capacitor C_3 at the output of the inverter. When $\phi_1 = 1$, the input data bit is transferred to C_1 and its complement to C_3. When $\phi_2 = 1$, the bit on C_3 is transferred to C_2 and its complement to C_4. The input data bit is now stored on C_4 which is the input of the next stage.

To see the need for C_3 consider again the transfer of the input data bit through the first inverter and through transmission gate $T4$ to C_2. Assume that the input bit is a **0** and that C_3 is not present. Then when ϕ_1 becomes $\phi_1 = 1$, the complement of the input data bit will appear at the output of the inverter. But when ϕ_1 returns to $\phi_1 = 0$, both $T3$ and $T2$ become nonconducting and the inverter output is isolated from its input. If, however, C_3 is present, then C_3 can store the input data complement. Then when ϕ_2 goes to $\phi_2 = 1$, the logic level **1** can be transferred to C_2 through $T4$. Since C_3 must charge C_2 and still maintain a voltage which is comfortably in the range of logic **1**, then, as a matter of fact it is required that $C_3 \gg C_2$. As can be verified, for the case where the input data bit is a **1**, the presence of C_3 is not required.

12.4 TWO-PHASE RATIOLESS SHIFT REGISTER

The shift-register stages of the preceding section involve, in every case, straight-forward inverter stages. As was discussed in Sec. 8.3, such stages are composed of a driver FET and a load FET. It is required that there be a large *ratio* between the channel resistance of the driver and the channel resistance of the load, that is, $(W/L)_D \gg (W/L)_L$, to ensure a narrow transition region. As was also noted, such ratio-type inverter stages are obtained by making the load-FET channel long and narrow in comparison with the driver-FET channel. Correspondingly, the load FET occupies a large area of the integrated-circuit chip in comparison with the area of the driver. Further, because of the high

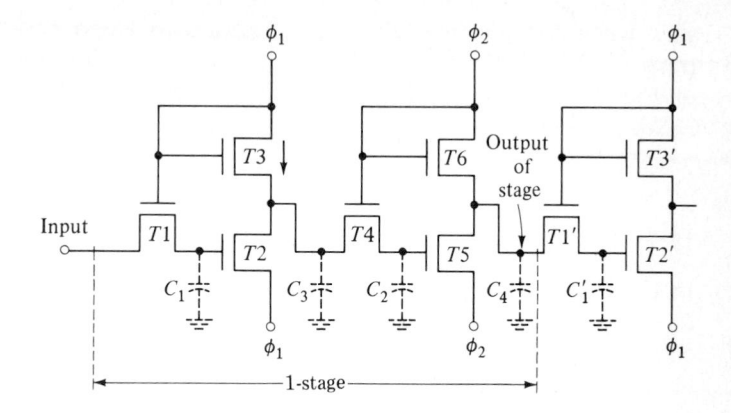

FIGURE 12.4-1
A ratioless dynamic shift-register stage.

resistance of the load channel, the speed of operation of the inverter is limited. For in any complete cycle of operation there will be an interval when some capacitance will have to charge through this high resistance. Both these disadvantages, in the matter of area and operating speed, can be relieved by devising a scheme which allows the inverter to be replaced by two low-resistance FETs of the same geometry.

Such a *ratioless* shift register stage is shown in Fig. 12.4-1. Observe that this circuit is similar to that shown in Fig. 12.3-4 except that both supply voltage and ground connections have been replaced by connections to the clock phases. In the ensuing discussion it will be convenient to think that $\phi = \mathbf{0}$ means that ϕ is at 0 V and that when $\phi = \mathbf{1}$, ϕ is at some positive voltage V_{DD} (say 10 V). Similarly, let the data to be transferred through the register stage have logic levels **0** and **1**, which are 0 and V_{DD}, respectively.

Now let the input data bit be at V_{DD}. Then when ϕ_1 goes to V_{DD}, C_1 will charge to V_{DD}. Since with ϕ_1 at V_{DD}, $T3$ will also be ON and, $T3$ will charge capacitor C_3 to V_{DD}. Note that during the operation of charging C_3 transistor $T2$ is OFF, since at no time does the gate voltage of $T2$ exceed the source voltage of $T2$ by the threshold voltage V_T. Thus the combination of $T2$ and $T3$ does not operate in the manner of an inverter.

When ϕ_1 returns to 0 V, $T3$ goes OFF and $T2$ goes ON because the charge stored on C_1 causes $V_{GS}(T2)$ to exceed the threshold voltage. As a result C_3 discharges through $T2$, and the capacitor C_3 is left at 0 V. The overall result is that *after* the clock pulse, ϕ_1 has returned to 0 V, the bit stored on C_3 will be the complement of the input bit. Hence, the net effect is the same as if the combination $T2$, $T3$ were indeed an inverter. It can similarly be verified that if the input data bit were a **0**, an inversion would again take place. In a similar way, when ϕ_2 goes through its cycle from **0** to **1** and back to **0**, it will have transferred to C_4 the complement of the bit on C_3. The ratioless register stage of Fig. 12.4-1 places a heavy capacitive burden on the clock signal source because the clock must supply current to charge the capacitors.

12.5   FOUR-PHASE RATIOLESS REGISTER STAGE

In the circuit of Fig. 12.4-1 (as in the circuit of Fig. 12.3-4) we require that $C_3 \gg C_2$ since C_2 must charge from C_3 without causing appreciable voltage change. The large capacitance of C_3 has a twofold disadvantage: (1) it places a limitation on the speed of operation and (2) it requires that, in fabrication, a relatively large area on the integrated-circuit chip be devoted to that capacitor. The scheme shown in Fig. 12.5-1a avoids this difficulty because it does not require that any capacitor take its charge from another capacitor. It has the inconvenience, however, of requiring a four-phase clock waveform. The clock

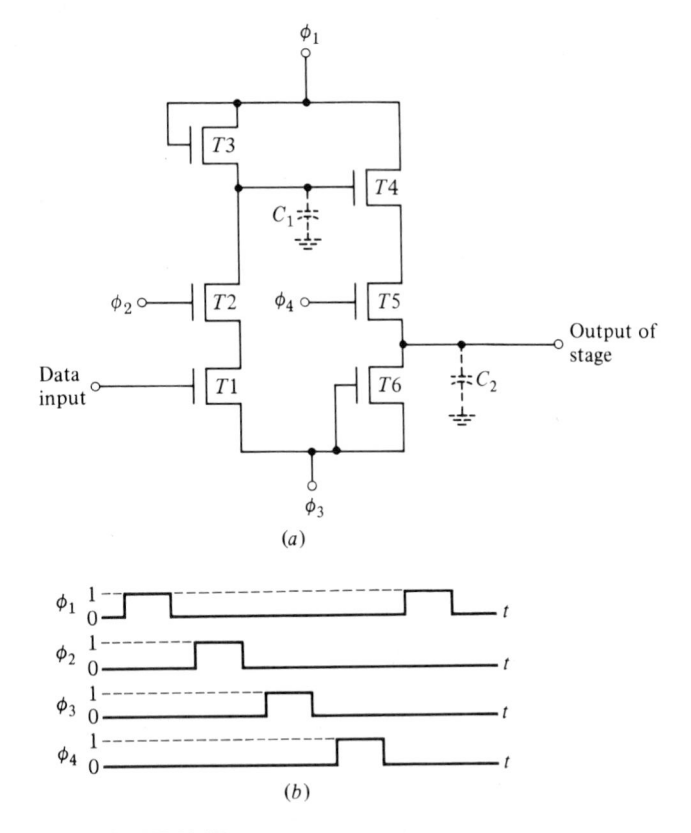

FIGURE 12.5-1
(a) A four-phase ratioless register stage. (b) Clocking waveforms.

waveforms are shown in Fig. 12.5-1b. During the interval when $\phi_1 = 1$, C_1 charges to logic **1**. Such is the case, no matter what the input, because $\phi_2 = 0$ and therefore $T2$ is OFF. This logic level **1** remains stored on C_1 after ϕ_1 returns to $\phi_1 = 0$. Assume now that the input is at logic **1**. Then, when ϕ_2 becomes **1** and turns $T2$ ON, C_1 will discharge through $T2$ and $T1$, leaving C_1 at logic **0**. Thus, we now have stored on C_1 the complement of the input bit. It may similarly be verified that if the input bit is **0**, then when ϕ_2 becomes **1**, $T1$ remains cut off and the bit stored on C_1 will again be the complement of the input bit. In a similar manner, through the operations of the clock phases ϕ_3 and ϕ_4, the complement of the level stored on C_1 will be transferred to C_2 and hence to the input of the next register stage. This circuit suffers from the disadvantage that ϕ_1 must supply the current to charge C_1 and ϕ_3 must charge C_2.

12.6 CMOS REGISTER STAGES

Like MOS register stages, CMOS register stages are constructed from inverter stages and transmission gates (taking advantage as well of stray capacitances). The CMOS inverter was discussed in detail in Secs. 1.15 and 8.4 and is shown again in Fig. 12.6-1. For the convenience of being specific, let us assume a supply voltage $V_{SS} = +10$ V. In such a system the input V_i and output V_o will be waveforms which make excursions between 0 V (ground) and 10 V. Let us then turn our attention to the matter of a transmission gate intended for use with voltages in this 0- to 10-V range. A clock waveform is required when a transmission gate is employed. We shall assume that the clock ϕ also makes excursions between 0 V and 10 V.

Consider, first, the MOS transmission-gate arrangement of Fig. 12.6-2a using a single N-channel device. The gate is being used to isolate the capacitor from, or to connect the capacitor to, an input V_i which is $V_i = 0$ V or $V_i = 10$ V. Assume that the threshold voltage of the transistor is $V_T = 2$ V. Starting with $V_i = 0$ V and the capacitor voltage $V_C = 0$ V, suppose that V_i becomes $V_i = 10$ V but that $V_G = 0$ V. Then the transistor will remain OFF since

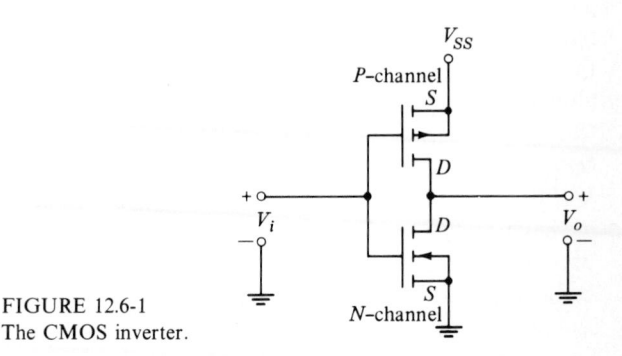

FIGURE 12.6-1
The CMOS inverter.

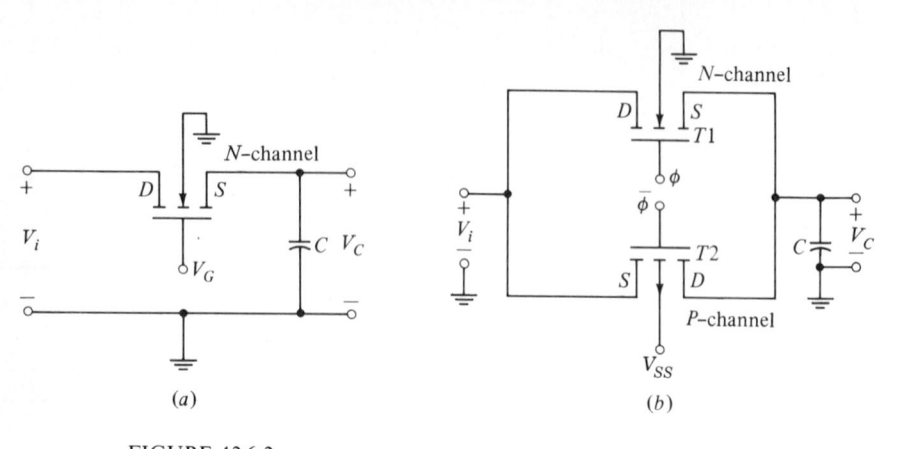

FIGURE 12.6-2
(a) An MOS transmission gate used to charge a capacitor from a voltage source.
(b) A CMOS transmission gate.

$V_{GS} < V_T$ and the capacitor will not charge. Now let V_G become $V_G = 10$ V. Then current will flow from left to right through the transistor. Since the transistor is an N-channel MOS, the current direction establishes that the left side of the transistor is the drain and the right side is the source, as indicated in the figure. Since the threshold voltage (which is the gate-to-source voltage V_{GS} at cutoff) is 2 V, the transistor will stop conducting when V_C becomes 8 V since at that point $V_{GS} = V_G - V_C = 10 - 8 = 2$ V. If now, with V_G still at $V_G = 10$ V, the input should become $V_i = 0$ V, the current direction will reverse, the source and drain will reverse, and the capacitor will be able to discharge completely to 0 V. Altogether we have the result that in this system with logic levels separated by 10 V, our single-transistor transmission gate is able to transmit only an 8-V change. Such a limitation applies to all the transmission gates employed in the dynamic MOS register stages discussed in the preceding section. This characteristic, of course, does not preclude the gate from being effective or the register stages from operating, but it does reduce noise margins.

The limitation of the MOS transmission gate is remedied in the CMOS gate, shown in Fig. 12.6-2b. Here the two transistors of a CMOS pair are paralleled, and the gating voltages applied are complementary. Return now to the case where $V_i = 10$ V and the capacitor is in the process of charging. Current through both transistors flows again from left to right. But since transistor $T2$ is a P-channel MOS, its source is at the left and its drain to the right, as indicated. When $\phi = 10$ V and $V_C = 8$ V, we have, as before, $V_{GS}(T1) = 10 - 8 = 2$ V, so that $T1$ is at the point of cutoff. However, for $T2$ we have $V_{SG}(T2) = 10 - \bar{\phi} = 10 - 0 = 10$ V. Since $V_T = 2$ V, transistor $T2$ will be ON as long as $\bar{\phi}$ is less than V_i by more than 2 V. Thus, the capacitor can charge to the full voltage $V_C = V_i = 10$ V since $T2$ remains on.

A CMOS register stage is shown in Fig. 12.6-3. It consists, like the most

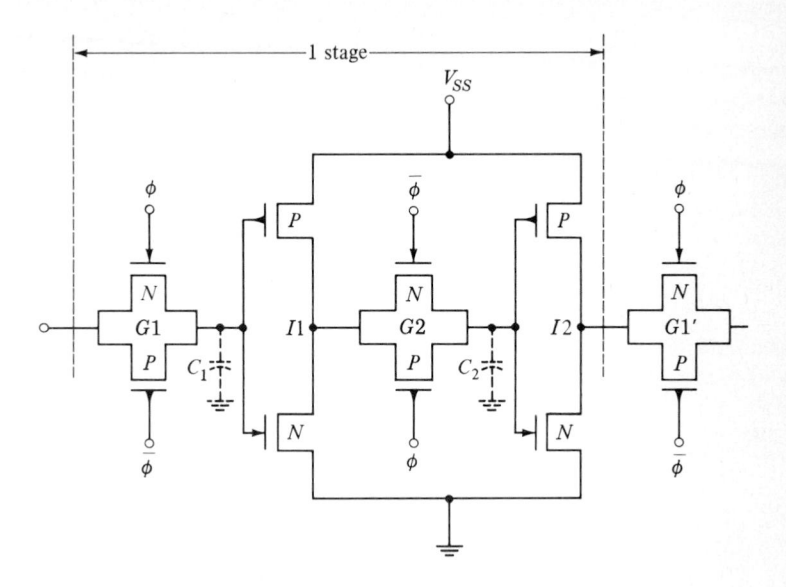

FIGURE 12.6-3
A CMOS register stage consisting of two transmission gates and two inverters.

elementary of the MOS register stages (Fig. 12.3-2), of two transmission gates $G1$ and $G2$, two inverters $I1$ and $I2$, and the two capacitors C_1 and C_2. In this figure we have again omitted the substrate connections for simplicity. Observe particularly the reversal of gate phases at alternate gates. The functional operation of this CMOS stage is so similar to the register stage of Fig. 12.3-2 that we need not repeat the description.

We noted that the MOS stage of Fig. 12.3-2 suffered disadvantages due to the fact that the inverters were ratioed and inverter stages would drain a direct current from the V_{SS} supply whenever the inverter gate was at logic **1**. Thereafter we considered a series of modifications designed to remedy these difficulties. We may, then, note that the CMOS inverter is "ratioless" and does not draw a steady current from the V_{SS} supply. On the other hand, a CMOS inverter and a CMOS gate use more area on a silion integrated-circuit chip than a MOS inverter or gate.

12.7 STATIC SHIFT-REGISTER STAGE

The dynamic register stages of the previous section will not operate if the clock interval is longer than the interval over which the stray capacitance is able to store a charge. When a shift register must operate at very low frequencies the dynamic stages must be replaced by static stages.

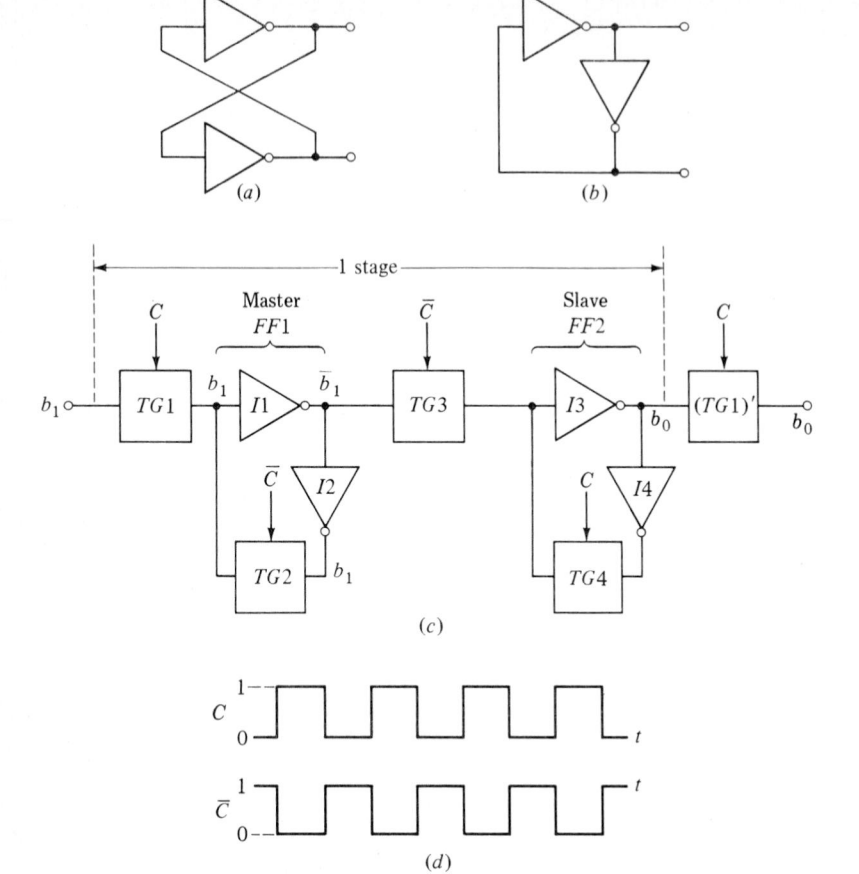

FIGURE 12.7-1
(a) A basic flip-flop with two cross-coupled inverters. (b) The basic flip-flop redrawn. (c) An MOS or CMOS static shift-register stage. (d) The clocking waveforms required in (c).

A conventional flip-flop reduced to its essentials is shown in Fig. 12.7-1a. It is redrawn in Fig. 12.7-1b to make it apparent that the circuit consists of a cascade of two inverters with the output fed back to the input. The circuit of a MOS or CMOS static shift-register stage is shown in Fig. 12.7-1c. Similar circuits were discussed in Sec. 9.21. Note that we have here a master and a slave flip-flop modified by the introduction of transmission gates ($TG2$ and $TG4$) in the feedback paths. Each shift-register stage has two additional transmission gates ($TG1$ and $TG3$) interposed between flip-flops. The circuit diagram of the inverter stages has been given for the CMOS in Fig. 12.6-1, and the diagram of the transmission gates for the CMOS is shown in Fig. 12.6-2b. A MOS trans-mission gate and inverter is shown in Fig. 12.3-1. Following conventional practice, the clocking waveform and its complement are here given as C and $\bar{C}$. These clocking waveforms are shown in Fig. 12.7-1d. The clock inputs of the gates have been labeled to make it clear that $TG1$ and $TG4$ open and close at the same time as $TG2$ and $TG3$. And, of course, when $TG1$ and $TG4$ allow transmission, $TG2$ and $TG3$ do not and vice versa. We shall assume that the clock pulses are applied to the gates in such manner that when $C = \mathbf{1}$ ($\bar{C} = \mathbf{0}$), the gates allow transmission (are short circuits between input and output) and do not allow transmission when $C = \mathbf{0}$. Viewing the gates as switches, we can say that the switch (gate) is CLOSED when $C = \mathbf{1}$ and the switch (gate) is OPEN when $C = \mathbf{0}$. Furthermore while only one clock pulse is shown clocking each transmission gate, it must be understood that in a CMOS system both C and $\bar{C}$ are present.

Let us start at a time when $C = \mathbf{1}$. Then the input data bit b_1 is being presented to $FF1$, but this bit will not be stored in $FF1$ because its feedback path is not closed. The feedback path of $FF2$ is closed, and there is a bit b_0 in storage there. This bit, whose logic level depends on the past history, is being presented through $TG1'$ to the first flip-flop of the next register stage. Note also that $TG3$ is open so that $FF2$ is not influenced by anything to the left of $TG3$. Now let C go to $C = \mathbf{0}$. Then $TG1$ opens. The gate $TG2$ closes, completing the flip-flop circuit and putting the bit b_1 in storage in $FF1$. The bit b_1 at the output of inverter $I2$ has been transmitted through $TG2$ and replaces the previous bit b_0, which was originally transmitted through $TG1'$. We may wonder whether any timing difficulty might arise because of the finite propagation time through $TG2$. No such problem arises because the inevitable stray capacitance at the input to $I1$ will hold the bit b_1 long enough to await the transmission through $TG2$. Note that here we depend on the stray capacitance to hold a charge only for an interval equal to the gate propagation time and not for the interval of a clock cycle. Hence the clock frequency has no bearing on our present discussion.

At the moment C becomes $C = \mathbf{0}$ and the bit b_1 is being stored in $FF1$, $TG3$ closes, and, after a propagation time (through $TG3$ and $I3$), the bit b_1 appears as a new bit at the input to $TG1'$. But at this time $TG1'$ is OPEN. When C returns to $C = \mathbf{1}$, the feedback path in $FF2$ will close and the bit will remain in storage in $FF2$.

The stage has two important properties relevant to our discussion of registers: (1) if the clock should stop at any point, the bit in the stage will remain in storage, either in $FF1$ or in $FF2$, depending on whether the clock should stop at $C = \mathbf{0}$ or $C = \mathbf{1}$; and (2) as is required generally when register stages are cascaded, the output of one stage changes at that point in the clock cycle when the input gate switch of the next stage is OPEN, thereby disallowing transmission.

12.8 A THREE-PHASE STATIC REGISTER STAGE

In addition to transmission gates, the static shift register of Fig. 12.7-1 has two flip-flops, $FF1$ and $FF2$, each involving two inverters. A static register stage still suitable for service in a shift register can be constructed which uses just a

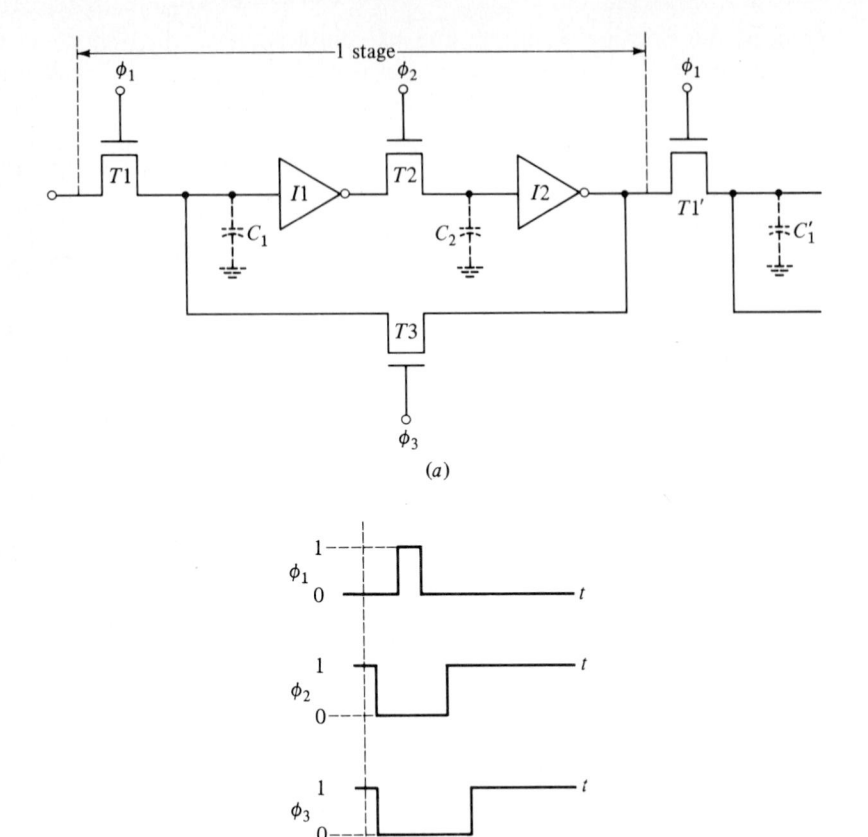

FIGURE 12.8-1
(a) A three-phase static register stage. (b) Clocking waveforms.

single flip-flop, albeit with some added complication in clocking. The circuit is shown in Fig. 12.8-1a. Here, for simplicity, we have shown the transmission gates as MOSFETS. However, CMOS gates can also be employed. The transistors $T1$, $T2$, and $T3$ in the present circuit are being used as transmission gates. As before, we assume that the gate allows transmission, i.e., the switch is CLOSED, when the corresponding clock signal is at logic **1** and the switch is OPEN when the clock is at logic **0**. A three-phase clock is required.

Consider, then, as shown in Fig. 12.8-1b, that at $t = 0$, $\phi_1 = 0$ while $\phi_2 = \phi_3 = 1$. Then the circuit loop, involving the inverters $I1$ and $I2$, is CLOSED and constitutes a flip-flop. Since the two transmission gates $T1$ and $T1'$ are open, this flip-flop is isolated from the preceding and succeeding stages. There will be a bit registered in the flip-flop; whether it is a **1** or **0** will be determined by the past history. Suppose that while ϕ_1 remains at $\phi_1 = 0$ we briefly set

$\phi_2 = \phi_3 = 0$. Then the bit in storage in the flip-flop will be preserved on capacitors C_1 and C_2. If we return $\phi_2 = \phi_3 = 1$ before the charges on C_1 and C_2 have changed appreciably, the flip-flop state will not have changed.

Now let ϕ_2 and ϕ_3 go to logic **0**, as in Fig. 12.8-1b. (In the figure it is indicated that ϕ_2 and ϕ_3 go to logic **0** simultaneously. This feature is not essential; nor is it important which clock signal goes to **0** first.) Suppose that shortly after ϕ_2 and ϕ_3 both become **0**, ϕ_1 becomes $\phi_1 = 1$. Then because of the voltage held on C_2 the state of the flip-flop, i.e., the logic level held at the output of I2, will be transferred to the capacitor C_1' of the next flip-flop. Similarly, the state of the preceding flip-flop will be transferred to C_1. Now let ϕ_1 return to **0** and bit ϕ_2 be *first* to go back to logic level **1**, thereby closing switch T2. Then whatever may have been the previous state of C_2 it will now go to the state complementary to C_1. Finally, when ϕ_3 goes back to $\phi_3 = 1$, the flip-flop circuit loop is again complete and the flip-flop will hold its new state indefinitely. The order of the reclosing of switches T2 and T3 is important. For, as can be verified, if ϕ_3 recloses first, the flip-flop will revert to its original state.

12.9 THE READ-ONLY MEMORY

A read-only memory (ROM) is a memory device intended to store information which is fixed. That is, there is an initial operation during which information is written into the memory and thereafter the memory is *read only* and is not again written into. Generally the information in the memory is placed there by the manufacturer of the device. There are, however, memories which allow the user to establish the store of information in the memory. Such memories are referred to as programmable memories (PROM). There are also ROMs in which the stored information can be changed. However, in such cases the writing operation is accomplished in a time which is many orders of magnitude larger than the time required to read. Such memory devices (described as *erasable*) are read-only memories in the sense that to change the stored information it is necessary to interrupt the digital processing in which the memory is involved.

A most important attribute of the ROM is that the information it stores will not be lost if the electric power that it uses to operate should be interrupted. Such memories are referred to as *nonvolatile*. In contrast, the sequential memories already discussed and the random-access memories to be discussed (beginning in Sec. 12.13) are both *volatile* memories.

The ROM has many applications in a digital system. It can be used to provide the realization of an arbitrary truth table. Whenever a truth table involves enough input and output logical variables for its physical implementation to require a large number of gates, a ROM may well be more economical in size, weight, and cost. ROMs are widely used in code conversion and in connection with alphanumeric displays. The ROM is also used to yield results

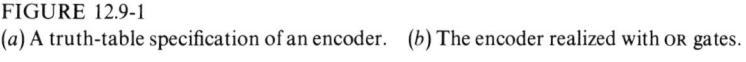

Z_0	Z_1	Z_2	W_0	W_1	W_2	W_3	W_4
0	0	1	1	0	1	1	1
0	1	0	0	1	1	0	1
1	0	0	1	1	1	1	0

(a)   (b)

FIGURE 12.9-1
(a) A truth-table specification of an encoder.   (b) The encoder realized with OR gates.

that would otherwise be achieved by a computation involving a sequence of arithmetic operations, e.g., multiplication, division, or evaluation of a trigonometric function (as sin x or ln x), or it may be used as a function generator.

A ROM is an *encoder*. An encoder is a logical-gate structure which has M inputs $Z_0, Z_1, \ldots, Z_{M-1}$ and K outputs $W_0, W_1, \ldots, W_{K-1}$. It is intended that at any time an individual input, say Z_i, is to be singled out, we set $Z_i = 1$ while all other inputs are at logic **0**. (Alternatively we may have $Z_i = 0$ while all other outputs are at logic **1**.) Corresponding to each input Z_i, which may be set to $Z_i = 1$, the K outputs will take on the logic levels $W_0(i)$, $W_1(i)$, $\ldots$, $W_{K-1}(i)$. That is, the encoder accepts the input Z_i (only) $= 1$ and identifies this situation through the *code* word $W_0(i)$, $W_1(i)$, $\ldots$, $W_{K-1}(i)$; or if the encoder is viewed in its application as a memory, the ith storage location in the memory is *addressed* by setting Z_i (only) $= 1$ and the ROM responds by presenting at its output the word stored in that location. A logical structure of an encoder is illustrated in Fig. 12.9-1. The truth-table specification is given in Fig. 12.9-1a, and its realization in logic gates is shown in Fig. 12.9-1b. The ROM illustrated stores three words, each of 5 bits.

Rather generally the address of a stored word is itself available in a digital system as a binary coded word. It is therefore necessary to interpose between the binary-coded address word and the ROM a device which takes note of the address and singles out a corresponding individual line. Such a device performs a function which is inverse to the function performed by the encoder and is hence called a *decoder*. A truth table for a decoder is given in Fig. 12.9-2a. The address word has 2 bits; hence four address words are available. A realization of this decoder with logical gates is shown in Fig. 12.9-2b. Note that in Fig. 12.9-1 only three of the four possible Z_i are being used.

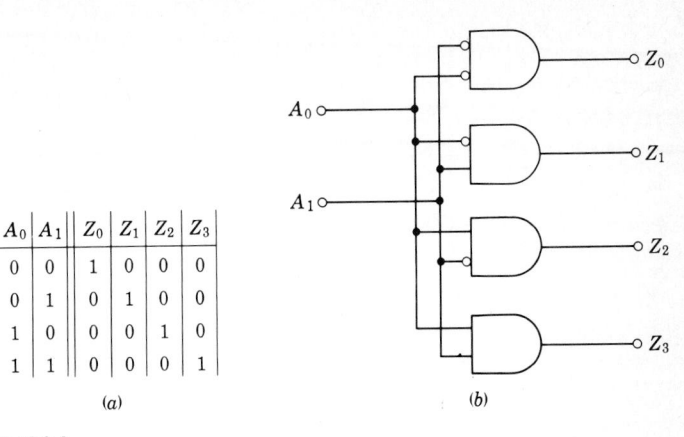

A_0	A_1	Z_0	Z_1	Z_2	Z_3
0	0	1	0	0	0
0	1	0	1	0	0
1	0	0	0	1	0
1	1	0	0	0	1

(a)   (b)

FIGURE 12.9-2
(a) A truth table for a four-word decoder. (b) The decoder realized with AND gates.

As a convenience to the user, manufacturers generally provide decoders as an integral adjunct of the ROM encoder. Such a memory device with decoder included, as indicated in Fig. 12.9-3, are generally referred to as a *decoded ROM*. In providing the decoder, the manufacturer does himself a service as well. For on an integrated-circuit chip there is generally a limitation to the number of pins which can conveniently be provided for external connection. And on an M-word memory without decoder, M pins would have to be provided for addressing purposes; while with the decoder N pins, with $2^N = M$, are adequate.

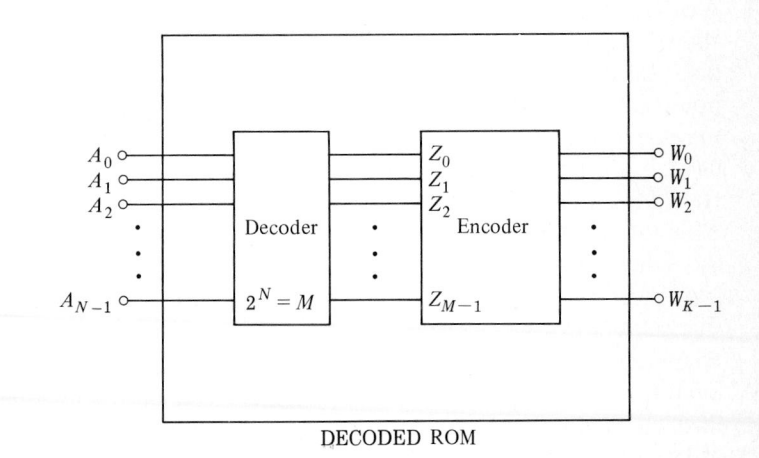

DECODED ROM

FIGURE 12.9-3
A ROM with decoder included.

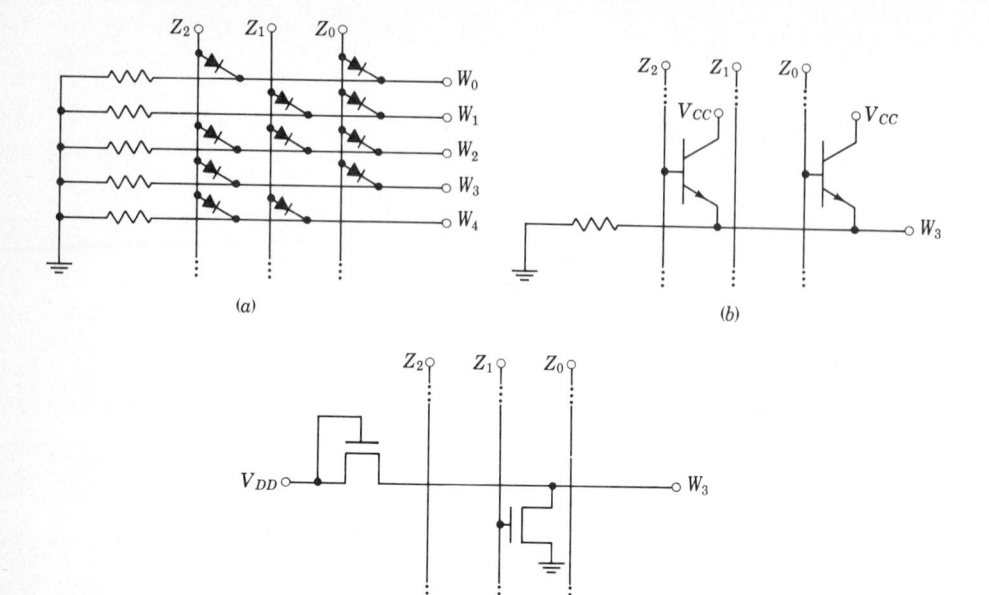

FIGURE 12.10-1

(a) A three-word, 5-bit ROM using diode connections between address and bit lines.
(b) Single-bit line of the ROM in (a) illustrating the use of bipolar transistors in
place of diodes. (c) The use of MOS transistors illustrated.

12.10 IMPLEMENTATION OF ROMs

A possible implementation of the ROM OR-gate encoder shown in Fig. 12.9-1b is
shown in Fig. 12.10-1a. It can be verified (Prob. 12.9-2) that the circuit shown
does indeed constitute precisely the OR-gate logic structure. The use of diodes,
however, has a drawback in that the inputs to the encoder must directly supply
the current which must be delivered at the outputs. This difficulty is relieved by
using bipolar transistors in place of diodes in the manner shown in Fig. 12.10-1b.
Here, for simplicity, we have indicated only the connections to bit output line W_3.
Now the output current may be supplied principally by the V_{CC} source. Since,
as we have noted, in bipolar technology, integrated-circuit diodes are generally
transistors with collector and base terminal joined, fabrication in the manner of
Fig. 12.10-1b is not substantially more complicated than the pattern in Fig.
12.10-1a. It is to be observed further that all the transistors connected to an
input line Z_i will have their collector terminals connected in common to V_{CC} and
their bases connected in common to the Z_i line. Hence such a vertical array
of transistors is not infrequently replaced by a single transistor with multiple
emitters. ROMs employing bipolar technology may have access times as low as
15 ns. The *access time* of a memory is the length of time that must elapse

after addressing before the addressed word is made available at the output of the memory.

The number of bits in a memory is defined to be equal to the product of the number of words and the number of bits per word. (In a ROM like that in Fig. 12.10-1a the number of diodes will generally be about one-half the number of bits, since we may expect 1s and 0s to occur with about equal frequency.) When the number of bits is large, say in excess of 1,000, MOS technology is preferred to bipolar technology since MOS devices are more economical of silicon-die real estate. The implementation of a single-bit line (again, for case of comparison, the $W3$ output bit line) is shown in Fig. 12.10-1c. Assuming positive logic as usual and assuming NMOS, we see that $W3$ is at logic **0** only when $Z_1 = 1$, as required by the truth table of Fig. 12.9-1a. Access times using MOS devices are typically about 400 ns.

12.11 PROGRAMMABLE AND ERASABLE ROMs

In a ROM, the address lines and the output word bit lines form a crossed array of lines, i.e., a grid structure. At each grid intersection is placed a device (diode, bipolar, or MOS transistor) or not, depending on whether the corresponding word bit is to be **1** or **0**. (In cases where there is no special interest in the type of device, the coupling between address line and bit line is often shown simply by a dot at the grid intersection.) In a programmable ROM (PROM) the manufacturer locates a connecting device at every grid intersection. However, in series with each such device there is provided a fusible link. Any particular fusible link is located at the intersection of some line Z_i and some line W_j. By making connection to Z_i and W_j and passing an adequately large current through the link, the link can be burned out. Thus, the user of such a PROM may burn out links as necessary, leaving transistors only on locations required to establish the memory storage desired.

One type of *erasable* or *alterable* ROM uses *floating-gate* PMOS transistors. These are transistors in which at normal operating voltage the gate is entirely insulated and isolated from electrical connection to any other part of the integrated-circuit chip. It turns out to be possible to establish a negative charge on these gates by the application of a high voltage between source and drain. The negative charge left on the gate by such treatment leaves the corresponding transistor with a conducting channel. The ROM can be erased by exposure to ultraviolet light, which serves to discharge any charged gate.

12.12 APPLICATIONS OF ROMs

A ROM multiplier Consider that we want to perform the arithmetic operation of multiplication. As we have seen in Sec. 11.16, multiplication can be performed by a sequence of shifting operations, i.e., multiplying by powers of 2, and a sequence of additions. On the other hand, we may view a multiplication table as a truth table. Thus, the entry in the multiplication table to effect the multiplication $11 \times 10 = 0110$, that is, $3 \times 2 = 6$, may be read to mean that a memory addressed by the input code $A_1 A_0 B_1 B_0 = 1110$ reads out the word $P_3 P_2 P_1 P_0 = 0110$.

When the multiplicands have many bits, straightforward multiplication by a ROM may get out of hand. For example, when the multiplicands each consist of 8 bits, the product contains 16 bits and the number of bits which need to be stored in the memory is 16 bits/word $\times 2^{16}$ words $\approx 10^6$. A saving in bit storage can generally be accomplished by using a number of ROMs in a manner now to be illustrated.

Let us consider that we want to multiply two 4-bit numbers $A_3 A_2 A_1 A_0$ and $B_3 B_2 B_1 B_0$. We write

$$A_3 A_2 A_1 A_0 = A_3 A_2 00 + 00 A_1 A_0 \tag{12.12-1a}$$

and

$$B_3 B_2 B_1 B_0 = B_3 B_2 00 + 00 B_1 B_0 \tag{12.12-1b}$$

the product is then

$$P = (A_3 A_2 00)(B_3 B_2 00) + (00 A_1 A_0)(B_3 B_2 00)$$
$$+ (A_3 A_2 00)(00 B_1 B_0) + (00 A_1 A_0)(00 B_1 B_0) \tag{12.12-2}$$

Note that the product P appears as the sum of four products. Four ROM multipliers are required, as indicated in Fig. 12.12-1. In these multiplications, however, the 0s in Eq. (12.12-1) are ignored. Thus, to effect the last product in Eq. (12.12-2) we use a ROM which accepts as input the address $A_1 A_0 B_1 B_0$ and yields one of 16 possible 4-bit output product words. We must now sum the outputs of the four multipliers, but in so doing we must take account of the differing numerical significance of the output bits of the different multipliers. As can be verified, the numerical significances are as given in the figure. Only one ROM yields bits of significance 2^1 and 2^0. These are brought out and become bits of the final product. Three ROMs provide bits of significance 2^2 and 2^3, and these are combined in two 2-bit adders. Similarly three ROMs provide bits of significance 2^4 and 2^5, and these are combined in two other 2-bit adders. It is left as a student exercise (Prob. 12.12-2) to show that the disposition of carry outputs from the various adders is proper to assure a correct product result.

Each of the ROMs in Fig. 12.12-1 stores 16×4 bits. The four ROMs store $16 \times 4 \times 4$ bits $= 2^8 = 256$ bits. If the multiplication had been effected by a single ROM, that ROM would need storage for 2^8 words $\times 8$ bits/word $= 2^8 \times 8$ bits. Hence the present scheme reduces the required storage capacity by a factor of 8. The saving is more impressive as the number of bits increases. When the factors to be multiplied each have 8 bits, there is a saving by a factor of $2^7 = 128$. Of course, some of this saving is dissipated on account of the need to provide five adders. In addition, the use of adders, even with a look-ahead carry feature, results in a significant increase in the time required to perform the multiplication.

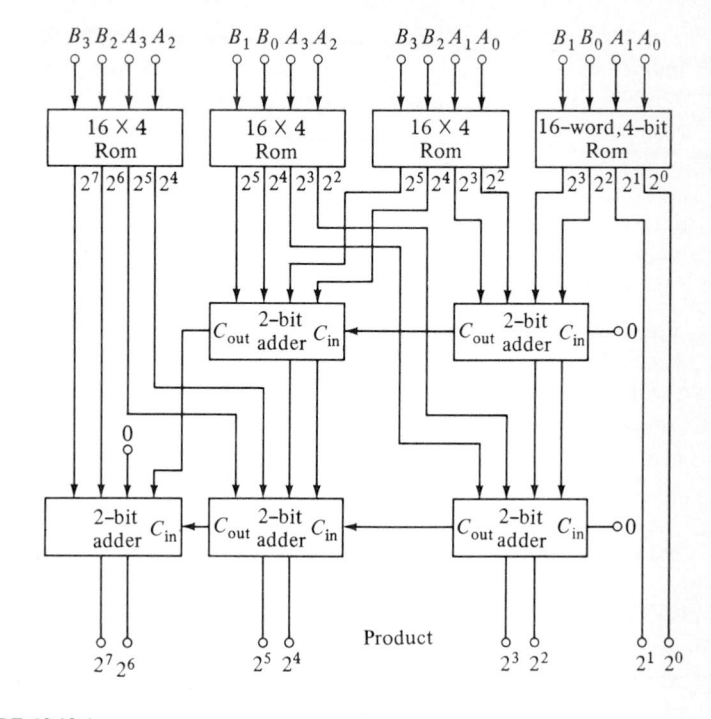

FIGURE 12.12-1
A ROM multiplier for two 4-bit numbers, illustrating a technique for reducing required storage capacity.

It is often required that a ROM multiplier produce an output having the same number of bits as each of the multiplicands. Thus, in the above example, where the inputs were each 4-bit words, the output would also be a 4-bit word. In this case, referring to Fig. 12.12-1, the output bits would consist of the four most significant bits. In this case it is possible to keep the maximum error between ± 8 by using additional circuitry (Prob. 12.12-1). If an error of 15 is permitted, the complexity of Fig. 12.12-1 can be reduced significantly to include only three ROMs and three adders.

The ROM as a *look-up* table ROMs are also widely used as look-up tables for mathematical functions such as logarithms, trigonometric functions, square roots, exponentials, etc. Techniques similar to that employed with the multiplier can be used effectively to reduce the storage capacity required of the memory. For example, consider a ROM for $\sin \theta$. If we wanted to provide for a resolution of 0.01°, then for the range $0 \le \theta \le 90°$, a 9,000-word memory would be required. We may, however, write $\theta = I + F$, where I is the integral part of θ and F is the fractional part. Then

$$\sin \theta = \sin (I + F) = \sin I \cos F + \cos I \sin F \qquad (12.12\text{-}3)$$

The implementation of Eq. (12.12-3) requires four smaller memories: two 90-word memories for sin I and cos I and two 100-word memories for cos F and sin F. Further simplifications are possible in the matter of bits per word through the recognition that cos $F \approx 1$ and sin $F \approx F$.

12.13 BIPOLAR-JUNCTION-TRANSISTOR RANDOM-ACCESS-MEMORY CELLS

The bit-storage cells in a *random-access memory* (RAM) may use BJT or may use MOS devices in either a static or dynamic mode. When high speed is required, the BJT is the RAM of choice inasmuch as it can be accessed in about 35 ns, in contrast to the MOS RAM, which has access times of approximately 400 ns. On the other hand, a BJT RAM is small, containing 1024 memory cells or less, while MOS RAMs are available with 4,096 memory cells.

In this section and in Sec. 12.14 we consider memory cells using the BJT. MOS cells are discussed beginning with Sec. 12.15.

The BJT RAM memory cell is simply a flip-flop. However some special features must be incorporated into it in order to facilitate *addressing* the cell, *writing* into it, and *reading* from it. Of course, given a flip-flop and the availability of gates as required, these three operations are easily effected. However, since useful RAMs may contain several hundred storage cells, these operations must be done with economy.

A RAM memory cell using multiple-emitter transistors is shown in Fig. 12.13-1. The cell itself is the flip-flop composed of $T1$ and $T2$. The remaining circuitry in the figure provides a mechanism for writing and reading data. This auxiliary circuitry may service additional memory cells as well, which may be bridged across the data lines; hence the dashed extensions of the data lines. Commercially available bipolar memories of the type being discussed operate with a supply voltage in the range 3.5 to 5.0 V. We may then reasonably consider that logic **0** represents any voltage less than about 0.3 V and logic **1** represents any voltage above about 3.0 V. As indicated in the figure, the signals X and Y make excursions between these logic levels. These signals are used to address the cell. As we shall now see, the cell is accessed for reading or writing when X and Y are simultaneously at logic **1**.

Let X and Y be at logic **0** (≤ 0.3 V), and let the read-write line be at logic **0** as well. Then $T3$ and $T4$ are ON, with collector voltages also at logic **0**, and diodes $D1$ and $D2$ do not conduct. If, say, the state of the flip-flop is such that $T1$ is conducting and $T2$ is OFF, then emitter current will flow in E_X and E_Y. A bias voltage of 0.5 V is applied through the resistor R_3 to emitter E_D. E_D is more positive than E_X and E_Y by at least 0.2 V, and hence E_D does not conduct. The transistors $T5$ and $T6$ are also OFF, and the $\overline{\text{data}}$ output terminal is at logic **1** and will persist at this logic level independently of the state of the flip-flop. Now let the cell be addressed by raising both X and Y to logic **1**. Then the currents through E_X and E_Y will be diverted to E_D. A component of this current

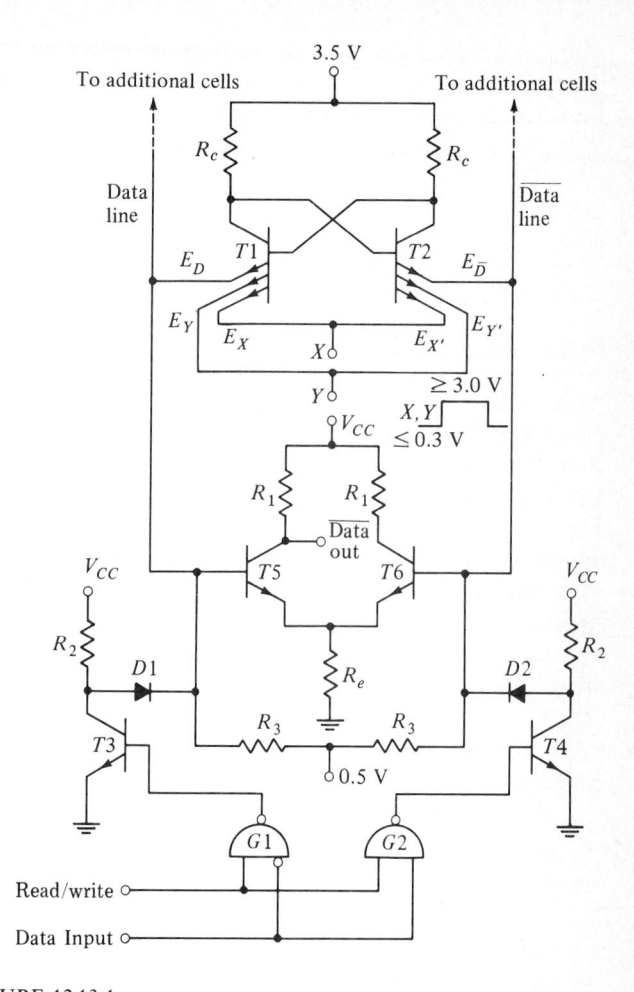

FIGURE 12.13-1
A RAM BJT memory cell. Provision is made for reading and writing.

will flow into the base of $T5$ and the $\overline{\text{data}}$ out terminal will assume the same logic level present at the collector of $T1$. Thus, with the read-write line at logic **0** (in which case gates $G1$ and $G2$ are disabled against the entry of data) the operation of addressing the cell provides a reading of the cell.

Now, again, let $X = Y = 1$, so that the flip-flop is using emitters E_D and $E_{\bar{D}}$, and suppose that the read-write line is at logic **1**. Now gates $G1$ and $G2$ are enabled, and if the data line is at logic **1**, $T3$ will remain ON but $T4$ will turn OFF. The collector of $T4$ will rise, pulling with it the emitter $E_{\bar{D}}$ of $T2$. Hence, no matter what the original state of the flip-flop, it will be forced to the state in which $T2$ does not conduct. Hence, the logic level at the collector of $T2$ will become the logic level of the data input. If, however, the cell had not been addressed, E_D and $E_{\bar{D}}$ would not have been carrying current and the flip-flop would not have responded to the writing operation. Of course, even if the cell is not addressed, the *data output* terminal will respond to the writing operation. Presumably, however, the memory is part of a system that "knows" that a writing operation is in process and will ignore the output until instructed to *read*.

The three-emitter transistor flip-flop cell of Fig. 12.13-1 is addressed when the logical product XY and $XY = 1$. When there is no need for the cell to perform this AND operation, one of the emitters may be omitted on each transistor.

12.14 OTHER BIPOLAR-TRANSISTOR MEMORY CELLS

An alternative bipolar memory cell is shown in Fig. 12.14-1. Schottky transistors ($T1$ and $T2$) and Schottky diodes ($D1$ and $D2$) are employed. In the standby condition (cell not addressed) the address line X is held at logic level **1**, which for the present cell is about 2.5 V. At this level, the flip-flop remains operational. The data lines are biased at 1.6 V so that when the cell is not addressed, the diodes ($D1$ and $D2$) are back-biased.

The cell is addressed by lowering the address line to logic **0** (≈ 0.3 V). Assume, then, that before being addressed the cell was in the state in which $T1$ was conducting and $T2$ was OFF. When X drops to 0.3 V, the base of $T1$ will drop to $0.3 + 0.75 = 1.05$ V. The forward voltage of a Schottky diode is about 0.45 V. Hence the $\overline{\text{data}}$ line will be pulled down slightly from 1.6 V to $1.05 + 0.45 = 1.5$ V. Allowing as usual, a voltage $V_{CE}(\text{sat}) = 0.2$ V across $T1$, the collector of $T1$ will be pulled down to $0.3 + 0.2 = 0.5$ V. If we neglect the

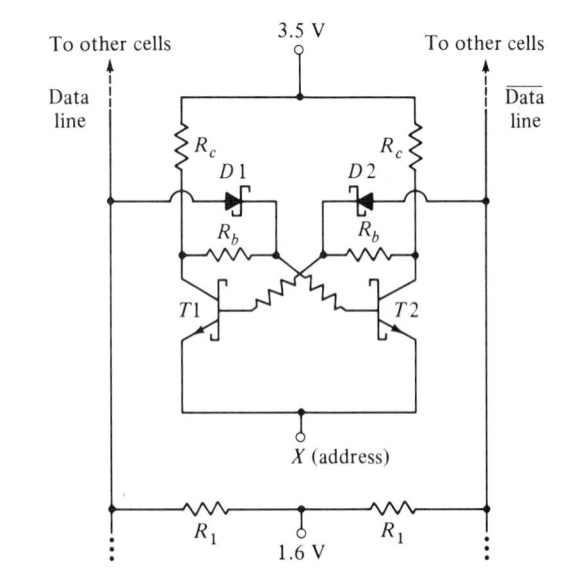

FIGURE 12.14-1
A Schottky diode bipolar-transistor memory cell.

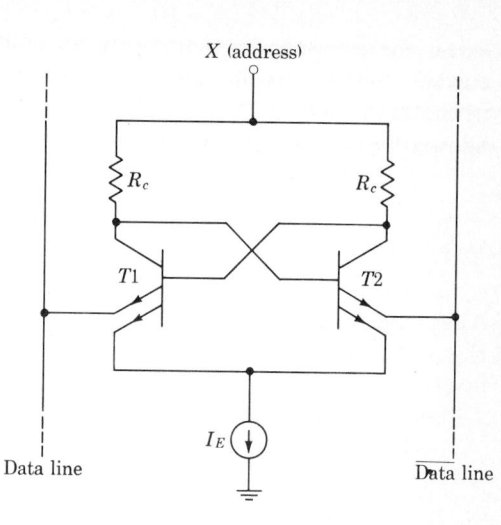

FIGURE 12.14-2
A bipolar-transistor memory cell which operates in the manner of emitter-coupled logic.

drop across the resistor R_b, we then find that the data line is pulled down rather substantially from 1.6 V to $0.5 + 0.45 = 0.95$ V. In practice, the resistor R_b is small enough (≈ 1 kΩ) to ensure that even when the drop across R_b is taken into account, the change in voltage on the data line is still much larger than the voltage change on the $\overline{\text{data}}$ line. The difference in the voltage change between the data line and the $\overline{\text{data}}$ line is used to read the cell.

It is important to remember that the data and $\overline{\text{data}}$ lines go to many other cells. When the cell shown in the figure is addressed, the other cells are not and diodes $D1$ and $D2$ of these other cells are cut off, disconnecting the other cells from the data and $\overline{\text{data}}$ lines.

To write into the cell we first address it by setting X to 0.3 V. Assume again that initially $T1$ is ON and that we want $T2$ to be ON. Then we force the data line to go to 2.8 V. As a result $T2$ will turn ON, and turning $T2$ ON will turn $T1$ OFF. The other cells, which have not been addressed, do not change state, for the voltage at the collector of each $T1$ of these other cells will be $2.5 + 0.2 = 2.7$ V, and to make diode $D1$ conduct it would have been necessary to raise the data line to $2.7 + 0.45 = 3.15$ V rather than 2.8 V.

The Schottky cell of Fig. 12.14-1 has some advantage compared with the multiple-emitter cell of Fig. 12.13-1. With the Schottky cell, the standby voltage across the cell is $3.5 - 2.5 = 1.0$ V. With the multiple-emitter cell this voltage is $3.5 - 0.3 = 3.2$ V. Hence, if the collector resistors R_c are the same in the two cases, the power dissipation in the Schottky cell will be appreciably smaller. Power dissipation can be reduced by increasing R_c. However, such increase in collector-circuit resistances would lower the speed. On the other hand, the Schottky cell does not have the two-address-line feature of the multiple-emitter cell.

Another memory cell is shown in Fig. 12.14-2. Here multiple-emitter transistors are again used. However, the addressing is done from the collector side of the transistors so that the common emitters can be returned to a constant-current source (possibly a large resistor). In this way the cell can be operated in the manner of emitter-coupled logic, where transistors $T1$ and $T2$ are not allowed to saturate.

12.15   MOS RAMs

A six-transistor static cell   A six-transistor static MOS memory cell is shown in Fig. 12.15-1a. A bit is stored in the flip-flop, which consists of two cross-coupled inverters. Transistors $T1$ and $T2$ are the driver and load of one inverter, and $T3$ and $T4$ serve correspondingly for the other inverter.

The memory cell shown is addressed by setting X and Y to logic **1**. Setting $X = 1$ connects the cell to the data line and the $\overline{\text{data}}$ line. To write into the cell we set $W = 1$. This connects the data input terminal to node D as $T5$, $T7$, and $T9$ are ON. If the data input is at logic **1**, this raises the gate of $T3$, turning it ON and making node $\overline{D} = 0$. If the data input is at logic **0**, then $T3$ would be turned OFF and $\overline{D}$ would be at **1**. To read the state of the flip-flop we set $R = 1$. This connects the $\overline{\text{data}}$ output terminal to $\overline{D}$ since now $T6$, $T8$, and $T10$ are ON. Thus, the complement of the data level written into the cell is read.

In general, in a RAM, there are many memory cells connected to the same input and output lines. Such a configuration is shown in Fig. 12.15-1b. In this figure there are $\alpha\beta$ memory cells, each containing six transistors. Note that there are β columns of cells, each with a different Y address, and α rows of cells, each with a different X address. There is however, a single input transistor $T9$ to connect the data input terminal to the selected memory cell when the *write* instruction is given and a single output transistor $T10$ to connect the selected memory cell to the $\overline{\text{data}}$ output terminal during the *read* operation.

A four-transistor dynamic cell   In the dynamic memory cell of Fig. 12.15-2 the transistor count is reduced from six to four with a corresponding saving of real estate on the silicon chip and a saving of power. The cell is composed of transistors $T1$, $T2$, $T5$, and $T6$. Transistors $T7$ and $T8$ as well as transistors $T11$ and $T12$ serve all cells having the same column Y address. Transistors $T9$ and $T10$ are common to all of the cells in the memory, as in Fig. 12.15-1b.

The state of the cell is stored on the stray capacitances C_1 and C_2, whose presence is essential. These capacitors become accessible to the data terminals when the transmission gates $T7$ and $T8$ as well as $T5$ and $T6$ are all made to conduct by simultaneously raising the X and Y addresses to logic **1**.

In one state of the cell, the voltage across C_1 is larger than the threshold voltage of $T1$, and $T1$ is ON. Correspondingly C_2 has zero voltage, and $T2$ is OFF. In the other state the voltages of C_1 and C_2 and the conducting states of $T1$ and $T2$ are reversed. The cell having been accessed, the state of the cell can be *read* by setting $R = 1$. We can *write* into the cell by setting $W = 1$.

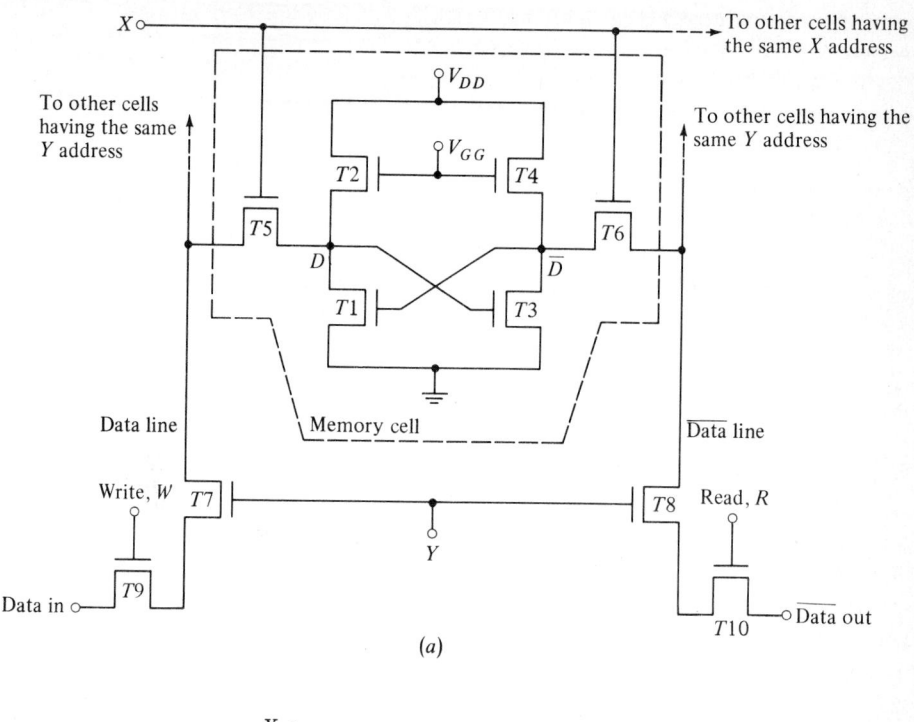

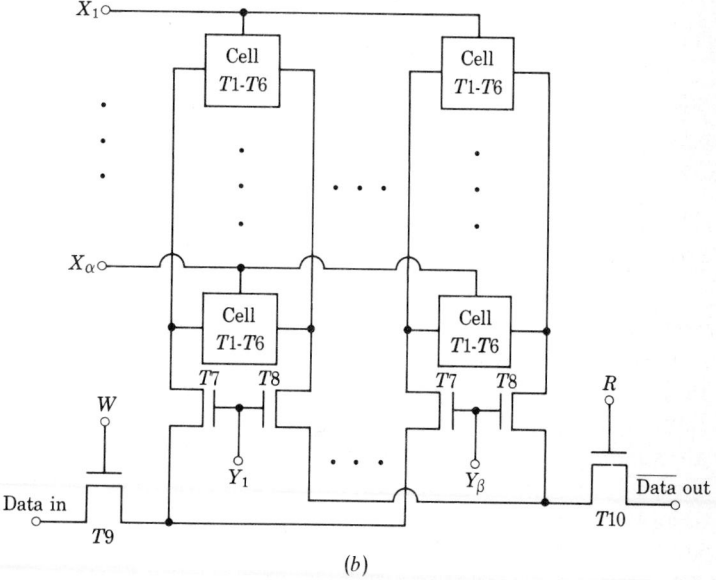

FIGURE 12.15-1

(a) A static MOS memory cell. (b) The interconnection of cells to form a RAM.

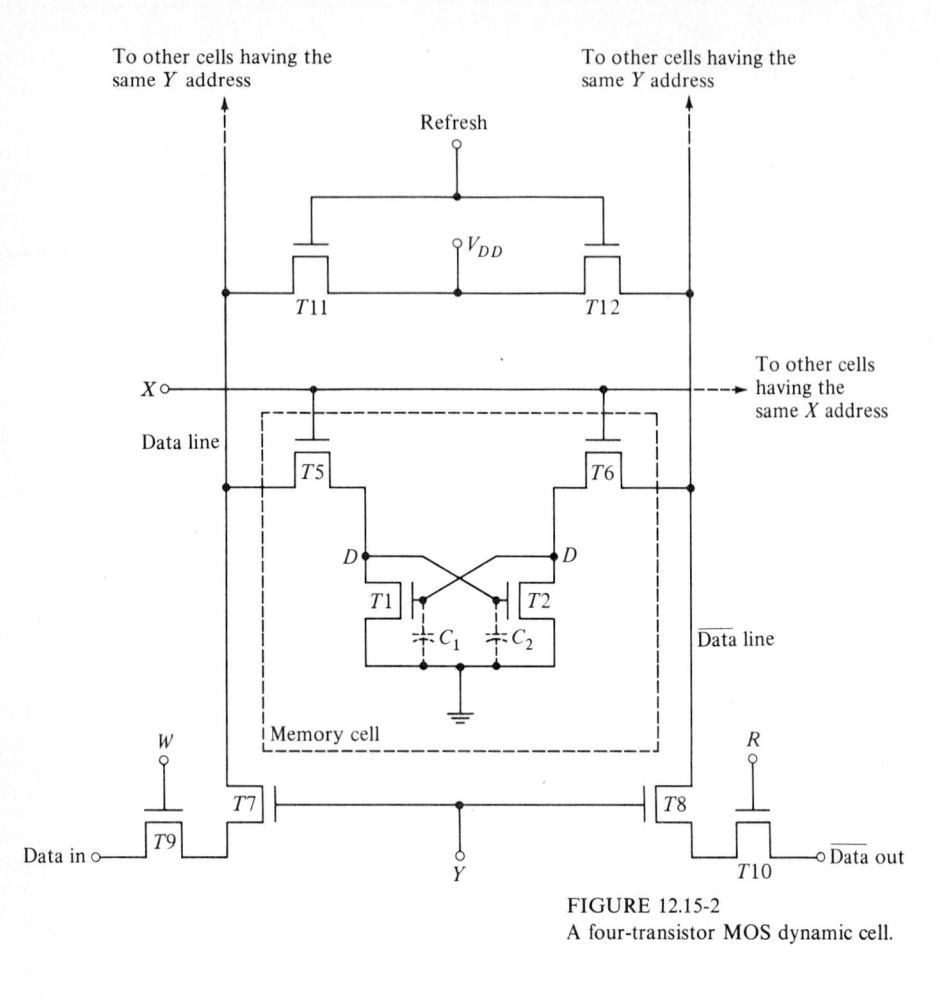

FIGURE 12.15-2
A four-transistor MOS dynamic cell.

If we have not performed a write operation for an extended time, then because of leakage of capacitor charge the information in the cell may be lost. It is therefore necessary to *refresh* the cell periodically. This refreshing operation is accomplished by allowing brief access from the supply voltage V_{DD} to the cell. This access becomes available when the X address and the refresh terminal are simultaneously at logic **1** so that $T5$ and $T6$ as well as $T11$ and $T12$ are thereby turned ON. Suppose now that initially $T1$ is ON, $T2$ is OFF, the voltage across C_1 is $V_{C_1} > V_T$, the threshold voltage, and $V_{C_2} = 0$ V. During the refresh interval V_{DD} is applied through $T12$ and $T6$ to C_1 paralleled by $T2$. However, $T2$ is OFF, and hence all the current from V_{DD} will be directed into C_1, allowing C_1 to replenish any charge lost due to leakage. Similarly V_{DD} is applied to C_2 which is in parallel to $T1$. But $T1$ is ON, and hence C_2 will not charge as rapidly as C_1. Observe that during the refresh interval $T6$ and $T12$ become a load for the driver transistor $T2$ and that $T5$ and $T11$ become a load for $T1$.

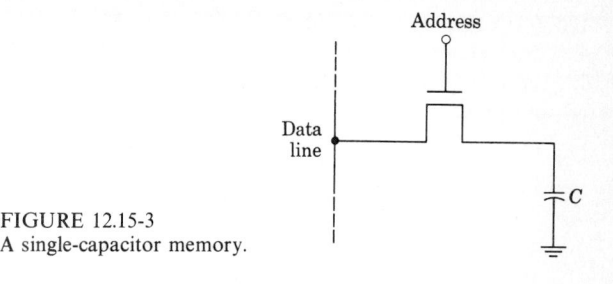

FIGURE 12.15-3
A single-capacitor memory.

Hence, during this refresh interval the cell becomes a conventional flip-flop consisting of two cross-coupled inverters. In any event it is clear that whatever the initial state of the flip-flop, during the refresh interval this initial state is reinforced.

A three-transistor memory cell It is apparent that if a bit is to be preserved through the charge stored on capacitors (as in Fig. 12.15-2), then a single capacitor is enough. Such a rudimentary memory is indicated· in Fig. 12.15-3. Of course, additional circuitry will be required to allow reading, writing, and refreshing, but each component of this extra circuitry will be able to service many memory cells. We are concerned principally with the components which must be duplicated in every cell.

An important difficulty associated with the simple one-transistor cell of Fig. 12.15-3 is that during a read operation the storage capacitor C will discharge into the data line. Unless the storage capacitance is large in comparison with the data-line capacitance, the read operation may well be destructive. That is, the process of reading a bit may well erase the bit from the memory. Suppose, on the other hand, that C is made quite large. Then during a write operation a large capacitor needs to be charged, and the interval which needs to be devoted to a write operation will thereby be extended.

One resolution of the difficulty is presented in Fig. 12.15-4, where we have provided separate access paths to the capacitor C for writing and reading. However, now a three-transistor memory cell is required. In this cell, transistor $T1$ is used during the write instruction, while transistors $T2$ and $T3$ are used when the logic state of capacitor C is to be read. Since the capacitor is the gate capacitor of $T2$, it is isolated from the output $\overline{data}$ line. A refresh circuit is also needed, of course, inasmuch as the charge on capacitor C can leak off through $T1$. Refreshing is accomplished using an inverter consisting of transmission gate $T9$ and the inverter amplifier formed by capacitor C_r and transistors $T10$ and $T11$. Note the similarity between this refresh amplifier and the inverter shown in Fig. 12.3-4, consisting of $T1$, $T2$, $T3$, and C_1.

To access the cell we set X and Y to **1**. As before, there are many cells having the same Y address. To write into the cell we disconnect the refresh circuit by setting $P = 0$. We then set $W = 1$. This connects the data-input

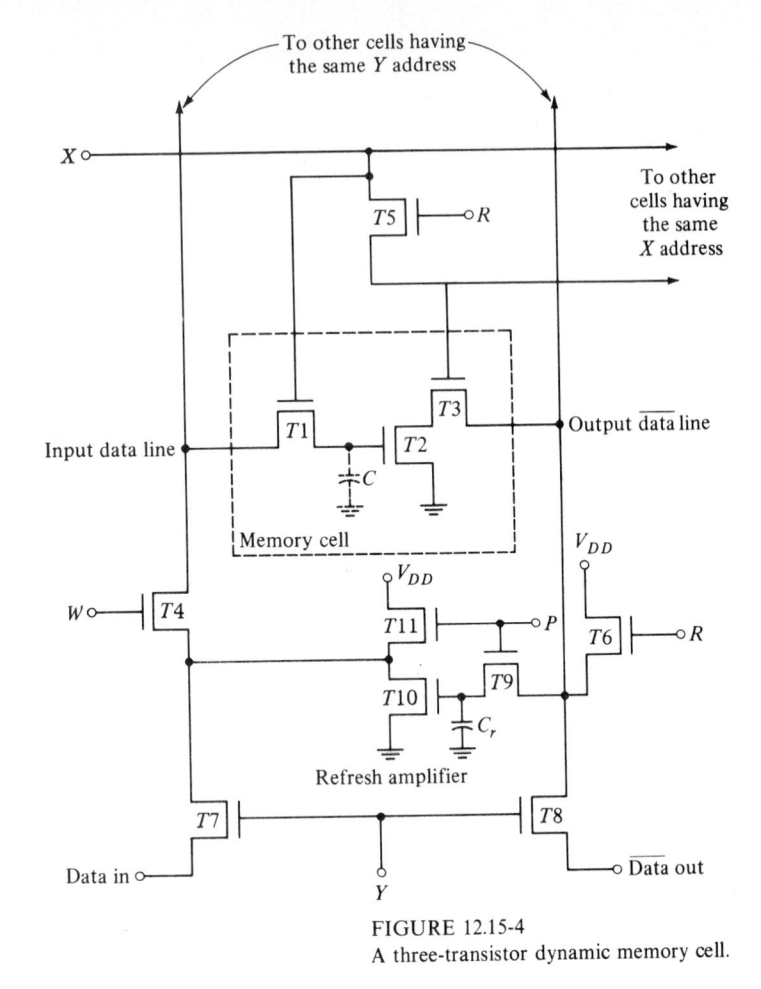

FIGURE 12.15-4
A three-transistor dynamic memory cell.

terminal across capacitor C since $T7$, $T4$, and $T1$ are ON. Capacitor C then charges to the state of the data input. To read the cell we set $R = 1$ $(W = 0)$, which turns $T6$ and $T3$ of the addressed cell ON. The $\overline{\text{data}}$-output line is then connected to the drain of $T2$ since $T3$ and $T8$ are ON. The complement of the level stored on C is read at the $\overline{\text{data}}$-output terminal. Note that transistor $T6$ acts as a load for $T2$ during the read operation so that $T2$ and $T6$ form an inverting amplifier.

To refresh the cell we set $Y = 0$, $X = 1$, $P = 1$, and $R = 1$. The data-input and the $\overline{\text{data}}$-output terminals are now disconnected from all the cells in the memory. The complement of the logic level of capacitor C is now transferred through $T9$, and this level is stored on capacitor C_r. The input terminal P, which permits the connection of C_r to the output $\overline{\text{data}}$ line, is called the *precharge input*. This term is used because when $P = 1$, C_r precharges to the complement of the level on C. After this precharging is accomplished, R is set to **0** and W is then set to **1**. The output of the inverter, $T10$ and $T11$, then refreshes the charge on capacitor C. Note that initially the refresh-amplifier output is disconnected from $T1$ and capacitor C. This is to ensure that $T11$ does not initially load capacitor C until after C_r has been precharged to the correct level. If this precaution were not taken, capacitor C might be discharged erroneously.

The memory cells having the same Y address are often refreshed sequentially, the cell having address X_1 being refreshed first, then the cell with address X_2, etc. For each column of cells there is a single refresh circuit; thus if there are β columns in the RAM (see Fig. 12.15-1b), there are β refresh circuits. The β cells having the same X address are refreshed simultaneously. Thus, if there are α rows, and if it takes a time T_r to refresh a single cell, the entire memory is refreshed in the time αT_r.

Note that although there are only three transistors in the memory cell shown in Fig. 12.15-4, many auxiliary transistors are needed to implement the reading, writing, and refreshing operations. However, these transistors are shared by other memory cells.

12.16 ORGANIZATION OF A RAM

The organization of a memory with storage facility for M words each of 3 bits is shown in Fig. 12.16-1. Cell (1, 1) stores the first bit of the first word, cell (1, 2) the second bit and so on. These first word-storage cells are addressed by raising the level of line Z_1. If the memory stores M words, M address lines Z_1, Z_2, ..., Z_M are required. The memory-storage location is presented here in coded form through the λ address bits $A_0 A_1 \cdots A_{\lambda-1}$ (where $M = 2^\lambda$). As required, then, a decoder has been interposed between the memory cells and the coded input address. As discussed in Sec. 12.9, such a decoder singles out one and only one of the address lines, the line so selected depending on the input address. Not all commercially available memories provide such decoding, and in such cases the decoders must be provided by the user. When a decoder is already part of a memory unit, the manufacturer will characterize his product as being "decoded" or "fully decoded."

All the cells ($C_{1,1}, C_{2,1}, ..., C_{M,1}$) intended to store first bits of words are connected to a common pair of data lines. All second-bit cells and all third-bit cells are similarly connected. The block marked I/O (input-output) represents all the circuitry shown in Fig. 12.13-1 or 12.15-4 with the exception of the memory-cell flip-flop itself. The bits b_{i1}, b_{i2}, and b_{i3} are input bits which will be written into the memory when the read/write (RW) line goes to logic **1**. The bits b_{01}, b_{02}, and b_{03} are the output bits which will be read out of the memory when $RW = \mathbf{0}$.

If the memory cell has two input address terminals, as in Figs. 12.13-1 and 12.15-4, we can arrange to make each cell separately and individually accessible. In such a case a memory with M cells can be used to provide

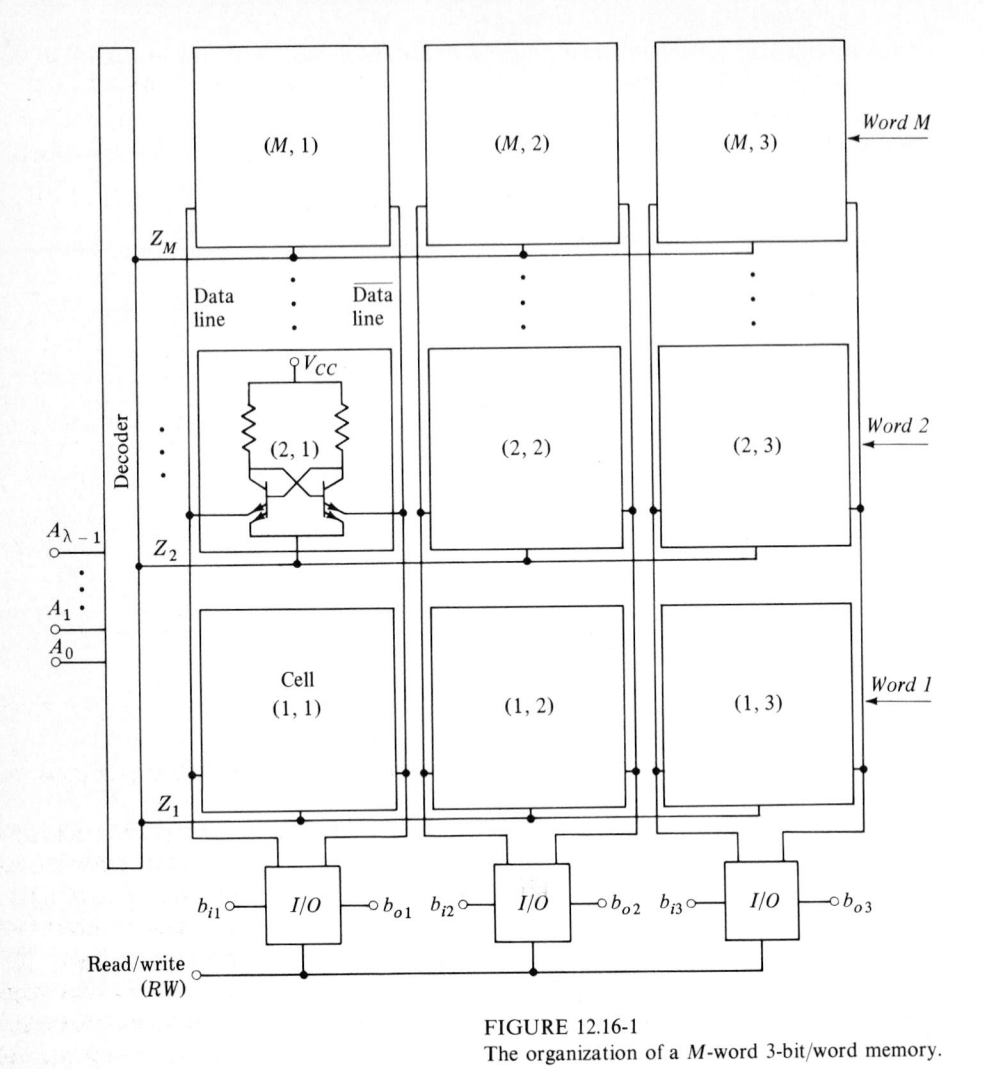

FIGURE 12.16-1
The organization of a M-word 3-bit/word memory.

storage for M words, each word having just 1 bit. The organization of such a memory is shown in Fig. 12.16-2. Note particularly that two decoders are required. At any time only one output line of the X decoder will be selected and only one line of the Y decoder will be selected. Correspondingly only that cell will be selected (addressed) which is at the intersection of these two selected lines. All the data lines of all the cells are paralleled, and only a single input-output stage is required to handle the 1-bit word.

This present arrangement allows the use of simpler decoders since some of the decoding is done in the memory cell itself. For example, suppose we need a memory with 256 (2^8) words each of 1 bit. If the cells had single address lines, we would need a decoder with 256 output lines. Suppose, however, the

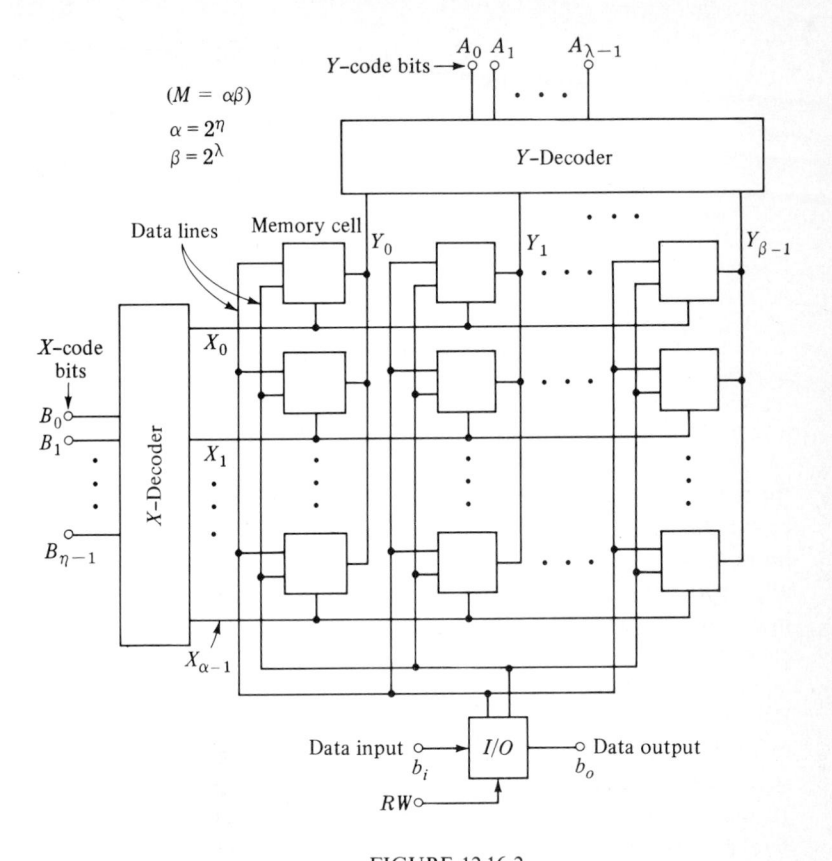

Y-code bits $\rightarrow$ A_0 A_1 $\cdots$ $A_{\lambda-1}$

$(M = \alpha\beta)$

$\alpha = 2^{\eta}$

$\beta = 2^{\lambda}$

Y-Decoder

Data lines   Memory cell Y_0 Y_1 $\cdots$ $Y_{\beta-1}$

X-code bits

X-Decoder

X_0

B_0
B_1

X_1

$B_{\eta-1}$

$X_{\alpha-1}$

Data input $\circ\!\!\rightarrow$ I/O $\circ$ Data output
b_i b_o

$RW\circ$

FIGURE 12.16-2
Organization of a memory with N 1-bit words.

cells were arranged, as in Fig. 12.16-2, in a square array ($\alpha = \beta = 16$). Then we should require two decoders, but each decoder would have only 4 input lines and 16 output lines.

If the memory cells employed in the RAM shown in Fig. 12.16-2 are dynamic, each column of cells has its own refresh circuit, as in Fig. 12.15-4. Thus, in a 1,024-bit memory containing 32 columns, 32 refresh circuits are used.

12.17 PARALLELING OF SEMICONDUCTOR MEMORY INTEGRATED-CIRCUIT CHIPS

On an integrated-circuit memory chip, the number of bits in a word is equal to the number of memory cells that have a common address line, and the number of words is equal to the number of separately addressable groups of bits. Manufacturers generally specify *word* and *bits per word* storage capacity through

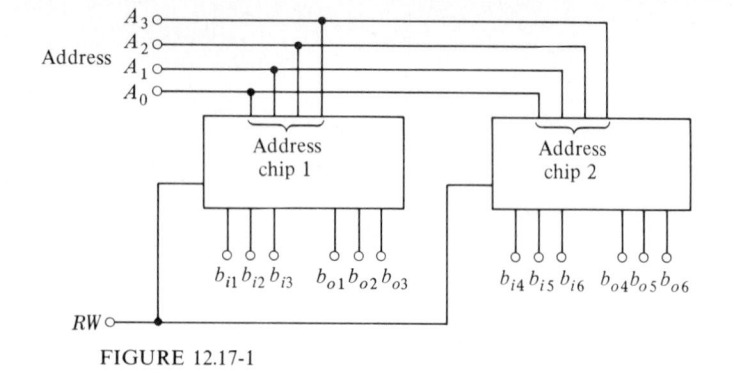

FIGURE 12.17-1
Paralleling memory chips to increase the number of bits per word.

the term *organization.* Thus, typically, a small memory may be characterized as having a capacity of 64 bits organized in 16 words of 4 bits each. (The number of words is invariably a power of 2 in order to put to effective use all the outputs of a decoder from a binary-code input.) As noted, the chip may also have a decoder and will have, as well, the required input-output circuitry.

The capacity of a memory system can be increased by paralleling chips. Paralleling chips in a manner which holds the number of words fixed but increases the number of bits per word is shown in Fig. 12.17-1. Here we represent the case of two identical chips, each of 48-bit capacity, organized in 16 words with 3 bits/word. The words are addressed by a 4-bit address code $A_3 A_2 A_1 A_0$. The address inputs of the two chips are paralleled, as is the read/write (RW) input. The 3 output bits of chip 1 (b_{01}, b_{02}, b_{03}) and the 3 output bits of chip 2 (b_{04}, b_{05}, b_{06}) become the 6 output bits $(b_{01}, \ldots, b_{06})$ of the new enlarged word. Similarly the input bits of the 6-bit word are $b_{i1} \cdots b_{i6}$. Additional chips, of course, may be paralleled to increase the number of bits per word further.

If the 1-bit word memories of the type shown in Fig. 12.16-2 are paralleled, the number of bits in a final output word is directly equal to the number of cells paralleled. The X-code input bits are applied to the X decoders of each chip, and the Y-code input bits to the Y decoders of each chip. This paralleling is illustrated in Fig. 12.17-2. Here we have drawn all the A and B address lines as single lines to simplify the figure.

Expansion of 1-bit per word memory chips Consider again that we have memory chips, each with M bits, in which each individual bit is separately addressable. We have already seen, as in Fig. 12.17-2, how N such chips may be organized into a memory of M words each of N bits. Suppose, however, that while N bits per word are adequate, we require more than M words. Let us say that we require $L \times M$ words. Such an expanded memory may be constructed by paralleling L groups of N chips, as shown in Fig. 12.17-3.

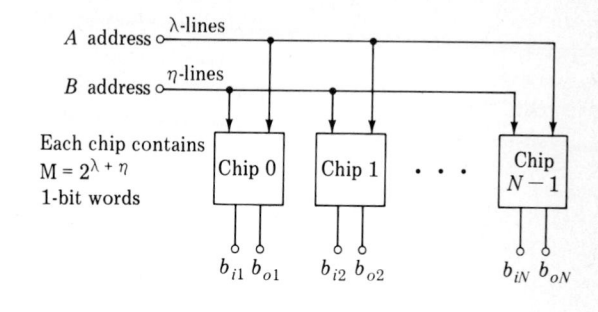

FIGURE 12.17-2
Paralleling chips to increase the number of bits per word. The special case in which each chip provides only one bit.

The chips in the top row store the first M words, the next M words are stored in the second row, and finally the Lth group of M words is stored in the bottom row. In order that it be possible to select a word in a particular row, the manufacturer makes available on the chip an additional address line called *chip select*. When the chip-select line is $CS = 0$, the chip decoder is disabled and no word on the chip is addressed. When the line is at $CS = 1$, the chip decoder is enabled and words are accessed in normal fashion.

In Fig. 12.7-3 the arrow marked A at each chip represents all the address

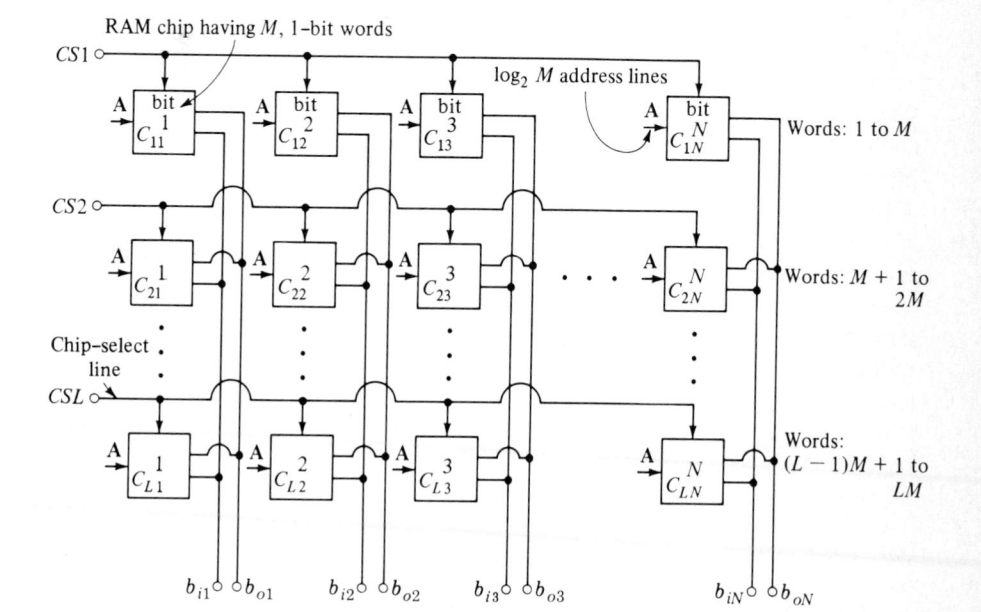

FIGURE 12.17-3
Connection of $M \times 1$ RAM chips to form an LM-word by N-bit memory.

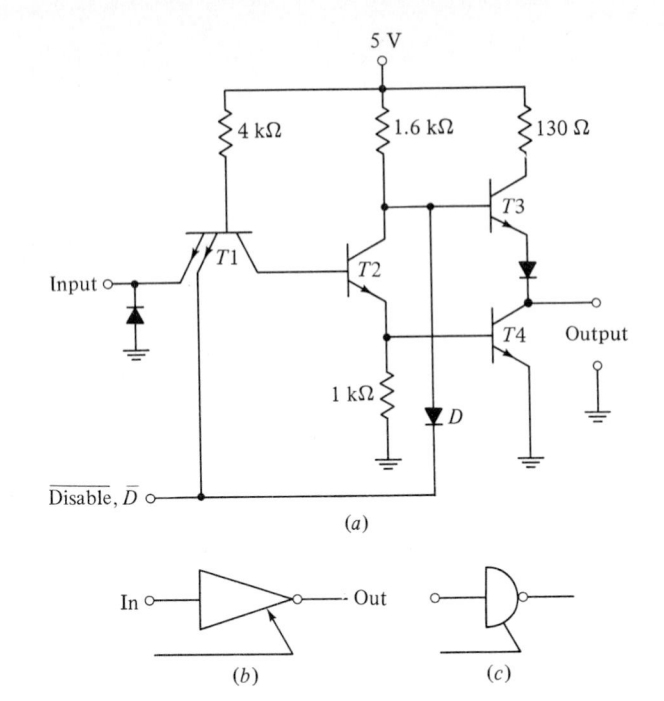

FIGURE 12.17-4

(a) A TTL tristate output stage; (b) and (c) circuit symbols for the stage.

input lines by which a single bit on the chip itself is addressed. The total input address applied to the memory, consisting of the address input lines A and the chip-select line CS, singles out a single chip-select line and the word is read from the corresponding row of chips. If we allow ourselves to compare reading from a memory with reading from a book, the *chip-select* selects a *page* and the address A then selects a word on the page.

Next we must take into account that when paralleling chips to increase the number of words it is necessary to connect in parallel the output bit lines of each of the other chips. Thus, for example, the output line of the parallel arrangement which bears the first bit of the word must be connected to the first-bit output terminals of each of the other chips; that is, $C_{11}, C_{21}, \ldots, C_{L1}$ are in parallel. It is therefore necessary to devise for a memory chip an output stage which allows the output logic level of a selected chip to be unaffected by connection of the output terminals of other chips which are not selected. The TTL output stage shown in Fig. 12.17-4a is very commonly used for this purpose. This stage is, of course, readily recognized as the standard TTL gate modified through the addition of a disabling input.

When the $\overline{\text{disable}}$ line is at logic **1**, diode D does not conduct nor does the emitter of $T1$, which is connected to this line. Hence with $\overline{D} = \mathbf{1}$ a straightforward output stage is provided with all of the usual advantages of active pull-up (see Chap. 5). However when $\bar{D} = 0$, both $T3$ and $T4$ will be OFF and the output terminal will be entirely isolated. The impedance seen looking back into the output terminal will be very high, being less than infinite only because of leakage and stray capacitance. The stage of Fig. 12.17-4 is often described as a *tristate* stage, the available states being logic **0**, logic **1**, and the *third state*, a nominally isolated high-impedance state. Symbols for the tristate stage are shown in Fig. 12.17-4*b* and *c*.

When memory chips are paralleled for the sake of increasing the number of words, the tristate output stage is enabled and disabled by the chip-select input. Only the output stages on the selected chip are enabled. CMOS gates are also available with a tristate output.

12.18 THE CHARGED-COUPLED DEVICE (CCD)

We have seen that an array of MOSFET devices fabricated on a silicon chip and forming an integrated-circuit dynamic-shift register may serve as a sequential memory. We consider now a different MOS dynamic-shift register sequential memory using the principle of the *charge-coupled device* (*CCD*). While the CCD memory will operate at about the same speed as the MOSFET memory, the CCD memory will dissipate appreciably less power. However, the most important relative advantage of the CCD memory is that it may be fabricated with a density of bits which is of the order of three times the density feasible with MOSFET memories. As we have noted, there is an advantage to using silicon chip "real estate" as economically as possible, since, as the area of the chip increases, there is a corresponding increase of the likelihood of encountering an imperfection in the chip. Hence the improved density possible with CCD devices results in better yields and lower cost. A still further advantage of the CCD device is that it requires a much simpler fabrication procedure involving many fewer operations than is required in MOSFET and in bipolar technology.

The structure of the CCD is represented in Fig. 12.18-1. The device is fabricated on a semiconductor substrate. We have indicated *n*-type silicon for the sake of being specific but *p*-type would serve as well. The substrate is covered with an insulating layer of silicon dioxide and on the oxide an array of closely spaced metal electrodes is arranged. The metal-oxide semiconductor sandwich is reminiscent of the MOSFET structure. Note however, the absence of *p*-type regions which in the MOSFET serve as source and drain.

Let us now consider initially that, while the bottom of the substrate is maintained at ground (0 V), all the metal electrodes are maintained at the same fixed negative voltage, $-V$. As a matter of practice, the spacing between electrodes is extremely small in comparison with their width. (The spacing in Fig. 12.18-1 is grossly exaggerated). Hence, in effect, the metalized layer can be considered to be an equipotential surface at the voltage $-V$. Since the base of the subtrate is an equipotential surface at 0 V, the equipotential surfaces generally are planes parallel to the faces of the structure.

If the voltage $-V$ on the metalization is large enough in magnitude so

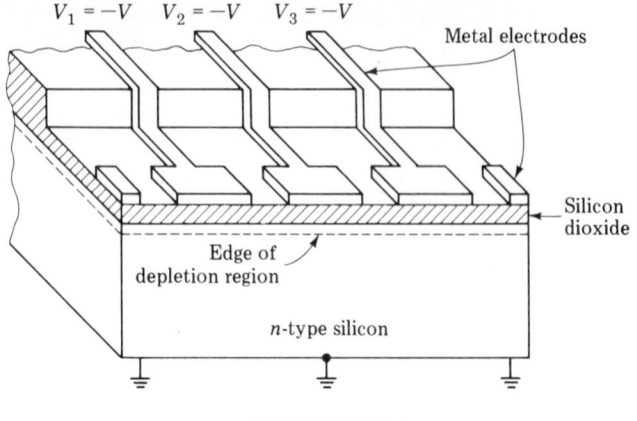

FIGURE 12.18-1
The structure of a charged-coupled device.

that it exceeds the threshold voltage V_T of the substrate, then there will form, under the oxide and adjacent to it, a depletion region. The depletion region is a region from which mobile carriers have been removed. In the present case, the relatively few minority holes present within the substrate will be drawn to the surface of the substrate under the oxide, while the majority carrier—electrons— will be pushed away from the surface. Since predominantly negative charge has been removed from the initially neutral region, the depletion region is left with a positive charge. Lines of electric field which originates from the negative charge on the metal will terminate on this positive charge. The boundary of the depletion region is established when enough positive charge has been provided to allow for the termination of all the electric-field lines. Like a typical equipotential surface, the depletion region boundary is a plane parallel to the structure surfaces (see Fig. 12.18-1).

Initially, the depletion region shown in Fig. 12.18-1 is devoid of mobile charges. However, as time progresses, holes present in the n-type substrate will diffuse into the depletion region. After a sufficient number of holes have entered the depletion region, there will be formed a conducting layer under the oxide precisely in the manner in which a channel is formed in an MOS device.

Depending on the nature and quantities of impurities in the semiconductor, the absence or presence of defects in the crystal structure of the semiconductor and other factors, the time that elapses between initial depletion and eventual channel generation will be in the range from tenths of seconds to tens of seconds or even longer. In any event, the time is long in comparison with the usual clock interval encountered in a digital system. This feature may seem surprising in view of our experience in connection with MOSFETS. There we noted that, when an appropriate voltage was applied to the gate, a channel formed immediately, i.e., in a matter of tens of nanoseconds. The difference arises from the fact that in the MOSFET, unlike the CCD device, there, there is a source of minority carriers immediately to the side of the depleted region. Thus in the MOSFET we have, at the outset, a voltage applied between the source and drain. As soon as a voltage is applied to the gate to generate a depleted region, minority carriers from the source and from the drain rush into the depleted region and convert it instead to a conducting channel.

12.19 STORAGE OF CHARGE

Consider now the situation represented in Fig. 12.19-1a. Here the voltage on one of the metal electrodes has been made substantially more negative than the voltage on its neighbors; that is, there is a larger negative voltage on electrode 2 than on the others. Accordingly, the depletion region under electrode 2 extends more deeply than under the adjacent electrodes. The boundary of the depleted region now has the form indicated by the dashed line in Fig. 12.19-1a.

Qualitative and approximate plots of electrical potential as a function of x, the distance measured horizontally through the substrate, at a number of levels are as shown in Fig. 12.19b. At any level which passes through the depletion region the potential, under electrode 2, is *depressed* relative to the potential under its neighboring electrodes. The potential depression is more pronounced for a level AA' which is located closer to the electrodes and is less pronounced for a level CC' located further from the electrodes. We have taken account, in drawing

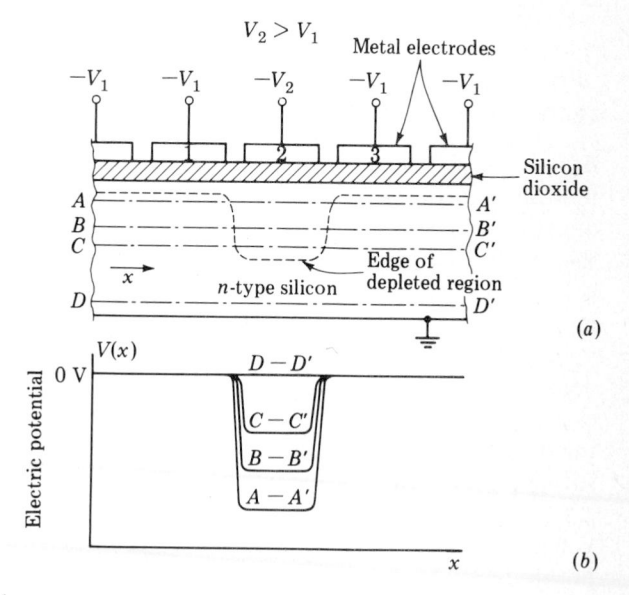

FIGURE 12.19-1

(a) The form of the depletion region when one electrode is made more negative that its neighbors. (b) The electric potential $V(x)$ through the substrate at various levels.

these plots, that outside the depletion region the substrate is a good conductor, being copiously supplied with majority carriers. Since no current flows, there is no electric field and, hence, no potential difference along the x axis within the n-type substrate.

The reader is now asked to recall that a charged particle of charge q in a field described by a (one-dimensional) potential $V(x)$ has exerted on it a force in the positive x direction given by $f = -q \, dV/dx$. Thus, in Fig. 12.19-1b where the potential is flat $(dV/dx = 0)$, there is no force. In the region between electrodes 1 and 2 where dV/dx is negative, the force is to the right, and in the region between electrodes 2 and 3, where dV/dx is positive, the force is to the left. Thus, a positive charge introduced into the depletion region at any level is free to move about only within the "potential well" shown in b.

Finally, then, it appears that the application to a particular electrode of a negative voltage larger in magnitude than is applied to its neighbors generates under that particular electrode an extended depletion region which penetrates more deeply into the substrate. If now we introduce into this region some positive charge (in a manner yet to be described), this charge will be trapped in position. Eventually, this introduced charge will lose its identity because of the diffusion of new carriers into the depletion region as described above. But as long as the charge is identifiable, the presence or absence of such a charge may be used to represent the two logic levels. As we shall now see, there are means by which such trapped charges may be transferred from a position under one electrode to a position under the next, and so on, the whole structure then serving as a shift register.

12.20 TRANSFER OF CHARGE

A mechanism by which charge may be transferred laterally in the CCD is illustrated in Fig. 12.20-1. The CCD device itself, extending laterally in the x direction, appears in Fig. 12.20-1b. We consider initially at time $t = t_0$ that the voltage on electrode k is at $V = -V_2$ while the voltage on all other electrodes is $V = -V_1$. Then there is an extended depletion region under electrode k and a minimal depletion region under all other electrodes. Then, at time $t = t_0$, the voltage $V(x)$ across the substrate, at any level that intersects the extended depletion region, will appear as shown in Fig. 12.20-1c. Here we have assumed that we have devised, by some means, to introduce some positive charge into the depletion region. Somewhat fancifully we have represented the charge as sitting at the bottom of the potential well.

Now, as is indicated in Fig. 12.20-1a, let us arrange that at time $t = t_1$ the voltage on electrode $k + 1$ should also drop to $V = -V_2$. Then the potential profile takes on the appearance shown in Fig. 12.20-1d. The potential barrier to the right of the charge in Fig. 12.20-1c has now been removed. Accordingly, as indicated by the arrow in Fig. 12.20-1d, the charge will diffuse to the right. No charge transport takes place, at least initially, in the reverse direction simply

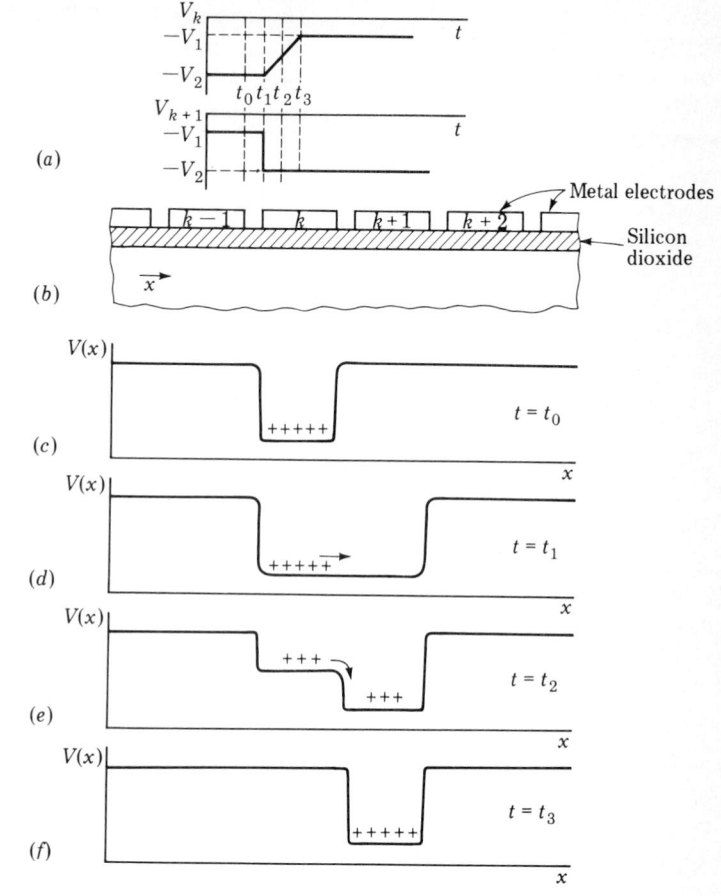

FIGURE 12.20-1
The mechanism of charge transfer. (a) The waveform V_k and V_{k+1}. (b) The CCD device. (c) through (f) The potential distributions at times t_0 through t_3 as identified in (a). The transfer of the charge is shown.

because there is no charge in the $k + 1$ region. If the voltage profile persisted as shown in Fig. 12.20-1d, then eventually the original charge would distribute itself evenly over the k and $k + 1$ region. However, we arrange, as shown by the waveforms of V_k and V_{k+1} in Fig. 12.20-1a, that the voltage V_k should now begin to return, relatively slowly, to the value $V_k = -V_1$.

As V_k rises, charge in the k region continues to move to the $k + 1$ region. However, now the mechanism of charge flow is due to a combination of diffusion and the presence of the electric field produced by the potential difference between the regions. Finally at $t = t_3$ and thereafter, as indicated in Fig. 12.20-1f, the charge is in the $k + 1$ region, having been transferred there from the k region.

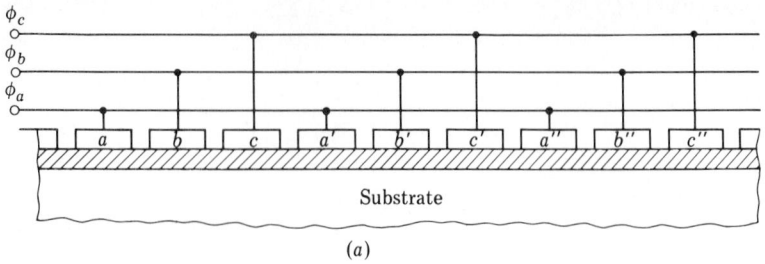

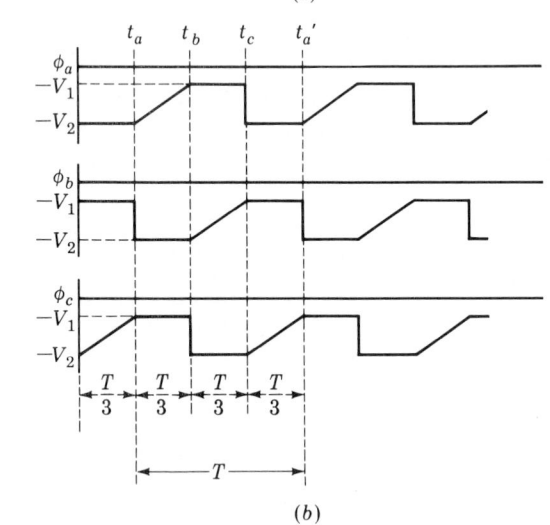

FIGURE 12.20-2
(*a*) The connection of a three-phase clocking waveform to the electrodes of a CCD device. (*b*) The waveforms of the clocking waveform phases.

During the interval from $t = t_1$ to $t = t_2$, while charge is being transferred from the k to the $k + 1$ region, it is necessary that the voltage at electrodes $k - 1$ and $k + 2$ be held at $V = -V_1$ so that no additional depletion region form under these electrodes. If a depletion region formed under electrode $k - 1$, some of the charge in the k region would be transported in the wrong direction. If a depletion region formed under $k + 2$, some of the charge would be transported in the forward direction by two steps rather by one step.

Taking account of the charge-transfer mechanism described, and taking account of the need to ensure that charge is not transferred backward or forward more than one step at a time, we may now consider how to effect a sequence of forward steps in the manner of a shift register. As shown in Fig. 12.20-2*a* every third electrode is connected to a common clocking bus and the three parts ϕ_a, ϕ_b, and ϕ_c of the composite clocking waveform is as in Fig. 12.20-2*b*. The composite clocking waveform has a period T. Each of the parts consists of three subintervals of duration $T/3$. In these subintervals the waveform is at $-V_1$, or at $-V_2$, or is rising from $-V_2$ to $-V_1$. At the time

t_a, ϕ_b has been at $-V_1$ for a full subinterval $T/3$ so that there is certainly no charge under the b electrodes (b, b', b'', etc.). Also ϕ_c has just completed its rise to $-V_1$, as a result of which any charge which might have been under the c electrodes has been completely dumped into the regions under the a electrodes. Accordingly, at $t = t_a$, any charge (or a lack of charge, depending on whether a logic **1** or a logic **0** is being represented) is under the a electrodes. Consider specifically that there is such a charge under the electrode a in Fig. 12.20-2a. Then, during the time interval from t_a to t_b, this charge will be transferred to the region under the b electrode, the process of transfer being completed at time $t = t_b$. At time $t = t_c$ the charge will have been transferred to the region under electrode c. Finally, at time $t = t'_a$, after a total time T the charge which was under a at $t = t_a$ will now be found under a'. Observe, most importantly, that, when the cycle has been completed, charge has been transferred from one region to a second region *three electrodes away*. Thus it takes *three electrodes* to provide for storage and transfer of *one bit*.

We have described a CCD device in which the geometry of the electrodes and insulating oxide layer are such that a three-phase clock is required. This geometry has the difficulty, in the matter of fabrication, that the separation between electrodes is required to be inconveniently small. Other more complicated geometries have been developed which require a four-phase clock, a two-phase clock, and even a single-phase clock.

12.21 INPUT AND OUTPUT ARRANGEMENT

We have seen how a charge trapped in the depletion region under an electrode may be transferred laterally in shift-register fashion. We need to consider briefly how a charge is introduced at the input side and finally detected at the output side.

One method suitable for use in shift-register application is shown in Fig. 12.21-1a. Here, under an opening in the insulating oxide layer we have diffused a p-type region in the n-type substrate. We have also added an additional gating electrode preceding the clocked transfer electrodes. Consider then that the clock phase is such that there is a depletion region under the electrode next to the gating electrode as shown. Then the similarity to the situation that prevails in a MOSFET is rather apparent. If at this time a negative gating voltage is applied, a depletion channel will open under the gate and minority p-type carriers will flow across the channel from the p-type region to the depletion region. When the depletion region has its required complement of charge, the gating voltage is removed. Thereafter the succeeding phases of the clocking waveform transfer this charge laterally, leaving this region under the first transfer electrode available for the next charge.

Correspondingly, the packets of positive charge may be detected, at the end of the register, in the manner indicated in Fig. 12.21-1b. Again a p-type region has been diffused under an opening in the oxide insulation. An external voltage

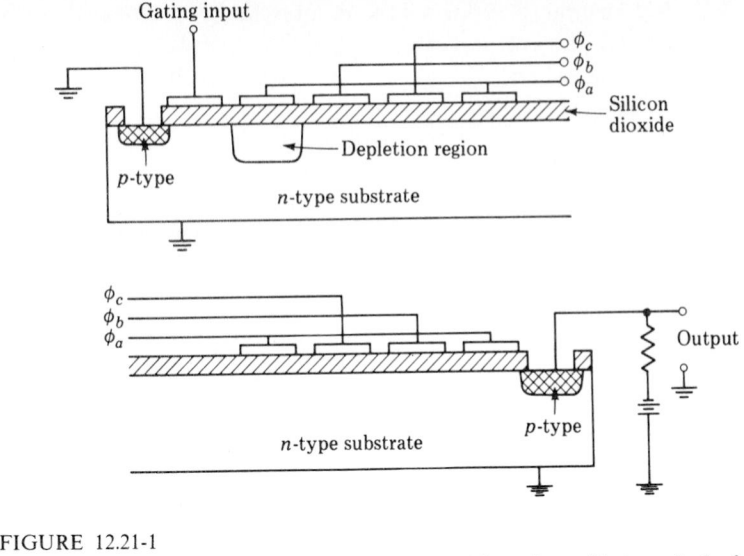

FIGURE 12.21-1

(a) A method of injecting a charge into a CCD shift register. (b) A method of detecting a charge at the end of a CCD shift register.

source reverse-biases the junction formed between the n-type substrate and the p-type diffusion. The minority carriers are transferred across this junction just as in the manner in which minority charges are collected by the collector junction of a conventional trasistor. An output voltage is developed across the resistor in series with the reverse-biasing voltage.

The methods described for injecting and collecting charge have the disadvantage, in the matter of fabrication, that high-temperature diffusions into the substrate are required. Other methods that require no such diffusions are also available.

Finally, we must note that the process of charge transfer is not completely efficient. At each transfer of charge, from under one electrode to the next, some charge remains behind. In a long register it is accordingly necessary periodically to provide for refreshing. Refresh amplifier stages are fabricated directly on the substrate of the CCD without involving an inordinate area of the chip.

An example of a CCD memory Typical of the CCD memories presently commercially available is the Intel type 2416 memory. This memory is organized as 64 recirculating shift registers each of 256 bits. We have noted that because of the gradual disappearance of depletion regions a bit cannot be stored in one position indefinitely. There is, accordingly, a minimum time between shifts which must be observed. In the type 2416 this minimum time is 9 μs. The maximum shift rate is 2 MHz. A four-phase clocking waveform is used. Average power dissipation is about 20 μW per bit.

REFERENCES

1 Seqrun, C. H., and M. F. Tompsett: "Charge Transfer Devices," Academic Press, 1975.
2 Boyle, W. S., and G. E. Smith: Charge-Coupled Semiconductor Devices, *Bell System Tech. J.*, vol. 49, April 1970, pp. 587–600.
3 Boyle, W. S., and G. E. Smith: Charge-Coupled Devices—A New Approach to MIS Device Structures, *IEEE Spectrum*, July 1971, pp. 18–27.
4 Altman, L.: The New Concept for Memory and Imaging: Charge Coupling, *Electronics*, June 21, 1971, pp. 50–59.
5 Kosonocky, W. F., and J. E. Carnes: Charge-Coupled Digital Circuits, *IEEE J. Solid State Circuits*, vol. SC-6, no. 5, October 1971, pp. 314–322.

13

ANALOG SWITCHES

Digital waveforms, ideally at least, make abrupt transitions between two separated ranges of values. One range represents logic level **1** while the other range represent logic level **0**. Within each range, the exact signal level is of no significance. In logical gates all inputs and outputs are digital signals.

Analog voltages, on the other hand, are voltages whose precise value is always significant. Such analog voltages may be fixed in value or may make excursions through a continuous range of values. There frequently occurs a need for switches in circuits and systems involving analog signals, in which the opening and closing of the switches are to be controlled by digital waveforms. Circuits of this type are variously called *analog gates, transmission gates, linear gates, time-selection circuits,* etc., depending on the purpose to which the circuit is put. The switch-control digital waveform is referred to as the *gating signal,* the *control signal,* or the *logic input.*

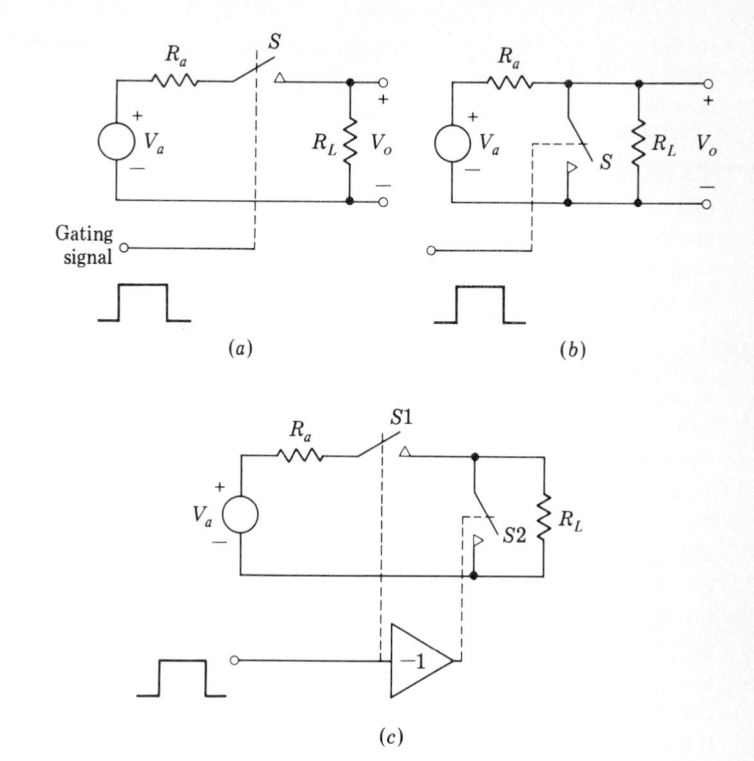

FIGURE 13.1-1
Switch configurations: (*a*) series switch; (*b*) shunt switch; (*c*) combination of series and shunt switch.

13.1 BASIC OPERATING PRINCIPLES OF ANALOG GATES

Several analog gates through which we can control the transmission of an analog signal V_a to a load R_L is shown in Fig. 13.1-1. A gating signal, which makes transitions between two ranges, controls the opening and closing of the switch S. The switch control is here displayed functionally. The exact mechanism of control will depend on the devices used to realize the switch. The gating signal is a digital signal; i.e., when the gating signal is at one logic level, the switch is OPEN, and when the gating signal is at the other logic level, the switch is CLOSED.

In Fig. 13.1-1*a* a series switch is used while in Fig. 13.1-1*b* a shunt switch is shown. In the first case V_a is transmitted to the load R_L when S is CLOSED; in the second case transmission occurs when S is OPEN. If the switches were ideal, there would be no basis for preferring either the series- or shunt-switch location. However, physical switches are not ideal in many respects. When the switch is CLOSED, the switch resistance is not zero and the volt-ampere characteristic of the switch device may not be linear. Hence in Fig. 13.1-1*a*, when S is CLOSED, some additional signal attenuation may be introduced and some signal distortion may occur. When S is opened, there will be some unintended signal transmission. Similar comments apply to Fig. 13.1-1b except for the alternate position of the switch. There will inevitably be some capacitance shunting the switch. As a result, when the switch is OPEN, in Fig. 13.1-1a, there will be some partial transmission of fast waveforms and, in Fig. 13.1-1b, there will be some distortion of fast waveforms, because of the shunt capacitance bridged across the load. Finally, there may well be some capacitive or other coupling between the analog-signal transmission path and the gating-input control signal. In such case the gating signal may, to some extent, appear superimposed at the output on the analog signal.

As a result of these switch imperfections, we may find, in particular cases, an advantage in the series- or shunt-switch arrangement. For example, consider a case with $R_a = 900\ \Omega$, $R_L = 1\ k\Omega$, and a switch which has an impedance when closed of $R_s(c) = 100\ \Omega$ and when open of $R_s(o) = 10\ M\Omega$. Then, as is easily verified, the series-switch arrangement is far more effective than the parallel arrangement. In the first case the ratio of the analog-signal output with the switch CLOSED to the switch OPEN is 5×10^3. In the shunt case the ratio of the analog-signal output with the switch OPEN to the switch CLOSED is only about 5. On the other hand, we may invent cases where, because of the resistor ratios involved, the shunt arrangement is better. Further, depending on the application and the type of analog signal involved, we may find that the various capacitances associated with the switch are of lesser consequence in one arrangement than the other. Finally, we may note, that especially in such cases where neither the shunt nor series arrangement has a clean-cut advantage, the arrangement of Fig. 13.1-1c has special merit. Here we use both a shunt and a series switch. The inverting amplifier indicated in the figure is one way of representing the fact that when one switch closes the other opens and vice versa. (Another common symbolism dispenses with the inverter and simply shows one switch in the closed position and one in the open position.)

13.2 APPLICATIONS OF SWITCHING CIRCUITS

Multiplexing Figure 13.2-1 shows an anlog switching circuit which allows us to connect a number of analog signals, one at a time, to a common load. This operation is called *multiplexing*. Four separate signals are indicated in the figure, but of course the number of signals which may be involved is, in principle, arbitrary. (The situation in Fig. 13.1-1a may be viewed as a special case in which only a single signal is involved.) An appropriate set of control-logic signals is easily generated in a four-stage ring counter.

The usefulness of the multiplexing arrangement indicated in Fig. 13.2-1 may be seen from the following considerations. Suppose we have a number of information-bearing signals (speech, music, data, etc.) which are to be transmitted over a communication channel (a pair of wires, a radio link, etc.) to a distant point. An entirely straightforward method of effecting such transmission is to use as many individual communication channels as there are signals.

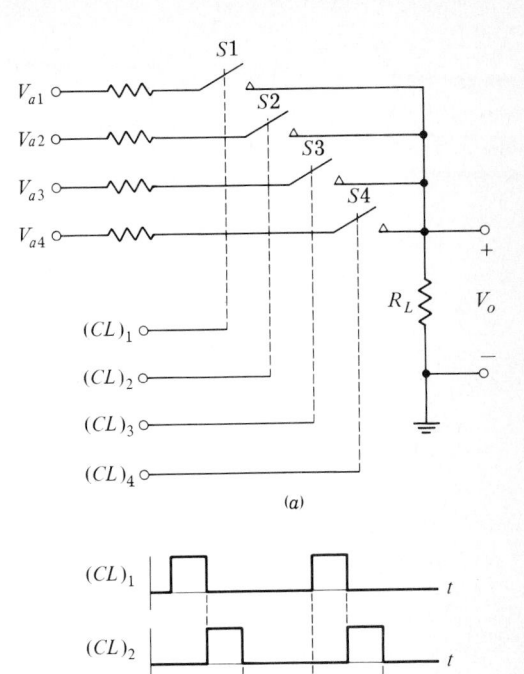

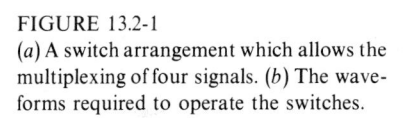

FIGURE 13.2-1

(a) A switch arrangement which allows the multiplexing of four signals. (b) The waveforms required to operate the switches.

Alternatively, we might *sample* each signal, arranging that the successive samples appear in turn, as in Fig. 13.2-1, across the input to a single communication channel. (In Fig. 13.2-1, the wires connected across R_L constitute this single communication channel.) It now turns out that at the receiving end of the communication channel it is possible to reconstruct each of the individual signals. To allow such reconstruction without error, it is necessary to be able at the receiving end to identify which sample corresponds to which signal. (These matters are discussed in more detail in Chap. 14).

Sample and hold (S/H)   A second analog switching application is indicated in Fig. 13.2-2. Here, an analog signal $M(t)$ is sampled at times separated by the sampling interval T_S. The sampling is controlled by the control waveform V_c, which closes and opens the switch. During the time T_C the switch is closed and the capacitor *charges* to $M(t_i)$. In the remaining time $T_H = T_S - T_C$ the sample value is *held* on capacitor C. The output waveform of a *sample-and-hold* circuit which performs such an operation is shown in Fig. 13.2-2d. The held sample values are indicated by solid levels. The output waveform between held samples

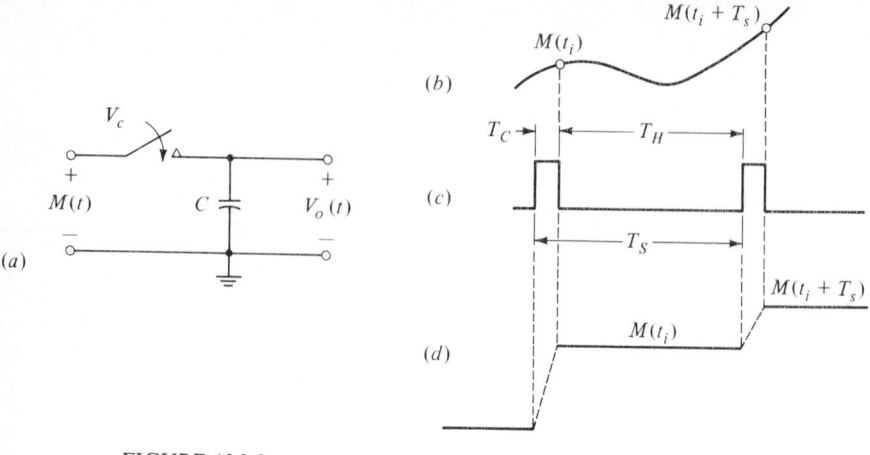

FIGURE 13.2-2
The basic sample-and-hold circuit: (a) the switching circuit, (b) the waveform to be sampled, (c) the gating waveform, V_c, and (d) the output of the sampling circuit V_o.

is of no special interest and is indicated by the dashed portion of the waveform.

One application of a sample-and-hold circuit is in *pulse-code modulation* (PCM). Here, the held waveform is converted from its analog voltage into a digital waveform. The process of analog-to-digital conversion is discussed in some detail in Chap. 14. A theoretical discussion of PCM and requirements relating to the sampling interval is found in Ref. 1.

A somewhat more detailed circuit is shown in Fig. 13.2-3. Note that when the switch S is kept in the closed position by the control logic waveform, the output voltage V_o follows the analog signal $M(t)$ as closely as allowed by the constraint imposed by the presence of the storage capacitor C. In principle, we should prefer that R_a be arbitrarily small so that V_o can follow $M(t)$ precisely when the switch is closed, and we should prefer that R_L be arbitrarily large so that the *held* value be held constant when S is opened. In practice, of course, we can require no such idealization since, at a minimum, R_a represents the

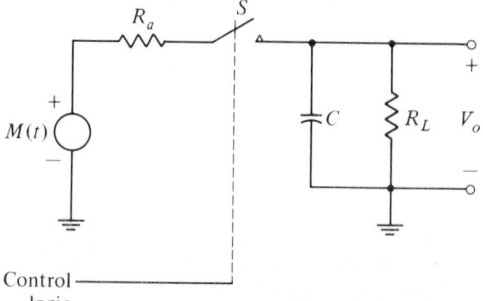

FIGURE 13.2-3
A sample-and-hold circuit with source resistor R_a and load resistor R_L.

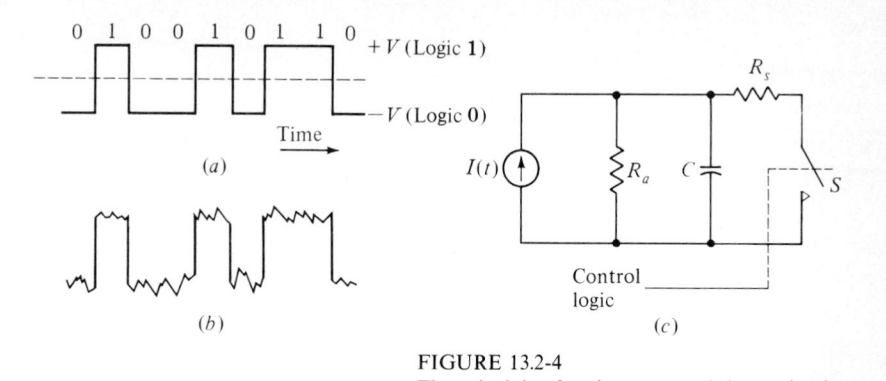

FIGURE 13.2-4
The principle of an integrate-and-dump circuit.

source impedance of $M(t)$ combined with the switch resistance and R_L represents the input impedance of whatever device (amplifier, etc.) is coupled across the capacitor to read the sampled value. We shall refer again (Sec. 13.8) to these and other departures from perfection of physical sample-and-hold circuits.

Integrate and dump To introduce the need for a third type of analog switch circuit, consider the situation represented in Fig. 13.2-4. In Fig. 13.2-4a there appears a digital signal. For simplicity we have assumed that logic **1** and logic **0** are represented respectively by $+V$ and $-V$. It may well be in the transmission of this signal through the communications channel to its destination that the signal has superimposed on it random erratic disturbances (noise). When received, the signal may appear as in Fig. 13.2-4b. As the waveform appears here, there is still no uncertainty within each bit interval as to whether the transmitted bit is a logic **1** or a logic **0**. Still we may well imagine that if the noise contamination were much more pronounced, we might well encounter difficulty in making such a determination. One way of making such a determination which suppresses the effect of the noise is to integrate the received signal over the interval of a bit. Such integration will yield an output which is proportional to the average signal value over the bit interval. Since the noise is random, going positive and negative with equal likelihood, we may expect that the noise average over a bit may well be small enough to allow the determination of bit value with little uncertainty. Indeed it can be established[1] that under some conditions this scheme for reading bits is, in principle, *optimum*; i.e., the number of errors made is a minimum. It must be noted, however, that no scheme, including this integration scheme can guarantee that no errors in determination will be made.

In the circuit of Fig. 13.2-4c the anlog-signal source is represented by a Norton's equivalent circuit. If the impedance R_a is adequately large, the analog current flowing into C will generate across C a voltage proportional to the integral of the analog signal over the interval during which the switch S is OPEN. In operation, the switch is OPEN during most of the bit duration and closes

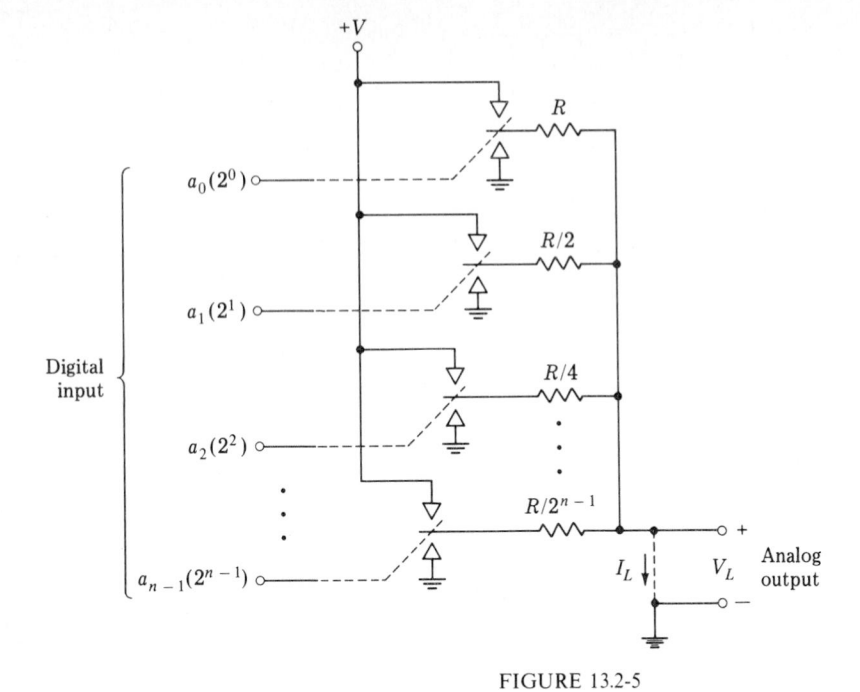

FIGURE 13.2-5
A digital-to-analog (D/A) converter.

briefly at the end of the bit. Just before the closing of the switch, a reading is taken of the voltage across C. The subsequent closing of the switch *dumps* the charge on C, that is, discharges the capacitor, and readies the capacitor to charge again during the next bit. Each reading of the capacitor voltage then indicates the estimate of the logic level received during the bit interval just passed, independently of the logic level of preceding bits. Ideally, of course, we should like R_s to be zero. For, the smaller R_s the shorter the interval during which S must be closed to effectively discharge C and the longer the portion of the bit interval during which integration can be carried out. It is intuitively obvious that the longer the integration time the greater the suppression of the effect of the noise.

Digital-to-analog conversion The use of logic-controlled switches to convert a digital to an analog signal is illustrated in Fig. 13.2-5. The binary digits, $a_0, a_1, \ldots, a_{n-1}$, assumed to be available simultaneously, i.e., in parallel rather than serially, are used to provide the control logic to operate the switches. A switch closes to the $+V$ side if a_k is at logic **1** and closes to ground if a_k is at logic **0**. As the figure shows, the resistor associated with the switch operated by the input bit of numerical significance 2^k has a resistance $R/2^k$.

The digital-to-analog converter (D/A) will operate as required with the analog output terminals open-circuited (in which case the output voltage is V_L)

or short-circuited (as indicated by the dashed short-circuit connection, in which case the output current is I_L). Let the digital input be $a_{n-1}, a_{n-2}, \ldots, a_0$, in which each a_k is $a_k = 1$ or $a_k = 0$. In the case of short-circuited output we see by inspection that

$$I_L = \sum_{k=0}^{k=n-1} \frac{V a_k}{R/2^k} = \frac{V}{R} \sum_{k=0}^{k=n-1} 2^k a_k \tag{13.2-1}$$

so that I_L, as required, is directly proportional to the numerical significance of the digital input. When the output terminals are open-circuited, we may note that the current through the kth resistor, flowing toward the node of common connection of all the resistors, is

$$I_k = \frac{a_k V - V_L}{R/2^k} \tag{13.2-2}$$

With the output open-circuited, the sum of all such currents must be zero. Hence

$$\sum_{k=0}^{k=n-1} I_k = \sum_{k=0}^{k=n-1} \frac{a_k V - V_L}{R/2^k} = 0 \tag{13.2-3}$$

Solving for V_L, we find, again as required, that

$$V_L = \left(\frac{V}{\sum_{k=0}^{n-1} 2^k} \right) \sum_{k=0}^{n-1} 2^k a_k \tag{13.2-4}$$

An alternative resistor arrangement for a D/A is shown in Fig. 13.2-6. Here again, the output may be a short circuit, in which case the output is the short-circuit current I_L, or the output may be terminated in a resistor $2R$ in which case the output is V_L. We shall compare the two arrangements in Sec. 14.5, where we discuss digital-to-analog conversion in more detail.

Chopper stabilization of amplifiers   Let us assume that we are required to amplify a very low frequency signal whose amplitude changes by a very small amount. The changes may be, say, of the order of microvolts, and if, by way of example, the signal is periodic, the period may be many minutes, hours, or even days. An ac amplifier with the customary capacitive coupling between stages would not be feasible, since these coupling capacitances would be impractically large. Instead it would be necessary to use direct coupling between stages. With such a dc amplifier we might not be able to distinguish between a change in output voltage as the result of a change in input voltage or as the result of a drift in some active device or component. If the amplifier has high gain, even small changes in the operating point of the first stage, amplified by the remaining stages, might cause a large change in output.

A method for circumventing this difficulty is shown in Fig. 13.2-7a. The slowly varying input signal V_s is shown in Fig. 13.2-7b. Assume that switch $S1$ is being driven so that it is alternately OPEN and CLOSED. Then the signal V_i at the amplifier input will appear as in Fig. 13.2-7c. When $S1$ is OPEN,

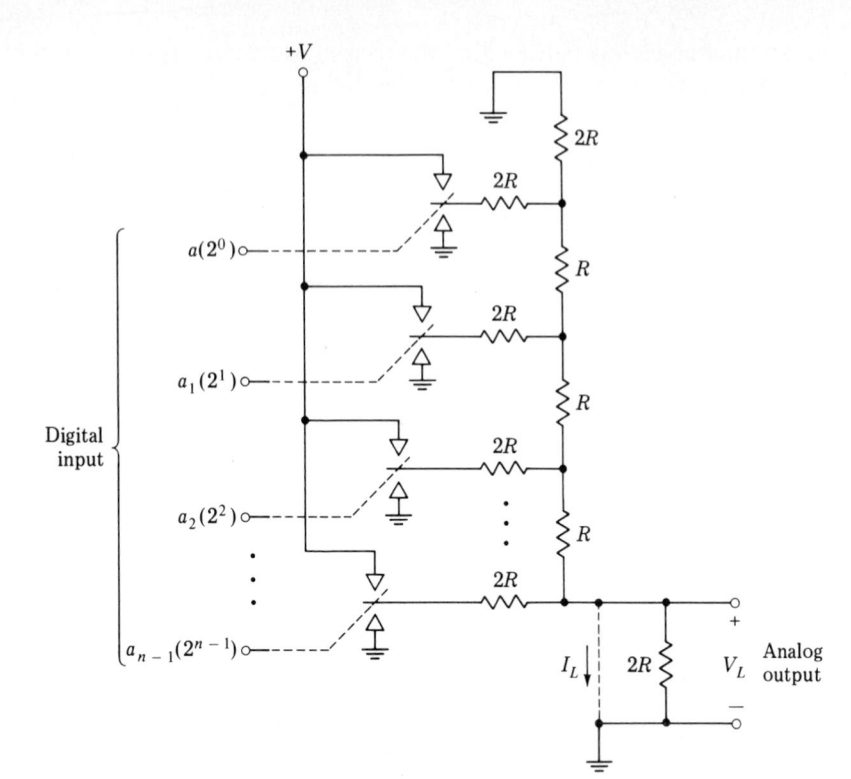

FIGURE 13.2-6
Another type of D/A converter.

$V_i = V_s$, and when $S1$ is CLOSED $V_i = 0$. Observe that the waveform V_i is a chopped version of the waveform V_s. It is for this reason that the circuit consisting of resistance R and switch $S1$, in the present application, is called a *chopper*.

In Fig. 13.2-7c we note that when the switch is OPEN the signal V_i reproduces the input signal V_s. As we have drawn the figure, a perceptible voltage change takes place in V_s during any interval when $S1$ is OPEN. Thus, when V_s is positive, the positive extremities of the waveform V_i are not at a constant voltage, and similarly for the negative extremities when V_s is negative. This feature is in no way essential to the operation. More customarily, the frequency of operation of the switch is very high (typically 100 times) compared with the maximum frequency of the signal V_s. Therefore no appreciable change takes place in V_s during the interval when $S1$ is OPEN. Accordingly, it is proper to describe the waveform V_i as a *square wave* of amplitude proportional to V_s and having an average value (shown dashed) that is also proportional to the signal V_s. Alternatively stated, the waveform V_i is a square wave at the switching frequency, amplitude-modulated by the input signal and superimposed

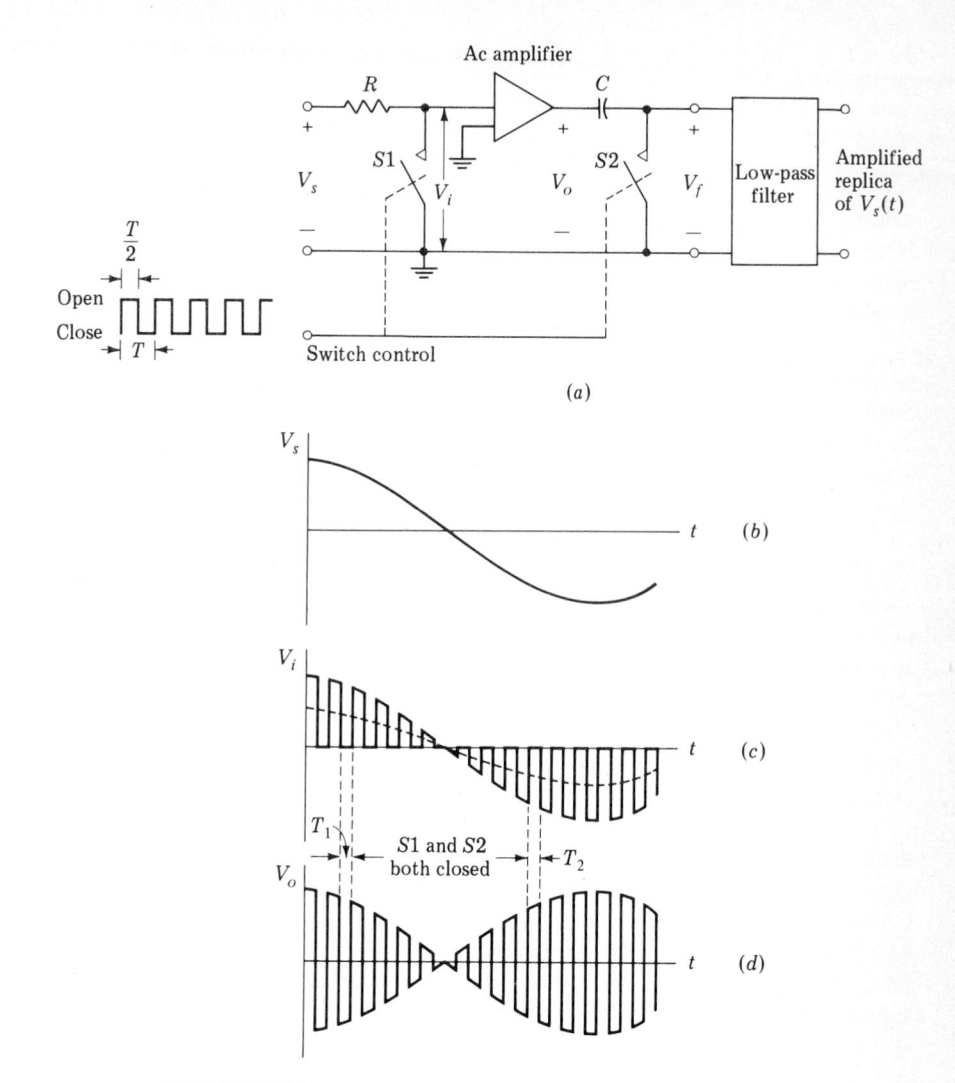

FIGURE 13.2-7
(a) A chopper-stabilized amplifier. (b) An input waveform. (c) The input to the
amplifier. (d) The amplifier output.

on a signal which is proportional to the input signal V_s. It is left for the problems
to show that $V_i(t)$ is given by

$$V_i(t) = \tfrac{1}{2}V_s(t) + V_s(t) \sum_{n=1}^{\infty} \left[\frac{\sin (n\pi/2)}{n\pi/2} \right] \cos \frac{2\pi nt}{T} \qquad (13.2\text{-}5)$$

The low-frequency cutoff of the ac amplifier shown in Fig. 13.2-7 is such
that the relatively high-frequency amplitude-modulated signals pass with small

distortion while the signal $\frac{1}{2}V_s(t)$ is well below the cutoff point. Consequently, at the output of the amplifier, the waveform V_o is given by

$$V_o(t) = A V_s(t) \sum_{n=1}^{\infty} \left[\frac{\sin{(n\pi/2)}}{n\pi/2} \right] \cos{\frac{2\pi nt}{T}} \qquad (13.2\text{-}6)$$

where A is the gain of the amplifier. The waveform is as shown in Fig. 13.2-7d. Here we are left with only the modulated waveform itself. Because of this process of modulation which has been accomplished, the chopper [or the chopper together with the ac amplifier to eliminate the signal $V_s(t)$] is often called a *modulator*.

The signal is recovered through the mechanism of the capacitor C and the switch $S2$. The switch $S2$ closes and opens in synchronism with $S1$. Thus, during the interval T_1, the negative extremity of V_o is held at zero, whereas during T_2 the positive extremity is held at zero. As a result, except for an increase in amplitude, the signal V_f across $S2$ assumes again the form of the signal V_s. If now this signal V_f is passed through a low-pass filter which rejects the frequency components of the square wave and transmits the frequency components of the signal $V_s(t)$, then at the filter output we shall have an amplified replica of the original signal. If $S2$ operates antisynchronously with $S1$ ($S2$ CLOSED while $S1$ is OPEN and vice versa), then at the output the signal will appear with reversed polarity. In either case the combination of the capacitor C, the switch $S2$, and the filter consitutes a *synchronous demodulator*. The amplifier of Fig. 13.2-7a is called a *chopper-stabilized amplifier*. Note that the amplifier is not stabilized by the choppers but that the choppers eliminate the necessity for a direct-coupled stabilized amplifier.

Currently available integrated-circuit op-amps have input offset-voltage drifts with temperature sensitivities of the order of 5 μV/°C. This sensitivity is too large for some applications, and chopper stabilization must be employed. Consider, for example, that an op-amp is to be used to amplify the very slowly varying output voltage of a thermocouple and that the op-amp must operate in nominally the same environment as the thermocouple. A thermocouple typically has an output which changes by about 30 μV for a temperature change

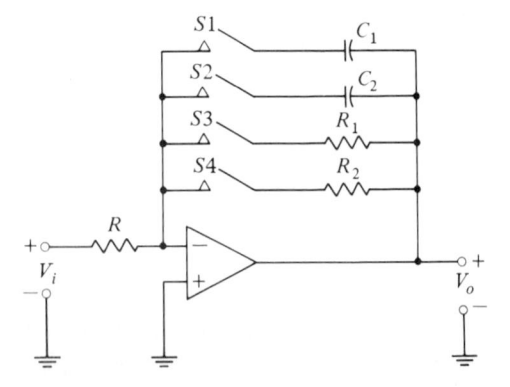

FIGURE 13.2-8
An active filter whose transfer function is determined by which of the switches are closed.

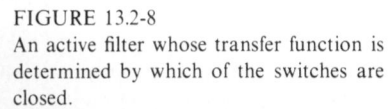

of 1°C.   To resolve temperature changes less than a degree we might then require a total amplifier drift of less than 10 μV.   If the range of temperature to be encountered is, say, 100°C, then we may allow an amplifier temperature drift of no more than 0.1 μV/°C; this figure is smaller by a factor of 50 than what is normally provided by an op-amp.

Parameter control   As a final example of logic-controlled analog switching we may note that such switching can be used very effectively to control the parameters associated with circuits intended to process analog signals or even their mode of operation.   Thus, in Fig. 13.2-8, depending on which switches are closed, we have an amplifier of adjustable gain, an integrator with adjustable time constant, or even a more complicated filter.

13.3   DIODE TRANSMISSION GATES

Diodes, bipolar transistors, and field-effect transistors (both junction and insulated-gate) can be used as analog switches.   Diode switches are faster than transistor switches since the storage times and capacitances associated with diodes are appreciably smaller than in transistors.   Such is particularly the case when hot-carrier, i.e., Schottky, diodes are used.   Diodes, on the other hand, being two-terminal devices, have no control electrode and hence generally lead to more awkward circuitry than transistors.   Still, where speed is at a premium, diodes are the active device of choice, and we shall now examine some diode analog gates.

One-diode gate   A one-diode gate is shown in Fig. 13.3-1a.   A control voltage V_c, as in Fig. 13.3-1b, opens and closes the gate to regulate the transmission of the analog signal V_a.   For a typical analog voltage, as shown in Fig. 13.3-1c, the output V_o appears as in Fig. 13.3-1d.   We note that the analog waveform, when transmitted through the gate, is superimposed on a *pedestal*, i.e., the base line for the output is different depending on whether the gate is open or closed.

This gate (like all gates) has a limited range of amplitudes (of V_a) over which it will operate.   With V_c at $V_c = -V_2$, the diode D is normally cut off. However, if the peak, positive-going excursion of the analog waveform is too large, there will be transmission of these peaks.   Similarly with $V_c = V_1$, a negative excursion of the analog waveform which is too large will cut the diode off. The allowable range of the analog input and also the attenuation of the analog signal depends on the values of the circuit components and is easily calculated (Probs. 13.3-1 to 13.3-3).

Finally, we note in Fig. 13.3-1e an alternate configuration of a one-diode gate in which the diode switch appears as a shunt element.   The operation of this device is left for the problems.

Two-diode gate   A principal limitation of the one-diode gate is the pedestal noted in Fig. 13.3-1d.   The pedestal is eliminated in the two-diode gate of Fig. 13.3-2.

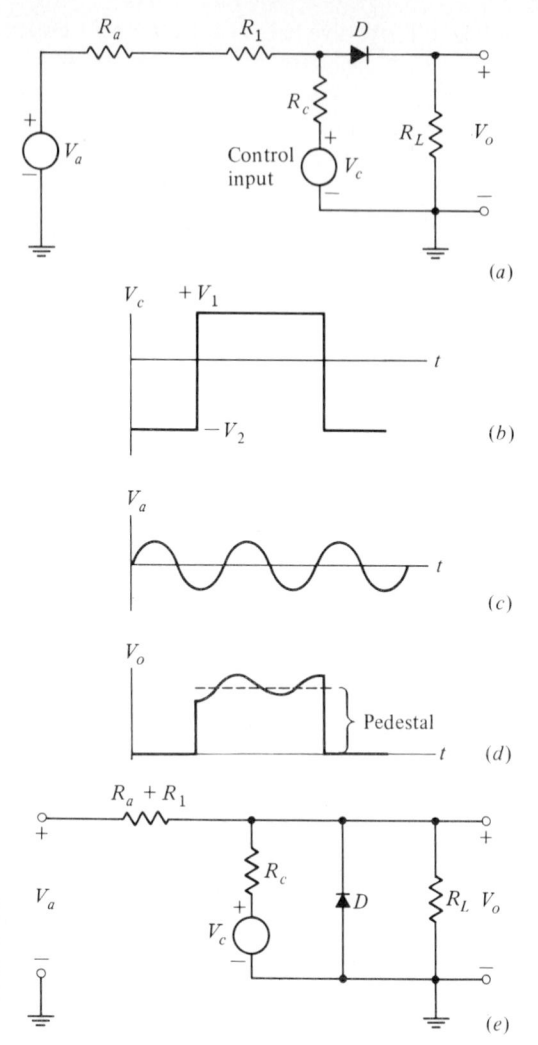

FIGURE 13.3-1
(a) A one-diode gate. (b) The control voltage. (c) An input waveform. (d) The output waveform. (e) A one-diode gate in which the diode appears as a shunt element.

Note that this two-diode gate is a combination of two one-diode gates of the type shown in Fig. 13.3-1a. Two symmetrical control voltages V_c and $-V_c$ are required. When the control voltages are at the levels V_1 and $-V_1$, respectively, the diodes conduct and the analog signal V_a is transmitted to the output. When the control waveforms are at the levels V_2 and $-V_2$, respectively, the diodes do not conduct and the gate is closed against transmission.

As a result of the symmetry of the circuit and the symmetry of the control waveforms, there will be no pedestal at the gate output. (We may note also that, because of the symmetry, no current flows through the analog-signal source in response to the control voltage sources.) We have been assuming that the control waveforms are rectangular in shape. However, unless it is necessary for the gate

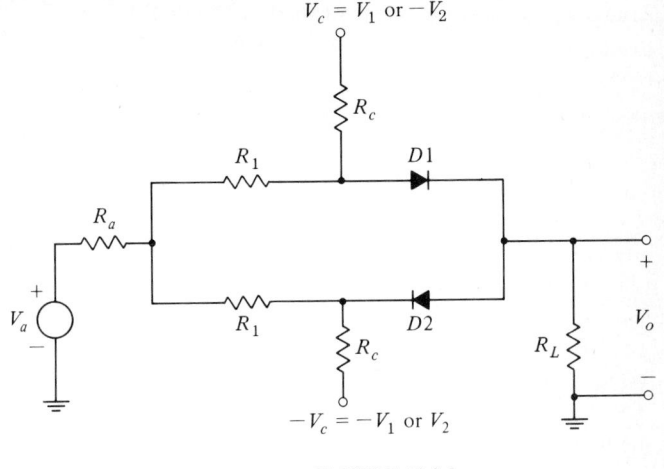

FIGURE 13.3-2
A two-diode bridge transmission gate.

to open and close rapidly, it is not necessary for the gating waveforms to make abrupt transitions between levels. Similarly, as long as the control-voltage levels are able to keep the diodes definitely conducting or definitely nonconducting (independently of V_a), there is no need for the control voltages to remain absolutely constant between transitions. What is important, with respect to balance, is that whatever the shape of the control voltages, they must be *identical in waveshape*, except of course for polarity. If, for example, the two control voltages have different rise times, spikes (called *glitches*) will appear at the output at the times of the opening and closing of the gates. One method which often provides an effective way of generating identically shaped control voltages is to get these waveforms from identical and bifilar-wound windings of a pulse transformer.

Balance also requires that the diodes have identical characteristics. With such considerations in mind, various manufacturers make available *matched-diode assemblies*. These diodes, consisting of a pair or a group of four diodes (quads) or even a group of six diodes, are matched with respect to forward current, reverse current, temperature dependence, and sometimes also frequency response. The quads and six-diode assemblies find application in the four- and six-diode gates, discussed below. The assemblies may consist of individual diodes, diodes in a common encapsulation, or diodes formed by integrated-circuit techniques for the purpose of ensuring a common temperature dependence.

As in the gate of Fig. 13.3-1, so here, in Fig. 13.3-2 the circuit component values and the range of excursion of the analog voltage will determine the required control-voltage levels V_1 and V_2. On the one hand, when the gate is to transmit, $V_c = V_1$ must be sufficiently positive ($-V_c = -V_1$ sufficiently negative) to maintain both diodes conducting over the full range of the analog voltage. Similarly, when the gate is to be closed against transmission, $V_c = -V_2$, which must be adequate to keep both diodes cut off over the full range of the analog voltage. These control-voltage levels and the gate gain V_o/V_a, where

$$\frac{V_o}{V_a} \approx \frac{R_L \left|\left| \frac{R_c}{2} \right.\right.}{R_a + \frac{R_1}{2} + \left(R_L \left|\left| \frac{R_c}{2} \right.\right. \right)} \tag{13.3-1}$$

are calculated in Probs. 13.3-7 and 13.3-8. To arrive at Eq. (13.3-1) we assumed that $V_c = V_1$ and that the diodes are ideal.

A difficulty associated with the gate of Fig. 13.3-2 is that each of the resistors R_1 interposed between the analog-signal source and output has the effect of reducing the gain of the gate. However, reducing the resistance of R_1 has an adverse effect on the control voltage level V_2, the level at which V_c must be held to keep the diodes cut off. As R_1 becomes smaller, the magnitude required of V_2 increases. In general (assuming that the diode cutoff voltage is $V_\gamma = 0$) the minimum allowable value of V_2 is (Prob. 13.3-8):

$$V_2(\text{min}) = \frac{(R_c/R_1)V_a(\text{max})}{1 + 2R_a/(R_1 + R_c)} \tag{13.3-2}$$

where $V_a(\text{max})$ is the maximum positive value of V_a. A quite reasonable situation is one in which R_a is very small in comparison with the sum of R_1 and R_c. In this case Eq. (13.3-2) becomes

$$V_2(\text{min}) = \frac{R_c}{R_1} V_a(\text{max}) \tag{13.3-3}$$

However, even in the general case, using Eq. (13.3-2), it is seen that $V_2 \to \infty$ as $R_1 \to 0$. Thus, the maximum possible gate gain is limited by R_1 through $V_2(\text{min})$.

Four-diode gate An improved diode gate, using four diodes, is shown in Fig. 13.3-3. Note that two *fixed* equal-magnitude voltages are applied to the resistors R_c while the gating voltages $+V_c$ and $-V_c$ are applied through two additional diodes, D3 and D4.

During transmission, with $V_c = V_1$, diodes D3 and D4 are cut off. The gate is then identical in form to the two-diode gate of Fig. 13.3-2 except that the current through diodes D1 and D2 is provided by the fixed reference voltages $+V_r$ and $-V_r$ rather than by the control voltage. In addition, referring to Fig. 13.3-3, we see that the output voltage can never exceed $\pm V_1$. For, if V_a should increase to a value which increases V_o above V_1, diode D3 would conduct, clamping point a to the voltage $V_1 + 0.7$. Similarly, b will be clamped to $-V_1 - 0.7$ if V_a is too negative a voltage.

During nontransmission, with $V_c = -V_2$, the points a and b are clamped by diodes D3 and D4 to $-V_2 + 0.7$ and $V_2 - 0.7$, respectively. Diodes D1 and D2 are therefore cut off and remain cut off for all values of V_a.

In this four-diode gate, resistors R_1 may have a small resistance, thereby increasing the gain of the gate. Their resistance values need only be large enough

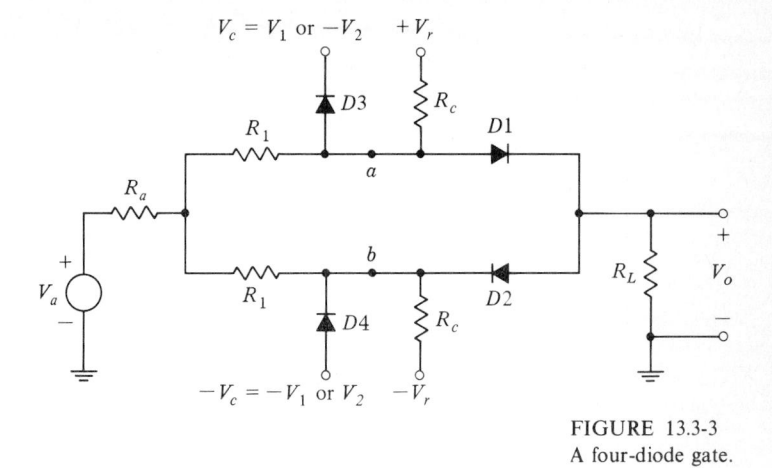

FIGURE 13.3-3
A four-diode gate.

to allow effective clamping so that when $V_c = -V_2$, the voltage at point a is negative and the voltage at point b positive. That is, resistances R_1 and R_c need only be large in comparison with the forward resistances of diodes $D3$ and $D4$. It can readily be verified that the gain of the four-diode gate is identical to Eq. (13.3-1) except that now R_1 can be significantly smaller.

As noted, a first advantage of the four-diode gate is that this gate makes it feasible to realize a gate gain close to unity (since R_1 can be small) without requiring an inconveniently large V_2. A second advantage is that during non-transmission the isolation between input and output is improved. This improvement results from the fact that during nontransmission not only are diodes $D1$ and $D2$ cut off but also points a and b are clamped to fixed voltages by diodes $D3$ and $D4$. A third advantage is to be seen in the fact that to eliminate an output pedestal it is not necessary for the control voltages $+V_c$ and $-V_c$ to be equal in magnitude. The suppression of the pedestal in the present case depends primarily on the symmetry of circuit components and the equal magnitude of the fixed voltages $+V_r$ and $-V_r$.

Alternate form of a four-diode gate   An alternate four-diode gate is shown in Fig. 13.3-4. Here the fixed-voltage sources $+V_r$ and $-V_r$ are not used. When $V_c = V_1$, all four diodes conduct and the gate transmits. When $V_c = -V_2$, the diodes are cut off and the gate is closed against transmission. During transmission the connection between input and output is made through the equivalent of a single forward-biased diode (actually two parallel paths of two series diodes each). During nontransmission, when all four diodes are cut off, the isolation between source and load is certainly better than in the two-diode gate of Fig. 13.3-2. In comparing the present gate with the four-diode gate of Fig. 13.3-3 we must recognize that in Fig. 13.3-3 we may operate with low values of resistance R_1 in order to keep the gain of the gate close to unity. In this case, the

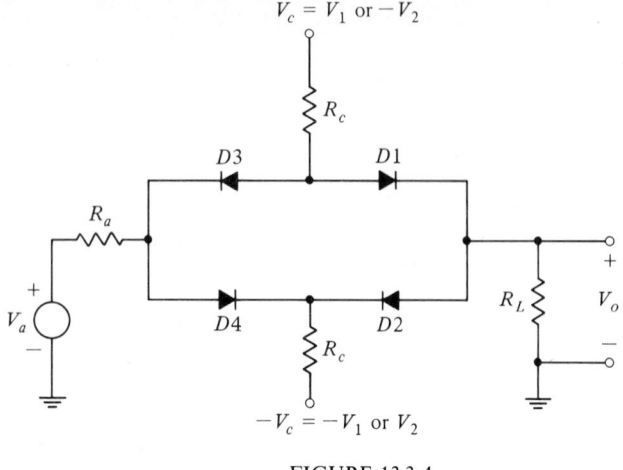

FIGURE 13.3-4
An alternate form of four-diode gate.

clamping action of diodes $D3$ and $D4$ will be of limited effectiveness, and again the isolation of the present four-diode gate will be superior.

Referring to Fig. 13.3-4, we see that if the analog signal V_a attains a peak value $V_a(\text{max})$, then, to keep the gate closed against transmission requires a voltage $V_2 = V_a(\text{max})$. On the other hand, during transmission we require that all four diodes be maintained in conduction. To satisfy this condition requires (Prob. 13.3-9) that the control level V_1 be approximately

$$V_1 \approx \frac{R_c V_a(\text{max})}{R_a + \left(R_L \,\|\, \dfrac{R_c}{2}\right)} = \frac{R_c V_a(\text{max})}{R_a + \dfrac{R_c R_L}{R_c + 2R_L}} \tag{13.3-4}$$

If, say, R_a is small in comparison with R_c and R_L, then Eq. (13.3-4) becomes

$$V_1 = \left(2 + \frac{R_c}{R_L}\right) V_a(\text{max}) \tag{13.3-5}$$

and for $R_c \approx R_L$, $V_1 = 3V_a(\text{max})$. Here then is a possible disadvantage of the present four-diode gate. That is, the required value of V_1 may become inconveniently high. This situation may become substantially worse if the diodes are not matched (see Prob, 13.3-11). On the other hand, suppose that the output of the switch is applied to the inverting terminal of an op-amp. Then the load R_L is effectively the virtual short circuit seen at the op-amp input. In this case $V_1 \approx R_c V_a(\text{max})/R_a$ and V_1 may even be smaller than $V_a(\text{max})$.

Six-diode gate A six-diode gate which combines the features of the four-diode gates of Figs. 13.3-3 and 13.3-4 is shown in Fig. 13.3-5. When the gate is closed against transmission, diodes $D5$ and $D6$ are conducting and serve as clamps,

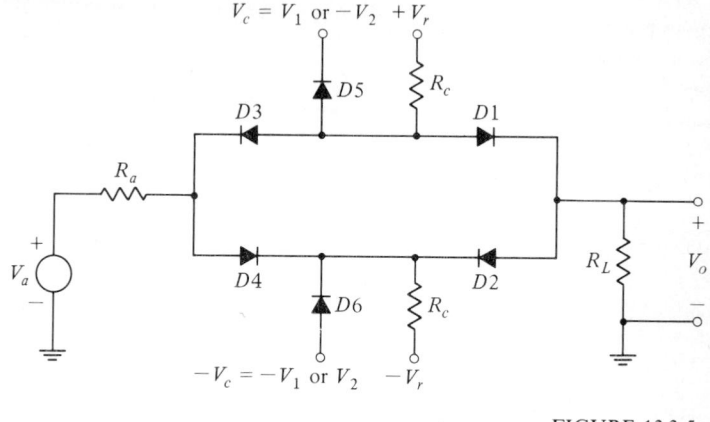

FIGURE 13.3-5
A six-diode gate.

back-biasing all the other diodes. During transmission diodes $D5$ and $D6$ are cut off while the other diodes are conducting.

During transmission, when diodes $D5$ and $D6$ are back-biased, the six-diode gate becomes equivalent to the four-diode gate of Fig. 13.3-4 except that the fixed voltages $+V_r$ and $-V_r$ replace the control-voltage levels V_1 and $-V_1$. Therefore Eqs. (13.3-4) and (13.3-5) apply in the six-diode case with V_1 replaced by V_r. We noted in the four-diode case of Fig. 13.3-4 a disadvantage in the possibly inconveniently large value required for V_1. The advantage of the six-diode case is apparent in the fact that here the large voltage need appear only as a fixed voltage and not as a control signal.

As in the gate of Fig. 13.3-3, so in the present case, the control signals need not be balanced. From Fig. 13.3-5 it appears that if the clamping diodes $D5$ and $D6$ are to remain back-biased for an analog signal of peak value $V_a(\text{max})$, then V_1 must be at least equal to $V_a(\text{max})$. On the other hand, it is also seen that in the nontransmission mode, where $V_c = -V_2$, diodes $D1$ and $D2$ remain cut off for all values of V_a. Thus, V_2 may be any nonnegative voltage. To summarize, we have the results that

$$V_1 = V_a(\text{max}), \qquad V_2 \geq 0$$

Commercially available diode-gate circuits The RCA type CA3019 integrated-circuit diode array consists of six diodes, four of which are internally connected in the manner of the four-diode gate of Fig. 13.3-4 while the other two are individually available. All the diodes are fabricated simultaneously on a single silicon chip and hence have nearly identical characteristics. Also their parameters track each other with small temperature variation as a result of their close proximity and the relatively good thermal conductivity of silicon. A typical six-diode gate using this integrated-circuit diode array is shown in Fig. 13.3-6. It will operate with analog signals up to about 0.5 V rms and at frequencies to about 170 MHz. The gating-signal amplitude required is in the

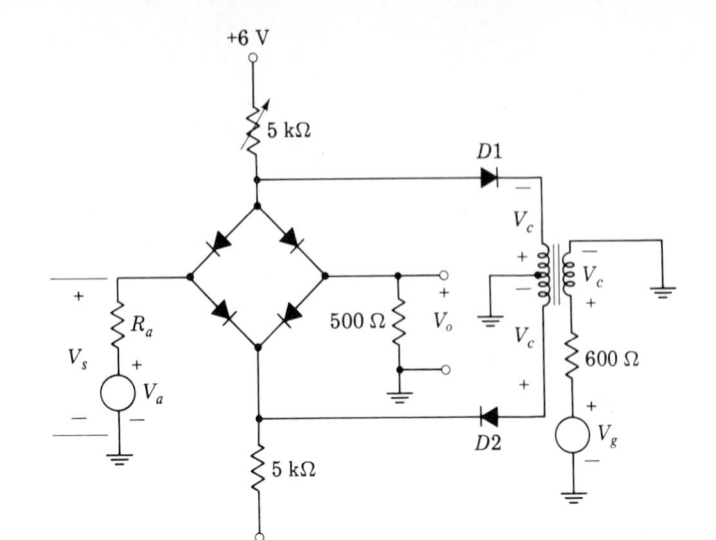

FIGURE 13.3-6
A six-diode gate using the RCA 3019 integrated-circuit diode array.

range 1 to 3 V rms, depending on the amplitude of the analog signal, and may range up to 500 kHz. A matter of principal interest is, of course, the absence of the pedestal at the output. The pedestal output may be observed directly by applying a gating signal and noting the relative amplitude of the gating signal at the output in the absence of an analog signal. In the present circuit, it is found that with V_g set at 2 V rms at 500 kHz and the adjustable 5-kΩ resistor set for minimum output, the output pedestal is 100 mV rms.

In our discussion of diode gates we have indicated the diode switch as a series element. It may of course be used equally well as a shunt element. An example[2] of such a shunt-switch diode gate, again using four of the diodes in the RCA array, is shown in Fig. 13.3-7. The diode $D1$, together with its series resistor, is not properly part of the gate and is not essential to its operation. It does, however, serve to keep the load seen by the control-signal source V_g more nearly constant over its entire cycle. For otherwise, when the polarity of the gating voltage is such that the four diodes of the gate are not conducting, the gating source looks nominally into an open circuit. Because of the higher resistor values involved, this gate will operate at signal frequencies only up to about 500 kHz and at gating frequencies up to about 100 kHz.

Finally, Fig. 13.3-8 shows a six-diode gate which is somewhat different from the gate of Fig. 13.3-6. In the present case no fixed voltages are used. The gate combines the use of diodes as series switch elements and also as shunt diode elements. During transmission, the gating waveform maintains point B positive with respect to point A. Diodes $D1$, $D2$, $D3$, and $D4$ conduct and

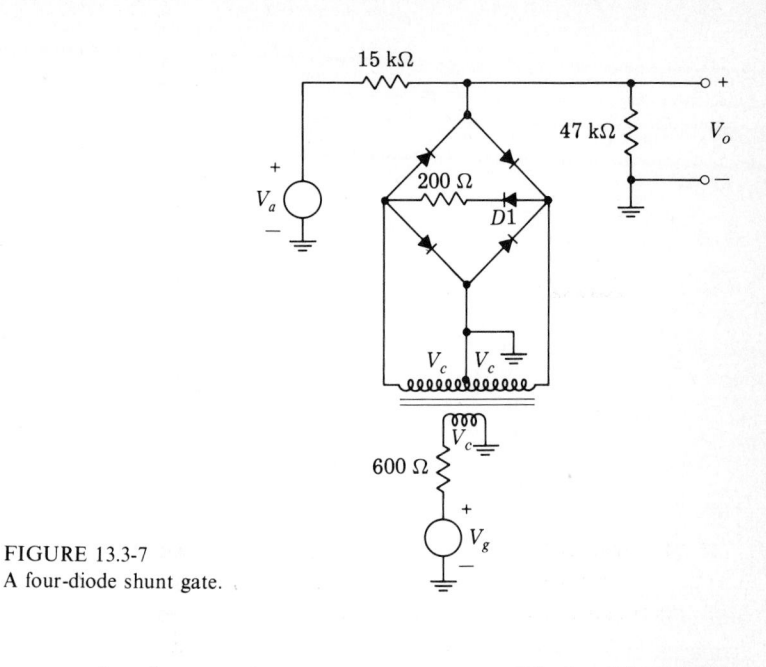

FIGURE 13.3-7
A four-diode shunt gate.

provide a series connection from analog source to output. Diodes *D5* and *D6* are back-biased. When the gate is closed against transmission, not only are the four bridge diodes back-biased, but also diodes *D5* and *D6* conduct, providing shunt paths to ground and thereby further assuring against signal transmission.

A very high speed S/H circuit In the four-diode gate considered in Fig. 13.3-4 we assumed a gating waveform V_c, which consisted of two signals equal in form but of opposite polarity, symmetrical with respect to ground. It is possible to avoid the requirement of two waveforms and to use a single control signal,

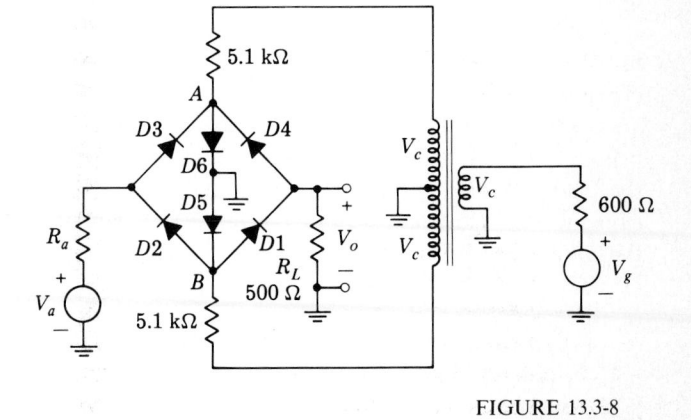

FIGURE 13.3-8
A series-shunt gate.

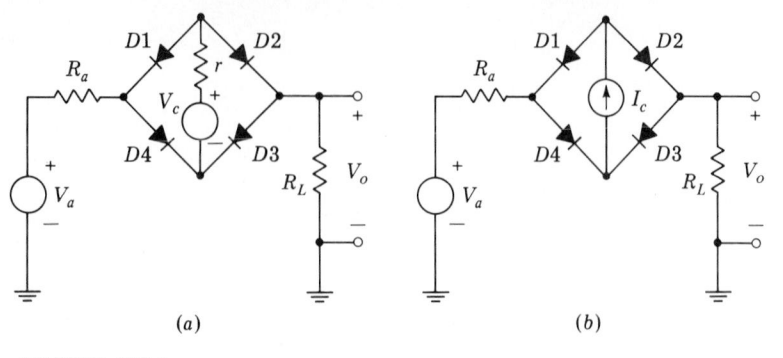

(a)   (b)

FIGURE 13.3-9

A four-diode gate in which the gating control source is not grounded: (a) voltage
source used for gating; (b) current source used for gating.

as we did in Figs. 13.3-6 to 13.3-8. The use of such a single control is
illustrated in Fig. 13.3-9a, where we have moved the control voltage into the
bridge. In Fig. 13.3-9a we see that during transmission the control voltage must
maintain a positive voltage level to keep the diodes conducting. In this case
the resistor r is required to limit the diode current. During nontransmission
the level of V_c is negative to keep the diodes OFF. Alternatively, we may use
the voltage source V_c in Fig. 13.3-9a only to keep the diodes back-biased (in
which case the resistor r may be omitted) and a current source as in Fig. 13.3-9b
to keep the diodes conducting.

The circuit configurations shown in Fig. 13.3-9a and b are employed in the
high-speed sample-and-hold circuit[3] shown in Fig. 13.3-10. This circuit is intended to operate with a charging time of the order of only about 15 ns. Charging
time is the time during which the switch is closed for transmission so that the
capacitor C can charge to the present value of the analog signal. To allow
such speeds, the diodes used are of the *Schottky barrier* type. To simplify the
figure, the Schottky symbol is not employed.

In Fig. 13.3-10 a current I is switched by transistors $T1$ and $T2$ to flow
in one or the other direction in the transformer primary. Thus, during the
charging interval, a nominally constant current flows through the diodes $D3i$
and $D4i$ and in the forward direction through the diodes of the four-diode
bridge. During this charging time diodes $D1$ and $D2$ are back-biased. With
the bridge diodes forward-biased, the input analog voltage is impressed across
capacitor C, which constitutes the load. Note that in this mode of operation
the four-diode bridge is being kept ON by a current source, as in Fig. 13.3-9b.

It is of some interest to estimate the magnetizing inductance required
of the transformer to keep the current in the secondary winding nominally
constant during the charging time, as required to ensure that the bridge diodes
remain fully conducting. A circuit that can be used to calculate the amount of
current decay is shown in Fig. 13.3-11. Figure 13.3-11a shows the secondary
current I_o impressed across the magnetizing inductance L_m and the $2N$ series

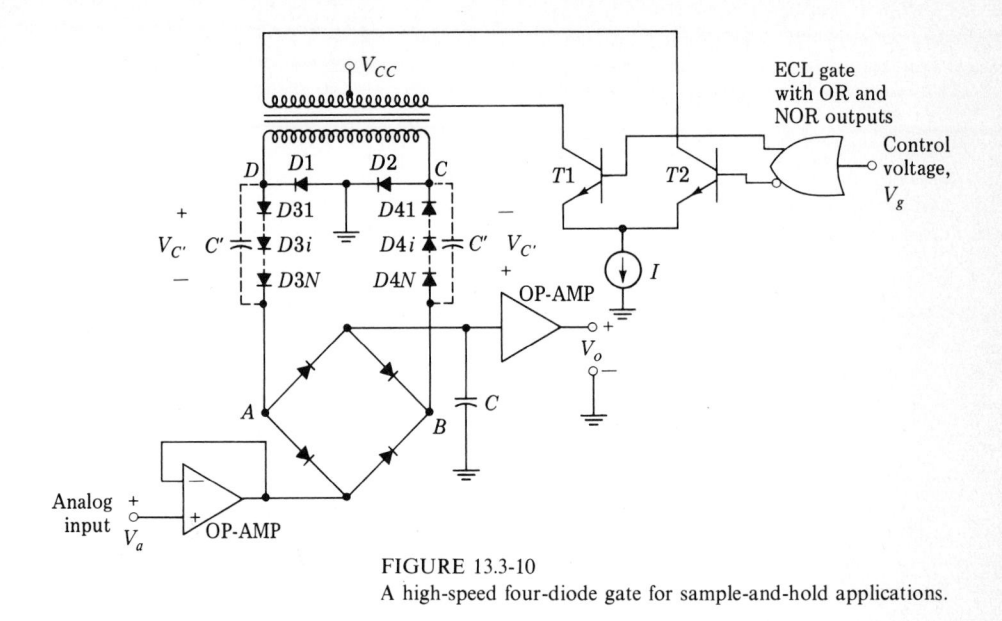

FIGURE 13.3-10

A high-speed four-diode gate for sample-and-hold applications.

diodes and the four bridge diodes. Let us assume that the voltage drop across each of these diodes is 0.7 V when the diode is conducting. The resulting circuit, simplified by combining the voltage drops across the diodes, is then as shown in Fig. 13.3-11b. Assume that at $t = 0$, the forward current I_0 is applied, thereby turning the diodes ON. Initially all the current I_0 flows through the diodes; however, since this current produces a voltage drop $V_{L_m}(t) = 2(N + 1)(0.7)$ across the inductance L_m, the current through L_m increases since

$$I_{L_m}(t) = \frac{1}{L_m} \int V_{L_m}(t) \, dt$$

Hence,

$$I_{L_m}(t) \approx \frac{2(N + 1)0.7}{L_m} t \tag{13.3-6}$$

Thus, the diode current $I_D(t)$ (Fig. 13.13-11b) decreases with time as

$$I_D(t) \approx I_o - \frac{1.4(N + 1)t}{L_m} \tag{13.3-7}$$

To ensure that the diode current remains almost constant during the charging time T_c, we shall design L_m so that the decrease in $I_D(t)$ is 1 percent. Then, with $t = T_c$, we have

$$L_m \approx \frac{140(N + 1)T_c}{I_o} \tag{13.3-8}$$

For example, if $I_o = 10$ mA, $T_c = 20$ ns, and $N = 5$, then $L_m = 840$ μH.

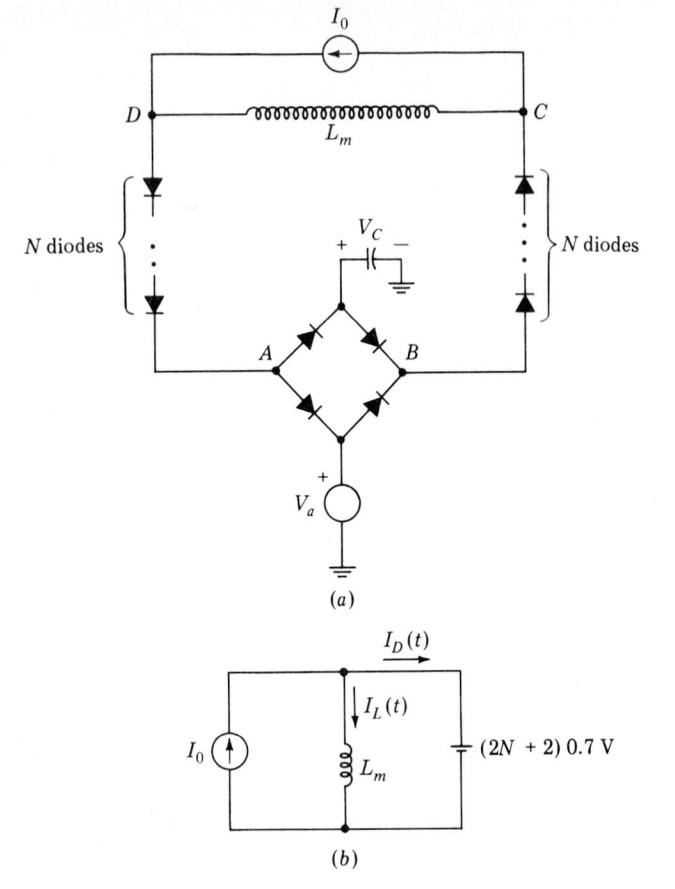

FIGURE 13.3-11
(a) Circuit to calculate the decay of the forward diode current. (b) Equivalent circuit with diodes combined.

Selection of the number of diodes N During the charging interval diodes $D1$ and $D2$ (Fig. 13.3-10) must be reverse-biased. Thus, the voltage at node C in Figs. 13.3-10 and 13.3-11a must be less than 0.65 V and the voltage at mode D must be greater than -0.65 V independently of the value of V_a. If the maximum value of V_a is $V_a(\max)$, then we have, from Fig. 13.3-11a that

$$0.65 \geq V \text{ (at node } C) = V_a - 0.7(N + 1)$$

Solving for N yields

$$N \geq 1.43 V_a(\max) - 1.93 \qquad (13.3\text{-}9)$$

For example, if $|V_a| \leq 5$ V, $N \geq 6$.

Operation during hold time   When the load capacitor C is to be isolated from the signal source, the transformer current is reversed.   Diodes $D1$ and $D2$ shown in Fig. 13.3-10 now become forward-biased.   Since the voltage $V_{C'}$ across the N diode capacitors in each leg (which we have represented by the capacitor C') cannot change instantaneously, the voltage at point B in the diode bridge rises instantaneously to $V_{C'} + V_{D2}$ while the voltage at point A falls to $-(V_{C'} + V_{D1})$. Assuming, again, that there are N diodes present, we see that $V_{BA} = (2N + 2)0.7$ V and therefore the diode bridge is instantaneously reverse-biased.

The voltage drop $V_{C'}$ across the stray capacitors C' rapidly decreases to zero, the discharge path being through the diodes $D3i$ and $D4i$.   After the discharge is completed, $V_{BA} = 1.4$ V and the bridge is maintained at cutoff.   It is readily verified that the bridge remains cut off, independently of V_a, until the transformer current is again reversed.

Note that in this mode of operation the diode bridge is maintained at cutoff by a voltage source as in Fig. 13.3-9a, the voltage source, in steady state, being the voltage drop across diodes $D1$ and $D2$.

13.4   BIPOLAR JUNCTION TRANSISTOR GATES

Diode gate circuits have the important advantage of speed, especially when *Schottky barrier diodes* are used.   Diodes, however, being two-terminal devices without control electrodes, provide no gain and lead to somewhat cumbersome circuits.   We are therefore led to inquire into the feasibility of transistor (bipolar or FET) gate circuits.

A simple transistor gate is shown in Fig. 13.4-1a.   A bipolar transistor is indicated, but an FET would serve as well.   The control voltage V_c may be used to carry the transistor into its active region, thereby allowing transmission of the analog signal V_a to the output; or V_c may carry the transistor into cutoff, thereby closing the gate.   This simple circuit suffers the disadvantage that a pedestal will appear at the output.   An attempt to suppress the pedestal is indicated in Fig. 13.4-1b.   Here, two transistors are used.   The two gating voltages of opposite polarity are designed to keep the quiescent load current the same, both when $T1$ is ON and $T2$ OFF (gate OPEN) and when $T1$ is OFF and $T2$ ON (gate CLOSED). The proposed arrangement can, in principle, be adjusted to eliminate the pedestal. However, as indicated in Fig. 13.4-1c, even with entirely symmetrical gating waveforms, objectionable spikes may appear at the times of opening and closing of the gate.   The difficulty arises because of the practical problems of trying to devise a technique whereby one transistor turns ON precisely at the same time and to the same extent that the other transistor is turned OFF.   The situation is different in the case of diode circuits, where the gating voltage drives (ON at the same time or OFF at the same time) all diodes through which the signal finds a path to the output.

If the rise and fall times of the gating voltages are short in comparison with the intervals during which the gates are kept OPEN or CLOSED, the spikes

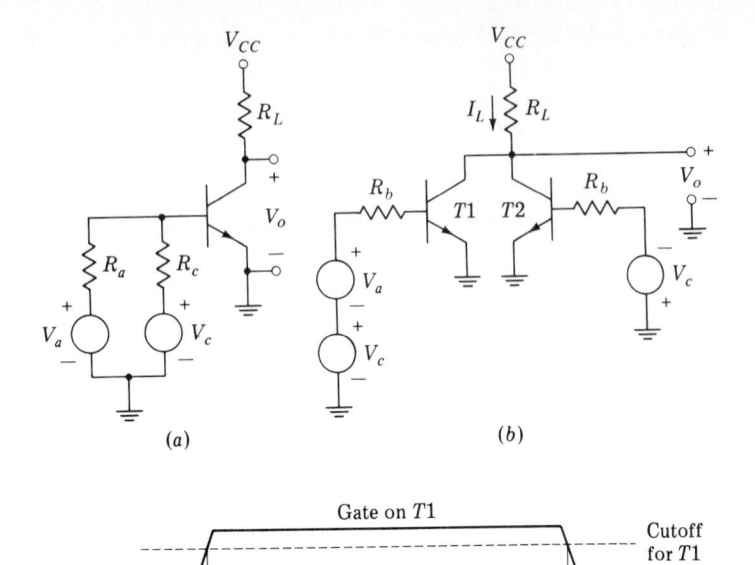

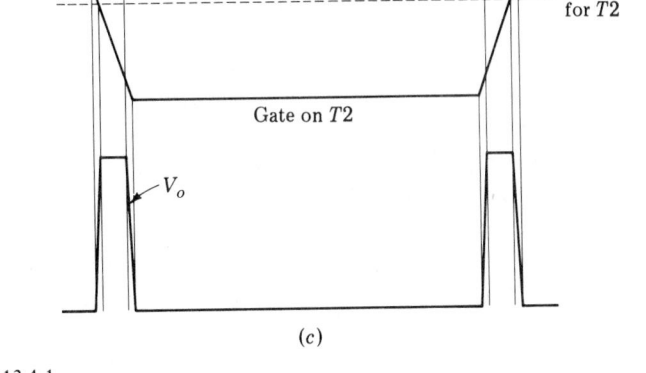

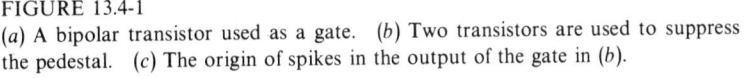

FIGURE 13.4-1
(a) A bipolar transistor used as a gate. (b) Two transistors are used to suppress the pedestal. (c) The origin of spikes in the output of the gate in (b).

described above may be relatively short enough not to be seriously objectionable. Nonetheless, because of these spikes and because of the difficulty of matching and keeping the two transistors matched well enough to suppress the pedestal, gates employing transistors in the amplifier mode are of limited applicability. Therefore, where transistors are used in a switching mode, they are used as illustrated in the three circuits of Fig. 13.1-1. With bipolar transistors the switch terminals are the collector and emitter, and the control voltage is applied to the base terminal. With FETs the switch terminals are source and drain, and the control voltage is applied to the gate.

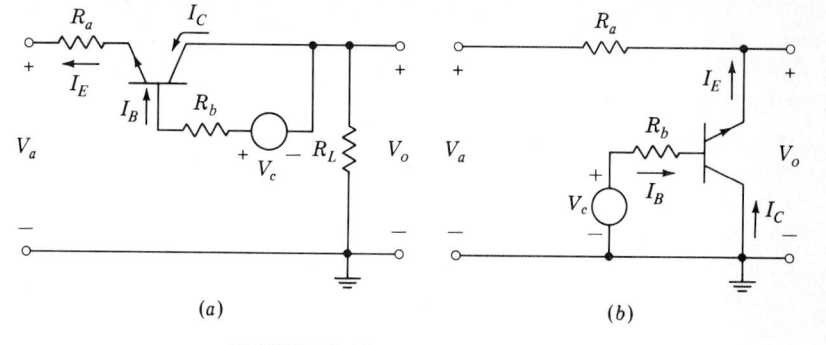

FIGURE 13.4-2
Transistor used as (a) a series element and (b) a shunt element.

In the transmission gates of Fig. 13.4-1 the transistors are used in the manner of amplifiers. An analog signal is applied to the base, and an output is taken at the collector. Alternative bipolar transistor gates are shown in Fig. 13.4-2. Here the transistors are used in the manner of variable-impedance elements. In Fig. 13.4-2a, to allow transmission, the transistor is driven to saturation by the control source V_c. To restrain transmission the transistor is driven to cutoff. In Fig. 13.4-2b transmission is allowed when the transistor is cut off and is not allowed when the transistor is in saturation. The circuit in Fig. 13.4-2b has some relative merit in that it allows the use of a control source, one terminal of which may be grounded in common with one terminal of the signal source. In Fig. 13.4-2a the control source will generally have to be applied through a transformer.

In Sec. 1.10 we noted that when a transistor is driven well into saturation ($\sigma \ll 1$), the collector-to-emitter saturation voltage is different, depending on whether the transistor is operating in the *normal* mode or in the *inverted* mode. In the normal mode, the base-drive source is connected between base and emitter, and we showed that

$$V_{CE}(\text{sat}) \approx V_T \ln \frac{1}{\alpha_I} \qquad (13.4\text{-}1)$$

With $\alpha_I \approx 0.1$ we find $V_{CE}(\text{sat}) \approx 60$ mV. On the other hand, in the inverted mode, where the base-drive source is connected between base and collector, we found that

$$V_{CE}(\text{sat}) = -V_T \ln \frac{1}{\alpha_N} \qquad (13.4\text{-}2)$$

With $\alpha_N = 0.98$, we have $V_{CE}(\text{sat}) = -1$ mV. The inverted-mode saturated transistor is clearly more nearly a closed ideal switch than the saturated transistor in the normal mode. It is for this reason that the transistors are driven in the inverted mode in the gates of Fig. 13.4-2.

The voltage across the saturated transistors in gate circuits is referred to as the *offset voltage.* In Fig. 13.4-2a, assuming $R_a \ll R_L$, the output V_o will be offset from V_a by the saturation voltage whenever the gate circuit allows transmission. Similarly in Fig. 13.4-2b the output V_o will be offset from its intended value (ground, 0 V) whenever the gate circuit is closed against transmission. The existence of an offset voltage is disadvantageous generally since it will generate a pedestal at the output. It is particularly disadvantageous when the gate is being used as a chopper where the input voltage is very small. In an ideal chopper circuit, as we have noted, we seek to generate at the gate output a square wave of amplitude equal to the magnitude of the input V_a. Consider then that in the circuits of Fig. 13.4-2, $V_a = 0$ V. Then there will be present at the output nonetheless a square wave of amplitude equal to the offset voltage. The smallest input voltage for which the chopper can effectively be used is limited by this residual square wave due to the offset voltage.

On account of the difficulties associated with the offset voltage and transistor nonlinearities in the saturation region, the gate circuits using bipolar transistor gates as transmission elements, as in Fig. 13.4-2, have fallen out of favor. The circuits are widely used, however, with FETs, as discussed in the next section.

13.5 FET GATES

A series and a shunt gate, each using an FET, are shown in Fig. 13.5-1a and b. The diagrams indicate MOS devices, but JFETs may be used as well, and, of course, the series and shunt switching may be combined in the manner of Fig. 13.1-1c to yield even more effective gating.

One important advantage of the FET switch over the bipolar transistor switch is that the FET switch exhibits no offset. That such is the case can be seen in Fig. 1.11-3. Here, we have reproduced the drain voltage-ampere characteristic of a typical JFET in the neighborhood of the origin. The voltage-ampere characteristic of a typical MOSFET is identical to the characteristic

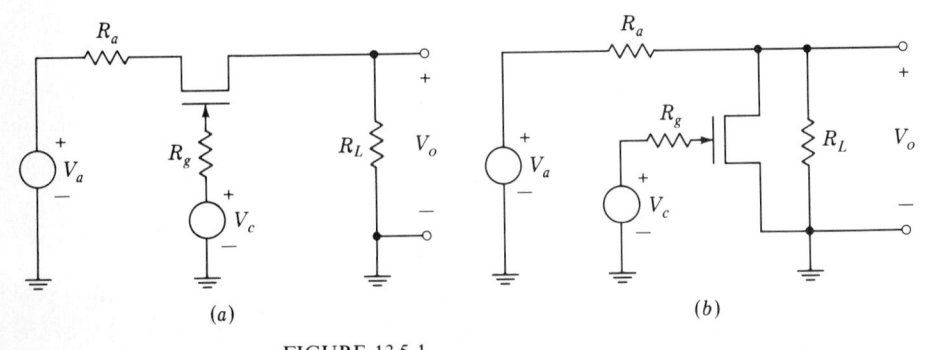

FIGURE 13.5-1
(a) An FET series-controlled gate. (b) An FET shunt-controlled gate.

shown. We note the precision with which the characteristics pass through the origin. We may note further that if voltage and current excursions are limited to a region quite close to the origin, the characteristics are those of a "linear" resistor albeit with a resistance which is a function of gate-to-source voltage.

A second advantage of the FET switch is its symmetry with respect to the origin. The FET has similar characteristics in its forward and inverse operating modes. The distinction between the FET and the BJT is easily seen by comparing Figs. 1.11-3 and 1.10-1. Figure 1.10-1 shows the voltage-ampere characteristic of the bipolar transistor when I_B is constant.

13.6 OPERATIONAL AMPLIFIERS

Operational amplifiers were discussed in Chap. 2. Since, however, we shall have occasion to include op-amps in gates to be discussed in succeeding sections in this chapter, we restate here some of the especially relevant points about such amplifiers. In Fig. 13.6-1a the amplifier is shown with the input signal and the feedback resistor R_2 directed toward the inverting ($-$) input while the noninverting input ($+$) is grounded. The amplifier has an open-loop gain A, an input impedance R_i, and an open-loop output impedance R_o, all of which are specified by the manufacturer. Then (for A and R_i large) the gain A_f and the output impedance Z_o are given very closely by

$$A_f \approx -\frac{R_2}{R_1} \tag{13.6-1}$$

and
$$Z_o = \frac{R_0}{A}\left(1 + \frac{R_2}{R_1}\right) \tag{13.6-2}$$

Assuming $R_1 = R_2$, we find $A_f = -1$, and assuming $R_o = 200 \ \Omega$ and $A = 5 \times 10^4$, we find $Z_o = 8 \ \text{m}\Omega$. The important point to note is that the output impedance is very low and hence such an amplifier is ideal for providing rapid charging and discharging for capacitors like those, for example, in sample-and-hold circuits.

A second noteworthy point about the circuit in Fig. 13.6-1a is that looking into the input terminals of the amplifier proper, we see a virtual ground. That is, the impedance r_i is very small. Applying Eq. (2.2-2), we find

$$r_i \approx \frac{R_2 + R_o}{1 + A} \approx \frac{R_2}{A} \tag{13.6-3}$$

if, as is typical, $R_2 \gg R_o$. Assuming $R_1 = R_2 = 10 \ \text{k}\Omega$, we find $r_i \approx 0.2 \ \Omega$. Since r_i is so small in comparison with R_1, we may consider that a virtual short circuit to ground exists at the amplifier input terminals. Hence an input voltage V_i generates an input current $I = V_i/R_1$. Since, however, the short circuit is only virtual, the current continues through R_2. The output is therefore $V_o = -IR_2/R_1$, giving a gain $A_f = V_o/V_i = -R_2/R_1$, as in Eq. (13.6-1). More importantly, for our purpose, is to note the following. Suppose, say with

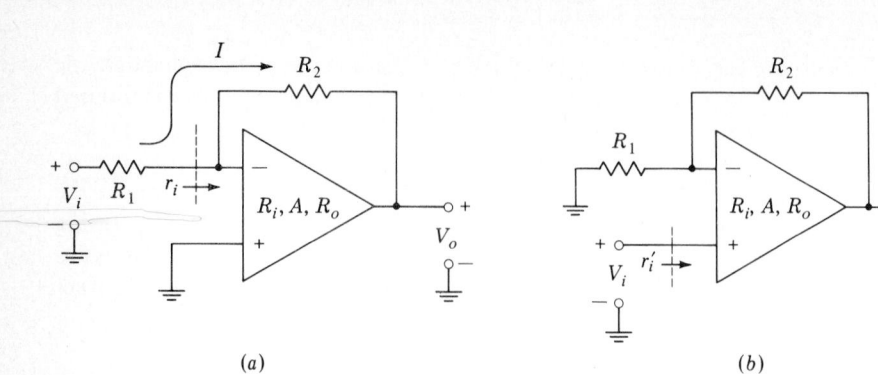

FIGURE 13.6-1

(*a*) An op-amp employing current error.   (*b*) An op-amp employing voltage error.

$R_1 = R_2$, V_i makes excursion $\pm M$; then V_o will make equal excursions $\pm M$, but at the amplifier input terminal the excursion will only be $\pm M/|A|$.   Thus with $M = 10$ V and $|A| = 5 \times 10^4$ we find that at the amplifier terminal the swing is only $\pm 10/5 \times 10^4 = \pm 0.2$ mV.

An amplifier can also be used in the manner of Fig. 13.6-1*b*, where the input is applied to the noninverting terminal.   Note that the feedback resistor R_2 is always connected to the *inverting* input.   In this case we find that the output impedance is again given as Eq. (13.6-2).   The gain is now (see Eq. (2.7-2))

$$A_f = 1 + \frac{R_2}{R_1} \qquad (13.6\text{-}4)$$

A gain $A_f = 1$ is obtained if R_2 is set at $R_2 = 0$.   In this case R_1 serves no purpose and may be removed.

If R_i is the input impedance of the amplifier proper, then the input impedance r_i' is (see Eq. (2.8-5))

$$r_i' = R_i \left(\frac{A}{1 + R_2/R_1} \right) \qquad (13.6\text{-}5)$$

This input impedance is generally very large.   Typically, with $R_2 = 0$ Ω, $A = 5 \times 10^4$, and $R_i = 10^5$ Ω, the input r_i' is $r_i' = 5,000$ MΩ.

13.7   AN FET GATE WITH AN OP-AMP LOAD

A simple, relatively inexpensive series gate is shown in Fig. 13.7-1.   The input analog signal V_a is transmitted to the output when the transistor $T1$ conducts. The control signal V_c, to turn $T1$ ON and OFF, is provided by the TTL driver. Ordinarily the voltage corresponding to logic **1** at the output of a TTL gate is in the neighborhood of 3.7 V (see Sec. 6.6).   In the present application it will be useful to arrange that logic **1** correspond to the full supply voltage of 5 V.

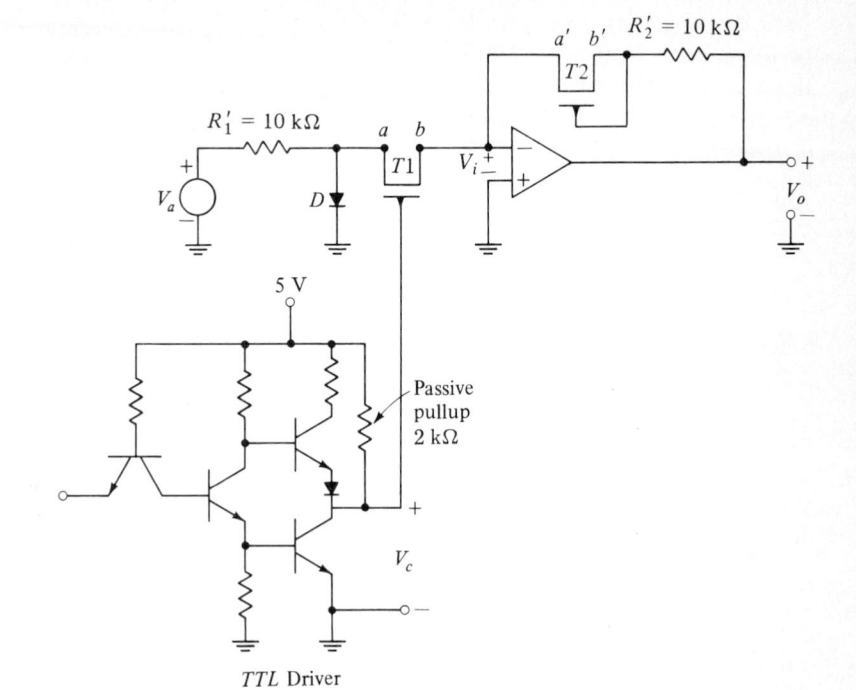

FIGURE 13.7-1
A series FET switch and its TTL driver.

For this reason there has been added the 2-kΩ passive pull-up resistor. (Such a resistor is frequently added when, as in the present case, it is not required that the voltage rise of V_c from 3.7 to 5 V be accomplished with the speed normally associated with TTL logic and when, in addition, the load on the driver is quite light.)

The transistors $T1$ and $T2$ are nominally identical p-channel devices and are of the depletion type. It will be recalled (see Sec. 1.12) that depletion transistors conduct when the gate-to-source voltage is zero. Further, in a p-channel transistor conduction increases as the source-to-gate voltage V_{SG} goes toward more positive voltages and decreases as V_{SG} goes toward more negative voltages. Typically the threshold voltage V_T is $V_{SG} = -4$ V, and we shall use this figure in our subsequent discussion.

Assume, now, that $V_c = +5$ V and let V_a go positive. In this case, if current were to flow through $T1$, it would flow from a to b. Such current direction would make a the source and b the drain. (Recall that the source is the terminal at which carriers, in this case holes, enter the transistor.) Because of the diode $D1$ the voltage at a will be clamped at 0.7 V. Hence, no matter how positive V_a may become, V_{SG} will never be greater than $-5 + 0.7 = -4.3$ V and hence $T1$ will stay OFF with a margin of safety of 0.3 V. Analogously to the situation which prevails in logic gates, this 0.3-V margin is called a *noise margin*.

Next, let V_a go negative. In this case, if $T1$ were to conduct, terminal b would be the source. Since the input terminal of the amplifier is virtually at ground, the 5 V on the gate keeps $T1$ OFF with a margin of $5 - 4 = 1$ V.

The transistor $T1$ is turned ON to allow transmission; and when the gate is transmitting, transistor $T2$ is intended to provide compensation for the temperature sensitivity and the possible slight nonlinearity of $T1$. The incremental gain (as well as the absolute gain) of the stage is given by

$$A = \frac{\Delta V_o}{\Delta V_i} = - \frac{R'_2 + R(T2)}{R'_1 + R(T1)} \tag{13.7-1}$$

Since, as can be noted in Fig. 13.7-1, R'_2 and R'_1 have been selected to be equal, if $R(T2) = R(T1)$, we shall have $A = -1$ independently of $R(T1)$.

In connection with our interest that $R(T2) = R(T1)$ the following points are pertinent. When V_c goes to logic **0** to turn $T1$ ON, we may well expect to find the gate of $T1$ nearly at ground since $V_c \approx 0.1$ V or even lower [rather than our customarily assumed $V_{CE}(\text{sat}) = 0.2$ V] on account of the very light load on the driver. We shall therefore make no serious error if we assume that just as the gate of $T2$ is connected to b', so also is the gate of $T1$ connected to b, since b is *virtually* at ground. Next, note that because b is at ground only *virtually* and not actually, no current flows into the amplifier input terminal and the transistor currents are equal and in the same direction. Hence, altogether, since the transistors are similarly connected and carry equal currents in the same direction, we may reasonably expect that $R(T2) = R(T1)$.

When the gate is transmitting, diode D does not conduct and serves no function. For assume $T1$ to have an ON resistance of 150 Ω. Then for a maximum 10-V input signal, we calculate that the voltage across the diode will be $10(150/(10,000 + 150)) \approx 150$ mV, which is less than the 0.65-V cut-in voltage of the diode.

In the present circuit the switch transistor $T1$ is located at the virtual-ground side of the 10-kΩ series resistor. In principle, $T1$ could be located at the input side of the resistor. The virtual-ground-side location has some merit. For, in this case, neither terminal of the switch is subjected to any appreciable voltage variation with respect to ground, and as a consequence the requirements on the gating voltage are thereby eased (see Prob. 13.7-2). Finally we may note that in this circuit the transistors are operating bilaterally, i.e., carrying current in one or the other direction depending on the polarity of the input. When V_a is positive, terminals a and a' are sources and b and b' are drains. When V_a is negative, source and drain interchange.

13.8 A SAMPLE-AND-HOLD CIRCUIT

The pattern of the circuit of Fig. 13.7-1 appears again in the sample-and-hold circuit shown in Fig. 13.8-1. When the FET is ON, the circuit is an amplifier with a dc gain equal to $-R_2/R_1$ and has a 3-dB frequency $f = 1/2\pi R_2 C$. In these expressions we have neglected the FET impedance. When the FET is

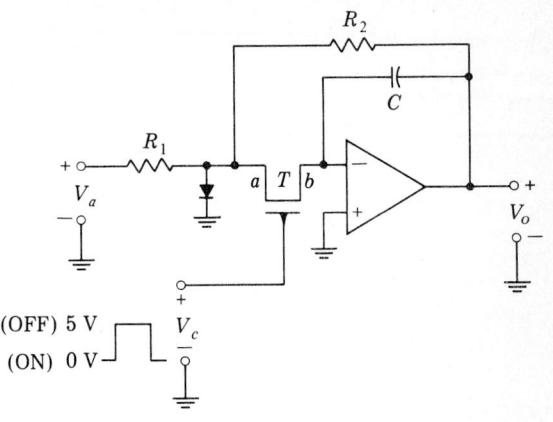

FIGURE 13.8-1
A sample-and-hold circuit.

turned OFF, the amplifier output voltage is "held" by the capacitor C. The capacitor will not hold its charge indefinitely. Instead, assuming that the leakage through the FET is negligible, the capacitor will gradually change its stored charge at a rate determined by the input bias current of the amplifier. In order to minimize this drift in the voltage to be held, it is advantageous to use a capacitor C as large as is consistent with the required frequency response of the amplifier. It is further advantageous to use an amplifier with small input bias current such as an op-amp with an FET input stage. Finally, we may note that the need to keep the discharge rate of the capacitor at a minimum accounts for the fact that the resistor R_2 is not bridged directly across C but is instead connected to the input of the FET switch. In this circuit, when the FET is open, R_2 is no longer in parallel with C.

Figure 13.8-2a shows a circuit combining the features of the circuits of Figs. 13.7-1 and 13.8-1. This circuit combines the sample-and-hold capability with the features of allowing the time multiplexing of three input signals, V_{a1}, V_{a2}, and V_{a3}.

In the circuits of Figs. 13.8-1 and 13.8-2a, as in other sample-and-hold circuits, we are required, for special applications, to take account of the capacitance in the switch device (in this case the junction FET) between the control electrode and the device terminal connected to the storage capacitance. This device capacitance (together with any parallel stray capacitance) gives rise to what is referred to as a *sample-to-hold offset error*. To illustrate the source of this error, consider the situation at the moment when the control voltage switches from the ON level to the OFF level. We require that after the transfer to the OFF level (hold) has been made, the charge on the storage capacitor shall remain the same as it was at the end of the charging interval just before the transfer to hold occurred. However, the abrupt change in control voltage will transfer a charge to or from the storage capacitor C, thereby generating an *offset error*. If C_d is the device

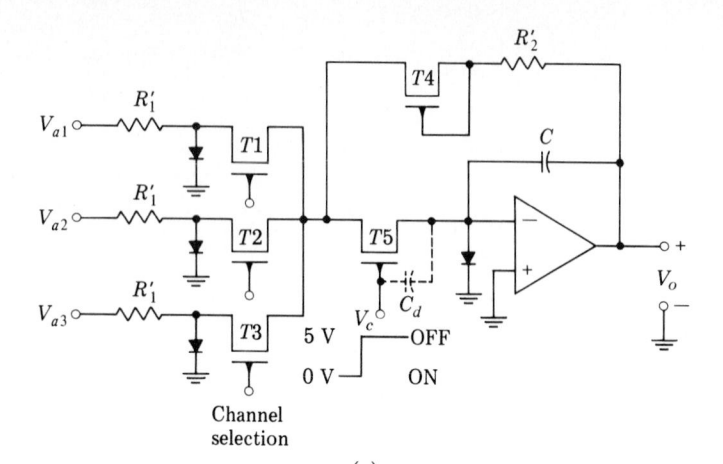

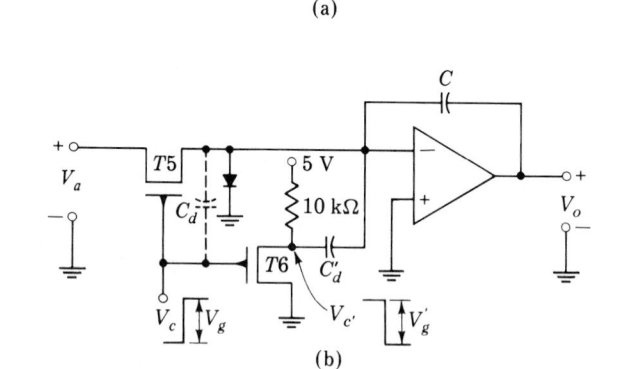

FIGURE 13.8-2
(*a*) A three-channel multiplexer, sample-and-hold amplifier.   (*b*) The transistor *T6* and capacitor C_d' are used to cancel the offset error.

capacitance (we ignore, for simplicity, the likelihood that C_d is a nonlinear capacitor), and if the *difference* in control voltage levels is V_g, then, as can be verified, the sample-to-hold offset error ΔV is given by

$$\Delta V = \frac{C_d}{C} V_g \tag{13.8-1}$$

If $V_g \approx 5$ V, $C_d \approx 10$ pF, and $C \approx 0.01$ μF, we find $\Delta V = 10$ mV.   When errors of this magnitude are not acceptable, it is necessary to provide some compensation for this *charge injection*.   A simple and straightforward method of compensation starts by making available a waveform similar to the switching waveform except of reversed polarity, as shown in Fig. 13.8-2*b*.   This reversed waveform of amplitude V_g' is now applied through a capacitor C_d' to the input of the storage capacitor, and a good measure of compensation is achieved by making $C_d' V_g' = C_d V_g$ (see Prob. 13.8-1).

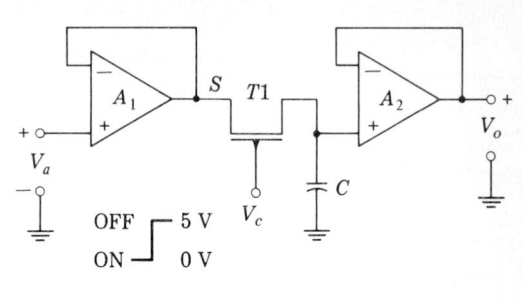

FIGURE 13.8-3
A sample-and-hold circuit using two op-
amps connected for high input impedance.

When offset-error compensation is required, the circuits of Figs. 13.8-1 and 13.8-2 have a special merit. The virtual ground of the operational amplifier keeps the input side of the storage capacitor at a fixed voltage. Hence, the offset error and the compensation required are constant, independently of the voltage of the analog signal.

The second type of error, called *glitching*, is also reduced by the presence of capacitor C_d'. A glitch is a pulse which occurs at the op-amp input of Fig. 13.8-2a when the control voltage V_c goes from logic **1** to logic **0**. Due to the stray capacitance C_d, this negative-going voltage step appears at the input to the op-amp. However, since switch $T5$ is closed when V_c is at logic **0**, the step decays rapidly (the time constant being the product of C_d and the effective resistance of the FET switch), forming a very sharp negative pulse (glitch). In Fig. 13.8-2b we see that as a result of the presence of $T6$ and C_d' a positive glitch is also presented to the input of the op-amp. This results in a significant reduction in glitch size.

Other sample-and-hold circuits[4]   In the circuit of Fig. 13.8-3, amplifier A_1 provides a high input impedance for the analog signal and a low output impedance to allow rapid charging and discharging of the storage capacitor C. Amplifier A_2 provides a buffer between the storage capacitor and the output. As in the previous circuit, the *drift rate*, i.e., the rate at which the storage capacitor voltage changes during the hold period, will be determined by the size of the storage capacitor, the input bias circuit current of the amplifier A_2, and the leakage through the FET switching device. We observe in the present case that the voltage with respect to ground at the switch terminals may range over the full excursions of the analog signal and that a larger-amplitude gate-control voltage will therefore be required than when the switch was located at a virtual ground.

The circuit of Fig. 13.8-3 suffers from an offset error which is different from the sample-to-hold offset error discussed earlier. This error (not confined to sample-and-hold circuits) simply reflects the characteristics of amplifiers generally. When an amplifier is adjusted so that the output voltage is zero when the input voltage is zero, such adjustment may not persist. With time, there will be a drift in operation of the amplifier so that there will develop an offset of the output with respect to the input. Similarly in the amplifiers of Fig. 13.8-3, whose

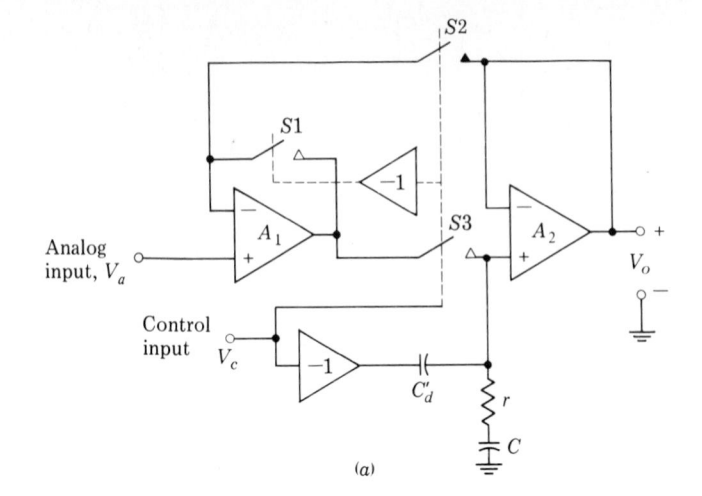

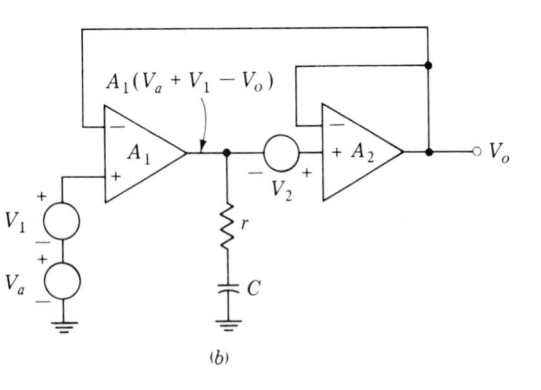

(b)

FIGURE 13.8-4
(a) The Intersil type 5110 sample-and-hold amplifier. (b) Equivalent circuit when S_1 is open and S_2 and S_3 closed, to show effect of offset voltages V_1 and V_2.

individual gains should be $+1$ with great precision, where an offset has developed, the output of the individual amplifiers will never quite equal the input, and so also for the operation of the entire circuit.

An interesting sample-and-hold circuit based on the circuit of Fig. 13.8-3 but incorporating a feature which suppresses somewhat the amplifier offset errors is shown in Fig. 13.8-4a. The circuit is that of the Intersil type 5110 general-purpose sample-and-hold amplifier. The switches, shown here functionally, are CMOS devices (see Sec. 13.10 for a discussion of the operation of the CMOS gate). The amplifiers A_1 and A_2 are op-amps, and the inverters are usually TTL inverters since the control input typically comes from a TTL driver. In this circuit, the storage capacitor C (≈ 0.01 to 0.1 μF) is in series with a resistor r (≈ 50 Ω). This resistor is in principle disadvantageous since it somewhat limits the speed with which C can charge. It serves, however, to suppress the inclination of A_1 to oscillate, a feature fairly characteristic of monolithic operational amplifiers driving large capacitances. Capacitor C_d' is used with a reversed-polarity switching waveform to reduce the offset error (see Fig. 13.8-2b). This compensation can be dispensed with where an offset error of the order of 10 mV is acceptable.

The operation of this gate is as follows. During sampling $S2$ and $S3$ are CLOSED and $S1$ is OPEN. In this mode, the present circuit is different from the circuit of Fig. 13.8-3 in that the feedback in Fig. 13.8-3 around A_1 is omitted and feedback around the entire amplifier cascade from output to input has been substituted. This overall feedback is shown in Fig. 13.8-4b, where V_1 and V_2 have been introduced to represent the input offset voltage of A_1 and A_2, respectively. The overall offset of the circuit will now be just the offset of A_1 because the offset of A_2 will be reduced as a result of the feedback, by the open-loop gain A_1 of the first amplifier. This result can be demonstrated by noting that the output of amplifier A_1 is $A_1(V_a + V_1 + V_o)$ and the output of amplifier A_2 is V_o, where

$$V_o = A_2[A_1(V_a + V_1 - V_o) + V_2 - V_o] \tag{13.8-2}$$

Hence

$$V_o = \frac{A_1 A_2 (V_a + V_1)}{1 + A_2 + A_1 A_2} + \frac{A_2 V_2}{1 + A_2 + A_1 A_2} \approx V_a + V_1 + \frac{1}{A_1} V_2 \tag{13.8-3}$$

The arrangement shown in Fig. 13.8-4a not only has the advantage of reducing the overall offset error but also permits using a single offset adjustment rather than two, one for each of the amplifiers. It is common practice to set $V_o = 0$ V by adjusting the input level of amplifier A_1 when $V_a = 0$.

The need for switch $S1$ can be seen from the following. During sampling, when the feedback is effective, the output of A_1 follows the analog signal very closely. During hold, when $S2$ and $S3$ are OPEN, if $S1$ were not CLOSED, there would be no feedback around amplifier A_1 and because of its high gain and capacitance, it would oscillate. Even if we were to neglect the possibility of oscillations, we would have other problems. For, the output of amplifier A_1 would be not V_a but $A_1 V_a$, at least up to the point where amplifier A_1 saturates. Hence, at the end of the hold interval, when $S2$ and $S3$ close again, there would be a delay before the output of A_1 changes from $A_1 V_a$ (or comes out of saturation) to V_a. Keeping switch $S1$ closed permits the output of A_1 to follow V_a all during the hold period.

The circuits of Figs. 13.8-3 and 13.8-4 use two amplifiers, one to provide a high input impedance for the analog signal and the second to isolate the storage capacitor C. The circuit of Fig. 13.8-5 is interesting in that it is a circuit wherein a single amplifier serves both functions. During sampling, transistors $T1$ and $T3$ are ON while $T2$ is OFF. The amplifier now provides a high input impedance to V_a and charges C from its low output impedance. During hold, $T1$ and $T3$ are OFF and $T2$ is ON. The amplifier is now a buffer between the capacitor and the output.

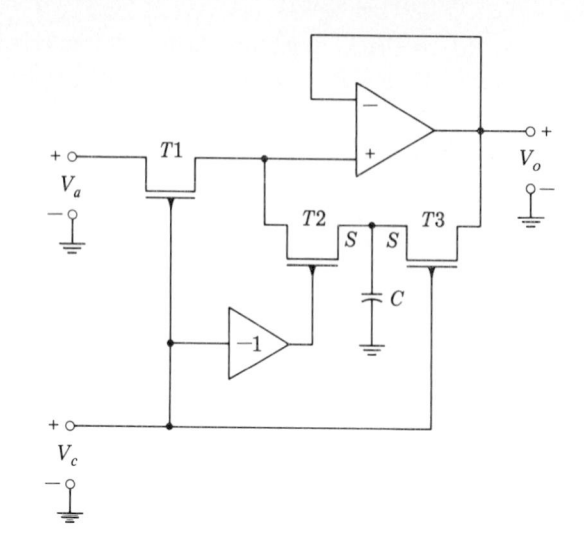

FIGURE 13.8-5

A sample-and-hold circuit patterned on the circuit of Fig. 13.8-4 but using only a single op-amp.

This circuit also has an advantage with respect to the sample-to-hold offset. The gate-source voltages present on $T2$ and $T3$ are of the same magnitude since the capacitor voltage appears at the source terminal of both $T2$ and $T3$, but since the gating voltages are of opposite polarity, the charge transfer through the gate-source capacitances of $T2$ and $T3$ to C will be equal but opposite.

13.9 FET GATE DRIVERS

The gate-to-source voltage of an FET determines whether the device is ON or OFF. (In a MOSFET, cutoff depends also, albeit with much less sensitivity, on the substrate-to-source voltage.) In many gate circuits, the source voltage varies with the analog signal, and account must be taken of this feature of circuit operation to determine how large an analog signal can be handled in a particular switching circuit. The situation would be otherwise if the gating voltage were applied directly between gate and source. However, such an arrangement generally requires the use of a transformer with its attendant limitations. As already noted, the matter is of minimal concern when the switch FET is located at the virtual-ground input of an op-amp. It is of serious concern, however, when the switch is located as in, say, Fig. 13.5-1, where the voltage excursion of the source is comparable to the excursion of the analog signal itself.

To be specific, consider the simple gate circuit of Fig. 13.9-1a using a type 3N126 n-channel junction FET. The FET is fully turned ON when the gate-to-source voltage is $V_{GS} = 0$ V and is fully turned OFF with $V_{GS} = -4$ V. Suppose

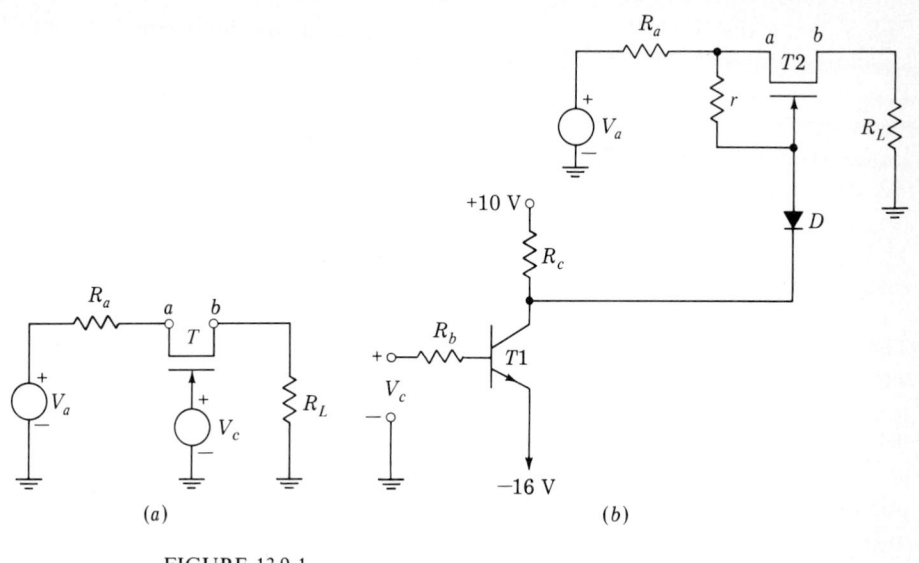

FIGURE 13.9-1
(*a*) A simple FET switch.   (*b*) An FET switch using a compensated control voltage.

then that we swing V_c between the two levels $V_c = 0$ V and $V_c = -4$ V.   Then the maximum allowable input-voltage swing will be determined by conditions that prevail when the switch is ON with $V_c = 0$ V.   For when V_a becomes negative, the *pn* diode formed between gate and source is being driven in the direction of forward bias.   Hence, the maximum allowable negative excursion of the input must be restricted to rather less than about 0.6 V, say about 0.4 V. When the input swings positively, V_{GS} goes negative and the switch begins to turn OFF.   It will be completely turned OFF with $V_{GS} = -4$ V but will begin to turn OFF with, say, $V_{GS} = -2$ V.   Hence, the allowable analog-input swing is about from $+2$ to -0.4 V.   (It is easily verified that the allowable swing when the FET is OFF is less restrictive.)

This restriction of the allowable input swing can obviously be remedied by devising a circuit whereby the gate voltage is not fixed but follows the source voltage.   This can be accomplished in a variety of ways (short of connecting the gating signal between gate and source).   One simple and effective method is indicated in Fig. 13.9-1*b*.   With this circuit, as we shall see, an input-voltage range of about ± 10 V is allowable.   When the bipolar transistor $T1$ is OFF, the diode D is OFF, since its *n* terminal is at $+10$ V.   If V_a is negative, then terminal *a* is the source and the gate-to-source voltage of the FET is at 0 V because of the connection through resistor *r*.   If V_a is positive, current flow through the FET reverses and therefore the source and drain terminals reverse. The gate-to-source voltage is now always slightly positive since the gate is now at the same potential as the drain.   Assuming $R_a \ll R_L$, the diode will remain OFF and the FET ON until the input voltage becomes $V_a = 10 + 0.65 = 10.65$ V,

0.65 V being the diode cut-in voltage. At that point diode D conducts, and (assuming that $R_c \ll r$) with increasing V_a the gate of $T2$ will be clamped at $V_a = 10 + 0.75 = 10.75$ V, 0.75 V being the ON voltage of the diode. Further increase in V_a, however, will increase the voltage at b, which is now the source. Hence, eventually the gate transistor will turn OFF. If we neglect the drop across $T2$, so long as it is ON, in comparison with the drop across R_L and assume also that $R_a \ll R_L$, cutoff of $T2$ will occur at $V_a = 10.75 + 4 = 14.75$ V. To allow a good margin of safety and to keep $T2$ well away from cutoff we might well restrict V_a to $V_a < +10$ V.

To turn OFF the FET, the bipolar transistor is turned ON and saturated. The n terminal of the diode is then approximately -15.8 V. In this case, even when V_a goes to $V_a = -10$ V, the gate-to-source voltage is -5.1 V (since there is a 0.7 V drop across the diode) and the FET is OFF. Positive excursions of V_a drive the FET further in the direction of cutoff since the gate voltage remains at -15.1 V and the drain and source reverse so that the source is at ground potential ($V_{R_L} = 0$). Hence the limitation on such positive excursions is the drain-to-gate breakdown voltage of the FET.

Gate drive for the MOSFET With a MOSFET, the problem of arranging for relatively large symmetrical analog-voltage swings in the gate circuit is different from the corresponding problem in the JFET. In the MOSFET the gate is insulated from source and drain. There are, however, junctions at the interface between source and substrate and between drain and substrate. These junctions

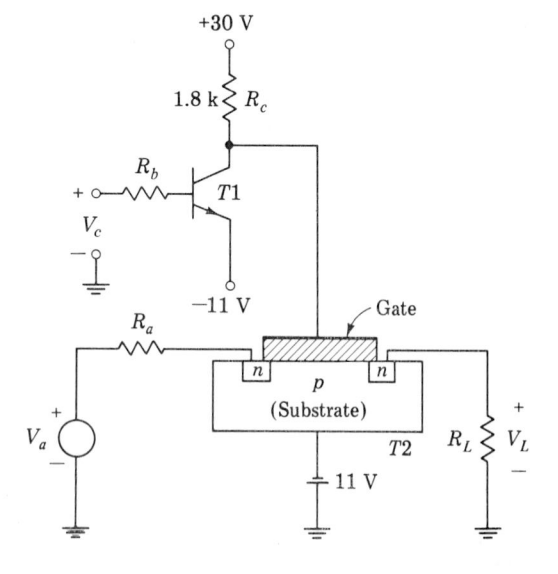

FIGURE 13.9-2
The substrate is returned to a voltage which assures that no pn junction is forward-biased.

must not be allowed to become forward-biased. One way to achieve this is simply to allow the substrate to *float*, i.e., to make no connection to the substrate lead of the device. In integrated circuitry, where the substrate is common to many individual devices, such floating is not acceptable because the substrate would then be a source of coupling between all the individual devices. In discrete circuitry such floating is feasible and may be resorted to. Even in discrete circuitry, however, the floating may be objectionable. For the substrate may be connected to the device case, and the substrate may then find itself coupled, through stray capacitance, to many signals from which it should be shielded.

Instead, as shown in Fig. 13.9-2, the substrate of the n-channel MOSFET is connected to the most negative point in the circuit. Thus, when $V_a = \pm 10$ V, the substrate remains back-biased with respect to the source and drain.

13.10 CMOS GATES

In gating operations, CMOS transistors have some advantages over simple MOS devices, as we shall now consider. We can see some limitations of the MOS gate by examining the situation represented in Fig. 13.10-1a. Here an analog source is connected to a load through an n-channel MOSFET. The load is represented as a parallel combination of R_L and C_L. If a sample-and-hold circuit is intended, the load is a relatively large capacitor C_L and R_L represents the input impedance of the op-amp connected to V_o (see Fig. 13.8-3). If an analog-signal switch is intended, R_L is the load resistance and C_L represents the small but unavoidable stray capacitance. We have omitted the analog-source output impedance as not being particularly relevant in the present discussion.

To be specific, let us consider that the analog signal makes excursions in the range ± 5 V. Then, as noted in Fig. 13.10-1, in order to assure that the junctions between substrate and source and drain are never forward-biased, we have returned the substrate to -5 V. Let us further assume that the FET is turned ON when the gate-to-source voltage is greater than or equal to the threshold voltage $V_T = +2$ V and that otherwise the FET is OFF. Then since the maximum possible voltage attainable at either switch terminal of the FET is $+5$ V, to assure that the FET is ON we shall have to supply a gate voltage of at least $+7$ V, as indicated on the figure. On the other hand, to keep the FET OFF even when the voltage at a switch terminal is -5 V we shall need a gate voltage of less than -3 V, again as noted on the figure. Actually, this gate-voltage range, from $+7$ to -3 V, is just barely enough to assure turn on and turnoff, and as matter of practice we would extend this range somewhat.

Now let us give some consideration to the supply voltages that would be required to operate the switch shown in Fig. 13.10-1a. Figure 13.10-1b shows a simple gate-driving circuit which will serve to furnish the gating voltage levels required in Fig. 13.10-1a. When the bipolar transistor is cut off, we shall have $V_c = +7$ V, and when the bipolar transistor is saturated, we shall have $V_c \approx -3$ V. We note in Fig. 13.10-1a that we require a supply voltage of -5 V for the substrate. Hence in Fig. 13.10-1b we might eliminate the -3-V supply and

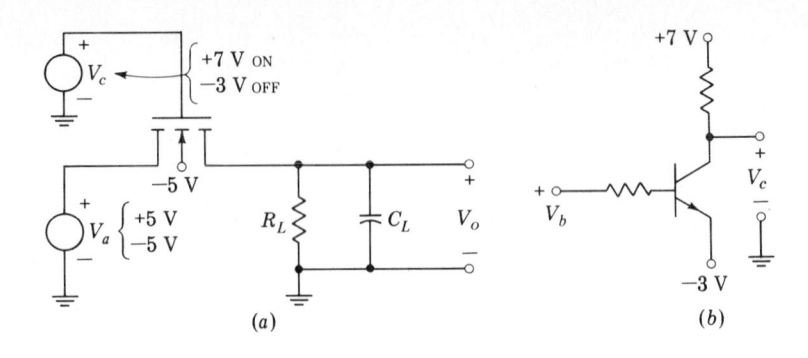

FIGURE 13.10-1

(a) An n-channel MOS gate.   (b) A bipolar transistor driver for the gate in (a).

return the emitter directly to -5 V, thereby incidentally providing an extra margin of assurance that the FET will be OFF when required.   There are two important things to notice: (1) to handle an analog signal whose total range extends over 10 V we need a power supply of -3 V and a supply of $+7$ V, a total range of 10 V, and (2) depending on V_a and V_o, the FET will sometimes be just barely ON and sometimes will be ON very substantially.   Even if we increase the gate voltage beyond $V_c = +7$ V and -3 V to avoid operating near cutoff, nonetheless the resistance between source and drain of the FET will vary widely, depending on V_a and V_o.   And whenever the FET resistance is large, the speed with which C_L can charge or discharge will be limited.

Next let us consider the circuit shown in Fig. 13.10-2, in which a *complementary-symmetry MOSFET* (CMOS) has been employed.   Note that the source and drain locations of the two transistors are reversed.   Thus, when a and b are respectively the source and drain of one transistor, the source and drain of the other transistor are b' and a', respectively.   As before, we consider that V_a makes excursions over the range ± 5 V.   As before, the substrate of the n-channel device is returned to -5 V, and, symmetrically, the substrate of the p-channel device is returned to $+5$ V.   Gating waveforms of opposite polarities are required, as indicated on the figure.   We shall again assume that to turn ON the n-channel device requires a gate-to-source voltage, $+2$ V.   Correspondingly, to turn ON the p-channel device requires a source-to-gate voltage, of 2 V.

Note that the gating-voltage levels are $+5$ and -5 V.   When V_c is $V_c = -5$ V, that is, -5 V at the gate of the n-channel and $+5$ V at the gate of the p-channel device, both FETS are substantially turned OFF no matter where in its range V_a is to be found.   ($V_c = -3$ V would just bring the devices to cutoff in the worst case.)   Next, consider that $V_c = +5$ V, and let us assume that the input voltage V_a is 0 V.   Now each FET is ON and $V_{GS}(T1) = V_{SG}(T2)$ is 3 V above cutoff.   Now let the input voltage change from 0 V.   Then one of the devices will be driven toward cutoff and will reach cutoff when the source changes by 3 V.   But at the same time the other device will be driven further

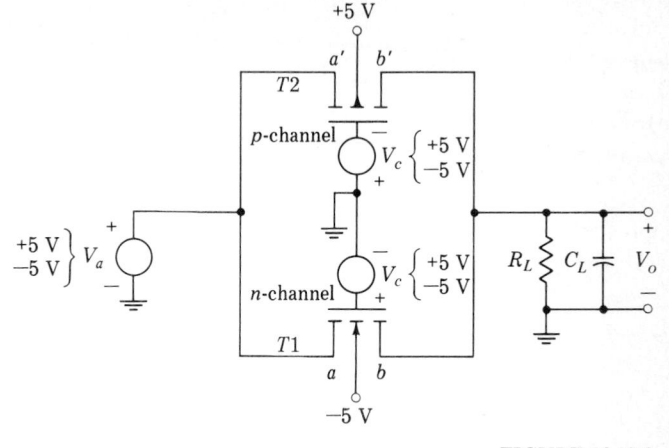

FIGURE 13.10-2
A CMOS gate.

into its ON region of operation. Over the entire range of input voltage to be encountered in the circuit, at least one FET will be ON.

In the present circuit we have the convenience that the required gating voltages extend over the range of excursion of the analog input. Further, the present circuit has an advantage over the single-MOS circuit in that in the CMOS circuit the resistance presented by the switch from source to drain is approximately constant. Such is the case because as the input voltage changes, the resistance of one FET increases while the resistance of the other FET decreases, the parallel resistance remaining approximately unchanged (see Prob. 13.10-2). In the single-MOSFET switch the resistance varies greatly. An important additional advantage of the CMOS transmission gate was discussed in Sec. 12.6.

13.11 APPLICATION OF ANALOG SWITCHES

Multiplexing Multiplexers (see Sec. 13.2) are used in most pulse-code-modulation (PCM) communication systems and have many other applications as well. Figure 13.11-1 shows a three-channel multiplexer with its associated gating drive and a ring counter (Sec. 10.16), which is used to turn switches $T1$ to $T3$ ON and OFF in sequence. Each of the four analog switches shown employs a p-channel depletion-mode JFET, as described in Sec. 13.7. However, MOS, CMOS, or diode gates can also be used. The gate drive is provided by the TTL gates.

Each gating amplifier $G1$, $G2$, $G3$ is driven by two signals. The first, V_B, provides *blanking*. For, when V_B is *low*, all the switches $T1$ to $T3$ are OPEN and the signals V_{a1} to V_{a3} are not transmitted to the output. The second input to each gate is the appropriate output of the ring counter. The counter is designed

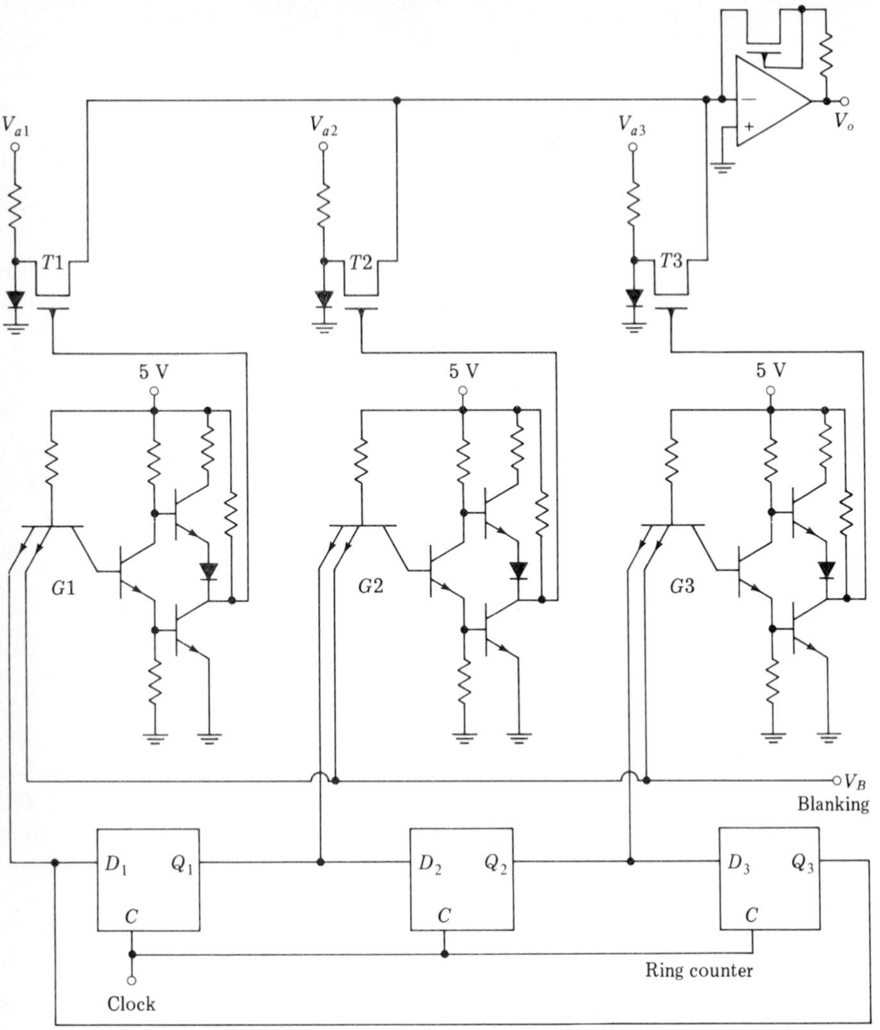

FIGURE 13.11-1
A three-channel multiplexer.

so that only one output of the counter is in the high state at any one time. Hence only one switch is CLOSED at a time. Thus, the output V_o represents the samples of $V_{a1}(t)$, $V_{a2}(t)$, and $V_{a3}(t)$ taken in succession.

Maximum-voltage decision making   In many digital communication systems it is necessary at regular intervals separated by a time T_s to select the largest of a number of available input voltages. This operation for four input signals is accomplished using the circuit shown in Fig. 13.11-2a. In this circuit the input voltages V_1, V_2, V_3, and V_4 are to be compared and the largest voltage is

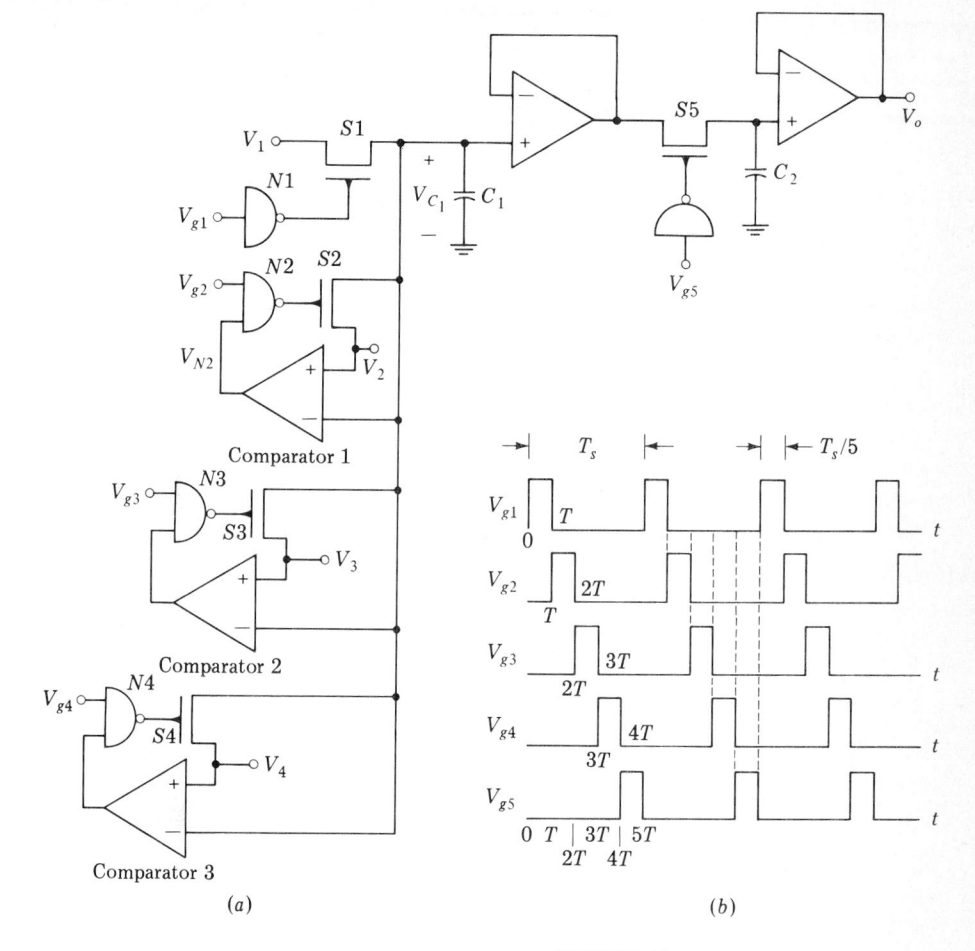

FIGURE 13.11-2
(*a*) Circuit diagram.   (*b*) Timing diagram.

presented at the output as V_o.   The procedure employed is to first compare V_1 and V_2 and select the largest, then compare the selected voltage with V_3, etc.

To compare V_2 with V_1 we first close switch $S1$, for the time interval 0 to T (see Fig. 13.11-2*b*), during which time the capacitor C_1 charges to V_1.   Then, during the time interval T to $2T$, V_2 is compared with V_1 in comparator 1. If V_2 is greater than $V_{C_1} = V_1$, the comparator output is positive, the output of NAND gate $N2$ goes to logic **0**, switch $S2$ closes, and capacitor C_1 charges to V_2. If, however, $V_2 < V_1$, switch $S2$ remains open and the capacitor voltage remains equal to V_{C_1}.   Note that the comparison of V_2 and the capacitor voltage V_{C_1} occurs throughout the time interval T_s and not just during the time T to $2T$. However, only during the time T to $2T$ can switch $S2$ be closed, as $V_{g2} = 1$ during this interval.

This procedure continues until $t = 4T$.   At this time $S5$ closes and capacitor

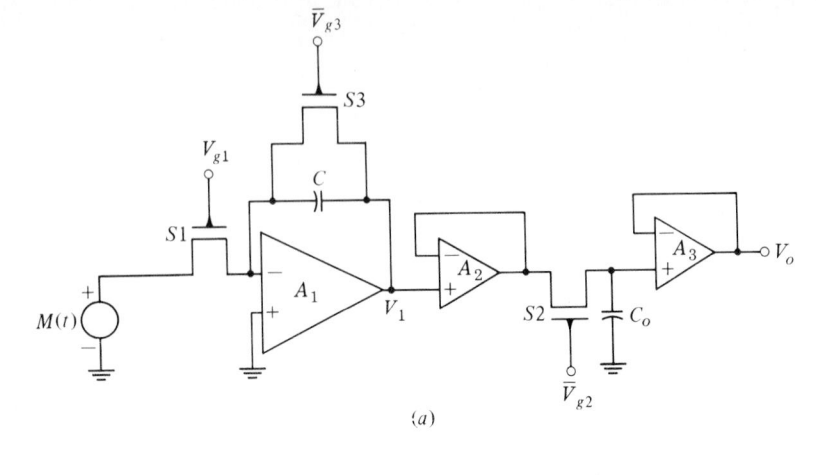

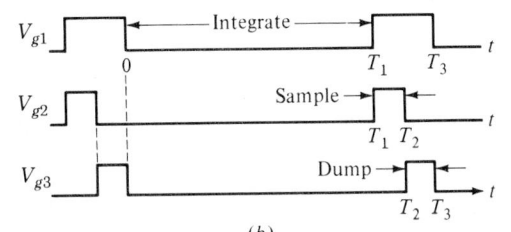

(a)

(b)

FIGURE 13.11-3
(a) Integrate-and-dump circuit. (b) Timing diagram.

C_2 charges to $V_{C_1}(\max)$. The output voltage holds this voltage until $t = 9T$, · at which time the voltage on C_2 changes to the new value of $V_{C_1}(\max)$.

A five-stage ring counter can be used to generate the gate pulses $V_{g1}, \ldots, V_{g5}$. This counter is not shown in the figure (see Sec. 10.16). By extending the concept illustrated in the circuit of Fig. 13.11-2a, we see that $N - 1$ comparators are required to compare N signals. It can be shown (Prob. 13.11-2) that a single comparator can be used if N input signals are time-division-multiplexed.

An integrate-and-dump circuit An *integrate-and-dump* circuit is shown in Fig. 13.11-3a. As explained in Sec. 13.2, this circuit is used to integrate the input waveform $M(t)$ for a time interval T_1, present the integrated voltage to the output, and then repeat this procedure. However, before each integration is begun, the capacitor which stores the integrated input voltage must be *dumped*.

The op-amp integrator shown in Fig. 13.11-3a is capable of dumping capacitor C. For when switch S1 is OPEN and switch S3 is CLOSED, the capacitor discharges through the FET switch. The integrated voltage V_1 is presented at the output V_o when switch S2 is closed.

Timing waveforms are shown in Fig. 13.11-3b. Note that the input voltage is integrated during the interval $t = 0$ to $t = T_1$. From $t = T_1$ to $t = T_2$ the integrator output V_1 is sampled and stored on capacitor C_o. Note that during this interval the input is disconnected from op-amp A_1. The capacitor C is discharged from T_2 to T_3. A new cycle begins at T_3.

13.12 MANUFACTURER'S SPECIFICATION OF S/H AMPLIFIERS

Manufacturers of sample-and-hold amplifiers generally provide the following information concerning their products.

Input voltage The maximum voltage range, typically ± 5 or ± 10 V.

Input impedance When the sampling switch is closed, i.e., during the charging time T_C, the input is typically 1 kΩ in series with a 100-pF capacitor. When the switch is open (during T_H) the input impedance is very high, usually $> 1,000$ MΩ.

Slew rate The *slew rate* is the maximum rate of change of voltage per unit time possible for the S/H amplifier. This is usually limited by the rise-and-fall-time capability of the op-amp. A typical number is 50 V/μs. To ensure that the S/H will follow the input signal we should make sure that the slew rate exceeds the maximum slope of the input signals. Note that if a 10-V step is applied at the input to a S/H amplifier with a slew rate of 50 V/μs, the system will reach 10 V in $10/50$ μs = 200 ns. This time to acquire a full-scale voltage step is called the *acquisition time*.

Settling (charge time) T_C The time required for the output to come to within a given percentage (usually 0.01 percent) of its steady-state value following a full-scale (10 V if ± 5-V input is allowed) step-input voltage; 5 μs is a typical number.

Aperture time This is the uncertainty between the time that the switch is "told" to open and the time that the switch has opened. This disconnect uncertainty is usually less than 10 ns.

Hold decay or droop During the hold time, the voltage on the capacitor will *decay* or droop. A typical value might be 1 mV/ms.

Switch (control) signal This specifies the logic level and voltage needed to turn the switch ON and OFF. This specification varies greatly with the manufacturer. For example, one manufacturer will *sample* with a **1** and *hold* on a **0** while another manufacturer will *sample* with a **0** and *hold* with a **1**.

REFERENCES

1 Taub, H., and D. L. Schilling: "Principles of Communication Systems," McGraw-Hill, New York, 1971.

2 Theriault, G. E., and R. Tipping: Applications of the RCA-CA 3019 Integrated-Circuit Diode Array, *RCA Appl. Note* 1, ICAN-5299.

3 Horna, O. A.: A 150 M bps A/D and D/A Conversion System, *COMSAT Tech. Rev.*, vol. 2, no. 1, pp. 39–72 (Spring 1972).

4 Burd, M., and B. Sear: High Performance Sample and Holds, *Electron. Eng.*, December 1967, p. 60.

14

ANALOG-TO-DIGITAL CONVERSIONS

14.1 INTRODUCTION

Digital signals, ideally at least, are represented by waveforms which make abrupt transitions between two values. Signals which may assume any value in a continuous range are called analog signals. When analog signals must be processed, there is often a great advantage in converting the signal to digital form so that the processing can be done digitally.

By way of example, an analog voltage may be a fixed (dc) voltage, and the required processing may consist of determining its value. An analog voltmeter will display the voltage value through the position of a pointer on a scale. A digital voltmeter will give a more convenient numerical indication on a display panel. Or the analog signal may be the time-varying output voltage $M(t)$ of a microphone. The required processing may then consist of transmitting $M(t)$ to the input of a distantly located loudspeaker so as to minimize the effect of noise (random unpredictable disturbance) which is invariably superimposed on the signal during transmission. A most effective way of suppressing noise is to transmit the signal digitally. A communication system which operates in this way, i.e., converts the analog signal to digital form and then reconstitutes the analog signal, is called a *pulse-code modulation* system. Or, again, given an input analog signal $M_i(t)$, the processing may consist of some digital operation such that when an output analog signal $M_o(t)$ is eventually reconstituted, the signals $M_i(t)$ and $M_o(t)$ are related by the transfer function of a *digital filter*.

At the input end of such a digital processing system, the overall process of converting an analog signal to a digital form involves a sequence of four individual processes, called *sampling, holding, quantizing*, and *encoding*. These processes are not necessarily performed as separate operations. Rather generally, as discussed in Chap. 13, sampling and holding are done simultaneously in a type of circuit referred to as a sample-and-hold (S/H) circuit, while quantizing and encoding are done simultaneously in a circuit referred to as an *analog-to-digital* (A/D) *converter*. After the digital processing is completed, the reconstitution of an analog output signal is accomplished by the operations of *digital-to-analog* (D/A) *conversion* followed by *filtering* or *smoothing*. In this chapter we shall discuss the circuits used to perform each of these functions.

14.2 THE SAMPLING THEOREM[1]

The validity of the entire process depends fundamentally on the *sampling theorem*, which we shall now state. This theorem is so well known and is discussed at such lengths in texts on communication theory and systems analysis that we shall not prove it here.

Let $M(t)$ be a signal which is band-limited such that its highest-frequency spectral component is f_m. Let the values of $M(t)$ be determined at regular intervals $T_s \leq 1/2f_m$; that is, the signal is to be sampled regularly every T_s or more frequently. Then these samples uniquely determine the signal, and the signal can be reconstructed from these samples with no error. The signal $M(t)$ can be reconstructed precisely by transmitting the samples through an ideal low-pass filter which has a flat response at least to f_m and cuts off at a frequency less than or equal to $f_s - f_m$, where $f_s = 1/T_s$.

The time T_s is called the *sampling time*. Note that the theorem requires that the *sampling rate* f_s be rapid enough for at least two samples to be taken during the course of the period corresponding to the highest-frequency spectral component of the signal $M(t)$.

The significance of the sampling theorem is illustrated in Fig. 14.2-1. In Fig. 14.2-1a the sampling of the band-limited signal is accomplished by the FET switch and its associated sampling signal. The sampling signal, which is the FET gate-control signal, consists, as shown, of a pulse sequence, the pulses having a duration τ and a period T_s. The gate allows transmission only during the interval τ. The signal, except for a multiplicative constant α is reconstituted at the output of the low-pass filter. A portion of a typical signal is shown in Fig. 14.2-1b, and the sampled waveform is shown in Fig. 14.2-1c. The sampled waveform $M_s(t)$ consists of pulses. The duration of each pulse is equal to the time interval during which the FET switch is closed. During each sampling

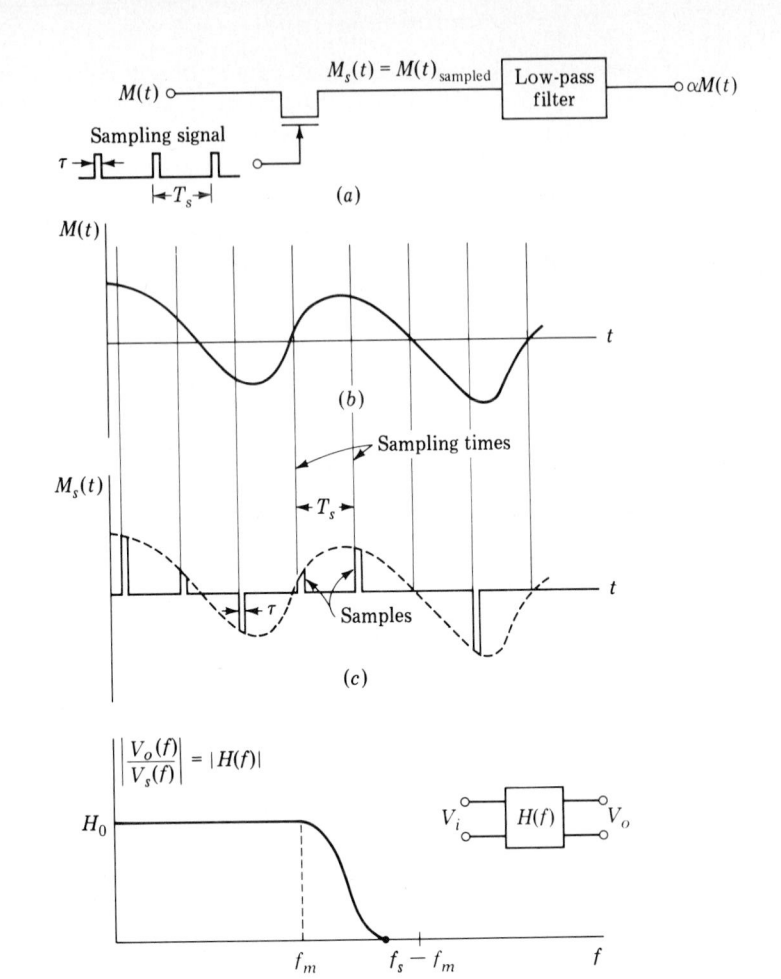

FIGURE 14.2-1

(a) A signal $M(t)$ is sampled and reconstructed.   (b) A signal $M(t)$.   (c) The signal $M(t)$ sampled.   (d) The transfer characteristic required of the filter.

interval $M_s(t) = M(t)$; that is, the top of each pulse follows the contour of $M(t)$. Outside the sampling interval $M_s(t) = 0$.

The characteristic required of the filter, for $M(t)$ to be reconstituted without error, is shown in Fig. 14.2-1d.   Here $H(f)$ is the transfer function of the filter; that is, $H(f) = V_o(f)/V_i(f)$.   We note that $H(f)$ must be flat at least up to the frequency f_m, which is the highest-frequency component of $M(t)$, and $H(f)$ must fall to zero before the frequency $f_s - f_m$.   By way of example, say $f_m = 1$ kHz; then the minimum allowable sampling frequency is $f_s = 2f_m = 2$ kHz.   To allow some latitude let us take $f_s = 2.5$ kHz.   Then $f_s - f_m = 2.5 - 1.0 = 1.5$ kHz.   It would then be required that $|H(f)|$ be constant, $|H(f)| = H_0$, from 0 to at least 1 kHz and then fall to zero at a frequency ≤ 1.5 kHz. If we had selected $f_s = 2f_m = 2$ kHz, we would have had $f_s - f_m = 2f_m - f_m = f_m$. In this case we would require that $|H(f)| = H_0$ up to f_m and then $|H(f)|$ should drop abruptly to zero.

While we have made no explicit reference to the phase characteristic of the filter, it should be noted that in the filter passband (0 to f_m) the phase characteristic must be linear. Such is the requirement in the passband of any filter if the filter is to pass, without distortion, any signal whose spectral components all lie in the filter's passband.

Suppose, referring to Fig. 14.2-1c, that we arrange that $\tau = T_s$, that is, that the sampling interval be as long as the interval between samples. In this case the FET switch will transmit *all the time*, and the amplitude of the reconstituted signal $M_o(t)$ is $M(t)$, that is, *all* of the signal. More generally, the FET switch transmits for only a fraction τ/T_s of the time. It seems reasonable, then, that in this situation the fraction of the original signal $M(t)$ available in the reconstituted signal should be $(\tau/T_s)M(t)$. The same result can be seen in another way. A low-pass filter is a time-averaging device. Thus the extent to which $M(t)$ is available in the sampled signal should be proportional to the width τ of the sample pulses in the sampled signal, the available signal being $M(t)$ itself when $\tau = T_s$. Finally, taking account of the gain H_0 of the low-pass filter itself, we would expect the recovered output signal to be

$$M_o(t) = \frac{\tau}{T_s} H_0 M(t) \qquad (14.2\text{-}1)$$

In Fig. 14.2-1a the distance between the signal source $M(t)$ and its eventual destination might be so large that it would be appropriate to refer to the connection between source and destination as a *communication channel*. The channel connection might be composed of wires, or it might consist of a radio link. The physical facility which constitutes the communication channel might well have been quite expensive to establish, and we would be interested in making maximum possible use of it. We note that in the transmission of the signal $M(t)$ the channel appears to be "in use" only a fraction τ/T_s of the available time, and we might inquire whether when the channel is not being used for $M(t)$ it might be used to transmit other signals. Such simultaneous transmission of signals is indeed possible and is referred to as *time-division multiplexing*.

14.3 TIME-DIVISION MULTIPLEXING

A technique by which we can take advantage of the sampling principle for the purpose of time-division multiplexing is illustrated in the idealized representation of Fig. 14.3-1. At the transmitting end on the left, a number of band-limited signals are connected to the contact points of a rotary switch. We assume that the signals are similarly band-limited. For example, they may all be voice signals, limited to 3.3 kHz. As the rotary arm of the switch swings

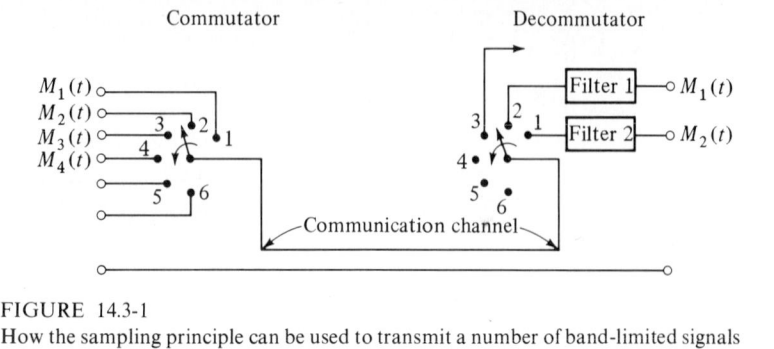

FIGURE 14.3-1

How the sampling principle can be used to transmit a number of band-limited signals over a single communications channel.

around, it samples each signal sequentially. The rotary switch at the receiving end is in synchronism with the switch at the sending end. The two switches make contact simultaneously at similarly numbered contacts. With each revolution of the switch, one sample is taken of each input signal and presented to the correspondingly numbered contact of the receiving-end switch. The train of samples at, say, terminal 1 in the receiver, passes through low-pass filter 1, and at the filter output the original signal $M_1(t)$ appears reconstructed. Of course, if f_m is the highest-frequency spectral component present in any of the input signals, the switches must make at least $2f_m$ revolutions per second.

When the signals to be multiplexed vary slowly with time, so that the sampling rate is correspondingly slow, mechanical switches, indicated in Fig. 14.3-1, can be used. When the switching speed required is outside the range of mechanical switches, electronic switching systems can be used. In either event, the switching mechanism, corresponding to the switch at the left in Fig. 14.3-1, which samples the signals, is called the *commutator*. The switching mechanism which performs the function of the switch at the right in Fig. 14.3-1 is called the *decommutator*. The commutator samples and combines the samples, while the decommutator separates samples belonging to individual signals so that these signals can be reconstructed.

The interlacing of the samples that allows multiplexing is shown in Fig. 14.3-2. Here, for simplicity, we have considered the multiplexing of just two signals, $M_1(t)$ and $M_2(t)$. Signal $M_1(t)$ is sampled regularly at intervals of T_s and at the times indicated in the figure. The sampling of $M_2(t)$ is similarly regular, but the samples are taken at a time different from the sampling time of $M_1(t)$. The input waveform to the filter numbered 1 in Fig. 14.3-1 is the train of samples of $M_1(t)$, and the input to the filter numbered 2 is the train of samples of $M_2(t)$. The timing in Fig. 14.3-2 has been deliberately drawn to suggest that there is room to multiplex more than two signals.

We observe that the train of pulses corresponding to the samples of each signal is modulated in amplitude in accordance with the signal itself. Accordingly,

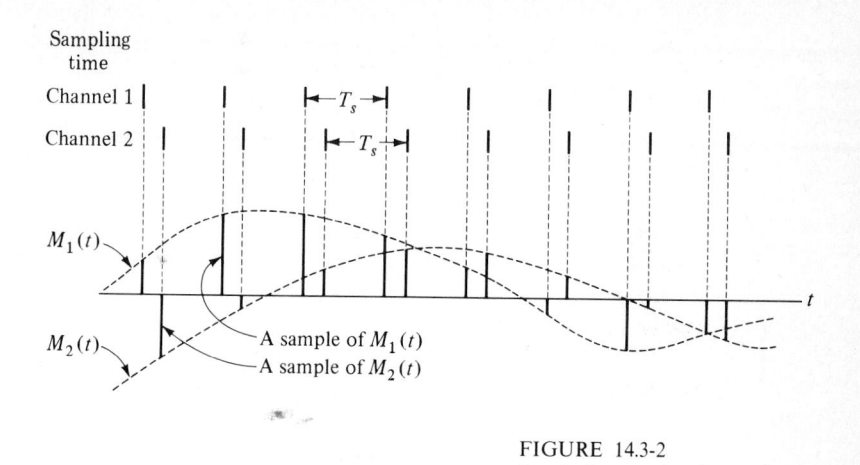

FIGURE 14.3-2
Interlacing two baseband signals.

the scheme of sampling is called *pulse-amplitude modulation* (PAM). We note, however, that these pulses may vary *continuously* in amplitude. Hence PAM is an *analog* and not a *digital* system. We have discussed it here simply as a necessary prelude to *pulse-code modulation*, which is a digital system.

14.4 QUANTIZATION

The validity of the sampling theorem makes it possible to transmit or to process an analog signal by digital means. For we need not take note of the analog signal at all times but only at the sampling times, and hence in the intervals between samplings we shall have time to convert each sample voltage to digital form. The samples are continuously varying analog voltages. In digital form the allowable variation is not continuous since sample values must differ, at a minimum, by the least significant of the digits used in the digital representation. Hence the process of digitizing samples involves making an approximation. This process of approximation is called *quantization*.

The operation of quantization is illustrated in Fig. 14.4-1. A signal $M(t)$ is shown in Fig. 14.4-1a. This signal is the waveform V_i applied to the quantizer input. The output of the quantizer is called V_o. The quantizer has the essential feature that its input-output characteristic has the staircase form shown in Fig. 14.4-1b. As a consequence, the output V_o, shown in Fig. 14.4-1c, is the quantized waveform $M_q(t)$. It is observed that while the input $V_i = M(t)$ varies smoothly over its range, the quantized signal $V_o = M_q(t)$ holds at one or another of a number of fixed levels ... M_{-2}, M_{-1}, M_0, M_1, M_2, ... Thus the signal $M_q(t)$ either does not change or changes abruptly by a quantum jump S called the *step size*.

The waveform $M'(t)$ shown dotted in Fig. 14.4-1c represents the output

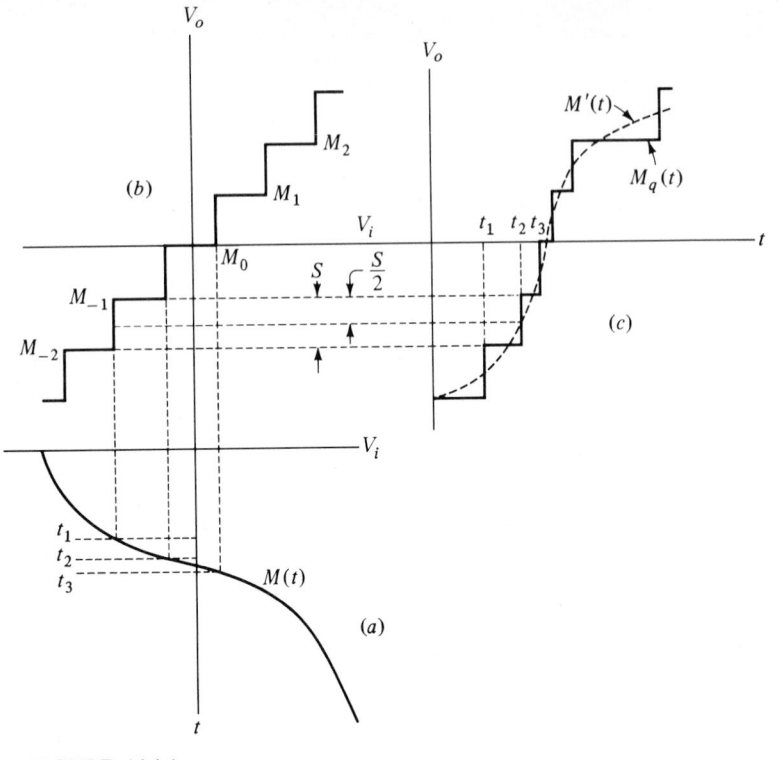

FIGURE 14.4-1
The operation of quantization. The step size is S. (a) The signal $M(t)$. (b) The input-output characteristic of the quantizer. (c) The quantizer output (solid line) response to $M(t)$. The dashed waveform $M(t)$ shows the waveform of the output signal for a linear characteristic.

waveform, assuming that the output is linearly related to the input. If the factor of proportionality is unity, $V_o = V_i$ and $M'(t) = M(t)$. We see then that the level held by the waveform $M_q(t)$ is the level to which $M'(t)$ is closest. The transition between one level and the next occurs at the instant when $M'(t)$ crosses a point midway between two adjacent levels.

We see, therefore, that the quantized signal is an approximation to the original signal. The quality of the approximation can be improved by reducing the size of the steps, thereby increasing the number of allowable levels. Eventually, with small enough steps, the human ear or eye will not be able to distinguish the original from the quantized signal. To give the reader an idea of the number of quantization levels required in a practical system, we note that 512 levels can be used to obtain the quality of commercial color TV, while 64 levels gives only fairly acceptable color TV performance.

If we propose to quantize a signal with peak-to-peak range R and to use Q quantization levels, the step size S is determined by the condition that $QS = R$.

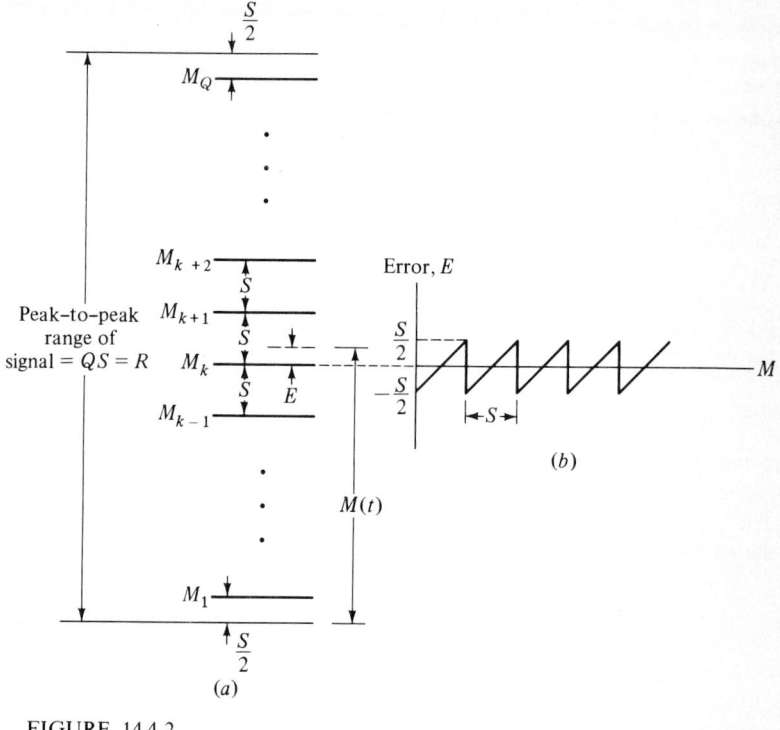

FIGURE 14.4-2
(*a*) A range of voltage over which a signal $M(t)$ makes excursions is divided into Q quantization ranges each of size S. The quantization levels are located at the center of the range. (*b*) The error voltage $E(t)$ as a function of the instantaneous value of the signal $M(t)$.

We would locate the quantization levels as indicated in Fig. 14.4-2*a*. In this way the maximum instantaneous quantization error would be $S/2$, as illustrated in Fig. 14.4-2*b*.

The complete process of digitizing an analog waveform is illustrated in Fig. 14.4-3. The signal $M(t)$ is regularly sampled at times indicated by the dots on the waveform. The anticipated peak-to-peak range R is 7 V extending from -3.5 to $+3.5$ V. We have allowed eight quantization levels located in such manner that maximum possible instantaneous quantization error is 0.5 V. Following common practice, we have assigned a set of binary digits to each level, using twos-complement representation (since there are eight levels, 3 bits are required).

The binary digits can now be transmitted or processed serially or in parallel. With serial processing and with three digits, as in the present case, the processing of each digit may occupy nominally one-third the interval between sampling times. With parallel processing, the entire interval is available for each bit.

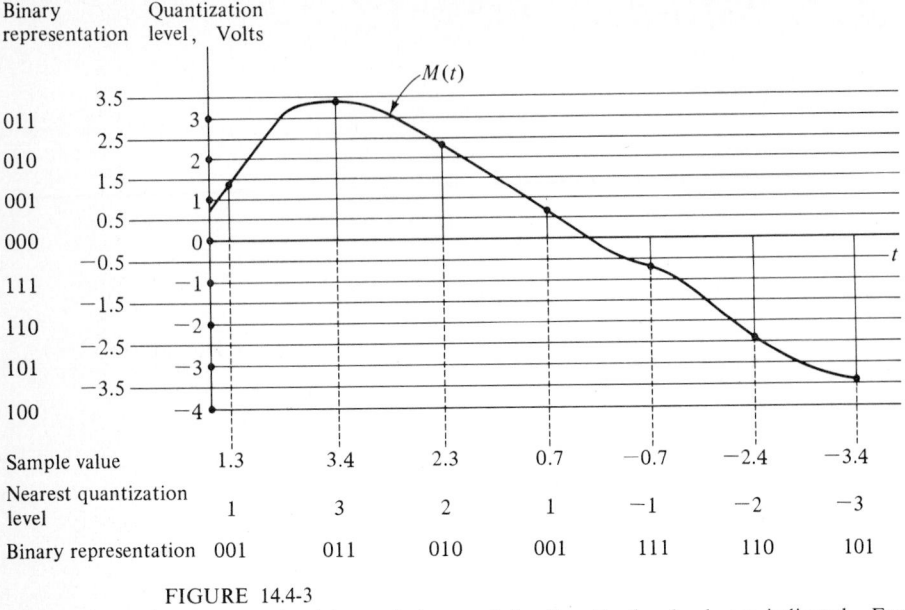

FIGURE 14.4-3

A message signal is regularly sampled. Quantization levels are indicated. For each sample the quantized value is given and its binary representation is indicated using twos-complement representation.

Beginning with the next section we shall consider circuits for converting digital representations to analog form. The D/A converters are rather simpler than the A/D converters. Further, a D/A converter is frequently used within the structure of A/D converters. For these reasons we shall consider D/A converters first.

14.5 THE WEIGHTED-RESISTOR D/A CONVERTER

The structure of a weighted-resistor D/A converter is shown in Fig. 14.5-1. The input (not shown in the figure) is an N-digit binary signal $V = V_{N-1}V_{N-2} \cdots V_0$ in which each V_k is a voltage whose level is such as to represent either logic 1 or logic 0. It is assumed that all the V_k are available simultaneously, i.e., in parallel on N lines. If the V_k appear serially, the bits must be entered into a shift register so that all bits can be made available simultaneously. The voltages V_k normally available to drive the converter need not be precisely fixed voltages but only identifiable as representing one logic level or the other. For this reason the V_k's are not applied directly to the converter but are used instead to operate the (electronic) switches $S_0, S_1, \ldots, S_{N-1}$.

When V_k corresponds to logic 1 or 0, the switch S_k is thrown to the 1 or 0 position connecting a resistor of resistance R_k to the precisely controlled voltage

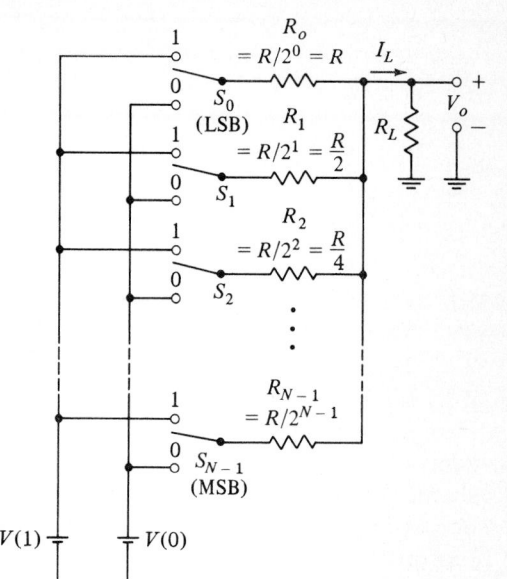

FIGURE 14.5-1
A weighted-resistor D/A converter.

source $V(1)$ or $V(0)$.   The least significant bit (LSB) V_0 operates switch S_0, which is connected to the output through the resistor of highest resistance $R_0 = R$.   The most significant bit (MSB) operates S_{N-1}.   Observe that the resistors R_0, R_1, ..., R_{N-1} are *weighted* so that their resistances are inversely proportional to the numerical significance of the corresponding binary digit.   The operation of the converter depends on the resistance values of successive resistors being related by a factor of 2 and does not depend on the absolute value of the resistors. Hence the parameter R in Fig. 14.5-1 is arbitrary.

For simplicity, let us assume initially that $V(1) = V_R$, a fixed reference voltage, that $V(0) = 0$, that is, all 0 switch positions are grounded, and that the load $R_L = 0$ (in which case $V_o = 0$).   Then the output current I_L is readily calculated in terms of the switch positions.   Let $S_k = 1$ and $S_k = 0$ represent respectively the situation in which S_k is thrown to the 1 or 0 position.   Then the load current I_L (which, being the current into a short circuit, we call here I_{LS}) is

$$I_{LS} = V_R \left(\frac{S_{N-1}}{R_{N-1}} + \frac{S_{N-2}}{R_{N-2}} + \cdots + \frac{S_0}{R_0} \right) \qquad (14.5\text{-}1a)$$

$$= \frac{V_R}{R} \left(S_{N-1} 2^{N-1} + S_{N-2} 2^{N-2} + \cdots + S_0 2^0 \right) \qquad (14.5\text{-}1b)$$

Thus we observe that I_{LS} has a numerical value which is directly proportional to the numerical value of the binary number $S = S_{N-1} S_{N-2} \cdots S_0$, the factor of proportionality being V_R/R.

FIGURE 14.5-2
The D/A converter resistor array of Fig.
14.5-1 replaced by a Norton's equivalent
circuit.

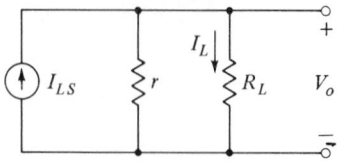

This proportionality persists, albeit with a different factor of proportionality, if R_L is not equal to zero. That such is the case is to be seen from Fig. 14.5-2. Here we have replaced the converter network (excluding R_L) by its Norton's equivalent circuit. This equivalent circuit consists of a current source I_{LS} shunted by a resistor r, which is the output impedance of the network. The output impedance r is equal in resistance to the parallel combination of all the resistors R_0, R_1, ..., R_{N-1} and, as can be verified (Prob. 14.5-1), is given by

$$r = \frac{R}{2^N - 1} \tag{14.5-2}$$

The new load current into R_L is

$$I_L = \frac{r}{r + R_L} I_{LS} \tag{14.5-3}$$

and the output voltage is, using Eqs. (14.5-1) to (14.5-3),

$$V_o = R_L I_L = \frac{R_L V_R}{R + (2^N - 1)R_L} (S_{N-1} 2^{N-1} + S_{N-2} 2^{N-2} + \cdots + S_0 2^0) \tag{14.5-4}$$

With $V(1) = V_R$ and $V(0) = 0$, the range of output voltage V_o extends from 0 V when $S = 0 \cdots 00$ to $V = R_L V_R/(r + R_L)$ when $S = 1 \cdots 11$. However, by selecting $V(1)$ and $V(0)$ appropriately we can offset the range of V_o to suit our convenience. Thus, for example, with $V(1) = V_R$ and $V(0) = -V_R$, the excursion of V_o will extent in the negative as well as in the positive direction, being symmetrical about 0 V. Most commonly an op-amp is employed to buffer the D/A resistor network. Such an arrangement is indicated in Fig. 14.5-3. For simplicity the switches have been omitted from this diagram, but their presence at the input sides of R_0, R_1, ..., R_{N-1} is to be understood. Because of the virtual ground at the inverting input terminal of the op-amp, the output is given by

$$V_o = -\frac{R_f V_R}{R_{N-1}} S_{N-1} - \frac{R_f V_R}{R_{N-2}} S_{N-2} - \cdots - \frac{R_f V_R}{R_0} S_0 \tag{14.5-5}$$

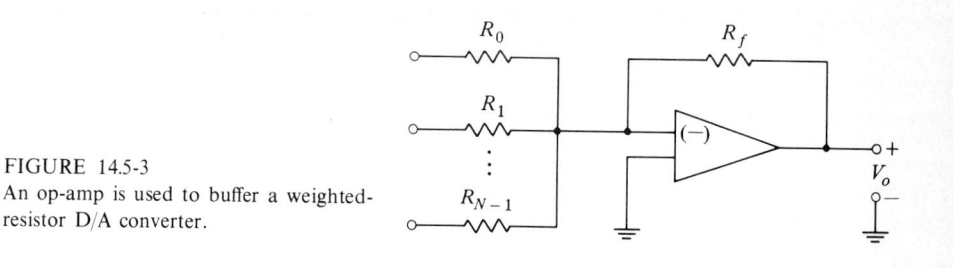

FIGURE 14.5-3
An op-amp is used to buffer a weighted-resistor D/A converter.

Taking account, as indicated in Fig. 14.5-1 of the fact that $R_{N-1} = R/2^{N-1}$, etc., we find

$$V_o = -\frac{V_R R_f}{R}\left(S_{N-1}2^{N-1} + S_{N-2}2^{N-2} + \cdots + S_0 2^0\right) \qquad (14.5\text{-}6)$$

It is of some interest to note that when an op-amp is used, if the op-amp were ideal, it would not be necessary to use double-pole switches at the input to the converter. Single-pole switches would serve as well. Each switch would connect the corresponding resistor to a voltage source V_R when the switch position was $S = 1$ and leave the resistor *floating*, i.e., disconnected at its input end, when $S = 0$. Such an arrangement is allowable, in principle, because of the virtual ground at the op-amp input. With this virtual ground the amplifier input current due to any switch being closed is independent of the settings of any of the other switches. However, as a matter of practice, a floating input terminal is often a source of spurious disturbances introduced into the system by unintended but often unavoidable couplings, and this procedure is not usually followed.

A difficulty associated with the weighted-resistor converter stems from the wide range of resistor values which must be used. Suppose, for example, that the resistor corresponding to the MSB is 2 kΩ and that the converter is to accommodate 13 bits. Then the resistor associated with the least significant bit is 2 kΩ $\times 2^{12} = 8.2$ MΩ. Resistors covering so large a range which have the required precision and which track over a wide temperature range are difficult to produce. It is especially difficult in a monolithic fabrication, such as would be of interest in integrated-circuit fabrication. We consider in the next section an alternative converter-resistor array which circumvents this difficulty.

14.6 THE $R - 2R$ LADDER D/A CONVERTER

The converter-resistor array of Fig. 14.6-1a uses resistors of only two sizes, R and $2R$. In this figure only 4 bits are indicated, but of course the array can be expanded to accommodate an arbitrary number of bits. Note that this array requires twice as many resistors for the same number of input bits as the weighted-resistor array. Again, for simplicity in Fig. 14.6-1a the details of the

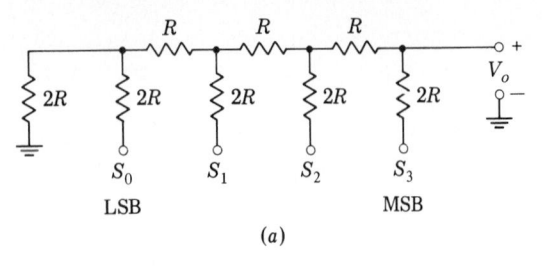

(a)

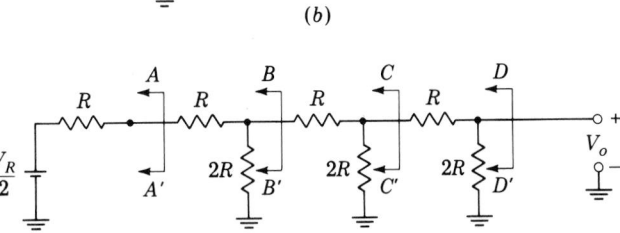

(b)

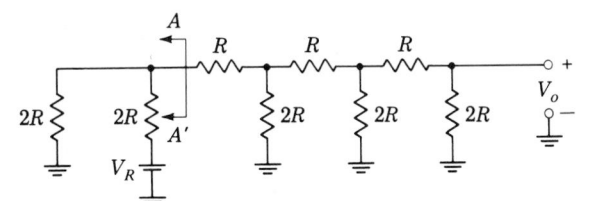

(c)

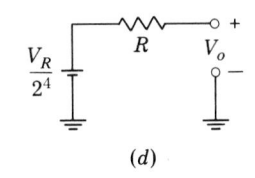

(d)

FIGURE 14.6-1
(a) The R-$2R$ ladder D/A converter. (b and c) Thevenin's theorem used to determine the output voltage V_o. (d) The final equivalent circuit as seen at the output.

switches have been omitted, but it is to be understood that when $S_k = 1$, the corresponding resistor is connected to a voltage V_R and when $S_k = 0$, the resistor input is grounded.

To see most simply the relative weight at the output of the individual switches, consider the situation in Fig. 14.6-1b, where we have set $S_0 = 1$ while $S_1 = S_2 = S_3 = 0$. In this example we find the voltage source V_R in series with a resistor $2R$. Applying Thevenin's theorem at AA', we find, as shown in Fig. 14.6-1c, that we now have a voltage source $V_R/2$ in series with R. We repeat the application of Thevenin's theorem at BB', CC', and DD'. We find that at each such application the voltage source is again divided by 2, while the Thevenin's equivalent output impedance remains constant at R. The final equivalent circuit for the output is shown in Fig. 14.6-1d. If we repeated these operations starting with $S_1 = 1, S_3 = S_2 = S_0 = 0$, we would find an equivalent circuit as in Fig. 14.6-1d except that the voltage source would be $V_R/2^3$, and so on for switches S_2 and S_3. Thus, at the output, each switch contributes its proper relative binary weight. For the arrangement in Fig. 14.6-1a we then have

$$V_o = V_R\left(\frac{S_3}{2^1} + \frac{S_2}{2^2} + \frac{S_1}{2^3} + \frac{S_0}{2^4}\right) \tag{14.6-1a}$$

and

$$V_o = \frac{V_R}{2^4}(S_3 2^3 + S_2 2^2 + S_1 2^1 + S_0 2^0) \tag{14.6-1b}$$

or, more generally, if there are N input digits and correspondingly N switches, we have

$$V_o = \frac{V_R}{2^N}(S_{N-1} 2^{N-1} + S_{N-2} 2^{N-2} + \cdots + S_0 2^0) \tag{14.6-2}$$

Equation (14.6-2) does not take into account a possible load placed across the output. Such a load will change the absolute value of the output, but, as with the weighted-resistor converter, it will not change the relative weight of the individual switches. For suppose that the load is R_L. Then, referring to Fig. 14.6-1d, the output when $S_0 = 1$ will be

$$V_o = \frac{V_R}{2^4} \frac{R_L}{R + R_L} \tag{14.6-3}$$

However, no matter what the position of the switches, the Thevenin's output impedance of the converter is R, and hence the effect on a load R_L is in every case simply to reduce the output by the factor $R_L/(R_L + R)$.

Usually the R-2R ladder converter array is constructed as shown in Fig. 14.6-2a. Note here the inclusion of an additional 2R load resistor to the right of the switch corresponding to the most significant digit. The feature (sometimes an advantage) which results from the inclusion of this extra resistor is that the impedances seen looking into each input of the array are the same. As indicated, this impedance is $3R$. A D/A converter using an R-2R ladder with the extra resistor and using an op-amp buffer is shown in Fig. 14.6-2b. It can be verified that the output V_o (except for sign) is given by Eq. (14.6-2).

The weighted-resistor converter assigns weights to digital input bits by using appropriately weighted resistors. In the R-2R ladder, bits are weighted by providing paths for current division with consequent successive attenuations for bits of lower significance. A compromise between these two schemes is possible. One such compromise circuit is shown in Fig. 14.6-3. As usual, S_0 is the least significant bit and S_7 the most significant bit. We have chosen to arrange the bits in groups of four. Within each group the bits are introduced through weighted resistors whose resistances are related by powers of 2, as required. It

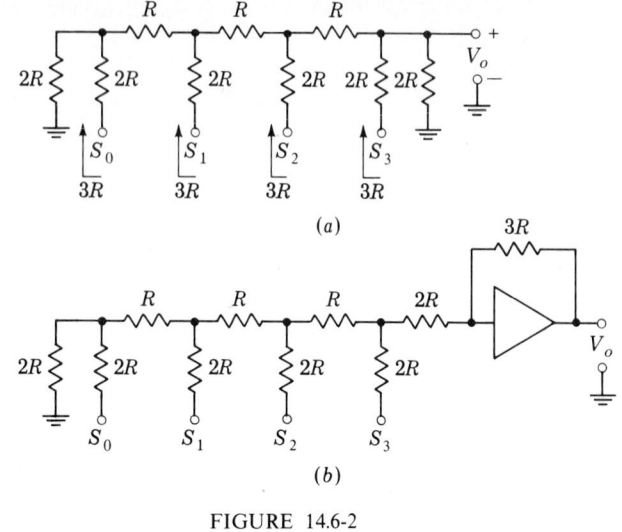

(a)

(b)

FIGURE 14.6-2
Connection of the R-$2R$ ladder to an op-amp.

is, however, required that there be an additional attenuation of 16 between the group $S_0 S_1 S_2 S_3$ and the group $S_4 S_5 S_6 S_7$. Thus when $S_3 = \mathbf{1}$, the consequent input current to the virtual ground of the op-amp should be one-sixteenth the current which results when $S_7 = \mathbf{1}$. Similarly this same relationship should hold between S_2 and S_6, etc. The resistor r has been inserted between the two groups to provide this attenuation. It can be verified that $r = 8R$ will reduce the current by the factor $\frac{1}{16}$, as required for a straight binary input format.

The arrangement of Fig. 14.6-3 is also very convenient to accommodate a BCD input format. In this format, the four bits $S_3 S_2 S_1 S_0$ represent a decimal digit 0 to 9 in the units position while $S_7 S_6 S_5 S_4$ represent a decimal digit in the tens position. In such cases the resistor r must be selected to provide an attenuation by 10 rather than by 16. It can be verified (Prob. 14.6-3) that such attenuation by 10 requires that $r = 4.8R$.

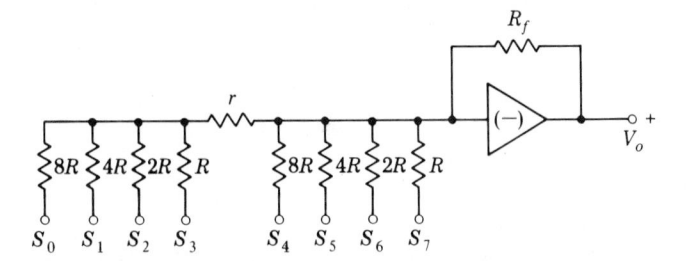

FIGURE 14.6-3
A D/A converter which effects a compromise between the use of weighted resistors and the use of a ladder network.

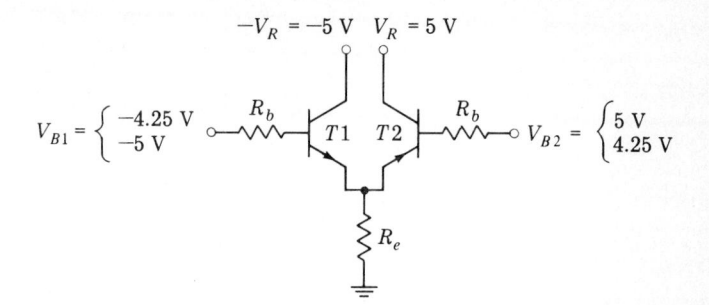

FIGURE 14.7-1
Bipolar transistors used in the inverse manner as switches in a D/A converter.

14.7 SWITCHES FOR D/A CONVERTERS

In Chap. 1 we discussed the use of diodes and transistors (both bipolar and FET) as switches.   In Chap. 13 we discussed the use of these devices in analog-signal switching circuits.   These devices can be used as well to serve as the switches in D/A converters.

Turned-ON FETs behave like simple resistors.   When such devices are used, either the resistors in the converter array must be large enough for the FET resistance to be ignored or the resistance of the FET must be taken into account.   In the latter case it may be necessary to provide some measure of temperature compensation (see Prob. 14.7-1) since the FET resistance is temperature-sensitive.   In saturation, bipolar transistors have negligible resistance, but, as noted, they have across them a small residual voltage, i.e., an offset voltage. Since the offset voltage is appreciably smaller when the transistor is used in the inverse direction (see Fig. 1.10-1), this inverse operation would normally be used.

There is, however, a difficulty associated with using bipolar transistor switches in the inverse direction in a converter circuit.   To appreciate the difficulty, consider the situation represented in Fig. 14.7-1.   Here the two transistors are to serve as the double-pole switch in, say, the converter circuit of Fig. 14.5-1, in which we have set $V(1) = V_R = 5$ V and $V(0) = -V_R = -5$ V.   The resistor R_e represents one of the converter resistors.   The side of R_e distant from the switches is considered to be at ground by virtue of its connection to the virtual ground at the input of an op-amp.   Each transistor is being used in the inverse direction with the normal collector junction serving here as the emitter junction.   We shall cut off a transistor by arranging that the voltage across its base-collector junction shall be 0 V, and we shall saturate the transistor by forward biasing its base-collector junction to the extent of 0.75 V.   Then the voltages indicated on the figure for V_{B1} and V_{B2} are appropriate to saturate one transistor while the other is OFF and vice versa.   In this way the ungrounded side of R_e is connected to -5 or $+5$ V respectively.

Now consider, say, that $V_{B1} = -5$ V and $V_{B2} = 4.25$ V, so that $T1$ is OFF and $T2$ is ON.   Then the voltage between the base and the emitter of $T1$ is very

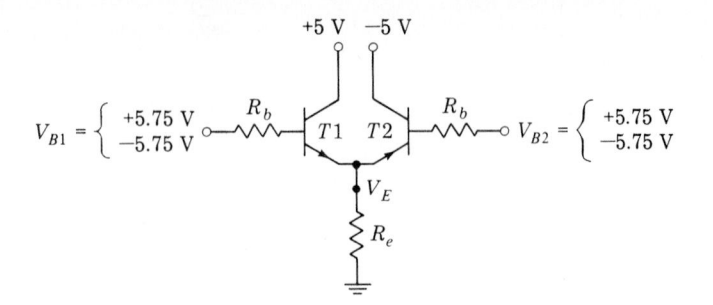

FIGURE 14.7-2
Overdriven emitter followers used as switches in a D/A converter.

nearly $V_{BE1} = -9.25$ V. This large a voltage between base and emitter of a transistor is larger than a transistor can normally sustain. For this reason the switching arrangement indicated in Fig. 14.7-1 is not acceptable.

An alternative switching arrangement which circumvents this voltage difficulty is indicated in Fig. 14.7-2. Here the transistors are used in the normal rather than in the inverted fashion, but the circuit configurations are those of emitter followers. As pointed out in Sec. 1.10, an overdriven emitter follower can have an offset voltage of 0 V since the saturation factor σ is negative. In Fig. 14.7-2 the bases are driven between about $+5.75$ and -5.75 V. In the first case $T1$ is in saturation and $V_E \approx 5$ V while $T2$ is OFF. The base-emitter voltages are the same with $V_{BE1} = V_{BE2} = 0.75$ V. In the second case $T2$ is ON, and $T1$ is OFF. The emitter voltage V_E is now $V_E \approx -5$ V. We note as a possible disadvantage of the present arrangement that large base-voltage swings (11.5 V) are required.

An example of a D/A converter As an example of a D/A converter using the switching scheme of Fig. 14.7-2 we consider the circuit of Fig. 14.7-3. Although a weighted-resistor converter is indicated, the scheme is suitable for a ladder converter as well. The reference voltages $\pm V_R$ are ± 5 V. Thus, neglecting the saturation voltage across $T4$ or $T3$, we have switch voltages of ± 5 V. To see that the collector-emitter voltage of $T4$ (and of $T3$) is indeed negligible assume that $T2$ is cut off. Then $T4$ is ON. The emitter voltage of $T4$ is therefore approximately 5 V, and the base voltage V_{B4} is approximately 5.75 V. Thus, the base current $I_{B4} \approx (12 - 5.75)/(4.4 \text{ k}\Omega) \approx 1.4$ mA. However, the emitter current $I_{E4} \approx 5/(10 \text{ k}\Omega) = 0.5$ mA. Note therefore that $I_{C4} \approx -0.9$ mA and $\sigma \equiv I_C/h_{FE} I_B \approx -0.6/h_{FE}$. It may be shown from Eq. (1.10-3) that when $\sigma \approx -0.6/h_{FE}$ that $V_{CE} \approx V_T \ln (1 + 0.4/h_{FE})$. If we assume that $h_{FC} = 0.1$ we then find that V_{CE} is only $V_{CE} \approx 40$ mV. The input driving voltage swings from about 0 V to about 3 V. With the input at 3 V, $T1$ and $T2$ are OFF, and $T4$ is driven to saturation by base current derived from the $+12$-V source and flowing through the 2.2-kΩ

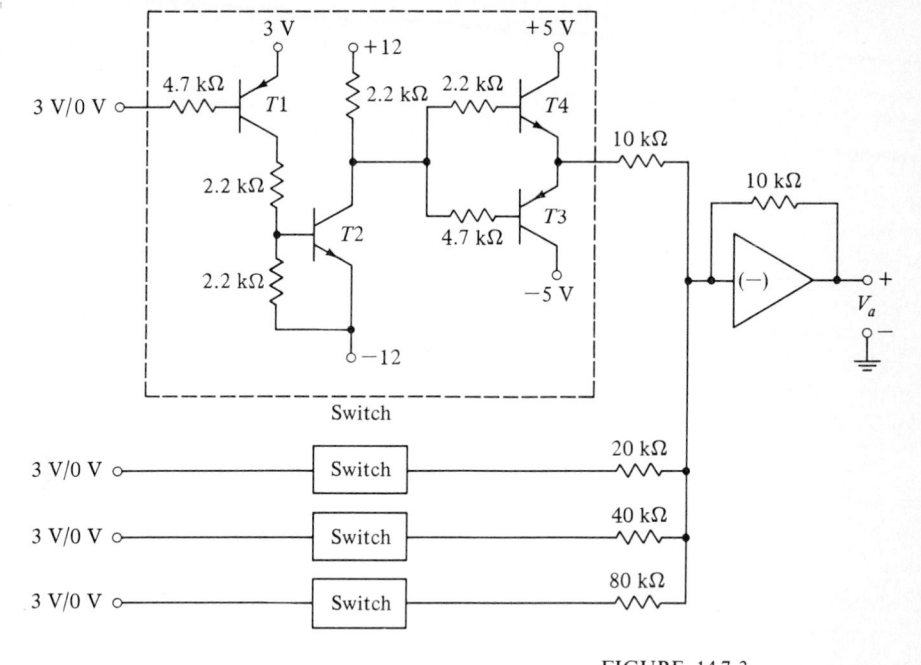

FIGURE 14.7-3
An example of a D/A converter.

collector resistor of $T2$ and the 2.2-kΩ base resistor of $T4$. While $T4$ is ON, $T3$ is OFF. When the input is at 0 V, both $T1$ and $T2$ saturate, as can be verified (Prob. 14.7-1). In this case, $T3$ is driven to saturation by base current derived from the -12-V supply and flowing through the saturated transistor $(T2)$ and through the 4.7-kΩ base resistor of $T3$. In this case, assuming that $V_{CE2} \approx 0.2$ V, we find $I_{B3} \approx 1.3$ mA and, hence, σ remains $\sigma \approx -0.6/h_{FE}$. Now the emitter voltages of $T3$ and $T4$ are again very nearly at -5 V.

14.8 A CURRENT-DRIVEN D/A CONVERTER

A D/A converter using the ladder array of resistors and using current sources is shown in Fig. 14.8-1. We assume that the switches are of the make-before-break type so that the current I need never be interrupted. The present network is equivalent to the former network of Fig. 14.6-2. In the former network a voltage source is applied, as required, in series with resistors of resistance $2R$, In the present network the series combinations of voltages and resistors are replaced by the combination of a current source in parallel with resistors of resistance $2R$.

The current-driven converter has a potential merit in comparison with the voltage-driven converter. The voltage-driven converter requires transistor

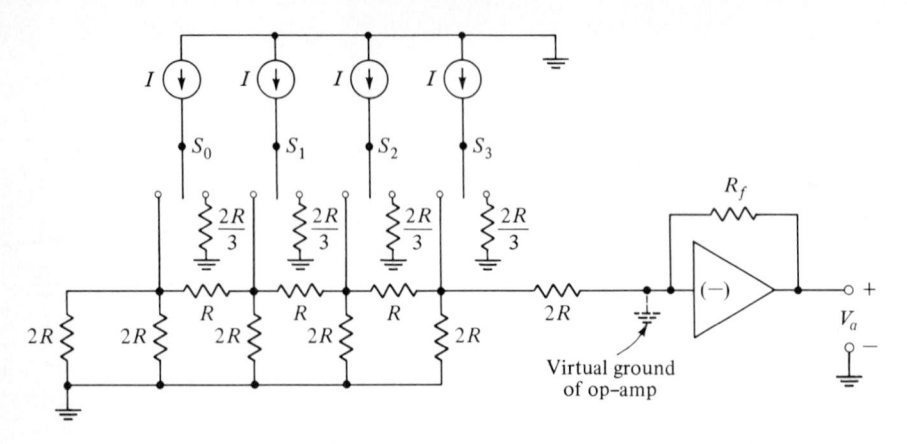

FIGURE 14.8-1
A current-driven ladder D/A converter.

switches which are driven to saturation. The time delay associated with bringing a transistor out of saturation often establishes an upper limit to the speed with which such voltage-driven converters can operate. On the other hand, in current-driven converters it is feasible to use a switching arrangement in which transistors are not driven into and out of saturation; instead, we can use an arrangement in which a current is switched into and out of the converter-resistor array, devising thereby a technique to drive transistors from the active region to cutoff and vice versa. We have already encountered such a switching scheme in the basic difference-amplifier configuration of emitter-coupled logic.

A current-driven ladder converter using the difference-amplifier switches encountered in emitter-coupled logic is shown in Fig. 14.8-2. Currents are switched into the ladder when the transistors $T0A$, $T1A$, $T2A$, or $T3A$ conduct, and the currents are diverted to transistors $T0B$, $T1B$, $T2B$, or $T3B$ when they conduct. The circuit shown in Fig. 14.8-2b provides the base-biasing voltage V_B for all the switches. The converter circuit is intended to be used in conjunction with ECL gates. At room temperatures these gates (see Sec. 7.4) have logic levels -0.76 V (logic **1**) and -1.58 V (logic **0**). Consequently, the base voltage V_B is set very nearly midway between these levels at $V_B = -1.15$ V. Thus, when, say, $V_0 = -1.58$ V, $T0B$ is OFF and $T0A$ is ON, and when $V_0 = -0.76$ V, $T0B$ is ON and $T0A$ is OFF.

We assume that at room temperature the voltage across diodes $D1$ and $D2$ as well as across the base emitter junction of a conducting transistor is 0.75. On this basis, it can readily be verified that because the supply voltage $-V_{EE}$ in the biasing circuit has been selected at $-V_{EE} = -8.1$ V, the base bias V_B turns out to be $V_B = -1.15$ V. For we have

$$V_B = -V_{EE} + V_Z + V_{D2} + V_{D1} - V_{BE}(TB) \tag{14.8-1}$$

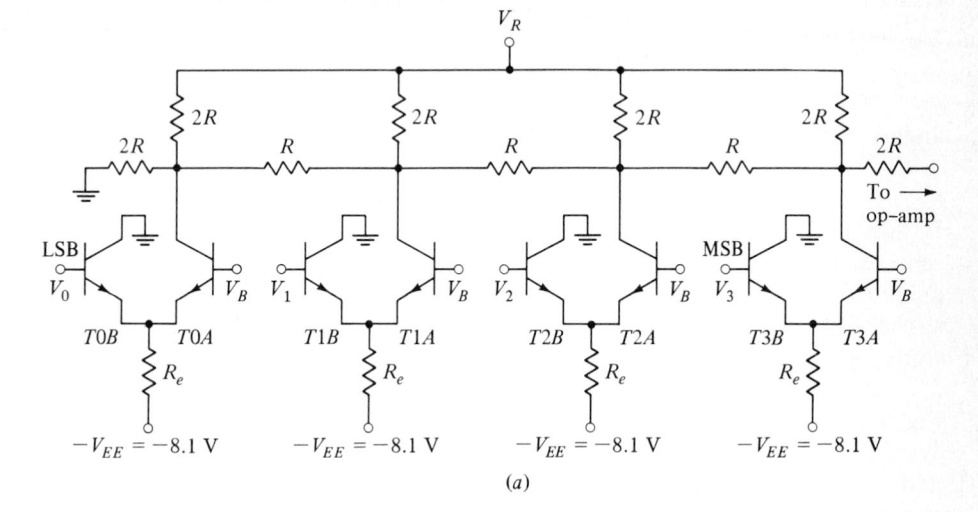

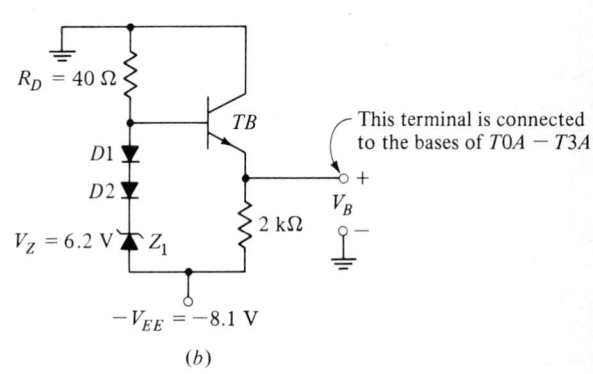

FIGURE 14.8-2

(a) A high-speed current-driven D/A converter. (b) Base-biasing circuit.

where V_Z = voltage across zener diode
V_{D1}, V_{D2} = voltages across diodes $D1$ and $D2$
$V_{BE}(TB)$ = base-to-emitter voltage of TB
We then find

$$V_B = -8.1 + 6.2 + 0.75 + 0.75 - 0.75 = -1.15 \text{ V} \qquad (14.8\text{-}2)$$

The circuitry also provides a measure of temperature compensation in that the current injected into the ladder at each switch is independent of the temperature. In each case this injected current is the collector current of the corresponding transistor. This collector current, in turn, except for a small and constant fraction due to base current, is equal to the emitter current. The voltage across the emitter resistor of, say, $T0$ is

$$V_{R_e} = V_B - V_{BE}(T0A) + V_{EE} \qquad (14.8\text{-}3)$$

Using Eq. (14.8-1) and assuming $V_{D1} = V_{D2} = V_{BE}(TB) = V_{BE}(T0A)$, we find V_{R_e} and the current I_{R_e} to be

$$V_{R_e} = V_Z \tag{14.8-4}$$

and

$$I_{R_e} = \frac{V_Z}{R_e} \tag{14.8-5}$$

In Sec. 1.4 it was pointed out that depending on the operating voltage of a zener diode, an operating current can generally be found at which the zener-diode voltage is temperature-independent. The zener diode in the bias circuit is so selected, and hence Eq. (14.8-5) indicates that the injected current is relatively independent of temperature.

We may consider now various details in connection with the circuit of Fig. 14.8-2, all of which combine to allow the circuit to operate at high speed. We have already noted that the transistor switches do not go to saturation and that storage-time delays are thereby avoided. Next we note that assuming the base-emitter voltage of a conducting transistor to be 0.75 V, we can readily verify that when either of the transistors in the switch pair ($T0A$ or $T0B$, for example) is OFF, it operates with a base-emitter voltage of 0.35 V. Since we have always assumed that cut-in occurs at about 0.65 V, the transistor is never deeply into cutoff and there need be no long delay in carrying the transistor from cutoff into the active region.

When a switch operates to inject current into a node of the ladder, the voltage at that node (and to a lesser extent at all other nodes) must change. There are the inevitable stray shunt capacitances present in the ladder network. These capacitances must charge and discharge, in response to the operation of the switches, with a consequent slowing of the operation of the converter. To minimize this speed limitation these stray capacitances must find themselves in circuits having a small time constant. For this reason the resistors of the ladder must be small. Resistors $R = 50\ \Omega$ and $2R = 100\ \Omega$ are not unusual.

There is, however, a problem associated with the circuit of Fig. 14.8-2a: the current injected by $T0A$ takes longer to reach the op-amp than the current injected by, say, $T3A$ since the ladder network appears to be a lossy transmission line at very high frequency. This time difference can result in a rather large spike in the op-amp output. To see how this can arise consider that the digital input is 1000 (8 V) at $t = 0$ and somewhat later switches to 0111 (7 V). We further assume that the switching occurs at the same time throughout the circuit. However, due to the propagation delay in the ladder network, we find that $T3A$ switches first, so that the 8-V level immediately drops to 0 V. The voltage then rises to 4, 6, and finally 7 V.

To reduce the duration of the spike, a very-high-speed D/A converter can arrange to apply the logic switch voltages to $T0B$ before $T1B$, etc. In a later section we discuss the need to offset the output of a D/A to encompass a range of voltages which are both positive and negative. The reference voltage V_R in Fig. 14.8-2a is used to provide for such an offset.

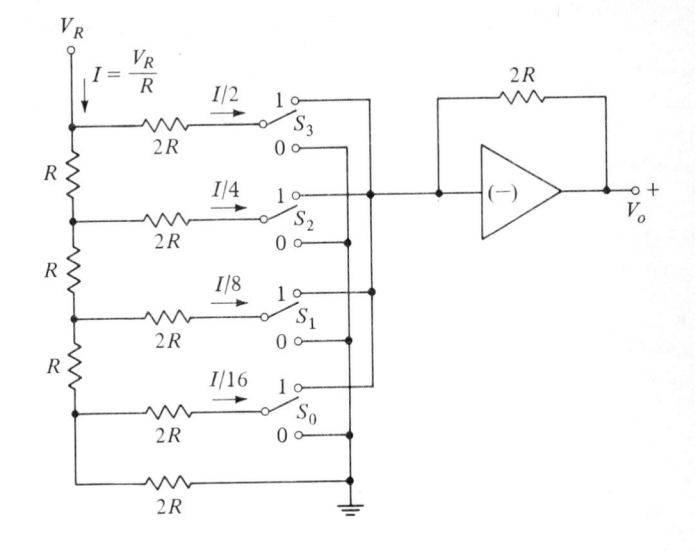

FIGURE 14.9-1
The inverted-ladder D/A converter.

14.9 THE INVERTED-LADDER D/A CONVERTER

We noted above that the D/A converter shown in Fig. 14.8-2 produces a spike during conversion as a result of the propagation delay suffered in the ladder. In the inverted ladder shown in Fig. 14.9-1 the switches are connected directly to the op-amp, thereby eliminating the propagation-delay-time problem.

In the inverted ladder the switches are located at the input to the op-amp rather than at the reference source V_R, as in Fig. 14.5-1. A 4-bit converter is indicated, but the extension to an arbitrary number of bits is obvious. Keeping in mind that at the input to the op-amp we see a virtual ground, it is clear that the currents that flow in the ladder are independent of the switch position. For in either position of the switch, the $2R$ resistors are connected to ground. We assume that the switches are of the make-before-break type, so that at no time is the switch arm disconnected from ground or virtual ground. It can now be readily verified that the current drawn from V_R is $I = V_R/R$ and that the currents in the individual $2R$ resistors are related by powers of 2, as shown. The switches serve to direct these currents into the op-amp or to direct the currents to ground. It can also be verified that the output of the amplifier V_o is given by

$$V_o = \frac{V_R}{2^3} (S_3 2^3 + S_2 2^2 + S_1 2^1 + S_0 2^0) \qquad (14.9\text{-}1)$$

The important merit of the inverted-ladder configuration is that the currents through the resistors of the ladder do not change with switching. Hence the

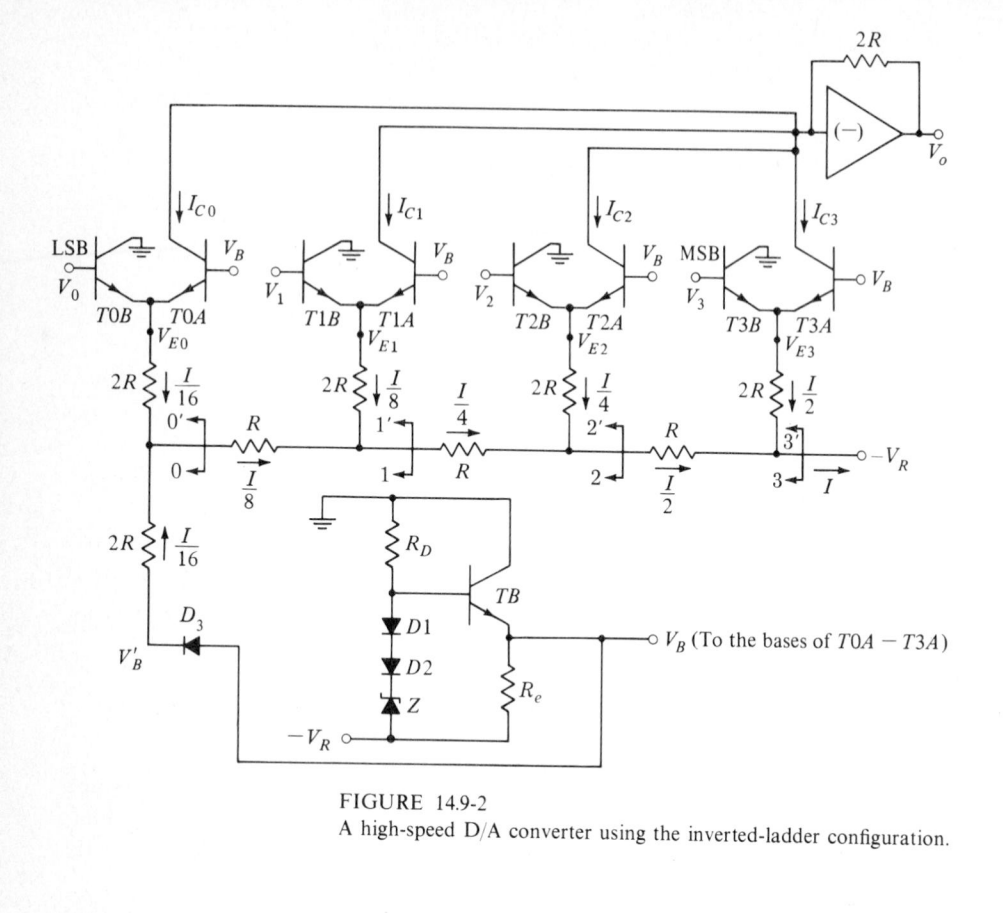

FIGURE 14.9-2
A high-speed D/A converter using the inverted-ladder configuration.

voltages across these resistors do not change, and there is no delay occasioned by the necessity for parasitic capacitors to charge or discharge. There is a second advantage. In the normal ladder the switches must switch the full reference voltage in and out of the circuit. In the inverted ladder the switches (thanks to their location) never have to sustain any appreciable voltage.

An example of an inverted-ladder converter is shown in Fig. 14.9-2. Here again the ECL-type difference amplifier has been used as a switch in order to avoid the delay associated with bringing a transistor out of saturation. We may note the correspondences and also the small differences between the circuits of Figs. 14.9-2 and 14.9-1. In Fig. 14.9-2 the reference voltage is $-V_R$ rather than V_R simply because *npn* transistors have been used. More important, we note that the last resistor in the ladder, which in Fig. 14.9-1 is connected to ground, is connected in the present circuit to the voltage V'_B. This connection, as we shall see, serves two functions. On the one hand it connects this last resistor to a voltage equal to the voltage $V_{E0} (= V_{E1} = V_{E2} = V_{E3})$ to which the other $2R$ resistors are connected. In Fig. 14.9-1 these resistors are all returned to the same point, namely, ground. The ground connection is required there since the switches are assumed ideal and this common point must be at the same potential as the virtual ground at the op-amp input. In Fig. 14.9-2 the switches are not perfect, and the common voltage is not the voltage at the op-amp input. The second function this connection provides is a measure of temperature compensation.

Let us consider that in each case in Fig. 14.9-2 the A transistor is ON. Then the B transistor is OFF, and the voltages at the common emitters are

$$V_{E0}\,(= V_{E1} = V_{E2} = V_{E3}) = V_B - V_{BE} \qquad (14.9\text{-}2)$$

in which V_{BE} is the voltage drop from base to emitter of a transistor ($T0A$, $T1A$, $T2A$, or $T3A$). We also find, referring to TB, that

$$V'_B = V_B - V_{D3} \qquad (14.9\text{-}3)$$

If we now assume that the voltage V_{D3} across the diode is the same as V_{BE}, then $V_{E0} = V'_B$, as required. On this basis the currents through the two leftmost resistors $2R$ in the circuit are the same, and we have assigned to them the value $I/16$. Following now the pattern of Fig. 14.9-1, i.e., noting that the circuits seen looking into 0–0', 1–1', 2–2', 3–3' each consist of an open-circuit voltage source V'_B in series with a resistor R, it is apparent that the currents assigned to the other resistors of the ladder are correct.

Now we can also verify that the currents in the ladder, and hence the currents injected into the op-amp, are independent of the temperature. These currents depend on the common voltage $V'_B = V_{E0} = V_{E1} = V_{E2} = V_{E3}$. Calculating V'_B, (see transistor TB) we find

$$V'_B = - V_R + V_Z + V_{D1} + V_{D2} - V_{BE}(TB) - V_{D3} \qquad (14.9\text{-}4)$$

Again assuming that the diode voltages are equal and equal also to V_{BE}, we find

$$V'_B = - V_R + V_Z \qquad (14.9\text{-}5)$$

which, as discussed in connection with the circuit of Fig. 14.8-2, can be adjusted to be temperature-independent.

There is a source of error in the circuit of Fig. 14.9-2. To explore this point assume initially that all the A transistors are ON. Now let V_0 increase in voltage in order to transfer current from $T0A$ to $T0B$. If $T0A$ is to turn OFF, the emitter voltage V_{E0} must rise. As a consequence, the current in the $2R$ emitter resistor of $T0$ must change, and thereby all the other currents (except in the emitter resistor of $T3$) will be disturbed. Since these currents are required to remain constant, an error will result. However, in a practical case the error need not be large enough to be serious. The calculation of this error in a typical case is left as an exercise (see Prob. 14.9-2).

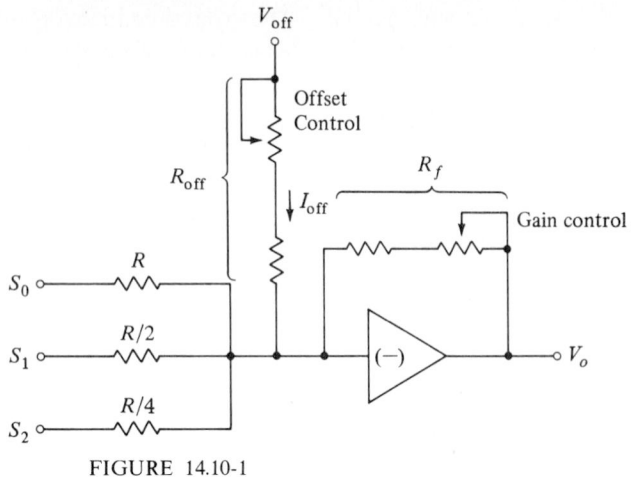

FIGURE 14.10-1
Circuit used to offset the output voltage of a D/A converter.

14.10   INPUT AND OUTPUT FORMATS OF A D/A CONVERTER

In the circuit arrangement of Fig. 14.5-1 suppose that we set $V(0) = 0$ V and deliver the current I_L to an operational amplifier, as in Fig. 14.5-3.   Then when $S = S_{N-1} \cdots S_0 = 0 \cdots 0$, that is, all zeros, V_o will be $V_o = 0$, and when $S = 1 \cdots 1$, that is, all ones, V_o will be given by Eq. (14.5-6).   If we take $V_R = 1$ V and $R = R_f$ and assume, say, a 3-bit converter, the output range will extend from 0 to -7 V.   We describe the output as *unipolar* since it swings in only one direction. Since the input bits have a straightforward binary significance (each bit has the numerical significance of some power of 2), the overall D/A converter is described as operating in the *binary unipolar format*.   If, however, $V(0) \neq 0$, we can *offset* the output swing.   For example, if $V(1) = \frac{1}{2}$ V and $V(0) = -\frac{1}{2}$ V, the output swing extends symmetrically about 0 V from -3.5 to $+3.5$ V.   In this case the format is called *bipolar binary* or *offset binary*.

An alternative method of offsetting the output range of a D/A converter is shown in Fig. 14.10-1.   Here we have made provision to inject an offset current I_{off} into the input of the amplifier.   Also included in the circuit is an adjustment of the gain of the amplifier.   It is to be noted that because of the virtual ground at the amplifier input, the adjustment of the gain control and the offset control are independent of each other.   Further, and again because of the virtual ground, the addition of the offset control has no effect on the input current of the amplifier which is due to the resistor array of the converter.   The offset produced in the output voltage V_o is $V_o = -(R_f/R_{\text{off}})V_{\text{off}}$.

As long as we do not intend to use the representation to perform arithmetic operations, the method of offsetting provides a useful and convenient representation of negative numbers.   For example, suppose, as above, that we start with a

3-bit converter which yields $V_o = 0$ V for an input **000** and $V_o = 7$ V for an input **111** [use $V(1) = -1$ V and $V(0) = 0$]. Now, suppose that we adjust the offset voltage so that $V_o = 0$ V for an input **100**. Then the digital input and analog output will be related as in Table 14.10-1. Here, 000 represents -4; 001 represents -3, and so on. Such a representation of negative numbers would be most convenient and acceptable if, say, the analog output were intended to deflect the pointer of a meter one way or another, depending on the sign of the input number to the D/A converter. Note that for N bits, 2^N is an even number. Hence the offset cannot be adjusted so that the output range is exactly symmetrical about 0 V if it is required that one digital input correspond to the analog output 0 V.

Suppose, on the other hand, the digital representation of negative numbers we must deal with is a representation more suitable for arithmetic computation, say the twos-complement representation. A procedure which will allow a D/A converter to read and convert such a number representation is given in Table 14.10-2.

As can be seen from Table 14.10-2 and taking Table 14.10-1 into account, to convert from a twos-complement format to an analog signal we apply to the

Table 14.10-1

Digital input	Analog output V_o
111	$+3$
110	$+2$
101	$+1$
100	0
011	-1
010	-2
001	-3
000	-4

Table 14.10-2

Decimal	Twos-complement representation	Offset binary format when **100** is adjusted for 0 V
$+3$	011	111
$+2$	010	110
$+1$	001	101
0	000	100
-1	111	011
-2	110	010
-3	101	001
-4	100	000

D/A converter the digital input with the most significant bit complemented and then adjust the offset so that **100** corresponds to 0 V.

If the digital input to the converter is presented in a ones-complement format, the appropriate conversion procedure is apparent from an examination of Table 14.10-3.

Observe that the ones-complement representations of the positive numbers $+0$ to $+3$ are identical to the representations of the unipolar binary numbers. Therefore the digital inputs representing positive numbers are applied directly to the D/A converter, which is set for a zero offset voltage.

If the ones-complement representation of the negative numbers **111** to **100** were applied directly to the converter, the output analog voltage would read $+7$ to 4 V. To shift these voltages we apply a -7-V offset voltage to the op-amp whenever the sign bit in the ones-complement format is **1**. For example, the number **110** (in ones complement) when applied to a D/A converter would read $+6$ V rather than -1 V. By applying a -7-V offset the correct analog voltage $+6 -7$ V $= -1$ V is read.

One way we can conveniently arrange for the correct offset is indicated in Fig. 14.10-2. Here we use symmetrical reference voltages $V(1) = -\frac{1}{2}$ V and $V(0) = +\frac{1}{2}$ V. We also use $V(1)$ and $V(0)$ as two offset voltages (V_{off} in Fig. 14.10-1) and add an additional switch, S_{off}. The offset resistor R_{off} is set to be equal to the parallel combination of R, $R/2$ and $R/4$, and finally we arrange that $S_{off} = \bar{S}_2$, so that when S_2, the switch corresponding to the most significant digit (the sign bit), is at logic **1**, S_{off} is at logic **0**, and vice versa. Because of the symmetrical input reference voltages V_o would range between $3\frac{1}{2}$ and $-3\frac{1}{2}$ V if the additional offset provided were ignored. Hence we start with an offset of $-3\frac{1}{2}$ V. When $S_2 = 0$, $S_{off} = 1$ and an additional offset of $+3\frac{1}{2}$ is

Table 14.10-3

Decimal	Ones-complement representation	Unipolar binary
$+7$		111
$+6$		110
$+5$		101
$+4$		100
$+3$	011	011
$+2$	010	010
$+1$	001	001
$+0$	000	000
-0	111	
-1	110	
-2	101	
-3	100	

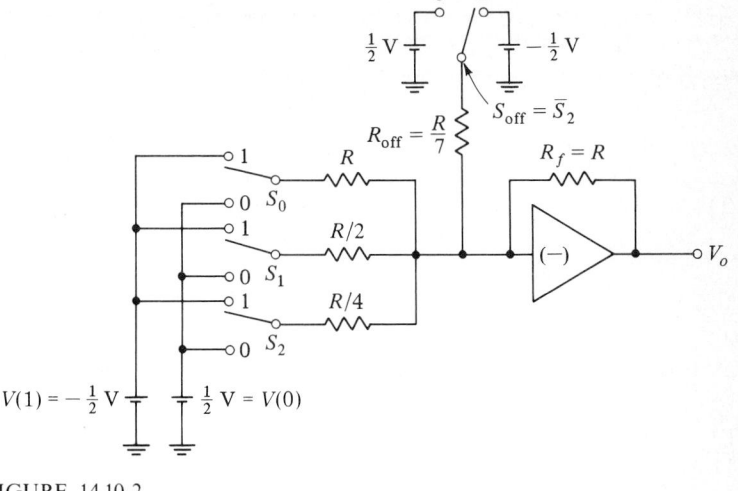

FIGURE 14.10-2
A D/A converter which accepts digital inputs in the ones-complement representation.

introduced, making the total offset zero. When $\bar{S}_2 = 1$, $S_{\text{off}} = \bar{S}_2 = 0$ and the offset voltages combine, making the total offset -7 V, as required.

It is left as a problem to show that the switch $S_{\text{off}} = \bar{S}_2$ can be eliminated by replacing S_2 by $\bar{S}_2$ and its resistance $R/4$ by $R/3$.

14.11 SPECIFICATIONS FOR D/A CONVERTERS

We consider now a number of the parameters which serve to describe the quality of performance of a D/A converter. These parameters are generally specified by manufacturers of converters.

Resolution This term specifies the number of bits the converter can accommodate and correspondingly the number of output voltages (or currents). For example, a converter which can accept 10 input bits is referred to as a converter with a 10-bit resolution. The number of possible output voltages is $2^{10} = 1,024$. Hence the smallest possible change in output voltage is $\frac{1}{1024}$ of the full-scale output range. Approximating 1,024 as 1,000, we can describe the resolution as being 1 part in 1,000, or 0.1 percent.

Linearity In an ideal D/A converter equal increments in the numerical significance of the digital-input should yield equal increments in the analog output. The *linearity* of a converter serves as a measure of the precision with which this requirement is satisfied. Linearity is measured in the manner suggested by Fig. 14.11-1. Assuming a unipolar binary input format, we have located the

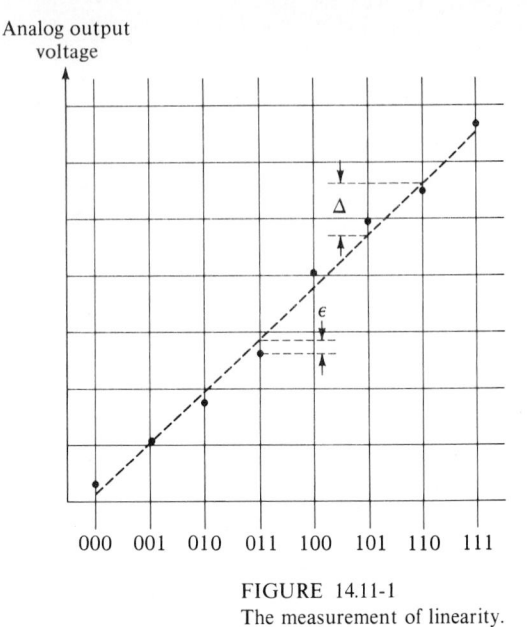

Analog output
voltage

000 001 010 011 100 101 110 111

FIGURE 14.11-1
The measurement of linearity.

input bit combinations with fixed-interval separations, in the order of numerical significance along the horizontal axis. Along the vertical axis we have indicated by dots the corresponding analog output voltage for each case, as might be encountered in a physical converter. If the converter were perfect, the dots would fall on a straight line. In Fig. 14.11-1 we have drawn a straight line best fit to the dots and in a typical case indicated by ϵ the linearity error. The voltage Δ is the nominal analog output change corresponding to a digital input change equivalent to a change in the least significant bit (LSB).

The linearity of a converter is generally specified by comparing ϵ to Δ. Thus, commonly, we find the linearity of a commercial unit specified as "less than $\pm\frac{1}{2}$ LSB," meaning that $|\epsilon| < \frac{1}{2}\Delta$. This is a very important specification. For suppose that at one digital input we find that ϵ is positive and $\epsilon > \frac{1}{2}\Delta$ while at the next higher digital input ϵ is negative and $|\epsilon| > \frac{1}{2}\Delta$. In this case the converter would have the unacceptable feature of not being *monotonic*; i.e., an increase in digital input would yield a decrease in analog output.

The linearity of a converter depends principally on the accuracy of the resistors. It depends as well on the precision with which the voltage drops across the switches are fixed. Since both resistors and switch voltages are temperature-dependent, linearity may be adversely affected by substantial temperature changes.

Accuracy The *accuracy* of a converter is a measure of the difference between the actual analog output voltage and what the output should be in the ideal case. Lack of linearity contributes to inaccuracy. Further limitations on

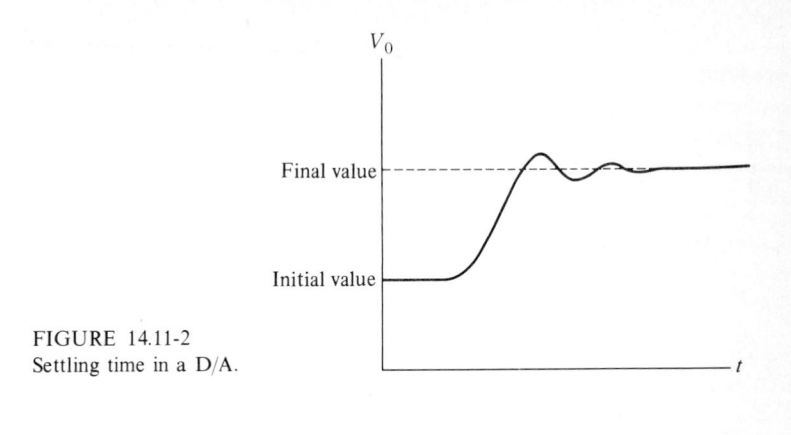

FIGURE 14.11-2
Settling time in a D/A.

accuracy are contributed by the uncertainty in the reference voltages, the amplifier gain, amplifier offset, etc. A typical manufacturer's specifications of a moderate-quality converter might read "0.2% of full scale $\pm \frac{1}{2}$ LSB."

Settling time When the digital input to a converter changes, switches open and close and abrupt voltage changes appear. Because of the inevitable stray capacitance and inductance present in the passive circuitry, the transients so initiated may persist for an appreciable time. Additional transients originate due to the characteristics of the active devices (switches, transistors, etc.). Typically a plot of the change in output voltage as a function of time might be as shown in Fig. 14.11-2. Note that not only is there a finite time required to reach the new output level, but also an oscillation may occur. The interval that elapses from the input change to the time when the output has come *close enough* to its final value is called the *settling time*. The settling time depends, among other things, on how we define "close enough." Typically a general purpose converter might have a settling time given as "500 ns to 0.2% of full scale."

Some of the largest transients (albeit, fortunately, generally of the shortest duration) are produced by the operation of the switches (Sec. 14.8). For example, suppose in a 4-bit unipolar converter the switches are at $S_3 S_2 S_1 S_0 = $ **0111**, corresponding to which the output is $+7$ V. Suppose an input change requires that now $S_3 S_2 S_1 S_0$ become **1000**, corresponding to 8 V. But suppose also that S_3 changes before the other switches have done so; then for a brief interval we shall have $S_3 S_2 S_1 S_0 = $ **1111** $= 15$ V. If the spikes produced in the D/A converter are objectionable, the D/A output voltage can be sampled and held. This new output is then low-pass-filtered. It is found in practice that the glitches formed by the S/H circuit contain significantly less energy than the typical D/A spike. Thus, the noise produced after filtering the S/H circuit is much less than the noise obtained after filtering the D/A.

Temperature sensitivity At any fixed digital input, the analog output will vary with temperature. This temperature sensitivity typically ranges from about

± 50 ppm/°C in a general-purpose converter to as low as ± 1.5 ppm/°C in a high-quality unit. The overall temperature sensitivity is due to the temperature sensitivities of the reference voltages, the resistors in the converter, the op-amp, and even the amplifier offset voltage.

14.12 A/D CONVERTERS: A PARALLEL-COMPARATOR TYPE

We turn now to systems employed to convert an analog input to a digital output. It should be noted at the outset that generally A/D converters are complex, sophisticated systems. Hence, unlike the situation which prevails with D/A converters, it is ordinarily not feasible to present a detailed schematic diagram. Instead we shall discuss A/D converters in terms of the basic digital and analog building blocks we have already described. These basic blocks include gates, flip-flops, registers, counters, D/A converters, comparators, etc. It is further to be noted that there are many possible schemes for A/D conversion and many variations possible within each scheme. The very large number of A/D converters described in the literature (a good number of which are available commercially) is a tribute to the persistence and ingenuity of engineers who have applied themselves to the problem. In the present and succeeding sections we shall describe a number of representative systems.

In the *comparator* A/D converter shown in Fig. 14.12-1 the range of analog input extends from 0 to V_0, and a 3-bit digital output is provided.

The relationship of the digital output to the analog input is presented in Fig. 14.12-2. The analog input is divided into eight ranges. Six of these ranges encompass an interval $S = V_0/7$. The other two ranges, at the ends, extend over the interval $S/2 = V_0/14$. When the analog input is anywhere in the lowest range from 0 to $V_0/14$, the A/D converter output is to be **000,** as indicated. As also indicated, if this digital output were in turn to be reconverted to an analog voltage, as by a D/A converter, the analog reading would be 0 V. Hence, in the A/D conversion, an error, the *quantization error*, has been introduced. In this lowest range the error is at most equal to $S/2 = V_0/14$. Similarly when the input is in the range S extending from $V_0/14$ to $(3\ V_0)/14$ the corresponding digital output will be **001**. This **001** will be interpreted as representing the analog voltage $V_0/7 = 2V_0/14$. Hence in this interval the quantization error will again never be larger than $S/2 = V_0/14$ no matter where in the range the input falls. It can now be seen that the reason for setting the intervals as in Fig. 14.12-2 is to arrange that throughout the input range from 0 to V_0 the maximum quantization error will be the same.

It will be recalled that a comparator (see Sec. 2.13) is a device with two inputs (a reference input and a signal input) and a single output. When the signal input is less than the reference-input voltage, the comparator output is at logic **0**. When the signal is higher than the reference, the output is at logic **1**. A comparator A/D converter with N output bits requires $2^N - 1$ comparators. For the 3-bit system of Fig. 14.12-1, seven comparators C_1 to C_7 are used.

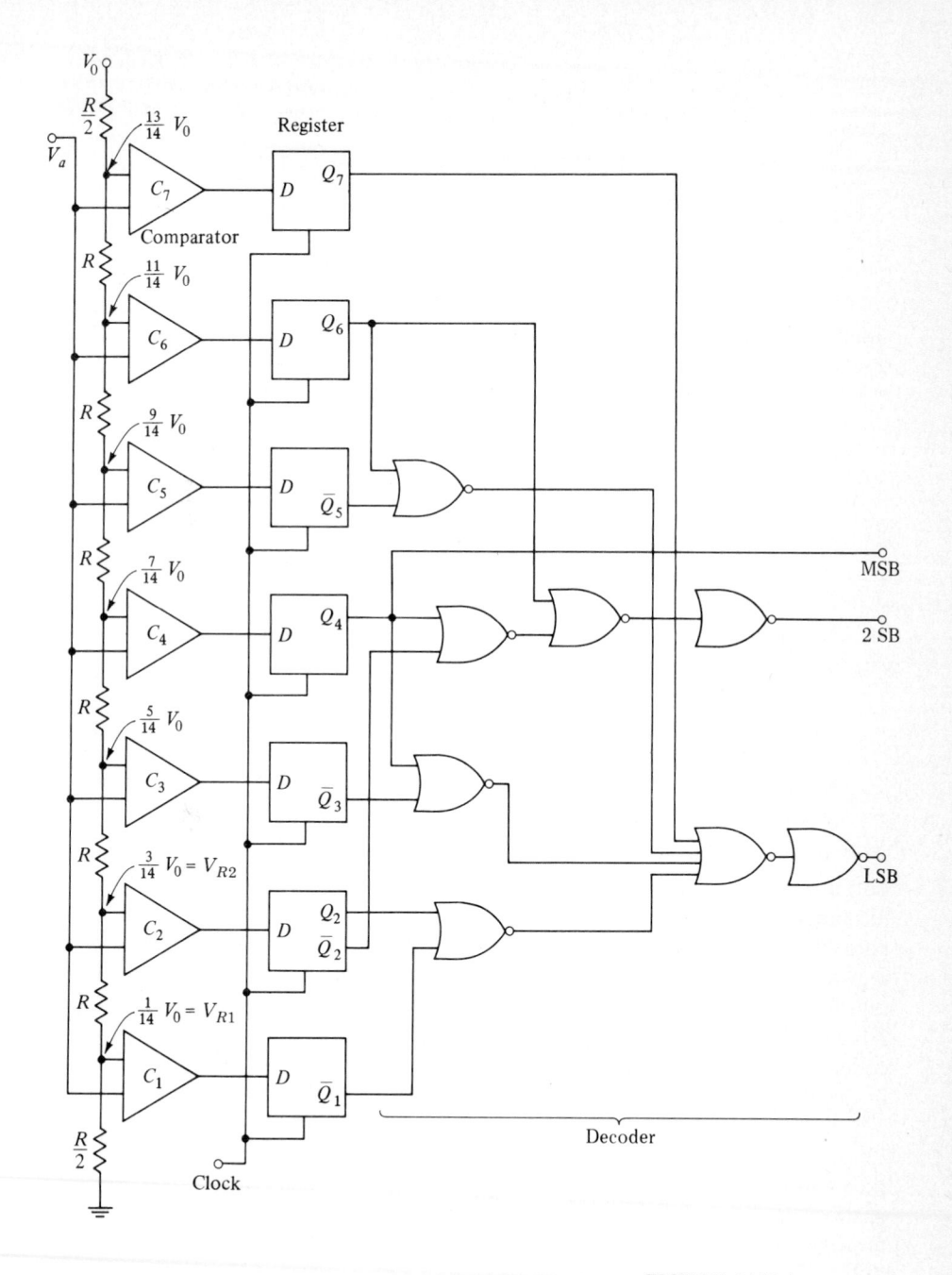

FIGURE 14.12-1
A comparator A/D converter.

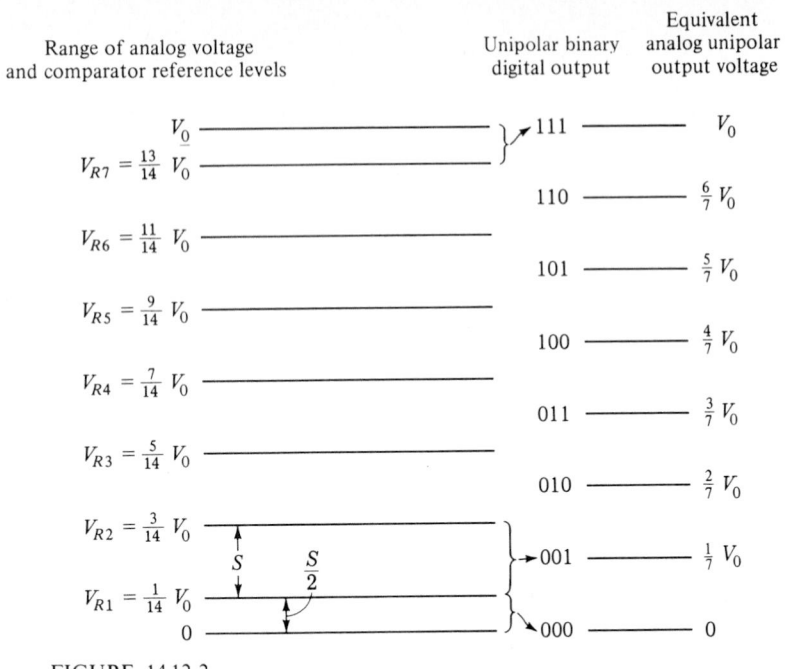

FIGURE 14.12-2
A unipolar analog voltage is divided into intervals and the intervals assigned digital representations in a manner assuring uniform maximum quantization error over the entire range.

If the analog voltage V_a is in the range 0 to $\frac{1}{14}V_0$, all the comparator output logic levels will be **0**, that is, $C_1 C_2 C_3 C_4 C_5 C_6 C_7 = 0000000$. If V_a is in the range $\frac{1}{14}V_0$ to $\frac{3}{14}V_0$, then $C_1 C_2 C_3 C_4 C_5 C_6 C_7 = 1000000$, etc. These comparator outputs will be transferred to the outputs of the seven flip-flops of the register at the occurrence of a clock pulse. Finally, as in Fig. 14.12-1, the register is followed by a decoder, which converts the register indications into a 3-bit unipolar binary code. It can be verified (see Prob. 14.12-1) that the decoder indicated in Fig. 14.12-1 does indeed assign the 3-bit output code words in the manner required in Fig. 14.12-2.

EXAMPLE 14.12-1 If the comparator A/D converter shown in Fig. 14.12-1 is to convert analog voltages varying from $-V_0$ to $+V_0$ into twos-complement arithmetic, determine the reference voltages at each comparator input. Illustrate the result using a chart similar to Fig. 14.12-2. Assume a 3-bit output.

SOLUTION When the output of the A/D converter is **000**, it will be read as 0 V. Hence, if the maximum quantization error is to be $S/2$, as in Fig. 14.12-2, the output indication **000** should be assigned to the analog range 0 V $\pm$ $S/2$, as shown in Fig. 14.12-3. A slight difficulty now arises because the analog range is symmetrical about 0 V

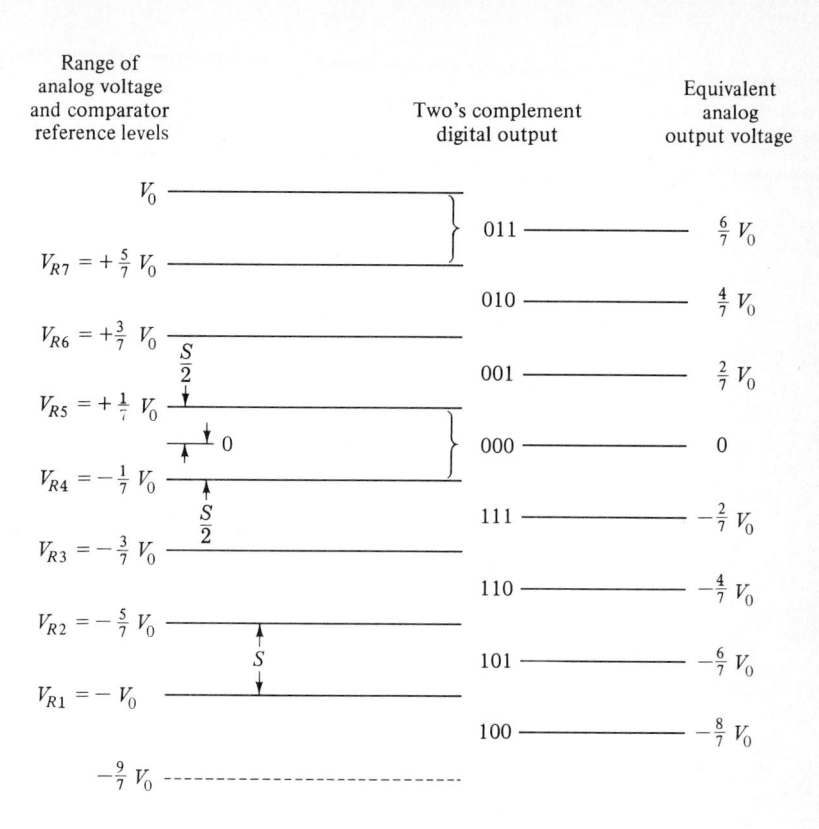

FIGURE 14.12-3

Reference levels and digital outputs for the comparator A/D converter using the twos-complement representation.

while in the twos-complement representation there is always one more negative number than positive number (see Sec. 11.9). If we ignore for the moment the most negative digital number (**100** $= -4$), seven digital outputs remain. Accordingly the analog range V_0 to $-V_0$ is divided into seven intervals, each of size $S = 2V_0/7$, and, as shown in Fig. 14.12-3, one digital output is assigned to each range.

We still have the extra digital output **100**, which will serve to represent the range $-\frac{8}{7}V_0 \pm S/2 = -\frac{8}{7}V_0 \pm \frac{1}{7}V_0$, that is, the range $-V_0$ to $-\frac{9}{7}V$. If we choose to use this output, the bottom of the resistor chain in Fig. 14.12-3 will have to be returned to $-9V_0/7$ and seven comparators will be required with reference voltages $-V_0$, $-\frac{5}{7}V_0$, etc., up to $\frac{5}{7}V_0$. If we choose to ignore this **100** output, the resistor chain will be returned to $-V_0$ and only six comparators will be needed. Most manufacturers choose to use the **100** output and specify their device as having a corresponding asymmetrical voltage range.

The comparator A/D converter is capable of great speed since the entire conversion process occurs simultaneously rather than sequentially. Its operation is fast enough to warrant the use of emitter-coupled logic for the flip-flops

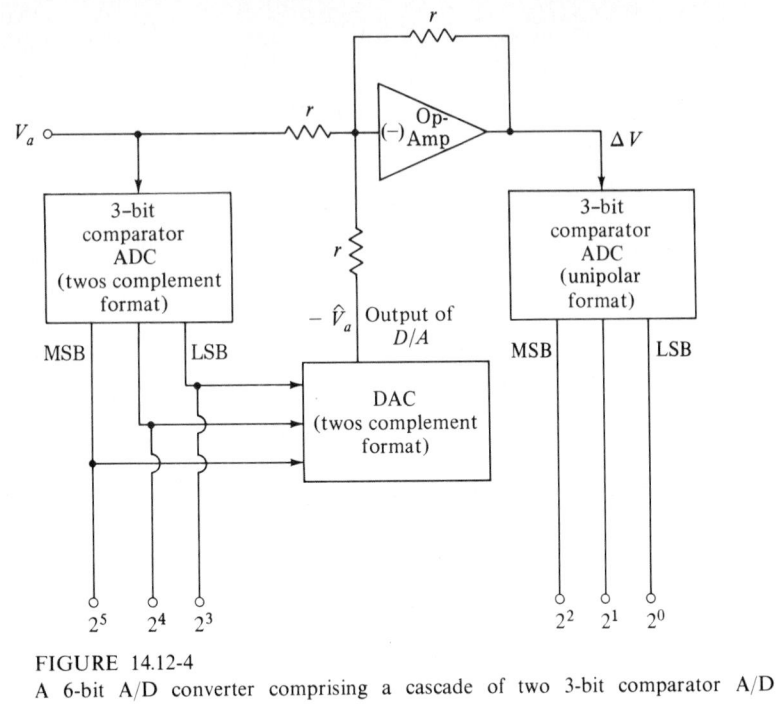

FIGURE 14.12-4

A 6-bit A/D converter comprising a cascade of two 3-bit comparator A/D converters.

and the decoder gates. Immediately on presentation of the analog input voltage and after only the short delay in the comparators, a digital representation of the analog voltage is available. We may note that the flip-flop register is not essential to the operation. However, the register is convenient, especially when the analog input is changing rapidly. The register permits us to hold the digital representation of the analog input until we are ready to accept a new sample.

A cascade-comparator A/D converter The comparator A/D converter has an inconvenient feature in that the hardware required nominally doubles for each additional output bit. Thus, while a 3-bit A/D converter requires seven comparators and seven flip-flops, a 4-bit A/D converter would require fifteen comparators, fifteen flip-flops, and a corresponding increase in decoding gates. At a sacrifice of speed of operation, it is possible to cascade comparator converters with a saving in hardware. Such a cascaded converter is shown in Fig. 14.12-4. The converter provides 6 output bits. Such a converter, if constructed as a single unit after the pattern of Fig. 14.12-1, would require $2^6 - 1 = 63$ comparators. In Fig. 14.12-4 however, we use two 3-bit converters involving $2(2^3 - 1) = 14$ comparators.

The first converter ADC-1 provides the three most significant bits while ADC-2 generates the three least significant bits. Let us assume for convenience that the scaling of the overall 6-bit converter is arranged so that the output reads directly in volts, (e.g., 001101 represents 13 V). In this case the step size of ADC-2 is $S_2 = 1$ V while the step size of ADC-1 is 8 V. The digital output of ADC-1 is applied to a D/A converter which yields an analog output $\hat{V}_a$. The difference $\Delta V = V_a - \hat{V}_a$ is in the quantization error range of ADC-1. This analog voltage difference is converted to digital form by the second converter ADC-2.

We note, most importantly that the bits added by ADC-2 can only *leave unchanged or increase* the numerical significance of the final 6-bit representation. It would then appear that we must arrange that ΔV always be zero or positive. We may assure that such be the case by setting the comparator reference levels in ADC-1 at ..., -8 V, 0 V, 8 V, 16 V, ... Thereafter we shall decode in such manner that when V_a is in the range -8 to 0 V the digital output be 111 (using twos complement representation), when V_a is in the range 0 to 8 V the output be 000, when V_a is in the range 8 to 16 V the output be 001, etc. Such setting of comparator levels and subsequent decoding would be nearly but not quite what is required. For, consider that V_a is infinitesimally smaller than 8 V. We shall require in this case that the 6-bit output read 001000 in order that the quantization error be no more than $\pm\frac{1}{2}$ LSB which in the present case is ± 0.5 V. However, with V_a just less than 8 V, the digital output of ADC-1 would be 000. Even if ADC-2 yielded an output 111 the 6-bit indication would be 000111 with an error of a full LSB. As is readily verified, the difficulty may be remedied by moving the comparator reference levels in the negative direction by $\frac{1}{2}$ LSB $= \frac{1}{2} S_2 = 0.5$ V. In this case ΔV may turn out to be negative for some range of V_a, but never by more than $\frac{1}{2}$ LSB and hence no difficulty results.

The difference ΔV applied to ADC-2 will now be in the range from -0.5 to 7.5 V. The unipolar format of comparator reference levels as displayed in Fig. 14.12-2 may be set to accommodate this range with quantization error ± 0.5 V. To accomplish this end we need only to set the comparator reference levels at 0.5, 1.5, ..., 6.5 V.

Sources of error   A quantization error is necessarily introduced when an analog signal is converted to digital form. This error is often specified as $\pm\frac{1}{2}$ LSB; that is, it has a magnitude which is one-half the numerical significance of the least significant bit of the digital output. For example, consider a straightforward (single-stage) converter intended to handle an input analog signal in the range 0 to 7 V with a 3-bit digital output from **000** to **111**. The reference level of the seven comparators would be set at 0.5, 1.5, ..., 5.5, 6.5 V. If, say, the analog voltage V were in the range $2.5 < V < 3.5$, then the converter output would read 011 = 3.0 V. In a physical situation, however, the error might be larger than $\pm\frac{1}{2}$ LSB. Such would be the case if the comparator reference levels were not precisely set and if the comparators operated less than ideally. Ideal operation of the comparator would require the comparator outputs to be at logic **1** or at logic **0** as the analog voltage is infinitesimally on one side or the other of the reference level, there being no range of uncertainty.

In the cascaded comparator converter we have sources of error not only in the comparators but also in the D/A converter, the mechanism used to form the difference $V - \hat{V}_a$, and the amplifier. These additional operations, none of which is present in the straightforward comparator converter, must be interposed between every stage of a cascaded converter.

High-speed operation   Although the cascaded A/D converter serves significantly to reduce the number of comparators required in the system, it also reduces the speed of operation by a factor of 2. For on the first clock pulse we convert the first 3 bits, and on the second clock pulse we convert the second 3 bits.

With some extra circuitry, the cascaded A/D converter can be operated at the same speed as the A/D converter shown in Fig. 14.12-1. We note that while the least significant 3 bits are being converted, the A/D converter for the most significant 3 bits is not performing any useful function. Thus, we can design circuits which enable this first converter to start operating on the next sample of V_a.

14.13   SUCCESSIVE-APPROXIMATION CONVERTER

The principle of the successive-approximation converter is set forth by the following example. Suppose that we have an object whose weight is unknown beyond the fact that it is in the range 0 to 1 kg. Suppose further that a balance is available and a set of known weights of $\frac{1}{2}$, $\frac{1}{4}$, $\frac{1}{8}$ kg, etc. These known weights are to be used in a succession of trials to determine the unknown weight. With the unknown weight W on one side of the balance we place the $\frac{1}{2}$ kg on the other side. If we find $W > \frac{1}{2}$ kg, we leave the $\frac{1}{2}$-kg weight on the scale and add the $\frac{1}{4}$-kg weight. If we find $W < \frac{1}{2}$ kg, we remove the $\frac{1}{2}$-kg weight and put on the $\frac{1}{4}$-kg weight. In this way we continue to try weights successively smaller by a factor of 2. Whenever the last trial weight added tilts the balance toward the side of the known trial weights, we remove the last weight added and try the next smaller weight. Thus, if we found that we could leave the $\frac{1}{2}$-kg weight, had to remove the $\frac{1}{4}$-kg weight, and could leave the $\frac{1}{8}$-kg weight, we would approximate the unknown weight as $1 \times \frac{1}{2}$ kg $+ 0 \times \frac{1}{4}$ kg $+ 1 \times \frac{1}{8}$ kg $= \frac{5}{8}$ kg. Assigning the numerical significance $\frac{1}{2}$ to the most significant binary digit, $\frac{1}{4}$ to the next, etc., we would have for the weight the binary designation **101**. It is clear that continuing this operation with successively smaller weights, we can establish the unknown weight to whatever precision we please.

If the number of allowable successive weighings is limitless, the procedure described above is acceptable. Suppose, however (as is normally the case), that the number of weighings is finite. Then for the sake of reducing the quantization error it is necessary to *offset* the scale, i.e., to bias or "tilt" the scale in favor of the unknown. The magnitude of the offset must be equal one-half the smallest weight. That such is the case is illustrated in Fig. 14.13-1. Here we assume two successive trials using weights of $\frac{1}{2}$ and $\frac{1}{4}$ kg to determine a

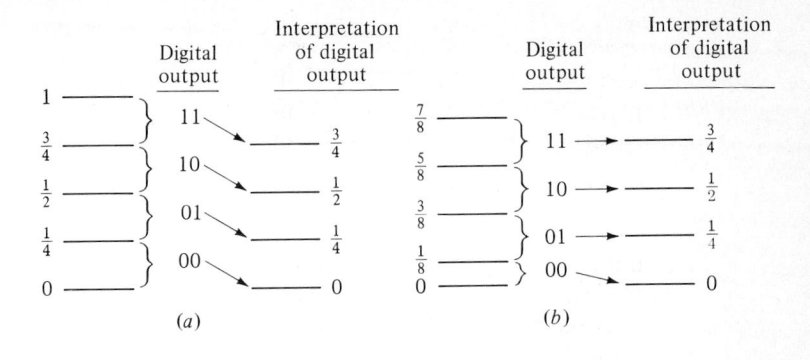

FIGURE 14.13-1
In successive approximation weighing it is necessary to offset the scale by one-half
the smallest trial weight.

weight assumed to lie in the range 0 to 1. In Fig. 14.13-1a, the range 0 to 1 is
divided into four intervals, and each interval is identified by its appropriate
digital representation. Also indicated is the interpretation given to each digital
representation. Now suppose that the unknown is infinitesimally smaller than
$\frac{1}{4}$. We would find that we could use neither the $\frac{1}{2}$ nor the $\frac{1}{4}$ trial weights and
the corresponding digital indication would be **00**. The quantization error is
then $\frac{1}{4}$. A similar maximum quantization error of $\frac{1}{4}$ will be encountered in every
other interval.

Now suppose we offset the scale by adding a weight of $\frac{1}{8}$ to the scale side
intended for the unknown. Then the four digital representations will be associated with weight ranges in the manner shown in Fig. 14.13-1b. An unknown
weight just less than $\frac{1}{4}$ will appear as a weight just less than $\frac{1}{8} + \frac{1}{4} = \frac{3}{8}$. This
weight falls in the range with digital indication **01**. This indication will be
interpreted as $\frac{1}{4}$ so that the quantization error is $\frac{1}{8}$. As can be seen in Fig. 14.13-1b,
the new range of the unknown is from 0 to $\frac{7}{8}$, and, as can be verified, any place in
this range the maximum quantization error to be encountered is $\frac{1}{8}$.

A 3-bit successive-approximation A/D converter   A diagram of a 3-bit successive-
approximation A/D converter is shown in Fig. 14.13-2. This converter is designed
to convert an analog waveform into binary code neglecting the sign bit. For
simplicity we have not been as economical of hardware as we might otherwise
have been. In this A/D converter we have allowed five (equal) time intervals
to accomplish a single A/D conversion. Three of these intervals are used to
determine the 3 digital bits; a fourth interval is used simply to read the digital
output; while a fifth interval is used to clear the converter in readiness for the
next conversion.

The five type-D flip-flops FFA to FFE are connected to form a modulo-5
ring counter (see Sec. 10.16). Such a counter provides at its outputs Q_A to Q_E,
five waveforms, only one of which is at logic level **1** at any time; the logical
level **1** is transferred from A to B to C, etc., with each successive clock cycle.

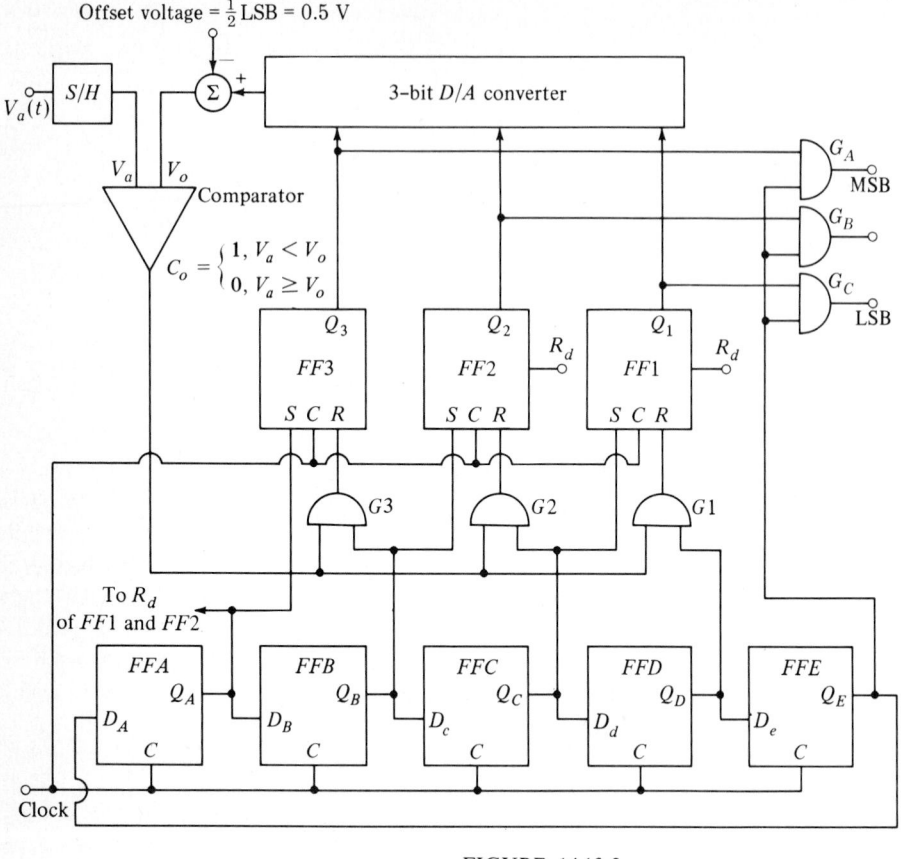

FIGURE 14.13-2

A 3-bit successive-approximation A/D converter.

The three flip-flops $FF1$, $FF2$, and $FF3$ are employed to register the digital bits, with $FF1$ corresponding to the LSB and $FF3$ to the MSB.

The conversion cycle begins with $Q_A = 1$, while $Q_B = Q_C = Q_D = Q_E = 0$. Then $FF3$ will be set while $FF2$ and $FF1$ will be reset. We then have $Q_3 = 1$ and $Q_2 = Q_1 = 0$. The input **100** is thus presented to the 3-bit D/A converter, which provides a corresponding analog output V_o. The comparator output C_o will then be $C_o = 0$ or **1** depending on whether $V_a \geq V_o$ or $V_a < V_o$. During the next clock interval $Q_B = 1$, while $Q_A = Q_C = Q_D = Q_E = 0$. With $Q_B = 1$ the AND gate $G3$ is enabled, and $FF3$ is reset if $C_o = 1$ and left in the set state if $C_o = 0$. Thus, altogether we have tentatively assigned a logic **1** to the most significant position, and at the beginning of the second clock interval this bit remains or is changed to logic **0** depending on the comparison of V_a and V_o.

During succeeding clock intervals the trial is repeated for the bits in the next two places.   The interval when $Q_E = 1$ is an interval when no comparisons are being made and we can read the digital output.   Thus, Q_E is used to strobe the output gates G_A, G_B, and G_C.

The point need hardly be belabored that during the sequence of operations leading finally to a digital output the sample value of the analog input must be held constant.   Therefore, the converter of Fig. 14.13-2 must be preceded by an S/H amplifier.   The sample operation must be synchronized to the operation of the converter.   The read-out interval when $Q_E = 1$ is an interval suitable for sampling, while during the interval from $Q_A = 1$ through $Q_D = 1$, that is, while $Q_E = 0$, the sampled signal is *held*.   Hence, the synchronization can be effected by using Q_E to operate the switches in the S/H circuit.

As the converter operates, the flip-flops must toggle back and forth, the D/A switches must open and close, and the comparator sees an input which switches abruptly from one level to another.   There are transients associated with all these switchings and abrupt changes, and the speed at which the converter can be driven depends on the decay time of these transients.   However, the accuracy of the converter does not depend on the flip-flop or on the gates.   The accuracy depends almost exclusively on the accuracy of the D/A converter and hence on the accuracy of the D/A resistors, etc.

At the outset of our discussion of the successive-approximation method of conversion we described the method in terms of a succession of weight comparisons on a scale.   Returning to that analogy, it is seen that the comparator shown in Fig. 14.13-2 is the scale, the analog input V_a is the unknown weight, and the A/D converter output V_o is the sum of all the trial weights which are "left on the scale."   Whenever a bit **1** causes the scale to tilt toward the side of the sum of the trial weights ($V_o > V_a$), the bit **1** is reset to a **0** and the next bit of lower significance is tried.

We noted earlier the need to tilt the scale in the direction of the unknown. In the circuit of Fig. 14.13-2 this tilt can be effected by adding a fixed voltage to the unknown V_a or, generally more conveniently, by offsetting V_o in the opposite direction.   (One arrangement for providing such an offset is shown in Fig. 14.10-1.)   Consider by way of example a 3-bit converter with outputs ranging from **000** to **111** intended for an analog voltage in the range 0 to 7 V. The least significant bit corresponds to an analog voltage of 1 V.   Hence the required offset of V_o is $\frac{1}{2}$ V.   Specifically the D/A converter must be offset such that when its digital input is **000**, its analog output must be -0.5 V, as shown in Fig. 14.13-3.   In this case the maximum quantization error will also be 0.5 V. Further, the maximum analog voltage which can be accommodated is 7.5 V. For when the output is **111**, the quantization error does not exceed 0.5 V until $V_a > 7.5$ V.

It is instructive to trace the steps needed to encode an analog voltage V_a. If, say, $V_a = 4.9$ V, we begin with $Q_A = 1$ and $Q_3 Q_2 Q_1 = 100$ so that V_o reads $4 - 0.5 = 3.5$ V.   Since $V_a > V_o$, when Q_B becomes **1**, Q_3 remains **1** and Q_2 also

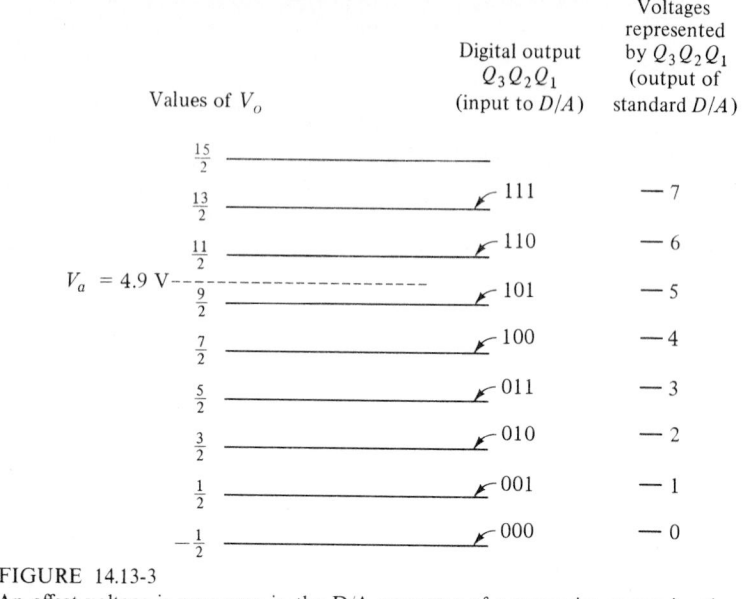

FIGURE 14.13-3

An offset voltage is necessary in the D/A converter of a successive approximation A/D converter.

becomes **1** so that $Q_3 Q_2 Q_1 = $ **110**.   Now $V_o = 6 - 0.5 = 5.5$ V $> V_a$.   Hence, when $Q_C = $ **1**, Q_2 becomes **0** and Q_1 goes to **1** yielding $Q_3 Q_2 Q_1 = $ **101**.   The result is $V_o = 5 - 0.5 = 4.5$ V $< V_a$.   Therefore Q_1 does not change when $Q_D = $ **1**, and the answer is $Q_3 Q_2 Q_1 = $ **101**.   This result is read when $Q_E = $ **1**.

14.14   THE COUNTING CONVERTER

A three-bit *counting* converter is shown in Fig. 14.14-1.   The three flip-flops are connected as a modulo-8 ripple counter.   The counter goes from state to state at each clock pulse when gate $G0$ is enabled.   Let us assume initially that the control line H (hold count) is at logic level **1**.   Then $G0$ is disabled and the counting does not proceed.   Let us assume further that the R_d (reset) line has been used to reset the counter to **000**, in which case the output will also read **000**.   Finally, let us consider that the H line is also used to control the operation of the S/H circuit, so that the waveform $V_a(t)$ is sampled when $H = $ **1** and is held when $H = $ **0**.

Initially, let $V_o = \frac{1}{2}$ LSB and let the comparator output be at logic **1**.   Now let H change to $H = $ **0**.   Then $G0$ is enabled, and the counting proceeds.   With each clock pulse the counter advances one count and the D/A converter output V_o jumps by one step.   Eventually, we shall have $V_o > V_a$.   At that time the comparator output C_o will become $C_o = $ **0**, and the counting will stop.   The count which has accumulated in the counter is the digital output and will be proportional (except for a quantization error) to the analog voltage $V_a(t)$.   Having allowed a counting time long enough to ensure that V_o has exceeded V_a, we can

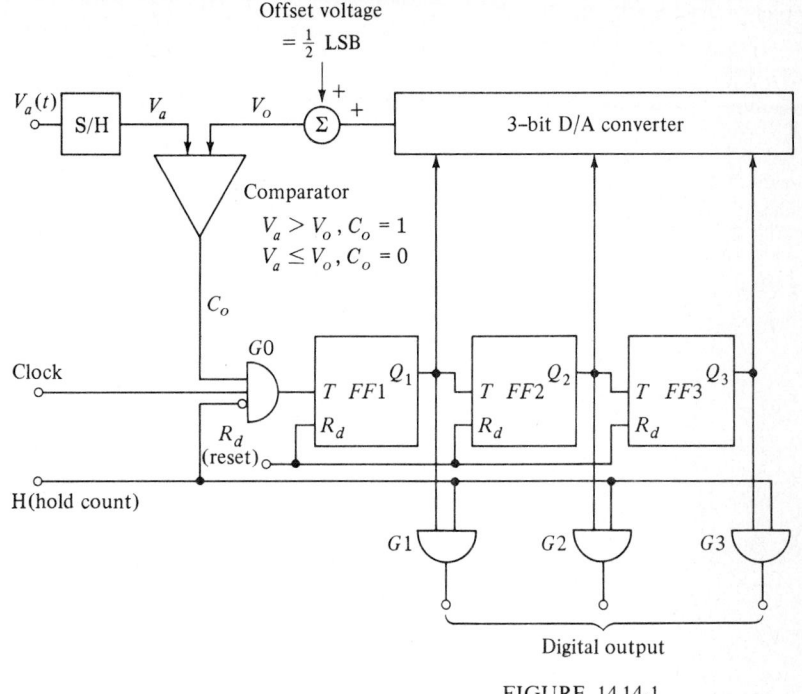

FIGURE 14.14-1
A 3-bit counting A/D converter.

raise H to $H = \mathbf{1}$, thereby allowing a reading of the digital output and allowing the S/H circuit to sample the input signal again. Before returning H to $H = \mathbf{0}$ we move the reset R_d briefly to $R_d = \mathbf{1}$ to clear the counter and then return R_d to $R_d = \mathbf{0}$. Now returning H to $H = \mathbf{0}$ will start a new conversion cycle.

In the counting A/D converter, as in the successive-approximation converter, it is necessary to offset the output V_o of the D/A converter. In the present case, however, the offset must be in the direction to *increase* rather than to decrease V_o by the voltage corresponding to $\frac{1}{2}$ LSB. To see that such is the case neglect the offset voltage and let the LSB equal 1.0 V, in which case the maximum quantization error is to be ± 0.5 V. Suppose now that V_a is infinitesimally higher than 0 V. Then at the start of the conversion we shall have $C_o = \mathbf{1}$, and the counter will advance by one count and then stop. But this one count will yield a digital output **001** with the interpretation 1.0 V. Correspondingly the quantization error will be 1.0 V. On the other hand, with an offset of 0.5 V the counter will not advance by one count until $V_a > 0.5$ V.

The converter waveforms are shown in Fig. 14.14-2. We assume as before a D/A converter which yields outputs (exclusive of the offset) of 0 and 7 V for digital inputs **000** and **111**, respectively. Because of the appearance of the waveform V_o, the counter converter is also referred to as the *digital-ramp converter*. Two conversion intervals are shown. In the first, the analog voltage

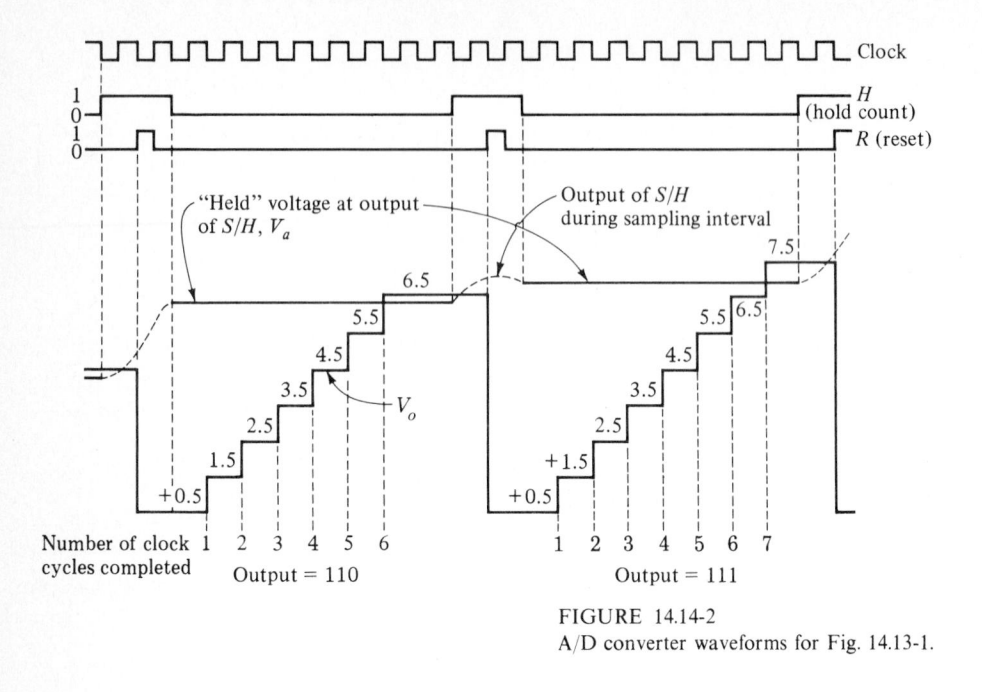

FIGURE 14.14-2
A/D converter waveforms for Fig. 14.13-1.

lies between 5.5 and 6.5 V, and the digital output is **110** = 6 V. In the second, V_a lies between 6.5 and 7.5, and the output is **111**. The largest analog input which can be accommodated with maximum quantization error 0.5 V is again 7.5 V.

For a given sampling rate and number of output bits the counter converter generally requires a much faster clock rate than the successive-approximation converter. In the counter converter, with N output bits, 2^N clock cycles are required for a conversion operation. In the successive-approximation converter the number required is N (or $N + 2$ if we include clock intervals for resetting and reading, as in Fig. 14.13-2). In any event the clock frequency increases exponentially with N in the counter converter and only linearly in the successive-approximation counter. The counter type of A/D converter is usually restricted to sampling frequencies which are less than 100 kHz, while with successive-approximation converters 1-MHz sampling rates are feasible.

At the expense of some increase in complexity the counter converter can be improved by substituting for the up counter of Fig. 14.14-1 an up-down counter (see Sec. 10.14). Such a converter is referred to as a *continuous-digital-ramp converter*, a *tracking converter*, or a *servo converter*. The counter is commanded to count up or down depending on whether the output of the comparator is at logic **1** or logic **0** and hence depending on whether V_o is larger or smaller than V_a. If initially $V_a > V_o$, the counter counts up until $V_o > V_a$. At this point the counter reverses. If after a single count down we find that $V_o < V_a$, the counter reverses again, and so on. The level of V_o will therefore

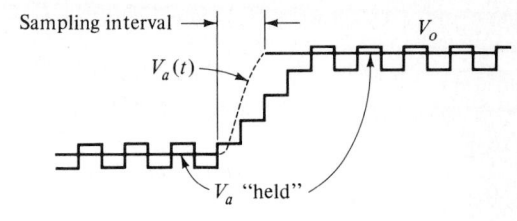

FIGURE 14.14-3
The behavior of a tracking converter.

hunt back and forth across V_a. The appearance of V_o during intervals when V_a is being held and during an intervening sampling interval is indicated in Fig. 14.14-3. The output of this converter is read, as before, at the end of the hold interval. Compared with the straight counting converter, in the servo converter, only half as many counts on the average will be required to complete a conversion. Hence a servo converter can operate at twice the speed.

Of the three types of converters just described which involve feedback through a D/A converter, the successive-approximation converter is by far the most popular.

14.15 THE DUAL-SLOPE CONVERTER

We consider now a type of converter which involves no feedback. This converter, shown in Fig. 14.15-1a and called the *dual-slope converter*, is often used in digital voltmeters. We now describe its principle of operation.

At the beginning of the conversion process, say at $t = 0$, the switch S_1 is connected to point A, and the held sample of the analog input V_a is applied to the analog integrator. If $\tau = RC$ is the time constant of the integrator, the integrator output is $V_o = -(t/\tau)V_a$. The waveform of V_o is shown in Fig. 14.15-1b. At the same time ($t = 0$) a clock waveform is applied to a counter which was initially clear. The counter counts until the counter flip-flops $FF0$ to $FF(N-1)$ simultaneously reset, so that $Q_0 = Q_1 = Q_{N-1} = \mathbf{0}$, at which time FFN is set, that is, $Q_N = \mathbf{1}$. This output Q_N controls the state of S_1, and when $Q_N = \mathbf{1}$, switch S_1 moves to point B. The output of the integrator now starts to move in the positive direction since the applied reference voltage is negative, that is, $-V_r$ (see Fig. 14.15-1b). The counter continues to count until the output voltage V_o becomes just barely positive. At this time the comparator output goes to the $\mathbf{0}$ state, gate $G1$ is disabled, and the counting stops.

We now show that the count recorded in the N-stage counter, $Q_{N-1}, \ldots, Q_1, Q_0$, is directly proportional to V_a and is independent of the time constant τ. The time T_1 required for the $N+1$ flip-flops to go from $\mathbf{00 \cdots 00}$ to $\mathbf{10 \cdots 00}$ is $2^N T_C$, where T_C is the time between clock pulses. At this time the output voltage V_o is

$$V_o = -\frac{V_a}{\tau} T_1 = -\frac{V_a}{\tau} 2^N T_C \qquad (14.15\text{-}1)$$

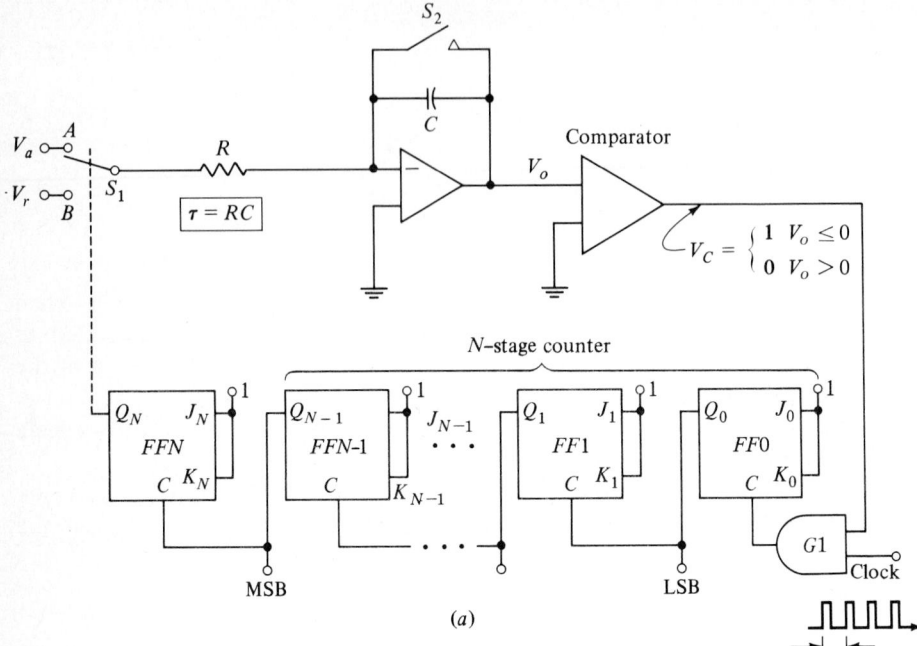

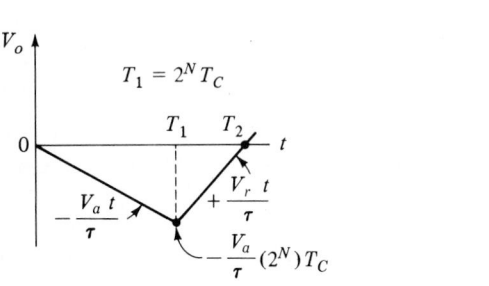

(b)

FIGURE 14.15-1
The dual-slope A/D converter.

Referring to Fig. 14.15-1b, we see that at the time T_2 V_o is again equal to 0 V and that therefore

$$\frac{V_r(T_2 - T_1)}{\tau} = \frac{V_a}{\tau} T_1 \qquad (14.15\text{-}2)$$

Hence, the time interval $T_2 - T_1$ is

$$T_2 - T_1 = \frac{V_a}{V_r} 2^N T_C \qquad (14.15\text{-}3)$$

If at time T_2 the count recorded in the first N flip-flops is λ, since the count was 0 at time T_1, we have

$$T_2 - T_1 = \lambda T_C = \frac{V_a}{V_r} 2^N T_C \qquad (14.15\text{-}4)$$

so that the count λ is

$$\lambda = \frac{V_a}{V_r} 2^N \qquad (14.15\text{-}5)$$

As long as $V_a < V_r$, the system operates as an A/D converter. Since $\lambda < 2^N$, the count is directly proportional to V_a and is a number which can be read from the counter. The converter can be made direct-reading if $V_r = 2^N$ V. Now $\lambda = V_a$, and the count recorded in the counter is numerically equal to the applied voltage V_a.

When the counting ceases, the outputs of $FF1$ to $FF(N-1)$ are recorded, all $N+1$ flip-flops are reset, and capacitor C is discharged using switch S_2. A new sample can now be converted. The logic circuitry needed to reset the system is left as a problem (Prob. 14.15-2).

14.16   A COMPARISON OF CONVERTER TYPES

The converters we have discussed represent devices whose speed of operation lies in three different ranges. Fastest is the comparator converter. In principle, except for the delay through the comparators, this converter makes available a digital output at the moment the analog input is applied. Hence, this converter is the system of choice where maximum speed is required. If the hardware requirements of a straightforward comparator converter become excessive, a cascaded arrangement can be used with some sacrifice in speed and accuracy.

Next in order of speed is the successive-approximation converter. Where a relatively fast converter of good quality is required, this comparator is by far the most popular. As we have noted, the time required to process a conversion increases linearly with the number of bits, requiring about as many clock pulses as bits. Counter converters are the slowest, requiring 2^N clock cycles per conversion, N being the number of bits. Very popular among counter converters is the dual-slope converter, which is widely used in such instruments as digital voltmeters, where conversion speed is not important.

Beyond the comparators we have discussed there are almost limitless other types, some differing in principle and some only in detail. We consider now briefly and in no great detail some additional converters which have the merit of simplicity and economy.

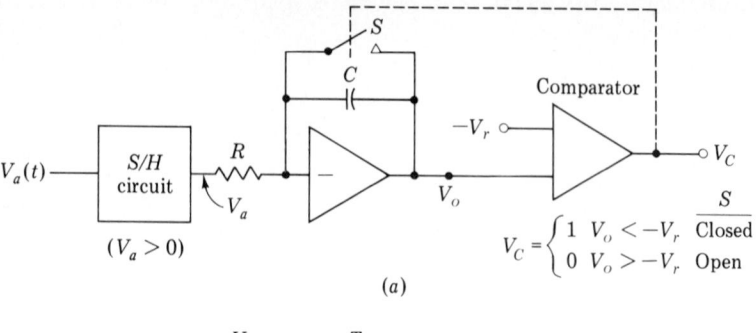

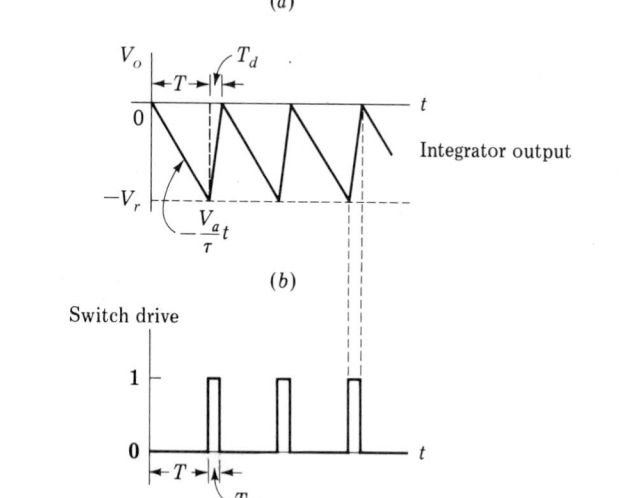

FIGURE 14.17-1
(a) A V/F converter. (b) The waveform V_o. (c) The waveform which drives the switch.

14.17 A CONVERTER USING VOLTAGE-TO-FREQUENCY CONVERSION

An A/D converter can be built using a counter and a device referred to as a *voltage-to-frequency* (V/F) converter. In rather simplified form, the V/F converter is shown in Fig. 14.17-1a. The principles of operation of this circuit are explained below.

An analog waveform $V_a(t)$ (assumed positive) is sampled and held to form the voltage V_a. This voltage is applied to an integrator, which is followed by a comparator. The other input to the comparator is a reference voltage $-V_r$. Initially the switch S bridging the integrating capacitor C is open, and the voltage V_o decreases linearly with time. If $\tau = RC$, we have $V_o = -V_a t/\tau$, which is shown in Fig. 14.17-1b. When V_o decreases to $-V_r$, after a time $t = T$, the comparator output V_C becomes positive for a small time interval T_d, during which time switch S closes, thereby discharging capacitor C and returning the

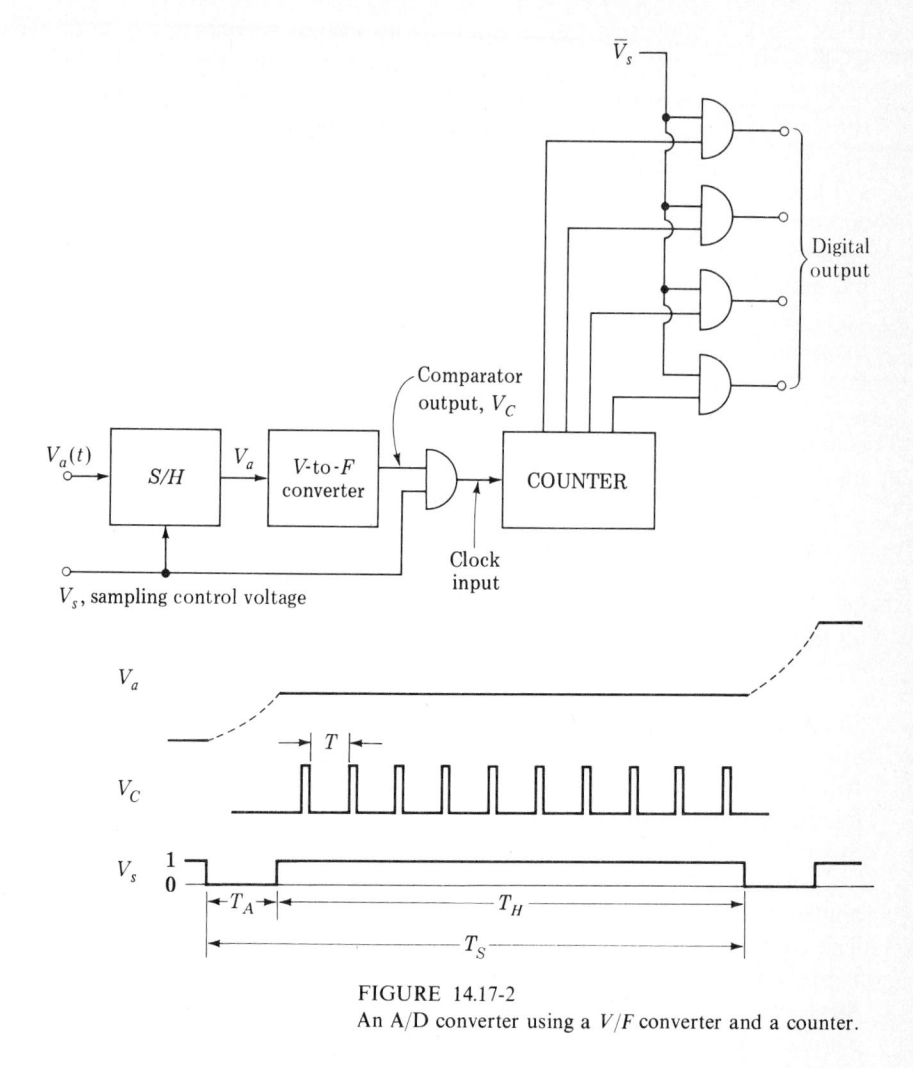

FIGURE 14.17-2
An A/D converter using a V/F converter and a counter.

integrator output V_o to approximately 0 V. The discharge rate is, of course, determined by the switch resistance. In an actual system, the comparator output might remain positive for too short a time to discharge the capacitor completely. In this case, a circuit called a monostable multivibrator (see Secs. 15.3–15.5), capable of using the narrow comparator pulse to form a pulse of width T_d, may be interposed between the comparator output and the switch.

After the time interval T_d the comparator voltage drops to the **0** state, switch S opens, and V_o starts to decrease once more. If the discharge time T_d is much less than the integration time T, the frequency of the waveforms V_o and V_C is

$$f = \frac{1}{T + T_d} \approx \frac{1}{T} = \frac{1}{\tau} \frac{V_a}{V_r} \qquad (14.17\text{-}1)$$

Thus, the V/F converter makes available an output waveform whose frequency is proportional to the input voltage. Such a waveform is said to be *frequency-modulated*. V/F converters, which operate in a manner similar to the device shown in Fig. 14.17-1a, are commercially available.

An A/D converter using a V/F converter is shown in Fig. 14.17-2. The V/F converter provides the clock input (which is the comparator output V_C) to a counter through an AND gate. A second input to the AND gate is the sampling voltage V_s, which holds the logic level **1** for a fixed time T_H. As long as $V_s = \mathbf{1}$, the sampled output V_a is held fixed at its value at the beginning of the interval T_H. With the AND gate enabled for the time T_H the counter reading will equal the number of cycles executed by the V/F converter output in the specified time interval. If we call the number read by the counter, λ, then, from Eq. (14.17-1), $\lambda = fT_H = T_H V_a/\tau V_r$, and the counter registration is proportional to V_a. The counter output is read when V_s is in the **0** state. In addition, during this time interval T_C, a new sample value of $V_a(t)$ is established.

14.18   A CONVERTER USING VOLTAGE-TO-TIME CONVERSION

In the previous section we discussed the principle of an A/D converter which operates by counting the cycles of a *variable-frequency* source for a *fixed period*. Alternatively, a converter may operate on the principle of counting the cycles of a *fixed-frequency* source for a *variable period*. Such an A/D converter is shown in some detail in Fig. 14.18-1.

Assuming a positive analog voltage, a negative reference voltage V_r is applied to an integrator, the output of which provides one input of a comparator. The comparator output is at logic level **1** when the integrator output V_i is less than the analog voltage V_a. A fixed-frequency clock V_{CL} is applied to a counter through an AND gate G. The circuit operates by devising to keep this AND gate enabled only during the time beginning at $t = 0$, when $V_i = 0$, to the time $t = T$, when $V_i = V_a$. Since $V_i = V_r t/\tau$,

$$T = \frac{\tau V_a}{V_r} \tag{14.18-1}$$

If f_{CL} is the clock frequency, during the interval T the count N registered will be

$$N = f_{CL} T = \frac{\tau f_{CL}}{V_r} V_a \tag{14.18-2}$$

The count N, as required, is proportional to V_a. Note that here, unlike the situation which prevails in the dual-slope converter, the calibration of the present converter does depend on the clock frequency and on the integrator time constant.

Waveforms for the A/D converter are shown in Fig. 14.18-1b. The

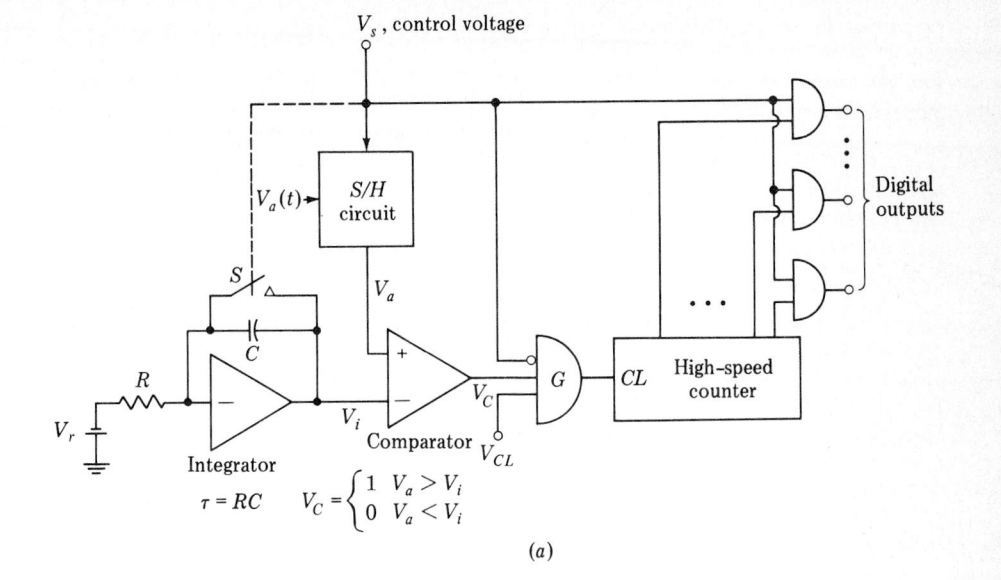

FIGURE 14.18-1
(*a*) An A/D converter using a voltage-to-time converter. (*b*) Waveforms.

sampling voltage V_s samples the positive input voltage $V_a(t)$ every T_S during the time interval T_A. The sampled voltage V_a is then held for a time T_H. During this time, switch S is open, and the integrator output is a voltage ramp. When $V_i < V_a$, the comparator output is in the **1** state. Gate G is enabled as long as V_s is in the **0** state and V_C is in the **1** state, i.e., for the time interval T. During the time interval T the clock voltage V_{CL} is transmitted by gate G to the high-speed counter. Thus the counter output is proportional to T. During the time interval T_A, the gate G is disabled and the counter read. In addition, switch S is closed, the capacitor is discharged, and V_i is reset to 0 V.

14.19 A/D CONVERTER SPECIFICATIONS

The specifications of an A/D converter which are normally supplied by commercial manufacturers include the following:

Analog input voltage This is the maximum allowable input-voltage range. Typical values are 0 to 10 V, ± 5 V, ± 10 V, etc.

Input impedance Values range from 1 kΩ to 1 MΩ, depending on the type of A/D converter. Input capacitance is in the range of tens of picofarads.

Accuracy The *accuracy* of an A/D converter includes quantization error, digital system noise, including that present in the reference voltage (used in the D/A converter), deviations from linearity, etc. Usually the quantization noise is specified as $\pm\frac{1}{2}$ LSB. The accuracy also includes the sum of all other error sources. Typical values are ± 0.02 percent of the full-scale (FS) reading; however, very high accuracy A/Ds can be purchased with an accuracy of 0.001 percent of the full-scale reading (FSR). The accuracy of a converter generally determines the number of bits which can usefully be provided. By way of example, consider a converter with an analog input range ± 10 V. If the accuracy is 0.02 percent of the FSR, the maximum error due to such accuracy limitation is 2 mV. For 9, 10, 11, and 12 bits the quantization error ($\frac{1}{2}$ LSB) are, respectively, 10, 5, 2.5, and 1.25 mV. There is accordingly an advantage in using 10 bits rather than 9 bits. We may even justify using 11 bits, but 12 bits is probably not warranted.

Stability System accuracy is generally temperature-dependent. Typical temperature coefficients of error are 20 ppm of the FSR per Celsius degree. For example, if a 10-V signal is applied at 75°C, an error of $(20 \times 10^{-6})(10\,\text{V})(75 - 25) = 10$ mV results. With a 10-bit A/D converter, the error limits the response to that of a 9-bit device.

Conversion time Typical conversion times vary from 50 μs for moderate speed units to 50 ns for a very high speed device.

Format An A/D converter can usually be obtained for any standard code: unipolar binary, offset binary, ones complement, and twos complement. In addition the output voltage levels are often adjusted so that direct connection is possible to some logic family (TTL, ECL, etc.).

14.20 INTERCONNECTING THE S/H AND THE A/D CONVERTER

The system employed to convert an analog signal into a digital bit stream consists of a S/H amplifier and an A/D converter. These units are operated in synchronism, the A/D converter "telling" the S/H amplifier when to sample and when to hold.

Figure 14.20-1*a* shows the S/H-A/D conversion system. Note that the complete A/D converter consists of two subsystems. The first is the A/D converter, such as the comparator converter shown in Fig. 14.12-1 or the successive-approximation converter shown in Fig. 14.13-2, etc. The second subsystem is a *timing circuit*. The input to the timing circuit is the sampling

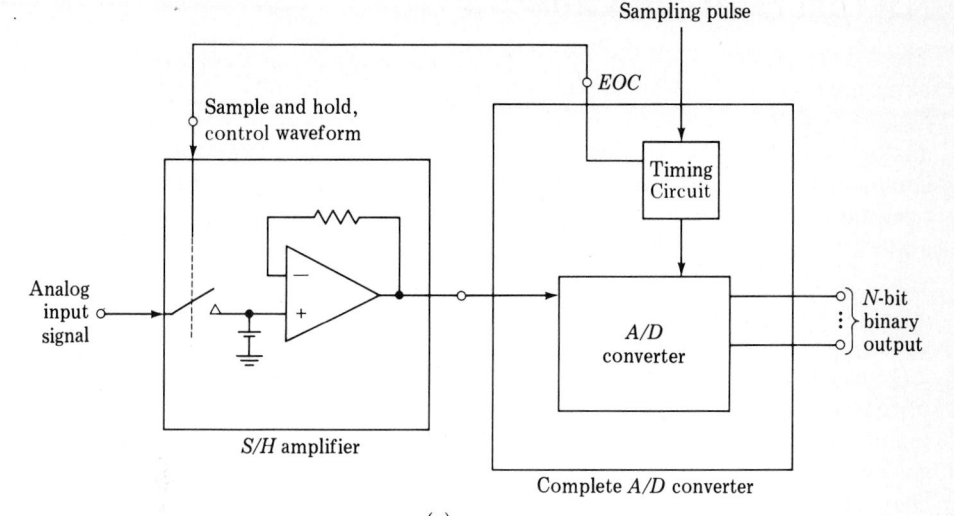

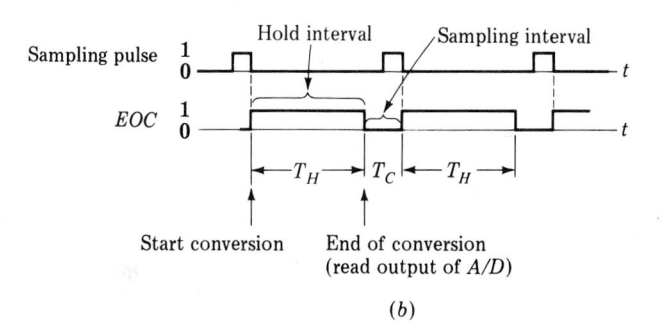

(b)

FIGURE 14.20-1
(a) Interconnection of the S/H circuit and A/D converter.   (b) Waveforms.

pulse train.   The timing circuit provides all the clock pulses needed for the A/D converter to convert the analog sample into an N-bit output signal.   In addition the timing circuit generates a timing waveform called the *end-of-conversion* (EOC) *waveform*, which tells the S/H circuit when to sample and when to hold (see Fig. 14.20-1b).

It is important to note that the sampling pulse train is applied to the A/D converter and not to the S/H circuit.   The control employed by the S/H circuit is the EOC waveform generated in the A/D converter.   This is done so that the output of the S/H circuit is held constant until the A/D converter has finished converting.   Then, while the converted output is displayed at the binary output terminals of the A/D converter, the S/H circuit is permitted to change its analog output level (this occurs during time T_A).   If the S/H circuit and A/D converter were not synchronized, one might well envision the S/H output changing during conversion, and such operation would lead to an incorrect digital output.

14.21 DELTA MODULATION[1]

The A/D converters discussed previously make available a digital representation of an analog signal. The fundamental characteristic common to each of these devices is that it produced an N-bit code word for each sample taken of the analog signal. Although the type of code may vary (offset binary, ones complement, twos complement, etc.), the generic term employed to describe the operation is *pulse-code modulation* (PCM). PCM has found wide-ranging applications, from computer-type arithmetic processing to data transmission of voice, video, and computer signals.

In the following sections we consider a different type of A/D converter called a *delta modulator* (DM). While applications of the DM are not nearly as large as for PCM, they are rapidly increasing. One of the major applications of the DM is the digital encoding of voice signals prior to transmission. It is found that similar clarity of voice, encoded using PCM and DM, can be achieved when the DM is operating at a *lower* bit rate than for PCM. Since channel bandwidth required increases with bit rate, and since bandwidth is usually at a premium, many digital voice systems now use DM.

A basic "linear" DM A block diagram of a DM is shown in Fig. 14.21-1a. Typical waveforms are shown in Fig. 14.21-1b, c, and d. The analog signal $M(t)$ and the estimate $\hat{M}(t)$ of $M(t)$ derived by the delta modulator are continually presented to a comparator. The output of the comparator can assume one of two possible voltage levels; logic **1** if $M(t) \geq \hat{M}(t)$ and logic **0** if $M(t) < \hat{M}(t)$. The comparator output waveform (not shown) is sampled once per clock cycle and held at the sampled value for the duration of the clock cycle. If at the sampling time kT_s (k an integer) the comparator output is at logic **0**, $E(t) = E(kT_s) = \mathbf{0}$ is transmitted; if, however, the comparator output is at logic **1**, $E(t) = E(kT_s) = \mathbf{1}$ is transmitted.

The estimate $\hat{M}(t)$ is formed using the following algorithm: if $E(kT_s) = \mathbf{1}$, which denotes that $M(kT_s) \geq \hat{M}(kT_s)$, *increase* $\hat{M}(t)$ by a step S_0 during the interval $kT_s + \tau < t < (k + 1)T_s + \tau$. If $E(kT_s) = \mathbf{0}$, which denotes that $M(kT_s) < \hat{M}(kT_s)$, *decrease* $\hat{M}(kT_s)$ by a step S_0. The time delay τ is the time required for the processor to convert changes in $E(t)$ into changes in $\hat{M}(t)$. The algorithm can be written in terms of the sampling times kT_s and $(k + 1)T_s$ as a recursive difference equation

$$\hat{M}[(k + 1)T_s] = \hat{M}(kT_s) + S_0 E(kT_s) \qquad \text{for all } k \qquad (14.21\text{-}1)$$

where $E(kT_s) = \begin{cases} +1 & \text{if comparator output is at logic level } \mathbf{1} \\ -1 & \text{if comparator output is at logic level } \mathbf{0} \end{cases}$

The application of this algorithm results in the estimate shown in Fig. 14.21-1c. The resulting bit stream, $E(t)$, is shown in Fig. 14.21-1d.

One way in which the processor shown in Fig. 14.21-1a can be realized is shown in Fig. 14.21-2. The processor consists of the up-down counter and the D/A converter. When at the sampling time it turns out that $E(kT_s) = \mathbf{1}$, the

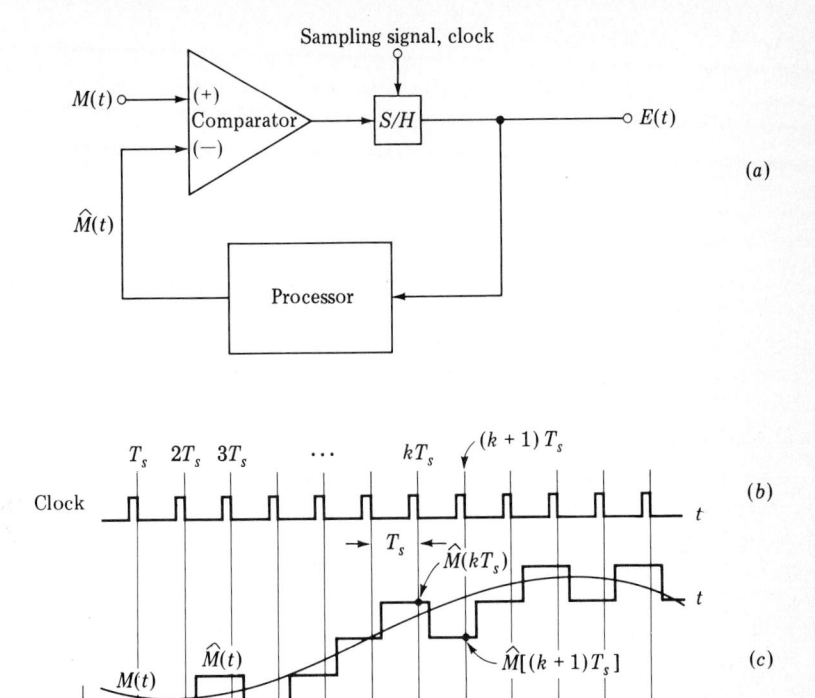

FIGURE 14.21-1
(a) A basic delta modulator. (b) Clock. (c) Signal and estimate. (d) Transmitted bit stream $E(t)$.

number stored in the counter is increased by **1**, and when $E(kT_s) = \mathbf{0}$, the number stored in the counter is decreased by **1**.

Another simple DM is shown in Fig. 14.21-3. Here the processor is not a digital device, consisting rather of an analog integrator. Here $E(kT_s)$ is $+V_0/4$ whenever $M(kT_s) < \hat{M}(kT_s)$ and $-V_0/4$ whenever $M(kT_s) > \hat{M}(kT_s)$. In this case the waveform $\hat{M}(t)$ consists not of a sequence of up or down jumps but a sequence of ramps with slopes $+V_0/4RC$ or $-V_0/4RC$. In the clock interval $T_s = 1/f_s$ the total change in $\hat{M}(t)$, which we again call S_0, is

$$S_0 = \frac{V_0 T_s}{4RC} \tag{14.21-2}$$

Reconstruction of the analog signal One of the great merits of delta modulation is the relative ease with which the quantized analog signal $\hat{M}(t)$ can be

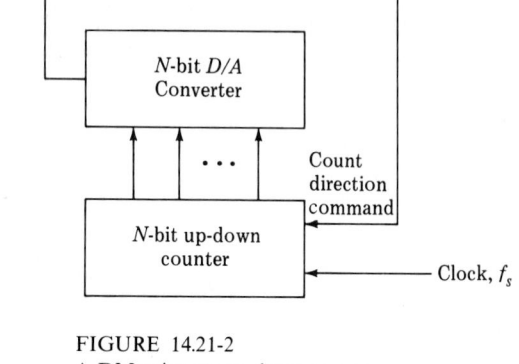

FIGURE 14.21-2
A DM using an up-down counter as a processor.

recovered from the digitized transmitted signal $E(t)$. This recovery process, of course, effects the D/A conversion, which is the inverse of the original A/D conversion. The recovery of $\hat{M}(t)$ is accomplished by passing the transmitted signal $E(t)$ through a processor such as is used at the transmitter. For, as we have noted, the processor generates $\hat{M}(t)$ from the waveform $E(t)$. However, no matter whether the transmitter processor is digital, as in Fig. 14.21-2, or analog, as in Fig. 14.21-3, the simple integrator processor is suitable for signal recovery.

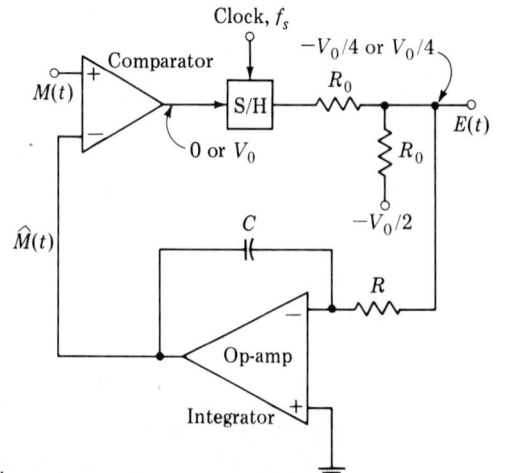

FIGURE 14.21-3
A simple realization of a DM.

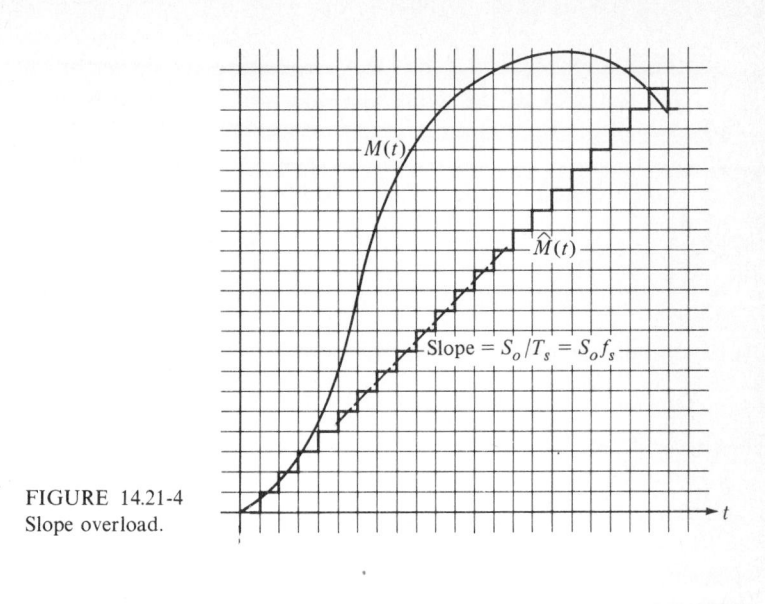

FIGURE 14.21-4
Slope overload.

Limitation of the linear DM The DM systems described in this section both have the characteristic that the estimate $\hat{M}(t)$ changes by $\pm S_0$, that is, by a constant amount, after each sampling pulse. Thus, if $E(t)$ should consist of a long stream of logic 1 bits, $\hat{M}(t)$ would resemble a voltage ramp or a stepwise approximation to a ramp. As a result of this "linear" change in $\hat{M}(t)$, the type of DM system described above is referred to as *linear delta modulation*.

The linear DM has the fundamental limitation that the magnitude of the "slope" of $\hat{M}(t)$ cannot exceed $S_o/T_s = S_o f_s$. If then the slope of the signal $M(t)$ should persist at a value greater than S_o/T_s for an extended time, the difference between $M(t)$ and $\hat{M}(t)$ may become quite large. This situation is illustrated in Fig. 14.21-4. The disparity between $M(t)$ and $\hat{M}(t)$ shown here is described as a *slope-overload error*.

To decrease the slope-overload error we can increase the step size S_0 with a corresponding increase in the quantization error, or we can increase the sampling rate f_s with a corresponding increase in bandwidth required to transmit the signal $E(t)$. As a result of this limitation linear DM is rarely used in practice. In the next section we discuss *adaptive DM*, which overcomes the slope-overload problem by generating different step sizes depending on the characteristics of $M(t)$.

14.22 ADAPTIVE DELTA MODULATION

Adaptive DM differs from linear DM in that the step size is no longer constant at S_0 but takes on other values, depending on the past history of the transmitted data, $E(t)$, that is, $E(kT_s)$, $E[(k-1)T_s]$, ..., $E(0)$. This type of DM is often used to digitally encode voice. The voice reproduction is found to be comparable to that obtained in PCM, even if the DM is operating at a lower bit rate. (The bit rate in DM is equal to the sampling rate. In PCM the bit rate is equal to the product of the sampling rate and the number of bits into which each sample is encoded.)

In this section we describe an adaptive DM (ADM), designed and constructed at the Communication Systems Laboratory of CCNY[2]; it can reproduce voice which has been band-limited to 2,400 Hz with a word intelligibility of 90 percent at 9.6 kb/s. This is equivalent to using PCM and encoding each sample into 2 bits.

Quite generally, in DM, we have

$$\hat{M}[(k + 1)T_s] = \hat{M}(kT_s) + S[(k + 1)T_s] \tag{14.22-1}$$

where $S[(k + 1)T_s]$ is the step size by which the estimate $M(t)$ changes at $t = kT_s + \tau$ (see Fig. 14.20-1c). In the linear DM, described above,

$$S[(k + 1)T_s] = S_0 E(kT_s) \tag{14.22-2}$$

leading to Eq. (14.21-1). In an ADM system the step size $S[(k + 1)T_s]$ is not of constant magnitude, as in Eq. (14.21-2). In the system described here $S[(k + 1)T_s]$ is given instead by

$$S[(k + 1)T_s] = |S(kT_s)| E(kT_s) + S_0 E[(k - 1)T_s] \tag{14.22-3}$$

Shown in Fig. 14.22-1 is a signal $M(t)$ and the estimate $\hat{M}(t)$ which results when $S[(k + 1)T_s]$ is generated through the algorithm of Eq. (14.22-3). To verify that $\hat{M}(t)$, as shown, is consistent with Eq. (14.22-3) it is well to keep in mind that $S[(k + 1)T_s]$ represents the jump in $\hat{M}(t)$ at the time $kT_s + \tau$, where τ is the processing delay time in the system. However $E(kT_s)$ refers to the processor input voltage just after the kth clock pulse. This is illustrated in Fig. 14.22-1. The signal $M(t)$ in Fig. 14.22-1 is one which would generate a large slope-overload error if linear DM were used. Observe, however, in the present case that as the condition $M(t) > \hat{M}(t)$ persists, the jumps in $\hat{M}(t)$ become progressively larger. The estimate $\hat{M}(t)$ therefore catches up with $M(t)$ sooner than it would with linear modulation. On the other hand, when, in response to a large slope in $M(t)$, $\hat{M}(t)$ develops large jumps, it may well require a large number of clock cycles for these jumps to decay in amplitude when they are no longer required. Such a situation is to be seen in $\hat{M}(t)$ in the neighborhood of the maximum of $M(t)$ in Fig. 14.22-1. Altogether, then, while this ADM system reduces slope error, it does so at the expense of increasing quantization error.

It turns out that in the matter of speech transmission the reduced slope error provides a net advantage in spite of the increased quantization error. The spurious frequency components introduced into the reconstructed signal by slope-overload error are principally in the low-frequency range, while quantization error introduces principally spurious high-frequency components. The power in speech is concentrated largely in the low-frequency components, and these low-frequency components are also the ones which must principally be reproduced without noise contamination in order to preserve intelligibility. Hence, if we

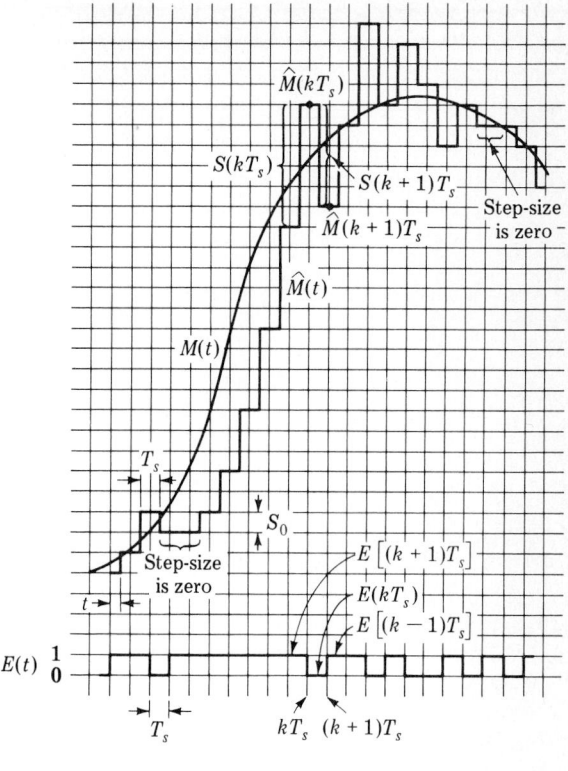

FIGURE 14.22-1
The ability of an ADM to approximate the same waveform as in Fig. 14.20-4. The minimum step size is the same as in Fig. 14.20-4.

pass the reconstructed signal through a low-pass filter, we shall be able to discriminate against the high-frequency noise due to quantization error without materially degrading the voice signal.

It is of incidental interest to note, in connection with $\hat{M}(t)$ in Fig. 14.22-1, that on occasion $\hat{M}(t)$ does not exhibit a jump in each clock interval. Two such instances are indicated by brackets. It can also be verified that in a steady-state condition when $M(t)$ remains constant (within an interval S_0), $\hat{M}(t)$ oscillates about $M(t)$ but the oscillation frequency is one-half the frequency that would be encountered in linear DM.

REFERENCES

1 Taub, H., and D. L. Schilling: "Principles of Communications Systems," McGraw-Hill, New York, 1971.
2 Song, C. L., J. Garodnick, and D. L. Schilling: A Robust Delta Modulator, *IEEE Trans. Commun. Technol.*, December 1971.

15

TIMING CIRCUITS

In this chapter we consider two timing circuits, the *monostable multivibrator* and the *astable multivibrator* and, also a general purpose *timer* capable of performing either operation as well as a wide variety of other timing functions.

The flip-flop circuit (also called a bistable multivibrator), it will be recalled, has two stable states, in either one of which it may remain permanently. The monostable multivibrator has only one permanently stable state and one quasi-stable state. In the monostable configuration, a triggering signal is required to induce a transition from the stable state to the quasi-stable state. The circuit may remain in its quasi-stable state for a time which is very long in comparison with the time of transition between states. Eventually, however, it will return from the quasi-stable state to its stable state, no external signal being required to induce this reverse transition.

Since, when it is triggered, the monostable circuit returns by itself to its original state after a time T, it is also known as a *one-shot*. Since it generates a rectangular waveform which can be used to gate other circuits, it is also called a *gating circuit*. Furthermore, since it generates a fast transition at a predetermined time T after the input trigger, it is also referred to as a *delay circuit*.

The astable multivibrator has two states, both of which are quasi-stable.

Without the aid of an external triggering signal the astable configuration will make successive transitions from one quasi-stable state to the other. Thus the astable circuit is an *oscillator* and is used as a generator of square waves or a clocking waveform.

15.1 CMOS MULTIVIBRATORS

CMOS gates are conveniently adaptable for use in monostable and astable multis. In this section we consider monostable circuits using such gates.

We discussed CMOS gates in Chap. 8. The diagram of a two-input NOR gate is given in Fig. 8.9-2. This NOR gate is shown again in Fig. 15.1-1. Note the presence of the input protective diodes (Sec. 1.16). These diodes are bridged between each input and the power-supply terminals and are used (operating in conjunction with the resistive impedance of the input driving sources) to restrain the voltages on the gates of the MOS devices to the voltage range nominally from ground to V_{SS}. This restraint is necessary to assure that the gate voltages will never exceed the breakdown voltage of the oxide layer which insulates the gate from the semiconductor. Actually, the breakdown voltage may be of the order of 100 V. Hence, in ordinary operation (using supply voltages V_{SS} in the range of about 10 V) it is not likely that the breakdown voltage would be exceeded. There is, however, a special difficulty which is encountered with MOS devices generally and which results from the extremely high insulation resistance ($\approx 10^{12}\ \Omega$) of the gate. As a result, when no dc path is provided to allow charge leakage from a gate, even a static charge inadvertently established on a gate (as by touching an isolated input-gate terminal) may cause breakdown. Such breakdown is precluded by the presence of the input diodes.

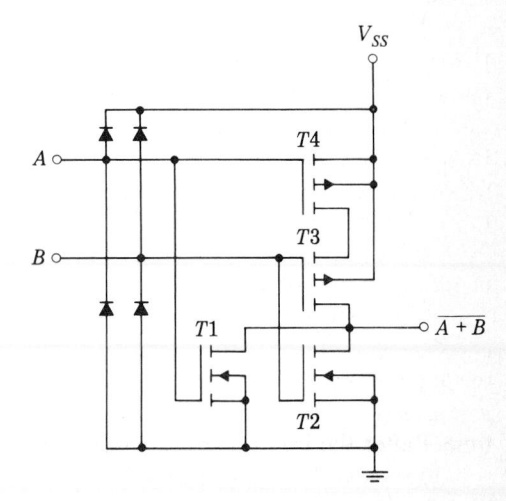

FIGURE 15.1-1
Schematic diagrams of two-input CMOS NOR gates, illustrating the input protective diodes (Motorola MC 14001).

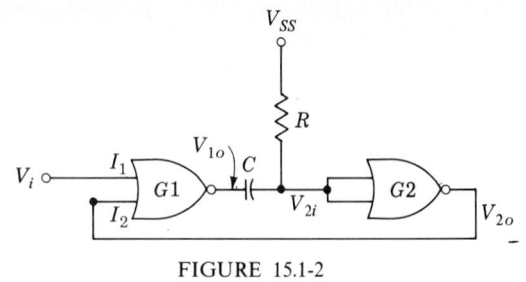

FIGURE 15.1-2
A CMOS NOR-gate monostable multi.

When these gates are used in straightforward gate circuits, the input diodes may actually never be called upon to conduct. On the other hand, in multi circuits some of the diodes are essential, and if they were not already incorporated in the gate itself, it would be necessary to add them externally.

A basic CMOS monostable multi using two NOR gates is shown in Fig. 15.1-2. As indicated, the resistor R is returned to the supply voltage V_{SS}, which powers the gates. The gate $G2$ has its inputs joined since this gate is being used simply as an inverter.

Some details of the waveforms associated with the CMOS multi depend on the precise form of the input-output characteristic of the CMOS gate. A typical characteristic is shown in Fig. 8.4-1b. However, the monostable multi is used essentially to establish a time interval. Matters relating to this time interval, its range, stability, etc., are of interest while the exact waveforms are of no serious consequence. Accordingly, to simplify the discussion we shall consider that the input-output characteristic of the CMOS may be represented as in Fig. 15.1-3. Here we assume an abrupt change in output between V_{SS} to 0 when the input is at the transition voltage V_T. With this assumption, the waveforms of the monostable multi appear as in Fig. 15.1-4.

In the stable state the input V_{2i} to gate $G2$ will be at logic level **1**; that is, $V_{2i} = V_{SS}$ (since MOS gates and hence $G2$ draw negligible input current).

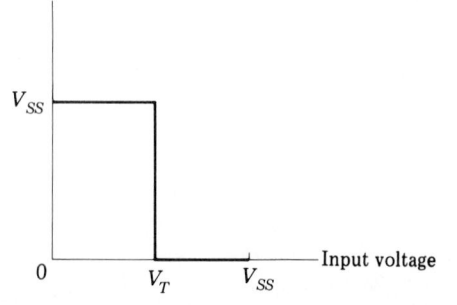

FIGURE 15.1-3
Idealized input-output characteristic of a
CMOS gate.

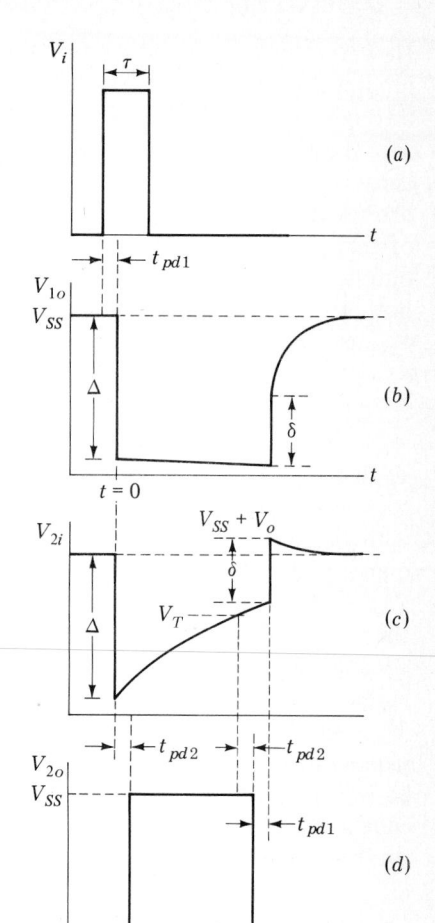

FIGURE 15.1-4
Waveforms of the CMOS astable multivibrator.

The output V_{2_o} will be approximately at 0 V, that is, at logic level **0**. The trigger input to initiate the quasi-stable state is applied at the input terminal I_1 of gate $G1$. With the input V_i and the input V_{2_o} at I_2 both at logic **0**, the output V_{1_o} of gate $G1$ will be at logic **1**, that is, at V_{SS}. Thus capacitor C is initially discharged.

Now let a short positive pulse be applied at I_1, as indicated in Fig. 15.1-4*a*, of amplitude adequate to carry I_1 to logic level **1**. Then after a propagation delay t_{pd1} due to $G1$, V_{1_o} will drop abruptly (Fig. 15.1-4*b*), and this abrupt drop will be transmitted through C to the input of gate $G2$. Assuming that the drop is great enough in magnitude to carry the voltage V_{2i} (Fig. 15.1-4*c*) to logic level **0**, the output V_{2_o} (Fig. 15.1-4*d*) will rise to logic **1** after a propagation delay t_{pd2}. The gate input I_2 will then be held at logic **1**, and the output of $G1$ will thereafter not be affected by the termination of the input trigger pulse V_i.

However, the trigger has served to initiate the quasi-stable state. Note that the minimum duration of the pulse V_i must exceed the sum of the propagation delays through gates $G1$ and $G2$, that is, $\tau > t_{pd1} + t_{pd2}$.

Initially, in the stable state, the current through R is zero and the voltage across C is zero. At the beginning of the quasi-stable state, when V_{1o} and V_{2i} drop, a current flows through R, continues through C, and sinks into the output of gate $G1$. If gate $G1$ were not being called upon to sink this current, the output V_{1o} would drop from its initial level V_{SS} down to ground. Suppose, however, that the output resistance of gate $G1$ is R_{o1}; then the drop Δ in V_{1o} and V_{2i}, as can be verified, will not be equal to V_{SS} but will instead be given by

$$\Delta = \frac{R}{R + R_{o1}} V_{SS} \qquad (15.1\text{-}1)$$

so that Δ will approach V_{SS} only if $R \gg R_{o1}$.

As C charges from V_{SS} through R and R_{o1}, the charging current decreases with time. Consequently the drop across R_{o1} decays as well, and during the quasi-stable interval T the waveform V_{1o} displays a downward tilt.

Starting at $t = 0$ (taken to be the time when V_{1o} and V_{2i} drop by amount Δ), the voltage V_{2i} rises from $V_{SS} - \Delta$ toward V_{SS} as an asymptotic limit. The time constant associated with this exponential waveform is $(R + R_{o1})C$. The quasi-stable state is terminated when V_{2i} reaches the transition voltage V_T. When V_{2i} passes V_T, V_{2o} returns to its initial value of 0 V and, as soon as the capacitor C has discharged the charge it has acquired, both V_{1o} and V_{2i} return as well to their initial values V_{SS}. The circuit will remain in this stable state indefinitely until again induced to make a transition to the quasi-stable state by an input pulse. The time of the quasi-stable state is (Prob. 15.1-1)

$$T = (R + R_{o1})C \ln \left(\frac{\Delta}{V_{SS} - V_T} \right) \approx (R_o + R_{o1})C \ln \frac{V_{SS}}{V_{SS} - V_T} \qquad (15.1\text{-}2)$$

In a typical case, with $V_T = V_{SS}/2$, we find $T = 0.7(R_o + R_{o1})C$.

The circuit which determines the charging and discharging of the timing capacitor C is shown in Fig. 15.1-5. Diode D represents the parallel combination of the diodes which connect the paralleled inputs of gate $G2$ to V_{SS}. Gate $G1$, as viewed from its output, is represented by the two switches S and S' and the "elements" R_{o1} and R'. In the stable state S' is closed, S is open, and the capacitor voltage V_C is $V_C = 0$ V. The quasi-stable state begins when S closes and S' opens and C charges from V_{SS} through R and R_{o1}. If R is large enough, then the voltage across the transistors ($T1$ and $T2$ in Fig. 15.1-1) will initially be low enough (and remain so as C charges) so that the transistors will operate in the triode region where they may reasonably be represented as a resistor R_{o1}.

The quasi-stable state ends when S opens and S' closes. During the quasi-stable state, the capacitor may acquire a voltage comparable to V_{SS}. At the end of the quasi-stable state C discharges through the diode D and through the "element" R'. The element R' is intended to represent the volt-ampere characteristic of the series combination of $T3$ and $T4$ in Fig. 15.1-1. Initially,

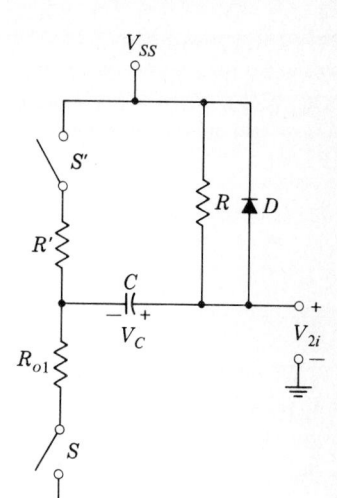

FIGURE 15.1-5
Circuit which determines the charging and
discharging cycles of the timing capacitor C.

when V_C is large, the transistors may be in the saturation region so that the capacitor will discharge at a nominally constant rate. As V_C decreases, the transistors may be in the triode region so that the capacitor discharge will continue exponentially. Finally, after the voltage V_C drops to the point where the diode voltage falls below its cut-in voltage V_D (≈ 0.65 V), the last part of the capacitor discharge will take place through R' and R.

When S opens, S' closes, and diode D starts to conduct, the voltage V_{2i} will become $V_{SS} + V_D$ (at this point, since the diode may well be carrying a substantial current, we may reasonably take V_D to be about 0.75 V). Hence, as is to be seen in Fig. 15.1-4c, at the end of the quasi-stable state V_{2i} jumps abruptly from V_T to $V_{SS} + V_D$. The size of the jump is $\delta = V_{SS} + V_D - V_T$. Since the output of $G1$ is coupled to the input of $G2$ by the capacitor, and since the voltage across the capacitor itself cannot change abruptly, there is an equal jump δ in the waveform of V_{1o}. Thereafter V_{2i} and V_{1o} decay to their steady-state value V_{SS}. As noted, R' is not a simple resistor and D does not conduct to the very end. Nonetheless, for simplicity, in Fig. 15.1-4b and c we have indicated the discharge of C as generating a simple exponential waveform.

EXAMPLE 15.1-1 In the monostable multivibrator circuit of Fig. 15.1-2, $V_{SS} = 10$ V, $V_T = 5$ V, $C = 0.01$ μF, $R = 10$ kΩ, $R_{o1} = 500$ Ω, and assume that R' is a "resistor" and let $R' = 1$ kΩ.

    (a) Find Δ, δ, and the voltage V_{1o} at the end of the quasi-stable state.

    (b) Find the time T of the quasi-stable state.

    (c) Estimate how long a time, after the end of the quasi-stable state, will be required for the capacitor to discharge to 0.1 V.

    (d) Suppose a second triggering pulse is applied at the time $V_C = 0.1$ V. How will the second timing interval compare with the first?

SOLUTION   (a) From Eq. (15.1-1), $\Delta = [10/(10 + 0.5)](10) = 9.5$ V; $\delta = V_{SS} + V_D - V_T = 10 + 0.75 - 5 = 5.7$ V; when $V_{2i} = V_T = 5$ V, the current through R and through R_{o1} is $(V_{SS} - V_T)/R$. The drop across R_{o1} is $R_{o1}(V_{SS} - V_T)/R = 0.5(5)/10 = 0.25$ V $= V_{1o}$.
(b) From Eq. (15.1-2),

$$T = (R + R_{o1})C \ln\left(\frac{V_{SS}}{V_{SS} - V_T}\right)$$

$$= (10 + 0.5) \times 10^3 \times 0.01 \times 10^{-6} \ln\left(\frac{10}{5}\right)$$

$$= 72 \ \mu s$$

(c) Assume that the diode maintains across itself a voltage 0.7 V (a compromise between 0.65 and 0.75) as long as the diode current is 0.1 mA or larger. Let $t = 0$ at the beginning of the capacitor discharge at which time the capacitor voltage is $V_C = 5 - 0.25 = 4.75$ V. Then the capacitor discharge current is

$$I_C = \frac{4.75}{R'} \epsilon^{-t/R'C} = 4.75\epsilon^{-10^5 t} \qquad \text{mA}$$

We find $I_C = 0.1$ mA at $t = 39$ μs. Just before the diode turns OFF the capacitor voltage V_C is 0.7 V $+ I_C R' = 0.7 + 0.1 = 0.8$ V. The decay from 0.8 to 0.1 V occurs as C discharges through the series combination of R and R'. This time is calculated from

$$V_C = 0.8\epsilon^{-t/(R + R')C} = 0.8\epsilon^{-1.1 \times 10^{-4}t}$$

We find that $V_C = 0.1$ V when $t = 230$ μs. The total time is $39 + 230 = 269$ μs. Observe that the time required, after the quasi-stable state, for the circuit to recover very nearly to its initial situation is rather long in comparison with the duration of the quasi-stable state itself.

(d) As calculated in (b) the capacitor charges from 0 to 5 V in 72 μs. Let us assume for simplicity that the capacitor charges at a constant rate. This is a rough approximation but, for the accuracy desired, it is sufficient. Hence to charge from 0.1 to 5 V will require a time $(4.9/5)72 = 70.6$ μs. The change in timing is 1.4 μs and the percentage change is about 2 percent.

In Fig. 15.1-3 we have suggested that the transition between logic levels in the CMOS gate occurs at an input voltage which is about one-half the supply voltage V_{SS}. In practice, the transition region of a CMOS gate may vary from sample to sample and may occur anywhere in the range from about 30 percent V_{SS} to 70 percent V_{SS}. Hence, if in the circuit of Fig. 15.1-2 the R and C are set for a particular time of the quasi-stable state, this time may vary very considerably from circuit to circuit with different samples of the gate [see Eq. (15.1-2)]. On the other hand, the input-output characteristic of a CMOS gate displays very little sensitivity to temperature (see Fig. 1.15-3), so that the timing with any particular CMOS sample will be rather temperature-insensitive.

A number of circuits are available[1, 2] which considerably reduce the dependence of timing on CMOS samples. These circuits depend on the fact that while there may be considerable variability from sample to sample as we go from one integrated-circuit chip to another, samples of gates fabricated on a single chip are quite well matched. An example is shown in Prob. 15.1-4.

15.2   THE CMOS ASTABLE MULTIVIBRATOR

The circuit diagram of a CMOS NOR-gate astable multi is shown in Fig. 15.2-1a. Again, in order that we may describe the essential features of the operation without going afield in a profusion of small details, we assume an input-output characteristic for the gates, as is shown in Fig. 15.1-3. We shall further neglect the output impedances of the gates and shall assume as well that the input protective diodes are ideal. Thus, we assume that the diodes cut in at zero volts and that when conducting their voltage drop is negligible.

With these simplifying assumptions, it is apparent that V and V_{2o} are complementary. When one is at V_{SS}, the other is at zero, and vice versa. Now let us assume that V_{1i} is above V_T. Then V is at zero and V_{2o} is at a *fixed* voltage V_{SS}. Hence V_{1i} is heading asymptotically toward zero. When V_{1i} reaches V_T, V will change abruptly to V_{SS} and V_{2o} will change abruptly to zero. This abrupt change in V_{2o} will be transmitted through C to V_{1i}. The consequent downward swing at V_{1i} will be limited at ground because of the clamping action of the protective diode at the input to gate $G1$. Now V_{1i} is below V_{tc} and is heading asymptotically toward V_{SS}, which is the voltage at V. Altogether there will be a periodic switching back and forth in V_{2o}, V, and V_{1i}, as shown by the idealized waveforms in Fig. 15.2-1b, c, and d. The circuit operation does not, of course, depend on V_T being equal to $V_{SS}/2$. If, however, $V_T \neq V_{SS}/2$, the waveform will not be symmetrical; i.e., we shall have $T_1 \neq T_2$. In general we shall have

$$T = T_1 + T_2 = RC \ln \left(\frac{V_{SS}}{V_{SS} - V_T} \cdot \frac{V_{SS}}{V_T} \right) \qquad (15.2\text{-}1)$$

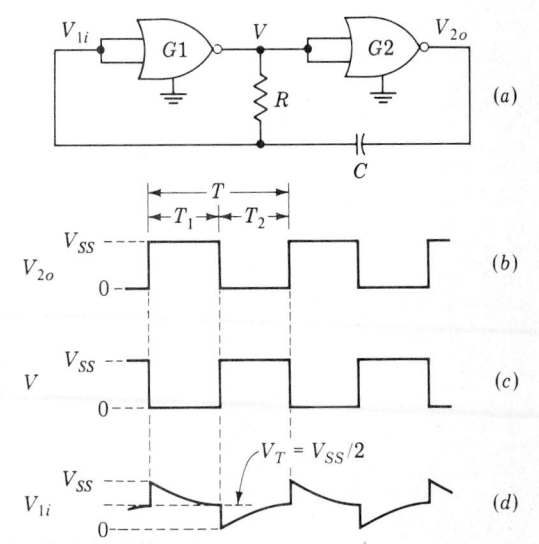

FIGURE 15.2-1
(a) A CMOS astable multi. (b to d) Idealized waveforms of the circuit.

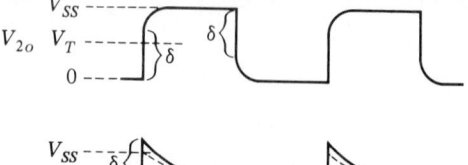

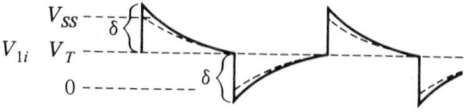

FIGURE 15.2-2
Waveforms of Fig. 15.2-1b and d taking
into account the finite output impedance
of gate G2 and the use of real diodes.

and for $T_1 = T_2$

$$T = 1.4RC \qquad (15.2\text{-}2)$$

We should note that the waveforms in Fig. 15.2-1 are actually inconsistent with one another. For we observe that on one side of the capacitor the voltage changes abruptly by amount V_{SS} while on the other side the abrupt changes are of magnitude $V_{SS}/2$. In a physical circuit the voltage $V_{SS}/2$, which does not appear in V_{1i}, actually develops across the output impedance of gate $G2$ and across the input diode of gate $G1$. That is, at each transition there is a brief interval during which V_{2o} does not attain V_{SS} or 0 and during that same interval V_{1i} extends above V_{SS} or below ground. During this brief interval the capacitor charges or discharges and thereby changes its voltage by amount $V_{SS}/2$. More realistic waveforms are illustrated in Fig. 15.2-2.

In Prob. 15.2-2 there is shown an alternative form of the astable CMOS multi which has the merit that its timing is less sensitive to variations of V_T and variations of supply voltage.

15.3 MONOSTABLE MULTIVIBRATORS USING ECL GATES

We turn our attention now to monostable circuits using integrated-circuit ECL NOR gates. A flip-flop using ECL gates is shown in Fig. 15.3-1. The flip-flop supplies the required gating functions, and the external r and C supply the required timing.

It will be recalled that ECL gates use transistors of high current gain and further that these transistors are always either in the active region or are cut off, i.e., never in saturation. Each input S (set) and R (reset) is applied directly to the base of a transistor in gate $G2$ and $G1$, respectively. The current into a base is zero when the transistor is cut off. And even when the transistor is ON, the input current may well be small enough for this input current not to have serious influence on the operation of the circuit. It is these features that make the ECL flip-flop particularly suitable for use in the configuration of Fig. 15.3-1.

In Fig. 15.3-1, when the input at S is at voltage $V(0)$ corresponding to logic **0**, the flip-flop will find itself in a permanently stable state with the flip-

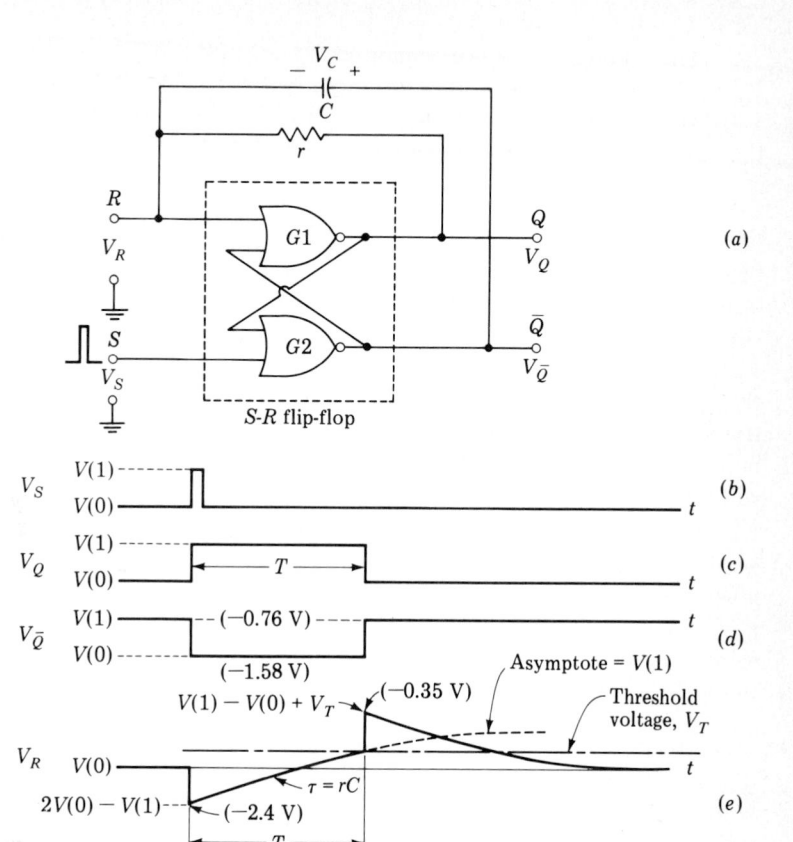

FIGURE 15.3-1
(a) A monostable multi using ECL NOR gates. (b to e) Waveforms of the circuit.

flop in the reset condition, that is, $Q = 0$, $\bar{Q} = 1$. The student may verify that such is the case by assuming that it is so and then showing that all logic levels at the inputs and outputs of the gates are consistent with one another. Assuming $Q = 1$, $\bar{Q} = 0$ leads to inconsistencies. In this stable reset condition $V_Q = V_R = V(0)$. The R input of gate $G1$ looks into the base of a transistor which is cut off. No current flows into this base, and the current through r is also zero. The voltage across the capacitor C is $V_C = V(1) - V(0)$.

Now let the S input rise briefly to the logic level corresponding to voltage $V(1)$, as shown in Fig. 15.3-1b. Then, as appears in Fig. 15.3-1c and d, the flip-flop will make a transition to the set state. The voltage $V_{\bar{Q}}$ will drop abruptly by amount $V(1) - V(0)$. This abrupt change will be transmitted through capacitor C to node R, so that R will now find itself at $V_R = V(0) - [V(1) - V(0)] = 2V(0) - V(1)$, as shown. Note that since the input R has been driven negative, the corresponding transistor is farther in cutoff so that again no current flows into the input R.

The voltage V_R cannot remain at $V_R = 2V(0) - V(1)$ because now we find V_Q at $V_Q = V(1)$. There is a voltage $V_r = V(1) - [2V(0) - V(1)] = 2[V(1) - V(0)]$ across r, and current flows through r, charging capacitor C. Accordingly, the voltage V_R increases exponentially with time constant

$$\tau = rC \tag{15.3-1}$$

heading asymptotically toward $V(1)$. [In writing Eq. (15.3-1) we have ignored the output impedance of gate $G2$ because in an ECL gate it is ordinarily very small in comparison with r.] As V_R rises, a point will be reached where $V_R = V_T$, V_T being the transition voltage of the flip-flop. We assume again, for simplicity, that there is a sharply defined voltage $V_R = V_T$ at which the flip-flop makes a transition between states, the voltage levels of the flip-flop being otherwise independent of V_R. At this point, a rapid reversal in the state of the flip-flop will take place back to the reset condition. The voltage $V_{\bar{Q}}$ will jump by amount $V(1) - V(0)$ and there will be an equal jump at point R. We shall then have, as shown, $V_R = V(1) - V(0) + V_T$. It may be verified that the time T of the quasi-stable state is

$$T = \tau \ln 2 \, \frac{V(1) - V(0)}{V(1) - V_T} \tag{15.3-2}$$

If the R input of gate $G1$ continued to draw no input current, V_R would decay from its peak value, $V(1) - V(0) + V_T$, toward its asymptotic limit $V(0)$ with the same time constant τ as prevailed during the quasi-stable state. In order to ensure that successive cycles would duplicate one another it would be necessary to allow a recovery time of several (3 or 4) time constants. As we may anticipate, T itself is of the order of τ. Hence we should have to allow a recovery time which is several times longer than the timing interval itself. This feature, which may well be quite inconvenient, can be corrected in a manner to be noted below. Actually, from the point where V_R reaches its peak value until it falls to V_T the gate input does draw some current, which is supplied through the capacitor C, and thereby the capacitor recharges more rapidly. However, as noted, the gate input current is small and makes no substantial improvement in the recovery time.

As noted in Chap. 7, the logic levels of an ECL gate operating with a supply voltage -5.2 V are $V(0) \approx -1.58$ V and $V(1) \approx -0.76$ V. The threshold voltage V_T is midway between these levels at $V_T \approx -1.17$ V. The other voltages indicated in Fig. 15.3-1 are

$$2V(0) - V(1) = -2.4 \text{ V} \qquad \text{and} \qquad V(1) - V(0) + V_T = -0.35 \text{ V}$$

Using Eq. (15.3-2), we find the quasi-stable-state duration T to be

$$T = rC \ln 2 \, \frac{V(1) - V(0)}{V(1) - [V(1) + V(0)]/2} = rC \ln 4 = 1.4rC \tag{15.3-3}$$

It is of interest to note that experimentally, using the Motorola MC 302 ECL flip-flop, T is found to be given by

$$T = 20 \text{ ns} + 1.4r(C + 5 \text{ pF}) \tag{15.3-4}$$

In Eq. (15.3-4), the 20 ns is due to propagation delay in the flip-flop, which we have not taken into account, and the 5 pF added to C is necessary because of the inevitable stray capacitance.

Finally, we may note that the circuit would perform equivalently if instead of as shown in Fig. 15.3-1, C were connected from Q to S and r connected from $\bar{Q}$ to S. In this case the triggering pulse would be applied instead to R. Other types of monostable circuits which employ clocked JK ECL flip-flops are also available commercially.

Triggering While we have not so indicated in the discussion above or in the waveforms of Fig. 15.3-1, the amplitude and duration of the triggering waveform V_S applied at S will have an effect on the duration of the quasi-stable state. Observe that in Fig. 15.3-1a the NOR outputs of the gate are being used. Referring to Fig. 7.4-1b, we note that when the ECL-gate output voltage enters the region corresponding to logic level **1**, the NOR output depends on how high the input voltage goes. Thus, more realistically, the circuit waveforms appear as in Fig. 15.3-2. The reflection of V_S in the waveform of V_R will clearly have an effect on the circuit timing. It is therefore advantageous to modify the circuit so that it is unresponsive to the triggering waveform after it has responded to its leading edge. Such a modification is shown in Fig. 15.3-3.

In this figure we have inserted an additional NOR gate $G3$ and we have inverted the polarity of the trigger. In the reset state, $Q = 0$ and the trigger level is high, so that the set input is at logic **0**. When the trigger drops to **0**, the S input goes to **1** and the flip-flop is set. As soon as Q becomes **1**, that is,

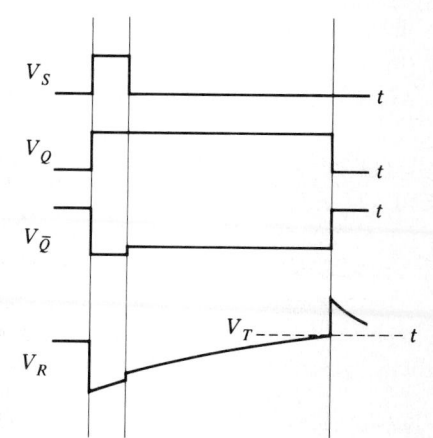

FIGURE 15.3-2
How the waveforms of Fig. 15.6-1 are modified because the NOR output of an ECL circuit varies with input when the input is in the range of logic level **1**.

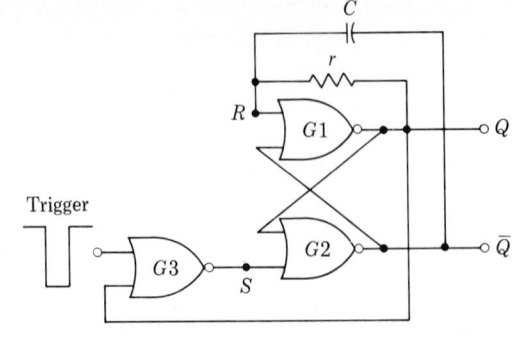

FIGURE 15.3-3
A modification of the circuit of Fig. 15.3-1 to ensure that the triggering waveform does not affect the timing of the multi.

after a time delay equal to the sum of the propagation delays in $G1$ and $G2$, gate $G3$ is disabled and its output is set to **0** independently of the trigger waveform. Note that the pulse width of the trigger must be at least equal to the sum of the propagation delays in $G1$ and $G2$. Observe that the S input voltage is now also independent of the trigger amplitude.

An alternate configuration of an ECL one-shot is shown in Fig. 15.3-4. Here we have replaced the two NOR gates $G1$ and $G2$ of Fig. 15.3-3 by an SR flip-flop, such as the Motorola MC 1014, and replaced gate $G3$ by an edge-triggered $J\overline{K}$ flip-flop, such as the MC 1013. The virtue of this circuit is that the trigger width can now exceed the pulse width of the monostable multi. It is left for the reader to show that the output-pulse width is independent of the trigger width in the circuit of Fig. 15.3-4 while the output-pulse width must exceed the trigger width when using the circuit shown in Fig. 15.3-3.

Note that triggering of the $J\overline{K}$ flip-flop occurs on the edge of the input waveform that has a positive slope. (Refer to the truth table of Fig. 9.18-2a.) If the trigger-pulse rise time is too slow, a Schmitt trigger (Sec. 2.17) must be inserted between the trigger and the JK flip-flop.

Recovery time   We have noted that the ECL monostable multi of Fig. 15.3-1 has a long recovery time. Shown in Fig. 15.3-5 are two circuit modifications which reduce recovery time. In Fig. 15.3-5a a diode D has been bridged across

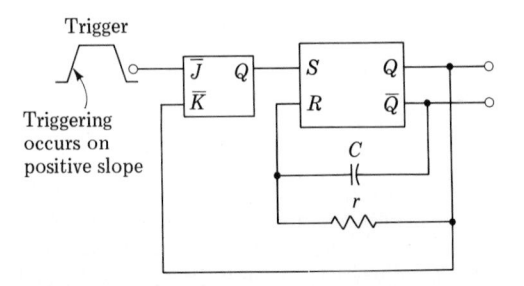

FIGURE 15.3-4
The use of a $J\overline{K}$ flip-flop rather than a NOR gate (Fig. 15.3-3).

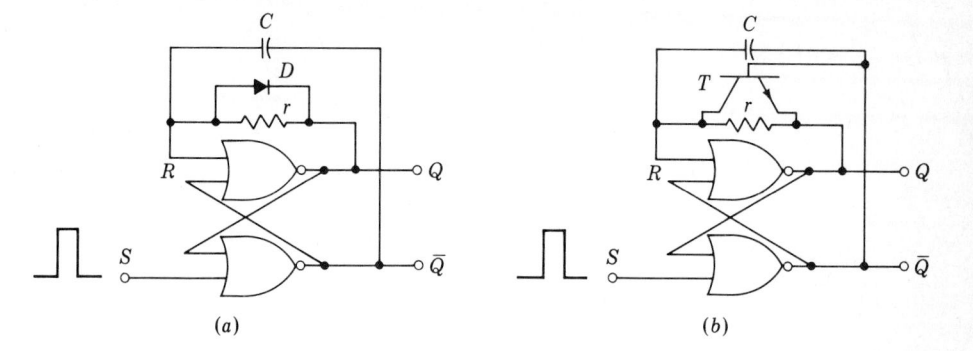

FIGURE 15.3-5
Two schemes to reduce the recovery time in an ECL monostable multi.

r. In the stable state the voltage across r is zero. During the timing interval the voltage across r is in the direction to back-bias the diode. During the recovery-time interval the diode is forward-biased, as can be verified from the waveforms of Fig. 15.3-1. When the diode is ON, its incremental resistance r_d will be much smaller than r and hence the capacitor may recharge much more rapidly than in the absence of the diode. However, the diode stops conducting when its voltage falls below about 0.65 V, while the recovery process is not complete until the voltage across r falls to zero. Consequently the diode reduces the recovery time by only about 50 percent.

A much more effective method of reducing recovery time is shown in Fig. 15.3-5b. During the timing interval, as can be verified from the waveform of Fig. 15.3-1, the transistor is cut off. However, during the recovery interval the transistor T is driven vigorously into saturation, thereby effectively short-circuiting the resistor r and discharging C.

15.4 MULTIVIBRATORS FOR SHORT TIMING INTERVALS

When a monostable multi is required to establish a very short timing interval (of the order of tens of nanoseconds), it is more practical to establish the timing interval with a delay line than with a resistor-capacitor circuit. Such delay-line multis generally use ECL logic gates whose propagation delays are comparably short.

A monostable circuit using ECL gates with their NOR and OR outputs is shown in Fig. 15.4-1. We shall assume a propagation-delay time t_{pd} for the gates themselves and assume as well that this delay is the same for the OR and the NOR outputs of gates GA and GB. The delay line itself has a delay time t_D.

The operation of the circuit is described by the logic waveforms in Fig. 15.4-1b. For the present we consider only the waveforms drawn in solid lines. We start with V_i at logic level **0** and the circuit in a stable steady state. Then

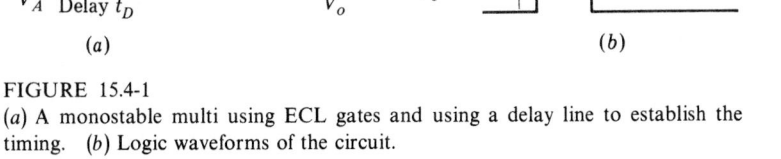

FIGURE 15.4-1

(a) A monostable multi using ECL gates and using a delay line to establish the timing. (b) Logic waveforms of the circuit.

it can be verified that the second input to gate GA (which is also the GB output $\overline{V}_o$) is also at logic **0**. (Assuming that $\overline{V}_o$ is at logic **1** leads to inconsistencies.) Now let V_i jump to logic **1**, as shown. Then, after a delay t_{pd}, V_A, which was at **0**, will jump to **1** and $\overline{V}_A$ will change to **0**. The logic level of V_{AD} is the same as V_D except for the time delay t_D. Finally, $\overline{V}_o$, which is $\overline{V}_o = 1$ whenever both $\overline{V}_A$ and V_{AD} are **0**, is as shown. In drawing $\overline{V}_o$ we have again taken account of the propagation delay t_{pd}, this time through gate GB. We now observe that we have established that $\overline{V}_o$ is $\overline{V}_o = 1$ for exactly the duration t_D. The interval during which $\overline{V}_o = 1$ begins after a delay $2t_{pd}$, but the duration of the interval is t_D entirely independently of t_{pd}.

In Fig. 15.4-1a with V_i a step, the feedback connection from the output of gate GB to the input of gate GA serves no purpose. For as long as $V_i = 1$ the output of GA is $\overline{V}_A = 0$ no matter what the logic level at the other input. The feedback connection, however, does permit us to replace the step by a pulse, as indicated by the dashed waveforms. For, in the presence of the feedback, V_i needs to be held at $V_i = 1$ only long enough for $\overline{V}_o$ to rise to $\overline{V}_o = 1$, that is, for a time equal to or longer than the sum of the delays of the two gates, that is, $2t_{pd}$. With a pulse input all the waveforms return eventually to their initial levels. As can be verified from the waveforms, if V_i changes to $V_i = 1$ at $t = 0$, the circuit will have completely returned to its initial state at a time $t = 2t_D + 3t_{pd}$. The *duty cycle* of the multi is defined as the maximum fraction of the interval between successive triggering inputs which can be taken up by the quasi-stable state of the multi. In the present case the duty cycle is seen to be $t_D/(2t_D + 3t_{pd})$, which is at most 50 percent if $t_D \gg t_{pd}$.

An alternative arrangement which allows an improved duty cycle is shown in Fig. 15.4-2. The circuit is recognized as a flip-flop (two crossed coupled inverters) in which the input V_i, directly or delayed, provides the set and reset

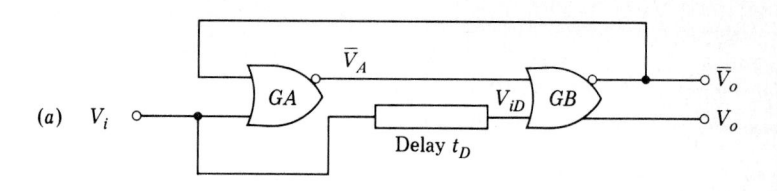

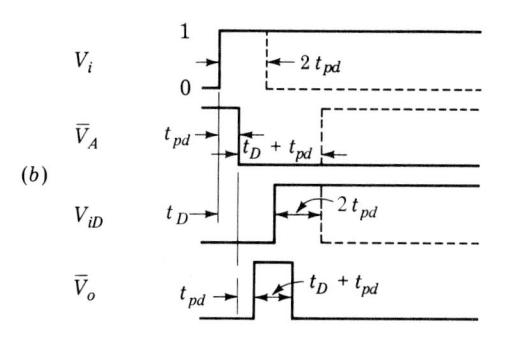

FIGURE 15.4-2
An alternative to the circuit of Fig. 15.4-1 which allows an improved duty cycle.
(*a*) Circuit diagram.   (*b*) Logic waveforms.

inputs.   Hence, initially, with $V_i = 0$ two states are possible.   We shall assume that, to start, the circuit is in the state in which $\bar{V}_o = 0$, $\bar{V}_A = 1$.   As before, we assume equal gate delays t_{pd}.

The solid waveforms in Fig. 15.4-2*b* correspond to the case in which V_i is a step from logic **0** to logic **1**.   We then observe that the timed waveform $\bar{V}_o$ has a duration $t_D - t_{pd}$ beginning at a time $2t_{pd}$ after the jump in V_i.   As in the previous case, so here, because of the feedback, the step in V_i can be replaced by a pulse of duration at least equal to $2t_{pd}$.   The waveforms for a pulse input are shown dashed.   As can be seen in the waveforms, the timed output $\bar{V}_o$ is the same for step or pulse input.   With a pulse input of minimum duration $2t_{pd}$ the circuit returns to its initial state at a time $t_D + 2t_{pd}$.   The duty cycle is therefore $(t_D - t_{pd})/(t_D + 2t_{pd})$, which may approach 100 percent for $t_D \gg t_{pd}$.

The ECL astable multivibrator   An ECL astable multivibrator can be constructed using the circuit shown in Fig. 15.2-1*a* by replacing the CMOS gates $G1$ and $G2$ by ECL NOR gates.

15.5   AN INTEGRATED-CIRCUIT TTL MONOSTABLE MULTIVIBRATOR

The 54121 TTL integrated-circuit monostable multi is shown in block-diagram form in Fig. 15.5-1*a*.   The circuit is seen to consist of two parts, a flip-flop (gates $G1$ to $G4$) and a trigger pulse-shaping circuit (gates $G5$ and $G6$).   In the actual circuit gates $G1$, $G3$, $G4$, and $G5$ are TTL gates, each with a logic **1** level of 3.5 V

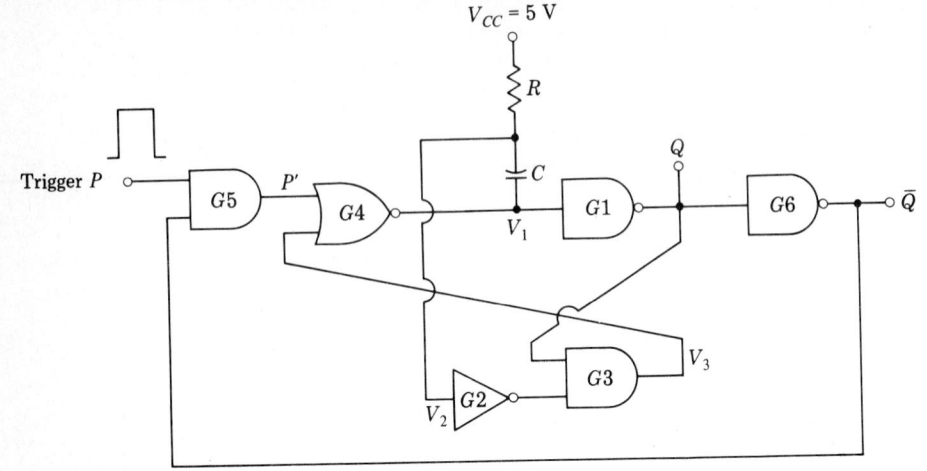

(a)

(b) Waveforms

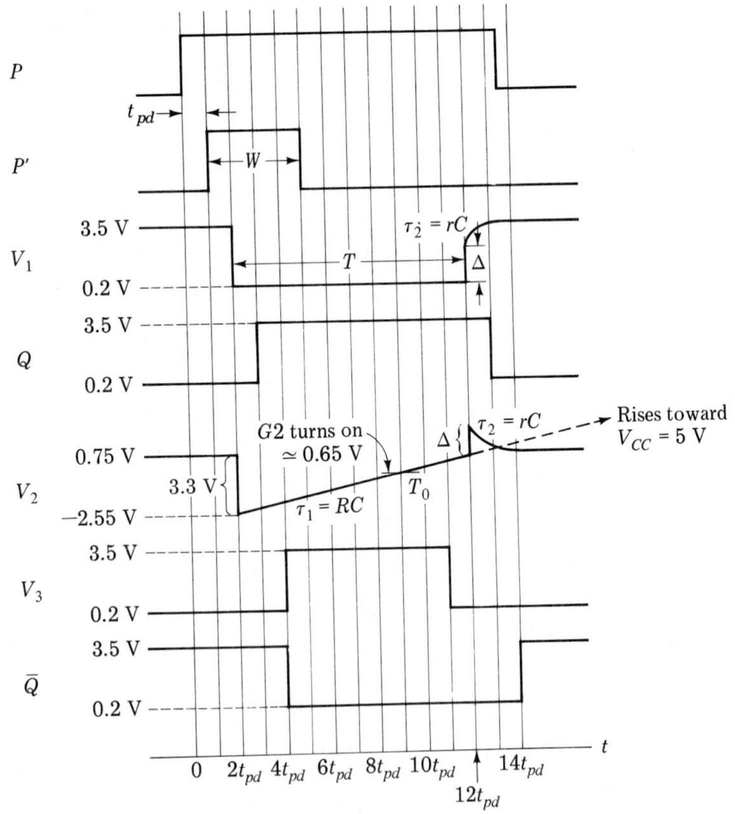

FIGURE 15.5-1
(a) Block diagram of monostable multivibrator. (b) Waveforms.

and a logic **0** level of 0.2 V, while gate $G2$ is a grounded emitter amplifier-inverter which saturates when its input voltage V_2 is 0.75 V and has a cut-in voltage of 0.65 V. In addition, in the actual circuit, the trigger input is first shaped by a Schmitt trigger (not shown) before application to gate $G5$.

In steady state $P = 0$, and therefore $P' = 0$. Furthermore, in steady state, there is no current flowing through capacitor C. Thus, any current in R flows into the base of the grounded emitter inverter, gate $G2$. The resistor R is small enough for $G2$ to be saturated, and $V_2 = 0.75$ V. Since $G2$ is saturated, V_3 is at logic **0**, and with P' also at **0**, the output of $G4$ is at logic **1**. We then see that $Q = 0$ and $\bar{Q} = 1$. Gate $G5$ is thereby enabled, so that P' will rise to **1** when P becomes **1**.

Now assume that a positive step is applied at P at time $t = 0$ (see Fig. 15.5-1b). Since $G5$ is enabled, P' goes high (the delay t_{pd} shown in Fig. 15.5-1b is the propagation delay, which we shall assume to be the same for each gate in the circuit). The output V_1 of gate $G4$ drops abruptly to 0.2 V. Since V_1 and V_2 are coupled by capacitor C, V_2 decreases by the same amount as V_1, an amount which is approximately $3.5 - 0.2 = 3.3$ V. Capacitor C now begins to charge, bringing V_2 from $0.75 - 3.3 = -2.55$ V asymptotically toward $V_{CC} = 5$ V. When V_2 rises to approximately 0.65 V, $G2$ turns ON and V_3 goes low. In Fig. 15.5-1b we have shown V_2 reaching 0.65 V at a time T_0. To simplify the figure we have caused T_0 to be a multiple of the propagation-delay time and have arbitrarily set $T_0 = 9t_{pd}$. The delay between V_2 reaching 0.65 V at time T_0 and V_3 falling is seen to be equal to the delay due to gates $G2$ and $G3$. This delay, $2t_{pd}$, is independent of the value of T_0.

The timing interval of the multi is completed at $t = T_0$. At this time, except for the propagation delays, the circuit returns to its permanently stable state. Suppose now that we want the circuit to respond again to a succeeding trigger and that we want the timing interval in response to this second trigger to be the same as to the first triggering signal. If such is the case, we must allow an interval between triggers for the charge on C to return to its initial value. The recharging of C gives rise to an overshoot in the waveform of V_2, and, correspondingly, the waveform of V_1 does not attain its logic **1** level of 3.5 V until this overshoot is completed. Since the voltages V_1 and V_2 appear at the two sides of a capacitor, the voltage jumps Δ in the two must be equal, as indicated in Fig. 15.5-1. The resistance in series with C, as C replaces its charge is, on the one side, the output impedance of $G4$ and, on the other side, the resistance R in parallel with the input resistance of $G2$. Let us call this total resistance r. Then the overshoot in V_2 and the approach of V_1 to 3.5 V take place with a time constant $\tau_2 = rC$ (a quantitative calculation of Δ and τ_2 is considered in Prob. 15.5-2).

We also note that when V_1 falls, Q rises and $\bar{Q}$ falls. This causes P' to return to the **0** state. The width W of P' is equal to the sum of the propagation delays of gates $G4$, $G1$, $G6$, and $G5$ and is independent of P. The width of P must exceed W, however. The virtue of $G5$ is that as soon as the multi has responded to the input trigger, gate $G5$ is disabled. The multi is thereby disconnected from the triggering source and is free to generate its timing interval without being influenced by the waveform at P. In particular the multi timing interval T can be less than the interval over which P is held at logic **1**.

Output pulse width The circuitry used in the 54121 is designed to minimize variations in output width due to changes in supply voltage and/or temperature. The output-pulse width is nominally

$$T = RC \ln 2 \approx 0.7RC \tag{15.5-1}$$

The output-pulse width T can vary from 40 ns when the external timing components R and C are not used to over 40 s. A duty cycle of 67 to 90 percent is obtainable.

If the supply voltage should vary from 5.25 to 4.75 V (± 5 percent), the pulse width will vary by less than ± 2 percent. In addition, temperature variations from 125 to $-55°C$ result in output-pulse width variations of from $+0.2$ to -0.8 percent, respectively.

15.6 AN INTEGRATED-CIRCUIT TIMER

In this section we consider an integrated circuit designed to serve in timing applications generally. This integrated-circuit device, type 555, can be used to generate stable time delays like a monostable multi, or it may be used as an oscillator like an astable multi. In the monostable mode, the timing is controlled by one external capacitor and one external resistor. In the astable mode the two operating features of interest, frequency and duty cycle of the output waveform, are controlled by one external capacitor and two external resistors.

The functional diagram of the type 555 timer is shown in Fig. 15.6-1. The names associated with the external terminals are those of the manufacturer. The unit is intended to be used with supply voltages in the range 5 to 15 V. The string of three resistors bridged between the supply voltage V_{CC} and ground provides reference voltages for the two comparators. The reference voltage for comparator 2 is $V_{CC}/3$, and the reference voltage for comparator 1 is $2V_{CC}/3$. As we shall see, these reference voltages have control over the timing. In applications where we may want to vary the timing electronically, we can do so by applying a modulation voltage to the *control-voltage* input terminal. In applications where no such modulation is intended, the manufacturers recommend that the control-voltage terminal be capacitively bypassed to ground (about 0.01 μF).

On a *negative-going* excursion of the *trigger* input and when the trigger voltage passes through the reference voltage $V_{CC}/3$, the output of comparator 2 *sets* the flip-flop. On a *positive-going* excursion of the *threshold* voltage and when the threshold voltage passes through the reference voltage $2V_{CC}/3$, the output of comparator 1 *resets* the flip-flop. The *reset* input provides a mechanism to reset the flip-flop in a manner which overrides the effect of any instruction to set which the flip-flop may have from comparator 2. This overriding reset will be in effect

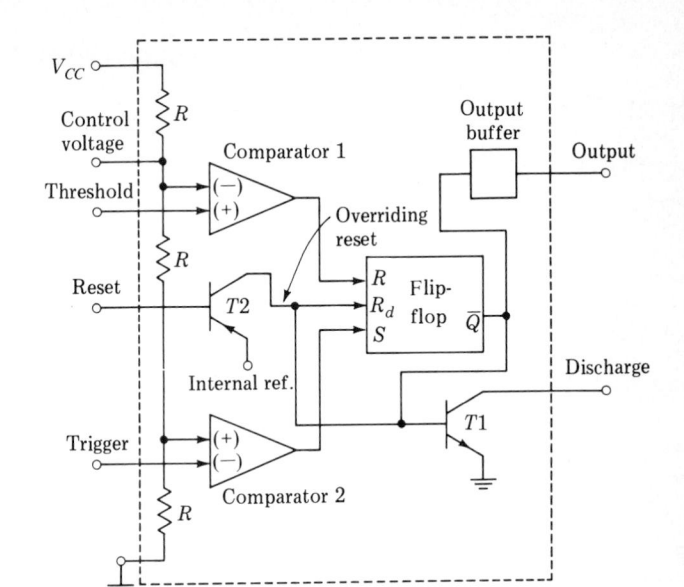

FIGURE 15.6-1
Functional diagram of the type 555 integrated-circuit timer.

whenever the reset input is less than about 0.4 V. When it is not intended that the overriding reset be used, the reset input is ordinarily returned to V_{CC}. The transistor $T2$ serves simply as a buffer to isolate the reset input from the flip-flop and the transistor $T1$. The output simply reflects the logic level at the output of the flip-flop. The block marked "output" is a buffer between the flip-flop and the output terminal. This output buffer stage is able to source currents as high as 200 mA and provides logic levels consistent with TTL logic.

An external timing capacitor (not shown in Fig. 15.6-1) may be bridged between the discharge terminal and ground. When the flip-flop is in the reset state, its output $\bar{Q}$ is at logic level **1**. Hence, the base of $T1$ is high, transistor $T1$ is in saturation, and the timing capacitor will be held in a discharged condition. A timing cycle will start when the flip-flop goes to the set state and transistor $T1$ is turned OFF. The external capacitor will then be free to charge through an external resistor (not shown). We can terminate this capacitor charging at any arbitrary time by dropping the voltage at the reset terminal below 0.4 V. The direct connection of the override-reset line output of $T2$ to the input of $T1$ is now seen as a means of turning ON $T1$ immediately and thereby circumventing the propagation delay through the flip-flop.

The type 555 timer connected for monostable operation is shown in Fig. 15.6-2. The timing capacitor C is charged from V_{CC} through R. The resistors R_1 and R_2 are rather arbitrary and are of resistance values in the range of many tens of thousands. They are selected to hold the trigger input comfortably above $V_{CC}/3$ in the absence of the negative trigger pulse applied through C'.

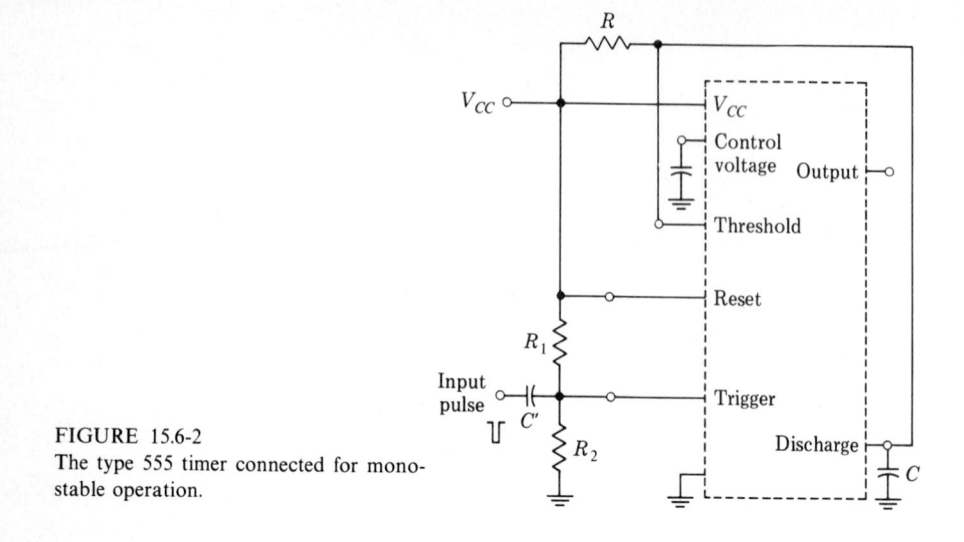

FIGURE 15.6-2
The type 555 timer connected for monostable operation.

In the stable state the capacitor C is held at discharge. The input triggering pulse turns $T1$ OFF (see Fig. 15.6-1) and allows the timing cycle to begin. (The circuitry of the 555 is such that once the timing cycle has begun, the voltage level at the trigger input has no effect until the cycle is completed, provided only that the voltage does not rise above V_{CC}.) The voltage across C is applied to the *threshold* input. When the capacitor voltage reaches $2V_{CC}/3$, comparator 1 responds, resets the flip-flop, and terminates the cycle by discharging C. The

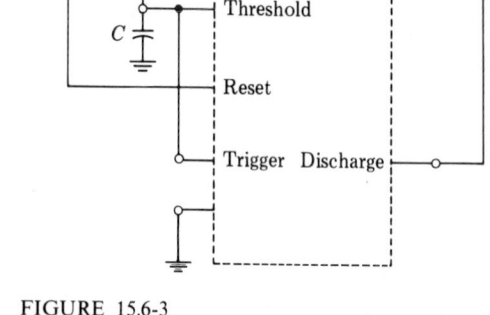

FIGURE 15.6-3
The type 555 timer connected for astable operation.

time T of a cycle is the time required for the capacitor to charge from nominally zero to $2V_{CC}/3$. This time, as can be verified, is $T = 1.1RC$. Note that T does not depend on V_{CC}. This independence from V_{CC} results from the fact that the threshold voltage is a fixed fraction of the voltage V_{CC}, which is the voltage toward which the capacitor voltage is charging asymptotically.

The type 555 timer connected for astable operation is shown in Fig. 15.6-3. The capacitor C charges through R_a and R_b from V_{CC} until the capacitor voltage just exceeds $2V_{CC}/3$. The discharge transistor $T1$ then goes ON, and C discharges through R_b. The discharge continues until the voltage across C is just less than $V_{CC}/3$. The charge and discharge times are $T_1 = 0.7(R_a + R_b)C$ and $T_2 = 0.7R_b C$. The total period is

$$T = T_1 + T_2 = 0.7(R_a + 2R_b)C \qquad (15.6\text{-}1)$$

The duty cycle of the present arrangement is necessarily different from 50 percent. It is left as a student exercise (Prob. 15.6-2) to modify the circuit to allow a 50 percent duty cycle.

REFERENCES

1. Motorola, "McMOS Handbook," Motorola, Inc., Phoenix, Ariz.
2. RCA, "COS/MOS Digital Integrated Circuits," RCA, Somerville, N.J.

APPENDIX
TRANSMISSION LINES

A.1 INTRODUCTION[1]

We present here a brief and simplified discussion of transmission lines. Our point is mainly to clarify the question of when a pair of connecting wires may be viewed as just that, i.e., a pair of connecting wires, and when, on the other hand, they must be viewed as a length of transmission line.

In Fig. A.1-1a is shown a source of fixed voltage V connected, when switch S is closed, to a load R. The connection is made through a pair of wires extending in length from $x = 0$ to $x = \ell$. Initially, we might say that if S were to be closed at time $t = 0$, the voltage V_x (the voltage drop from the top wire to the bottom) at any distance x, as well as the voltage across the load, would *immediately* also attain the value V. Actually, however, the effect of the closing of S does not make itself felt everywhere immediately but propagates from source to load with a finite velocity.

Let us assume that the lines connecting source to load are such that the geometry of the lines in cross section at any value of x is everywhere the same. That is, we assume that the line has a *uniform* cross section. In this case, the velocity of propagation u is given by

$$u = \frac{1}{\sqrt{LC}} \tag{A.1-1}$$

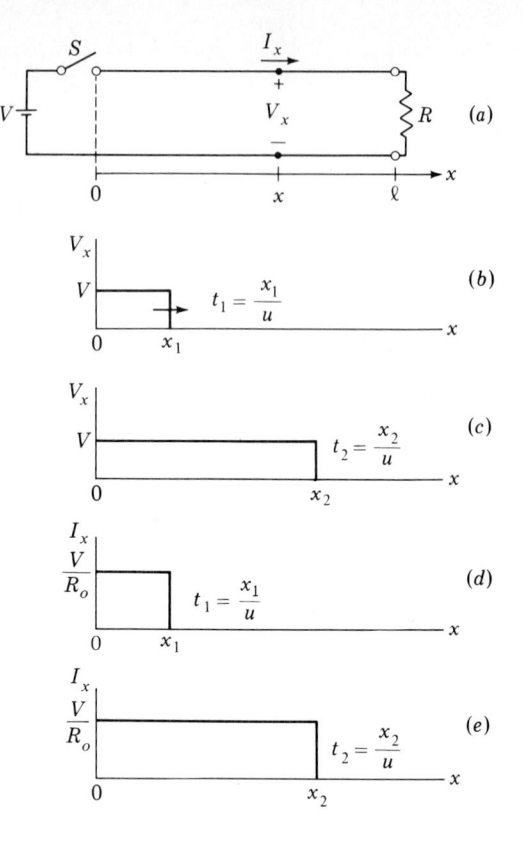

FIGURE A.1-1
(a) A source of voltage V connected to a load R through a transmission line of length ℓ. (b) Voltage V_{x_1} at time $t_1 = x_1/\mu$. (c) Voltage V_{x_2} at time $t_2 = x_2/\mu$. (d) Current I_{x_1} at time $t_1 = x_1/\mu$. (e) Current I_{x_2} at time $t_2 = x_2/\mu$.

where L and C are respectively the inductance and capacitance per unit length of the lines. Equation (A.1-1) suggests that u is a function of the geometry of the cross section since both L and C depend on the geometry. If, for example, the transmission system consists of two parallel wires, then as the spacing between the wires is reduced, L decreases and C increases. It turns out, however, that the product LC is constant independently of the geometry and has the value $u = 3 \times 10^8$ m/s when the medium between the wires is air. This velocity u is a fundamental constant of the physical universe generally referred to as the *velocity of light*. This result, that the propagation velocity does not depend on geometry, is not surprising when we take account of the fact that actually the power being transferred from source to load is not carried by the wires. Instead the power is transferred through the free space between and surrounding the wires, the wires serving only as *guides* to direct the power flow from source to load.

The distribution of voltage along the line at times $t_1 = x_1/u$ and $t_2 = x_2/u$ are shown in Fig. A.1-1b and c. At t_1, as shown in Fig. A.1-1b, the line voltage is V from $x = 0$ to $x = x_1$ and is zero for $x > x_1$. The *front* of voltage travels to the right with a velocity u so that at time $t_2 > t_1$ the voltage has propagated to $x = x_2$.

As the voltage front moves down the line, the capacitance of the line charges to voltage V. For such capacitance charging to take place it is necessary that a front of *current* accompany the voltage front. The magnitude of this current is easily calculated.

When the front moves a distance dx, the additional capacitance that is charged to voltage V is $C\,dx$. The charge required is $dQ = VC\,dx$. We now have [using Eq. (A.1-1)]

$$I = \frac{dQ}{dt} = VC\frac{dx}{dt} = VCu = VC\frac{1}{\sqrt{LC}} = V\sqrt{\frac{C}{L}} \equiv \frac{V}{R_0} \qquad (A.1\text{-}2)$$

in which we have introduced the parameter $R_0 \equiv \sqrt{L/C}$, called the *characteristic impedance* of the line. The current I in Eq. (A.1-2) is the magnitude of the current which flows on that part of the line on which current is flowing, i.e., from $x = 0$ up to the point where the voltage front is located. To the right of the voltage front, the current is zero. We shall use I_x to represent the current as a function of x. We adopt the sign convention that I_x is positive when current flows to the *right* on the upper wire and correspondingly to the *left* on the lower wire, as shown in Fig. A.1-1a.

To recapitulate, it now appears that at the closing of switch S, a front of voltage V moves to the right on the line with a velocity u. Accompanying this voltage front is a current front I. The distribution of current on the line at times t_1 and t_2 is shown in Fig. A.1-1d and e. At all points on the line up to the voltage and current front we have

$$\frac{V_x}{I_x} = \frac{V}{I} = R_0 = \sqrt{\frac{L}{C}} \qquad (A.1\text{-}3)$$

Suppose, now, that in Fig. A.1-1a we interchanged the locations of source and load. Then, at the closing of S, a voltage and current front would start moving toward the left. If, however, we maintain the *same sign convention* as before for V_x and I_x, we would then have

$$\frac{V_x}{I_x} = -R_0 = -\sqrt{\frac{L}{C}} \qquad (A.1\text{-}4)$$

A.2 THE CHARACTERISTIC IMPEDANCE

The characteristic impedance $R_0 = \sqrt{L/C}$ is a function of the geometry of the cross section of the line. For example, if the transmission line consists of two parallel wires, R_0 decreases as the spacing between the wires is reduced. Such is the case since, with reduced spacing, C increases and L decreases. If a dielectric is introduced between the wires, C will increase while L will remain unchanged so that R_0 will decrease. (Such a dielectric will therefore also reduce the propagation velocity.)

In Fig. A.2-1 are shown the cross sections of three common lines, and the expressions

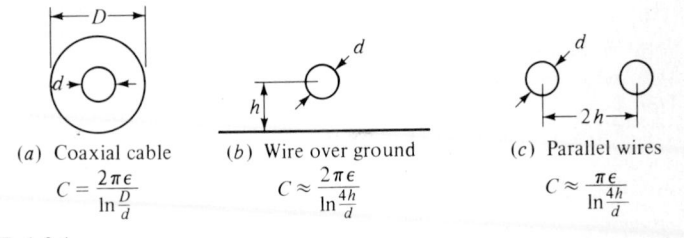

(a) Coaxial cable	(b) Wire over ground	(c) Parallel wires
$C = \dfrac{2\pi\epsilon}{\ln\dfrac{D}{d}}$	$C \approx \dfrac{2\pi\epsilon}{\ln\dfrac{4h}{d}}$	$C \approx \dfrac{\pi\epsilon}{\ln\dfrac{4h}{d}}$

FIGURE A.2-1
Several types of transmission lines, together with expressions for their capacitance per meter. The approximate expressions for the capacitance of the wire lines in (b) and (c) are correct to better than 5 percent if $h/d \geq 1$.

for their capacitance are given. It is also true that the impedance depends on the logarithm of a dimension ratio. Since the logarithm is a very slowly varying function of its argument, it is generally not feasible to make very large changes in R_0 by reasonable changes in dimensions. For example, consider the case of the coaxial line. When attenuation in the line results principally from ohmic losses in the conductors, the loss (for a fixed D) is a minimum for $D/d = 3.6$, a reasonable ratio. For this ratio and with a relative dielectric constant $\epsilon_r = 2.3$, $R_0 = 51 \ \Omega$. On the other hand to attain an impedance $R_0 = 1 \ k\Omega$ would require that $D/d = 10^{11}$! Most commercially available coaxial lines have impedances which are less than $100 \ \Omega$. Parallel-wire lines may have impedances up to several hundred ohms.

A.3   REFLECTIONS

At the moment of the closing of switch S in Fig. A.1-1 the source applies a voltage V to the line and delivers a current V/R_0. Hence, the impedance seen by the source in looking into the line is the characteristic impedance R_0. If the line were infinitely long so that the fronts of voltage and current never reached the end, the fronts would continue indefinitely, maintaining between themselves the ratio $V/I = R_0$ and there would be a constant impedance R_0 seen looking into the input of the line.

Now suppose that the line is not infinitely long but of finite length and that at its termination, there is bridged across the line a resistor $R = R_0$. This termination, being equal to the characteristic resistance of the line, would appear as an infinite extension of the line. Hence, beginning with the closing of the switch a front of voltage V and current V/R_0 would sweep to the right. After a time $t = \ell/u$ the fronts will have reached the line termination. From that time forward the voltage across the line at any position, as well as the voltage across the load, will be V while the current at all points on the line and in the load will be V/R_0. In short, after the fronts reach the termination, nothing further happens.

Now consider that the terminating resistor $R \neq R_0$. When the fronts arrive at the end of the line where $x = \ell$, they will be related by $V_\ell/I_\ell = R_0$. However, since at $x = \ell$ the resistor is R, it is required that $V_\ell/I_\ell = R$. This potential inconsistency is resolved by virtue of the fact that when $R \neq R_0$, a *reflection* develops at the termination. That is, at the moment the *incident* fronts (started down the line at the closing of S) reach the termination, there immediately develop *reflected* fronts, which start moving toward the left. These reflected fronts are related by Eq. (A.1-4). The amplitude and polarity of the reflected fronts will be as required so that the *total* voltage V_ℓ (the sum of incident and reflected voltage at $x = \ell$) and the total current I_ℓ will be related by $V_\ell/I_\ell = R$. Thus if the incident voltage front has an amplitude V, the reflected voltage front will have an amplitude ρV. The parameter ρ is called the *reflection coefficient* and is to be determined. The incident current is V/R_0. The reflected current is $-\rho V/R_0$ [the reason for the minus sign is explained in the discussion immediately preceding Eq. (A.1-4)]. The ratio of the total voltage to the total current is R, and hence

$$\frac{V + \rho V}{V/R_0 - \rho V/R_0} = R \tag{A.3-1}$$

Solving for ρ, we find

$$\rho = \frac{R/R_0 - 1}{R/R_0 + 1} \tag{A.3-2}$$

The reflection coefficient lies in the range -1 to $+1$; $\rho = 0$ when $R = R_0$, $\rho = 1$ when the end of the line is open-circuited, and $\rho = -1$ when the line is terminated in a short circuit.

A.4 MULTIPLE REFLECTIONS

Consider the situation shown in Fig. A.4-1 in which a line of impedance R_0 is terminated at its receiving end in $R \neq R_0$ and at its sending end in a source resistance $R_s \neq R_0$. Let V_i be a step of amplitude V applied at $t = 0$. Until such time as the line input becomes aware of the fact that the receiving termination does not *match* the line imped- ance, the input to the line appears to be a resistance R_0. Hence, at $t = 0+$ the voltage step at the line input at $x = 0$ will be

$$V'_i = \frac{R_0}{R_s + R_0} V_i \qquad (A.4\text{-}1)$$

This step in voltage will travel down the line and a reflection will develop at the receiving end. On arriving at the line input the reflected front will again be reflected and so on. Each time a front arrives at a terminating resistor, at either end of the line, the reflected front will be smaller than the incident front so that eventually a steady-state situation will be established. In the special cases in which each termination is an open or short circuit the reflection coefficients will be $+1$ or -1 and (excluding the effect of attenuation, as a front moves down the line) a steady state will never be attained.

EXAMPLE A.4-1   A line of impedance $R_0 = 100\,\Omega$ and of one-way delay time t_d is terminated at its receiving end in a resistor $R = 900\ \Omega$ and at its sending end in $R_s = 15\ \Omega$. The input is a step of voltage $V_i = 12$ V. Draw the waveform at the receiving end.

SOLUTION   The reflection coefficient at the receiving end ρ_r is, from Eq. (A.3-2),

$$\rho_r = \frac{\frac{900}{100} - 1}{\frac{900}{100} + 1} = 0.8$$

The reflection coefficient at the sending end can similarly be calculated to be $\rho_s = -0.75$. A 12-V input step will appear at time $t = 0$ at the line input, as a step of amplitude

$$V'_i = V_1 = \frac{R_0}{R_s + R_0} V = \frac{100}{15 + 100} \times 12 \simeq 10 \text{ V}$$

The front V_1 will arrive at the termination at time $t = t_d$, where there will immediately appear a reflected front $V_2 = \rho_r V_1 = 0.8(10) = 8$ V. Hence, at $t = t_d$ the receiving-end voltage V_ℓ will jump from zero to $V_\ell = 10 + 8 = 18$ V. V_2 goes back to the input end and

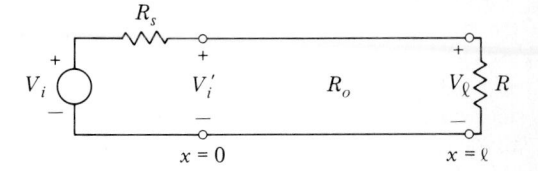

FIGURE A.4-1
Voltage source V_i with source impedance R_s is connected to load resistor R by means of a transmission line with characteristic impedance R_0.

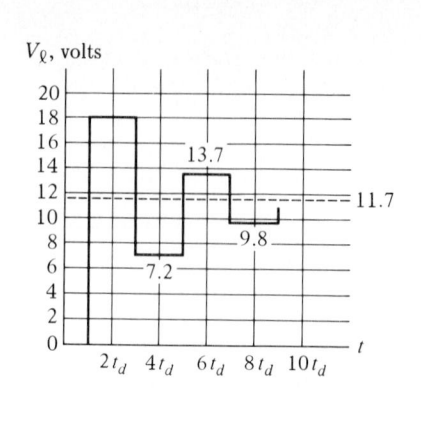

FIGURE A.4-2

Voltage waveform at $x = \ell$ due to reflections.

is reflected as $V_3 = -0.75V_2 = -0.75(8) = -6$ V. When V_3 gets back to the line end at time $t = t_d + 2t_d = 3t_d$, it immediately gives rise to $V_4 = 0.8V_3 = 0.8(-6) = -4.8$ V. Hence, at time $t = 3t_d+$ the total receiving-end voltage is $V_1 + V_2 + V_3 + V_4 = 10 + 8 - 6 - 4.8 = 7.2$ V. Continuing in this manner, we find the waveform is as shown in Fig. A.4-2. It oscillates back and forth about an asymptotic limit. This limiting value, which is the same along the entire length of the line, is

$$V_x \text{ (at all } x \text{ and at } t = \infty) = V_i \frac{R}{R + R_s} = 12 \frac{900}{900 + 20} = 11.7 \text{ V}$$

A.5 EFFECT OF WAVEFORM RISE TIME

Suppose, in connection with the circuit of Fig. A.4-1 considered in Example A.4-1, that we had ignored the transmission-line character of the connection from source to load. In such a case, we would have anticipated that the voltage response across R due to a 12-V step input would simply be a step of amplitude 11.7 V. We see that actually, as in Fig. A.4-2, there are rather violent oscillations about the level 11.7 V. The amplitude of these oscillations does not depend on the delay time t_d. However, this lack of dependence results from the fact that we have assumed an applied input of zero rise time. When the rise time is not zero, the situation is different, as we shall now see.

To explore the matter most simply, let us consider the special case in which a line of one-way delay t_d is terminated in its receiving end in an open circuit and at its sending end in a short circuit; i.e., in Fig. A.4-1, $R = \infty$ and $R_s = 0$. The voltage V_ℓ across the load is shown for two cases in Fig. A.5-1. In Fig. A.5-1a the input V_i is a step, as shown by the dashed waveform, and, as can be verified, the load voltage V_ℓ oscillates with a peak-to-peak amplitude which is twice the amplitude of the input. In Fig. A.5-1b the input has the same amplitude as in Fig. A.5-1a but is a linear ramp which rises *slowly in comparison with* t_d, the rise occupying an interval of $10t_d$. Observe how much less pronounced the oscillations in Fig. A.5-1b are compared with those in Fig. A.5-1a. In Fig. A.5-1b, except for a delay t_d, the waveform V_ℓ does reasonably follow the input. Of course, in both cases, if there were terminating resistors R and R_s, the oscillations would decay with time.

Quite generally, when waveforms are such that they make sensible changes only over intervals which are very long in comparison with the time t_d, the transmission-line character of a connection does not make itself particularly apparent. As these rise

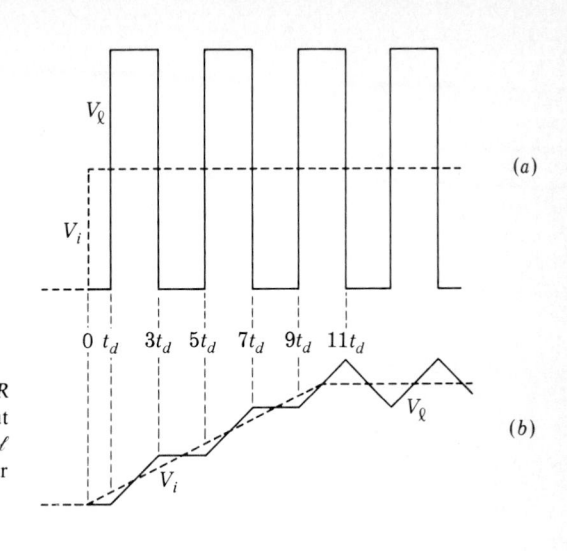

FIGURE A.5-1

(a) Voltage at $x = \ell$ when load resistor R is infinite and source resistor $R_s = 0$; input voltage is a step. (b) Voltage at $x = \ell$ when R is infinite, $R_s = 0$, and V_i is a linear voltage ramp.

times get progressively longer in comparison with t_d, the lines may be replaced by lumped-circuit equivalents, i.e., stray capacitance or stray inductance or both, with progressively better accuracy. On the other hand, where rise times are very fast, the delay-line character of the connections must be taken into account and some considerations given to proper line terminations to suppress reflections. By way of example, in ECL gates, rise times are of the order of nanoseconds and connecting wires extending only over a length of some tens of centimeters become transmission lines.

REFERENCE

1 Millman, J., and H. Taub: "Pulse, Digital, and Switching Waveforms," McGraw-Hill, New York, 1965, chap. 3.

Chapter 1

1.1-1 By measurement, a diode is found to draw 1 mA of current when the voltage drop across the diode is 0.75 V.

(a) If the volt-ampere characteristic of the diode is given by Eq. (1.1-1), Find I_0.

(b) Plot the volt-ampere characteristic of the diode.

1.1-2 (a) A diode is operating in a circuit where the diode current is I_D and the voltage across the diode is V_D. If the current in the diode is doubled, calculate the increase in the diode voltage V_D. Assume $V_T = 25$ mV.

(b) If the diode voltage in (a), at the lower current, was 750 mV, show that the doubling of the current increases the diode voltage by only 2 percent.

1.1-3 (a) Diode D1 has $I_0 = I_{01}$, while diode D2 has $I_0 = I_{02}$. Show that when the diodes carry equal currents, the voltages across them are different by amount $V_1 - V_2 = V_T \ln (I_{02}/I_{01})$.

(b) Let $I_{02} = 10I_{01}$. Diode D1 is carrying a small enough current for us to consider it at cutoff, and we find $V_1 = 0.65$ V. What would be the voltage across D2 if it were carrying the same small current?

1.2-1 The dependence of I_0 on temperature T is given approximately by

$$I_0 = KT^2\epsilon^{-V_g/V_T}$$

where K is a constant of proportionality, $V_g \approx 1.12$ V for silicon, and $V_T = kT/e$.

(a) Obtain an expression for $\ln I_0$. Differentiate $\ln I_0$ with respect to temperature T to show that

$$\frac{d}{dT}(\ln I_0) = \frac{2}{T} + \frac{V_g}{TV_T}$$

(b) Show that when $V \gg V_T$ Eq. (1.1-1) can be written

$$\ln I = \ln I_0 + \frac{V}{V_T}$$

so that

$$\frac{1}{I}\frac{dI}{dT} = \frac{d}{dT}(\ln I_0) + \frac{1}{V_T}\frac{dV}{dT} - \frac{V}{TV_T}$$

(c) Show that, at constant current,

$$\frac{dV}{dT} = \frac{V - V_g}{T} - \frac{2k}{e}$$

(d) Calculate dV/dT at $T = 300$ °K when $V = 0.65$ V and when $V = 0.75$ V. Note that the answer is not -2 mV/°C, and also note that the answer depends on the diode voltage V.

(e) The differential equation in part (c) can be solved by the change of variables $\mu = V/T$. From the result in (c) show that if $V(T_1)$ and $V(T_0)$ are the diode voltages at some fixed current the two temperatures T_1 and T_0, then

$$\frac{V(T_1)}{T_1} - \frac{V(T_0)}{T_0} = V_g\left(\frac{1}{T_1} - \frac{1}{T_0}\right) - \frac{2k_T}{e}\ln\frac{T_1}{T_0}$$

If $V(T_0) = 0.75$ V when $T_0 = 25$ °C, calculate $V(T_1)$ when $T_1 = 125$ and -55 °C.

1.2-2 Two identical diodes $D1$ and $D2$ are connected in parallel and are reverse-biased, as shown in Fig. P1.2-2. If $D2$ is operating at 125 °C and $D1$ is operating at 25 °C, determine the current in each diode. Assume that I_0 doubles for each 10 °C temperature increase.

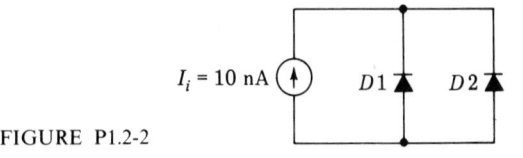

$I_i = 10$ nA

$D1$ $D2$

FIGURE P1.2-2

1.3-1 For the 1N916 diode, assuming that $C_T \approx C_0/(1 + V)^n$, find values for C_0 and n suitable to make C_T agree with the experimentally determined capacitance given in Fig. 1.3-1.

1.4-1 Three diodes are connected in series with a 9-V zener diode, and the series arrangement is used at 5 mA to provide a reference voltage of $9 + 3(0.75) = 11.25$ V. From Fig. 1.4-2b show that the temperature sensitivity of the arrangement is nearly zero.

1.4-2 A single forward-biased diode is to be connected in series with a zener diode. The pair is to be operated at 5 mA as a reference source. Use Fig. 1.4-2b to estimate what the operating voltage of the zener diode should be if the pair is to display no temperature sensitivity.

1.4-3 From Fig. 1.4-2a plot V_Z as a function of I_Z for zero temperature coefficient.

1.7-1 Define the symbols I_E, I_C, and I_B in such manner that they are positive when the emitter, collector, and base currents, respectively, are in the directions of normal current flow in a *pnp* transistor. Write the Ebers-Moll equations (1.7-3) and (1.7-4) using these current symbols and making whatever other changes are appropriate for a *pnp* transistor.

1.7-2 For an *npn* transistor the base current I_B may be written as the sum

$$I_B = I_{B1}(\epsilon^{V_{BE}/V_T} - 1) + I_{B2}(\epsilon^{V_{BC}/V_T} - 1)$$

Find I_{B1} and I_{B2}. Use the results of Prob. 1.7-1 to find a similar expression for the base current of a *pnp* transistor.

1.7-3 Using Eq. (1.7-4), find V_{CE} when $I_C = 0$. Assume $V_{BE} \gg V_T$ and $V_{CE} \gg V_T$.

1.7-4 Compare the base current of a transistor having $V_{BE} = 0.7$ V and $V_{CE} = 0.7$ V to the base current in a transistor having $V_{BE} = 0.7$ V and $V_{CE} = 0.1$ V. Let $\alpha_N = 0.9$ and $\alpha_I = 0.1$.

1.8-1 (a) Assume that a transistor is at the edge of cutoff so that I_E is given by Eq. (1.8-2). Compare the cut-in voltages of two transistors, the first with $\alpha_N = 0.9$ and $\alpha_I = 0.1$ and the second with $\alpha_N = 0.9$ and $\alpha_I = 0.01$.
   (b) Repeat part (a) if the collector is open-circuited so that $I_C \equiv 0$.

1.8-2 (a) Show that for $I_E = 0$ and also for $I_C = 0$, the Ebers-Moll equations reduce to the simple diode equation.
   (b) Show that for equal diode voltages the ratio of the diode currents is equal to the ratio α_N/α_I.

1.10-1 Verify Eqs. (1.10-3) to (1.10-5) and (1.10-9).

1.10-2 Refer to Fig. P1.10-2. If $h_{FE} = 50$ and $h_{FC} = 0.1$ find (a) the value of I_B needed for $V_{CE} = 0.3$ V, (b) the value of I_B needed for $V_{CE} = 0.1$ V. (c) Can V_{CE} equal 0 V with this configuration?

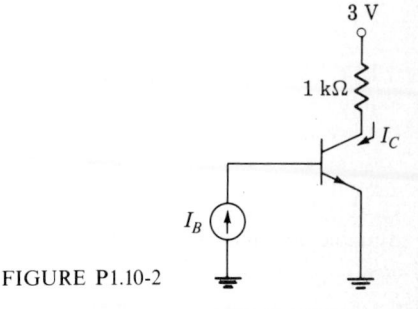

FIGURE P1.10-2

1.10-3 Refer to Fig. P1.10-3. Each transistor has $h_{FE} = 50$ and $h_{FC} = 0.1$.

(a) If $I_{B1}(T1) = 0$ and $V_{BE}(T1) = V_\gamma = 0.65$ V, calculate $I_{E0}(T0)$ and V_i. Assume $V_{BE}(T0) = 0.75$ V.

(b) When $V_i = 5$ V, T1 is saturated. Calculate $I_{B1}(T1)$ and $V_{CE}(T1)$. *Hint:* Use Fig. 1.10-1.

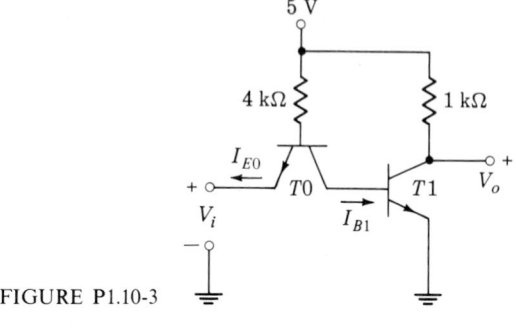

FIGURE P1.10-3

1.10-4 (a) Verify Eq. (1.10-12), which applies to the configuration of Fig. 1.10-3c.

(b) Find I_B, I_C, and I_E when $V_{CC} = 3$ V, $R_L = 1$ kΩ, $h_{FE} = 50$, and $h_{FC} = 0.1$.

1.10-5 (a) Show that in Fig. P1.10-5 I_C is given approximately by

$$I_C = \frac{(1 + I_{E0} R/V_T) I_{C0}}{1 - \alpha_N \alpha_I + (1 - \alpha_N) I_{E0} R/V_T}$$

Assume that the collector junction is reverse-biased and that the emitter junction is slightly forward-biased.

(b) Assume $\alpha_N = 0.98$, $I_{C0} = 10$ nA, $\alpha_I = 0.1$. Calculate I_C for $R = 0$ and $R = \infty$.

(c) What value of R will give a current midway between the currents corresponding to $R = 0$ and $R = \infty$?

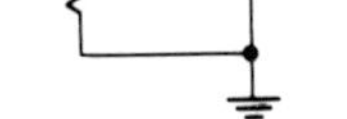

FIGURE P1.10-5

1.12-1 An inverter yields a high output voltage when the input is low and a low output when the input is high. One measure of the usefulness of an inverter in logic applications is its ability to maintain this large difference between input and output indefinitely, as many stages of the inverter are cascaded. Consider an inverter which uses the MOSFET with characteristics as in Fig. 1.12-2. Assume the load to be a resistor of resistance 2 kΩ (as might be encountered in a bipolar inverter). The supply voltage is $V_{DD} = 20$ V. Estimate the number of such inverters which can be cascaded before the difference between input and output becomes uncomfortably small.

1.13-1 In an integrated-circuit MOSFET inverter, as in Fig. 1.13-1*b*, $V_{DD} = 12$ V, $V_{GG} = 15$ V, and $V_T = 3$ V. Assume that the parameter C in Eq. (1.12-3) is $C = 1.2$. Find the inverter output voltage when the drive transistor is not conducting.

1.15-1 Idealize the plot of I_D in Fig. 1.15-5. Assume that I_D rises linearly from $I_D = 0$ at $V_i = 3.0$ V to 1.4 mA at $V_i = 8.0$ V and falls linearly to zero at $V_i = 12$ V. Let V_i have the waveform shown in Fig. P1.15-1. Neglecting the effect of capacitances, calculate the power dissipation in the gate.

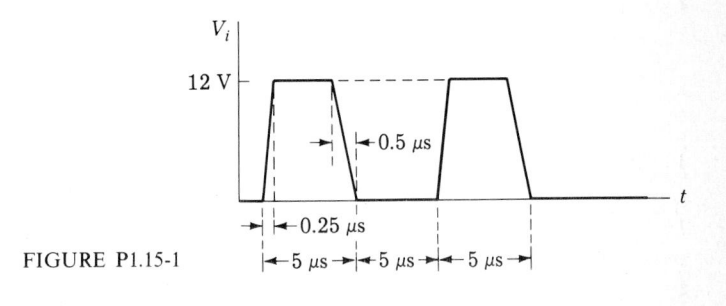

FIGURE P1.15-1

1.18-1 The excess minority-carrier charge Q in any volume of a semiconductor will decay because the minority charge recombines with the majority charge of opposite sign. The rate of decay of charge is proportional to Q, that is,

$$\frac{dQ}{dt} = -\frac{1}{\tau} Q$$

where $1/\tau$ is a factor of proportionality. (τ has the dimensions of time and is called the *storage time constant*.) Further if a current I flows out of the volume, there is an additional decrease at a rate $dQ/dt = I$. Altogether

$$\frac{dQ}{dt} + \frac{Q}{\tau} = I$$ 1

In the steady state, where $dQ/dt = 0$, $I = Q/\tau$.

(*a*) A diode has a storage time constant $\tau = 1$ μs. A voltage of 10 V is applied to the diode through a 10-kΩ resistor in the forward direction for a time long enough for a steady state to have been reached. Assuming that the forward voltage of the diode is zero, calculate the excess minority-carrier charge stored in the diode.

(*b*) At time $t = 0$ and after the situation described in part (*a*) the applied voltage is abruptly reversed to -5 V. Assume that the diode voltage remains at zero until the minority charge is removed and calculate the time at which the diode will be turned off.

(*c*) The combined diode transition capacitance and external shunt capacitance across the diode is 20 pF. Draw the waveform of the diode voltage and of the current through the 10-kΩ resistor from the time before $t = 0$ until the diode voltage becomes -5 V.

1.20-1 The input voltage V_i in Fig. P1.20-1 is initially $V_i = 5$ V, and the transistor is in saturation. The total stored base charge is 1 nC. At $t = 0$, V_i drops abruptly to 0 V. The capacitor C is included so that the base charge can be removed abruptly.

   (a) Calculate the minimum value of C required to allow removal of all base charge.

   (b) Draw the base waveform for C selected as in part (a) and also for the case where C is twice as large as in part (a). (Assume that at saturation $V_{BE} = 0.75$ V and that when $V_{BE} < 0.75$, the base current is negligible.)

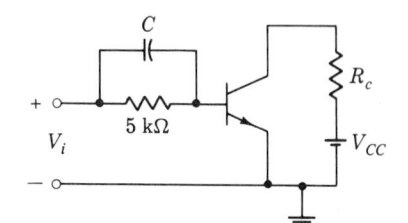

FIGURE P1.20-1

Chapter 2

2.2-1 (a) Show that the voltage gain of the op-amp of Fig. 2.2-1 is

$$A_f \equiv \frac{V_o}{V_i} = -\frac{A - R_o/Z_f}{1 + R_o/Z_f}$$

   (b) Typical values of A, R_o, and Z_f are $A = 5 \times 10^4$, $R_o = 100$ Ω, and $Z_f = 10$ kΩ. If the maximum value of V_o before overload is $V_o(\max) = 10$ V, find the maximum value of V_i which the amplifier will accommodate.

   (c) If $R_i = 100$ kΩ and V_i is the value found in part (b), calculate the current in R_i.

2.2-2 Refer to Fig. 2.2-1.

   (a) Compute the ratio of the current in R_i to the current in Z_f in terms of A, R_i, R_o, and Z_f.

   (b) If $R_i = 10$ kΩ, $Z_f = 10$ kΩ, $R_o = 100$ Ω, and $A = 10^4$, calculate the ratio of the currents.

2.2-3 Refer to Fig. 2.4-1.

   (a) Obtain an expression for the ratio V_i/V_s in terms of R, R_i, R_f, R_o, and A.

   (b) If $R = R_i = R_f = 10$ kΩ, $R_o = 100$ Ω, and $A = 5 \times 10^4$, evaluate this ratio.

   (c) Comment on why the input to the op-amp is considered a virtual ground.

2.3-1 Refer to Fig. 2.3-1b. If the capacitor C has a resistor R_c in parallel with it, determine the transfer function $V_o(s)/V_i(s)$. At what frequency does a pole occur?

2.3-2 An *all-pass network* provides phase shift but no attenuation over the entire frequency band. In the op-amp circuit shown in Fig. P2.3-2 $V_o = A(V_2 - V_1)$. Show that the circuit is an all-pass network by finding the transfer function $V_o(\omega)/V_s(\omega)$.

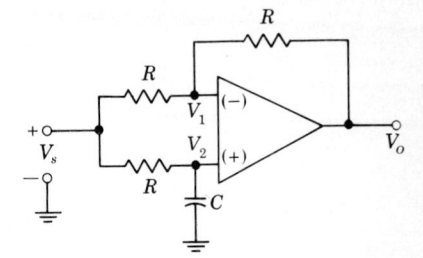

FIGURE P2.3-2

2.4-1 Verify Eq. (2.4-3) by calculating Z_o according to the following procedure (see Fig. P2.4-1): apply a voltage generator V_o to the output terminals of the op-amp and calculate the current I_o supplied to the generator. The output impedance $Z_o = V_o/I_o$.

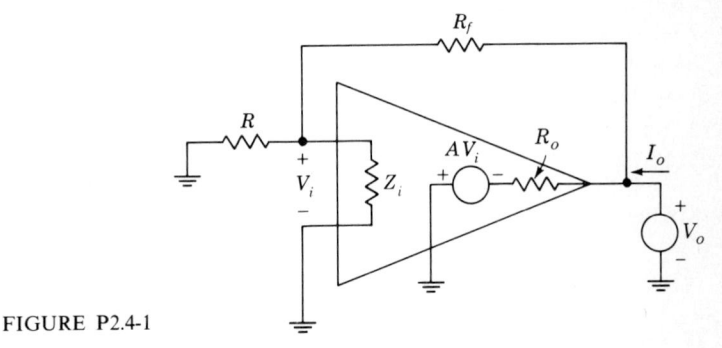

FIGURE P2.4-1

2.5-1 Obtain an expression for the input impedance between the bases of transistors $T1a$ and $T2a$ in Fig. P2.5-1.

(*a*) Show that the result is

$$R_i = 2h_{ie}(Ta) + 2h_{FE}\, h_{ie}(Tb)$$

(*b*) Since $I_{B1b} = h_{FE} I_{B1a}$, show that

$$R_i = 4h_{ie}(Ta)$$

(*c*) If $h_{FE} = 100$ and $I_{E1b} = I_{E2b} = 1$ mA, calculate R_i.

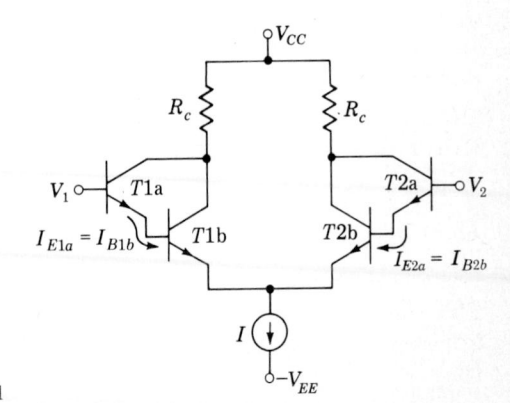

FIGURE P2.5-1

2.7-1 (a) Show that the gain V_o/V_s of the op-amp shown in Fig. 2.7-1, when R_i is infinite, is

$$\frac{V_o}{V_s} = \frac{A}{1 + RA/(R + R_f + R_0)}\left(\frac{R + R_f}{R + R_f + R_0}\right)$$

(b) If $A = 5 \times 10^4$, $R = R_f = 10$ kΩ, and $R_o = 200$ Ω, calculate the gain V_o/V_s.
(c) Show that when A is very large, the result in part (a) reduces to Eq. (2.7.-2).

2.7-2 (a) In Fig. 2.7-1, assume for simplicity that $R_o = 0$. Calculate the ratio V_i/V_s and V_i/V_o. Observe that these ratios are very small and that the virtual short seen looking into the op-amp actually appears between the input terminals.
(b) Let $R_f = R = 1$ kΩ, $A = 10^5$, and $V_s = 10$ V. Calculate V_i and V_o.

2.8-1 (a) Calculate the input impedance of the op-amp configuration shown in Fig. 2.7-1 when the amplifier is designed for unity gain, i.e., when $R = \infty$ and $R_f = 0$. Assume $A = 5 \times 10^4$ and $R_i = 50$ kΩ.
(b) Calculate the input impedance of the amplifier of part (a) when $R = R_f = 1$ kΩ.

2.11-1 Consider that the current source in Fig. 2.5-1 is a real, not ideal, current source to be represented by an ideal current generator shunted by a resistor R_e.
(a) Obtain the small-signal equivalent circuit of the difference amplifier. *Hint:* Modify Fig. 2.5-3 appropriately. Show that with $h_{ib} = h_{ie}/(h_{FE} + 1)$

$$\Delta I_{E1} = \frac{1}{2h_{ib}}(\Delta V_1 - \Delta V_2) + \frac{1}{h_{ib} + 2R_e}\frac{\Delta V_1 + \Delta V_2}{2}$$

and

$$\Delta I_{E2} = \frac{1}{2h_{ib}}(\Delta V_2 - \Delta V_1) + \frac{1}{h_{ib} + 2R_e}\frac{\Delta V_1 + \Delta V_2}{2}$$

(b) Assume that $h_{ib} \ll R_e$. Note that ΔI_{E1} and ΔI_{E2} and therefore ΔV_{C1} and ΔV_{C2} of the difference amplifier respond to the difference voltage, $\Delta V_1 - \Delta V_2$, and also to the common-mode voltage, $(\Delta V_1 + \Delta V_2)/2$. Note also that if R_e becomes infinite, the current due to the common-mode voltage goes to zero.

2.11-2 The CMRR of a difference amplifier is 10,000 (80 dB). If the common-mode signal is $V_c = 10$ V, find the smallest difference voltage V_d that can be amplified before the output voltage due to the common mode exceeds 1 percent of the output voltage due to the difference signal.

2.12-1 The slew rate of the op-amp shown in Fig. 2.7-1 is 50 V/μs. The input signal V_s is a ramp which rises from 0 to 5 V at a rate which is very fast in comparison with the slew rate, so that the input can be considered to be a voltage step. If $R_f = R = 1$ kΩ, sketch (a) the output waveform, (b) the voltage $V_1(t)$, (c) $V_i(t)$. (d) For how long will the amplifier be overloaded?

2.12-2 When a low-frequency sinusoidal waveform is applied to an input of the op-amp shown in Fig. 2.7-1, the amplifier responds linearly over an output range from -10 to $+10$ V. If $R = R_f$ and the slew rate of the amplifier is 50 V/μs, what is the maximum allowable frequency of an input sinusoid if the output signal swing is to be maintained from -10 to $+10$ V without distortion?

2.14-1 In Fig. 2.14-1 $T3$ is intended as a constant-current source. Use a small-signal circuit representation to calculate the resistance seen looking down into the collector of $T3$. Use $h_{ie} = 1$ kΩ, $h_{FE} = 50$, and $1/h_{oe} = 100$ kΩ.

2.15-1 Calculate V_{C5} in Fig. 2.15-1.

2.15-2 Refer to Fig. 2.14-1. Assume that $T5$ is saturated and $V_{CE5}(\text{sat}) = 0.2$ V. (*a*) Calculate I_{C5}, (*b*) show that $V_{B6} \approx 8.1$ V, (*c*) calculate I_{C4} assuming $D2$ is off, (*d*) determine $V_{B4} - V_{B5}$, and (*e*) calculate V_i.

2.15-3 (*a*) Find the value of V_i needed to produce an output voltage $V_o = 0$ V in Fig. 2.14-1. Note that $T5$ is not saturated at this output voltage.
(*b*) Is $D2$ cut off?

2.15-4 Verify Eq. (2.15-11).

2.17-1 Figure P2.17-1 shows a regenerative comparator (Schmitt circuit) in which the capacitor C has been included to acknowledge that any physical amplifier has a limited frequency response. Initially switch S is open, and the comparator is in an initial equilibrium state with $V_s = V_2 = V_o = 0$. Now introduce a small perturbation into the system so that the system is *constrained* (switch S is closed) to reside in the new state $V_s = 0$, $V_o = \Delta V$, $V_2 = \beta \Delta V$.

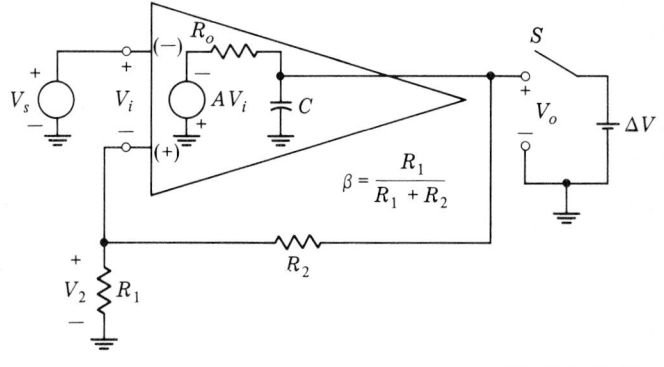

FIGURE P2.17-1

(*a*) Calculate the response of the system when the constraint is removed, that is, S is opened. Show that the system does not return to its initial state but departs farther from its initial state and at a rate which becomes progressively faster. Write an expression for V_o as a function of time.

(*b*) Apply the result of part (a) to the situation represented in Fig. 2.17-3. Let $V_s = V_{s3}$ and assume that the system is in equilibrium. Let V_s be changed abruptly to $V_{s3} + \Delta V$. Describe qualitatively the response of the circuit. What will be the final stable equilibrium state of the system?

2.17-2 Use Eq. (2.17-2) and Fig. 2.17-3 to verify Eq. (2.17-3).

2.17-3 The Schmitt trigger circuit shown in Fig. 2.17-1 has $A_o = 1000$ and $R_1 = R_2 = 1$ kΩ. The output voltage has the levels $V_H = 5$ V and $V_L = 0$ V.
(*a*) Sketch V_o versus V_s.
(*b*) Calculate the hysteresis.

2.17-4 The basic Schmitt trigger circuit of Fig. 2.17-1 has been adapted as in Fig. P2.17-4 to operate as a square-wave generator. Assume that at $t = 0$, $V_C = 0$ V and $V_o = 5$ V.

(a) Sketch $V_o(t)$ and $V_C(t)$ as a function of time.

(b) Note that $V_o(t)$ is not a symmetrical square wave. What should V_R be to make it so?

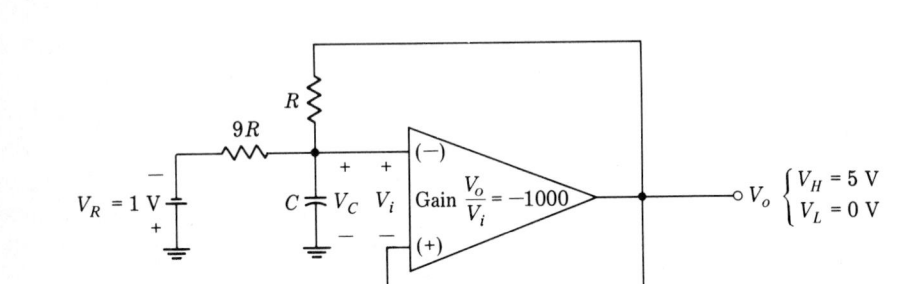

FIGURE P2.17-4

2.18-1 Verify Eq. (2.18-2).

2.18-2 (a) For the circuit of Fig. 2.18-1 find how V_R must be related to the ratio R_1/R_2 for the hysteresis range to be symmetrical with respect to $V_s = 0$ V; that is, in Fig. 2.17-3 $|V_{s4}| = |V_{s5}|$.

(b) If $A_o = 1,000$ and $R_1 = R_2 = 1$ kΩ, calculate the value of the hysteresis for V_R equal to the value found in part (a).

Chapter 3

3.3-1 Draw a circuit showing a lamp that can be lit only if three switches S_1, S_2, and S_3 are simultaneously closed.

3.3-2 Draw a circuit showing a lamp that can be lit only if switch S_1 is closed and switch S_2 is open.

3.4-1 Draw a circuit showing a lamp that can be lit only if switch S_1 or S_2 or both are closed.

3.5-1 Verify Eq. (3.5-5), which states that the NAND operation is not associative. *Hint:* Use a truth table, and let $X = (\overline{AB})C$ and $Y = A(\overline{BC})$. Show that $X \neq Y$.

3.5-2 Verify Eq. (3.5-9).

3.6-1 Show how the switch circuit of Fig. P3.6-1 can be used to perform the EXCLUSIVE-OR operation. *Hint:* The switches must be ganged so that there will be only two independent logical variables to represent the switch positions.

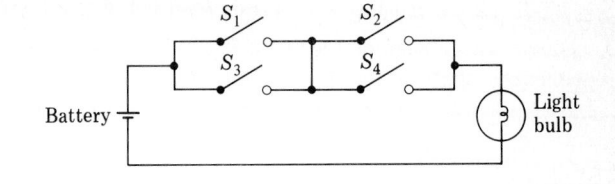

3.6-2 Show that the EXCLUSIVE-OR function can be synthesized using NAND gates.

3.6-3 Show that Z in Eq. (3.6-3) is true if an *odd* number of variables are true and that Z is false if an even number of variables are true.

3.7-1 (*a*) For two independent variables A and B, make a truth table showing all possible functions $Z = f(A, B)$.

(*b*) Show that two of these "functions" are actually not functions at all, being independent of A and B.

(*c*) Show that four of these "functions" are actually functions of only a single one of the variables.

(*d*) The function

A	B	Z
F	F	T
F	T	T
T	F	F
T	T	T

is written $A \supset B$ and is read "A implies B." What is the motivation for this terminology? Show that among the 16 functions are to be found $\overline{A \supset B}$, $\overline{B \supset A}$, and $B \supset A$.

3.8-1 John will go to the movies if Alice will go with him and if he can use the family car. However, Alice has decided to go to the beach if it is not raining and if the temperature is above 80 °F. John's father has made plans to use the car to visit friends if it rains or if the temperature is above 80 °F. Under what conditions will John go to the movies? Construct a special-purpose computer using NOT, AND, and OR gates, switches, battery, and a light bulb to solve the problem. The bulb should light if John can go to the movies.

3.8-2 Buses leave the terminal every hour on the hour unless there are fewer than 10 passengers or if the driver is late. If there are fewer than 10 passengers, the bus will wait 10 min or until the number of passengers increases to 10. If the bus leaves on time, it can travel at 60 mi/h. If the bus leaves late, or if it rains, it can travel at only 30 mi/h. Under what conditions will the bus travel at 60 mi/h? Construct a special-purpose computer using AND, OR, and NOT gates, switches, battery, and a light bulb. The bulb should light if the bus can travel at 60 mi/h.

3.8-3 Seven switches operate a lamp in the following way: if switches 1, 3, 5, and 7 are closed and switch 2 is open, or if switches 2, 4, and 6 are closed and switch 3 is open, or if all seven switches are closed, the lamp will light.

Use NOT, AND, and OR gates to show how the switches must be connected.

3.8-4 Verify that the lamp in Fig. 3.8-1 will light as specified in Example 3.8-1.

3.8-5 The circuit in Fig. P3.8-5 does not make efficient use of logic gates. Find a different circuit having the same transfer function but using fewer gates.

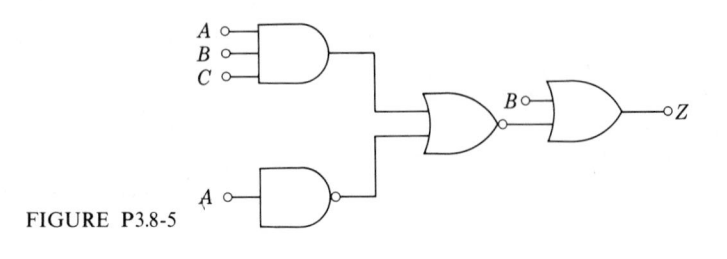

FIGURE P3.8-5

3.8-6 An "economist" proposed the following technique for making money in the stock market:

> *1* If the dividends paid on a stock exceed those paid on a bond, buy the stock.
> *2* If the dividends paid on a bond exceed those paid on a stock, buy the bond unless the growth rate of the stock has been at least 25 percent annually for the past 5 years, in which case the stock should be purchased.

The economist designed a special-purpose computer to tell him what to buy. The computer requires three switches (a higher dividend in the stock, a higher dividend in the bond, growth rate greater than 25 percent) and two lamps, one to light if the stock is selected and the other to light if a bond is selected. Design the computer using NOR logic.

3.10-1 (*a*) Design an EXCLUSIVE-OR circuit using NAND and NOR gates.
(*b*) Repeat using only NOR gates.

3.10-2 (*a*) Design a circuit to light the lamp in Example 3.8-1 using only NAND gates.
(*b*) Repeat using NOR gates.

3.11-1 Verify Eqs. (3.11-6*a*) and (3.11-6*b*) by substituting **0** or **1** for each of the three variables. Note that there are eight possible combinations.

3.11-2 Using Eq. (3.11-6*a*), show that

$$X + WY + UVZ = (X + WY + UV)(X + WY + Z)$$
$$= (X + U + W)(X + U + Y)(X + V + W)$$
$$(X + V + Y)(X + Z + W)(X + Z + Y)$$

3.11-3 Verify Eq. (3.11-9*b*) (*a*) by simplifying the expression and (*b*) by direct substitution of **0** or **1** for each variable.

3.11-4 Verify Eq. (3.11-11*a*).

3.11-5 Verify Eqs. (3.11-12*a*) and (3.11-12*b*).

3.12-1 Simplify the following expressions:
(*a*) $(A + \bar{B})\bar{A}\bar{B}\bar{C}$
(*b*) $A\bar{B}C + \bar{A}\bar{C}D + \bar{C}A$
(*c*) $AB + \bar{A}C\bar{D}E + \bar{B}C\bar{D}$
(*d*) $\bar{A}\bar{B} + AC + \bar{B}\bar{C}\bar{D} + B\bar{C}E + \bar{B}CF + BC\bar{G}$

3.12-2 Complement and simplify the following expressions:
  (a) $\bar{A}B$
  (b) $(\bar{A} + B)(\bar{C}\bar{D})$
  (c) $(A + \bar{A}B)(C + \bar{D})$

3.13-1 Find the binary equivalent for (a) 13, (b) 5, (c) 18, (d) 30. Each binary number should have four digits.

3.13-2 Find the binary numbers for (a) 0.625, (b) 0.278, (c) 18.454.

3.14-1 Write the binary equivalent of the decimal numbers 0 to 31 using the Grey code.

3.15-1 Reduce the expressions given below to a minimum sum of products:
  (a) $(A + \bar{B} + C)(\bar{A} + \bar{B} + \bar{C})$
  (b) $(A + B)(\bar{B} + \bar{A})$
  (c) $AB(C + D)E + (C + D)AC$

3.15-2 Write the expressions of Example 3.15-1 as a standard sum of products.

3.16-1 Find the standard product of sums for
  (a) $AB + \bar{A}C\bar{D}E + \bar{B}CD$
  (b) $AB\bar{C}D + B\bar{C}\bar{E} + AD$
  (c) $\bar{A}\bar{B}C + \bar{A}\bar{B}D + CD$

3.16-2 Write the expressions of Prob. 3.16-1 as a minimum product of sums.

3.17-1 Simplify:
  (a) $f(A, B, C) = \sum m(0, 1, 2, 4)$
  (b) $f(A, B, C) = \sum m(1, 3, 5, 6, 7)$
  (c) $f(A, B, C) = \sum m(0, 1, 2, 3, 4, 5, 6)$

3.17-2 Simplify:
  (a) $f(A, B, C) = \prod M(0, 1, 2, 4)$
  (b) $f(A, B, C) = \prod M(3, 5, 6, 7)$

3.17-3 A function $f = f(A, B, C, D)$. If $f = \sum m(0, 1, 2, 3)$, express f in terms of maxterms.

3.17-4 Show that $\sum m(0, 3, 4, 5)$ and $\prod M(0, 3, 4, 5)$ are complements, while $\sum m(0, 3, 4, 5) = \prod M(1, 2, 6, 7)$.

3.23-1 Minimize the following expressions using the K-map method:
  (a) $\bar{A}\bar{B}C + AD + \bar{D}(B + C) + A\bar{C} + \bar{A}\bar{D}$
  (b) $\sum m(0, 1, 2, 3, 4, 9, 10, 12, 13, 14, 15)$
  (c) $\prod M(0, 1, 2, 3, 4)$
  (d) $B\bar{C} + \bar{A}B + \bar{A}\bar{B}\bar{D} + BCD + A\bar{B}\bar{C}\bar{D}$

3.25-1 If $F = C + \bar{A}B$, plot $\bar{F}$ on a K map. Show that $\bar{F} = B\bar{C} + A\bar{C}$.

3.25-2 Verify that Fig. 3.25-1d represents Eq. (3.25-3b).

3.25-3 Using a minimum number of
  (a) Two-input NAND gates
  (b) Two-input NOR gates
  (c) Three-input NAND gates
  (d) Three-input NOR gates
synthesize the following functions:
  (1) $f = \sum m(0, 1, 2, 3, 8, 9, 10, 11)$
  (2) $f = \sum m(0, 1, 8, 9, 10)$

(3) $f = \prod M(5, 7, 13, 15)$

(4) $f = \prod M(1, 3, 9, 10, 11, 14, 15)$

3.26-1 Using a minimum number of

(a) Two-input NAND gates

(b) Two-input NOR gates

(c) Three-input NAND gates

(d) Three-input NOR gates

synthesize the following functions:

(1) $f(A, B, C, D) = \sum m(0, 1, 4, 5, 9, 11, 14, 15) + \sum d(10, 13)$

(2) $f(A, B, C, D) = \sum m(0, 13, 14, 15) + \sum d(1, 2, 3, 9, 10, 11)$

Chapter 4

4.2-1 Draw circuit diagrams to show how DCTL gates can be connected to perform the logic operations AND and NOT.

4.2-2 Refer to Fig. 4.2-1. Let $V_{CC} = 3$ V and $R_c = 1$ kΩ. Determine the number of transistors N that gate $G0$ can saturate if 0.1 mA of base current is needed to saturate a transistor.

4.2-3 Plot the input-output characteristic of a DCTL gate, $G11$, shown in Fig. 4.2-1, at the temperatures $T = -55$ and 125 °C. (Manufacturers often specify gate characteristics at these temperatures as well as at room temperature.)

4.4-1 Refer to the RTL gate of Fig. 4.1-1. If $h_{FE} = 200$ and transistors $T2$ through TN are OFF, determine approximately the minimum value of V_1 required to obtain an output $V_o = 0.3$ V. Assume that $V_{BE1} = 0.75$ V. Use Fig. 1.10-1 with $h_{FC} = 0.1$.

4.4-2 Refer to the RTL gate of Fig. 4.1-1. Assume that $T2$ through TN are OFF. Suppose that $T1$ may have h_{FE} in the range $50 < h_{FE} < 200$ and that $0.01 \leq h_{FC} \leq 0.1$. Assume that $T1$ is in saturation with $V_{BE1} = 0.75$ V, $V_o = 0.2$ V, and $V_{CC} = 3.0$ V. What is the allowable V_1 which will ensure that $T1$ will be saturated in the worst case? Use Eq. 1.10-3.

4.4-3 Repeat Prob. 4.4-2 for $R_b = 1.5$ kΩ and $R_c = 3.6$ kΩ.

4.4-4 Refer to the gate of Fig. 4.1-1 and to the input-output characteristic of Fig. 4.4-1. Let $T2$ through TN be OFF.

(a) Obtain an expression for the slope of the input-output characteristic $\Delta V_o / \Delta V_1$ as a function of I_C when $T1$ is in the active region. Assume that in the active region the base current is related to the base-emitter voltage by the diode equation.

(b) If $V_{CC} = 3$ V and $h_{FE} = 100$, evaluate the slope when $V_o = 1.6$ V.

4.5-1 Refer to Fig. 4.5-3. Let $R_b = 1.5$ kΩ, $R_c = 3.6$ kΩ, and $V_{CC} = 3$ V. Transistors $T1$ and $T2$ are identical, with parameters $h_{FE} = 50$ and $h_{FC} = 0.1$.

(a) If $T2$ is cut off, find I_{C1} and I_{B1} so that $V_{CE1} = V_{CE2} = 0.2$ V.

(b) If I_{B1} is maintained at the value specified in part (a) and $T2$ is turned on, find the value of I_{B2} required for V_{CE1} to drop to 0.1 V. Obtain the ratio I_{B1}/I_{B2}.

4.5-2 (a) In Fig. 4.5-3, $R_b = 1.5$ kΩ, $R_c = 3.6$ kΩ, $V_{CC} = 3$ V, $T1$ and $T2$ are identical, and $h_{FE} = 50$, while $h_{FC} = 0.1$. $T2$ is OFF and I_{B1} is adjusted to saturate $T1$ with $V_{CE} = 0.2$ V. Find I_{B1}.

(b) Set $I_{B1} = I_{B2}$. Find $I_{B1} = I_{B2}$ for $V_{CE} = 0.2$ V.

(c) Obtain the ratio of I_{B1} for parts (a) and (b). Comment on the result.

4.6-1 (a) Calculate the input-output characteristic of an RTL gate operating with a fan-out of 1 at $T = -55, 25,$ and $125\ °C$. Let $R_b = 450\ \Omega$, $R_c = 640\ \Omega$, $V_{CC} = 3$ V, $h_{FE} = 50$, and $h_{FC} = 0.1$. Let the input voltage V_i increase to $\sigma = 0.1$. Show on the plot the temperature dependence of $V_{CE}(\text{sat})$.

(b) For each temperature determine the noise margin $\Delta 1$ and $\Delta 0$.

4.6-2 Repeat the calculation of the input-output characteristic of Prob. 4.6-1 with $R_b = 1.5\ \text{k}\Omega$ and $R_c = 3.6\ \text{k}\Omega$.

4.6-3 Repeat Prob. 4.6-2 for a fan-out of 5.

4.6-4 Verify Eq. (4.5-4).

4.7-1 (a) Plot the input-output characteristic of the RTL buffer shown in Fig. 4.7-1. Assume a fan-out $N = 1$. Let the temperature $T = -55, 25,$ and $125\ °C$. Make the usual approximations that the transistors turn on at $V_{BE} = 0.65$ and go to saturation at $V_{BE} = 0.75$ and that $V_{CE}(\text{sat}) = 0.2$ V. Assume $h_{FE} = 100$.

(b) Determine the $\Delta 0$ and $\Delta 1$ noise immunities at each temperature.

4.7-2 Plot the output volt-ampere characteristic $(-I_o$ versus $V_o)$ of the RTL buffer in Fig. 4.7-1 when V_i is at logic **0**.

4.7-3 Refer to Fig. 4.7-2. Let $N = 25$ and $V_\sigma = 0.75$ V and assume that $T2$ is saturated. Calculate (a) I_o, (b) the collector current I_C, and (c) the base current I_B.

4.7-4 Verify Eqs. (4.7-1) and (4.7-2).

4.8-1 The circuit of $T2$ and $T3$ of the EXCLUSIVE-OR circuit shown in Fig. 4.8-1 is redrawn in Fig. P4.8-1.

(a) If $V_A = 3$ V, sketch qualitatively the base voltage of $T3$ as V_B varies from 0 to 3 V.

(b) If $V_A = 0$ V, show that $T3$ is off, independently of V_B.

(c) Obtain a truth table to show that $V_o = \overline{V_A \overline{V_B}}$.

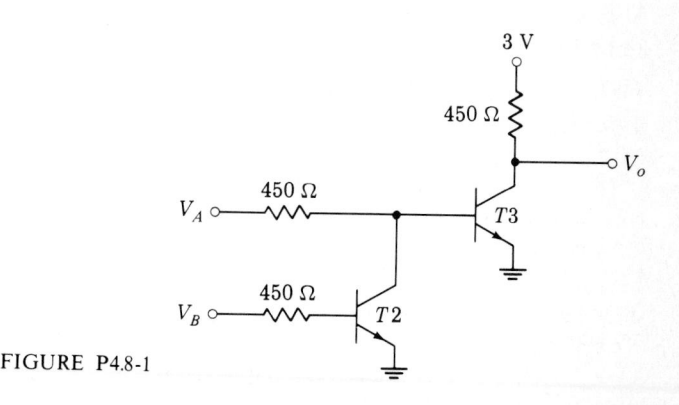

FIGURE P4.8-1

4.9-1 A particular RTL gate has the following characteristics. The maximum input current $I_{in} = 400\ \mu$A, the minimum output current is 2 mA, and the leakage current of each transistor is $I_x = 100\ \mu$A. Each gate has four inputs. Assuming that all gates have the same characteristics, what fan-out can a gate accommodate?

4.10-1 (a) Two identical RTL gates, each with two inputs, are paralleled. Assume that both gates are on the same chip. When operated individually, each gate has associated with it an input loading factor of 1 and an output loading factor of 5. Calculate the input and output loading factors of the combined gate.

(b) Two identical RTL gates, located on different chips, are paralleled. Each gate has two inputs, an input loading factor of 1, and an output loading factor of 5. Calculate the input and output loading factors of the combined gate.

4.10-2 Consider that two buffers, each as shown in Fig. 4.7-1, are paralleled. Let the input to buffer A be in the **1** state and the input to buffer B be in the **0** state. Calculate the current flowing in $T3$ of buffer A.

4.11-1 An RTL gate has the worst-case voltages listed below:

Temp., °C	V_{OH}	V_{IH}	V_{IL}	V_{OL}
−55	1.014	1.01	0.718	0.710
25	0.844	0.815	0.565	0.300
125	0.673	0.67	0.325	0.320

Calculate the worst case $\Delta 0$ and $\Delta 1$ noise immunities.

4.12-1 The input voltage V_1 in Fig. P4.12-1 is an abrupt positive 3-V step. Gates $G1$, $G2$, and $G3$ have propagation delays of 1, 2, and 4 μs, respectively. Draw plots of the waveforms V_1, V_2, V_3, and V_4 on a common time scale.

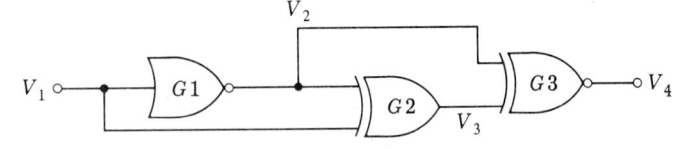

FIGURE P4.12-1

4.13-1 Verify that Fig. 4.13-1 can be redrawn as shown in Fig. 4.13-2.

4.13-2 Verify that Fig. 4.13-2 can be reduced to Fig. 4.13-3.

4.15-1 Redraw Fig. 4.15-1 showing the pnp and npn transistors. Show the eight outputs ABC, $AB\overline{C}$, ..., $\overline{A}\overline{B}\overline{C}$ in your figure and the three inputs A, B, and C.

Chapter 5

Unless otherwise specified, assume $h_{FE} = 50$ and $h_{FC} = 0.1$. Refer to Fig. 1.10-1.

5.1-1 Plot the input-output characteristic of the DTL gate shown in Fig. 5.1-1. Assume that the diode voltage is 0.75 V at a current of 1.5 mA. Plot your results for the temperatures (a) 25 °C, (b) −55 °C, and (c) 125 °C.

5.1-2 When transistor $T2$ of Fig. 5.1-1 is cut off, its input circuit can be approximated by a capacitor $C = 5$ pF. (a) Calculate the time required for the base of transistor $T2$ to reach the cut-in voltage $V_\gamma = 0.65$ V if at $t = 0$, V_i suddenly rises from 0 to 5 V. Assume that diodes $D1$ and $D2$ can be represented by 0.7-V batteries.

5.2-1 Refer to Fig. 5.2-1. Suppose that we specify the extent to which $T2$ is to be driven into saturation by specifying the value of the parameter $\sigma \equiv I_C/h_{FE}I_B$ (see Sec. 1.10) at which this transistor is to operate. Show that the fan-out N that can be accommodated is given by

$$N = \sigma h_{FE} \frac{V_{CC} - 3V_\sigma}{V_{CC} - V_\sigma - V_{CE}(\text{sat})} - \frac{R}{R_c} \frac{V_{CC} - V_{CE}(\text{sat})}{V_{CC} - V_{CE}(\text{sat}) - V_\sigma}$$

where V_σ is the voltage across a forward-biased diode and across the base-emitter voltage of a saturated transistor, that is, $V_\sigma = 0.75$ at 25 °C. Show that for V_{CC} very large $N = \sigma h_{FE} - R/R_c$.

5.2-2 Plot the input-output characteristic of the loaded DTL gate shown in Fig. 5.2-1. Assume that $N = 10$. Let $T = 25$ °C. Assume diode voltage drops as in Prob. 5.1-1.

5.3-1 Refer to Fig. 5.3-1. (*a*) Compare the power dissipated in transistors $T1$ and $T2$ when $\rho = 0$ and $\rho = 1$, (that is, $T1$ is replaced by a diode). Assume $R = 3.75$ kΩ.

(*b*) Calculate the total current supplied by the V_{CC} supply to the IC DTL gate when $V_i = 0$ V.

(*c*) Repeat part (*b*) when $V_i = 5$ V.

5.4-1 (*a*) Calculate the current in diode DA, in Fig. 5.4-1, when $V_i = 0.2$ V. Assume that the voltage drop $V_{DA} = 0.75$ V.

(*b*) Calculate the current in diode DA when $V_P = 2.05$ V, corresponding to the point where $T2$ is just at cut-in.

5.4-2 When V_i in Fig. 5.4-1 is at $V_i = 1.6$ V, DA is cut off and, as indicated in Fig. 5.4-2, $T2$ is in saturation. Show that $T2$ will saturate even before DA cuts off by making the following calculations:

(*a*) Find I_{C2} such that $V_{C2} = 0.2$ V.

(*b*) Use Fig. 1.10-1 to find I_{B2}.

(*c*) Calculate I_D through diode $D2$.

(*d*) Calculate V_P.

(*e*) Determine current I_{DA} and estimate the corresponding diode voltage V_{DA}. Find V_i and note that V_i is less than 1.6 V.

5.4-3 Repeat the calculations of Prob. 5.4-3 for temperatures of -55 and 125 °C.

5.4-4 The Western Electric version of the DTL NAND gate is shown in Fig. P5.4-4.

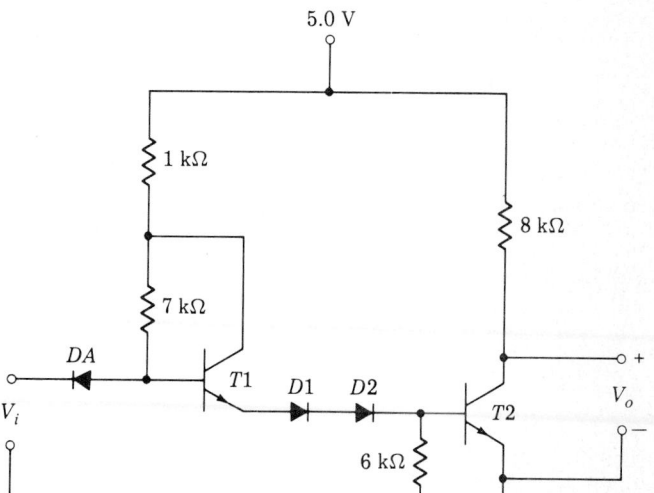

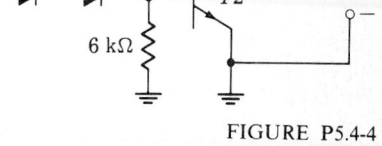

FIGURE P5.4-4

For simplicity only one input is shown. Note that in this circuit two diodes $D1$ and $D2$ are employed. Also note that $\rho = \frac{1}{8}$.

(a) Plot the input-output characteristic of the gate at $T = -55$, 25, and 125 °C.

(b) Determine the $\Delta 0$ and $\Delta 1$ noise margins at the three temperatures -55, 25, and 125 °C.

(c) Determine an approximate relationship between the fan-out N of this gate and ρ corresponding to Eq. (5.3-7) which applies to the circuit of Fig. 5.3-2.

5.4-5 The Western Electric DTL buffer, in which $T5$ is an active pull-up and $T1$ serves as an input diode, is shown in Fig. P5.4-5.

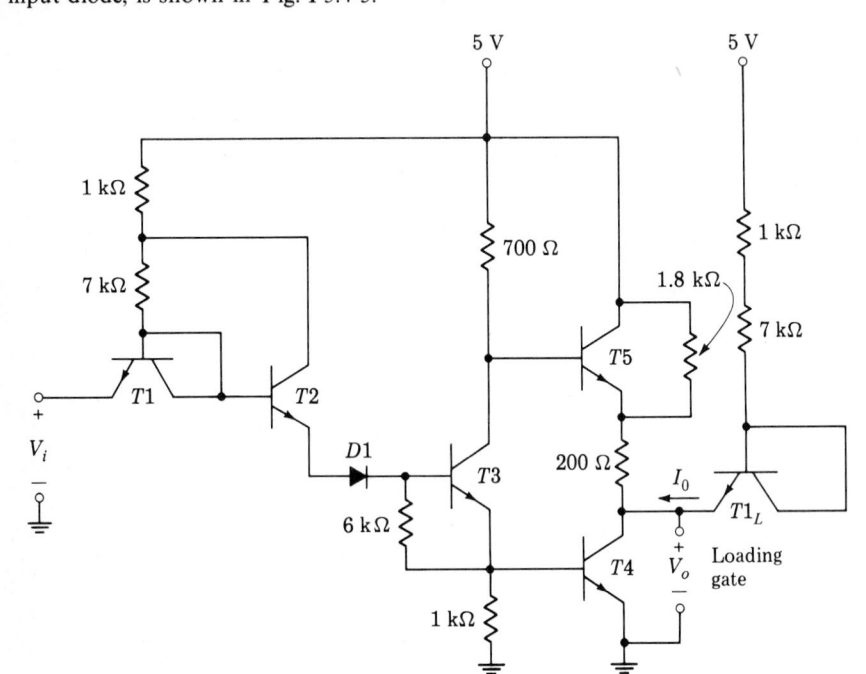

FIGURE P5.4-5

(a) At what voltage V_{B2} will $T2$ just cut in?

(b) At what voltage V_{B2} will $T3$ just cut in?

(c) What are the voltages V_i corresponding to the situation in part (a) and the situation in part (b)?

(d) What, qualitatively, is the purpose of the active pull-up $T5$?

5.6-1 Suppose, in Fig. 5.6-3 that $T1$ is in saturation with $V_{CE}(\text{sat}) = 0.2$ V while all other transistors $T2$, ..., TK are off.

(a) Find I_{C1}, and use Fig. 1.10-1 to find I_{B1}.

(b) Now let both $T1$ and $T2$ be in saturation each with base current as in part (a). Estimate the collector currents and the collector-emitter voltage.

5.8-1 Refer to Fig. 5.7-1. Assume that $T2$ is at cut-in, that is, $V_{BE}(T2) = V_\gamma = 0.65$ V. Let $V_Z = 6.9$ V. Calculate:

(a) $I_E(T1)$.

(b) I_{DA}.

(c) Assume that with the value of current calculated in part (b) that the voltage $V_{DA} \approx 0.75$ V. Calculate the input voltage V_i.

5.8-2 When transistor $T2$ of Fig. 5.7-1 is at the edge of saturation, $V_{CE}(T2) \approx 0.2$ V and $\sigma = 0.85$. Calculate (a) I_{B2} and (b) $I_E(T1)$. With this value of $I_E(T1)$ we can assume that $V_{BE}(T1) \approx 0.7$ V. Calculate (c) I_{DA}. (d) With this value of I_{DA} we can assume that $V_{DA} \approx 0.75$ V. Calculate V_i.

5.8-3 Plot the input-output characteristic of the HTL gate shown in Fig. 5.7-1 at $T = -30$ and 75 °C.

5.9-1 An HTL buffer is shown in Fig. P5.9-1. The buffer is intended to improve the speed of operation of the gate when the gate is heavily loaded capacitatively. Discuss qualitatively the operation of the circuit. Through what path does the load capacitance charge? Through what path does it discharge?

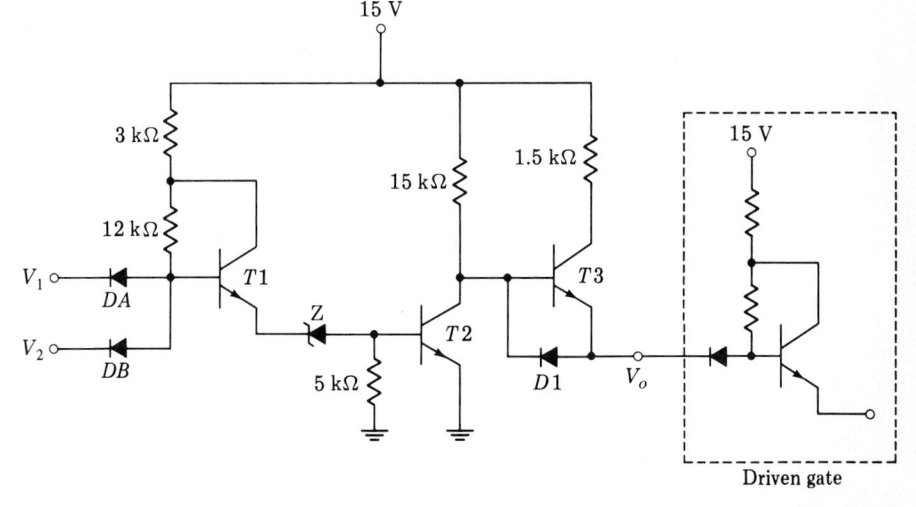

Driven gate

FIGURE P5.9-1

Chapter 6

Unless otherwise indicated, use $h_{FE} = 50$ and $h_{FC} = 0.1$ in all problems. Refer to Fig. 1.10-1.

6.3-1 In Fig. 6.3-2 with the input voltages at logic **1**, $T3$ is in saturation. The stored saturation base charge Q_{Bs} and the stored active base charge Q_{BA}, as shown in Fig. 1.20-1d, are respectively 100 and 150 pC. If one of the inputs is abruptly dropped to 0 V, estimate the time for transistor $T3$ to go from saturation to the edge of cutoff.

6.4-1 Assume in Fig. 6.4-1 that $T2$ and $T3$ are saturated simultaneously. If $V_{CE}(\text{sat}) = 0.2$ V and $V_{BE3} = V_{BE2} = V_D = 0.75$ V, calculate (a) the current in R_{c2} and (b) V_{c4} and V_{E4}. Is $T4$ saturated?

6.5-1 Refer to Fig. 6.5-1. Plot V_o versus V_{B4} as V_{B4} increases from 0 to 1.5 V. Use $R_{c4} = 1$ and 10 kΩ, all other components are as shown. Assume a fan-out of 1.

6.5-2 Refer to Fig. 6.5-1. Plot V_o versus V_{B4} at the temperatures -55 and 125 °C. Assume a fan-out of 1.

6.5-3 Assume that the gate of Fig. 6.5-1 is operating without load.
  (a) At what base current I_{B4} will $T3$ just reach cut-in?
  (b) At what base current I_{B4} will $T2$ just turn off? (Assume here that $T2$ is carrying some minute current.)

(c) Find V_{CE4} under the condition given in part (b).

6.5-4 Using Fig. 6.5-1, calculate and plot I_{E2} as a function of I_{B4}. Assume a fanout of 1.

6.8-1 Calculate and sketch the input volt-ampere characteristic of the TTL gate shown in Fig. 6.5-1 when $T = -55$ and 125 °C.

6.9-1 Calculate and sketch the output volt-ampere characteristic of the TTL gate shown in Fig. 6.5-1 when $T = -55$ and 125 °C when (a) $T3$ is cut off; (b) $T3$ is saturated.

6.10-1 A TTL gate has the following specifications: $V_{IL} = 0.8$ V, $V_{IH} = 1.8$ V, $V_{OL} = 0.4$ V, $V_{OH} = 2.4$ V. Calculate the $\Delta 0$ and $\Delta 1$ noise margins.

6.10-2 The TTL gates shown in Fig. 6.5-1 are operating at $T = 25$ °C and have the specifications given by Figs. 6.10-2 to 6.10-4.
   (a) Estimate the fan-out N if V_o is not to rise above 0.2 V.
   (b) Estimate the fan-out N if V_o is not to fall below 3.5 V.

6.10-3 A TTL gate is guaranteed to sink 10 mA without exceeding an output voltage $V_{OL} = 0.4$ V and to source 5 mA without dropping below $V_{OH} = 2.4$ V. If $I_{IH} = 100$ μA at 2.4 V and $I_{IL} = 1$ mA at 0.4 V, calculate the **0**-state and **1**-state fan-outs.

6.11-1 Refer to Fig. 6.5-1.
   (a) Calculate the power-supply current drain when $V_i = 3.5$ V.
   (b) Repeat part (a) when $V_i = 0.2$ V.
   (c) What is the peak current drawn from the power supply during the brief interval when $T2$ and $T3$ are simultaneously saturated?

6.12-1 In the circuit of Fig. 6.12-1, let each of the inputs be at logic **0** so that $T3$ and $T4$ are OFF. Assume that when a transistor is in the active region its base current may be neglected. Sketch the volt-ampere characteristic seen looking back into the output by determining corresponding source currents I_s and output voltages V_o as follows:
   (a) Find I_s when $V_o = 0$ V.
   (b) Find V_o when I_s just becomes 0 V. At this point $T2$ is at cut-in, that is, $V_{BE2} = V_\gamma$.
   (c) Find V_o and I_s when $V_C(T6) = 4.7$ V. Note that $T6$ is not saturated.
   (d) Find V_o and I_s when $I_{C6} = 0$. Note that $T6$ is saturated.
   (e) Verify that $T2$ never saturates.

6.12-2 If the output voltage of the TTL gate shown in Fig. 6.12-1 is to be $V_o = 2.4$ V, calculate the output current available when $T3$ is off.

6.13-1 Consider that the TTL gate shown in Fig. 6.5-1 has an open collector output; i.e., transistor $T2$ and diode D are removed. Now let two such gates be connected in the WIRED-AND configuration and a passive pull-up resistor R_o be connected between the collectors and the 5 V supply.
   (a) Calculate the base current supplied to the output transistor $T3$ of one of the gates when the inputs to that gate are high and the inputs to the other gate are low.
   (b) If both output transistors are off, calculate R_o so that 4 mA will flow into a load when $V_o = 2.4$ V.
   (c) If the gate drives N gates, each requiring 100 μA at 2.4 V, calculate the fan-out N if $I_{OH} = 4$ mA at 2.4 V.
   (d) If an input to one of the two WIRED-AND gates is low while the inputs to the other gate are high, the output V_o will be low. If $V_{OL} = 0.4$ V, when $I_{OL} = 10$ mA, the leakage current in transistor $T3$ of the "off" gate is 100 μA and it is necessary to sink 1 mA from each drain gate. Find R_o to ensure a fan-out of 5.

6.14-1 Sketch, qualitatively, the input-output characteristic of the TTL AND gate shown in Fig. P6.14-1.

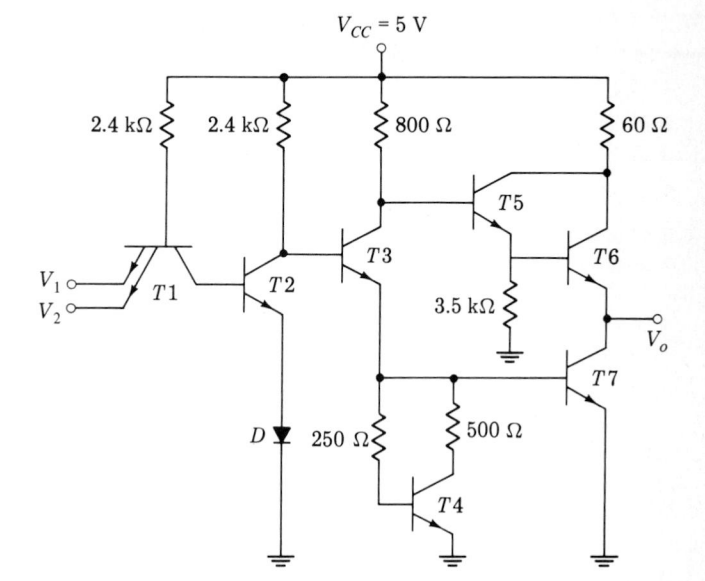

FIGURE P6.14-1

6.14-2 Write an expression which gives the logic level at the output in terms of the logic levels of the input V_1, V_2, V_3, and V_4 in Fig. P6.14-2.

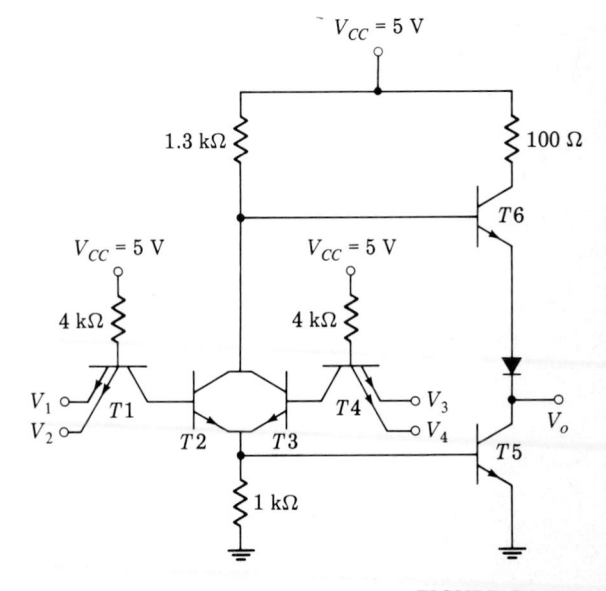

FIGURE P6.14-2

6.14-3 Sketch, qualitatively, the input-output characteristic of the TTL NOR gate shown in Fig. P6.14-3.

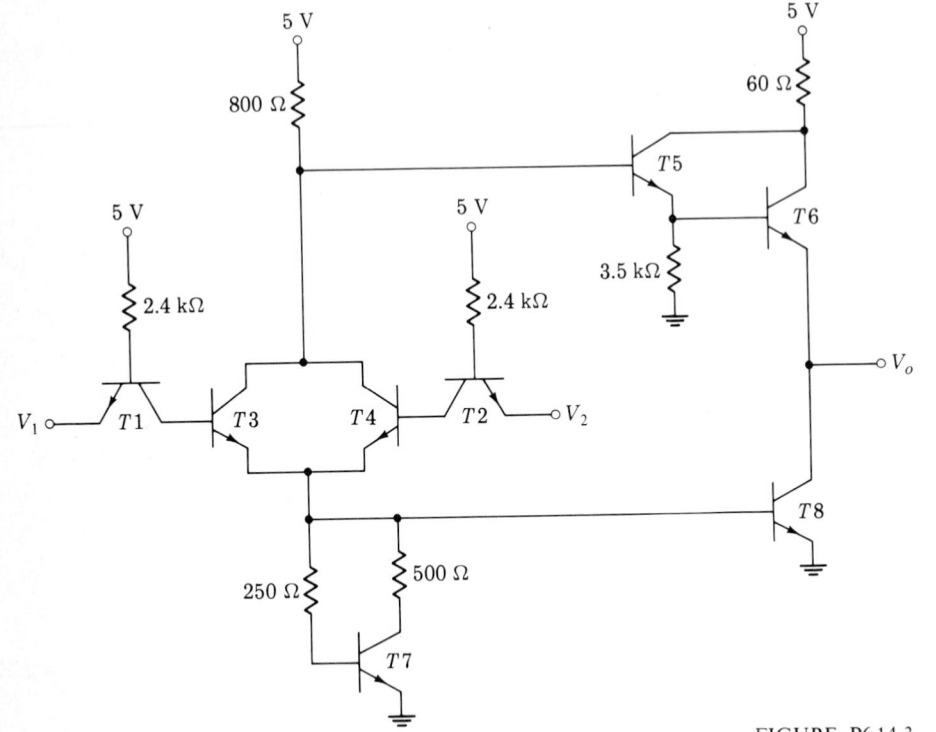

FIGURE P6.14-3

6.14-4 Sketch, qualitatively, the input-output characteristic of the TTL OR gate shown in Fig. P6.14-4.

Chapter 7

Unless otherwise indicated use $h_{FE} = 50$ and $h_{FC} = 0.1$ in all problems.   Refer to Fig. 1.10-1.

7.2-1 Refer to Fig. 7.1-1. Find the output voltage V_{o2} as a function of V_R, V_{CC}, R_{c2} and R_e when (a) $V_i \gg V_R$ and (b) $V_i \ll V_R$.

7.4-1 Verify that as shown in Fig. 7.4-1 the values of V_{o2} at the edges of the transition region are -0.79 and -1.54 V, corresponding to V_i equal to -1.1 and -1.25 V.

7.4-2 A MECL II gate has $R_{c1} = 150\ \Omega$, $R_{c2} = 170\ \Omega$, and $R_e = 650\ \Omega$. Obtain the transfer function V_{o2} versus V_i for a MECL II OR gate. Does the characteristic differ from Fig. 7.4-1a?

7.5-1 (a) Plot the NOR-gate input-output characteristic of the MECL II gate described in Prob. 7.4-2.

(b) Determine the $\Delta 0$ and $\Delta 1$ noise margins of the gate.

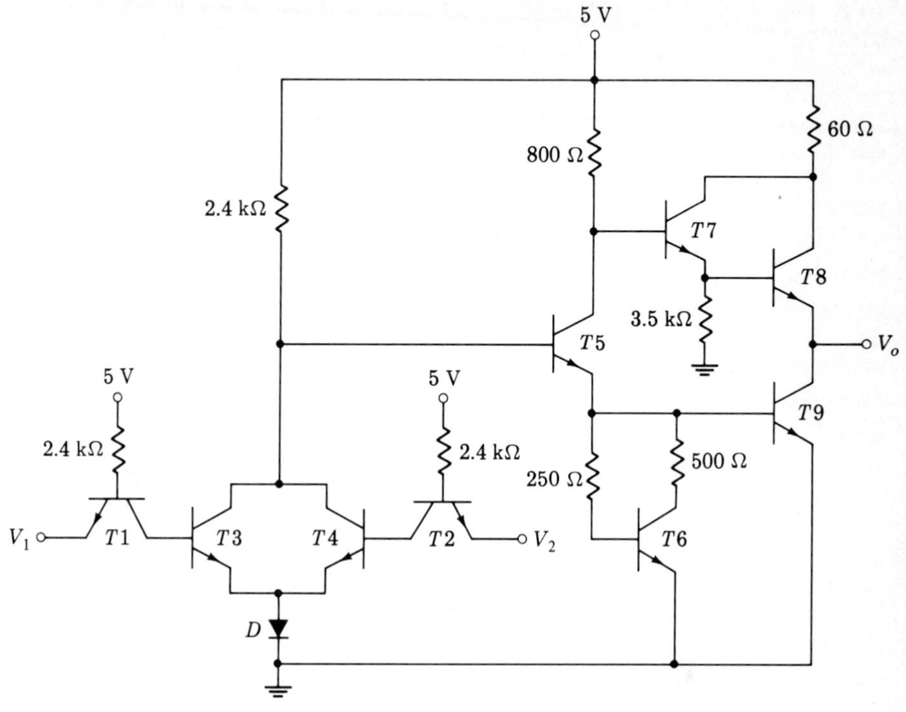

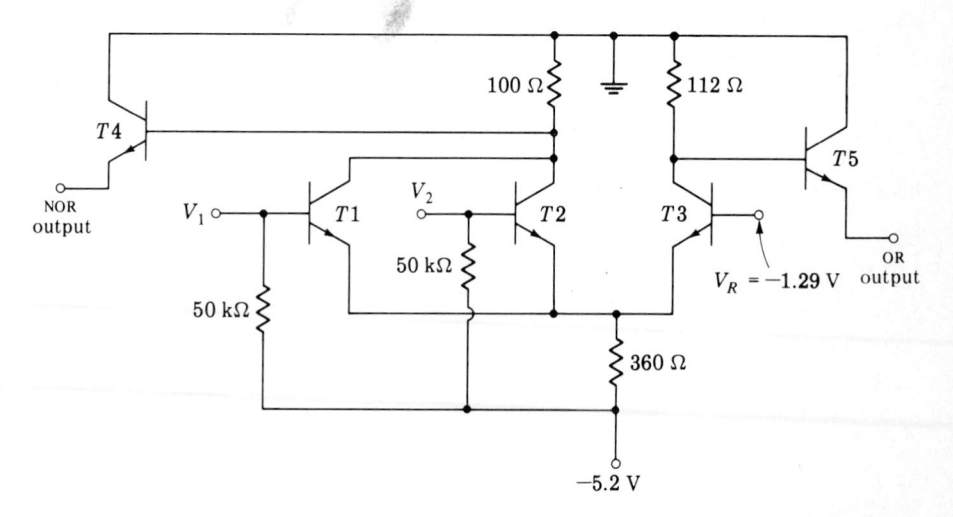

FIGURE P6.14-4

7.6-1 (*a*) Plot the input-output characteristic of the MECL III OR gate shown in Fig. P7.6-1. Assume that a fan-out of 1 is employed.

(*b*) Calculate the Δ**0** and Δ**1** noise margins.

FIGURE P7.6-1

7.6-2 Repeat Prob. 7.6-1 for the MECL III NOR gate shown in Fig. P7.6-1.

7.7-1 Determine the fan-out of the ECL gate shown in Fig. 7.7-1 if the $\Delta 1$ noise margin is allowed to fall to only 0.1 V.

7.9-1 Consider that the ECL NOR gate shown in Fig. 7.2-1 is using the bias circuit shown in Fig. 7.9-1.

    (a) Sketch the input-output characteristic of the ECL NOR gate at $T = -55$ and 125 °C.

    (b) Determine the $\Delta 0$ and $\Delta 1$ noise margin at each temperature.

7.9-2 (a) Sketch the input-output characteristic of the ECL OR gate shown in Fig. 7.2-1 at $T = -55$ and 125 °C.

    (b) Determine the $\Delta 0$ and $\Delta 1$ noise margins at each temperature. Compare your results with the $\Delta 0$ and $\Delta 1$ noise margins found when $T = 25$ °C.

7.9-3 The temperature-compensated bias circuit for the MECL II gate of Fig. P7.6-1 is shown in Fig. P7.9-3. Assume a fan-out of 5. Calculate, for the ECL bias circuit and OR gate of Fig. P7.9-3 and for a temperature change ΔT, the following changes:

    (a) ΔV_R.

    (b) $\dfrac{\Delta V_{o2}(1) + \Delta V_{o2}(0)}{2}$.

    (c) Compare (a) and (b) at $T = -55$ and 125 °C.

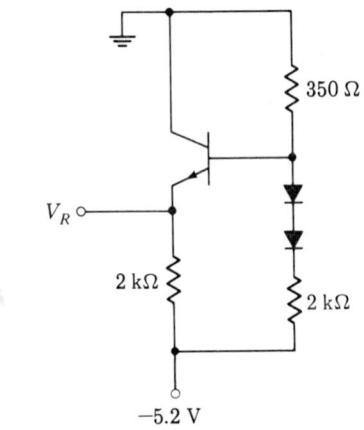

FIGURE P7.9-3   −5.2 V

7.9-4 Repeat Prob. 7.9-3 for the ECL NOR gate of Prob. 7.9-3.

7.9-5 Consider the gate of Fig. 7.2-1 using the bias circuit of Fig. 7.9-1. Assume that $T = 25$ °C but V_{EE} can vary by an amount ΔV_{EE}.

    (a) Calculate ΔV_R as a function of ΔV_{EE}.

    (b) Calculate $\dfrac{\Delta V_{o2}(1) + \Delta V_{o2}(0)}{2}$ as a function of ΔV_{EE}.

    (c) Compare (a) and (b) by plotting the ratio

$$\lambda = \frac{2\Delta V_R}{\Delta V_o(1) + \Delta V_o(2)}$$

as a function of ΔV_{EE} as ΔV_{EE} varies from −1 to 1 V. Comment on the result.

7.11-1 In Sec. 7.11 we note that in ECL logic, noise introduced in series with the supply voltage appears at the output attenuated to a greater extent if the positive side of the supply is grounded than if the negative side is grounded. The intent of this problem is to show that in RTL logic no such advantage occurs.

Consider the two RTL gates shown in Fig. P7.11-1a. Let $V_{CC} = 4$ V, $R_b = 1$ kΩ, and $R_c = 2$ kΩ. If $V_1 = V_2 = 0$ V, $T3$ is saturated.

(a) Calculate the value by which V_{CC} must change for $T3$ to be brought to the edge of cut-in.

(b) Refer to Fig. P7.11-1b. Let $V_{EE} = 4$ V, $R_b = 1$ kΩ, and $R_c = 2$ kΩ. If $V_1 = V_2 = -4$ V, $T3$ is saturated. Calculate the value by which V_{EE} must change for $T3$ to be at cut-in.

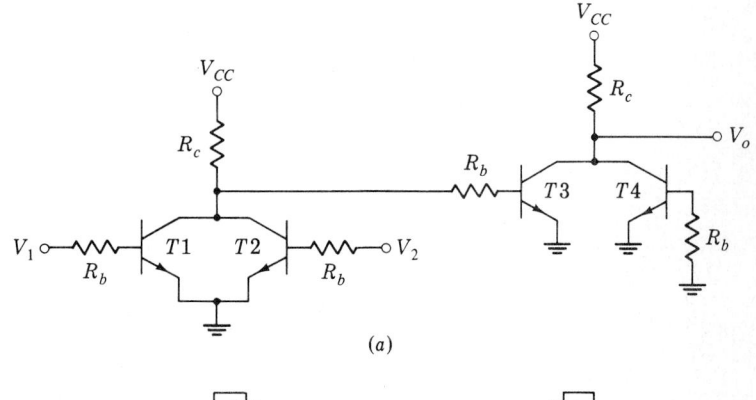

(a)

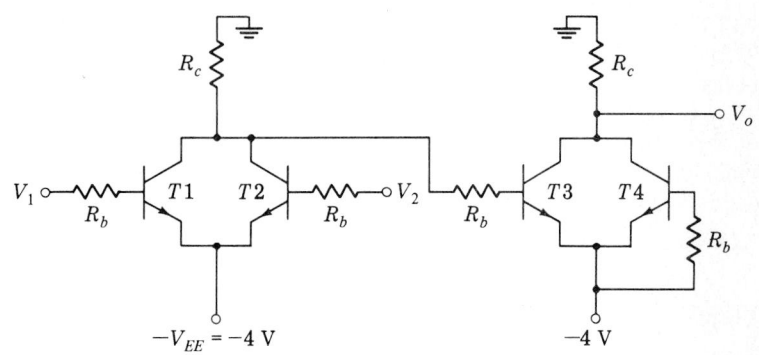

(b)

FIGURE P7.11-1a

7.11-2 Calculate V_o and V_o', in Fig. 7.11-1b, due to V_n when $T1$ is OFF and $T2$ is ON. Assume $R_e = 1.8$ kΩ and $V_R = -1.175$ V.

7.12-1 (a) Refer to Fig. 7.12-1. Assume $V_2 = 5$ V. Let $V_1 = 3.5$ V, corresponding to logic **1** in TTL. Calculate V_{B1} and V_o.

(b) Find V_{B1} and V_o when $V_1 = 0.2$ V (logic **0**).

7.12-2 The TTL to ECL translator shown in Fig. P7.12-2 is used with MECL II.

(a) Show that logically $V_{ECL} = V_1 \cdot V_2$.

(b) For $V_2 = 5\,V$ (logic **1**) determine and sketch the transfer characteristic V_{ECL} vs V_1.

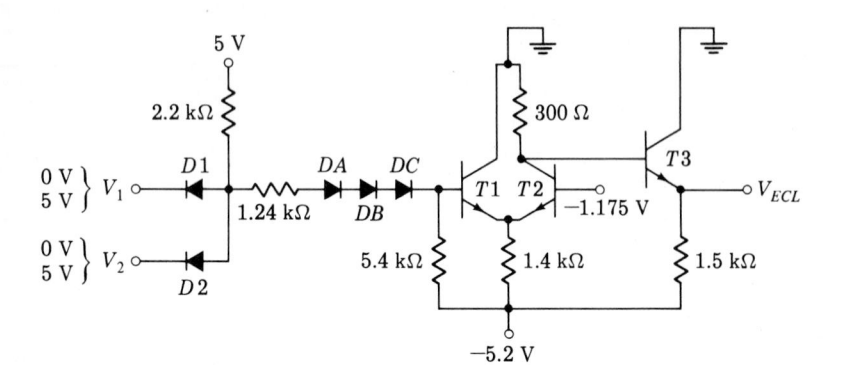

FIGURE P7.12-2

7.12-3 What logic operation is performed by the circuit shown in Fig. 7.12-3?

7.12-4 (a) Plot the input-output characteristic for the translator shown in Fig. 7.12-3. Let $T2$ and $T4$ be cut off and $V_3 = -1.175\,V$; sketch V_o as a function of V_1.

(b) What are the $\Delta 0$ and $\Delta 1$ noise margins?

7.13-1 Refer to Fig. 7.13-1b. Calculate the voltage levels of $V_o(t)$ if $V_i(t)$ is a voltage step of 1 V amplitude, $R_o = 50\,\Omega$, $R_s = 5\,\Omega$, and $R = 100\,k\Omega$.

7.13-2 Referring to Fig. 7.13-4, consider that the characteristic impedance of the line is 50 Ω, the output impedance of the driving amplifier is 5 Ω, and the input impedance of the drive stage can be approximated by an open circuit.

(a) Sketch the input voltage to the driven gate if the open-circuit output voltage of the driving gate is a 1-V pulse of duration τ less than the one-way propagation time of the line.

(b) Repeat part (a) for a driving voltage that is a unit step.

7.13-3 Verify the waveforms shown in Fig. 7.13-5b and c.

Chapter 8

8.1-1 Plot I_{DS} as a function of V_{DS} from Eqs. (8.1-2) and (8.1-3) using V_{GS} as a parameter. Let $k = 1\,mA/V^2$ and $V_T = 2\,V$. Let $V_{GS} = 2$, 3, and 4 V.

8.1-2 (a) What is the drain-to-source voltage V_{DS} of an n-channel enhancement-mode FET if the gate is connected to the drain and $I_{DS} \simeq 0$?

(b) Show that, quite independently of the value of I_{DS}, the FET is in saturation if $V_{GD} = 0$.

8.1-3 The NMOSFET shown in Fig. P8.1-3 has the parameters $k = 1\,mA/V^2$ and $V_T = 2\,V$.

(a) Sketch V_o as a function of V_{DD} as V_{DD} varies from 0 to 20 V.

(b) Repeat part (a) if $R_L = 100\,\Omega$ rather than 1 kΩ.

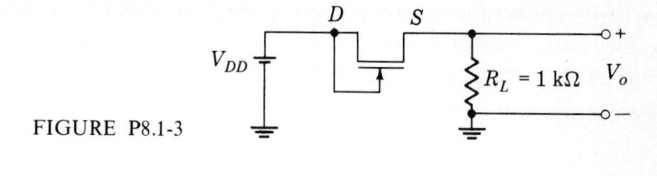

FIGURE P8.1-3

8.1-4 Repeat Prob. 8.1-3 for an NMOS depletion-mode FET having a threshold voltage $V_T = -2$ V.

8.1-5 A depletion-mode NMOSFET is operating in the circuit shown in Fig. P8.1-3 but is reconnected so that the gate is tied to the source rather than to the drain. Sketch V_o and I_{DS}, each as a function of V_{DD}, as V_{DD} varies from 0 to 20 V. Assume $V_T = -2$ V and $k = 1$ mA/V^2.

8.1-6 A PMOSFET and an NMOSFET are connected in parallel as shown in Fig. P8.1-6.

(*a*) Sketch V_i as a function of I_i if the FETs are enhancement-mode devices each with a threshold voltage $V_T = 2$ V and with the same value of $k = 1$ mA/V^2. Let V_i vary from -20 to 20 V, and let $V_{GG} = 5$ V.

(*b*) Estimate the effective resistance seen looking into the circuit.

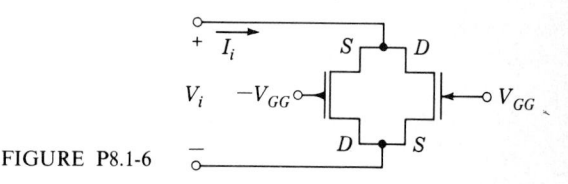

FIGURE P8.1-6

8.1-7 The enhancement-mode NMOS FETs shown in Fig. P8.1-7*a* and *b* are identical. The circuits in which they are employed differ since the substrate of the FET in (*a*) is connected to the source while the substrate of the FET in (*b*) is connected to ground. The threshold voltage is $V_T = 2 + \sqrt{|V_{BS}|}$, where V_{BS} is the voltage drop from substrate to source. If $k = 1$ mA/V^2,

(*a*) Plot V_o versus V_i for both circuits (Fig. P8.1-7*a* and *b*).

(*b*) Compare the sketches obtained in part (*a*).

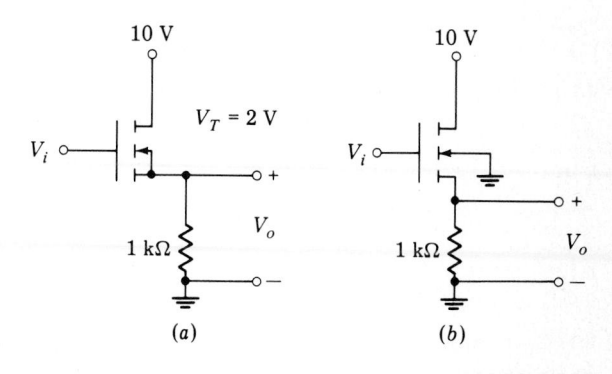

FIGURE P8.1-7

8.3-1 Obtain equations and plot the input-output characteristic of the inverter shown in Fig. 8.3-1a. Let $k_1 = 40k_2$.

8.3-2 Repeat Prob. 8.3-1 for the inverter shown in Fig. 8.3-1b. Prove that the load transistor is in saturation.

8.5-1 Verify Eq. (8.5-8).

8.5-2 Verify Eq. (8.5-9).

8.6-1 Verify the truth tables of Figs. 8.6-1 and 8.6-2.

8.6-2 What would the logic operations be if in Figs. 8.6-1 and 8.6-2 the NMOSFETs were replaced by PMOSFETs?

8.6-3 Consider stacking two BJT transistors as shown in Fig. P8.6-3. V_1 and V_2 are each either 0 or 5 V. If the saturation voltage of each transistor is 0.2 V, how many transistors can be stacked? Remember that if V_o is to be at logic level **0**, its voltage must be less than 0.65 V.

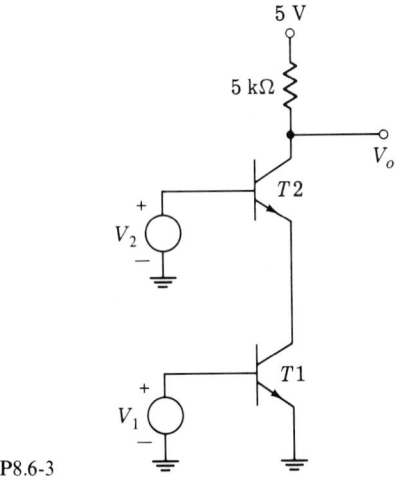

FIGURE P8.6-3

8.6-4 Determine the logic operations that would be performed by the gates shown in Fig. 8.6-3a and c if NMOSFETs were employed instead of PMOSFETs.

8.7-1 Verify Eq. (8.7-3).

8.8-1 Verify Eq. (8.8-4).

8.10-1 Verify that for a CMOS gate the values of the rise time and the fall time are the same. Find these times for $C_L = 5\,\text{pF}$, $k_L = k_D = 1\,\text{mA/V}^2$, $V_T = 2\,\text{V}$, and $V_{SS} = 10\,\text{V}$. Assume an ideal step input.

8.12-1 Design a circuit which will translate the output logic levels of CMOS to the logic levels required at the input to ECL.

8.12-2 Repeat Prob. 8.12-1 for ECL driving a CMOS.

8.12-3 Verify Eqs. (8.12-1).

Chapter 9

9.1-1 Consider a cross-coupled gate configuration as in Fig. 9.1-1 except that OR gates are used.

(*a*) Show that when the data input terminals enable both gates, the two output terminals may both remain permanently at logic **0** or at logic **1** and that hence the circuit constitutes a flip-flop, i.e., a device with two permanent states.

(*b*) Show that by manipulation of the data terminals it is possible to induce a state transition in one direction but not in the other.

(*c*) Repeat parts (*a*) and (*b*) for AND gates.

(*d*) If one gate is an AND gate and the other is an OR gate, show that transitions either way can be induced.

9.4-1 Design an *RS* flip-flop using NAND gates, where the input terminals are *S* and *R* (not $\bar{S}$ and $\bar{R}$) and which satisfies the truth table shown in Fig. 9.4-1*c*. How many NAND gates are needed?

9.4-2 (*a*) Refer to Fig. 9.1-1*a*. Assume that $R = 0$ and that a pulse of width τ appears at the *S* terminal so that $S = 1$ for a time τ. If the propagation delay time in each gate is t_{pd}, sketch the output *Q* as a function of time if *Q* is initially **0** for:

(1) $\tau > 2t_{pd}$

(2) $t_{pd} < \tau < 2t_{pd}$

(3) $\tau < t_{pd}$

(*b*) Repeat part (*a*) for the NAND gate shown in Fig. 9.4-1*a*.

9.6-1 Design a clocked *SR* flip-flop using NOR gates such that the direct inputs S_d and R_d override the clock.

9.7-1 Show that if two master-slave flip-flops, each of which is as shown in Fig. 9.7-2, are cascaded, the input to the first flip-flop will appear at the output of the second flip-flop after two clock intervals have elapsed.

9.7-2 If, in Fig. 9.7-3, $\bar{S}_d = 1$ and $\bar{R}_d = 0$, show that $Q = 0$ and $\bar{Q} = 1$ regardless of the logic state of the inputs *S*, *R*, and *C*.

9.7-3 Obtain a truth table for the flip-flop shown in Fig. 9.7-3 if the direct inputs $\bar{R}_d$ and $\bar{S}_d$ are removed from the master flip-flop. That is, with $\bar{R}_d$ and $\bar{S}_d$ applied only to the slave flip-flop, find *Q* for all combinations of $\bar{R}_d$ and $\bar{S}_d$ and for $C = 1$ and **0**. Include all possibilities.

9.8-1 (*a*) Refer to Fig. 9.1-1 and consider that ECL gates are being employed. Assume initially that $R = S = 0$ so that $R = S = -1.58$ V (see Fig. 7.4-1). Estimate to what voltage one or the other of the data terminals must be raised in order to effect a flip-flop transition.

(*b*) Next refer to Fig. 9.8-1. Let one of the dynamic inputs be derived from the output of a driving ECL gate. Assume that this input changes from logic **0** to logic **1** linearly in a time T. Let the *rC* time constant be 1 ns. Neglect the propagation delays in the flip-flops and ignore any loading at the *R* and *S* inputs. What is the maximum rise time *T* for which the flip-flop will respond?

9.8-2 The flip-flop in Fig. 9.8-1 uses ECL gates with logic levels -1.58 and -0.76 V. Assume that the flip-flop will make a transition when the inputs *R* or *S* pass through the threshold voltage -1.175 V. The gate propagation delays are 1 ns. The inputs at $\bar{R}_D$ or $\bar{S}_D$ are abrupt steps from logic **0** to logic **1**. Find the minimum time constants

rC to assure that R or S will remain above the threshold long enough to make it certain that the flip-flop will respond.

9.8-3 Consider that DTL NAND gates are used in Fig. 9.8-3. Then $\bar{S}$ (and $\bar{R}$) must go from 5 V to some voltage less than, say, 1.3 V to ensure a change of state in the output. Let S_D decrease linearly from 5 V to 0 V in a time T. Let $rC = 1$ ns, and find the maximum value of T for the flip-flop to respond. Neglect the propagation delay time of the gates.

9.9-1 Verify the truth tables given in Fig. 9.9-1b, c, and d.

9.10-1 Refer to Fig. 9.10-1. The logic levels are 0 and 5 V. The time constant for the charging of a capacitor is 100 ns, and for discharging is 40 ns.

(*a*) Let $S = 1$ and $C = 0$ initially. At $t = 0$, C rises abruptly to logic **1**. How long must $\bar{C}$ remain at logic **1** if S' is to rise to 2 V (2 V and above is logic **1**)?

(*b*) Consider that C has been at logic level **1** for a time sufficiently long to ensure that S has reached 5 V. Assume that S' must remain above 2 V so that when $\bar{C} = 1$, the output of gate 2A will be at logic level **0**. What is the maximum allowable propagation delay time of the gate to ensure an output transition when clock $\bar{C}$ returns to logic **1**? Assume that all gates have the same propagation delay times.

9.11-1 Refer to Fig. 9.11-1a. Let A vary from 0 V (logic **0**) to 5 V (logic **1**) linearly in a time T_R. Let the propagation delay time of gates 1 and 2 be t_{pd}, and let the output of gate 1 decrease linearly from 5 to 0 V in a time T_F. Assume that to ensure that the output of gate 2 goes to logic **1**, inputs A and C must simultaneously exceed 2 V. Obtain an expression for the time T_o during which the two inputs to gate 2 simultaneously remain above 2 V in terms of T_F, T_R, and t_{pd}.

9.11-2 Verify the truth table given in Fig. 9.11-2b.

9.12-1 A JK toggle flip-flop is to be constructed using the ac-coupled edge-triggered flip-flop shown in Fig. 9.9-1. Redraw the figure to show the connection of Q and $\bar{Q}$ to the input gates 1A and 1B.

9.12-2 Repeat Prob. 9.12-1 using the capacitance-storage flip-flop shown in Fig. 9.10-1.

9.12-3 Repeat Prob. 9.12-1 using the propagation-delay flip-flop shown in Fig. 9.11-2.

9.13-1 In the circuit of Fig. 9.13-2 replace every gate by a NOR gate and change $\bar{S}$ and $\bar{R}$ at the inputs to gates 3A and 3B to S and R, respectively. Show that the circuit remains a type D flip-flop by verifying that:

(*a*) When $C = 0$ or when $C = 1$, changes in D have no effect on the output.

(*b*) When $D = 0$, a transition of C from **1** to **0** results in Q going to or remaining at $Q = 0$.

(*c*) When $D = 1$, a transition of C from **1** to **0** results in Q going to or remaining at $Q = 1$.

(*d*) A transition of C from **0** to **1** results in no output change.

9.15-1 Show that the flip-flop shown in Fig. 9.15-5 can function as a master-slave flip-flop.

9.17-1 If, in Fig. 9.17-2, $S = R = -1.58$ V and $Q = 1$ while $\bar{Q} = 0$, calculate voltage values for Q, $\bar{Q}$, V_{E1}, and V_{E2}.

9.17-2 A high-speed ECL flip-flop capable of toggling at 85 MHz is shown in Fig. P9.17-2. In this flip-flop circuit, diodes $D1$, $D2$, their associated resistors, and transistors TRA and TRB provide reference voltage V_R (not available in the flip-flop shown in Fig. 9.17-2).

Calculate:

    (a) V_R

    (b) The output voltage levels at Q and $\bar{Q}$ with $S = R = 0$

    (c) V_{E1} and V_{E2} at $S = R = 0$

    (d) V_{E1} and V_{E2} when $S = 1\ (-0.76\ \text{V})$ and $R = 0\ (-1.58\ \text{V})$

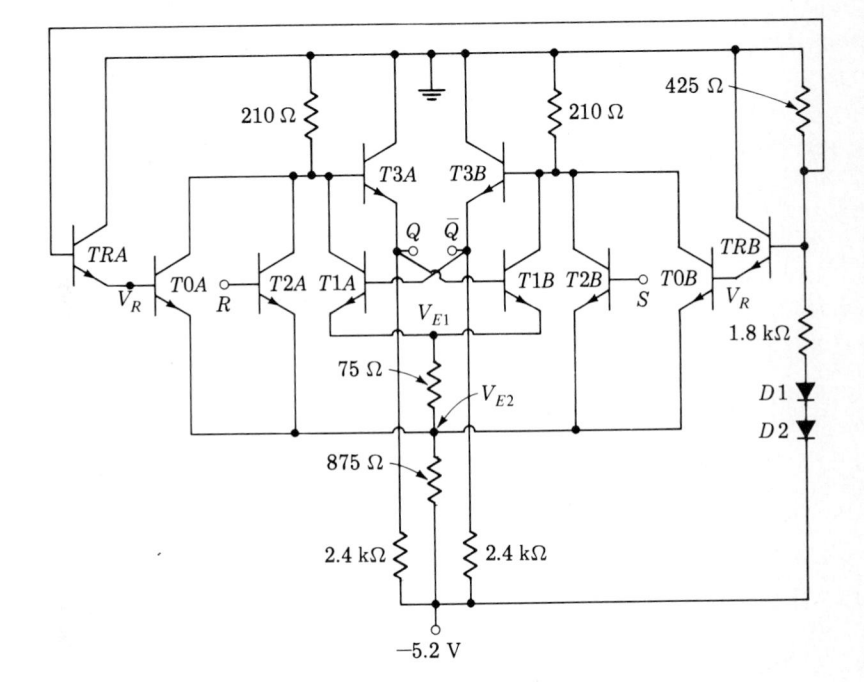

FIGURE P9.17-2

9.18-1 Demonstrate that the flip-flop shown in Fig. 9.18-1 corresponds to the gate structure of the flip-flop of Fig. 9.9-1.

9.18-2 Show that the truth tables given in Fig. 9.18-2 describe the operation of the ECL flip-flop shown in Fig. 9.18-1.

9.18-3 Verify the quiescent voltage levels given in Fig. 9.18-3 for $\bar{J} = \bar{K} = 0$, $\bar{Q} = 1$, and $Q = 0$.

9.20-1 Referring to Fig. 9.20-2, show that $T4A$ is at cut-in when the voltage at the emitter of $T2B$ is approximately 1 V.

9.20-2 Referring to Fig. 9.20-2 with $J = K = \bar{C} = \bar{Q} = 1$, determine the approximate base voltage and base current of $T4B$ at steady state, i.e., when the current in capacitor C_A is zero.

9.20-3 A TTL edge-triggered JK flip-flop capable of toggling at a 50-MHz rate is shown in Fig. P9.20-3. The circuit does not involve capacitors. Show that nonetheless it is an edge-triggered flip-flop. Derive the truth table for the circuit.

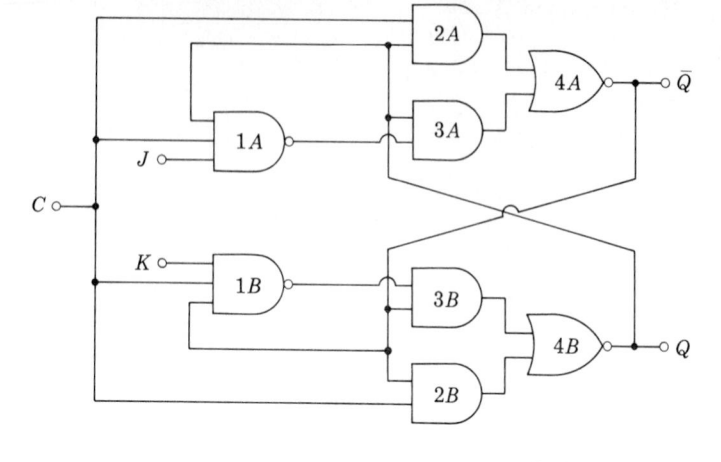

FIGURE P9.20-3

9.21-1 Using Fig. 9.21-3, verify that the input to switch S_z is $Z = Q\overline{K} + J\overline{Q}$.

9.21-2 Show that the input gate configuration in Fig. 9.21-3 with inputs Q, J, and K and output Z can be realized as in Fig. P9.21-2.

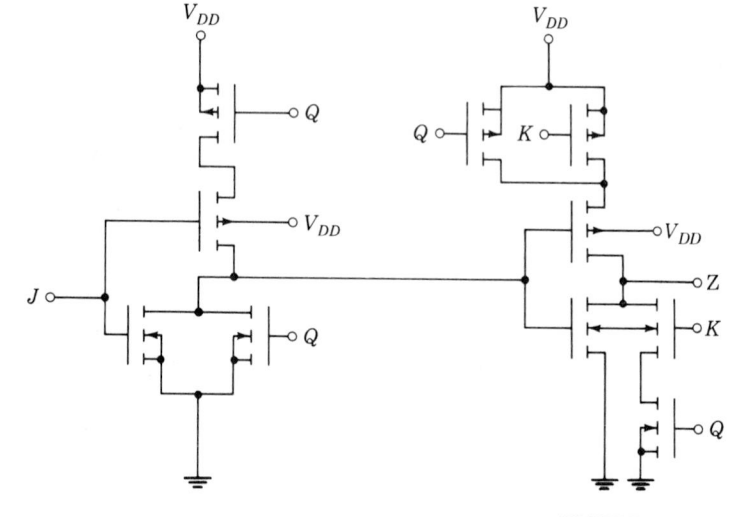

FIGURE P9.21-2

9.21-3 Verify the JK truth table in Fig. 9.21-3b.

Chapter 10

10.1-1 Consider the 4-bit shift register shown in Fig. 10.1-1a. If an input sequence **1010110** is applied to input D_o synchronously with the clock, sketch the output of each flip-flop. Assume each $Q = 0$ initially.

10.1-2 Consider the 4-bit register shown in Fig. 10.1-1b. Assume that initially $Q_0 = 1$, $Q_1 = 0$, $Q_2 = 1$, $Q_3 = 1$. Sketch the output of each flip-flop if an input sequence **101101** is applied to D_0 synchronously with the clock.

10.4-1 Figure P10.4-1 shows a three-stage shift register with end-around carry. The system is designed so that the input signal is disconnected from the register input D_0 and the output Q_2 is connected to D_0 after the third clock pulse. New information can be inserted into the register only after applying a *direct-reset* pulse to the counter, $FF3$ and $FF4$. After applying the reset pulse, 3 new bits of data can be fed into the register.

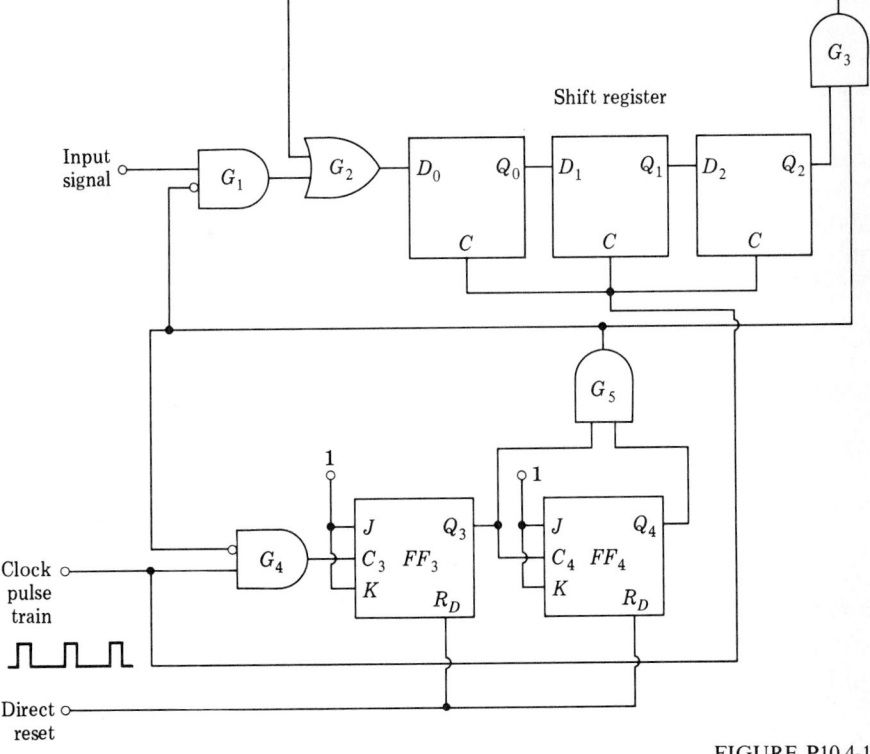

FIGURE P10.4-1

(a) Show that if Q_3 and Q_4 are both in the **0** state initially, the output of gate $G5$ will become **1** after three clock pulses.

(b) Show that once the output of $G5$ is **1**, the clock pulses are inhibited from the JK flip-flops $FF3$ and $FF4$.

(c) Show that after the third clock pulse, the input signal is inhibited from the register and Q_2 is connected into input D_0.

(d) Redesign the system assuming that only NAND gates are available.

10.5-1 Each steering circuit shown in Fig. 10.5-1 consists of two AND gates and an OR gate, say gates $1A$, $1B$, and $1C$.

(a) If only NAND gates are available, redesign the steering circuit.

(b) Repeat if only NOR gates are available.

(c)   When using CMOS gates both NAND and NOR gates may be used together. Is any economy achieved by using both NAND and NOR gates to construct a steering circuit? Show your design.

10.6-1   In Fig. 10.6-2 if the strobe is always in the **1** state, sketch the "real" and "ideal" outputs K_1 to K_7. The waveforms for Q_0, Q_1, and Q_2 are shown in Fig. 10.6-3. Assume that the propagation delay time of each flip-flop is 20 percent of the clock period.

10.6-2   A ripple counter is to operate at a maximum frequency of 10 MHz. If the propagation delay time of each flip-flop in the counter is 10 ns and the strobing time is 50 ns, how many stages can the counter have?

10.6-3   A 10-stage ripple counter is constructed using individual flip-flops, each having a propagation delay time of 5 ns. The strobing time is to be 50 ns. What is the maximum frequency of the counter which will still allow correct reading?

10.7-1   The propagation delay time of an ECL gate is 1 ns and of an ECL flip-flop is 2 ns. If the strobing time is 5 ns and a five-stage counter is desired, compare the maximum counting frequency of a ripple counter and a synchronous serial-carry counter.

10.7-2   Consider the synchronous-parallel counter shown in Fig. 10.7-2. Assume that the relationship between the propagation delay time and the fan-out of any of the flip-flops used is

$$t_{pd} = t_{pd0} + \lambda F$$

where λ is a constant of proportionality and F is the required fan-out. If an N-stage counter is to be designed with a strobing time T_s, show that the maximum clock frequency is

$$f_{\max} = \frac{1}{t_{pd0} + \lambda(N - 1) + T_s}$$

In particular, if $t_{pd0} = 8$ ns, $\lambda = 1$ ns, $N = 5$, and $T_s = 20$ ns, compute $f_{\max}$.

10.9-1   Verify the truth table for the JK flip-flop given in Table 10.9-2.

10.9-2   (a)   Verify the K maps shown in Fig. 10.9-1.
(b)   Show that when the K maps are read, Eq. (10.9-1) results.
(c)   Show that another possible solution is $J_0 = \bar{Q}_1$, $K_0 = \bar{Q}_1$, $J_1 = Q_0$, $K_1 = \bar{Q}_0$.

10.9-3   Two synchronously operated flip-flops are interconnected using the connections $J_0 = \bar{Q}_1$, $K_0 = \bar{Q}_1$, $J_1 = Q_0$, and $K_1 = \bar{Q}_0$. Derive the resulting counter configuration and tabulate its counting sequence. What is the modulo of the counter?

10.10-1   Verify the K maps shown in Fig. 10.10-1.

10.10-2   (a)   Show that a reading of Fig. 10.10-1 is $J_0 = \bar{Q}_2$, $K_0 = \bar{Q}_2$, $J_1 = Q_0$, $K_1 = Q_0$, $J_2 = Q_0 Q_1$, and $K_2 = \bar{Q}_0$.
(b)   Find the resulting mod-5 counter configuration.

10.10-3   (a)   Obtain K maps for the J and K inputs of the three flip-flops used to construct the mod-5 counter described by the state-assignment table given in Table 10.10-2.
(b)   Verify Eq. (10.10-2).

10.10-4 Consider the following state assignment table for a mod-5 counter:

S	Q_2	Q_1	Q_0
0	0	0	0
1	0	0	1
2	0	1	1
3	1	1	1
4	1	1	0

Derive a counter configuration.

10.11-1 Show that *lockout* will not be encountered with any of the unused states of the counter of Fig. 10.10-2.

10.11-2 Design a mod-5 counter using the state table given by Prob. 10.10-4. Your design should include circuitry to ensure that if we end up in an unused state, the next clock pulse will reset the counter to $Q_2 Q_1 Q_0 = \mathbf{000}$.

10.11-3 Design a mod-7 counter. Incorporate in your design a provision to ensure that if an unused state occurs, the next clock pulse will reset the counter to $Q_2 Q_1 Q_0 = \mathbf{000}$, which is to be a used state.

10.12-1 Design a decade counter by cascading a mod-5 and a mod-2 counter. Let the mod-5 counter be connected to the input and the mod-2 counter follow at the end of the cascade.

   (*a*) Show the decoding logic needed for the configuration.

   (*b*) Which decoding logic requires fewer gates, the logic needed in this system or in the system shown in Fig. 10.12-1?

10.12-2 Design a mod-6 counter using a mod-2 counter and a mod-3 counter. Show the decoding logic required.

10.12-3 Design a mod-9 counter by cascading two mod-3 counters. Show the required decoding logic.

10.13-1 Design a mod-3 counter using the technique outlined in Sec. 10.13, that is, by using the direct set and/or reset inputs.

10.13-2 Design a mod-5 counter using the technique outlined in Sec. 10.13.

10.13-3 Design a mod-9 counter using the technique outlined in Sec. 10.13.

10.13-4 Ripple-type counters of arbitrary modulo may be constructed by adding gates to a counter array of flip-flops in a manner to cause the counter to bypass some states. There is no special design procedure so that some ingenuity is required. Examples are given in Fig. P10.13-4. All J and K terminals not explicitely indicated are fixed at logic level **1**.

   (*a*) Verify that the counter in (*a*) has a modulo of 5 and that the counter in (*b*) has a modulo of 9.

   (*b*) Try your hand at a modulo-7 counter.

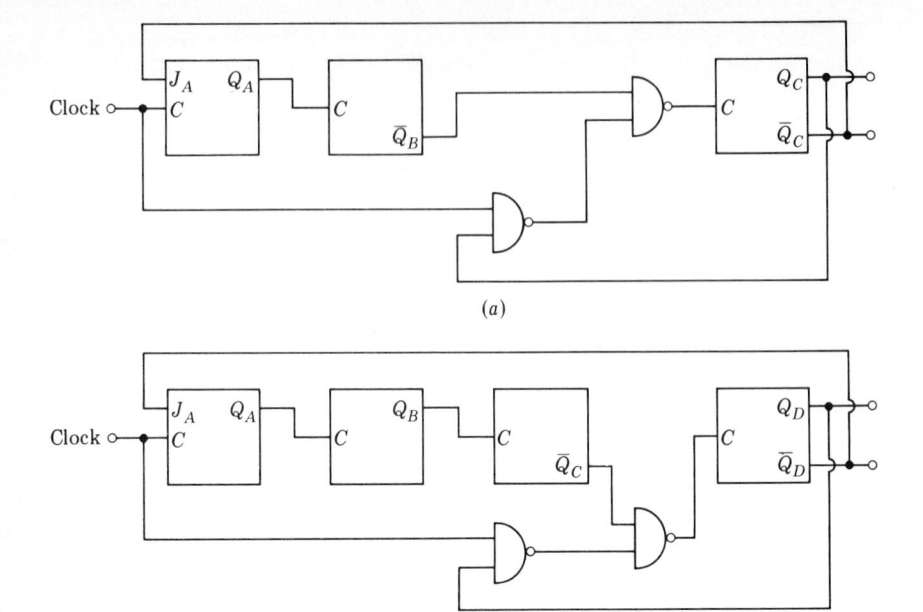

(a)

(b)

FIGURE P10.13-4

10.13-5 In the counter circuit of Fig. P10.13-5 the flip-flops are $\bar{J}\bar{K}$ flip-flops such as are encountered in ECL. A clock input is made available by joining one of the $\bar{J}$ inputs with one of the $\bar{K}$ inputs. The flip-flops respond to the clock when the clock makes a transition from **0** to **1**. They have the additional feature that, if all of the $\bar{J}$ and $\bar{K}$ terminals are at logic **0** and if then any one $\bar{K}$ input should make a transition from **0** to **1**, the flip-flop will reset (or stay reset). Similarly, a transition of a $\bar{J}$ input from **0** to **1** will set the flip-flop (or leave it set). In the counter shown all $\bar{J}$ and $\bar{K}$ inputs not shown are at logic **0**.

(a) Verify that, if the connection to $\bar{J}_0$ is ignored, the first N flip-flops constitute a forward-counting modulo-2^N counter.

(b) Show that by virtue of the addition of the last flip-flop and the feedback connection to $\bar{J}_0$ the counter has a modulo $2^N + 1$.

(c) Using the principle illustrated, and using flip-flops only (no external gates) construct a modulo-3 counter and a modulo-5 counter.

(d) Construct a modulo-7 counter [*Hint:* $7 = 2 \times (2 + 1) + 1$], a modulo-11 counter [$11 = 2 \times (2^2 + 1) + 1$], and a modulo-25 counter [$25 = 8 \times (2 + 1) + 1$].

(e) Construct an algorithm which will lead to a counter of arbitrary modulo.

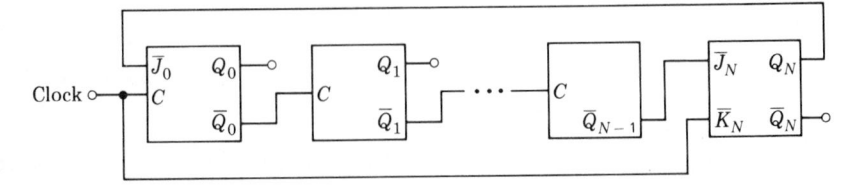

FIGURE P10.13-5

10.14-1 Refer to the four-stage forward-ripple counter shown in Fig. 10.6-1. Show that if C_1 is connected to $\bar{Q}_0$, C_2 to $\bar{Q}_1$, and C_3 to $\bar{Q}_2$, the counter outputs $Q_3 Q_2 Q_1 Q_0$ will count down. That is, if initially we set $Q_3 Q_2 Q_1 Q_0$ to **0000**, the next count will be **1111**, then **1110**, etc.

10.14-2 Refer to the UP-DOWN ripple counter shown in Fig. 10.14-1. Assume that we have been counting in the forward direction and after the kth input signal the count is $Q_3 Q_2 Q_1 Q_0 = $ **0101**. At this time the counting direction is reversed; that is, M becomes $M = \mathbf{0}$.

    (a) Find the value recorded by the counter just after M becomes $M = \mathbf{0}$ and before the next counting input arrives.

    (b) Find the value recorded by the counter after the $(k + 1)$st input.

10.14-3 Show that the UP-DOWN counter in Fig. 10.14-2 can be reversed at any count without error. As an example, consider that after the kth forward count, the counter reads $Q_3 Q_2 Q_1 Q_0 = $ **0101**. At this time the counting direction is reversed so that $M = \mathbf{0}$.

    (a) Find the counter reading just after M has changed to $M = \mathbf{0}$ but before the $(k + 1)$st input pulse.

    (b) Find the counter reading after the $(k + 1)$st input pulse.

10.15-1 Verify Eq. (10.15-2).

10.15-2 Design an UP-DOWN synchronous decade counter.

10.15-3 Design an UP-DOWN mod-9 counter. Use two mod-3 counters in cascade.

10.16-1 (a) Verify the truth table given in Fig. 10.16-2b.

    (b) Verify the decoding logic tabulated for this decade counter.

10.16-2 Assume that because of noise the decade ring counter of Fig. 10.16-2 finds itself in the state $Q_5 Q_4 Q_3 Q_2 Q_1 = $ **10101**. Determine the resulting counter states during the next 10 input pulses. Show that lockout results.

10.16-3 Refer to Fig. 10.16-3. The addition of the AND gate makes $J_1 = \bar{Q}_4 \cdot \bar{Q}_5$.

    (a) Starting with $Q_5 Q_4 Q_3 Q_2 Q_1 = $ **00000**, show that the counter states given in the truth table of Fig. 10.16-2b result.

    (b) If, because of noise, the counter finds itself in the state $Q_5 Q_4 Q_3 Q_2 Q_1 = $ **10101**, show that the counter will eventually return to an allowable state.

    (c) There are 22 unused states for the counter shown in Fig. 10.16-3. Show by direct manipulation that we shall always eventually return to a used state. For example, if $Q_5 Q_4 Q_3 Q_2 Q_1 = $ **01000**, the next state will be $Q_5 Q_4 Q_3 Q_2 Q_1 = $ **10000**.

10.17-1 (a) Design a maximal-length sequence generator having a sequence length $S = 31$.

    (b) Show that the correlation function of any flip-flop output, say $Q_1(k)$, is

$$R_{Q_1(i)} = \sum_{k=1}^{31} Q_1(k) Q_1(k + i) = \begin{cases} 31 & i = 0 \\ -1 & i \neq 0 \end{cases}$$

To obtain this result let 1 V represent logic **1** and -1 V represent logic **0**.

10.17-2 Design a sequence generator to generate the sequence **101101110110** at Q_1.

    (a) What sequence appears at D_1?

    (b) What sequence appears at Q_2?

    (c) How many flip-flops are needed to uniquely determine D_1 as a function of $Q_1, Q_2, \ldots$?

    (d) Determine the logical expression for $D_1 = f(Q_1, Q_2, \ldots)$.

10.17-3 Design a sequence generator to have a length $L = 10$. Such a generator can be used as a decade counter. Show the decoding logic needed if the generator is to perform as a counter. Note that a counter can always be designed by constructing the appropriate sequence generator. Such a sequence generator need not be *maximal length*.

Chapter 11

11.3-1 The serial adder of Fig. 11.3-1 has the numbers **01101** and **00111** stored in the two 5-bit addend and augend registers, respectively. Make timing charts of the operation of the adder through the interval of six clock cycles. Start by drawing the clock waveform, and underneath on the same time scale draw the waveforms of the full-adder output, the type-D flip-flop output, and the bits stored in the sum register.

11.4-1 (*a*) Add the numbers $A = 01101$ and $B = 00111$ using the parallel adder shown in Fig. 11.4-1. Determine the sum $S = C_{n-1}S_{n-1}S_{n-2} \cdots S_0$.
  (*b*) If the propagation delay time of each 1-bit full adder is τ, calculate the time delay between the presentation of the numbers to be added and the appearance of the complete sum at the output.

11.5-1 The numbers **0010, 0011, 0101** are presented, in sequence, in synchronism with a clock waveform to the accumulator in Fig. 11.5-1. In each clock cycle specify the registration of each of the three registers.

11.5-2 Verify Eq. (11.5-2).

11.5-3 The number $X_2(0) X_1(0) X_0(0) = 011$ has been placed in the register of Fig. 11.5-2 during the zeroth clock cycle. Thereafter the number $X_2(1)X_1(1)X_0(1)$ **101** is presented to the inputs and is to be added to the first number during the first clock cycle. Show that the completed sum $S = S_3 S_2 S_1 S_0$ does not appear until after all the carries have rippled through the accumulator.

11.6-1 Design a combinatorial circuit (gates only, no flip-flops) which accepts two 2-bit numbers and yields the sum. Assume that the bits and their complements are available. Note that in such a scheme there is no carry ripple.

11.6-2 Repeat Prob. 11.6-1 for three 2-bit numbers.

11.6-3 Verify Eq. (11.6-1).

11.6-4 Refer to Fig. 11.6-1. If the propagation delay in each full adder is T_A and in gates G_1 to G_4, G_0, and G_A it is T_G, calculate the maximum propagation delay between C_{in} and C_{out}.

11.6-5 Refer to Fig. 11.6-2b. Repeat Prob. 11.6-4 and find the maximum delay between C_0 and C_8.

11.6-6 In a 24-bit adder use a bypass circuit for each 4 bits of addition, as in Fig. 11.6-2. Calculate the worst-case delay time in the adder. Assume that the delay in each bypass is comparable to the delay in each 1-bit full adder.

11.7-1 Repeat Prob. 11.6-1 for subtraction.

11.7-2 Repeat Prob. 11.6-2 for subtraction.

11.8-1 Numbers are to be stored in an eight-place register. Negative numbers are to be represented by the twos-complement representation.

(a) Indicate the registration in the register when the decimal numbers are 14, 20, 67, −1.

(b) Repeat part (a) for the case where negative numbers represented in the ones-complement representation.

11.9-1 Subtract the number $B = 0011$ from $A = 1010$ using the mod-16 circular register shown in Fig. 11.9-1d. Why is the result incorrect?

(b) Repeat using twos-complement arithmetic; use Fig. 11.9-1c.

11.9-2 Subtract the number $B = 0111$ from $A = 1010$ using ones-complement arithmetic:

(a) Find B_1 the ones complement of B.

(b) Show that the difference $D = A - B = A + B_1 + 1$.

11.10-1 (a) Design a circuit to convert a 4-bit number expressed in ones complement into a 4-bit number expressed in twos complement.

(b) Design a circuit to convert a 4-bit number which is in twos complement into a 4-bit number in ones complement.

11.11-1 Using 4-bit arithmetic, an extra sign bit, and ones-complement arithmetic, add

(a) 10 and 5

(b) 10 and −5

(c) −10 and 5

(d) −10 and −5

11.13-1 Modify the adder shown in Fig. 11.5-1 so that a sequence of positive and negative numbers, expressed in ones complement, can be accumulated.

11.13-2 Show that when a sequence of positive and negative numbers is added, the result will be correct if the final result is within the range of the adder, even though overflow may have occured during the sequence.

(a) Add the numbers 6, 4, and −7 using 4-bit registers and ones-complement arithmetic. Repeat using twos-complement arithmetic.

(b) Refer to the circular register shown in Fig. 11.3-1c. Let each number be in twos complement. Show that $6 + 4 - 7 = +3 + 0011$, as expected.

(c) Repeat part (b) using ones complement (Fig. 11.9-1d).

11.14-1 Prove, in general, that if overflow occurs as a result of adding two positive numbers, the sign bit of the sum will be **1**. Repeat for a sum of two negative numbers to show that the sign bit of the sum will be **0**.

11.14-2 Derive the saturation logic needed for an adder using ones-complement arithmetic. Show the required circuitry.

11.14-3 (a) Design a circuit having an input X and an output $|X|$. X is given in ones complement format.

(b) Repeat part (a) if X is in twos-complement format.

11.15-1 If $Y = 2X$ and X and Y are expressed using 4 bits, form a table showing X, Y, and the error E caused by overflow as X varies from **1000** to **0111**. Assume X is in a twos-complement format.

11.15-2 The error obtained due to scaling can be reduced by using saturation logic. Let $Y = 4X$, where X and Y are each expressed using 4 information bits and a sign bit. X and Y may be positive or negative and are expressed using twos-complement arithmetic. Design a system and an input X and an output Y which satisfies the equation

$$Y = \begin{cases} 15 & X \geq 4 \\ 4X & |X| \leq 3 \\ -15 & X \leq -4 \end{cases}$$

as X varies from -15 to 15.

11.15-3 Using twos-complement arithmetic, a number X is stored in a 5-bit register. The number is scaled to replace the registration of X by Y, where $Y = X/4$. The error caused by discarding the two least significant bits may be larger than one-half the numerical significance of the least significant remaining bit. For example, if $X = \mathbf{01111} = 15$, then $Y = \mathbf{00011} = 3$, while the correct result is $\frac{15}{4} = 3.75$. Rounding off the correct answer should yield $4 = \mathbf{00100}$.

Design a logic system which rounds off Y to the nearest integer.

11.15-4 Repeat Prob. (11.15-2) if ones-complement arithmetic is used.

11.15-5 Repeat Prob. (11.15-3) if ones-complement arithmetic is used.

11.15-6 Verify Fig. 11.15-2a and b.

11.15-7 Verify Fig. 11.15-2c and d.

11.15-8 Design a system having an input X and output $Y = 2X + 3$. X and Y each use 7 information bits and a sign bit and employ twos-complement arithmetic. Saturation logic is to be used. (See Prob. 11.15-2.)

11.15-9 Repeat Prob. (11.15-8) using ones-complement arithmetic.

11.15-10 In a synchronous system $X(k)$ is the input during the kth clock cycle. The output $Y(k)$ is given by the equation

$$Y(k) = X(k) + \tfrac{1}{2}Y(k-1)$$

X and Y are each expressed using twos-complement arithmetic, and each has 4 information bits and a sign bit. Saturation logic is to be used. Ignore errors which result from truncation due to scaling. (This system is a recursive first-order digital filter.)

11.15-11 Design a system where the input X is related to the output Y by the equation

$$Y = \tfrac{3}{4}X$$

X and Y each have 4 information bits and a sign bit.

11.17-1 Design a divider, using repeated subtraction. Employ twos-complement arithmetic.

Chapter 12

12.2-1 The registers shown in Figs. 12.2-1 and 12.2-2 are *synchronous registers*; i.e., the bits in the register shift only when a clock pulse occurs. It is often important to shift quickly to a new word N words advanced from the word initially present at the register output. The register shown in Fig. P12.2-1 is capable of shifting 1, 2, or 3 bits in an asynchronous manner. Instructions concerning the number of bits to be shifted are provided by logic levels at inputs S_1 and S_2. (By providing additional inputs S_3, etc., the number of bits can be increased.) The time required to shift N bits depends not on

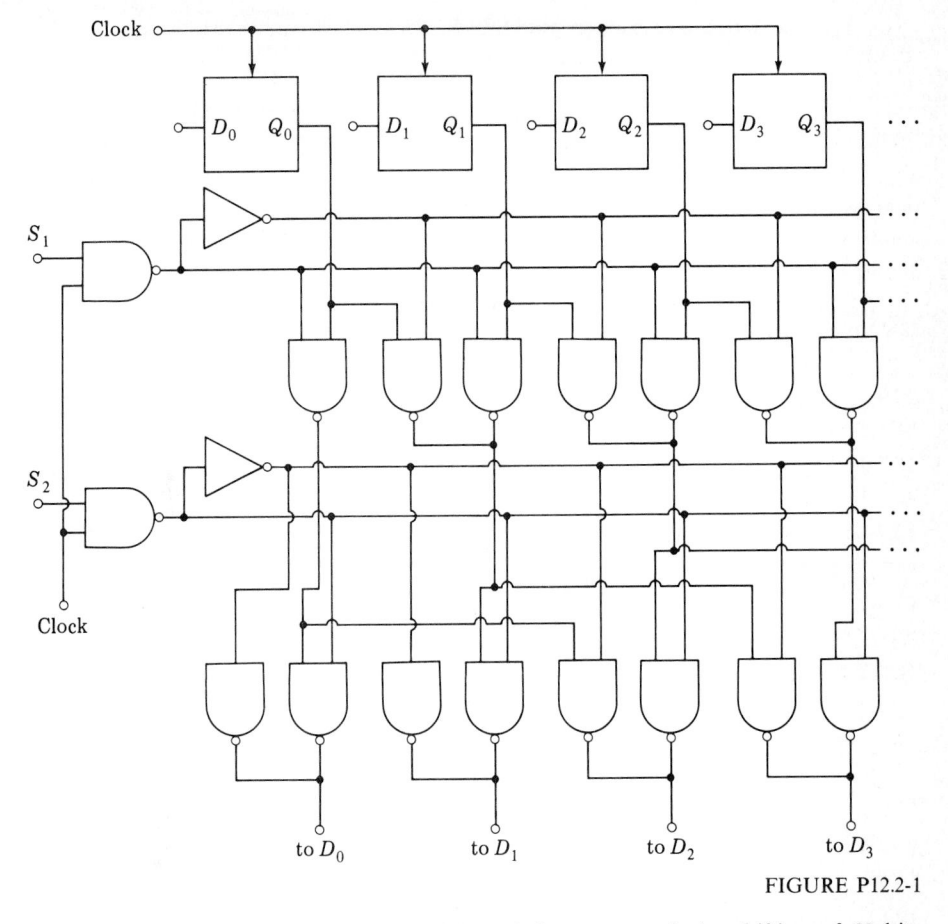

a clock pulse but on the propagation delay of the gates, and the shifting of N bits occurs within one clock interval. In the figure, where gate outputs are tied together, a wired-OR connection is intended.

(a) Explain how the circuit operates.

(b) Expand the system to permit shifts of up to 7 bits.

(c) Determine the propagation time required to effect a shift of 7 bits if the propagation delay time of each gate is 2 ns.

(d) Redesign the system to use only NOR gates.

12.3-1 The leakage current in $T1$ of Fig. 12.3-1 can be considered to be approximately 10 nA. If capacitor $C = 5\,\text{pF}$, how long does it take the voltage on the gate of $T2$ to decrease from 5 to 3 V?

12.3-2 In the shift-register stage shown in Fig. 12.3-2 the leakage current in an OFF transistor is 10 nA. Each capacitor is 5 pF. The load transistors $T3$ and $T6$ are equivalent to 5-kΩ resistors, while the driver and switch transistors have a resistance of about 100 Ω when turned ON. $V_{DD} = 5$ V.

(a) If $T4$ is kept ON for a time sufficient to charge C_2 from 0 to 4 V, calculate the minimum time that $\phi 2$ must be high.

(b) As a result of leakage the voltage on C_2 will eventually decrease to 0 V. If, after C_2 has reached 4 V, ϕ_2 goes to the **0** state, when must ϕ_2 return to the **1** state to assure that the voltage across C_2 does not drop below 3 V?

(c) Using the results obtained in parts (a) and (b), determine the minimum clock frequency.

12.3-3 In Fig. 12.3-4, when the input is at logic **0**, C_3 will charge to V_{DD}. When $T4$ goes ON, C_2 will charge from C_3.

(a) If $V_{DD} = 5$ V and we require that C_2 charge to 4 V, what must be the ratio C_3/C_2?

(b) If the resistance of $T4$ is 100 Ω, calculate the time required for C_2 to reach the level 90 percent of the asymptotic limit if $C_2 = 1$ pF.

(c) Verify that the requirement $C_3 \gg C_2$ does not apply when the input is at logic **1**.

12.4-1 Describe the operation of the shift register of Fig. 12.4-1 when the input is at logic **0**.

12.7-1 Design a shift-left shift-right register using the shift register stage shown in Fig. 12.7-1. Show all circuits.

12.7-2 Modify the shift register stage shown in Fig. 12.7-1 so that it becomes an asynchronous-parallel-input–synchronous-serial-output shift register.

12.8-1 If the register shown in Fig. 12.8-1 is to operate properly, verify that clock ϕ_2 must return to the **1** state before ϕ_3 returns to the **1** state.

12.9-1 Design an encoder using NAND gates satisfying the truth table given in Fig. 12.9-1a.

12.9-2 An encoder provides N outputs.

(a) How many OR gates are required?

(b) What is the maximum number of inputs that can be accommodated?

12.9-3 Redesign the decoder shown in Fig. 12.9-2b using only (a) NAND gates and (b) NOR gates.

12.10-1 Repeat Prob. 12.9-1 using (a) NOR gates and (b) diode gates.

12.12-1 ROMs and adders are to be used to multiply two 4-bit numbers. The output of the multiplier is to be a 4-bit number consisting of the 4 most significant bits. Design the multiplier.

12.12-2 Referring to Fig. 12.12-1 show that the disposition of the adder carry outputs is proper to assure a correct product result.

12.12-3 (a) Design a ROM system to yield an output Y equal to sin θ, where θ is the input. Let θ vary from 0 to 90° in 0.1° steps. Make use of Eq. (12.12-3) in your design by letting $I = 0, 1, \ldots, 90°$ and F have values between -0.5 and $0.5°$. Also assume $\sin F = F$ and $\cos F = 1$. The output Y is to yield results correct to the nearest hundredth. Specify the size of the ROM.

(b) If the system is designed so that a single ROM is employed with input θ and output Y, specify the size of the ROM.

12.12-4 Design a ROM which accepts the digits 0 to 9 in BCD code and generates outputs which can be used to drive a seven-segment light-emitting diode (LED) array display.

12.13-1 (a) Sketch a four-cell RAM by adding three cells to the RAM shown in Fig. 12.13-1.

(b) To address a cell in a four-cell RAM requires a 2-bit input address. Design the decoding circuit.

12.14-1 (a) Sketch a four-cell RAM using the MOS cell shown in Fig. 12.15-1a and the configuration of Fig. 12.15-1b. Show the individual transistors.

(b) To address one of the four cells requires a 2-bit address $A_1 A_2$. Design the decoder.

12.15-1 Repeat Prob. 12.14-1 using the cell and auxiliary circuits shown in Fig. 12.15-2. Show all the transistors.

12.15-2 Repeat Prob. 12.14-1 using the cell and auxiliary circuits shown in Fig. 12.15-4.

12.16-1 The memory cell shown in Fig. 12.15-4 is used in a 4,096-bit RAM organized as in Fig. 12.16-2.

(a) How large (number of bits) are the X and the Y addresses?

(b) If the time to refresh a single cell is 400 ns, how long does it take to refresh the entire memory?

12.17-1 A RAM chip contains 256 1-bit words. Following the pattern of Fig. 12.17-3, sketch a 1,024-word 8-bit system. Show decoding logic for the chip-select inputs.

12.17-2 A RAM chip contains 64 words of 4 bits. Following the pattern of Fig. 12.17-3 sketch a 256 word, 8-bit word system. Show decoding logic for the chip select inputs.

Chapter 13

13.2-1 In Fig. 13.2-3 $M(t) = 5$ V, $R_a = 100 \, \Omega$, $R_L = 1$ MΩ, and $C = 0.01 \, \mu f$. Assume that the switch is ideal.

(a) If the capacitor is initially uncharged, calculate the time required for V_o to reach 99 percent of $M(t)$ after the switch is closed.

(b) After the switch is opened, the capacitor voltage decreases. Calculate the time required for V_o to decrease by 1 percent.

13.2-2 When switch S is open, the circuit of Fig. 13.2-4 approximates an integrator.

(a) If $I(t) = I_0$ for a time T, comment on the ratio $T/R_a C$ if the capacitor voltage $V_c(t)$ is to approximate the integral of $I(t)$.

(b) If $I_0 = 10$ mA, $R_a = 1$ MΩ, and $C = 0.1 \, \mu f$, calculate $V_c(t)$ when $t = T = 1$ ms. Let switch S be open.

(c) At time T switch S closes. Calculate the time required for V_c to decrease to 1 percent of its maximum value. Assume $R_s = 100 \, \Omega$ and the switch to be ideal.

13.2-3 The input voltage V_s in Fig. 13.2-7 is a dc voltage of 50 mV. The ac amplifier passes all signals, with frequencies above 1 kHz with a gain of 10 and completely attenuates signals with frequencies less than 1 kHz. The switching rate is 2 kHz.

(a) Sketch $V_i(t)$.

(b) Sketch $V_o(t)$ on the same time scale used in part (a).

(c) Sketch $V_f(t)$ on the same time scale as part (a). Assume switch S2 closes synchronously with S1.

(d) Repeat part (c) assuming that S2 closes when S1 opens and S2 opens when S1 closes.

13.2-4 The filter components used in Fig. 13.2-8 are $R = 1$ kΩ, $R_1 = 1$ kΩ, $R_2 = 2$ kΩ, $C_1 = 0.01 \, \mu f$, and $C_2 = 0.02 \, \mu f$. If V_i is a pulse which is zero for $t < 0$, 1 V for

$0 \leq t \leq 10\,\mu$s, and zero for $t > 10\,\mu$s, plot $V_o(t)$ as a function of time. Assume that at $t = 0$ switch $S1$ closes, at $t = 5\,\mu$s switch $S2$ closes additionally, at $t = 10\,\mu$s switch $S3$ also closes, and at $t = 15\,\mu$s all four switches are closed.

13.3-1 In Fig. 13.3-1a diode D can be approximated by a 0.7-V battery in series with an ideal diode. When $V_c = V_1$, the diode is to pass the input signal $V_a(t)$.

(a) Obtain an expression for the most negative value $V_a(t)$ can have before diode D cuts off, in terms of R_a, R_1, R_2, and R_c.

(b) Let $V_1 = +5\,$V, $R_a + R_1 = 100\,\Omega$, $R_c = 1\,$kΩ, and $R_L = 1\,$kΩ. The input $V_a(t) = 5\sin \omega t$. Draw a plot of the output $V_o(t)$.

13.3-2 In Fig. 13.3-1a, diode D can be approximated by a 0.7-V battery in series with an ideal diode. When $V_c = -V_2$ the diode is to be cut off and $V_o(t)$ is to be equal to 0 V independently of $V_a(t)$. Obtain an expression for the maximum allowable value of $V_a(t)$ to ensure that diode D remains cut off. Your answer should be expressed as a function of V_2, R_a, R_1, R_2, and R_c.

13.3-3 Repeat Prob. 13.3-1 for the gate shown in Fig. 13.3-1e.

13.3-4 Repeat Prob. 13.3-2 for the gate shown in Fig. 13.3-1e.

13.3-5 In the two-diode gate shown in Fig. 13.3-2 diodes $D1$ and $D2$ can each be represented by a 0.7-V battery in series with an ideal diode.

(a) If $V_c = V_1$ and $V_a = V_A \sin \omega t$, calculate the largest value V_A the gate can transmit without either diode being cutoff.

(b) If $V_c = V_1$ and $V_A < V_A(\max)$, calculate the gain V_o/V_A.

(c) Verify that as long as the diodes conduct, the voltage at the junction of the resistors R_1 is the same as the voltage at the junction of the diodes.

(d) If $V_a = 1\,$V, $V_c = V_1 = 5\,$V, $R_a = 0$, $R_c = 1\,$kΩ, $R_1 = 1\,$kΩ, and $R_L = 1\,$kΩ, determine the currents in diodes $D1$ and $D2$. Since the currents differ, you would expect the diode voltages to differ and not be 0.7 V for each diode. We might expect that for a real diode $V_{D1} > V_{D2}$. Estimate $V_{D1} - V_{D2}$. What is the relevance of this result to your answer to part (c)?

13.3-6 In the two-diode gate shown in Fig. 13.3-2, $V_c = -V_2$ to open the gate and thereby stop transmission. If diodes $D1$ and $D2$ are each represented by a 0.7-V battery, find the maximum permissible value of V_A, where $V_a(t) = V_A \sin \omega t$, such that the gate remains open, neither diode conducting.

13.3-7 Repeat part (b) of Prob. 13.3-5 assuming diodes $D1$ and $D2$ are ideal, thereby verifying Eq. (13.3-1).

13.3-8 Repeat Prob. 13.3-6 assuming diodes $D1$ and $D2$ are ideal, thereby verifying Eq. (13.3-2).

13.3-9 Equation (13.3-4) applies to the four-diode gate shown in Fig. 13.3-4 when the diodes are assumed to be ideal.

(a) Verify Eq. (13.3-4).

(b) Obtain an expression for V_1 when the diodes are represented by a 0.7-V battery in series with an ideal diode.

13.3-10 The four-diode gate shown in Fig. 13.3-3 is used to gate the voltage V_a, where $|V_a| \leq 5\,$V. Assume $R_a = 0$ and that the diodes are ideal. $R_L = 1\,$kΩ. The voltages ± 5 and ± 10 V are available. Also available is facility, as required, to switch V_c from one of these available voltages to another. Design the four-diode gate by specifying R_1, R_c, V_1, V_2, and V_r.

13.3-11 Consider that for the four-diode gate shown in Fig. 13.3-4 diodes $D1$ and $D3$ can each be represented by a 0.7-V battery in series with an ideal diode while diodes $D2$ and $D4$ are each represented by a 0.75-V battery in series with an ideal diode. Assuming $R_a = 0$, obtain an expression for V_1 as a function of $V_a(\max)$, R_c, and R_L.

13.3-12 (a) Obtain an expression for the reference voltage V_r in terms of $V_a(\max)$, R_a, R_c, and R_L, for the six-diode gate shown in Fig. 13.3-5. Assume that each diode can be represented by a 0.7-V battery in series with an ideal diode.

(b) If $V_a = 5$ V, $V_r = 10$ V, $R_c = 1$ kΩ, $R_L = 1$ kΩ, and $R_a = 0$, calculate the current in diode $D1$ and diode $D3$.

(c) Using the result of part (b) and the fact that the diode voltages actually differ from one another since the diode currents differ, calculate V_o. Compare this result with the result obtained when the diode-voltage difference is ignored.

13.3-13 Obtain an expression for the value of V_a required just to cut off diodes $D1$ and $D3$ in Fig. 13.3-9b. Assume that each diode can be represented by a 0.7-V battery in series with an ideal diode. Your answer should be a function of I_c, R_a, and R_L.

13.3-14 Assume that in the high-speed gate shown in Fig. 13.3-11a the capacitive load is replaced by a 2-kΩ resistor. If $V_a = 5$ V, $I_0 = 10$ mA, $T_c = 20$ ns, and $N = 6$, find the minimum value of L_m to ensure that each diode in the four-diode bridge remains ON. Assume each diode is represented by a 0.7-V battery in series with an ideal diode.

13.7-1 If, in Fig. 13.7-1, $T1$ is replaced by an enhancement-type NMOSFET having a threshold voltage $V_T = 2$ V, find:

(a) The level of V_c needed to open switch $T1$ and to close switch $T1$. Assume $V_a = 0$ V.

(b) The maximum positive and most negative values of V_a which can be used if the switch is to operate properly. For the purpose of this calculation assume a margin of safety equal to zero.

13.7-2 The series FET switch shown in Fig. P13.7-2 differs from the switch shown in Fig. 13.7-1 inasmuch as the respective placement of $T1$ and R'_1 has been reversed. If $|V_a(t)| \leq 5$ V, find the values of V_c needed to turn $T1$ ON and OFF.

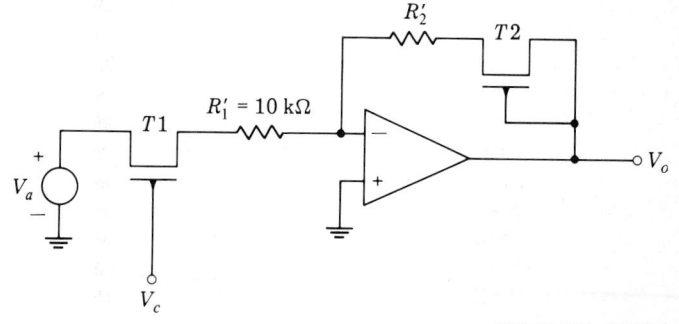

FIGURE P13.7-2

13.8-1 Use the circuit of Fig. 13.8-2b to show that if $C_d V_g/V'_g = C'_g$, the offset error is zero.

13.8-2 The op-amps of Fig. 13.8-3 have an input offset voltage of 1 mV; that is, when operated without feedback, the output voltage of each op-amp is 0 V when the input voltage to the op-amp is 1 mV. If $V_a = 0$ V and $T1$ is ON, calculate V_o.

13.8-3 The op-amps shown in Fig. 13.8-4 have the same characteristics as the op-amps described in Prob. 13.8-2. What voltage V_a must be applied to make $V_o = 0$ V?

13.10-1 Refer to Fig. 13.10-1. The FET shown has a threshold voltage of 2 V, and $k_n = 400$ μA/V^2. Let $V_c = 7$ V, and let V_a vary from -5 to 5 V.
(a) Plot the resistance of the NMOSFET as a function of V_a. (Refer to Sec. 8.1.)
(b) Assume $R_L = 1$ kΩ and neglect C_L; plot V_o as a function of V_a over the range $|V_a| \le 5$ V.

13.10-2 Refer to Fig. 13.10-2. The FETs shown each have a threshold voltage of 2 V and $k_n = k_p = 400$ μA/V^2. Let $V_c = 5$ V, and let V_a vary from -5 to 5 V.
(a) Plot the resistance of the CMOS gate as a function of V_a.
(b) Assume $R_L = 1$ kΩ and neglect C_L; plot V_o as a function of V_a over the range $|V_a| \le 5$ V. (Refer to Sec. 8.1.)

13.11-1 Design a circuit which accepts two independent input dc voltages simultaneously. The circuit has two outputs A and B. The larger of the inputs is routed to A and the smaller of the inputs to B.

13.11-2 Show that Fig. 13.11-2 can be constructed using a single comparator. Sketch the required circuit.

Chapter 14

14.2-1 Verify the sampling theorem by proceeding as follows:
1 Let there be a sampling waveform $S(t)$ which consists of a regular sequence of pulses of unit amplitude of arbitrarily small time duration Δt and separated by intervals T_s. Write out the Fourier series for this periodic pulse train.
2 Let $M(t)$ be an arbitrary waveform. Write the product $S(t)M(t)$, which then represents the waveform $M(t)$ sampled at intervals T_s. Assume that the signal $M(t)$ has a spectral range which extends from 0 to f_M Hz. Find the spectral range of each of the terms in the product $S(t)M(t)$.
3 Show that if $1/T_s = f_s > 2f_M$, the spectral ranges of the terms in $S(t)M(t)$ do not overlap. Note, then, that one of the terms in $S(t)M(t)$ is $(M(t)\,\Delta t)/T_s$ and that hence $M(t)$ is recoverable.

14.3-1 The time-division multiplexing system shown in Fig. 14.3-1 is used to multiplex the four signals $M_1(t) = \cos \omega_o t$, $M_2(t) = 0.5 \cos \omega_o t$, $M_3(t) = 2 \cos 2\omega_o t$, and $M_4(t) = \cos 4\omega_o t$.
(a) What is the minimum allowable sampling rate for each signal?
(b) Adapt the rotating commutator system of Fig. 14.3-1 so that each of the four signals can be sampled at its required rate. What is the speed of revolution of the commutator arm?

14.4-1 When a sample of a signal is quantized, the quantization introduces an error. If the step size between quantization levels is S, the error lies in the range $-S/2$ to $S/2$ and any error in the range is equally likely. The mean-square *quantization error* is defined to be the average value of the square of the error. Show that the mean-square quantization error is $S^2/12$.

14.4-2 The bandwidth allowed for voice transmission over a telephone line is 3,000 Hz. A number of telephone conversations are to be multiplexed over a single communications channel by time-division multiplexing using pulse-code modulation. It is found in a particular system that for reliable identification of bits at the receiving end, an interval of at least 1 μs must be allowed for each bit. If the signals are to be quantized into 16 levels, estimate the number of voice signals that can be multiplexed.

14.5-1 Verify Eq. (14.5-2).

14.5-2 (a) A digital-to-analog converter accepts 12 input bits, i.e., has a resolution of 12 bits, and provides an output which is 10 V maximum. Suppose that as a result of drift of component values, etc., the output may be in error by an amount ΔV. How large can ΔV be before the least significant bit would no longer be significant?

(b) Repeat for a 16-bit converter.

14.5-3 A 10-bit digital-to-analog weighted-resistor converter, as in Fig. 14.5-1, has $R_L = 0$ (the analog output is the current I_L). The error in output due to resistor tolerance is not to exceed one-half the change in the output corresponding to a change in the least significant input bit.

(a) If only the resistor R_{N-1} corresponding to the most significant bit is in error, how large a percentage change in R_{N-1} can be tolerated?

(b) If only the resistor R_0 corresponding to the least significant bit is in error, how large a percentage change in R_0 can be tolerated?

14.5-4 A 6-bit digital-to-analog converter delivers its output current to the virtual ground of an op-amp, as shown in Fig. 14.5-3. The input voltage levels are $V_i(1) = 10$ V and $V_i(0) = 0$ V. The resistor R_0 corresponding to the least significant bit is 320 kΩ and the feedback resistor is $R_f = 10$ kΩ. What is the output V_o corresponding to an input **110101**?

14.6-1 Verify that the output V_o in the digital-to-analog converter of Fig. 14.6-2b is given by Eq. (14.6-2).

14.6-2 Consider an R-2R digital-to-analog converter ladder as in Fig. 14.6-1a with provision for N bits. Suppose that all resistors are of resistance value R or 2R as required except that the 2R resistors associated with the most significant bit has drifted in value to $2R(1 + \delta)$. Write an expression for V_o corresponding to Eq. (14.6-2) which applies in this case. If $N = 10$, how large can δ become before the least significant bit is no longer significant?

14.6-3 Refer to Fig. 14.6-3.

(a) Verify that if a straight binary 8-bit converter is intended, we require that $r = 8R$.

(b) If the arrangement is intended as a two-decimal-digit BCD converter, verify that it is required that $r = 4.8R$.

14.7-1 Refer to Fig. 14.7-3. Verify that when the input is at 0 V, the transistors $T1$ and $T2$ both saturate.

14.7-2 In Fig. 14.7-3, starting from the top, the input voltages are 3, 0, 3, and 0 V. What is the corresponding value of the analog voltage output V_a?

14.8-1 Refer to Fig. 14.8-1.

(a) Verify that the resistance seen looking into any ladder tap position is $2R/3$.

(b) Assuming $R = 1$ kΩ, $R_f = 2R$, and $I = 10$ mA, write an expression for the output voltage V_o in terms of the logic level associated with the switch; that is, $S = 1$ for switch connected to the ladder tap.

14.8-2 In Fig. 14.8-2, $R = 100$ Ω and the op-amp (not shown) has a feedback resistor of $2R$. Find R_e if the full-scale output voltage of the op-amp is to be 1 V.

14.8-3   Refer to Fig. 14.8-2.   Verify that when either of the transistors in a switch pair (say $T0A$ and $T0B$) is OFF, its base-to-emitter voltage is $+0.35$ V.

14.9-1   Verify Eq. (14.9-1), which applies to the inverted ladder circuit of Fig. 14.9-1.

14.9-2   In Fig. 14.9-2 let the bias-voltage circuit have values as in Fig. 14.8-2, that is, $V_Z = 6.2$ V and $-V_R = -8.1$ V, so that V_B has the value $V_B = -1.15$ V as in the former circuit.   Let $R = 1$ kΩ.

    (a)   Calculate the currents I_{C0}, I_{C1}, I_{C2}, and I_{C3} when all the B transistors are OFF.

    (b)   Let V_0 be increased to the point where $T0A$ just turns OFF.   Estimate the consequent changes in the collector currents.

14.10-1   Four lines present digitally the numbers 7, 6, ..., 0, -1, ..., -7, -8 in the twos-complement format.   Design a digital-to-analog converter which accepts this input and provides an output voltage ranging from 7 to -8 V.

14.10-2   (a)   Refer to Fig. 14.10-2.   Verify that the operation of the circuit will be unaltered if the offset switch is eliminated, the switch S_2 replaced by a switch $\bar{S}_2$, and its resistor $R/4$ replaced by a resistor $R/3$.

    (b)   Adapt the circuit of Fig. 14.10-2 to four inputs.   Draw two circuits, one which employs an offset switch and one which does not.

14.12-1   Refer to Fig. 14.12-1.   Verify that the array of gates (decoder) following the flip-flop register properly assigns the 3-bit output code words in the manner required in Fig. 14.12-2.

14.12-2   A comparator analog-to-digital converter is to provide a 3-bit twos-complement output representation.

    (a)   Assume that the extra negative number available in twos-complement is to be ignored and that reference voltages V_0 and $-V_0$ are available.   Draw the resistor chain that will provide comparator reference voltages.

    (b)   Repeat part (a) for the case where the extra negative number is to be used.

14.12-3   Design decoders for the twos-complement comparator-type 3-bit analog-to-digital converter.   Consider separately (a) that the extra negative output is ignored and (b) that it is used.

14.12-4   Design a 3-bit comparator analog-to-digital converter in which the output is in ones-complement representation.   Draw the resistor chain to provide comparator references and design the output decoder.

14.13-1   Refer to the successive-approximation converter of Fig. 14.13-2.   Assume that the analog input sample is of such magnitude that it corresponds to the digital output **001**.   Draw a clock waveform, and draw on the same time scale the waveforms of Q_A through Q_E, the waveforms Q_3, Q_2, and Q_1, and the waveform of the comparator output C_o.

14.13-2   The 3-bit successive-approximation converter of Fig. 14.13-2 is to be used for analog input samples in the range 0 to 10 V.   What should the offset be to assure that the maximum quantization error will be the same in each range.

14.14-1   (a)   Draw the block diagram of a 2-bit counting analog-to-digital converter.

    (b)   The converter is to be used for analog samples in the range 0 to 10 V. Design the digital-to-analog converter providing the required offset and arranging that the maximum quantization error will be the same in each range.

    (c)   Draw waveforms as in Fig. 14.14-2 for analog sample inputs equal to 4 and to 9 V.

14.14-2 The sampling rate of a sampled analog signal presented to the input of a counting analog-to-digital converter is 5 kHz. The flip-flops in the converter have a maximum guaranteed toggling rate of 2 MHz. What is the best possible resolution of the converter?

14.14-3 A 10-kHz sinusoidal waveform of peak value 10 V is presented to a servo-type converter with 8-bit resolution. If the conversion error is to be kept to ± 1 least significant bits, what is the minimum acceptable clock frequency?

14.15-1 The discussion of the converter of Fig. 14.15-1 and the waveform in Fig. 14.15-1b assumes that when switch S_2 is closed $V_o = 0$ V and that the comparator output makes its transition between logic levels precisely as V_o passes through $V_o = 0$ V. These conditions would be hard to maintain in practice. Show that the need to satisfy these conditions can be circumvented by an appropriate adjustment of the comparator reference input and that this adjustment has no effect on the accuracy of the converter.

14.15-2 (*a*) In the dual-slope converter of Fig. 14.15-1 the clock rate is 100 kHz, and the converter has a resolution of 10 bits. What is the maximum rate at which samples can be converted?

(*b*) Design logical circuitry, which, allowing adequate time to convert a sample, performs all the operations required for reading the converter and resetting it so that it can convert the next sample.

14.17-1 In the circuit of Fig. 14.17-1, $-V_r = -10$ V, $R = 10$ kΩ, and $C = 0.01$ μF. Neglect T_d, and make a plot of the frequency as a function of the analog input V_a for values of V_a between 0 and 10 V.

14.21-1 (*a*) In the delta modulator of Fig. 14.21-1, $M(t)$ is the sinusoidal waveform $M(t) = 0.1 \sin (2\pi \times 10^3)t$. The clock rate is 2×10^4 Hz, and the step size is $S = 20$ mV. Plot $M(t)$, $\hat{M}(t)$, and $E(t)$. Assume that at $t = 0$ and $M(t) - \hat{M}(t) = S$.

(*b*) Repeat for $S = 4$ mV and for 60 mV. Note that the step size $S = 20$ mV yields the "best" estimate $\hat{M}(t)$. It can be shown that the best step size is $S_{opt} \approx 4A/(f_s/f_m)$, where A is the peak amplitude of the sinewave, f_s is the clock rate, and f_m is the frequency of the sinewave. Why is this result correct?

14.21-2 (*a*) Repeat part (*a*) of Prob. 14.21-1 but let the clock rate be 10 kHz and $S = 40$ mV.

(*b*) Which delta modulator, this one, in which $S = 40$ mV and $f_s = 10$ kHz, or the system in part (*a*) of Prob. 14.21-1, in which $f_s = 20$ kHz and $S = 20$ mV, yields the better estimate? *Note*: It can be rigorously proved that providing the step size is optimally chosen, the higher sampling rate yields the better estimate.

14.21-3 (*a*) In a delta modulator let $M(t) = M_0 \sin 2\pi f_0 t$, and let the step size $S > 2M_0$. Show that step-size limiting occurs; i.e., the waveform $\hat{M}(t)$ is no longer responsive to changes in $M(t)$.

(*b*) The step size is S, and the clock-sampling frequency is f_s. Show that if slope overload is to be avoided, the largest allowable amplitude of the sinusoid is $M_0 = Sf_s/2\pi f_0$.

(*c*) Show that if both slope overload and step-size limiting are to be avoided, it is additionally required that $f_s > 3f_0$.

14.22-1 In the adaptive delta-modulator system characterized by Eq. (14.22-3) assume that $M(t)$ remains constant to within the step size S_0. Show that in this case $\hat{M}(t)$ oscillates about $M(t)$ at one-half the frequency that would be encountered in linear delta modulation.

14.22-2 A waveform $M(t)$ applied to the adaptive delta modulator [Eq. (14.22-3)] rises linearly from 0 to 9.7 V in the time $10T_s$, remains at 9.7 V for $20T_s$, and thereafter falls linearly back to 0 V in $5T_s$. The minimum step size is $S_0 = 0.5$ V. Let $\hat{M}(t) = 0$ at $t = 0$, and plot $M(t)$ and $\hat{M}(t)$ on the same set of axes for $\tau = 0$ to $30T_s$.

Chapter 15

15.1-1 Consider an RC series circuit. If the initial voltage across the capacitor is V_i and the final asymptotic value of the voltage is V_f:

    (a) Show that if $RC = \tau$, then the voltage across the capacitor V_C is

$$V_C(t) = V_f - (V_f - V_i)\,\epsilon^{-t/\tau}$$

    (b) Find an expression for $V_R(t)$, that is, the voltage across the resistor.

    (c) Show that, in general, for a single-time-constant circuit, $V(t)$ the voltage between any two terminals in that circuit can be written in the form given in part (a).

    (d) Let V_T be some "critical" voltage of some special interest which is between V_i and V_f. Show that the time T required for $V(t)$, starting at V_i, to reach V_T is

$$T = \tau \ln \frac{V_f - V_i}{V_f - V_T}$$

15.1-2 In the circuit of Fig. 15.1-2, $V_{SS} = 12$ V, $V_T = 7$ V, $R = 5\ \text{k}\Omega$, and $R_{o1} = 200\ \Omega$.

    (a) Draw the waveforms V_{1o}, V_{2i} and V_{2o} and label all voltage levels.

    (b) If $C = 0.001\ \mu$F, what is the duration of the quasi-stable state?

15.1-3 (a) Assume that for a supply voltage of 10 V the input-output characteristic of the CMOS gate is as appears in Fig. P15.1-3a. (This characteristic is intended as a better approximation of a real characteristic, as seen in Fig. 8.4-1, than the idealization of Fig. 15.1-3.) For $V_{SS} = 10$ V, $R = 10\ \text{k}\Omega$, $R_{o1} = 500\ \Omega$ redraw the waveform of Fig. 15.1-4 labeling all voltage levels.

    (b) Assume that the input-output characteristic is as appears in Fig. P15.1-3b. Show that in this case the quasi-stable state will terminate when V_{1o} reaches 3.5 V. (Neglect the loading of R on gate $G1$.)

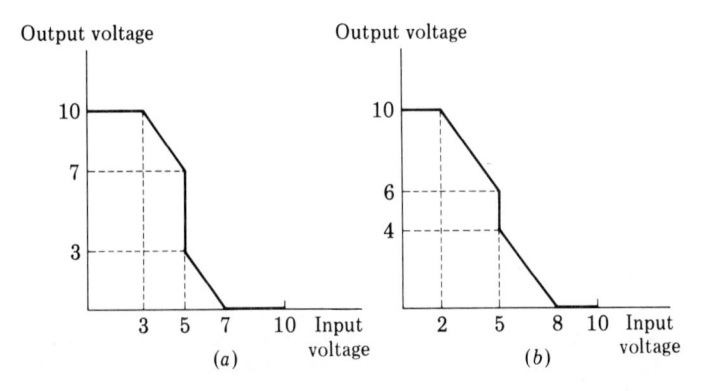

FIGURE P15.1-3a   FIGURE P15.1-3b

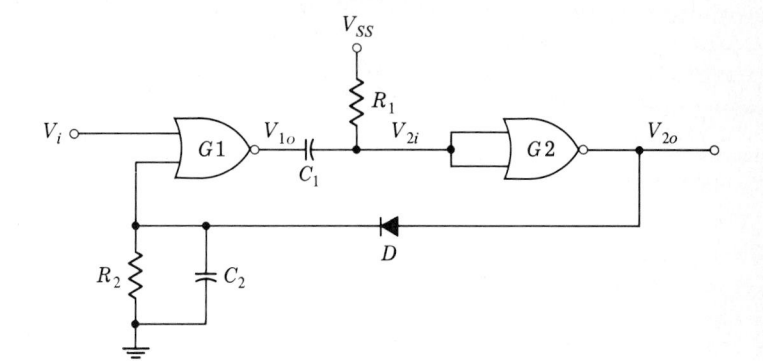

FIGURE P15.1-4

15.1-4 In the monostable multi circuit of Fig. P15.1-4 assume that the input-output characteristic of a CMOS gate appears as in Fig. 15.1-3. Assume as well that, as viewed from the output, the gate can be represented by the two-switch arrangement S and S', as in Fig. 15.1-5, except that $R_{o1} = R' = 0$. Finally assume that diode D is ideal with cut-in voltage $V_\gamma = 0$.

    (a) With V_i at logic level **0**, what are the voltage levels at V_{1o}, V_{2i}, and V_{2o}?

    (b) Let the circuit be triggered into its quasi-stable state by a short positive pulse at V_i. The transition voltage V_T of each of the gates is the same. Draw the waveforms of V_i, V_{1o}, V_{2i}, and V_{2o} from the beginning of the quasi-stable state until the circuit returns to its initial stable state.

    (c) Consider that $R_1 C_1 = R_2 C_2 = \tau$. Calculate the time T of the quasi-stable state as a function of V_{SS}, τ, and V_T.

    (d) Make a plot of T/τ as a function for V_T for $V_{SS} = 10$ V for $V_T = 1, 2, 3, \ldots, 9$. Compare the plot with the corresponding plot for the single RC circuit arrangement of Fig. 15.1-2.

15.1-5 A monostable multi employing discrete transistors is shown in Fig. P15.1-5a. A short negative trigger at V_i starts the timing. The waveforms of the circuit are shown in Fig. P15.1-5b. It is assumed that the components are such that each transistor, when conducting, is in saturation. When $T2$ goes back ON at $t = T$, its base current is so large that it is necessary to recognize that its base-emitter voltage will exceed the saturation voltage $V_{BE}(\text{sat}) = 0.75$ V. This excess of $V_{BE}(\text{sat})$ over 0.75 V accounts for the overshoot in V_{B2} at $t = T$ (see Fig. P15.1-5b). To account for this overshoot while assuming that $V_{BE}(\text{sat}) = 0.75$ V we can include a small resistor r in series with the base of $T2$. (This resistor r takes account of the ohmic resistance of the base as well as the incremental input resistance.)

    (a) Describe the operation of the circuit qualitatively, thereby verifying that the waveforms shown are correct.

    (b) Use $V_{CC} = 10$ V, $R_{c1} = 2$ kΩ, $R = R_{b1} = 50$ kΩ, $r = 50$ Ω, $C = 0.01$ μF, and $h_{FE} = 50$. Compute all voltage levels on the waveforms not already given. What is the size of the jump δ? Calculate the time constants of the exponential portions of the waveforms.

    (c) Calculate the time T.

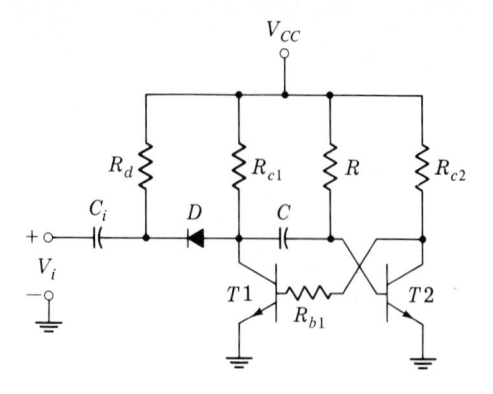

FIGURE P15.1-5a

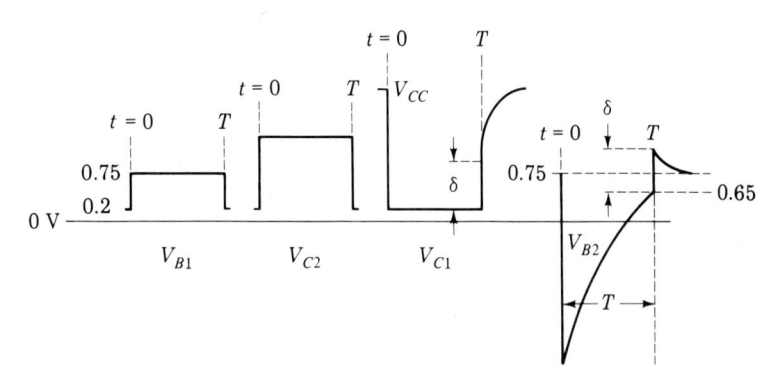

FIGURE P15.1-5b

15.2-1 In the astable CMOS multi of Fig. 15.2-1, represent the gates as viewed from their output by the two-switch arrangement of Fig. 15.1-5, in which $R_{o1} = 500\ \Omega$ and $R' = 1,000\ \Omega$. Take into account the fact that the cut-in voltage of the input protective diodes is 0.75 V. Let $V_{SS} = 12$ V, and assume R very large in comparison with R_{o1} and R'.

(a) Draw the waveforms V, V_{2o}, and V_{1i}.

(b) If $R = 50$ kΩ, $C = 0.01\ \mu$F, and $V_T = 7$ V, calculate the times T_1 and T_2.

15.2-2 In the astable CMOS multi of Fig. 15.2-1 add a large resistor R_s in series with the input to gate $G1$. Let this added resistor be large enough for the waveform at the junction of R and C to be unaffected by the protective diodes at the input to $G1$. Represent the gates by the two-switch arrangement of Fig. 15.1-5, in which $R_{o1} = R' = 0$.

(a) Draw the waveforms V_{1i}, V, V_{2o} and the waveform at the junction of R_s, R, and C. Label the voltages on the waveform in terms of V_{SS}, V_T, and the protective diode cut-in voltages $V_{\gamma D}$.

(b) Show that the total period of the circuit is

$$T = RC\left(\ln \frac{V_{SS} + V_T}{V_T} + \ln \frac{2V_{SS} - V_T}{V_{SS} - V_T} \right)$$

15.2-3   Let the transition voltage V_T of the gates used in the astable circuit of Fig. 15.2-1 be in the range from $0.3V_{SS}$ to $0.7V_{SS}$.   Use Eq. (15.2-1) to determine the corresponding range of T.

15.2-4   (a)   An astable multi using discrete transistors is shown in Fig. P15.2-4.   Draw the waveform at a collector and at a base.   (See comments and waveforms in Prob. 15.1-5.)

   (b)   Label the voltage levels and time constants in the waveform in part (a) for the case $V_{CC} = 12$ V, $R_c = 3$ kΩ, $R = 50$ kΩ, $r = 50$ Ω, and $C = 0.01$ μF.   Calculate the total timing interval.

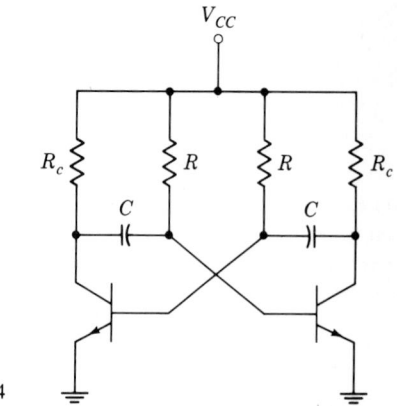

FIGURE P15.2-4

15.2-5   Consider that in Fig. 15.2-1a, gates $G1$ and $G2$ each have a propagation delay t_{pd}.   Redraw Fig. 15.2-1a, b, and c, taking this propagation delay into account.

15.3-1   (a)   Show that in the circuit of Fig. 15.3-3 the pulse waveform at the Q output must exceed the width of the triggering pulse.

   (b)   Show that in the circuit of Fig. 15.3-4 the output pulse waveform need have no relationship to the width of the triggering pulse.

15.3-2   Assume that in the circuit of Fig. 15.3-5a the diode maintains across itself a fixed voltage of 0.75 V when conducting.   Taking the diode into account, redraw the waveforms of Fig. 15.3-1.   Label all voltage levels.

15.3-3   In Fig. 15.3-1, each of the NOR gates has a propagation delay t_{pd}.
   (a)   Sketch the response at Q due to an input trigger at S when $t_{pd} = 0.25rC$.
   (b)   Express the minimum trigger duration W as a function of t_{pd} and the time constant $\tau = rC$.

15.3-4   Repeat Prob. 15.3-3 for the monostable multi shown in Fig. 15.3-3.

15.3-5   The propagation delay times of the flip-flops shown in Fig. 15.3-4 are each t_{pd}.   Express the minimum trigger duration W as a function of T_{pd} and $\tau = rC$.

15.4-1   In the gate-delay-line multi of Fig. 15.4-1, the propagation delay of GA is $t_{pd}(A) = 5$ ns, the delay of GB is $t_{pd}(B) = 8$ ns, and the delay line has a delay $t_D = 20$ ns.   Draw the waveforms for the cases where the input V_i is (a) a step, (b) a pulse of duration 15 ns, (c) a pulse of duration 10 ns, and (d) a pulse of duration 15 ns following by a second pulse 50 ns later.

15.4-2 Repeat Prob. 15.4-1 for the circuit of Fig. 15.4-2. Compare the duty cycle of this circuit with the duty cycle of the circuit in Fig. 15.4-1.

15.5-1 For the TTL monostable multi shown in Fig. 15.5-1*a*:
(*a*) Calculate the minimum duration of the input trigger *P* as a function of the propagation delay times of the gates. Assume for simplicity that each gate has the same propagation delay t_{pd}.
(*b*) Calculate the maximum input-trigger *pulse-repetition frequency*, i.e., the maximum rate of firing the monostable multi. Express your answer in terms of t_{pd}, τ_1, and τ_2.

15.5-2 In connection with the overshoot developed in the waveform of V_2 in Fig. 15.5-1*b* make the following estimates. Assume that the gate propagation times are very small in comparison with the time of the quasi-stable state so that the overshoot occurs when V_2 reaches 0.65 V. Assume that at the overshoot time, looking into *G2*, we see a 0.75-V battery in series with a 25-Ω resistor. Neglect the current into gate *G1*. Replace *G4*, as viewed looking into the output by the usual active pull-up of the TTL gate (as in Fig. 6.4-1). Assume that *R* is large enough for its presence to be neglected. Calculate Δ and the time constant τ_2 with which the overshoot decays.

15.6-1 (*a*) Explain qualitatively how the circuit of Fig. P15.6-1 can serve as an astable multi in which the two parts of the timing cycle can be made equal, i.e., the circuit can have a 50 percent duty cycle.

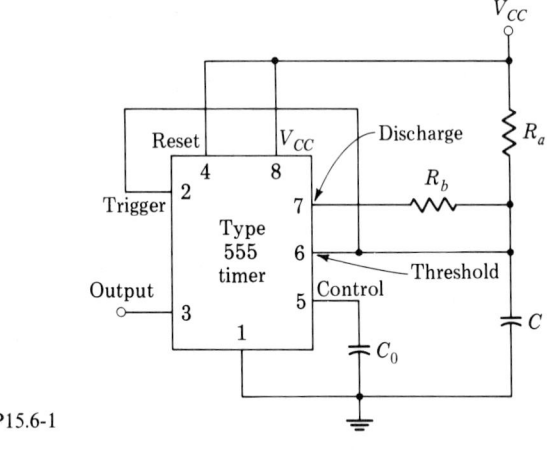

FIGURE P15.6-1

(*b*) Verify that the time of the complete cycle is $T = T_1 + T_2$, where $T_1 = 0.69 R_a C$ and T_2 is given by

$$T_2 = \frac{R_a R_b}{R_a + R_b} C \ln \frac{R_b - 2R_a}{2R_b - R_a}$$

(*c*) If $R_a = 50$ kΩ, find R_b so that $T_1 = T_2$.
(*d*) Verify from the expression for T_2 that R_b cannot be permitted to exceed $R_a/2$. Explain physically the reason for this restriction.

15.6-2 The timer circuit of Fig. P15.6-2 is used to generate a voltage across the capacitor *C* which varies linearly with time, i.e., a linear ramp. The transistor *T*

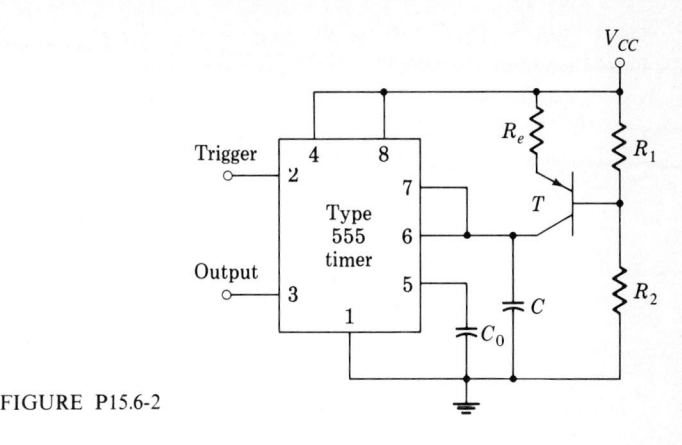

FIGURE P15.6-2

provides a constant charging current to the capacitor. The ramp is initiated by the application of a brief negative pulse at the trigger input. When the capacitor voltage reaches a critical level, the capacitor discharges abruptly.

(a) Assuming that the time duration of the ramp is shorter than the interval between input triggers draw waveforms on a common time scale of the triggering waveform, of the output at pin 3 and of the voltage across C.

(b) Assume that the ramp starts at 0 V and verify that the ramp continues to rise for a time

$$T = \frac{\frac{2}{3}V_{CC}\,R_e(R_1 + R_2)\,C}{R_1 V_{CC} - V_{EB}(R_1 + R_2)}$$

where $V_{EB}(\approx 0.7\,\text{V})$ is the base-emitter voltage of the transistor.

15.6-3 (a) Describe qualitatively how the type-555 timer can be used as a pulse frequency divider. A frequency divider accepts a train of input pulses of fixed frequency f_1 and makes available as an output a train of pulses of frequency $f_2 = f_1/n$, in which n is an integer.

15.6-4 An op-amp is connected to form a monostable multi as in Fig. P15.6-4.

(a) What polarity pulse must V_i be in order to fire the multi? What is the purpose of the -0.5-V reference?

(b) Sketch the responses $V_o(t)$ and $V_2(t)$ after an input trigger pulse. Take account in your sketch of the propagation delay t_{pd} of the op-amp. Assume that the op-amp

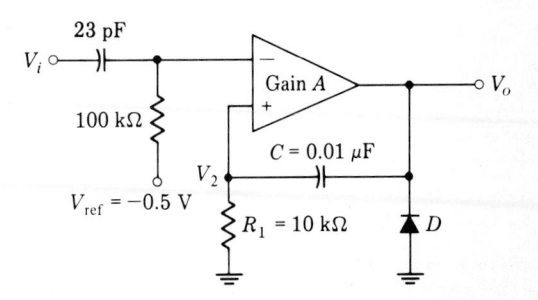

FIGURE P15.6-4

has a gain A from -5 to $5\,V$ and a gain of zero elsewhere, an infinite input impedance, and a negligibly small output impedance. Assume that diode D clamps when its voltage drop is 0.7 V.

(c) Obtain a general expression for the timing period of the monostable multi shown, in terms of V_{ref}, R_1, C and V_{max}, the most positive value that V_f can attain, and the op-amp gain A.

REPRESENTATIVE
MANUFACTURERS' SPECIFICATIONS

FAIRCHILD SUPER HIGH SPEED TTL/SSS • 9S20 (54S20/74S20)

DUAL 4-INPUT NAND GATE

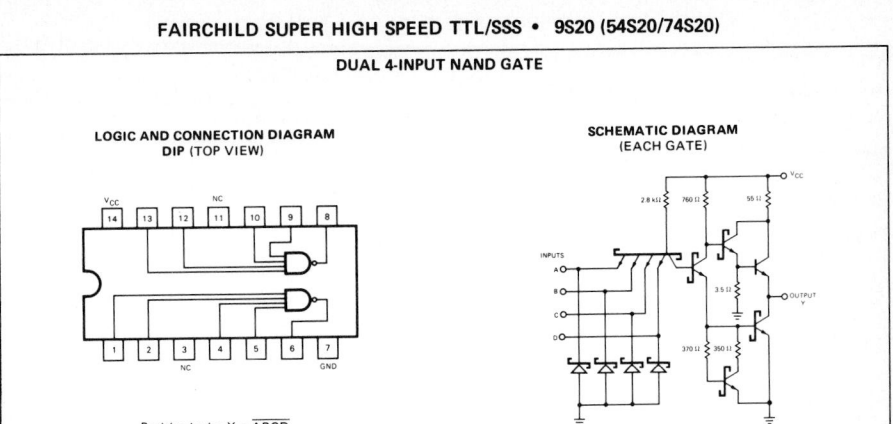

LOGIC AND CONNECTION DIAGRAM
DIP (TOP VIEW)

SCHEMATIC DIAGRAM
(EACH GATE)

Positive logic: Y = $\overline{ABCD}$

NC — No internal connection.

Component values shown are typical.

RECOMMENDED OPERATING CONDITIONS

PARAMETER	9S20XM (54S20XM)			9S20XC (74S20XC)			UNITS
	MIN.	TYP.	MAX.	MIN.	TYP.	MAX.	
Supply Voltage V_{CC}	4.5	5.0	5.5	4.75	5.0	5.25	V
Operating Free-Air Temperature Range	−55	25	125	0	25	75	°C
Input Loading for Each Input			1.25			1.25	U.L.

X = package type; F for Flatpak, D for Ceramic Dip, P for Plastic Dip. See Packaging Information Section for packages available on this product.

DC CHARACTERISTICS OVER OPERATING TEMPERATURE RANGE (Unless Otherwise Noted)

SYMBOL	PARAMETER		LIMITS			UNITS	TEST CONDITIONS (Note 1)
			MIN.	TYP. (Note 2)	MAX.		
V_{IH}	Input HIGH Voltage		2.0			V	Guaranteed Input HIGH Voltage
V_{IL}	Input LOW Voltage				0.8	V	Guaranteed Input LOW Voltage
V_{CD}	Input Clamp Diode Voltage			−0.65	−1.2	V	V_{CC} = MIN., I_{IN} = −18 mA
V_{OH}	Output HIGH Voltage	XM	2.5	3.4		V	V_{CC} = MIN., I_{OH} = −1.0 mA, V_{IN} = 0.8 V
		XC	2.7	3.4			
V_{OL}	Output LOW Voltage			0.35	0.5	V	V_{CC} = MIN., I_{OL} = 20 mA, V_{IN} = 2.0 V
I_{IH}	Input HIGH Current			1.0	50	µA	V_{CC} = MAX., V_{IN} = 2.7 V	Each Input
					1.0	mA	V_{CC} = MAX., V_{IN} = 5.5 V
I_{IL}	Input LOW Current			−1.4	−2.0	mA	V_{CC} = MAX., V_{IN} = 0.5 V Each Input
I_{OS}	Output Short Circuit Current (Note 3)		−40	−65	−100	mA	V_{CC} = MAX., V_{OUT} = 0 V
I_{CCH}	Supply Current HIGH			5.4	8.0	mA	V_{CC} = MAX., V_{IN} = 0 V
I_{CCL}	Supply Current LOW			12.6	18.0	mA	V_{CC} = MAX., Inputs Open

AC CHARACTERISTICS: T_A = 25°C

SYMBOL	PARAMETER	LIMITS			UNITS	TEST CONDITIONS	TEST FIGURES
		MIN.	TYP.	MAX.			
t_{PLH}	Turn Off Delay Input to Output	2.0	3.0	4.5	ns	V_{CC} = 5.0 V	A
t_{PHL}	Turn On Delay Input to Output	2.0	3.0	5.0	ns	C_L = 15 pF	

NOTES:
1. For conditions shown as MIN. or MAX., use the appropriate value specified under recommended operating conditions for the applicable device type.
2. Typical limits are at V_{CC} = 5.0 V, +25°C.
3. Not more than one output should be shorted at a time.

FAIRCHILD SUPER HIGH SPEED TTL/SSI • 9S74 (54S74/74S74)

DUAL D-TYPE EDGE TRIGGERED FLIP-FLOP

DESCRIPTION — The 9S74 (54S74/74S74) dual edge-triggered flip-flops utilize Schottky TTL circuitry to produce very high speed D-type flip-flops. Each flip-flop has individual clear and preset inputs, and also complementary Q and $\bar{Q}$ outputs.

Information at input D is transferred to the Q output on the positive-going edge of the clock pulse. Clock triggering occurs at a voltage level of the clock pulse and is not directly related to the transition time of the positive-going pulse. When the clock input is at either the HIGH or LOW level the D input signal has no effect.

These circuits are fully compatible for use with most TTL or DTL circuits. Maximum clock frequency is 100 MHz with a typical power dissipation of 66 mW per flip-flop.

LOGIC AND CONNECTION DIAGRAM
DIP (TOP VIEW)

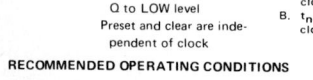

Positive logic: LOW input to preset sets
Q to HIGH level
LOW input to clear resets
Q to LOW level
Preset and clear are independent of clock

SYNCHRONOUS TRUTH TABLE
(EACH FLIP-FLOP)

t_n		t_{n+1}	
INPUT		OUTPUT	
D		Q	$\bar{Q}$
L		L	H
H		H	L

ASYNCHRONOUS TRUTH TABLE
(EACH FLIP-FLOP)

INPUT		OUTPUT	
Preset	Clear	Q	$\bar{Q}$
L	L	H	H
L	H	H	L
H	L	L	H
H	H	No Change	

H = HIGH level
L = LOW level
D = Data

NOTES:
A. t_n = bit time before clock pulse
B. t_{n+1} = bit time after clock pulse

LOGIC DIAGRAM
(EACH FLIP-FLOP)

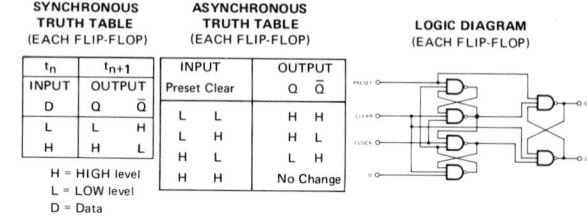

RECOMMENDED OPERATING CONDITIONS

PARAMETER	9S74XM (54S74XM)			9S74XC (74S74XC)			UNITS
	MIN.	TYP.	MAX.	MIN.	TYP.	MAX.	
Supply Voltage V_{CC}	4.5	5.0	5.5	4.75	5.0	5.25	V
Operating Free-Air Temperature Range	−55	25	125	0	25	75	°C

X = package type; F for Flatpak, D for Ceramic Dip, P for Plastic Dip. See Packaging Information Section for packages available on this product.

DC CHARACTERISTICS OVER OPERATING TEMPERATURE RANGE (Unless Otherwise Noted)

SYMBOL	PARAMETER		LIMITS			UNITS	TEST CONDITIONS (Note 1)
			MIN.	TYP. (Note 2)	MAX.		
V_{IH}	Input HIGH Voltage		2.0			V	Guaranteed Input HIGH Voltage
V_{IL}	Input LOW Voltage				0.8	V	Guaranteed Input LOW Voltage
V_{CD}	Input Clamp Diode Voltage			−0.65	−1.2	V	V_{CC} = MIN., I_{IN} = −18 mA
V_{OH}	Output HIGH Voltage	XM	2.5	3.4		V	V_{CC} = MIN., I_{OH} = −1.0 mA
		XC	2.7	3.4			
V_{OL}	Output LOW Voltage			0.35	0.5	V	V_{CC} = MIN., I_{OL} = 20 mA
I_{IH}	Input HIGH Current at	D		1.0	50	μA	V_{CC} = MAX., V_{IN} = 2.7 V
		Preset or Clock		2.0	100		
		Clear		3.0	150		
I_{IL}	Input LOW Current at	D		−1.4	−2.0	mA	V_{CC} = MAX., V_{IN} = 0.5 V
		Preset or Clock		−2.8	−4.0		
		Clear		−4.2	−6.0		
I_{OS}	Output Short Circuit Current (Note 3)		−40	−65	−100	mA	V_{CC} = MAX., V_{OUT} = 0V
I_{CC}	Supply Current			30	50	mA	V_{CC} = MAX. (Note 4)

NOTES:
(1) For conditions shown as MIN. or MAX., use the appropriate value specified under recommended operating conditions for the applicable device type.
(2) Typical limits are at V_{CC} = 5.0 V, 25°C.
(3) Not more than one output should be shorted at a time.
(4) I_{CC} is measured with clock and data inputs grounded and either preset or clear inputs grounded.

FAIRCHILD SUPER HIGH SPEED TTL/SSI • 9S74 (54S74/74S74)

AC CHARACTERISTICS: $T_A = 25°C$

SYMBOL	PARAMETER		LIMITS			UNITS	TEST CONDITIONS
			MIN.	TYP.	MAX.		
f_{max}	Maximum Clock Frequency		75	110		MHz	
t_{PLH}	Turn Off Delay Clear or Preset to Output			4	6	ns	
t_{PLH}	Turn Off Delay Clock to Output			6	9	ns	$V_{CC} = 5.0$ V
t_{PHL}	Turn On Delay Clock to Output			8	11	ns	$C_L = 15$ pF
t_{PHL}	Turn On Delay Clear or Preset to Output	CP = HIGH		9	13.5	ns	
		CP = LOW		6	8		

SWITCHING CHARACTERISTICS

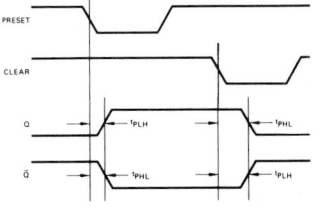

*Includes all probe and jig capacitance.

PULSE GENERATOR SETTINGS

CLOCK

$f \approx 1$ MHz
$t_f = t_r = 2.5$ ns
Amp = 0 to 3 V
Duty cycle = 50%

D

$f \approx 500$ kHz
$t_f = t_r = 2.5$ ns
Amp = 0 to 3 V
t_{set-up} (HIGH) = 7 ns (Typ. = 4 ns)
t_{set-up} (LOW) = 5 ns (Typ. = 2 ns)
Duty cycle = Adjust pulse width
to attain t_{set-up} (HIGH) and
t_{set-up} (LOW) relative to clock
as shown in waveforms.

DIRECT SET, CLEAR

$f \approx 1$ MHz
$t_f = t_r = 2.5$ ns
Amp = 0 to 3 V
Duty cycle = Adjust pulse
width and synch to attain
waveforms shown.

SWITCHING WAVEFORMS

CLOCK TO OUTPUT DELAY[†]

SWITCH IN POSITION 1

**DIRECT SET, AND CLEAR
TO OUTPUT DELAY**

SWITCH IN POSITION 2

[†]Direct set and clear inputs connected to
V_{CC} thru 2 kΩ resistor during test.

FAIRCHILD SUPER HIGH SPEED TTL/SSI • 9S114 (54S114/74S114)

DUAL JK EDGE TRIGGERED FLIP-FLOP

DESCRIPTION — The 9S114 (54S114/74S114) offer common clock and common clear inputs and individual J, K, and preset inputs. These monolithic dual flip-flops are designed so that when the clock goes HIGH, the inputs are enabled and data will be accepted. The logic level of the J and K inputs may be allowed to change when the clock pulse is HIGH and the bistable will perform according to the truth table as long as minimum set-up times are observed. Input data is transferred to the outputs on the negative-going edge of the clock pulse.

LOGIC AND CONNECTION DIAGRAM
DIP (TOP VIEW)

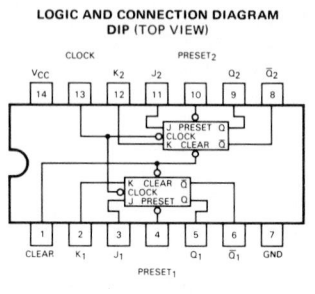

Positive logic: LOW input to preset sets Q to HIGH level.
LOW input to clear resets Q to LOW level.
Preset and clear are independent of clock.

TRUTH TABLES

t_n		t_n+1
J	K	Q
L	L	Q_n
L	H	L
H	L	H
H	H	$\bar{Q}_n$

Preset	Clear	Q	$\bar{Q}$
L	L	H	H
L	H	H	L
H	L	L	H
H	H	No Charge	

NOTES:
A. t_n = Bit time before clock pulse.
B. t_n+1 = Bit time after clock pulse.

LOGIC DIAGRAM
(Each Flip-Flop)

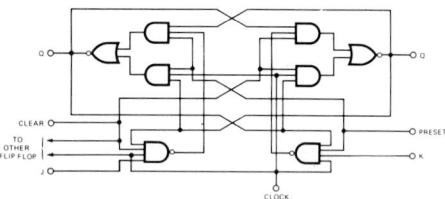

RECOMMENDED OPERATING CONDITIONS

PARAMETER	9S114XM (54S114XM)			9S114XC (74S114XC)			UNITS
	MIN.	TYP.	MAX.	MIN.	TYP.	MAX.	
Supply Voltage V_{CC}	4.5	5.0	5.5	4.75	5.0	5.25	V
Operating Free-Air Temperature Range	−55	25	125	0	25	75	°C

X = package type; F for Flatpak, D for Ceramic Dip, P for Plastic Dip. See Packaging Information Section for packages available on this product.

FAIRCHILD SUPER HIGH SPEED TTL/SSI • 9S114 (54S114/74S114)

DC CHARACTERISTICS OVER OPERATING TEMPERATURE RANGE (Unless Otherwise Noted)

SYMBOL	PARAMETER		LIMITS			UNITS	TEST CONDITIONS (Note 1)
			MIN.	TYP. (Note 2)	MAX.		
V_{IH}	Input HIGH Voltage		2.0			V	Guaranteed Input HIGH Voltage
V_{IL}	Input LOW Voltage				0.8	V	Guaranteed Input LOW Voltage
V_{CD}	Input Clamp Diode Voltage			−0.65	−1.2	V	V_{CC} = MIN., I_I = −18 mA
V_{OH}	Output HIGH Voltage	XM	2.5	3.4		V	V_{CC} = MIN., I_{OH} = −1.0 mA
		XC	2.7	3.4			
V_{OL}	Output LOW Voltage			0.35	0.5	V	V_{CC} = MIN., I_{OL} = 20 mA
I_{IH}	Input HIGH Current at	JK		1.0	50	µA	V_{CC} = MAX., V_{IN} = 2.7 V
		Clock		4.0	200		
		Preset		2.0	100		
		Clear		4.0	200		
I_{IL}	Input LOW Current at	JK		−0.96	−1.6	mA	V_{CC} = MAX., V_{IN} = 0.5 V
		Clock		−5.6	−8.0		
		Preset		−4.9	−7.0		
		Clear		−9.8	−14		
I_{OS}	Output Short Circuit Current (Note 3)		−40	−65	−100	mA	V_{CC} = MAX., V_{OUT} = 0V
I_{CC}	Supply Current			30	50	mA	V_{CC} = MAX. (Note 4)

AC CHARACTERISTICS: T_A = 25°C

SYMBOL	PARAMETER	LIMITS			UNITS	TEST CONDITIONS
		MIN.	TYP.	MAX.		
f_{max}	Maximum Clock Frequency	80	125		MHz	V_{CC} = 5.0 V C_L = 15 pF
t_{PLH}	Turn Off Delay Clear or Preset to Output			7.0	ns	
t_{PHL}	Turn On Delay Clear or Preset to Output			7.0	ns	
t_{PLH}	Turn Off Delay Clock to Output			7.0	ns	
t_{PHL}	Turn On Delay Clock to Output			7.0	ns	

NOTES:
1. For conditions shown as MIN. or MAX., use the appropriate value specified under recommended operating conditions for the applicable device type.
2. Typical limits are at V_{CC} = 5.0 V, 25°C.
3. Not more than one output should be shorted at a time.
4. I_{CC} is measured with outputs open, clock grounded and J, K, preset and clear at 4.5 V.

signetics

TRIPLE EXCLUSIVE OR/NOR GATE 10107

10107B,F: −30 to +85°C

DIGITAL 10,000 SERIES ECL

DESCRIPTION

The 10107 is a triple high speed 2-input Exclusive OR/Exclusive NOR gate. The 10107 is optimized for high speed comparator and parity functions, and has an excellent speed power product for this function. All inputs are terminated with a 50 kΩ resistor to V_{EE} which eliminates the need to tie unused inputs low. The high impedance inputs and high output fanout are ideal for a transmission line environment. The 10107 contains a temperature tracking internal bias which insures that the threshold point remains in the center of the transition region over temperature. The 10107 has complementary outputs.

FEATURES

- **FAST PROPAGATION DELAY**
  - **2.0 ns TYP (INPUTS 4, 9, 14)**
  - **−2.8 ns TYP (INPUTS 5, 7, 15)**
- **LOW POWER DISSIPATION = 115 mW/PACKAGE TYP (NO LOAD)**
- **VERY HIGH FANOUT CAPABILITY**
  - **CAN DRIVE SIX 50 Ω LINES**
- **HIGH Z INPUTS — INTERNAL 50 kΩ PULLDOWNS**
- **HIGH IMMUNITY FROM POWER SUPPLY VARIATIONS:** V_{EE} = −5.2 V ±5% RECOMMENDED
- **COMPLEMENTARY OR/NOR OUTPUTS**
- **OPEN EMITTERS FOR BUSSING AND LOGIC CAPABILITY**

LOGIC DIAGRAM

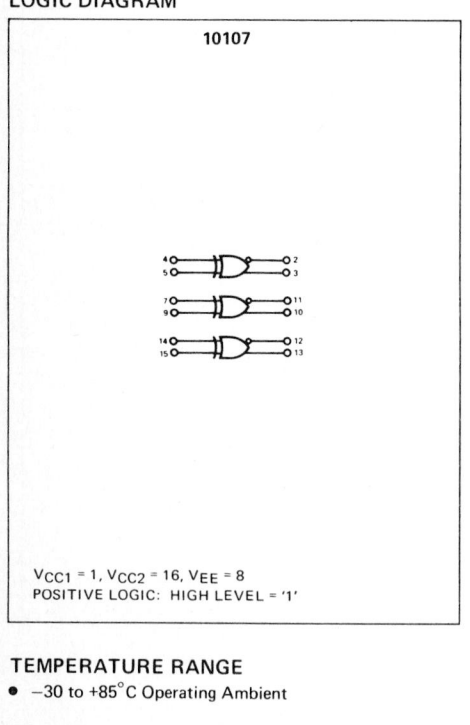

10107

V_{CC1} = 1, V_{CC2} = 16, V_{EE} = 8
POSITIVE LOGIC: HIGH LEVEL = '1'

CIRCUIT SCHEMATIC

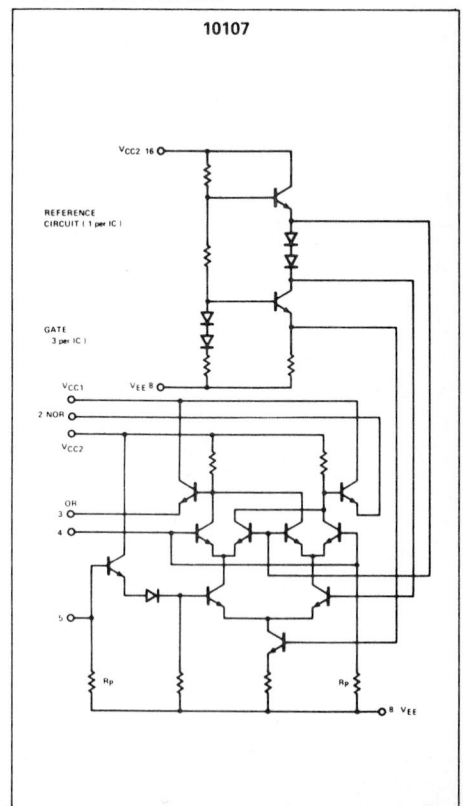

10107

TEMPERATURE RANGE

- −30 to +85°C Operating Ambient

PACKAGE TYPE

B: 16-Pin Silicone DIP
F: 16-Pin CERDIP

SIGNETICS TRIPLE EXCLUSIVE OR/NOR GATE ■ 10107

ELECTRICAL CHARACTERISTICS
(at Listed Voltages and Ambient Temperatures).

TEST VOLTAGE VALUES (Volts)

@ Test Temperature	$V_{IH\,max}$	$V_{IL\,min}$	$V_{IHA\,min}$	$V_{ILA\,max}$	V_{EE}
−30°C	−0.890	−1.890	−1.205	−1.500	−5.2
+25°C	−0.810	−1.850	−1.105	−1.475	−5.2
+85°C	−0.700	−1.825	−1.035	−1.440	−5.2

Characteristic	Symbol	Pin Under Test	−30°C Min	−30°C Max	+25°C Min	+25°C Max	+85°C Min	+85°C Max	Unit	$V_{IH\,max}$	$V_{IL\,min}$	$V_{IHA\,min}$	$V_{ILA\,max}$	V_{EE}	(V_{CC}) Gnd
Power Supply Drain Current	I_E	8	−	−	−	28	−	−	mAdc	All Inputs	−	−	−	8	1,16
	I_{inH}	4,9,14	−	* −	−	265	−	−	µAdc	*	−	−	−	8	1,16
		5,7,15	−	−	−	220	−	−	µAdc		−	−	−	8	1,16
	I_{inL}	*	−	−	0.5	−	−	−	µAdc	−	*	−	−	8	1,16
Logic "1" Output Voltage	V_{OH}	2	−1.060	−0.890	−0.960	−0.810	−0.890	−0.700	Vdc	4,5	−	−	−	8	1,16
		2	−1.060	−0.890	−0.960	−0.810	−0.890	−0.700	Vdc	−	4,5	−	−	↓	↓
		3	−1.060	−0.890	−0.960	−0.810	−0.890	−0.700		4	5	−	−	↓	↓
		3	−1.060	−0.890	−0.960	−0.810	−0.890	−0.700		5	4	−	−	↓	↓
Logic "0" Output Voltage	V_{OL}	2	−1.890	−1.675	−1.850	−1.650	−1.825	−1.615	Vdc	4	5	−	−	8	1,16
		2	−1.890	−1.675	−1.850	−1.650	−1.825	−1.615		5	4	−	−	↓	↓
		3	−1.890	−1.675	−1.850	−1.650	−1.825	−1.615		4,5	−	−	−	↓	↓
		3	−1.890	−1.675	−1.850	−1.650	−1.825	−1.615		−	4,5	−	−	↓	↓
Logic "1" Threshold Voltage	V_{OHA}	2	−1.080	−	−0.980	−	−0.910	−	Vdc	5	−	4	−	8	1,16
		2	−1.080	−	−0.980	−	−0.910	−		↓	−	−	4	↓	↓
		3	−1.080	−	−0.980	−	−0.910	−		−	−	4	−	↓	↓
		3	−1.080	−	−0.980	−	−0.910	−		−	−	5	−	↓	↓
Logic "0" Threshold Voltage	V_{OLA}	2	−	−1.655	−	−1.630	−	−1.595	Vdc	−	−	4	−	8	1,16
		2	−	−1.655	−	−1.630	−	−1.595		−	−	5	−	↓	↓
		3	−	−1.655	−	−1.630	−	−1.595		5	−	4	−	↓	↓
		3	−	−1.655	−	−1.630	−	−1.595		−	−	−	4	↓	↓

Switching Times† (50-ohm load)

Characteristic	Symbol		−30°C Min	−30°C Max	+25°C Min	+25°C Typ	+25°C Max	+85°C Min	+85°C Max	Unit	+1.1 V		Pulse In	Pulse Out	−3.2 V	+2.0 V
Propagation Delay	t++	Inputs 4,9,or 14 to either Output	1.0	3.8	1.1	2.0	3.7	1.1	4.0	ns	5,7,15	−	Input 4,9, or 14	Corresponding Ex-OR/Ex-NOR Outputs	8	1,16
	t+−					↓					↓	−			↓	↓
	t−+					↓					↓	−			↓	↓
	t−−										↓	−			↓	↓
	t++	Inputs 5,7, or 15 to either Output				2.8					4,9,14	−	Input 5,7, or 15	Corresponding Ex-OR/Ex-NOR Outputs	↓	↓
	t+−					↓					↓	−			↓	↓
	t−+					↓					↓	−			↓	↓
	t−−										↓	−			↓	↓
Rise Time (20% to 80%)	t+	**	1.1	3.5	2.5		3.5		3.8		4,9,14	−	Any Input	Corresponding Ex-OR/Ex-NOR Outputs	↓	↓
Fall Time (20% to 80%)	t−	**	1.1	3.5	2.5		3.5		3.8		4,9,14	−	Any Input		↓	↓

*Individually test each input applying V_{IH} or V_{IL} to input under test.

**Any Output

†Unused outputs connected to a 50-ohm resistor to ground.

SWITCHING TIME TEST CIRCUIT

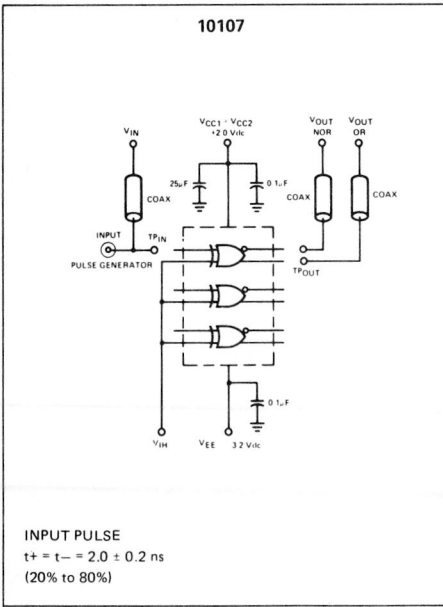

10107

INPUT PULSE
t+ = t− = 2.0 ± 0.2 ns
(20% to 80%)

PROPAGATION DELAY WAVEFORMS @ 25°C

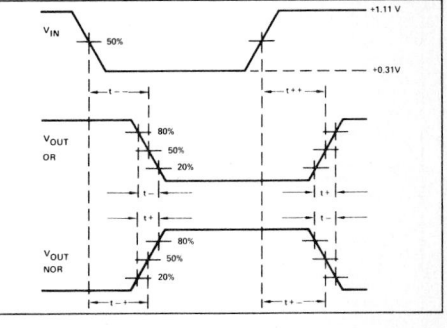

NOTES:

1. Each ECL 10,000 series device has been designed to meet the DC specifications shown in the test table, after thermal equilibrium has been established. The circuit is in a test socket or mounted on a printed circuit board and transverse air flow greater than 500 linear fpm is maintained. Voltage levels will shift approximately 4 mV with an air flow of 200 linear fpm. Outputs are terminated through a 50-ohm resistor to 2.0 volts.

2. For AC tests, all input and output cables to the scope are equal lengths of 50-ohm coaxial cable. Wire length should be < 1/4 inch from TP_{in} to input pin and TP_{out} to output pin. A 50-ohm termination to ground is located in each scope input. Unused outputs are connected to a 50-ohm resistor to ground.

3. Test procedures are shown for only one input or set of input conditions. Other inputs are tested in the same manner.

4. All voltage measurements are referenced to the ground terminal. Terminals not specifically referenced are left electrically open.

absolute maximum ratings (Note 1)

Voltage at Any Pin	$-0.3V$ to $V_{CC} + 0.3V$
Operating Temperature Range	
MM54C30	$-55^\circ C$ to $+125^\circ C$
MM74C30	$-40^\circ C$ to $+85^\circ C$
Storage Temperature Range	$-65^\circ C$ to $+150^\circ C$
Package Dissipation	500 mW
Operating V_{CC} Range	3.0V to 15V
Absolute Maximum V_{CC}	16V
Lead Temperature (Soldering, 10 seconds)	$300^\circ C$

dc electrical characteristics

Min/max limits apply across temperature range, unless otherwise noted.

PARAMETER	CONDITIONS	MIN	TYP	MAX	UNITS
CMOS TO CMOS					
Logical "1" Input Voltage ($V_{IN(1)}$)	$V_{CC} = 5.0V$	3.5			V
	$V_{CC} = 10V$	8.0			V
Logical "0" Input Voltage ($V_{IN(0)}$)	$V_{CC} = 5.0V$			1.5	V
	$V_{CC} = 10V$			2.0	V
Logical "1" Output Voltage ($V_{OUT(1)}$)	$V_{CC} = 5.0V$, $I_O = -10\mu A$	4.5			V
	$V_{CC} = 10V$, $I_O = -10\mu A$	9.0			V
Logical "0" Output Voltage ($V_{OUT(0)}$)	$V_{CC} = 5.0V$, $I_O = +10\mu A$			0.5	V
	$V_{CC} = 10V$, $I_O = +10\mu A$			1.0	V
Logical "1" Input Current ($I_{IN(1)}$)	$V_{CC} = 15V$, $V_{IN} = 15V$		0.005	1.0	μA
Logical "0" Input Current ($I_{IN(0)}$)	$V_{CC} = 15V$, $V_{IN} = 0V$	-1.0	-0.005		μA
Supply Current (I_{CC})	$V_{CC} = 15V$		0.01	15	μA
CMOS/LPTTL INTERFACE					
Logical "1" Input Voltage ($V_{IN(1)}$)	54C, $V_{CC} = 4.5V$	$V_{CC}-1.5$			V
	74C, $V_{CC} = 4.75V$	$V_{CC}-1.5$			V
Logical "0" Input Voltage ($V_{IN(0)}$)	54C, $V_{CC} = 4.5V$			0.8	V
	74C, $V_{CC} = 4.75V$			0.8	V
Logical "1" Output Voltage ($V_{OUT(1)}$)	54C, $V_{CC} = 4.5V$, $I_O = -360\mu A$	2.4			V
	74C, $V_{CC} = 4.75V$, $I_O = -360\mu A$	2.4			V
Logical "0" Output Voltage ($V_{OUT(0)}$)	54C, $V_{CC} = 4.5V$, $I_O = 360\mu A$			0.4	V
	74C, $V_{CC} = 4.75V$, $I_O = 360\mu A$			0.4	V
OUTPUT DRIVE (See 54C/74C Family Characteristics Data Sheet)					
Output Source Current (I_{SOURCE}) (P-Channel)	$V_{CC} = 5.0V$, $V_{OUT} = 0V$, $T_A = 25^\circ C$	-1.75	-3.3		mA
Output Source Current (I_{SOURCE}) (P-Channel)	$V_{CC} = 10V$, $V_{OUT} = 0V$, $T_A = 25^\circ C$	-8.0	-15		mA
Output Sink Current (I_{SINK}) (N-Channel)	$V_{CC} = 5.0V$, $V_{OUT} = V_{CC}$, $T_A = 25^\circ C$	1.75	3.6		mA
Output Sink Current (I_{SINK}) (N-Channel)	$V_{CC} = 10V$, $V_{OUT} = V_{CC}$, $T_A = 25^\circ C$	8.0	16		mA

ac electrical characteristics $T_A = 25^\circ C$, $C_L = 50$ pF, unless otherwise specified.

PARAMETER	CONDITIONS	MIN	TYP	MAX	UNITS
Propagation Delay Time to Logical "1"	$V_{CC} = 5.0V$		125	180	ns
or "0" (t_{pd})	$V_{CC} = 10V$		55	90	ns
Input Capacitance (C_{IN})	(Note 2)		4.0		pF
Power Dissipation Capacitance (C_{pd})	(Note 3) Per Gate		26		pF

INDEX